Lecture Notes in Computer Science 4978

Commenced Publication in 1973
Founding and Former Series Editors:
Gerhard Goos, Juris Hartmanis, and Jan van Leeuwen

Manindra Agrawal Dingzhu Du
Zhenhua Duan Angsheng Li (Eds.)

Theory and Applications of Models of Computation

5th International Conference, TAMC 2008
Xi'an, China, April 25-29, 2008
Proceedings

Springer

Volume Editors

Manindra Agrawal
Indian Institute of Technology (IIT)
Kanpur 208016, India
E-mail: manindra@iitk.ac.in

Dingzhu Du
University of Texas
Dallas, TX, USA
E-mail: dzdu@utdallas.edu

Zhenhua Duan
Xidian University
Xi'an 710071, China
E-mail: zhhduan@mail.xidian.edu.cn

Angsheng Li
Chinese Academy of Sciences
Beijing 100080, China
E-mail: angsheng@ios.ac.cn

Library of Congress Control Number: 2008924623

CR Subject Classification (1998): F.1.1-2, F.2.1-2, F.4.1, I.2.6, J.3

LNCS Sublibrary: SL 1 – Theoretical Computer Science and General Issues

ISSN 0302-9743
ISBN-10 3-540-79227-9 Springer Berlin Heidelberg New York
ISBN-13 978-3-540-79227-7 Springer Berlin Heidelberg New York

Springer is a part of Springer Science+Business Media

springer.com

Printed in Germany

Typesetting: Camera-ready by author, data conversion by Scientific Publishing Services, Chennai, India
Printed on acid-free paper SPIN: 12257317 06/3180 5 4 3 2 1 0

Preface

Theory and Applications of Models of Computation (TAMC) is an international conference series with an interdisciplinary character, bringing together researchers working in computer science, mathematics (especially logic), and the physical sciences. This crossdisciplinary character, together with its focus on algorithms, complexity, and computability theory, gives the conference a special flavor and distinction.

TAMC 2008 was the fifth conference in the series. The previous four meetings were held during May 17–19, 2004 in Beijing, May 17–20, 2005 in Kunming, May 15–20, 2006 in Beijing, and May 22–25, 2007 in Shanghai. TAMC 2008 was held in Xi'an, during April 25–29, 2008.

At TAMC 2008 we had two plenary speakers, Bernard Chazelle and Cynthia Dwork, giving one-hour talks each. Bernard spoke on "Why Algorithms Matter" and Cynthia on "Differential Privacy: A Survey of Results." Their respective papers accompanying the talks are included in the proceedings.

In addition, there were two special sessions organized by Barry Cooper and Ying Jiang on "Models of Computation" and by Jianer Chen on "Algorithms and Complexity." The invited speakers in the first session were Jose Felix Costa, Vincent Danos, Luke Ong, Mingsheng Ying, Miklos Santha, and Gilles Dowek. Invited speakers in the second session were Daniel Brown, Dieter Kratsch, Xiaotie Deng, and Jianer Chen.

The TAMC conference series arose naturally in response to important scientific developments affecting how we compute in the twenty-first century. At the same time, TAMC is already playing an important regional and international role, and promises to become a key contributor to the scientific resurgence seen throughout China and other parts of Asia.

For TAMC 2008, we received 192 submissions from all over the world, one of which was withdrawn. The Program Committee finally selected 50 papers for presentation at the conference and inclusion in this LNCS volume.

We are very grateful to the Program Committee, and the many outside referees they called on, for their hard work and expertise during the difficult selection process. We also wish to thank all those authors who submitted their work for our consideration. The Program Committee could have accepted many more submissions without compromising standards, and were only restrained by the practicalities of timetabling so many talks, and by the inevitable limitations on the size of this proceedings volume.

Finally, we would like to thank the members of the Editorial Board of *Lecture Notes in Computer Science* and the Editors at Springer for their encouragement and cooperation throughout the preparation of this conference.

Of course TAMC 2008 would not have been possible without the support of our sponsors, and we therefore gratefully acknowledge their help in the realization of this conference.

April 2008

Manindra Agrawal
Dingzhu Du
Zhenhua Duan
Angsheng Li

Organization

The conference was organized by the Institute of Computing Theory and Technology, Xidian University, Xi'an, China.

Steering Committee

Manindra Agrawal (IIT Kanpur, India)
Jin-Yi Cai (University of Wisconsin-Madison, USA)
S. Barry Cooper (University of Leeds, UK)
Angsheng Li (Chinese Academy of Sciences, China)

Conference Chair

Dingzhu Du (University of Texas at Dallas, USA)

Program Committee Co-chairs

Manindra Agrawal (IIT Kanpur, India)
Dingzhu Du (University of Texas at Dallas, USA)
Zhenhua Duan (Xidian University, China)
Angsheng Li (Chinese Academy of Sciences, China)

Program Committee

Manindra Agrawal (IIT Kanpur, India)
Giorgio Ausiello (Rome, Italy)
Paola Bonizzoni (University of Milano-Bicocca, Italy)
Jin-Yi Cai (University of Wisconsin-Madison, USA)
Cristian S. Calude (University of Auckland, Australia)
Jianer Chen (Texas A&M University, USA)
Francis Chin (University of Hong Kong)
S. Barry Cooper (University of Leeds, UK)
Decheng Ding (Nanjing University, China)
Rod Downey (Wellington, New Zealand)
Dingzhu Du (University of Texas at Dallas, USA)
Zhenhua Duan (Xidian University, China)
Rudolf Fleischer (Fudan University, China)
Hiroshi Imai (University of Tokyo, Japan)
Kazuo Iwama (Kyoto University, Japan)
Ying Jiang (Chinese Academy of Sciences, China)

Valentine Kabanets (Simon Fraser University, Canada)
Maciej Koutny (Newcastle University, UK)
Andrew Lewis (University of Leeds, UK)
Angsheng Li (Chinese Academy of Sciences, China)
Xingwu Liu (Chinese Academy of Sciences, China)
Satyanarayana Lokam (Microsoft Research, India)
Giuseppe Longo (Paris, France)
Janos Makowsky (Israel Institute of Technology, Israel)
Jaikumar Radhakrishnan (Tata Institute of Fundamental Research, India)
Rudiger Reischuk (Universitat zu Lubeck, Germany)
Helmut Schwichtenberg (Mathematisches Institut der Universität München, Germany)
Xiaoming Sun (Tsinghua University, China)
Luca Trevisan (UC Berkeley, USA)
Christopher Umans (CalTech, USA)
Hanpin Wang (Beijing University, China)
Osamu Watanabe (Tokyo Institute of Technology, Japan)
Mingsheng Ying (Tsinghua University, China)
Shengyu Zhang (California Institute of Technology, USA)
Ting Zhang (Microsoft Research Asia)
Wenhui Zhang (ISCAS)
Yunlei Zhao (Fudan University, China)
Hong Zhu (Fudan University, China)

Organizing Committee

Dingzhu Du (University of Texas at Dallas, USA)
Zhenhua Duan (Xidian University, Xi'an, China)
Angsheng Li (Institute of Software, Chinese Academy of Sciences, China)

Sponsoring Institutions

Institute of Computing Theory and Technology of Xidian University
Xidian University
The National Natural Science Foundation of China based on NSFC Grant No. 60433010

Table of Contents

Differential Privacy: A Survey of Results

Cynthia Dwork

Microsoft Research
dwork@microsoft.com

Abstract. Over the past five years a new approach to privacy-preserving data analysis has born fruit [13, 18, 7, 19, 5, 37, 35, 8, 32]. This approach differs from much (but not all!) of the related literature in the statistics, databases, theory, and cryptography communities, in that a formal and *ad omnia* privacy guarantee is defined, and the data analysis techniques presented are rigorously proved to satisfy the guarantee. The key privacy guarantee that has emerged is *differential privacy.* Roughly speaking, this ensures that (almost, and quantifiably) no risk is incurred by joining a statistical database.

In this survey, we recall the definition of differential privacy and two basic techniques for achieving it. We then show some interesting applications of these techniques, presenting algorithms for three specific tasks and three general results on differentially private learning.

1 Introduction

Privacy-preserving data analysis is also known as statistical disclosure control, inference control, privacy-preserving datamining, and private data analysis. Our principal motivating scenario is a *statistical database.* A statistic is a quantity computed from a sample. Suppose a trusted and trustworthy curator gathers sensitive information from a large number of respondents (the sample), with the goal of learning (and releasing to the public) statistical facts about the underlying population. The problem is to release statistical information without compromising the privacy of the individual respondents. There are two settings: in the *noninteractive* setting the curator computes and publishes some statistics, and the data are not used further. Privacy concerns may affect the precise answers released by the curator, or even the set of statistics released. Note that since the data will never be used again the curator can destroy the data (and himself) once the statistics have been published.

In the *interactive* setting the curator sits between the users and the database. Queries posed by the users, and/or the responses to these queries, may be modified by the curator in order to protect the privacy of the respondents. The data cannot be destroyed, and the curator must remain present throughout the lifetime of the database. Of course, any interactive solution yields a non-interactive solution, provided the queries are known in advance: the curator can simulate an interaction in which these known queries are posed, and publish the resulting transcript.

There is a rich literature on this problem, principally from the satistics community (see, *e.g.*, [10, 14, 27, 28, 29, 38, 40, 26, 39] and the literature on controlled

M. Agrawal et al. (Eds.): TAMC 2008, LNCS 4978, pp. 1–19, 2008.

release of tabular data, contingency tables, and cell suppression), and from such diverse branches of computer science as algorithms, database theory, and cryptography, for example as in [3, 4, 21, 22, 23, 33, 34, 41, 48], [1, 24, 25, 31], and [6, 9, 11, 12, 13, 18, 7, 19]; see also the survey [2] for a summary of the field prior to 1989.

This survey is about *differential privacy.* Roughly speaking, differential privacy ensures that the removal or addition of a single database item does not (substantially) affect the outcome of any analysis. It follows that no risk is incurred by joining the database, providing a mathematically rigorous means of coping with the fact that distributional information may be disclosive.

We will first describe three differentially private algorithms for specific, unrelated, data analysis tasks. We then present three general results about computational learning when privacy of individual data items is to be protected. This is not usually a concern in the learning theory literature, and signals the emergence of a new line of research.

2 Differential Privacy

In the sequel, the randomized function $\mathcal{K}$ is the algorithm applied by the curator when releasing information. So the input is the data set, and the output is the released information, or *transcript.* We do not need to distinguish between the interactive and non-interactive settings.

Think of a database as a set of rows. We say databases D_1 and D_2 *differ in at most one element* if one is a proper subset of the other and the larger database contains just one additional row.

Definition 1. *A randomized function* $\mathcal{K}$ *gives* ϵ-differential privacy *if for all data sets* D_1 *and* D_2 *differing on at most one element, and all* $S \subseteq Range(\mathcal{K})$,

$$\Pr[\mathcal{K}(D_1) \in S] \leq \exp(\epsilon) \times \Pr[\mathcal{K}(D_2) \in S] \qquad (1)$$

The probability is taken is over the coin tosses of $\mathcal{K}$.

A mechanism $\mathcal{K}$ satisfying this definition addresses concerns that any participant might have about the leakage of her personal information: even if the participant removed her data from the data set, no outputs (and thus consequences of outputs) would become significantly more or less likely. For example, if the database were to be consulted by an insurance provider before deciding whether or not to insure a given individual, then the presence or absence of that individual's data in the database will not significantly affect her chance of receiving coverage.

Differential privacy is therefore an *ad omnia* guarantee. It is also a very strong guarantee, since it is a statistical property about the behavior of the mechanism and therefore is independent of the computational power and auxiliary information available to the adversary/user.

Differential privacy is not an absolute guarantee of privacy. In fact, Dwork and Naor have shown that any statistical database with any non-trivial utility compromises a natural definition of privacy [15]. However, in a society that has decided that the benefits of certain databases outweigh the costs, differential privacy ensures that only a limited amount of additional risk is incurred by participating in the socially beneficial databases.

Remark 1.
1. The parameter ϵ in Definition 1 is public. The choice of ϵ is essentially a social question and is beyond the scope of this paper. That said, we tend to think of ϵ as, say, 0.01, 0.1, or in some cases, $\ln 2$ or $\ln 3$. If the probability that some bad event will occur is very small, it might be tolerable to increase it by such factors as 2 or 3, while if the probability is already felt to be close to unacceptable, then an increase by a factor of $e^{0.01} \approx 1.01$ might be tolerable, while an increase of e, or even only $e^{0.1}$, would be intolerable.
2. Definition 1 discusses the behavior of the mechanism $\mathcal{K}$, and is independent of any auxiliary knowledge the adversary, or user, may have about the database. Thus, a mechanism satisfying the definition protects the privacy of an individual row in the database even if the adversary knows every other row in the database.
3. Definition 1 extends to group privacy as well (and to the case in which an individual contributes more than a single row to the database). A collection of c participants might be concerned that their collective data might leak information, even when a single participant's does not. Using this definition, we can bound the dilation of any probability by at most $\exp(\epsilon c)$, which may be tolerable for small c. Of course, the point of the statistical database is to disclose aggregate information about large groups (while simultaneously protecting individuals), so we should expect privacy bounds to disintegrate with increasing group size.

3 Achieving Differential Privacy in Statistical Databases

We will presently describe an interactive mechanism, $\mathcal{K}$, due to Dwork, McSherry, Nissim, and Smith [19], for the case of continuous-valued queries. Specifically, in this section a *query* is a function mapping databases to (vectors of) real numbers. For example, the query "Count P" counts the number of rows in the database having property P.

When the query is a function f, and the database is X, the *true answer* is the value $f(X)$. The mechanism $\mathcal{K}$ adds appropriately chosen random noise to the true answer to produce what we call the *response*. The idea of preserving privacy by responding with a noisy version of the true answer is not new, but this approach is delicate. For example, if the noise is symmetric about the origin and the same question is asked many times, the responses may be averaged, cancelling out the noise[1]. We must take such factors into account.

[1] We do not recommend having the curator record queries and their responses so that if a query is issued more than once the response can be replayed: If the query language is sufficiently rich, then semantic equivalence of two syntactically different queries is undecidable; even if the query language is not so rich, the devastating attacks demonstrated by Dinur and Nissim [13] pose completely random and unrelated queries.

Definition 2. *For $f : \mathcal{D} \rightarrow R^k$, the* sensitivity *of f is*

$$\Delta f = \max_{D_1, D_2} \|f(D_1) - f(D_2)\|_1 \tag{2}$$

for all D_1, D_2 differing in at most one element.

In particular, when $k = 1$ the sensitivity of f is the maximum difference in the values that the function f may take on a pair of databases that differ in only one element.

For many types of queries Δf will be quite small. In particular, the simple counting queries discussed above ("How many rows have property P?") have $\Delta f = 1$. Our techniques work best – introduce the least noise – when Δf is small. Note that sensitivity is a property of the function alone, and is independent of the database. The sensitivity essentially captures how great a difference (between the value of f on two databases differing in a single element) must be hidden by the additive noise generated by the curator.

The scaled symmetric exponential distribution with standard deviation $\sqrt{2}\Delta f/\epsilon$ denoted $\text{Lap}(\Delta f/\epsilon)$, has mass at x proportional to $\exp(-|x|(\epsilon/\Delta f))$. More precisely, let $b = \Delta f/\epsilon$. The probability density function is $p(x) = \exp(-|x|/b)/2b$ and the cumulative distribution function is $D(x) = (1/2)(1 + \text{sgn}(x)(1 - \exp(|x|/b)))$.

On query function f the privacy mechanism $\mathcal{K}$ responds with

$$f(X) + (\text{Lap}(\Delta f/\epsilon))^k$$

adding noise with distribution $\text{Lap}(\Delta f/\epsilon)$ independently to each of the k components of $f(X)$. Note that decreasing ϵ, a publicly known parameter, flattens out the $\text{Lap}(\Delta f/\epsilon)$ curve, yielding larger expected noise magnitude. When ϵ is fixed, functions f with high sensitivity yield flatter curves, again yielding higher expected noise magnitudes.

For simplicity, consider the case $k = 1$. The proof that $\mathcal{K}$ yields ϵ-differential privacy on the single query function f is straightforward. Consider any subset $S \subseteq Range(\mathcal{K})$, and let D_1, D_2 be any pair of databases differing in at most one element. When the database is D_1, the probability mass at any $r \in Range(\mathcal{K})$ is proportional to $\exp(-|f(D_1) - r|(\Delta f/\epsilon))$, and similarly when the database is D_2. Applying the triangle inequality in the exponent we get a ratio of at most exp $(-|f(D_1) - f(D_2)|(\Delta f/\epsilon))$. By definition of sensitivity, $|f(D_1) - f(D_2)| \leq \Delta f$, and so the ratio is bounded by $exp(-\epsilon)$, yielding ϵ-differential privacy.

It is easy to see that ϵ-differential privacy can be achieved for any (adaptively chosen) query sequence $f_1, \ldots, f_d$ by running $\mathcal{K}$ with noise distribution $\text{Lap}(\sum_i \Delta f_i/\epsilon)$ on *each* query. In other words, the quality of each answer deteriorates with the sum of the sensitivities of the queries. Interestingly, it is sometimes possible to do better than this. Roughly speaking, what matters is the maximum possible value of $\Delta = \|(f_1(D_1), f_2(D_1), \ldots, f_d(D_1)) - (f_1(D_2), f_2(D_2), \ldots, f_d(D_2))\|_1$. The precise formulation of the statement requires some care, due to the potentially adaptive choice of queries. For a full treatment see [19]. We state the theorem here for the non-adaptive case, viewing the (fixed) sequence of queries $f_1, f_2, \ldots, f_d$, with respective arities $k_1, \ldots, k_d$, as a single $k = \sum_{i=1}^{d} k_i$-ary query f, and recalling Definition 2 for the case of arbitrary k.

Theorem 1 ([19]). *For $f : \mathcal{D} \to R^k$, the mechanism $\mathcal{K}_f$ that adds independently generated noise with distribution* $\mathrm{Lap}(\Delta f/\epsilon)$ *to each of the k output terms enjoys ϵ-differential privacy.*

The mechanism $\mathcal{K}$ described above has excellent accuracy for insensitive queries. In particular, the noise needed to ensure differential privacy depends only on the sensitivity of the function and on the parameter ϵ. Both are independent of the database and the number of rows it contains. Thus, if the database is very large, the errors for many typical queries introduced by the differential privacy mechanism is relatively quite small.

We can think of $\mathcal{K}$ as a differential privacy-preserving interface between the analyst and the data. This suggests a general approach to privacy-preserving data analysis: find algorithms that require few, insensentitive, queries. See, *e.g.*, [7, 8, 32]. Indeed, even counting queries are extremely powerful, permitting accurate and differentially private computations of many standard datamining tasks including principal component analysis, k-means clustering, perceptron learning of separating hyperplanes, and generation of an ID3 decision tree [7], as well as (nearby) halfspace learning [8] (see Section 4.3 below).

Among the many applications of Theorem 1, of particular interest is the class of *histogram* queries. A histogram query is an arbitrary partitioning of the domain of database rows into disjoint "cells," and the true answer is the set of counts describing, for each cell, the number of database rows in this cell. Although a histogram query with k cells may be viewed as k individual counting queries, the addition or removal of a single database row can affect the entire k-tuple of counts in at most one location (the count corresponding to the cell to (from) which the row is added (deleted); moreover, the count of this cell is affected by at most 1, so by Definition 2, every histogram query has sensitivity 1. Many data analyses are simply histograms; it is thus particularly encouraging that complex histograms, rather than requiring large variance in each cell, require very little.

3.1 When Noise Makes No Sense

In some tasks, the addition of noise makes no sense. For example, the function f might map databases to strings, strategies, or trees. In a recent paper McSherry and Talwar address the problem of optimizing the output of such a function while preserving ϵ-differential privacy [35]. Assume the curator holds a database X and the goal is to produce an object y. In a nutshell, their *exponential mechanism* works as follows. There is assumed to be a *utility function $u(X,y)$* that measures the quality of an output y, given that the database is X. For example, if the database holds the valuations that individuals assign a digital good during an auction, $u(X, y)$ might be the revenue, with these valuations, when the price is set to y. Auctions are a good example of where noise makes no sense, since an even slightly too high price may prevent many bidders from buying.

McSherry and Talwar's *exponential mechanism* outputs y with probability proportional to $\exp(-\epsilon u(X, y)/2)$. This ensures $\epsilon\Delta u$-differential privacy, or

ϵ-differential privacy whenever $\Delta u \leq 1$. Here Δu is defined slightly differently from above; it is the maximum possible change to the value of u caused by *changing* the data of a single row (as opposed to removing or adding a row; the notions differ by at most a factor of two); see [35].

With this approach McSherry and Talwar obtain approximately-truthful auctions with nearly optimal selling price. Roughly speaking, this says that a participant cannot dramatically reduce the price he pays by lying about his valuation. Interestingly, they show that the simple composition of differential privacy can be used to obtain auctions in which no cooperating group of c agents can significantly increase their utility by submitting bids other than their true valuations. This is analagous to the situation of Remark 1 above, where composition is used to obtain privacy for groups of c individuals,

4 Algorithms for Specific Tasks

In this section we describe differentially private algorithms for three unrelated tasks.

4.1 Statistical Data Inference

The results in this Section are due to Dwork and Nissim [18].

Consider a setting in which each element in the database is described by a set of k Boolean attributes $\alpha_1, \ldots \alpha_k$, and the rows are independently sampled from some underlying distribution on $\{0,1\}^k$. Let $1 \leq \ell \leq k/2$ be an integer. The goal here is to use information about the incidence of settings of any ℓ attribute values to learn the incidence of settings of any 2ℓ attribute values.

Although we use the term "queries," these will all be known in advance, and the mechanism will be non-interactive. From something like $\binom{k}{\ell}2^\ell$ pieces of released information it will be possible to compute approximations to the incidence of all $\binom{k}{2\ell}2^{2\ell}$ minterms. This will allow the data analyst to approximate the probabilities of all $2^{\binom{k}{2\ell}2^{2\ell}}$ subsets of 2ℓ-ary minterms of length 2ℓ, provided the initial approximations are sufficiently accurate.

We will identify probability with incidence, so the probability space is over rows in the database. Fix any set of ℓ attributes. The incidence of all possible settings of these attribute values is described by a histogram with 2^ℓ cells, and histograms have sensitivity 1, so we are going to be working with a query sequence of overall sensitivity proportional to $\binom{k}{\ell}$ (in fact, it will be worse than this by a factor t, discussed below).

Let α and β be attributes. We say that α *implies* β *in probability* if the conditional probability of β given α exceeds the unconditional probability of β. The ability to measure implication in probability is crucial to datamining. Note that since $\Pr[\beta]$ is simple to estimate well using counting queries, the problem of measuring implication in probability reduces to obtaining a good estimate of $\Pr[\beta|\alpha]$. Moreover, once we can estimate $\Pr[\beta|\alpha]$, $\Pr[\beta]$, and $\Pr[a]$, we can use Bayes' Rule and de Morgan's Laws to determine the statistics for any Boolean function of attribute values. For example, $\Pr[\alpha \wedge \beta] = \Pr[\alpha]\Pr[\beta|\alpha]$, so if we have estimates of the two multiplicands, within an additive η, we have an estimate for the products that is accurate within 3η.

As a step toward the non-interactive solution, consider the interactive case and assume that we have a good estimate for $\Pr[\alpha]$ and $\Pr[\beta]$. The key to determining $\Pr[\beta|\alpha]$ is to find a *heavy set for* α, that is, a set $q \subseteq [n]$ such that the incidence of α is at least, say, a standard deviation higher than expected, and then to determine whether the incidence of β on this heavy set is higher than the overall incidence of β. More specifically, one can test whether this conditional incidence is higher than a given threshold, and then use binary search to find the "right" threshold value. Finding the heavy set is easy because a randomly chosen subset of $[n]$ has constant probability of exceeding the expected incidence of α by at least one standard deviation.

To "simulate" the interactive case, the curator chooses some number of random subsets and for each one releases (noisy) estimates of the incidence of α and the incidence of β within this subset. With high probability (depending on t), at least one of the subsets is heavy for α.

Putting the pieces together: for t random subsets of $[n]$ the curator releases good approximations to the incidence of all $m = \binom{k}{\ell}2^\ell$ conjunctions of ℓ literals. Specifically, we require that with probability at least $1 - \delta/m^2$ a computed implication in distribution $\Pr[\alpha|\beta]$ is accurate to within $\eta/3m^2$, where α and β are now minterms of ℓ literals. This ensures that with probability least $1 - \delta$ all computed implications in distribution are accurate to within $\eta/3m^2$, and so all estimated probabilities for minterms of 2ℓ literals are accurate to within η/m^2. The number t is rather large, and depends on many factors, including the differential privacy parameter ϵ as well as η, δ, k and ℓ. The analysis in [18] shows that, when η and δ are constant, this approach reduces the number of queries from $\binom{k}{2\ell}$ (one histogram for each 2ℓ-tuple of variables (not literals!)), to $O(2^{4\ell}k^\ell\ell^2\log k)$. Note the interesting tradeoff: we require accuracy that depends on m^2 in order to avoid making m^2 queries. When the database is sufficiently large this tradeoff can be accomplished.

4.2 Contingency Table Release

The results in this Section are due to Barak, Chaudhuri, Dwork, Kale, McSherry, and Talwar [5]. A contingency table is a table of counts. In the context of a census or other survey, we think of the data of an individual as a *row* in a database. We do not assume the rows are mutually independent. For the present, each row consists of k bits describing the values of k binary attributes $a_1, \ldots, a_k$.[2] Formally, the contingency table is a vector in $\mathbb{R}^{2^k}$ describing, for each setting of the k attributes, the number of rows in the database with this setting of the attribute values. In other words, it is a histogram with 2^k cells.

Commonly, the contingency table itself is not released, as it is likely to be sparse when k is large. Instead, for various subsets of attributes, the data curator

[2] Typically, attributes are non-binary. Any attribute with m possible values can be decomposed into $\log(m)$ binary attributes.

releases the projection of the contingency table onto each such subset, *i.e.*, the counts for each of the possible settings of the restricted set of attributes. These smaller tables of counts are called marginals, each marginal being named by a subset of the attributes. A marginal named by a set of j attributes, $j \leq k$, is called a *j-way* marginal. The data curator will typically release many sets of low-order marginals for a single contingency table, with the goal of revealing correlations between many different, and possibly overlapping, sets of attributes.

Since a contingency table is a histogram, we can add independently generated noise proportional to ϵ^{-1} to each cell of the contingency table to obtain an ϵ-differentially private (non-integer and not necessarily non-negative) table. We will address the question of integrality and non-negativity later. For now, we simply note that any desired set of marginals can be computed directly from this noisy table, and consistency among the different marginals is immediate. A drawback of this approach, however, is that while the noise in each cell of the contingency table is relatively small, the noise in the computed marginals may be large. For example, the variance in the 1-way table describing attribute a_1 is $2^{k-1}\epsilon^{-2}$. We consider this unacceptable, especially when $n \ll 2^k$.

Marginals are also histograms. A second approach, with much less noise in the (common) case of low-order marginals, but not offering consistency between marginals, works as follows. Let C be the set of marginals to be released. We can think of a function f that, when applied to the database, yields the desired marginals. Now apply Theorem 1 with this choice of f, (adding noise to each cell in the collection of tables independently), with sensitivity $\Delta f = |C|$. When n (the number of rows in the database) is large compared to $|C|/\epsilon$, this also yields excellent accuracy. Thus we would be done if the small table-to-table inconsistencies caused by independent randomization of each (cell in each) table are not of concern, and if the user is comfortable with occasionally negative and typically non-integer cell counts.

We have no philosophical or mathematical objection to these artifacts – inconsistencies, negativity, and non-integrality – of the privacy-enhancing technology, but in practice they can be problematic. For example, the cell counts may be used as input to other, possibly off-the-shelf, programs that anticipate positive integers, giving rise to type mismatch. Inconsistencies, not to mention negative values, may also be confusing to lay users, such as casual users of the American FactFinder website.

We now outline the main steps in the work of Barak *et al* [5].

Move to the Fourier Domain. When adding noise, two natural solutions present themselves: adding noise to entries of the source table (this was our first proposal; accuracy is poor when k is large), or adding noise to the reported marginals (our second proposal; consistency is violated). A third approach begins by transforming the data into the Fourier domain. This is just a change of basis. Were we to compute all 2^k Fourier coefficients we would have a non-redundant encoding of the entire consistency table. If we were to perturb the Fourier coefficients and then convert back to the contingency table domain, we would get a (different, possibly non-integer, possibly negative) contingency table, whose "distance" (for example,

ℓ_2 distance) from the original is determined by the magnitude of the perturbations. The advantage of moving to the Fourier domain is that if only a set C of marginals is desired then we do not need the full complement of Fourier coefficients. For example, if C is the set of all 3-way marginals, then we need only the Fourier coefficients of weight at most 3, of which there are $\binom{k}{3} + \binom{k}{2} + k + 1$. This will translate into a much less noisy set of marginals.

The Fourier coefficients needed to compute the marginals C form a model of the dataset that captures everything that can be learned from the set C of marginals. Adding noise to these coefficients as indicated by Theorem 1 and then converting back to the contingency table domain yields a procedure for generating *synthetic datasets* that ensures differential privacy and yet to a great (and measurable) extent captures the information in the model. This is an example of a concrete method for generating synthetic data with provable differential privacy.

The Fourier coefficients exactly describe the information required by the marginals. By measuring exactly what is needed, Barak *et al.* add the least amount of noise possible using the techniques of [19]. Moreover, the Fourier basis is particularly attractive because of the natural decomposition according to sets of attribute values. Even tighter bounds than those in Theorem 4 below can be placed on sub-marginals (that is, lower order marginals) of a given marginal, by noting that no additional Fourier coefficients are required and fewer noisy coefficients are used in computing the low-order marginal, improving accuracy by reducing variance.

Use Linear Programming and Rounding. Barak *et al.* [5] employ linear programming to obtain a non-negative, but likely non-integer, data set with (almost) the given Fourier coefficients, and then round the results to obtain an integer solution. Interestingly, the marginals obtained from the linear program are no "farther" (made precise in [5]) from those of the noisy measurements than are the true marginals of the raw data. Consequently, the additional error introduced by the imposition of consistency is no more than the error introduced by the privacy mechanism itself.

Notation and Preliminaries. Recall that, letting k denote the number of (binary) attributes, we can think of the data set as a vector $x \in \mathbb{R}^{2^k}$, indexed by attribute tuples. For each $\alpha \in \{0,1\}^k$ the quantity x_α is the number of data elements with this setting of attributes. We let $n = \|x\|_1$ be the total number of tuples, or rows, in the data set.

For any $\alpha \in \{0,1\}^k$, we use $\|\alpha\|_1$ for the number of non-zero locations. We write $\beta \preceq \alpha$ for $\alpha, \beta \in \{0,1\}^k$ if every zero location in α is also a zero in β.

The Marginal Operator. Barak *et al.* describe the computation of a set of marginals as the result of applying a *marginal operator* to the contingency table vector x. The operator $C^\alpha : \mathbb{R}^{2^k} \to \mathbb{R}^{2^{\|\alpha\|_1}}$ for $\alpha \in \{0,1\}^k$ maps contingency tables to the marginal of the attributes that are positively set in α (there are $2^{\|\alpha\|_1}$ possible settings of these attributes). Abusing notation, $C^\alpha x$ is only defined at those locations β for which $\beta \preceq \alpha$: for any $\beta \preceq \alpha$, the outcome of $C^\alpha x$ at position β is the sum over those coordinates of x that agree with β on the coordinates described by α:

$$(C^\alpha(x))_\beta = \sum_{\gamma:\gamma\wedge\alpha=\beta} x_\gamma \tag{3}$$

Notice that the operator C^α is linear for all α.

Theorem 2. *The f^α form an orthonormal basis for $\mathbb{R}^{2^k}$.*

Consequently, one can write any marginal as the small summation over relevant Fourier coefficients:

$$C^\beta x = \sum_{\alpha\preceq\beta} \langle f^\alpha, x\rangle C^\beta f^\alpha \ . \tag{4}$$

The coefficients $\langle f^\alpha, x\rangle$ are necessary and sufficient data from x for the computation of $C^\beta x$.

Theorem 3 ([5]). *Let $B \subseteq \{0,1\}^k$ describe a set of Fourier basis vectors. Releasing the set $\phi_\beta = \langle f^\beta, x\rangle + Lap(|B|/\epsilon 2^{k/2})$ for $\beta \in B$ preserves ϵ-differential privacy.*

Proof: Each tuple contributes exactly $\pm 1/2^{k/2}$ to each output coordinate, and consequently the L_1 sensitivity of the set of $|B|$ outputs is at most $|B|/2^{k/2}$. By Theorem 1, the addition of symmetric exponential noise with standard deviation $|B|/\epsilon 2^{k/2}$ gives ϵ-differential privacy.

Remark: To get a sense of scale, we could achieve a similar perturbation to each coordinate by randomly adding or deleting $|B|^2/\epsilon$ individuals in the data set, which can be much smaller than n.

Putting the Steps Together. To compute a set A of marginals, we need all the Fourier coefficients f^β for β in the downward closure of A uner $\preceq$.

Marginals($A \subseteq \{0,1\}^k, D$):

1. Let B be the downward closure of A under $\preceq$.
2. For $\beta \in B$, compute $\phi_\beta = \langle f^\beta, D\rangle + Lap(|B|/\epsilon 2^{k/2})$.
3. Solve for w_α in the following linear program, and round to the nearest integral weights, w'_α.

$$\begin{aligned} &\text{minimize} \quad b \\ &\text{subject to:} \\ &\qquad w_\alpha \geq\ 0\ \forall\alpha \\ &\phi_\beta - \sum_\alpha w_\alpha f^\beta_\alpha \leq\ b\ \forall\beta\in B \\ &\phi_\beta - \sum_\alpha w_\alpha f^\beta_\alpha \geq -b\ \forall\beta\in B \end{aligned}$$

4. Using the contingency table w'_α, compute and return the marginals for A.

Theorem 4 ([5]). *Using the notation of* ***Marginals****(A), with probability* $1-\delta$, *for all* $\alpha \in A$,

$$\|C^{\alpha}x - C^{\alpha}w'\|_1 \leq 2^{\|\alpha\|_1} 2|B| \log(|B|/\delta)/\epsilon + |B| \,. \tag{5}$$

When k is Large. The linear program requires time polynomial in 2^k. When k is large this is not satisfactory. However, somewhat surprisingly, non-negativity (but not integrality) can be achieved by adding a relatively small amount to the first Fourier coefficient before moving back to the data domain. No linear program is required, and the error introduced is pleasantly small. Thus if polynomial in 2^k is an unbearable cost and one can live with non-integrality then this approach serves well. We remark that non-integrality was a non-issue in a pilot implementation of this work as counts were always converted to percentages.

4.3 Learning (Nearby) Halfspaces

We close this Section with an example inspired by questions in learning theory, appearing in a forthcoming paper of Blum, Ligett, and Roth [8]. The goal is to give a non-interactive solution to half-space queries. At a high level, their approach is to publish information that (approximately) answers a large set of "canonical" queries of a certain type, with the guarantee that for any (possibly non-canonical) query of the given type there is a "nearby" canonical query. Hence, the data analyst can obtain the answer to a query that in some sense is close to the query of interest.

The queries in [8] are *halfspace* queries in $\mathbb{R}^d$, defined next. Throughout this section we adopt the assumption in [8] that the database points are scaled into the unit sphere.

Definition 3. *Given a database* $D \subset \mathbb{R}^d$ *and unit length* $y \in \mathbb{R}^d$, *a halfspace query* H_y *is*

$$H_y(D) = \frac{|\{x \in D : \sum_{i=1}^{d} x_i \cdot y_i \geq 0\}|}{|D|}.$$

Note that a halfspace query can be estimated from two counting queries: "What is $|D|$?" and "What is $|\{x \in D : \sum_{i=1}^{d} x_i \cdot y_i \geq 0\}|$?" Thus, the halfspace query has sensitivity at most 2.

The *distance* between halfspace queries H_{y_1} and H_{y_2} is defined to be the sine of the angle between them, $\sin(y_1, y_2)$. With this in mind, the algorithm of Blum, Ligett, and Roth, ensures the following notion of utility:

Definition 4 ([8]). *A database mechanism* A *is* $(\epsilon, \delta, \gamma)$-useful *for queries in class* C *according to some metric* d *if with probability* $1-\delta$, *for every* $Q \in C$ *and every database* D, $|Q(A(D)) - Q'(D)| \leq \epsilon$ *for some* $Q' \in C$ *such that* $d(Q, Q') \leq \gamma$.

Note that it is the queries that are close, not (necessarily) their answers.

Given a halfspace query H_{y_1}, the algorithm below will output a value v such that $|v - H_{y_2}(D)| < \epsilon$ for some H_{y_2} that is γ-close to H_{y_1}. Equivalently, the algorithm arbitrarily counts or fails to count points $x \in D$ such that $\cos(x, y_1) \leq \gamma$. Blum *et al.* note that γ plays a role similar to the notion of margin in machine learning, and that even if H_{y_1} and H_{y_2} are γ-close, this does not imply that true answers to the queries $H_{y_1}(D)$ and $H_{y_2}(D)$ are close, unless most of the data points are outside a γ margin of H_{y_1} and H_{y_2}.

Definition 5 ([8]). *A halfspace query H_y is* b-discretized *if for each $i \in [d]$, y_i can be specified with b bits. Let C_b be the set of all b-discretized halfspaces in $\mathbb{R}^d$.*

Consider a particular $k < d$ dimensional subspace of $\mathbb{R}^d$ defined by a random $d \times k$ matrix M with entries chosen independently and uniformly from $\{-1, 1\}$. Consider the projection $P_M(x) = (1/\sqrt{k})x \cdot M$, which projects database points into the subspace and re-scales them to the unit sphere. For a halfspace query H_y, the projection $P_M(H_y)$ is simply the k-dimensional halfspace query defined by the projection $P_M(y)$. The key fact is that, for a randomly chosen M, projecting a database point x and the halfspace query specifier y is very unlikely to significantly change the angle between them:

Theorem 5 (Johnson-Lindenstrauss Theorem). *Consider a projection of a point x and a halfspace H_y onto a random k-dimensional subspace as defined by a projection matrix M. Then*

$$\Pr[|\cos(x, H_y) - \cos(P_M(x), H_{P_M(y)})| \geq \gamma/4] \leq 2e^{-((\gamma/16)^2 - (\gamma/16)^3)k/4}.$$

The dimension k of the subspace is chosen such that the probability that projecting a point and a halfspace changes the angle between them by more than $\gamma/4$ is at most $\epsilon_1/4$. This yields

$$k \geq \frac{4\ln(8/\epsilon_1)}{(\gamma/16)^2 - (\gamma/16)^3}.$$

Thus, the answer to the query H_y can be estimated by a privacy-preserving estimate of the answer to the projected halfspace query, and overall accuracy could be improved by choosing m projection matrices; the angle between x and y would be estimated by the median of the angles induces by the m resulting pairs of projections of x and y.

Of course, if the goal were to respond to a few half-space queries there would be no point in going through the projection process, let alone taking several projections. But the goal of [8] is more ambitious: an$(\epsilon, \delta, \gamma)$-*useful* non-interactive mechanism for (non-discretized) halfspace queries; this is where the lower dimensionality comes into play.

The algorithm chooses m projection matrices, where m depends on the discretization parameter b, the dimension d, and the failure probability δ (more specifically, $m \in O(\ln(1/\delta)+\ln(bd))$). For each random subspace (defined by a projection matrix M), the algorithm selects a *net* N_M of "canonical" halfspaces (defined by canonical vectors in the subspace) such that for every vector $y \in R^k$ there is a nearby canonical vector, specifically, of distance (induced sine) at most $(3/4)\gamma$.

The number of canonical vectors needed is $O(1/\gamma^{k-1})$. For each of these, the curator publishes a privacy-preserving estimate of the projected halfspace query. The mechanism is non-interactive and the curator will play no further role.

To handle an arbitrary query y, the analyst begins with an empty multiset. For each of the m projections M, the analyst finds a vector $\hat{y} \in N_M$ closest to $P_M(y)$, adding to the multiset the answer to that halfspace query. The algorithm outputs the median of these m values.

Theorem 6 ([8]). *Let*

$$n \geq \frac{\log(1/\delta) + \log m + (k-1)\log(1/\gamma) + mO(1/\gamma)^{k-1}}{\epsilon_2 \alpha}$$

Then the above algorithm is $(\epsilon, \gamma, \delta)$*-useful while maintaining* α*-differential privacy for a database of size* n. *The algorithm runs in time poly*$(\log(1/\delta), 1/\epsilon, 1/\alpha, b, d)$ *for constant* γ.

5 General Learning Theory Results

We briefly outline three general results regarding what can be learned privately in the interactive model.

We begin with a result of Blum, Dwork, McSherry and Nissim, showing that anything learnable in the statistical queries learning model can also be efficiently privately learned interactively [7]. We then move to results that ignore computational issues, showing that the exponential mechanism of McSherry and Talwar [35] can be used to

1. Privately learn anything that is PAC learnable [32]; and
2. Generate, for any class of functions C with polynomial VC dimension, a differentially private "synthetic database" that gives "good" answers to any query in C [8].

The use of the exponential mechanism in this context is due to Kasiviswanathan, Lee, Nissim, Raskhodnikova, and Smith [32].

5.1 Emulating the Statistical Query Model

The Statistical Query (SQ) model, proposed by Kearns in [10], is a framework for examining statistical algorithms executed on samples drawn independently from an underlying distribution. In this framework, an algorithm specifies predicates $f_1, \ldots, f_k$ and corresponding accuracies $\tau_1, \ldots, \tau_k$, and is returned, for $1 \leq i \leq k$, the expected fraction of samples satisfying f_i, to within additive error τ_i. Conceptually, the framework models drawing a sufficient number of samples so that the observed count of samples satisfying each f_i is a good estimate of the actual expectation.

The statistic/al queries model is most commonly used in the computational learning theory community, where the goal is typically to learn a (in this case, Boolean) *concept*, that is, a predicate on the data, to within a certain degree of accuracy. Formally, an algorithm δ-learns a concept c if it produces a predicate such that the probability of misclassification under the latent distribution is at most $1 - \delta$.

Blum, Dwork, McSherry, and Nissim have shown that any concept that is learnable in the statistical query model is privately learnable using the equivalent algorithm on a so-called "SuLQ" database [7]; the proof is not at all complicated. Reformulting these results using the slightly different technology presented in the current survey the result is even easier to argue.

Assume we have an algorithm in the SQ model that makes the k statistical queries $f_1, \ldots, f_k$, and let $\tau = \min\{\tau_1, \ldots, \tau_k\}$ be the minimum required tolerance in the SQ algorithm. Assume that n, the size of the database, is known. In this case, we can compute the answer to f_i by asking the predicate/counting query corresponding to f_i, call it p_i, and dividing the result by n. Thus, we are dealing with a query squence of sensitivity at most k.

Let $b = k/\epsilon$. Write $\tau = \rho/n$, for ρ to be determined later, so that if the noise added to the true answer when the counting query is p_i has magnitude bounded by ρ, the response to the statistical query is within tolerance τ.

We want to find ρ so that the probability that a response has noise magnitude at least ρ is bounded by δ/k, when the noise is generated according to $\mathrm{Lap}(k/\epsilon)$. The Laplace distribution is symmetric, so it is enough to find $x < 0$ such that the cumulative distribution function at x is bounded by $\delta/2k$:

$$\frac{1}{2}e^{-|x|/b} < \frac{\delta}{2k}$$

By a straightforward calculation, this is true provided $|x| > b\ln(k/\delta)$, ie, when $|x| > (k/\epsilon)\ln(k/\delta)$. We therefore set $\rho > (k/\epsilon)\ln(k/\delta)$. So long as $\rho/n < \tau$, or, more to the point, so long as $n > \rho/\tau$, we can emulate the SQ algorithm.

This analysis only takes into account noise introduced by $\mathcal{K}$; that is, it assumes $\frac{1}{n}\sum_{i=1}^{n} f_j(d_i) = \Pr_{x \in_R \mathcal{D}}[f_j(x)]$, $1 \leq j \leq k$, where $\mathcal{D}$ is the distribution on examples. The results above apply, *mutatis mutandis*, when we assume that the rows in the database are drawn iid according to $\mathcal{D}$ using the well known fact that taking $n > \tau^{-2}\log(k/\delta)$ is sufficient to ensure tolerance τ with probability at least $1 - \delta$ for all f_j, $1 \leq j \leq k$, simultaneously. Replacing τ by $\tau/2$ everywhere and finding the maximum lower bound on n handles both types of error.

5.2 Private PAC Learning

The results in this section are due to Kasiviswanathan, Lee, Nissim, Raskhodnikova, and Smith [32]. We begin with an informal and incomplete review of the concept of probably-appproximately-correct (PAC) learning, a notion due to Valiant [47].

Consider a concept $t : X \longrightarrow Y$ that assigns to each *example* taken from the domain X a *label* from the range Y. As in the previous section, a learning algorithm is given labeled examples drawn from a distribution $\mathcal{D}$, labeled by a target concept; the goal is to produce an *hypothesis* $h : X \longrightarrow Y$ from a specified hypothesis class, with small *error*, defined by:

$$\text{error}(h) = \Pr_{x \in_R \mathcal{D}}[t(x) \neq h(x)].$$

A *concept class* is a set of concepts. Learning theory asks what kinds of concept classes are learnable. Letting α, β denote two error bounds, if the target concept belongs to C, then the goal is to minimize error, or at least to ensure that with probability $1 - \beta$ the error is bounded by α. This is the setting of traditional PAC learning. If the target concept need not belong to C, the goal is to produce an hypothesis that, intuitively, does almost as well as any concept in C: Letting $\text{OPT} = \min_{c \in C}\{\text{error}(c)\}$, we want

$$\Pr[\text{error}(h) \leq \text{OPT} + \alpha] \geq 1 - \beta,$$

where the probability is over the samples seen by the learner, and the learner's randomness. This is known as *agnostic learning.*[3]

Following [32] we will index concept classes, domains, and ranges by the length d of their binary encodings. For a target concept $t : X_d \to Y_d$ and a distribution $\mathcal{D}$ over X_d, let $Z \in D^n$ be a database containing n labeled independent draws from $\mathcal{D}$. That is, Z contains n pairs (x_i, y_i) where $y_i = t(x_i)$, $1 \leq i \leq n$. The goal of the data analyst will be to agnostically learn a hypothesis class C; the goal of the curator will be to ensure ϵ-differential privacy.

Theorem 7 ([32]). *Every concept class C is ϵ-differentially pivately agnostically learnable using hypothesis class $\mathcal{H} = C$ with $n \in O(\log C_d + \log(1\beta) \cdot \max\{\frac{1}{\epsilon\alpha}, \frac{1}{\alpha^2}\})$. The learner may not be efficient.*

The theorem is proved using the exponential mechanism of McSherry and Talwar, with utility function

$$u(Z, h) = -|\{i : t(x_i) \neq h(x_i)\}|$$

for $Z \in (X \times Y)^n, h \in \mathcal{H}_d$. Note that u has sensitivity 1, since changing any element in the database can change the number of misclassifications by at most 1. The (inefficient) algorithm outputs $c \in \mathcal{H}_d$ with probability proportional to $\exp(\epsilon u(Z, c)/2)$. Privacy follows from the properties of the exponential mechanism and the small sensitivity of u. Accuracy (low error with high probablity) is slightly more difficult to argue; the proof follows, intuitively, from the fact that outputs are produced with probability that falls exponentially in the number of misclassifications.

5.3 Differentially Private Queries of Classes with Polynomial VC Dimension

Our last example is due to Blum, Ligett, and Roth [8] and was inspired by the result of the previous section, This learning result again ignores computational

[3] The standard agnostic model has the input drawn from an arbitrary distribution over labeled examples (x, y) (that is, the label need not be a deterministic function of the example). The error of a hypothesis is defined with respect to the distribution (i.e. probability that $y \neq h(x)$). The results (and proofs) of Kasiviswanathan *et al.* stay the same in this more general setting.

efficiency, using the exponential mechanism of McSherry and Talwar. The object here is to release a "synthetic dataset" that ensures "reasonably" accurate answers to *all* queries in a specific class C. The reader is assumed to be familiar with the Vapnick-Chervonenkis (VC) dimension of a class of concepts. Roughly speaking, it is a measure of how complicated a concept in the class can be.

Definition 6 ([8]). *A database mechanism A is (γ, δ)-*useful *for queries in class C if with probability $1-\delta$, for every $Q \in C$ and every database D, for $\widehat{D} = A(D)$, $|Q(\widehat{D}) - Q(D)| \le \gamma$.*

Let C be a fixed class of queries. Given a database $D \in (\{0,1\}^d)^n$ of size n, where n is sufficiently large (as a function of the VC dimension of C, as well as of ϵ and δ), the goal is to produce produce a synthetic dataset $\hat{D}$ that is (γ, δ)-useful for queries in C, while ensuring ϵ-differential privacy.

The synthetic dataset will contain $m = O(\mathrm{VC}\ \dim(C)/\gamma^2)$ d-tuples. It is chosen according to the exponential mechanism using the utility function

$$u(D, \hat{D}) = -\max_{h \in C} \left| h(D) - \frac{n}{m} h(\hat{D}) \right| .$$

Theorem 8 ([8]). *For any class of functions C, and any database $D \subset \{0,1\}^d$ such that*

$$|D| \ge O\left(\frac{d \cdot \mathrm{VC}\ \dim(C)}{\gamma^3 \epsilon} + \frac{\log(1/\delta)}{\epsilon\gamma}\right)$$

we can output an (γ, δ)-useful database $\widehat{D}$ that preserves ϵ-differential privacy. Note that the algorithm is not necessarily efficient.

Blum *et al.* note that this suffice for (γ, δ)-usefulness because the set of all databases of this size forms a γ-cover with respect to C of the set of all possible databases. One can resolve the fact that, since $|\hat{D}| < |D|$, the number of database entries matching any query will be proportionately smaller by considering the fraction of entries matching any query.

6 Concluding Remarks

The privacy mechanisms discussed herein add an amount of noise that grows with the complexity of the query sequence applied to the database. Although this can be ameliorated to some extent using Gaussian noise instead of Laplacian, an exciting line of research begun by Dinur and Nissim [13] (see also [17, 20]) shows that this increase is essential. To a great extent, the results of Dinur and Nissim drove the development of the mechanism $\mathcal{K}$ and the entire interactive approach advocated in this survey. A finer analysis of *realistic* attacks, and a better understanding of what failure to provide ϵ-differential privacy can mean in practice, are needed in order to sharpen these results – or to determine this is impossible, in order to understand the how to use these techniques for all but very large, "internet scale," data sets.

Acknowledgements. The author thanks Avrim Blum, Katrina Ligett, Aaron Roth and Adam Smith for helpful technical discussions, and Adam Smith for excellent comments on an early draft of this survey. The author is grateful to all the authors of [32, 8] for sharing with her preliminary versions of their papers.

References

[1] Achugbue, J.O., Chin, F.Y.: The Effectiveness of Output Modification by Rounding for Protection of Statistical Databases. INFOR 17(3), 209–218 (1979)

[2] Adam, N.R., Wortmann, J.C.: Security-Control Methods for Statistical Databases: A Comparative Study. ACM Computing Surveys 21(4), 515–556 (1989)

[3] Agrawal, D., Aggarwal, C.: On the Design and Quantification of Privacy Preserving Data Mining Algorithms. In: Proceedings of the 20th Symposium on Principles of Database Systems (2001)

[4] Agrawal, R., Srikant, R.: Privacy-Preserving Data Mining. In: Proceedings of the ACM SIGMOD Conference on Management of Data, pp. 439–450 (2000)

[5] Barak, B., Chaudhuri, K., Dwork, C., Kale, S., McSherry, F., Talwar, K.: Privacy, Accuracy, and Consistency Too: A Holistic Solution to Contingency Table Release. In: Proceedings of the 26th Symposium on Principles of Database Systems, pp. 273–282 (2007)

[6] Beck, L.L.: A Security Mechanism for Statistical Databases. ACM TODS 5(3), 316–338 (1980)

[7] Blum, A., Dwork, C., McSherry, F., Nissim, K.: Practical Privacy: The SuLQ framework. In: Proceedings of the 24th ACM SIGMOD-SIGACT-SIGART Symposium on Principles of Database Systems (June 2005)

[8] Blum, A., Ligett, K., Roth, A.: A Learning Theory Approach to Non-Interactive Database Privacy. In: Proceedings of the 40th ACM SIGACT Symposium on Thoery of Computing (2008)

[9] Chawla, S., Dwork, C., McSherry, F., Smith, A., Wee, H.: Toward Privacy in Public Databases. In: Proceedings of the 2nd Theory of Cryptography Conference (2005)

[10] Dalenius, T.: Towards a methodology for statistical disclosure control. Statistik Tidskrift 15, 222–429 (1977)

[11] Denning, D.E.: Secure Statistical Databases with Random Sample Queries. ACM Transactions on Database Systems 5(3), 291–315 (1980)

[12] Denning, D., Denning, P., Schwartz, M.: The Tracker: A Threat to Statistical Database Security. ACM Transactions on Database Systems 4(1), 76–96 (1979)

[13] Dinur, I., Nissim, K.: Revealing Information While Preserving Privacy. In: Proceedings of the Twenty-Second ACM SIGACT-SIGMOD-SIGART Symposium on Principles of Database Systems, pp. 202–210 (2003)

[14] Duncan, G.: Confidentiality and statistical disclosure limitation. In: Smelser, N., Baltes, P. (eds.) International Encyclopedia of the Social and Behavioral Sciences, Elsevier, New York (2001)

[15] Dwork, C.: Differential Privacy. In: Bugliesi, M., Preneel, B., Sassone, V., Wegener, I. (eds.) ICALP 2006. LNCS, vol. 4052, pp. 1–12. Springer, Heidelberg (2006)

[16] Dwork, C., Kenthapadi, K., McSherry, F., Mironov, I., Naor, M.: Our Data, Ourselves: Privacy Via Distributed Noise Generation. In: Vaudenay, S. (ed.) EUROCRYPT 2006. LNCS, vol. 4004, pp. 486–503. Springer, Heidelberg (2006)

[17] Dwork, C., McSherry, F., Talwar, K.: The Price of Privacy and the Limits of LP Decoding. In: Proceedings of the 39th ACM Symposium on Theory of Computing, pp. 85–94 (2007)
[18] Dwork, C., Nissim, K.: Privacy-Preserving Datamining on Vertically Partitioned Databases. In: Franklin, M. (ed.) CRYPTO 2004. LNCS, vol. 3152, pp. 528–544. Springer, Heidelberg (2004)
[19] Dwork, C., McSherry, F., Nissim, K., Smith, A.: Calibrating Noise to Sensitivity in Private Data Analysis. In: Proceedings of the 3rd Theory of Cryptography Conference, pp. 265–284 (2006)
[20] Dwork, C., Yekhanin, S.: New Efficient Attacks on Statistical Disclosure Control Mechanisms (manuscript, 2008)
[21] Evfimievski, A.V., Gehrke, J., Srikant, R.: Limiting Privacy Breaches in Privacy Preserving Data Mining. In: Proceedings of the Twenty-Second ACM SIGACT-SIGMOD-SIGART Symposium on Principles of Database Systems, pp. 211–222 (2003)
[22] Agrawal, D., Aggarwal, C.C.: On the design and Quantification of Privacy Preserving Data Mining Algorithms. In: Proceedings of the 20th Symposium on Principles of Database Systems, pp. 247–255 (2001)
[23] Agrawal, R., Srikant, R.: Privacy-Preserving Data Mining. In: Proceedings of the ACM SIGMOD International Conference on Management of Data, pp. 439–450 (2000)
[24] Chin, F.Y., Ozsoyoglu, G.: Auditing and infrence control in statistical databases. IEEE Trans. Softw. Eng. SE-8(6), 113–139 (1982)
[25] Dobkin, D., Jones, A., Lipton, R.: Secure Databases: Protection Against User Influence. ACM TODS 4(1), 97–106 (1979)
[26] Fellegi, I.: On the question of statistical confidentiality. Journal of the American Statistical Association 67, 7–18 (1972)
[27] Fienberg, S.: Confidentiality and Data Protection Through Disclosure Limitation: Evolving Principles and Technical Advances. In: IAOS Conference on Statistics, Development and Human Rights (September 2000), http://www.statistik.admin.ch/about/international/fienberg_final_paper.doc
[28] Fienberg, S., Makov, U., Steele, R.: Disclosure Limitation and Related Methods for Categorical Data. Journal of Official Statistics 14, 485–502 (1998)
[29] Franconi, L., Merola, G.: Implementing Statistical Disclosure Control for Aggregated Data Released Via Remote Access. In: United Nations Statistical Commission and European Commission, joint ECE/EUROSTAT work session on statistical data confidentiality, Working Paper No.30 (April 2003), http://www.unece.org/stats/documents/2003/04/confidentiality/wp.30.e.pdf
[30] Goldwasser, S., Micali, S.: Probabilistic Encryption. J. Comput. Syst. Sci. 28(2), 270–299 (1984)
[31] Gusfield, D.: A Graph Theoretic Approach to Statistical Data Security. SIAM J. Comput. 17(3), 552–571 (1988)
[32] Kasiviswanathan, S., Lee, H., Nissim, K., Raskhodnikova, S., Smith, S.: What Can We Learn Privately? (manuscript, 2007)
[33] Lefons, E., Silvestri, A., Tangorra, F.: An analytic approach to statistical databases. In: 9th Int. Conf. Very Large Data Bases, October-November 1983, pp. 260–274. Morgan Kaufmann, San Francisco (1983)
[34] Machanavajjhala, A., Gehrke, J., Kifer, D., Venkitasubramaniam, M.: l-Diversity: Privacy Beyond k-Anonymity. In: Proceedings of the 22nd International Conference on Data Engineering (ICDE 2006), p. 24 (2006)

[35] McSherry, F., Talwar, K.: Mechanism Design via Differential Privacy. In: Proceedings of the 48th Annual Symposium on Foundations of Computer Science (2007)

[36] Narayanan, A., Shmatikov, V.: How to Break Anonymity of the Netflix Prize Dataset, http://www.cs.utexas.edu/~shmat/shmat_netflix-prelim.pdf

[37] Nissim, K., Raskhodnikova, S., Smith, A.: Smooth Sensitivity and Sampling in Private Data Analysis. In: Proceedings of the 39th ACM Symposium on Theory of Computing, pp. 75–84 (2007)

[38] Raghunathan, T.E., Reiter, J.P., Rubin, D.B.: Multiple Imputation for Statistical Disclosure Limitation. Journal of Official Statistics 19(1), 1–16 (2003)

[39] Reiss, S.: Practical Data Swapping: The First Steps. ACM Transactions on Database Systems 9(1), 20–37 (1984)

[40] Rubin, D.B.: Discussion: Statistical Disclosure Limitation. Journal of Official Statistics 9(2), 461–469 (1993)

[41] Shoshani, A.: Statistical databases: Characteristics, problems and some solutions. In: Proceedings of the 8th International Conference on Very Large Data Bases (VLDB 1982), pp. 208–222 (1982)

[42] Samarati, P., Sweeney, L.: Protecting Privacy when Disclosing Information: k-Anonymity and its Enforcement Through Generalization and Specialization, Technical Report SRI-CSL-98-04, SRI Intl. (1998)

[43] Samarati, P., Sweeney, L.: Generalizing Data to Provide Anonymity when Disclosing Information (Abstract). In: Proceedings of the Seventeenth ACM SIGACT-SIGMOD-SIGART Symposium on Principles of Database Systems, p. 188 (1998)

[44] Sweeney, L.: Weaving Technology and Policy Together to Maintain Confidentiality. J. Law Med Ethics 25(2-3), 98–110 (1997)

[45] Sweeney, L.: k-anonymity: A Model for Protecting Privacy. International Journal on Uncertainty, Fuzziness and Knowledge-based Systems 10(5), 557–570 (2002)

[46] Sweeney, L.: Achieving k-Anonymity Privacy Protection Using Generalization and Suppression. International Journal on Uncertainty, Fuzziness and Knowledge-based Systems 10(5), 571–588 (2002)

[47] Valiant, L.G.: A Theory of the Learnable. In: Proceedings of the 16th Annual ACM SIGACT Symposium on Theory of computing, pp. 436–445 (1984)

[48] Xiao, X., Tao, Y.: M-invariance: towards privacy preserving re-publication of dynamic datasets. In: SIGMOD 2007, pp. 689–700 (2007)

On the Complexity of Measurement in Classical Physics

Edwin Beggs[1], José Félix Costa[2,3,⋆], Bruno Loff[2,3], and John Tucker[1]

[1] Department of Mathematics and Department of Computer Science
University of Wales Swansea, Singleton Park, Swansea, SA3 2HN
Wales, United Kingdom
E.J.Beggs@Swansea.ac.uk, J.V.Tucker@swansea.ac.uk

[2] Department of Mathematics, Instituto Superior Técnico
Universidade Técnica de Lisboa
Lisboa, Portugal
fgc@math.ist.utl.pt, bruno.loff@gmail.com

[3] Centro de Matemática e Aplicações Fundamentais do Complexo Interdisciplinar
Universidade de Lisboa
Lisbon, Portugal

Abstract. If we measure the position of a point particle, then we will come about with an interval $[a_n, b_n]$ into which the point falls. We make use of a *Gedankenexperiment* to find better and better values of a_n and b_n, by reducing their relative distance, in a succession of intervals $[a_1, b_1] \supset [a_2, b_2] \supset \ldots \supset [a_n, b_n]$ that contain the point. We then use such a point as an oracle to perform relative computation in polynomial time, by considering the succession of approximations to the point as suitable answers to the queries in an oracle Turing machine. We prove that, no matter the precision achieved in such a *Gedankenexperiment*, within the limits studied, the Turing Machine, equipped with such an oracle, will be able to compute above the classical Turing limit for the polynomial time resource, either generating the class $P/poly$ either generating the class $BPP//log*$, if we allow for an arbitrary precision in measurement or just a limited precision, respectively. We think that this result is astonishingly interesting for Classical Physics and its connection to the Theory of Computation, namely for the implications on the nature of space and the perception of space in Classical Physics. (Some proofs are provided, to give the flavor of the subject. Missing proofs can be found in a detailed long report at the address `http://fgc.math.ist.utl.pt/papers/sm.pdf`.)

1 Introducing a *Gedankenexperiment*

If a physical experiment were to be coupled with algorithms, would new functions and relations become computable or, at least, computable more efficiently?

⋆ Corresponding author. The authors are listed in alphabetic order.

M. Agrawal et al. (Eds.): TAMC 2008, LNCS 4978, pp. 20–30, 2008.

To pursue this question, we imagine using an experiment as an oracle to a Turing machine, which on being presented with, say, x_i as its i-th query, returns y_i to the Turing machine. In this paper we will consider this idea of coupling experiments and Turing machines in detail. The first thing to note is that choosing a physical experiment to use as an oracle is a major undertaking. The experiment comes with plenty of theoretical baggage: concepts of equipment, experimental procedure, instruments, measurement, observable behaviour, etc.

In earlier work [5], an experiment was devised to measure the position of the vertex of a wedge to arbitrary accuracy, by scattering particles that obey some laws of elementary Newtonian kinematics. Let *SME* denote this *Scatter Machine Experiment.* The Newtonian theory was specified precisely and the *SME* was put under a theoretical microscope: theorems were proved that showed that *the experiment was able to compute positions that were not computable by algorithms.* Indeed, the *SME* could, in principle, measure *any* real number. Thus, [5] contains a careful attempt to answer the question above, in the positive; it does so using a methodology developed and applied in earlier studies [3,4]. To address the question, here we propose to use the *SME* as an oracle to a Turing machine and to classify the computational power of the new type of machine, which we call an *analogue-digital scatter machine.* Given the results in [5], we expect that the use of *SME* will enhance the computational power and efficiency of the Turing machine.

To accomplish this, we must establish some principles that do not depend upon the *SME.* In a Turing machine, the oracle is normally specified very abstractly by a set. Here we have to design a new machine where the oracle is replaced by a specification of some physical equipment and a procedure for operating it. The design of the new machine depends heavily upon the interface and interaction between the experiment and the Turing machine.

Following some insights provided by the work of Hava T. Siegelmann and Eduardo Sontag [6], we use non uniform complexity classes of the form $\mathcal{B}/\mathcal{F}$, where $\mathcal{B}$ is the class of computations and $\mathcal{F}$ is the advice class. Context and proofs are, however, different. Examples of interest for $\mathcal{B}$ are P and BPP; examples for $\mathcal{F}$ are $poly$ and log. The power of the machines correspond with the choice of different $\mathcal{B}/\mathcal{F}$.

2 The Scatter Machine

Experiments with scatter machines are conducted exactly as described in [5], but, for convenience and to use them as oracles, we need to review and clarify some points. The *scatter machine experiment (SME)* is defined within the Newtonian mechanics, comprising of the following laws and assumptions: (a) point particles obey Newton's laws of motion in the two dimensional plane, (b) straight line barriers have perfectly elastic reflection of particles, i.e., kinetic energy is conserved exactly in collisions, (c) barriers are completely rigid and do not deform on impact, (d) cannons, which can be moved in position, can project

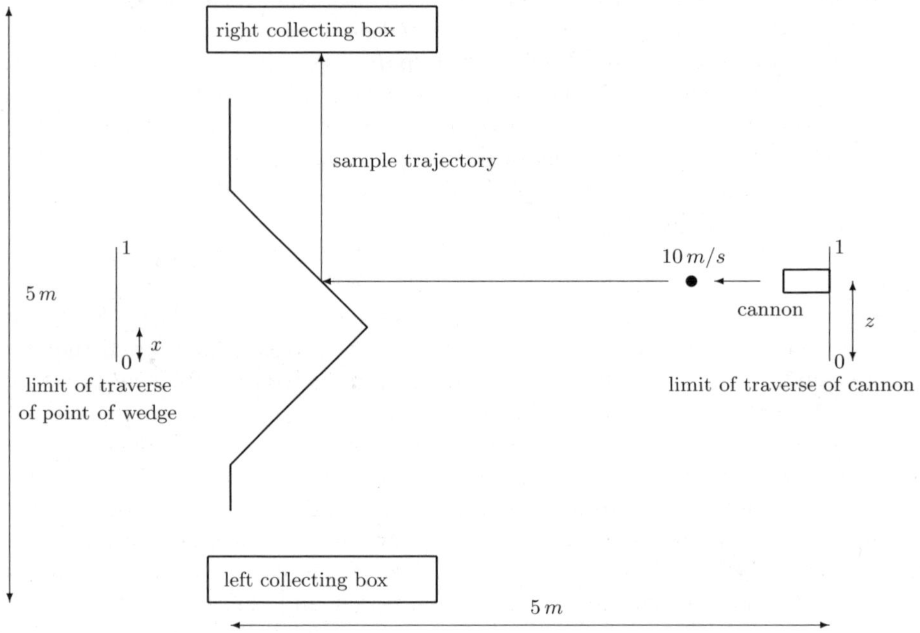

Fig. 1. A schematic drawing of the scatter machine

a particle with a given velocity in a given direction, (e) particle detectors are capable of telling if a particle has crossed a given region of the plane, and (f) a clock measures time.

The machine consists of a cannon for projecting a point particle, a reflecting barrier in the shape of a wedge and two collecting boxes, as in Figure 1.

The wedge, our motionless point particle, can be at *any* position. But we will assume it is fixed for the duration of all the experimental work. Under the control of a Turing machine, the cannon will be moved and fired repeatedly to find information about the position of the wedge. Specifically, the way the *SME* is used as an oracle in Turing machine computations, is this: a Turing machine will set a position for the canon as a query and will receive an observation about the result of firing the cannon as a response. For each input to the Turing machine, there will be finitely many runs of the experiment.

In Figure 1 the parts of the machine are shown in bold lines, with description and comments in narrow lines. The double headed arrows give dimensions in meters, and the single headed arrows show a sample trajectory of the particle after being fired by the cannon. The sides of the wedge are at 45° to the line of the cannon, and we take the collision to be perfectly elastic, so the particle is deflected at 90° to the line of the cannon, and hits either the left or right collecting box, depending on whether the cannon is to the left or right of the point of the wedge. Since the initial velocity is 10 m/s, the particle will enter one of the two boxes within 1 second of being fired. Any initial velocity $v > 0$

will work with a corresponding waiting time. The wedge is sufficiently wide so that the particle can only hit the 45° slopping sides, given the limit of traverse of the cannon. The wedge is sufficiently rigid so that the particle cannot move the wedge from its position. We make the further assumption, without loss of generality (see the report mentioned in the abstract) that the vertex of the wedge is *not* a dyadic rational.

Suppose that x is the arbitrarily chosen, but non dyadic and fixed, position of the point of the wedge. For a given cannon position z, there are two outcomes of an experiment: (a) one second after firing, the particle is in the right box — conclusion: $z > x$ —, or (b) one second after firing, the particle is in the left box — conclusion: $z < x$. The *SME* was designed to find x to arbitrary accuracy by altering z, so in our machine $0 \leq x \leq 1$ will be fixed, and we will perform observations at different values of $0 \leq z \leq 1$.

Consider the precision of the experiment. When measuring the output state the situation is simple: either the ball is in one tray or in the other tray. Errors in observation do not arise. Now consider some of the non-trivial ways in which precision depends on the positions of the cannon. There are different postulates for the precision of the cannon, and we list some in order of decreasing strength:

Definition 2.1. *The* SME *is error-free if the cannon can be set exactly to any given dyadic rational number. The* SME *is error-prone with arbitrary precision if the cannon can be set only to within a non-zero, but arbitrarily small, dyadic precision. The* SME *is error-prone with fixed precision if there is a value* $\varepsilon > 0$ *such that the cannon can be set only to within a given precision* ε.

The Turing machine is connected to the *SME* in the same way as it would be connected to an oracle: we replace the query state with a *shooting state* (q_s), the "yes" state with a *left state* (q_l), and the "no" state with a *right state* (q_r). The resulting computational device is called the *analog-digital scatter machine*, and we refer to the *vertex position* of an analog-digital scatter machine when mean to discuss the vertex position of the corresponding *SME*.

In order to carry out a scatter machine experiment, the analog-digital scatter machine will write a word z in the query tape and enter the shooting state. This word will either be "1", or a binary word beginning with 0. We will use z indifferently to denote both a word $z_1 \ldots z_n \in \{1\} \cup \{0s : s \in \{0,1\}^*\}$ and the corresponding dyadic rational $\sum_{i=1}^{n} 2^{-i+1} z_i \in [0,1]$. In this case, we write $|z|$ to denote n, i.e., the size of $z_1 \ldots z_n$, and say that the analog-digital scatter machine is *aiming* at z. The Turing machine computation will then be interrupted, and the *SME* will attempt to set the cannon at z. The place where the cannon is actually set at depends on whether the *SME* is error-free or error-prone. If the *SME* is error-free, the cannon will be placed exactly at z. If the *SME* is error-prone with arbitrary precision, then the cannon will be placed at some point in the interval $[z - 2^{-|z|-1}, z + 2^{-|z|-1}]$ *with a uniform probability distribution over this interval.* This means that for different words representing the same dyadic rational, the longest word will give the highest precision. If the *SME* is error-prone with fixed dyadic precision ε, then the cannon will be placed somewhere in the interval $[z - \varepsilon, z + \varepsilon]$, again with a uniform probability distribution.

After setting the cannon, the *SME* will fire a projectile particle, wait one second and then check if the particle is in either box. If the particle is in the right collecting box, then the Turing machine computation will be resumed in the state q_r. If the particle is in left box, then the Turing machine computation will be resumed in the state q_l. With this behaviour, we obtain three distinct analog-digital scatter machines.

Definition 2.2. *An* error-free analog-digital scatter machine *is a Turing machine connected to an error-free* SME. *In a similar way, we define an* error-prone analog-digital scatter machine with arbitrary precision, *and an* error-prone analog-digital scatter machine with fixed precision.

The error-free analog-digital scatter machine has a very simple behaviour. If such a machine, with vertex position $x \in [0, 1]$, aims at a dyadic rational $z \in [0, 1]$, we are certain that the computation will be resumed in the state q_l if $z < x$, and that it will be resumed in the state q_r when $z > x$. We define the following decision criterion.

Definition 2.3. *Let $A \subseteq \Sigma^*$ be a set of words over Σ. We say that an error-free analog-digital scatter machine $\mathcal{M}$* decides *A if, for every input $w \in \Sigma^*$, w is accepted if $w \in A$ and rejected when $w \notin A$. We say that $\mathcal{M}$* decides A in polynomial time, *if $\mathcal{M}$ decides A, and there is a polynomial p such that, for every $w \in \Sigma^*$, the number of steps of the computation is bounded by $p(|w|)$.*

The error-prone analog-digital scatter machines, however, do not behave in a deterministic way. If such a machine aims the cannon close enough to the vertex position, and with a large enough error, there will be a positive probability for both the particle going left or right. If the vertex position is x, and the error-prone analog-digital scatter machine $\mathcal{M}$ aims the cannon at z, then the probability of the particle going to the left box, denoted by $\mathbb{P}(\mathcal{M} \leftarrow \textit{left})$, is given by:

$$\mathbb{P}(\mathcal{M} \leftarrow \textit{left}) = \begin{cases} 1 & \text{if } z < x - \varepsilon \\ \frac{1}{2} + \frac{x-z}{2\varepsilon} & \text{if } x - \varepsilon \leq z \leq \tilde{x} + \varepsilon \\ 0 & \text{if } z > \tilde{x} + \varepsilon \end{cases}$$

The value ε will be either $2^{-|z|-1}$ or a fixed value, depending on the type of error-prone analog-digital scatter machine under consideration. The probability of the particle going to the right box is $\mathbb{P}(\mathcal{M} \leftarrow \textit{right}) = 1 - \mathbb{P}(\mathcal{M} \leftarrow \textit{left})$. We can thus see that a deterministic decision criteria is not suitable for these machines. For a set $A \subseteq \Sigma^*$, an error-prone analog-digital scatter machine $\mathcal{M}$, and an input $w \in \Sigma^*$, let the *error probability* of $\mathcal{M}$ for input w be either the probability of $\mathcal{M}$ rejecting w, if $w \in A$, or the probability of $\mathcal{M}$ accepting w, if $w \notin A$.

Definition 2.4. *Let $A \subseteq \Sigma^*$ be a set of words over Σ. We say that an error-prone analog-digital scatter machine $\mathcal{M}$* decides *A if there is a number $\gamma < \frac{1}{2}$, such that the error probability of $\mathcal{M}$ for any input w is smaller than γ. We say* *that* $\mathcal{M}$ decides A in polynomial time, *if* $\mathcal{M}$ *decides* A*, and there is a polynomial* p *such that, for every input* $w \in \Sigma^*$*, the number of steps in every correct computation is bounded by* $p(|w|)$*.*

3 The Relevant Computational Classes

We will see, in the sections that follow, that the non-uniform complexity classes give the most adequate characterisation of the computational power of the analog-digital scatter machine. Non-uniform complexity classifies problems by studying families of finite machines (e.g., circuits) $\{\mathcal{C}_n\}_{n\in\mathbb{N}}$, where each $\mathcal{C}_n$ decides the restriction of some problem to inputs of size n. It is called *non-uniform*, because for every $n \neq m$ the finite machines $\mathcal{C}_n$ and $\mathcal{C}_m$ can be entirely unrelated, while in uniform complexity the algorithm is the same for inputs of every size. A way to connect the two approaches is by means of *advice classes*: one assumes that there is a unique algorithm for inputs of every size, which is aided by certain information, called *advice*, which may vary for inputs of different sizes. The advice is given, for each input w, by means of function $f : \mathbb{N} \to \Sigma^*$, where Σ^* is the input alphabet.

Definition 3.1. *Let* $\mathcal{B}$ *be a class of sets and* $\mathcal{F}$ *a class of functions. The advice class* $\mathcal{B}/\mathcal{F}$ *is the class of sets* A *for which some* $B \in \mathcal{B}$ *and some* $f \in \mathcal{F}$ *are such that, for every* w*,* $w \in A$ *if and only if* $\langle w, f(|w|)\rangle \in B$*.*

$\mathcal{F}$ is called the *advice class* and f is called the *advice function*. Examples for $\mathcal{B}$ are the well known classes P, or BPP (see [1, Chapter 6]). We will be considering two instances for the class $\mathcal{F}$: *poly* is the class of functions with polynomial size values, i.e., *poly* is the class of functions $f : \mathbb{N} \to \Sigma^*$ such that, for some polynomial p, $|f(n)| \in O(p(n))$; log is the class of functions $g : \mathbb{N} \to \Sigma^*$ such that $|g(n)| \in O(\log(n))$. We will also need to treat prefix non-uniform complexity classes. For these classes we may only use prefix functions, i.e., functions f such that $f(n)$ is always a prefix of $f(n+1)$. The idea behind prefix non-uniform complexity classes is that the advice given for inputs of size n may also be used to decide smaller inputs.

Definition 3.2. *Let* $\mathcal{B}$ *be a class of sets and* $\mathcal{F}$ *a class of functions. The prefix advice class* $\mathcal{B}/\mathcal{F}*$ *is the class of sets* A *for which some* $B \in \mathcal{B}$ *and some prefix function* $f \in \mathcal{F}$ *are such that, for every length* n *and input* w*, with* $|w| \leq n$*,* $w \in A$ *if and only if* $\langle w, f(n)\rangle \in B$*.*

For the non-deterministic classes we need a refined definition:

Definition 3.3. $BPP//poly$ *is the class of sets* A *for which a probabilistic polynomial Turing machine* $\mathcal{M}$*, a function* $f \in poly$*, and a constant* $\gamma < \frac{1}{2}$ *exist such that* $\mathcal{M}$ *rejects* $\langle w, f(|w|)\rangle$ *with probability at most* γ *if* $w \in A$ *and accepts* $\langle w, f(|w|)\rangle$ *with probability at most* γ *if* $w \notin A$*. A similar definition applies to the class* $BPP//\log*$*.*

It can be shown that $BPP//poly = BPP/poly$, but it is unknown whether $BPP/\log * \subseteq BPP//\log *$. It is important to notice that the usual non-uniform complexity classes contain undecidable sets, e.g., $P/poly$ contains the *halting set.*

4 Infinite Precision

FIRST GEDANKENEXPERIMENT: the cannon can be placed at some dyadic rational with infinite precision. I.e., we began by investigating the error-free machine, which gives the simplest situation.

For every $n \in \mathbb{N}$ and $x \in \mathbb{R}$, let $x\restriction_n$ be the rational number obtained by truncating x after the first n digits in its binary expansion. Then we showed the following result:

Proposition 4.1. *Let $\mathcal{M}$ be an error-free analog-digital scatter machine with the vertex placed at the position x. Let $\tilde{\mathcal{M}}$ be the same machine, but with the vertex placed at the position $\tilde{x} = x\restriction_t$. Then, for any input w, $\mathcal{M}$ and $\tilde{\mathcal{M}}$, after t steps of computation, make the same decisions.*

We may now sketch the proof of the following main theorem of this section.

Theorem 4.1. *The class of sets decided by error-free analog-digital scatter machines in polynomial time is exactly $P/poly$.*

The sketch of the proof is done by the way of polynomial advice. Let A be a set in $P/poly$, and, by definition, let $B \in P$, $f \in poly$ be such that $w \in A \iff \langle w, f(|w|)\rangle \in B$. Let $\tilde{f} : \mathbb{N} \to \Sigma^*$ be a function, also in $poly$, such that, if the symbols of $f(n)$ are $\xi_1\xi_2\ldots\xi_{p(n)}$, then $\tilde{f}(n) = 0\xi_1 0\xi_2 \ldots 0\xi_{p(n)}$ We can create an error-free analog-digital scatter machine which also decides A, setting the vertex at the position $x = 0.\tilde{f}(1)11\tilde{f}(2)11\tilde{f}(3)11\ldots$ Given any input w of size n, the error-free analog-digital scatter machine $\mathcal{S}_x$ can use the bisection method to obtain $f(n)$ in polynomial time. Then the machine uses the polynomial-time algorithm which decides B, and accepts if and only if $\langle w, f(n)\rangle$ is in B. Thus we have shown that an error-free analog-digital scatter machine can decide any set in $P/poly$ in polynomial time.

As for the converse, let C be any set decided in polynomial time by an error-free analog-digital scatter machine with the vertex at the position x. Proposition 4.1 ensures that to decide on any input w, the machine only makes use of $p(|w|)$ digits of x, where p is a polynomial. Thus we can see that the set must be in $P/poly$, using the advice function $g \in poly$, given by $g(n) = x\restriction_{p(n)}$.

We then conclude that *measuring the position of a motionless point particle in Classical Physics, using a infinite precision cannon,* [1] *in polynomial time, we are deciding a set in $P/poly$. Note that, the class $P/poly$ includes the* Halting Set. But, most probably, if we remove the *infinite precision criterion for the cannon,* we will loose accuracy of observations, and we will loose the computational power of our physical oracle...

[1] Remember that by infinite precision we mean the cannon to be settle at a rational point with infinite precision.

5 From Infinite to Unlimited Precision

SECOND GEDANKENEXPERIMENT: the cannon can be placed at some dyadic rational z up to some dyadic arbitrary precision ε, let us say $\varepsilon = 2^{-|z|-1}$. I.e., we will be investigating the error-prone machine, which gives the next simplest situation. We will conclude that the computational power of these machines is not altered by considering such a small error, since they decide exactly $BPP//poly = BPP/poly = P/poly$ in polynomial time. As in Section 4, we showed that only a number of digits of the vertex position linear in the length of computation influences the decision of such an error-prone analog-digital scatter machine. Notice that the behaviour of the error-prone analog-digital scatter machines is probabilistic, because if the machine shoots close enough to the motionless point particle position — the vertex —, and with a large enough error ε, the projectile can go both left or right with a non-zero probability. Thus, we can not ensure that when we truncate the vertex position to $O(t(n))$ digits, the state of the machine will be exactly the same after $t(n)$ steps. Instead we showed that if a machine decides in time $t(n)$, then by truncating the vertex position to $O(t(n))$ digits, the machine will decide the same set for inputs up to size n. In the following statement, note that if a set is decided by an error-prone analog-digital scatter machine with error probability bounded by γ, we may assume without loss of generality that $\gamma < \frac{1}{4}$.

Proposition 5.1. *Let $\mathcal{M}$ be an error-prone analog-digital scatter machine with arbitrary precision, with the vertex at x, deciding some set in time $t(n)$ with error probability bounded by $\gamma < \frac{1}{4}$. Let $\tilde{\mathcal{M}}$ be an error-prone machine with arbitrary precision, with the same finite control as $\mathcal{M}$ and with the vertex placed at $\tilde{x} = x\restriction_{5t(n)}$. Then $\mathcal{M}$ and $\tilde{\mathcal{M}}$ make the same decision on every input of size smaller or equal to n.*

This allows us to show the following.

Proposition 5.2. *Every set decided by an error-prone analog-digital scatter machine with arbitrary precision in polynomial time is in $BPP//poly$.*

Let A be a set decided by a precise error-free analog-digital scatter machine $\mathcal{M}$ in polynomial time p, and with a error probability bounded by $\frac{1}{4}$. Let x be the position of the vertex of $\mathcal{M}$. We use the advice function $f \in poly$, given by $f(n) = x\restriction_{5p(n)}$, to construct a probabilistic Turing machine $\tilde{\mathcal{M}}$ which decides A in polynomial time.

Given any dyadic rational $\tilde{x} \in [0, 1]$, the machine $\tilde{\mathcal{M}}$ can carry out a Bernoulli trial X with an associated probability $\mathbb{P}(X = 1) = \tilde{x}$. If $\tilde{x}$ has the binary expansion $\xi_1 \dots \xi_k$, the machine $\tilde{\mathcal{M}}$ tosses its balanced coin k times, and constructs a word $\tau_1 \dots \tau_k$, where τ_i is 1 if the coin turns up heads and 0 otherwise. The Bernoulli trial will have the outcome 1 if $\xi_1 \dots \xi_k < \tau_1 \dots \tau_k$, and 0 otherwise, and this will give the desired probability.

The probabilistic machine $\tilde{\mathcal{M}}$ will decide if $w \in A$ by simulating $\mathcal{M}$ on the input w with the vertex placed at the position $\tilde{x} = x \restriction 5p(n)$. In order to mimic the shooting of the cannon from the position z, which should have an error $\varepsilon = 2^{-|z|-1}$, the machine will carry out a Bernoulli trial X with an associated dyadic probability.

Then $\tilde{\mathcal{M}}$ will simulate a left hit when $X = 1$ and a right hit when $X = 0$. As we have seen in the previous Proposition 5.1, $\mathcal{M}$ will, when simulated in this way, decide the same set in polynomial time and with bounded error probability.

In a way similar to the first part of the proof of Theorem 4.1 we can prove the statement:

Proposition 5.3. *An error-prone analog-digital scatter machine with arbitrary precision can obtain n digits of the vertex position in $O(n^2)$ steps.*

The conclusion for this section comes with the following theorem.

Theorem 5.1. *The class of sets decided by error-prone analog-digital scatter machines with arbitrary precision in polynomial time is exactly $BPP//poly = P/poly$.*

It seems that measurement in more reasonable conditions do not affect the computational power of a motionless point particle position taken as oracle. I.e., making measurements in Classical Physics with incremental precision decide languages above the Turing limit, namely the *halting set*. But, surely, the computational power of a fuzzy motionless point particle position, i.e., a point to which we have access only with a finite *a priori* precision in measurement, will drop above the Turing limit...

6 *A Priori* Finite Precision

THIRD GEDANKENEXPERIMENT: the cannon can be placed at some dyadic rational z up to some dyadic fixed precision ε. We will show that such machines may, in polynomial time, make probabilistic guesses of up to a logarithmic number of digits of the position of the vertex. We will then conclude that these machines decide exactly $BPP//\log *$.

Proposition 6.1. *For any real value $\delta < \frac{1}{2}$, prefix function $f \in \log$, there is an error-prone analog-digital scatter machine with fixed precision which obtains $f(n)$ in polynomial time with an error of at most δ.*

The proof will take two steps. First we show that if f is a prefix function in log, then there is a real value $0 \le r \le 1$ such that it is possible to obtain the value $f(n)$ from a logarithmic number of digits of r. Then we will show that by carefully choosing the vertex position, we can guess a logarithmic number of digits of r with an error rate δ. If $f(n)$ is ultimately constant, then the result is trivial, and so we assume it is not so. After the work of [2], we can assume, without loss of generality, that there exist $a, b \in \mathbb{N}$ such that $|f(n)| = \lfloor a \log n + b \rfloor$.

Since f is a prefix function, we can consider the infinite sequence φ which is the limit of $f(n)$ as $n \to \infty$. Let φ_n be the n-th symbol (1 or 0) in this sequence, and set $r = \sum_{n=1}^{\infty} \varphi_n 2^{-n}$. Then, since f is not ultimately constant, the digits in the binary expansion of r are exactly the symbols of φ. Then the value $f(n)$ can be obtained from the first $\lfloor a \log n + b \rfloor$ digits of r.

The real number r is a value strictly between 0 and 1. Now suppose that ε is the error when positioning the cannon. We then set the vertex of our analog-digital scatter machine at the position $x = \frac{1}{2} - \varepsilon + 2r\varepsilon$. Our method for guessing digits of r begins by commanding the cannon to shoot from the point $\frac{1}{2}$ a number z of times. If the scatter machine experiment is carried out by entering the shooting state after having written the word 01 in the query tape, the cannon will be placed at some point in the interval $[\frac{1}{2} - \varepsilon, \frac{1}{2} + \varepsilon]$, with a uniform distribution. Then we conclude that the particle will go left with a probability of r and go right with a probability of $1 - r$.

After shooting the cannon z times in this way, we count the number of times that the particle went left, which we denote by L. The value $\tilde{r} = \frac{L}{z}$ will be our estimation of r. In order for our estimation to be correct in the sufficient number of digits, it is required that $|r - \tilde{r}| \leq 2^{-|f(n)|-1}$. By shooting the cannon z times, we have made z Bernoulli trials. Thus L is a random variable with expected value $\mu = zr$ and variance $\nu = zr(1-r)$. By Chebyshev's inequality, we conclude that, for every Δ,

$$\mathbb{P}(|L - \mu| > \Delta) = \mathbb{P}(|z\tilde{r} - zr| > \Delta) = \mathbb{P}\left(|\tilde{r} - r| > \frac{\Delta}{z}\right) \leq \frac{\nu}{\Delta^2}.$$

Choosing $\Delta = z2^{-|f(n)|-1}$, we get

$$\mathbb{P}(|\tilde{r} - r| > 2^{-|f(n)|-1}) \leq \frac{r(1-r)2^{2|f(n)|+2}}{z}$$

And so, the probability of making a mistake can be bounded to δ by making $z > \delta^{-1} r(1-r) 2^{2a \log n + 2b + 2} \in O(n^{2a})$ experiments.

The proposition above will guarantee us that for every fixed error ε we can find a vertex position that will allow for an *SME* to extract information from this vertex position. It does not state that we can make use of any vertex position independently of the fixed error ε. It can be shown, however, that if ε is a dyadic rational, then we may guess $O(\log n)$ digits of the vertex position in polynomial time.

Proposition 6.2. *The class of sets decided in polynomial time by error-prone analog-digital scatter machines with* a priori *fixed precision is exactly* $BPP//\log *$.

Since the class $BPP//\log *$ include non-recursive sets, measurements in this more realistic condition still decide super-Turing languages.

We can even think about changing the (uniform) probability distribution in the last two sections, making experiments closer and closer to reality...

7 Conclusion

We have seen that every variant of the analogue-digital scatter machine has a hypercomputational power. For instance, if $K = \{0^n :$ the Turing machine coded by n halts on input $0\}$, then K can be decided, in polynomial time, either by an error-free analog-digital scatter machine, or by an error-prone analog-digital scatter machine with arbitrary precision. It is obvious that the hypercomputational power of the analog-digital scatter machine arises from the precise nature of the vertex position. If we demand that the vertex position is a computable real number, then the analog-digital scatter machine can compute no more than the Turing machine, although it can compute, in polynomial time, faster than the Turing machine, namely $REC \bigcap P/poly$, the recursive part of $P/poly$.

In order to use the scatter machine experiment as an oracle, we need to assume that the wedge is sharp to the point and that the vertex is placed on a precise value x. Without these assumptions, the scatter machine becomes useless, since its computational properties arise exclusively from the value of x. The existence of an arbitrarily sharp wedge seems to contradict atomic theory, and for this reason the scatter machine is not a valid counterexample to the physical Church–Turing thesis. If this is the case, then what is the relevance of the analog-digital scatter machine as a model of computation? The scatter machine is relevant when it is seen as a *Gedankenexperiment*. In our discussion, we could have replaced the barriers, particles, cannons and particle detectors with any other physical system with this behaviour. So the scatter machine becomes a tool to answer the more general question: *if we have a physical system to measure an answer to the predicate $y \leq x$, to what extent can we use this system in feasible computations?*

As an open problem, besides a few other aspects of the measurement apparatus that we didn't cover up in this paper, we will study a point mass in motion, according to some physical law, like a Newtonian gravitation field, and we will apply instrumentation to measure its position and velocity.

References

1. Balcázar, J.L., Días, J., Gabarró, J.: Structural Complexity I. Springer, Heidelberg (1988)
2. Balcázar, J.L., Hermo, M.: The structure of logarithmic advice complexity classes. Theoretical Computer Science 207(1), 217–244 (1998)
3. Beggs, E., Tucker, J.: Embedding infinitely parallel computation in newtonian kinematics. Applied Mathematics and Computation 178(1), 25–43 (2006)
4. Beggs, E., Tucker, J.: Can Newtonian systems, bounded in space, time, mass and energy compute all functions? Theoretical Computer Science 371(1), 4–19 (2007)
5. Beggs, E., Tucker, J.: Experimental computation of real numbers by Newtonian machines. Proceedings of the Royal Society 463(2082), 1541–1561 (2007)
6. Siegelmann, H.: Neural Networks and Analog Computation: Beyond the Turing Limit. Birkhäuser (1999)

Quantum Walk Based Search Algorithms*

Miklos Santha[1,2]

[1] CNRS–LRI, Université Paris–Sud, 91405 Orsay, France
[2] Centre for Quantum Technologies, Nat. Univ. of Singapore, Singapore 117543

Abstract. In this survey paper we give an intuitive treatment of the discrete time quantization of classical Markov chains. Grover search and the quantum walk based search algorithms of Ambainis, Szegedy and Magniez et al. will be stated as quantum analogues of classical search procedures. We present a rather detailed description of a somewhat simplified version of the MNRS algorithm. Finally, in the query complexity model, we show how quantum walks can be applied to the following search problems: Element Distinctness, Matrix Product Verification, Restricted Range Associativity, Triangle, and Group Commutativity.

1 Introduction

Searching is without any doubt one of the major problems in computer science. The corresponding literature is tremendous, most manuals on algorithms include several chapters that deal with searching procedures [22,14]. The relevance of finite Markov chains (random walks in graphs) to searching was recognized from early on, and it is still a flourishing field. The algorithm of Aleliunas et al. [4] that solves s–t connectivity in undirected graphs in time $O(n^3)$ and in space $O(\log n)$, and Schöning's algorithm [33] that provides the basis of the currently fastest solutions for 3-SAT are among the most prominent examples for that.

Searching is also a central piece in the emerging field of quantum algorithms. Grover search [16], and in general amplitude amplification [11] are well known quantum procedures which are provably faster than their classical counterpart. Grover's algorithm was used recursively by Aaronson and Ambainis [2] for searching in grids.

Discrete time quantum walks were introduced gradually by Meyer[27,28] in connection with cellular automata, and by Watrous in his works related to space bounded computations [37]. Different parameters related to quantum walks and possible speedups of classical algorithms were investigated by several researchers [30,8,3,29,19,32].

The potential of discrete time quantum walks with respect to searching problems was first pointed out by Shenvi, Kempe, and Whaley [34] who designed a quantum walk based simulation of Grover search. Ambainis, in his seminal paper [7],

* Research supported by the European Commission IST Integrated Project Qubit Applications (QAP) 015848, and by the ANR Blanc AlgoQP grant of the French Research Ministry.

M. Agrawal et al. (Eds.): TAMC 2008, LNCS 4978, pp. 31–46, 2008.

used quantum walks on the Johnson graphs to settle the query complexity of the Element Distinctness problems. Inspired by the work of Ambainis, Szegedy [35] designed a general method to quantize classical Markov chains, and developed a theory of quantum walk based search algorithms. A similar approach for the specific case of searching in grids was taken by Ambainis, Kempe and Rivosh [9]. The frameworks of Ambainis and Szegedy were used in various contexts to find algorithms with substantial complexity gains over simple Grover search [26,12,23,15]. In a recent work, Magniez, Nayak, Roland and Santha [25] proposed a new quantum walk based search method that expanded the scope of the previous approaches. The MNRS search algorithm is also conceptually simple, and improves various aspects of many walk based algorithms.

In this survey paper we give an intuitive (though formal) treatment of the quantization of classical Markov chains. We will be concerned with discrete time quantum walks, the continuous case will not be covered here. Grover search and the quantum walk based search algorithms of Ambainis, Szegedy and Magniez et al. will be stated as quantum analogues of classical search procedures. We present a rather detailed description of a somewhat simplified version of the MNRS algorithm. Finally, in the query complexity model, we show how quantum walks can be applied to the following search problems: Element Distinctness, Matrix Product Verification, Restricted Range Associativity, Triangle, and Group Commutativity. For a detailed introduction to quantum walks the reader is referred to the excellent surveys of Kempe [20] and Ambainis [5]. Another survey on quantum search algorithms is also due to Ambainis [6].

2 Classical Search Algorithms

At an abstract level, any search problem may be cast as the problem of finding a marked element from a set X. Let $M \subseteq X$ be the set of marked elements, and let ε be a known lower bound on $|M|/|X|$, the fraction of marked elements, whenever M is non-empty. If no further information is available on M, we can choose ε as $1/|X|$. The simplest approach, stated in **Search Algorithm 1**, to solve this problem is to repeatedly sample from X uniformly until a marked element is picked, if there is any.

Search Algorithm 1
Repeat for $t = O(1/\varepsilon)$ steps
1. Sample $x \in X$ according to the uniform distribution.
2. If x is in M then output it and stop.

More sophisticated approaches might use a Markov chain on the state space X for generating the samples. In that case, to generate the next sample, the resources expended for previous generations are often reused.

Markov chains can be viewed as random walks on directed graphs with weighted edges. We will identify a Markov chain with its transition matrix

$P = (p_{xy})$. A chain is *irreducible* if every state is accessible from every other state. An irreducible chain is *ergodic* if it is also aperiodic. The eigenvalues of a Markov chain are at most 1 in magnitude. By the Perron-Frobenius theorem, an irreducible chain has a unique stationary distribution $\pi = (\pi_x)$, that is a unique left eigenvector π with eigenvalue 1 and positive coordinates summing up to 1. If the chain is ergodic, the eigenvalue 1 is the only eigenvalue of P with magnitude 1. We will denote by $\delta = \delta(P)$ the *eigenvalue gap* of P, that is $1 - |\lambda|$, where λ is an eigenvalue with the second largest magnitude. It follows that when P is ergodic then $\delta > 0$.

The *time-reversed Markov chain* P^* of P is defined by the equations $\pi_x p_{xy} = \pi_y p^*_{yx}$. The Markov chain P is said to be *reversible* if $P^* = P$. Reversible chains can be viewed as random walks on undirected graphs with weighted edges, and in these chains the probability of a transition from a state x to another state y in the stationary distribution is the same as the probability of the transition in the reverse direction. The Markov chain P is *symmetric* if $P = P^t$ where P^t denotes the transposed matrix of P. The stationary distribution of symmetric chains is the uniform distribution. They can be viewed as random walks on regular graphs, and they are time-reversible.

We consider two search algorithms based on some ergodic and symmetric chain P. **Search Algorithm 2** repeatedly samples from approximately stationary distributions, and checks if the element is marked. To get a sample the Markov chain is simulated long enough to mix well. **Search Algorithm 3** is a greedy variant: a check is performed after every step of the chain.

Search Algorithm 2

1. Initialize x to a state sampled from the uniform distribution over X.
2. Repeat for $t_2 = O(1/\varepsilon)$ steps
   (a) If the element reached in the previous step is marked then output it and stop.
   (b) Simulate $t_1 = O(1/\delta)$ steps of P starting with x.

Search Algorithm 3

1. Initialize x to a state sampled from the uniform distribution over X.
2. Repeat for $t = O(1/\varepsilon\delta)$ steps
   (a) If the element reached in the previous step is marked then output it and stop.
   (b) Simulate one step of P starting with x.

We state formally the complexity of the three algorithms to clarify their differences. They will maintain a data structure d that associates some data $d(x)$ with every state $x \in X$. Creating and maintaining the data structure incurs a certain cost, but the data $d(x)$ can be helpful to determine if $x \in M$. We distinguish three types of cost.

Setup cost S**:** The cost to sample $x \in X$ according to the uniform distribution, and to construct $d(x)$.
Update cost U**:** The cost to simulate a transition from x to y according to P, and to update $d(x)$ to $d(y)$.
Checking cost C**:** The cost of checking if $x \in M$ using $d(x)$.

The cost may be thought of as a vector listing all the measures of complexity of interest, such as query and time complexity. The generic bounds on the efficiency of the three search algorithms can be stated in terms of the cost parameters.

Proposition 1. *Let P be an ergodic and symmetric Markov chain on X. Then all three algorithms find a marked element with high probability if there is any. The respective costs incurred by the algorithms are of the following order:*

1. **Search Algorithm 1***:* $(\mathsf{S}+\mathsf{C})/\varepsilon$,
2. **Search Algorithm 2***:* $\mathsf{S}+(\mathsf{U}/\delta+\mathsf{C})/\varepsilon$,
3. **Search Algorithm 3***:* $\mathsf{S}+(\mathsf{U}+\mathsf{C})/\delta\varepsilon$.

The generic bound of $O(1/\delta\varepsilon)$ in **Search Algorithm 3** on the hitting time is not always optimal, which in some cases, for example in the 2-dimensional grid, can be significantly smaller.

3 Quantum Analogue of a Classical Markov Chain

We define a quantum analogue of an arbitrary irreducible Markov chain P as it is given by Magniez et al. [25]. This definition is based on and slightly extends the concept of quantum Markov chain due to Szegedy [35]. The latter was inspired by an earlier notion of quantum walk due to Ambainis [7]. We also point out that a similar process on regular graphs was studied by Watrous [37].

The quantum walk may be thought of as a walk on the *edges* of the original Markov chain, rather than on its vertices. Thus, its state space is a vector subspace of $\mathcal{H} = \mathbb{C}^{X\times X} \cong \mathbb{C}^X \otimes \mathbb{C}^X$. For a state $|\psi\rangle \in \mathcal{H}$, let $\Pi_\psi = |\psi\rangle\langle\psi|$ denote the orthogonal projector onto $\mathsf{Span}(|\psi\rangle)$, and let $\mathrm{ref}(\psi) = 2\Pi_\psi - \mathrm{Id}$ denote the reflection through the line generated by $|\psi\rangle$, where Id is the identity operator on $\mathcal{H}$. If $\mathcal{K}$ is a subspace of $\mathcal{H}$ spanned by a set of mutually orthogonal states $\{|\psi_i\rangle : i \in I\}$, then let $\Pi_\mathcal{K} = \sum_{i\in I} \Pi_{\psi_i}$ be the orthogonal projector onto $\mathcal{K}$, and let $\mathrm{ref}(\mathcal{K}) = 2\Pi_\mathcal{K} - \mathrm{Id}$ be the reflection through $\mathcal{K}$. Let $\mathcal{A} = \mathsf{Span}(|x\rangle|p_x\rangle : x \in X)$ and $\mathcal{B} = \mathsf{Span}(|p^*_y\rangle|y\rangle : y \in X)$ be vector subspaces of $\mathcal{H}$, where

$$|p_x\rangle = \textstyle\sum_{y\in X} \sqrt{p_{xy}}\,|y\rangle \quad \text{and} \quad |p^*_y\rangle = \textstyle\sum_{x\in X} \sqrt{p^*_{yx}}\,|x\rangle.$$

Definition 1 (Quantum walk). *The unitary operation $W(P) = \mathrm{ref}(\mathcal{B})\cdot\mathrm{ref}(\mathcal{A})$ defined on $\mathcal{H}$ by is called the* quantum walk *based on the classical chain P.*

Let us give some motivations for this definition. Classical random walks do not quantize in the space of the vertices. The standard way advocated by several papers (see the survey [20]) is to extend the vertex space X by a coin space C, and define the state space of the walk as $X \times C$. Then a step of the walk is defined as the product of two unitary operations. The first one is the *flip* operation F controlled by the vertex state, which means that for every $x \in X$, it performs a unitary coin flip F^x on the states $\{|x, c\rangle : c \in C\}$. For d-regular undirected graphs, C can be taken as the set $\{1, \ldots, d\}$, and in that case the coin flip F^x is independent from x. The second one is the *shift* operation S which is controlled by the coin state, and takes a vertex to one if its neighboring vertices. For d-regular graphs the simplest way to define it is via a labeling of the directed edges by the numbers between 1 and d such that for every $1 \leq i \leq d$, the directed edges labeled by i form a permutation. Then, if the coin state is i, the new vertex is the ith neighbor according to the labeling. For general walks it is practical to take the coin space also to be X, then the state space of the walk corresponds naturally to the directed edges of the graph. In this case there is a symmetry between the two spaces, and the shift operation simply exchanges the vertices, that is $S|x, y\rangle = |y, x\rangle$, for every $x, y \in X$.

Let us pause here for a second and consider how a classical walk defined by some Markov chain P can be thought of as a walk on the directed edges of the graph (instead of the vertices). Let's think about an edge (x, u) as the state of the walk P being at x, where the previous state was u. According to this interpretation, in one step the walk on edges should move from state (x, u) to state (y, x) with probability p_{xy}. This move can be accomplished by the stochastic flip operation F controlled by the left end-point of the edge, where $F^x_{uy} = p_{xy}$ for all $x, u, y \in X$, followed by the shift S defined previously. If we define the flip F' as F but controlled by the right end-point of the edge, then it is not hard to see that $SFSF = F'F$. Therefore one can get rid of the shift operations, and two steps of the walk can be accomplished by two successive flips where the control and the target registers alternate.

Coming back to the quantization of classical walks, we thus want to find unitary coin flips which mirror the walk P, and which alternately mix the right end-point of the edges over the neighbors of the left end-point, and then the left end-point of the edges over the neighbors of the new right end-point. The reflections $\text{ref}(\mathcal{A})$ and $\text{ref}(\mathcal{B})$ are natural choices for that. They are also generalizations of the Grover diffusion operator [16]. Indeed, when the transition to each neighbor is equally likely, they correspond exactly to Grover diffusion. In Szegedy's original definition the alternating reflections were $\text{ref}(\mathcal{A})$ and $\text{ref}(\mathcal{B}')$ with $\mathcal{B}' = \mathsf{Span}(|p_y\rangle|y\rangle : y \in X)$, mirroring faithfully the classical edge based walk. The reason why the MNRS quantization chooses every second step a reflection based on the reversed walk P^* is explained now.

The eigen-spectrum of the transition matrix P plays an important role in the analysis of a classical Markov chain. Similarly, the behaviour of the quantum process $W(P)$ may be inferred from its spectral decomposition. The reflections through subspaces $\mathcal{A}$ and $\mathcal{B}$ are (real) orthogonal transformations, and so is their product $W(P)$. An orthogonal matrix may be decomposed into a direct sum of the identity, reflection through the origin, and two-dimensional rotations over orthogonal vector subspaces [17, Section 81]. These subspaces and the corresponding eigenvalues are revealed by the singular value decomposition of the product $\Pi_{\mathcal{A}}\Pi_{\mathcal{B}}$ of the orthogonal projection operators onto the subspaces $\mathcal{A}$ and $\mathcal{B}$. Equivalently, as done by Szegedy, one can consider the singular values of the *discriminant* matrix $D(P) = (\sqrt{p_{xy}p^*_{yx}})$. Since $\sqrt{p_{xy}p^*_{yx}} = \sqrt{\pi_x}p_{xy}/\sqrt{\pi_y}$, we have

$$D(P) \quad = \quad \text{diag}(\pi)^{1/2} \cdot P \cdot \text{diag}(\pi)^{-1/2},$$

where $\text{diag}(\pi)$ is the invertible diagonal matrix with the coordinates of the distribution π in its diagonal. Therefore $D(P)$ and P are similar, and their spectra are the same. When P is reversible then $D(P)$ is symmetric, and its singular values are equal to the absolute values of its eigenvalues. Thus, in that case we only have to study the spectrum of P.

Since the singular values of $D(P)$ all lie in the range $[0,1]$, they can be expressed as $\cos\theta$, for some angles $\theta \in [0, \frac{\pi}{2}]$. The following theorem of Szegedy relates the singular value decomposition of $D(P)$ to the spectral decomposition of $W(P)$.

Theorem 1 (Szegedy [35]). *Let P be an irreducible Markov chain, and let $\cos\theta_1, \ldots, \cos\theta_l$ be an enumeration of those singular values (possibly repeated) of $D(P)$ that lie in the open interval $(0,1)$. Then the exact description of the spectrum of $W(P)$ on $\mathcal{A}+\mathcal{B}$ is:*

1. *On $\mathcal{A}+\mathcal{B}$ those eigenvalues of $W(P)$ that have non-zero imaginary part are exactly $e^{\pm 2i\theta_1}, \ldots, e^{\pm 2i\theta_l}$, with the same multiplicity.*
2. *On $\mathcal{A}\cap\mathcal{B}$ the operator $W(P)$ acts as the identity* Id. *$\mathcal{A}\cap\mathcal{B}$ is spanned by the left (and right) singular vectors of $D(P)$ with singular value* 1.
3. *On $\mathcal{A}\cap\mathcal{B}^\perp$ and $\mathcal{A}^\perp\cap\mathcal{B}$ the operator $W(P)$ acts as* $-$Id. *$\mathcal{A}\cap\mathcal{B}^\perp$ (respectively, $\mathcal{A}^\perp\cap\mathcal{B}$) is spanned by the left (respectively, right) singular vectors of $D(P)$ with singular value* 0.

Let us now suppose in addition that P is ergodic and reversible. As we just said, reversibility implies that the singular values of $D(P)$ are equal to the absolute values of the eigenvalues of P. From the ergodicity it also follows that $D(P)$ has a unique singular vector with singular value 1. We have therefore the following corollary.

Corollary 1. *Let P be an ergodic and reversible Markov chain. Then, on $\mathcal{A}+\mathcal{B}$ the spectrum of $W(P)$ can be characterized as:*

$$|\pi\rangle = \sum_{x\in X} \sqrt{\pi_x}\,|x\rangle|p_x\rangle = \sum_{y\in X} \sqrt{\pi_y}\,|p^*_y\rangle|y\rangle$$

is the unique 1-eigenvector, $e^{\pm 2i\theta}$ are eigenvalues for every singular value $\cos\theta \in (0,1)$ of $D(P)$, and all the remaining eigenvalues are -1.

The *phase gap* $\Delta(P) = \Delta$ of $W(P)$ is defined as 2θ, where θ is the smallest angle in $(0, \frac{\pi}{2}]$ such that $\cos\theta$ is a singular value of $D(P)$. This definition is motivated by the previous theorem and corollary: the angular distance of 1 from any other eigenvalue of $W(P)$ on $\mathcal{A}+\mathcal{B}$ is at least Δ. When P is ergodic and reversible, there is a quadratic relationship between the phase gap Δ of the quantum walk $W(P)$ and the eigenvalue gap δ of the classical Markov chain P, more precisely $\Delta \geq 2\sqrt{\delta}$. Indeed, let $\delta \in (0, \frac{\pi}{2}]$ such that $\delta = 1 - \cos\theta$ and $\Delta = 2\theta$. The following (in)equalities can easily be checked: $\Delta \geq \left|1 - e^{2i\theta}\right| = 2\sqrt{1-\cos^2\theta} \geq 2\sqrt{\delta}$. The origin of the quadratic speed-up due to quantum walks may be traced to this phenomenon.

4 Quantum Search Algorithms

As in the classical case, the quantum search algorithms look for a marked element in a finite set X. We suppose that the elements of X are coded by binary strings and that $\bar{0}$, the everywhere 0 string is in X. A data structure attached to both vertex registers is maintained during the algorithm. Again, three types of cost will be distinguished, generalizing those of the classical search. In all quantum search algorithms the overall complexity is of the order of these specific costs, which justifies their choices. The operations not involving manipulations of the data will be charged at unit cost. For the sake of simplicity, we do not formally include the data into the description of the unitary operations defining the costs. The initial state of the algorithm is explicitly related to the stationary distribution π of P.

(Quantum) Setup cost S**:** The cost for constructing the state $\sum_{x\in X}\sqrt{\pi_x}|x\rangle|\bar{0}\rangle$ with data.
(Quantum) Update cost U**:** The cost to realize any of the unitary transformations and inverses with data

$$\begin{array}{rcl} |x\rangle|\bar{0}\rangle & \mapsto & |x\rangle\sum_{y\in X}\sqrt{p_{xy}}|y\rangle, \\ |\bar{0}\rangle|y\rangle & \mapsto & \sum_{x\in X}\sqrt{p^*_{yx}}|x\rangle|y\rangle. \end{array}$$

(Quantum) Checking cost C**:** The cost to realize the unitary transformation with data, that maps $|x\rangle|y\rangle$ to $-|x\rangle|y\rangle$ if $x \in M$, and leaves it unchanged otherwise.

In the checking cost we could have included the cost of the unitary transformation which realizes a phase flip also when $y \in M$, our choice was made just for simplicity. Observe that the quantum walk $W(P)$ with data can be implemented at cost $4\mathsf{U}+2$. Indeed, the reflection $\mathrm{ref}(\mathcal{A})$ is implemented by mapping states $|x\rangle|p_x\rangle$ to $|x\rangle|\bar{0}\rangle$, applying $\mathrm{ref}(\mathbb{C}^X \otimes |\bar{0}\rangle)$, and undoing the first transformation. In our accounting we charge unit cost for the second step since it does not depend on the database. Therefore the implementation of $\mathrm{ref}(\mathcal{A})$ is of cost $2\mathsf{U}+1$. The reflection $\mathrm{ref}(\mathcal{B})$ may be implemented similarly.

Let us now describe how the respective algorithms of Grover, Ambainis and Szegedy are related to the classical search algorithms of Section 2. We suppose that ε, a lower bound on the proportion of marked elements is known in advance, though the results remain true even if it is not the case. Grover search (which we discuss soon in detail) is the quantum analogue of **Search Algorithm 1**, and it doesn't involve my specific data structure.

Theorem 2 (Grover [16]). *There exists a quantum algorithm which with high probability finds a marked element, if there is any, at cost of order* $\frac{\mathsf{S}+\mathsf{C}}{\sqrt{\varepsilon}}$.

In the original application of Grover's result to unordered search there is no data structure involved, therefore $\mathsf{S}+\mathsf{C}=O(1)$, and the cost is of order $\frac{1}{\sqrt{\varepsilon}}$.

The algorithm of Ambainis is the quantum analogue of **Search Algorithm 2** in the special case of the walk on the Johnson graph and for some specific marked sets. Let us recall that for $0 < r \leq n/2$, the vertices of the Johnson graph $J(n,r)$ are the subsets of $[n]$ of size r, and there is an edge between two vertices if the size of their symmetric difference is 2. In other words, two vertices are adjacent if by deleting an element from the first one and adding a new element to it we get the second vertex. The eigenvalue gap δ of the symmetric walk on $J(n,r)$ is $n/r(n-r) = \Theta(1/r)$. If the set of marked vertices in $J(n,r)$ is either empty, or it consists of vertices that contain a fixed subset of constant size $k \leq r$ then $\varepsilon = \Omega(\frac{r^k}{n^k})$.

Theorem 3 (Ambainis [7]). *Let P be the random walk on the Johnson graph $J(n,r)$ where $r = o(n)$, and let M be either empty, or the class of vertices that contain a fixed subset of constant size $k \leq r$. Then there is a quantum algorithm that finds, with high probability, the k-subset if M is not empty at cost of order* $\mathsf{S} + \frac{1}{\sqrt{\varepsilon}}(\frac{1}{\sqrt{\delta}}\mathsf{U} + \mathsf{C})$.

Szegedy's algorithm is the quantum analogue of **Search Algorithm 3** for the class of ergodic and symmetric Markov chains. His algorithm is therefore more general than the one of Ambainis with respect to the class of Markov chains and marked sets it can deal with. Nonetheless, the approach of Ambainis has its own advantages: it is of smaller cost when C is substantially greater than U, and it also finds a marked element.

Theorem 4 (Szegedy [35]). *Let P be an ergodic and symmetric Markov chain. There exists a quantum algorithm that determines, with high probability, if M is non-empty at cost of order* $\mathsf{S} + \frac{1}{\sqrt{\delta\varepsilon}}(\mathsf{U}+\mathsf{C})$.

The MNRS algorithm is a quantum analogue of **Search Algorithm 2** for ergodic and reversible Markov chains. It generalizes the algorithms of Ambainis an Szegedy, and it combines their benefits in terms of being able to find marked elements, incurring the smaller cost of the two, and being applicable to a larger class of Markov chain.

Theorem 5 (Magniez et al. [25]). *Let P be an ergodic and reversible Markov chain, and let $\varepsilon > 0$ be a lower bound on the probability that an element chosen from the stationary distribution of P is marked whenever M is non-empty. Then, there exists a quantum algorithm which finds, with high probability, an element of M if there is any at cost of order* $\mathsf{S} + \frac{1}{\sqrt{\varepsilon}}(\frac{1}{\sqrt{\delta}}\mathsf{U} + \mathsf{C})$.

There is an additional feature of Szegedy's algorithm which doesn't fit into the MNRS algorithmic paradigm. In fact, the quantity $\frac{1}{\sqrt{\delta\varepsilon}}$ in Theorem 4 can be replaced by the square root of the classical hitting time [35]. The search algorithm for the 2-dimensional grid obtained this way, and the one given in [9] have smaller complexity than what follows from Theorem 5.

5 The MNRS Search Algorithm

We give a high level description of the MNRS search algorithm. Assume that $M \neq \emptyset$. Let $\mathcal{M} = \mathbb{C}^{M \times X}$ denote the marked subspace, that is the subspace with marked items in the first register. The purpose of the algorithm is to approximately transform the initial state $|\pi\rangle$ to the target state $|\mu\rangle$, which is the normalized projection of $|\pi\rangle$ onto $\mathcal{M}$:

$$|\mu\rangle = \frac{\Pi_{\mathcal{M}}|\pi\rangle}{\|\Pi_{\mathcal{M}}|\pi\rangle\|} = \frac{1}{\sqrt{\varepsilon}} \sum_{x \in M} \sqrt{\pi_x}\, |x\rangle |p_x\rangle,$$

where $\varepsilon = \|\Pi_{\mathcal{M}}|\pi\rangle\|^2 = \sum_{x \in M} \pi_x$ is the probability of the set M of marked states under the stationary distribution π. Let us recall that Grover search [16] solves this problem via the iterated use of the rotation $\mathrm{ref}(\pi) \cdot \mathrm{ref}(\mu^\perp)$ in the two-dimensional real subspace $\mathcal{S} = \mathsf{Span}(|\pi\rangle, |\mu\rangle)$, where $|\mu^\perp\rangle$ is the state in $\mathcal{S}$ orthogonal to $|\mu\rangle$ making some acute angle φ with $|\pi\rangle$. The angle φ is given by $\sin\varphi = \langle\mu|\pi\rangle = \sqrt{\varepsilon}$. Then $\mathrm{ref}(\pi) \cdot \mathrm{ref}(\mu^\perp)$ is a rotation by 2φ within the space $\mathcal{S}$, and therefore $\mathrm{O}(1/\varphi) = \mathrm{O}(1/\sqrt{\varepsilon})$ iterations of this rotation, starting with $|\pi\rangle$, approximates well $|\mu\rangle$. The MNRS search basically follows this idea.

Restricted to the subspace $\mathcal{S}$, the operator $\mathrm{ref}(\mu^\perp)$ is identical to $-\mathrm{ref}(\mathcal{M})$. Therefore, if the state of the algorithm remains close to the subspace $\mathcal{S}$ throughout, it can be implemented at the price of the checking cost. The reflection $\mathrm{ref}(\pi)$ is computationally harder to perform. The idea is to apply the phase estimation algorithms of Kitaev [21] and Cleve at al. [13] to $W(P)$. Corollary 1 implies that $|\pi\rangle$ is the only 1-eigenvector of $W(P)$, and all the other eigenvectors have phase at least Δ. Phase estimation approximately resolves any state $|\psi\rangle$ in $\mathcal{A}+\mathcal{B}$ along the eigenvectors of $W(P)$, and thus distinguishes $|\pi\rangle$ from all the others. Therefore it is possible to flip the phase of all states with a non-zero estimate of the phase, that is simulate the effect of the operator $\mathrm{ref}(\pi)$ in $\mathcal{A} + \mathcal{B}$. The following result of [25] resumes this discussion:

Theorem 6. *There exists a uniform family of quantum circuits $R(P)$ that uses $O(k \log(\Delta^{-1}))$ additional qubits and satisfies the following properties:*

1. *It makes $O(k\Delta^{-1})$ calls to the controlled quantum walk* c$-W(P)$ *and its inverse.*
2. *$R(P)|\pi\rangle = |\pi\rangle$.*
3. *If $|\psi\rangle \in \mathcal{A} + \mathcal{B}$ is orthogonal to $|\pi\rangle$, then $\|(R(P) + \mathrm{Id})|\psi\rangle\| \leq 2^{-k}$.*

The essence of the MNRS search algorithm is the following simple procedure that satisfies the conditions of Theorem 5, but with a slightly higher complexity, of the order of $\mathsf{S} + \frac{1}{\sqrt{\varepsilon}}(\frac{1}{\sqrt{\delta}} \log \frac{1}{\sqrt{\varepsilon}} \mathsf{U} + \mathsf{C})$. Again, we suppose that ε is known.

Quantum Search(P)
1. Start from the initial state $|\pi\rangle$.
2. Repeat $\mathrm{O}(1/\sqrt{\varepsilon})$–times:
   (a) For any basis vector $|x\rangle|y\rangle$, flip the phase if $x \in M$.
   (b) Apply circuit $R(P)$ of Theorem 6 with $k = \mathrm{O}(\log(1/\sqrt{\varepsilon}))$.
3. Observe the first register and output if it is in M.

To see the correctness, let $|\phi_i\rangle$ be the result of i iterations of $\mathrm{ref}(\pi) \cdot \mathrm{ref}(\mu^\perp)$ applied to $|\pi\rangle$, and let $|\psi_i\rangle$ be the result of i iterations of step (2) in **Quantum Search**(P) applied to $|\pi\rangle$. It is not hard to show by induction on i, using a hybrid argument as in [10,36], that $\||\psi_i\rangle - |\phi_i\rangle\| \leq O(i2^{-k})$. This implies that $\||\psi_k\rangle - |\phi_k\rangle\|$ is an arbitrarily small constant when k is chosen to be $\mathrm{O}(\log(1/\sqrt{\varepsilon}))$ and therefore the success probability is arbitrarily close to 1.

The cost of the procedure is simple to analyze. Initialization costs $\mathsf{S} + \mathsf{U}$, and in each iteration the single phase flip costs C. In the circuit $R(P)$, the controlled quantum walk and its inverse can be implemented, similarly to $W(P)$, at cost $4\mathsf{U} + 2$, simply by controlling $\mathrm{ref}(\mathbb{C}^X \otimes |\bar{0}\rangle)$ and $\mathrm{ref}(|\bar{0}\rangle \otimes \mathbb{C}^Y)$. The number of steps of the controlled quantum walk and its inverse is $\mathrm{O}((1/\Delta)\log(1/\sqrt{\varepsilon}))$. Since $\Delta \geq 2\sqrt{\delta}$, this finishes the cost analysis. Observe that the $\log(1/\sqrt{\varepsilon})$-factor in the update cost was necessary for reducing the error of the approximate reflection operator. In [25] it is described how it can be eliminated by adapting the recursive amplitude amplification algorithm of Høyer et al. [18]

6 Applications

We give here a few examples where the quantum search algorithms, in particular the MNRS algorithm can be applied successfully. All examples will be described in the query model of computation. Here the input is given by an oracle, a query can be performed at unit cost, and all other computational steps are free. A formal description of the model can be found for example in [23] or [26]. In fact, in almost all cases, the circuit complexity of the algorithms given will be of the order of the query complexity, with additional logarithmic factors.

6.1 Grover Search

As a first (and trivial) application, we observe that Grover's algorithm [16] for the unordered search problem is a special case of Theorem 5.

UNORDERED SEARCH
Oracle Input: A boolean function f defined on $[n]$.
Output: An element $i \in [n]$ such that $f(i) = 1$.

Theorem 7. UNORDERED SEARCH *can be solved with high probability in quantum query complexity* $O((n/k)^{1/2})$, *where* $|\{i \in [n] \ : \ f(i) = 1\}| = k$.

Proof. Consider the symmetric random walk in the complete graph on n vertices, where an element v is marked if $f(v) = 1$. The eigenvalue gap of the walk is $1 - \frac{1}{n-1}$, and the probability ε that an element is marked is k/n. There is no data structure involved in the algorithm, the setup, update and checking costs are 1.

6.2 Johnson Graph Based Algorithms

All these examples are based on the symmetric walk in the Johnson graph $J(n, r)$, with eigenvalue gap $\Theta(1/r)$.

Element Distinctness. This is the original problem for which Ambainis introduced the quantum walk based search method [7].

> ELEMENT DISTINCTNESS
> *Oracle Input:* A function f defined on $[n]$.
> *Output:* A pair of distinct elements $i, j \in [n]$ such that $f(i) = f(j)$ if there is any, otherwise reject.

Theorem 8. ELEMENT DISTINCTNESS *can be solved with high probability in quantum query complexity* $O(n^{2/3})$.

Proof. A vertex $R \subseteq [n]$ of size r is marked if there exist $i \neq j \in R$ such $f(i) = f(j)$. The probability ε that an element is marked, if there is any, is in $\Omega((r/n)^2)$. For every R, the data is defined as $\{(v, f(v)) : v \in R\}$. Then the setup cost is in $O(r)$, the update cost is $O(1)$, and the checking cost is 0. Therefore the overall cost is $O(r + n/r^{1/2})$ which is $O(n^{2/3})$ when $r = n^{2/3}$. This upper bound is tight, the $\Omega(n^{2/3})$ lower bound is due to Aaronson and Shi [1].

Matrix Product Verification. This problem was studied by Buhrman and Spalek [12], and the algorithm in the query model is almost identical to the previous one.

> MATRIX PRODUCT VERIFICATION
> *Oracle Input:* Three $n \times n$ matrices A, B and C.
> *Output:* Decide if $AB = C$ and in the negative case find indices i, j such that $(AB)_{ij} \neq C_{ij}$.

Theorem 9. MATRIX PRODUCT VERIFICATION *can be solved with high probability in quantum query complexity* $O(n^{5/3})$.

Proof. For an $n \times n$ matrix M and a subset of indices $R \subseteq [n]$, let $M|_R$ denote the $|R| \times n$ submatrix of M corresponding to the rows restricted to R. The submatrices $M|^R$ and $M|_R^R$ are defined similarly, when the restriction concerns the columns and both the rows and columns. A vertex $R \subseteq [n]$ is marked if there exist $i, j \in R$ such that $AB_{ij} \neq C_{ij}$. The probability of being marked, if there is such an element, is in $\Omega((r/n)^2)$. For every R, the data is defined as the set of entries in $A|_R, B|^R$ and $C|_R^R$. Then the setup cost is in $O(rn)$, the update cost is $O(n)$, and the checking cost is 0. Therefore the overall cost is $O(n^{5/3})$ when $r = n^{2/3}$. The best known lower bound is $\Omega(n^{3/2})$, and in [12] an $O(n^{5/3})$ upper bound was also proven for the time complexity with a somewhat more complicated argument.

Restricted Range Associativity. The problem here is to decide if a binary operation $\circ$ is associative. The only algorithm known for the general case is Grover search, but Dörn and Thierauf [15] have proved that when the range of the operation is restricted, Theorem 5 (or Theorem 3) give a non-trivial bound. A triple (a, b, c) is called *non associative* if $(a \circ b) \circ c \neq a \circ (b \circ c)$.

RESTRICTED RANGE ASSOCIATIVITY
Oracle Input: A binary operation $\circ : [n] \times [n] \to [k]$ where $k \in O(1)$.
Output: A non associative triple (a, b, c) if there is any, otherwise reject.

Theorem 10. RESTRICTED RANGE ASSOCIATIVITY *can be solved with high probability in quantum query complexity* $O(n^{5/4})$.

Proof. We say that $R \subseteq [n]$ is marked if there exist $a, b \in R$ and $c \in [n]$ such that (a, b, c) is non associative. Therefore ε is in $\Omega((r/n)^2)$, if there is a marked element. For every $R \subseteq [n]$, the data structure is defined as $\{(a, b, a \circ b) : a, b \in R \cup [k]\}$. Then the setup cost is $O((r + k)^2 = O(r^2)$ and the update cost is $O(r + k) = O(r)$. Observe that if $b \in R$ and $c \in [n]$ are fixed, then computing $(a \circ b) \circ c$ and $a \circ (b \circ c)$ for all $a \in R$ requires at most $k+1$ queries with the help of the data structure. Thus using Grover search to find b and c, the checking cost is $O(k\sqrt{rn}) = O(\sqrt{rn})$. The overall complexity is then $O(r^2 + \frac{n}{r}(\sqrt{r}r + \sqrt{rn}))$ which is $O(n^{5/4})$ when $r = \sqrt{n}$. The best lower bound known both in the restricted range and the general case is $\Omega(n)$.

Triangle. In an undirected graph G, a complete subgraph on three vertices is called a *triangle*. The algorithm of Magniez et al. [26] for finding a triangle uses the algorithm for ELEMENT DISTINCTNESS in the checking procedure.

TRIANGLE
Oracle Input: The adjacency matrix f of a graph G on vertex set $[n]$.
Output: A triangle if there is any, otherwise reject.

Theorem 11. TRIANGLE *can be solved with high probability in quantum query complexity* $O(n^{13/10})$.

Proof. We show how to find the edge of a triangle, if there is any, in query complexity $O(n^{13/10})$. This implies the theorem since given such an edge, Grover search finds the third vertex of the triangle with $O(n^{1/2})$ additional queries.

An element $R \subseteq [n]$ is marked if it contains a triangle edge. The probability ε that an element is marked is in $\Omega((r/n)^2)$, if there is a triangle. For every R, the data structure is the adjacency matrix of the subgraph induced by R. defined as $\{(v, f(v)) : v \in R\}$. Then the setup cost is $O(r^2)$, and the update cost is $O(r)$. The interesting part of the algorithm is the checking procedure, which is a quantum walk based search itself, and the claim is that it can be done at cost $O(\sqrt{n} \times r^{2/3})$.

To see this, let R be a set of r vertices such that the graph G restricted to R is explicitly known, and for which we would like to decide if it is marked. Observe that R is marked exactly when there exists a vertex $v \in [n]$ such that v and an edge in R form a triangle. Therefore for every vertex v, one can define a secondary search problem on R via the boolean oracle f_v, where for every $u \in R$, by definition $f_v(u) = 1$ if $\{u, v\}$ is an edge. The output of the problem is by definition positive if there is an edge $\{u, u'\}$ such that $f_v(u) = f_v(u') = 1$. To solve the problem we consider the Johnson graph $J(r, r^{2/3})$, and look for a subset which contains such an edge. In that search problem both the probability of being marked and the eigenvalue gap of the underlying Markov chain are in $\Omega(r^{-2/3})$. The data associated with a subset of R is just the values of f_v at its elements. Then the setup cost is $r^{2/3}$, the update cost is $O(1)$, and the checking cost is 0. Therefore by Theorem 5 the cost of solving a secondary search problem is in $O(r^{2/3})$. Finally the checking procedure of the original search problem consists of a Grover search for a vertex v such that the secondary search problem defined by f_v has a positive outcome. Putting things together, the problem can be solved in quantum query complexity $O(r^2 + \frac{n}{r}(\sqrt{r} \times r + \sqrt{n} \times r^{2/3}))$ which is $O(n^{13/10})$ when $r = n^{3/5}$. The best known lower bound for the problem is $\Omega(n)$.

6.3 Group Commutativity

The problem here is to decide if a group multiplication is commutative in the (sub)group generated by some set of group elements. It was defined and studied in the probabilistic case by Pak [31], the quantum algorithm is due to Magniez and Nayak [23].

> GROUP COMMUTATIVITY
> *Oracle Input:* The multiplication operation $\circ$ for a finite group whose base set contains $[n]$.
> *Output:* A non commutative couple $(i, j) \in [n] \times [n]$ if G, the (sub)group generated by $[n]$, is non-commutative, otherwise reject.

Theorem 12. GROUP COMMUTATIVITY *can be solved with high probability in quantum query complexity* $O(n^{2/3} \log n)$.

Proof. For $0 < r < n$ let $S(n, r)$ be the set of all r-tuples of distinct elements from $[n]$. For $u = (u_1, \ldots, u_r)$ in $S(n, r)$, we set $\bar{u} = u_1 \circ \cdots \circ u_r$. We define a random walk over $S(n, r)$. Let $u = (u_1, \ldots, u_r)$ be the current vertex. Then with probability 1/2 stay at u, and with probability 1/2 pick uniformly random $i \in [r]$ and $j \in [n]$. If $j = u_m$ for some m then exchange u_i and u_m, otherwise set $u_i = j$. The random walk P at the basis of the quantum algorithm is over $S(n, r) \times S(n, r)$, and it consists of two independent simultaneous copies of the above walk. The stationary distribution of P is the uniform distribution, and it is proven in [23] that its eigenvalue gap is $\Omega(1/(r \log r))$.

A vertex (u, v) is marked if $\bar{u} \circ \bar{v} \neq \bar{v} \circ \bar{u}$. It is proven again in [23] that when G is non-commutative and $r \in o(n)$, then the probability ε that an element is marked is $\Theta(r^2/n^2)$. For $u \in S(n, r)$ let T_u be the balanced binary tree with r leaves that are labeled from left to right by $u_1, \ldots, u_r$, and where each internal node is labeled by the product of the labels of its two sons. For every vertex (u, v) the data consists of (T_u, T_v). Then the setup cost is r, and the update cost is $O(\log r)$ for recomputing the leaf–root paths. The checking cost is simply 2 for querying $\bar{u} \circ \bar{v}$ and $\bar{v} \circ \bar{u}$. Therefore the query complexity to find a marked element is $O(r + \frac{n}{r}(\sqrt{r \log r} \log r + 1))$ which is $O(n^{2/3} \log n)$ when $r = n^{2/3} \log n$. Once a marked element is found, Grover search yields a non-commutative couple at cost $O(r)$. In [23] an $\Omega(n^{2/3})$ lower bound is also proven. And, it turns out that a Johnson graph based walk can be applied to this problem too [24], yielding an algorithm of complexity $O((n \log n)^{2/3})$.

Acknowledgment

I would like to thank Frédéric Magniez, Ashwin Nayak and Jérémie Roland, my coauthors in [25], for letting me to include here several ideas which were developed during that work, and for numerous helpful suggestions.

References

1. Aaronson, S., Shi, Y.: Quantum lower bounds for the collision and the element distinctness problems. Journal of the ACM, 595–605 (2004)
2. Aaronson, S., Ambainis, A.: Quantum search of spatial regions. Theory of Computing 1, 47–79 (2005)
3. Aharonov, D., Ambainis, A., Kempe, J., Vazirani, U.: Quantum walks on graphs. In: Proc. of the 33rd ACM Symposium on Theory of Computing, pp. 50–59 (2001)
4. Aleliunas, R., Karp, R., Lipton, R., Lovász, L., Rackoff, C.: Random Walks, Universal Traversal Sequences, and the cComplexity of Maze Problems. In: Proc. of the 20th Symposium on Foundations of Computer Science, pp. 218–223 (1979)
5. Ambainis, A.: Quantum walks and their algorithmic applications. International Journal of Quantum Information 1(4), 507–518 (2003)
6. Ambainis, A.: Quantum search algorithms. SIGACT News 35(2), 22–35 (2004)
7. Ambainis, A.: Quantum Walk Algorithm for Element Distinctness. SIAM Journal on Computing 37, 210–239 (2007)
8. Ambanis, A., Bach, E., Nayak, A., Vishwanath, A., Watrous, J.: One-dimensional quantum walks. In: Proc. of the 33rd ACM Symposium on Theory of computing, pp. 37–49 (2001)
9. Ambainis, A., Kempe, J., Rivosh, A.: Coins make quantum walks faster. In: Proc. of the 16th ACM-SIAM Symposium on Discrete Algorithms, pp. 1099–1108 (2005)
10. Bennett, C., Bernstein, E., Brassard, G., Vazirani, U.: Strengths and weaknesses of quantum computing. SIAM Journal on Computing 26(5), 1510–1523 (1997)

11. Brassard, G., Høyer, P., Mosca, M., Tapp, A.: Quantum amplitude amplification and estimation. In: Lomonaco Jr., S.J., Brandt, H.E. (eds.) Quantum Computation and Quantum Information: A Millennium Volume, American Mathematical Society. Contemporary Mathematics Series, vol. 305, pp. 53–74 (2002)
12. Buhrman, H., Spalek, R.: Quantum verification of matrix products. In: Proc. of the 17th ACM-SIAM Symposium on Discrete Algorithms, pp. 880–889 (2006)
13. Cleve, R., Ekert, A., Macchiavello, C., Mosca, M.: Quantum algorithms revisited. In: Proc. of the Royal Society A: Mathematical, Physical and Engineering Sciences, vol. 454(1969), pp. 339–354 (1998)
14. Cormen, T., Leiserson, C., Rivest, R., Stein, C.: Introduction to Algorithms, 2nd edn. The MIT Press and McGraw-Hill (2001)
15. Dörn, S., Thierauf, T.: The Quantum Query Complexity of Algebraic Properties. In: Proc. of the 16th International Symposium on Fundamentals of Computation Theory, pp. 250–260 (2007)
16. Grover, L.: A fast quantum mechanical algorithm for database search. In: Proc. of the 28th ACM Symposium on the Theory of Computing, pp. 212–219 (1996)
17. Halmos, P.: Finite-dimensional vector spaces. Springer, Heidelberg (1974)
18. Høyer, P., Mosca, M., de Wolf, R.: Quantum search on bounded-error inputs. In: Baeten, J.C.M., Lenstra, J.K., Parrow, J., Woeginger, G.J. (eds.) ICALP 2003. LNCS, vol. 2719, pp. 291–299. Springer, Heidelberg (2003)
19. Kempe, J.: Discrete Quantum Walks Hit Exponentially Faster. In: Proc. of the International Workshop on Randomization and Approximation Techniques in Computer Science, pp. 354–369 (2003)
20. Kempe, J.: Quantum random walks – an introductory survey. Contemporary Physics 44(4), 307–327 (2003)
21. Kitaev, A.: Quantum measurements and the abelian stabilizer problem. Electronic Colloquium on Computational Complexity (ECCC) 3 (1996)
22. Knuth, D.: Sorting and Searching. The Art of Computer Programming, vol. 3. Addison-Wesley, Reading (1973)
23. Magniez, F., Nayak, A.: Quantum complexity of testing group commutativity. Algorithmica 48(3), 221–232 (2007)
24. Magniez, F., Nayak, A.: Personal communication (2008)
25. Magniez, F., Nayak, A., Roland, J., Santha, M.: Search via quantum walk. In: Proc. of the 39th ACM Symposium on Theory of Computing, pp. 575–584 (2007)
26. Magniez, F., Santha, M., Szegedy, M.: Quantum Algorithms for the Triangle Problem. SIAM Journal of Computing 37(2), 413–427 (2007)
27. Meyer, D.: From quantum cellular automata to quantum lattice gases. Journal of Statistical Physics 85(5-6), 551–574 (1996)
28. Meyer, D.: On the abscence of homogeneous scalar unitary cellular automata. Physical Letter A 223(5), 337–340 (1996)
29. Moore, C., Russell, A.: Quantum Walks on the Hypercube. In: Proc. of the 6th International Workshop on Randomization and Approximation Techniques in Computer Science, pp. 164–178 (2002)
30. Nayak, A., Vishwanath, A.: Quantum walk on the line. Technical Report quant-ph/0010117, arXiv (2000)
31. Pak, I.: Testing commutativity of a group and the power of randomization (2000), Electronic version `http://www-math.mit.edu/~pak/research.html`
32. Richter, P.: Almost uniform sampling via quantum walks. New Journal of Physics (to appear)
33. Schöning, U.: A Probabilistic Algorithm for k -SAT Based on Limited Local Search and Restart. Algorithmica 32(4), 615–623 (2002)

34. Shenvi, N., Kempe, J., Whaley, K.B.: Quantum random-walk search algorithm. Physical Review A 67(052307) (2003)
35. Szegedy, M.: Quantum Speed-Up of Markov Chain Based Algorithms. In: Proc. of the 45th IEEE Symposium on Foundations of Computer Science, pp. 32–41 (2004)
36. Vazirani, U.: On the power of quantum computation. Philosophical Transactions of the Royal Society of London, Series A 356, 1759–1768 (1998)
37. Watrous, J.: Quantum simulations of classical random walks and undirected graph connectivity. Journal of Computer and System Sciences 62(2), 376–391 (2001)

Propositional Projection Temporal Logic, B*ü*chi Automata and ω-Regular Expressions*

Cong Tian and Zhenhua Duan

Institute of Computing Theory and Technology
Xidian University
{ctian,zhhduan}@mail.xidian.edu.cn

Abstract. This paper investigates the language class defined by Propositional Projection Temporal Logic with star (PPTL with star). To this end, B*ü*chi automata are first extended with stutter rule (SBA) to accept finite words. Correspondingly, ω-regular expressions are also extended (ERE) to express finite words. Consequently, by three transformation procedures between PPTL with star, SBA and ERE, PPTL with star is proved to represent exactly the full regular language.

Keywords: Propositional Projection Temporal Logic, B*ü*chi automata, ω-regular expression, expressiveness.

1 Introduction

Temporal logic is a useful formalism for describing sequences of transitions between states in a reactive system. In the past thirty years, many kinds of temporal logics are proposed within two categories, linear-time and branching-time logics. In the community of linear-time logics, the most widely used logics are Linear Temporal Logic (LTL) [?] and its variations. In the propositional framework, Propositional LTL (PLTL) has been proved to have the expressiveness of star-free regular expressions [12,16]. Considering the expressive limitation of PLTL, extensions such as Quantified Linear Time Temporal Logic (QLTL) [13], Extended Temporal Logic (ETL) [9,14] and Linear mu-calculus (νTL) [15] etc, were introduced to PLTL to express the full regular language. Nevertheless, results [17,18,19,20] have shown that temporal logic needs some further extensions in order to support a compositional approach to the specification and verification of concurrent systems. These extensions should enable modular and compositional reasoning about loops and sequential composition as well as about concurrent ones. Therefore, kinds of extensions are proposed. Prominently, one of the important extensions is the addition of the *chop* operator. The work in [9] showed that process logic with both *chop* operator and its reflexive-transitive closure (*chop star*), which is called *slice* in process logic, is strictly more expressive. The resulting logic is still decidable and in fact has the expressiveness of full regular expressions.

* This research is supported by the NSFC Grant No.60433010.

M. Agrawal et al. (Eds.): TAMC 2008, LNCS 4978, pp. 47–58, 2008.

Interval Temporal Logic (ITL) [4] is an easily understood temporal logic with *next, chop* and a projection operator *proj*. In the two characteristic operators, *chop* implements a form of sequential composition while *proj* yields repetitive behaviors. ITL without projection has similar expressiveness as Rosner and Pnueli's choppy logic [3]. Further, addition of the *proj* operator will brings more powerful expressiveness, since repetitive behaviors are allowed. However, no systematic proofs have been given in this aspect. Projection Temporal Logic (PTL) [5] is an extension of ITL. It extends ITL to include infinite models and a new *projection* construct, $(P_1, ..., P_m)\ prj\ Q$, which is much more flexible than the original one. However, in the propositional case[1], the *projection* construct needs further to be extended to *projection star*, $(P_1, ..., (P_i, ..., P_j)^{\circledast}, ..., P_m)\ prj\ Q$, so that it can subsume *chop*, *chop star*, and the original projection (*proj*) in [4]. This extension makes the underlying logic more powerful without the lose of decidability [22].

Within PTL, plenty of logic laws have been formalized and proved [5], and a decision procedure for checking the satisfiability of Propositional Projection Temporal Logic (PPTL) formulas are given in [6,7]. Based on the decision procedure, a model checking approach based on SPIN for PPTL is proposed [8]. Further, in [22], *projection star* is introduced to PPTL, and the satisfiability for PPTL with star formulas is proved to be still decidable. Instinctively, PPTL with star is powerful enough to express the full regular expression. Thus, by employing PPTL with star formulas as the property specification language, the verification of concurrent systems with the model checker SPIN will be completely automatic. This will overcome the error-prone hand-writing of a never claim in the original SPIN since some properties cannot be specified by PLTL formulas. Further, since PPTL with star can subsume *chop* construct, compositional approach for the specification and verification of concurrent systems with SPIN will be allowed. Therefore, we are motivated to give a systematic proof concerning the expressiveness of PPTL with star formulas. To this end, stutter B*ü*chi automata and extended ω-regular expressions are introduced first. Subsequently, by three transformation procedures beteen PPTL with star, SBA, and ERE, PPTL with star is proved to represent exactly the full regular language.

The paper is organized as follows. The syntax and semantics of PPTL with star are briefly introduced in the next section. Section 3 and Section 4 present the definition of stutter B*ü*chi automata and extended regular expressions respectively. Section 5 is devoted to proving the expressiveness of PPTL with star formulas. Precisely, three transformations between PPTL with star, SBA and ERE are given. Finally, conclusions are drawn in Section 6.

2 Propositional Projection Temporal Logic with Star

Our underlying logic is propositional projection temporal logic with star. It extends PPTL to include *projection star*. It is an extension of propositional interval temporal Logic (PITL).

[1] In the first order case, *projection star* is a derived formula.

Syntax: Let *Prop* be a countable set of atomic propositions. The formula P of PPTL with star is given by the following grammar:

$$P ::= p \mid \bigcirc P \mid \neg P \mid P \vee Q \mid (P_1, ..., P_m)\ prj\ Q \mid (P_1, ..., (P_i, ..., P_j)^{\circledast}, ..., P_m)\ prj\ Q$$

where $p \in Prop$, $P_1, ..., P_m$, P and Q are all well-formed PPTL formulas. $\bigcirc$ (*next*), *prj* (*projection*) and $prj^{\circledast}$ (*projection star*) are basic temporal operators.

The abbreviations *true*, *false*, $\wedge$, $\rightarrow$ and $\leftrightarrow$ are defined as usual. In particular, $true \stackrel{\text{def}}{=} P \vee \neg P$ and $false \stackrel{\text{def}}{=} P \wedge \neg P$. In addition, we have the following derived formulas.

$$\begin{array}{rcl@{\qquad}rcl}
empty & \stackrel{\text{def}}{=} & \neg \bigcirc true & more & \stackrel{\text{def}}{=} & \neg empty \\
\bigcirc^0 P & \stackrel{\text{def}}{=} & P & \bigcirc^n P & \stackrel{\text{def}}{=} & \bigcirc(\bigcirc^{n-1} P), n \geq 1 \\
len(0) & \stackrel{\text{def}}{=} & empty & len(n) & \stackrel{\text{def}}{=} & \bigcirc^n\ empty, n \geq 1 \\
skip & \stackrel{\text{def}}{=} & len(1) & \odot P & \stackrel{\text{def}}{=} & empty \vee \bigcirc P \\
P;Q & \stackrel{\text{def}}{=} & (P,Q)\ prj\ empty & \Diamond P & \stackrel{\text{def}}{=} & true; P \\
\Box P & \stackrel{\text{def}}{=} & \neg \Diamond \neg P & P^0 & \stackrel{\text{def}}{=} & empty \\
P^1 & \stackrel{\text{def}}{=} & P & P^n & \stackrel{\text{def}}{=} & (P^{(n)})\ prj\ empty, n \geq 1 \\
P^* & \stackrel{\text{def}}{=} & (P^{\circledast})\ prj\ empty & P^+ & \stackrel{\text{def}}{=} & (P^{\oplus})\ prj\ empty
\end{array}$$

$$\begin{array}{rl}
& (P_1, ..., P_{i-1}, (P_i, ..., P_j)^{(0)}, P_{j+1}, ..., P_m)\ prj\ Q \\
\stackrel{\text{def}}{=} & (P_1, ..., P_{i-1}, P_{j+1}, ..., P_m)\ prj\ Q \\
& (P_1, ..., P_{i-1}, (P_i, ..., P_j)^{(1)}, P_{j+1}, ..., P_m)\ prj\ Q \\
\stackrel{\text{def}}{=} & (P_1, ..., P_{i-1}, P_i, ...P_j, P_{j+1}, ..., P_m)\ prj\ Q \\
& (P_1, ..., P_{i-1}, (P_i, ..., P_j)^{(n)}, P_{j+1}, ..., P_m)\ prj\ Q \\
\stackrel{\text{def}}{=} & (P_1, ..., P_{i-1}, P_i, ...P_j, (P_i, ..., P_j)^{n-1}, P_{j+1}, ..., P_m)\ prj\ Q, n \geq 1 \\
& (P_1, ..., P_{i-1}, (P_i, ..., P_j)^{\oplus}, P_{j+1} ..., P_m)\ prj\ Q \\
\stackrel{\text{def}}{=} & (P_1, ..., P_{i-1}, P_i, ...P_j, (P_i, ..., P_j)^{\circledast}, P_{j+1}, ..., P_m)\ prj\ Q
\end{array}$$

where $\odot$ (weak next), $\Box$ (always), $\Diamond$ (sometimes), ; (chop), $prj^{\oplus}$ (projection plus), $*$ (chop star) and $+$ (chop plus) are derived temporal operators; *empty* denotes an interval with zero length, and *more* means the current state is not the final one over an interval.

Semantics: Following the definition of Kripke's structure [2], we define a state s over *Prop* to be a mapping from *Prop* to $B = \{true, false\}$, $s : Prop \longrightarrow B$. We will use $s[p]$ to denote the valuation of p at state s. An interval σ is a non-empty sequence of states, which can be finite or infinite. The length, $|\sigma|$, of σ is ω if σ is infinite, and the number of states minus 1 if σ is finite. To have a uniform notation for both finite and infinite intervals, we will use extended integers as indices. That is, we consider the set N_0 of non-negative integers and ω, $N_\omega = N_0 \cup \{\omega\}$, and extend the comparison operators, $=, <, \leq$, to N_ω by considering $\omega = \omega$, and for all $i \in N_0$, $i < \omega$. Moreover, we define $\preceq$ as $\leq -\{(\omega, \omega)\}$. To simplify definitions, we will denote σ by $< s_0, ..., s_{|\sigma|} >$, where $s_{|\sigma|}$ is undefined if σ is infinite. With such a notation, $\sigma_{(i..j)}$ $(0 \leq i \preceq j \leq |\sigma|)$ denotes the sub-interval $< s_i, ..., s_j >$

and $\sigma^{(k)}$ $(0 \leq k \preceq |\sigma|)$ denotes $< s_k, ..., s_{|\sigma|} >$. Further, the concatenation $(\cdot)$ of two intervals σ and σ' is defined as follows,

$$\sigma \cdot \sigma' = \begin{cases} \sigma, \text{ if } |\sigma| = \omega \\ < s_0, ..., s_i, s_{i+1}, ... > \text{ if } \sigma =< s_0, ..., s_i >, \sigma' =< s_{i+1}, ... >, i \in N_0 \end{cases}$$

And the fusion of two intervals σ and σ' is also defined as below,

$$\sigma \circ \sigma' = \begin{cases} \sigma, \text{ if } |\sigma| = \omega \\ < s_0, ..., s_i, ... > \text{ if } \sigma =< s_0, ..., s_i >, \sigma' =< s_i, ... >, i \in N_0 \end{cases}$$

Moreover, $\sigma^{\cdot\omega}$ means infinitely many interval σ are concatenated, while $\sigma^{\circ\omega}$ denotes infinitely many σ are fused together.

Let $\sigma =< s_0, s_1, \ldots, s_{|\sigma|} >$ be an interval and $r_1, \ldots, r_h$ be integers $(h \geq 1)$ such that $0 \leq r_1 \leq r_2 \leq \ldots \leq r_h \preceq |\sigma|$. The projection of σ onto $r_1, \ldots, r_h$ is the interval (namely projected interval)

$$\sigma \downarrow (r_1, \ldots, r_h) =< s_{t_1}, s_{t_2}, \ldots, s_{t_l} >$$

where $t_1, \ldots, t_l$ is obtained from $r_1, \ldots, r_h$ by deleting all duplicates. That is, $t_1, \ldots, t_l$ is the longest strictly increasing subsequence of $r_1, \ldots, r_h$. For instance,

$$< s_0, s_1, s_2, s_3, s_4 > \downarrow (0, 0, 2, 2, 2, 3) =< s_0, s_2, s_3 >$$

We need also to generalize the notation of $\sigma \downarrow (r_1, ..., r_m)$ to allow r_i to be ω. For an interval $\sigma =< s_0, s_1, ..., s_{|\sigma|} >$ and $0 \leq r_1 \leq r_2 \leq ... \leq r_h \leq |\sigma|$ $(r_i \in N_\omega)$, we define $\sigma \downarrow () = \varepsilon$, $\sigma \downarrow (r_1, ..., r_h, \omega) = \sigma \downarrow (r_1, ..., r_h)$. This is convenient to define an interval obtained by taking the endpoints (rendezvous points) of the intervals over which $P_1, \ldots, P_m$ are interpreted in the projection construct.

An interpretation is a tuple $\mathcal{I} = (\sigma, k, j)$, where σ is an interval, k is an integer, and j an integer or ω such that $k \preceq j \leq |\sigma|$. We use the notation $(\sigma, k, j) \models P$ to denote that formula P is interpreted and satisfied over the subinterval $< s_k, ..., s_j >$ of σ with the current state being s_k. The satisfaction relation $(\models)$ is inductively defined as follows:

I − prop $\mathcal{I} \models p$ iff $s_k[p] = true$, for any given proposition p
I − not $\mathcal{I} \models \neg P$ iff $\mathcal{I} \not\models P$
I − or $\mathcal{I} \models P \vee Q$ iff $\mathcal{I} \models P$ or $\mathcal{I} \models Q$
I − next $\mathcal{I} \models \bigcirc P$ iff $k < j$ and $(\sigma, k+1, j) \models P$
I − prj $\mathcal{I} \models (P_1, ..., P_m)\ prj\ Q$, if there exist integers $r_0 \leq r_1 \leq ... \leq r_m \leq j$ such that $(\sigma, r_0, r_1) \models P_1, (\sigma, r_{l-1}, r_l) \models P_l, 1 < l \leq m$, and $(\sigma', 0, |\sigma'|) \models Q$ for one of the following σ' :
(a) $r_m < j$ and $\sigma' = \sigma \downarrow (r_0, ..., r_m) \cdot \sigma_{(r_m+1..j)}$ or
(b) $r_m = j$ and $\sigma' = \sigma \downarrow (r_0, ..., r_h)$ for some $0 \leq h \leq m$
I − prj$^{\circledast}$ $\mathcal{I} \models (P_1, ..., (P_i, ..., P_j)^{\circledast}, ..., P_m)\ prj\ Q$ iff $\exists n \in N_0, \mathcal{I} \models (P_1, ..., (P_i, ..., P_j)^n, ..., P_m)\ prj\ Q$ or there exist integers $r_0 \leq r_1 \leq ... \leq r_i \leq ... \leq r_j \leq r_{2i} \leq ... \leq r_{2j} \leq r_{3i} ... \leq r_k \preceq \omega, lim_{k \to \omega} r_k = \omega$, such that $(\sigma, r_{l-1}, r_l) \models P_l, 0 < l < i, (\sigma, r_{l-1}, r_l) \models P_t, l \geq i, t = i + (l\ mod\ (j - i + 1))$, and $(\sigma', 0, |\sigma'|) \models Q$, where $\sigma' = \sigma \downarrow (r_0, ..., r_k, \omega)$, $l\ mod\ 1 = 0$.

Satisfaction and Validity: A formula P is satisfied by an interval σ, denoted by $\sigma \models P$, if $(\sigma, 0, |\sigma|) \models P$. A formula P is called satisfiable if $\sigma \models P$ for some σ. A formula P is valid, denoted by $\models P$, if $\sigma \models P$ for all σ.

3 Büchi Automata with Stutter

Definition 1. A Büchi automaton is a tuple $B = (Q, \Sigma, I, \delta, F)$, where,

- $Q = \{q_0, q_1, ..., q_n\}$ is a finite, non-empty set of locations;
- $\Sigma = \{a_0, a_1, ..., a_m\}$ is a finite, non-empty set of symbols, namely alphabet;
- $I \subseteq Q$ is a non-empty set of initial locations;
- $\delta \subseteq Q \times \Sigma \times Q$ is a transition function;
- $F \subseteq Q$ is a set of accepting locations.

An infinite word w over Σ is an infinite sequence $w = a_0 a_1...$ of symbols, $a_i \in \Sigma$. A run of B over an infinite word $w = a_0 a_1...$ is an infinite sequence $\rho = q_0 q_1...$ of locations $q_i \in Q$ such that $q_0 \in I$ and $(q_i, a_i, q_{i+1}) \in \delta$ holds for all $i \in N_0$. In this case, we call w the word associated with ρ, and ρ the run associated with w. The run ρ is an accepting run iff there exists some $q \in F$ such that $q_i = q$ holds for infinitely many $i \in N_0$. The language $L(B)$ accepted by a Büchi automaton B is the set of infinite words for which there exists some accepting run ρ of B.

Similar to the approach adopted in SPIN [10] for modeling finite behaviors of a system with a Büchi automaton, the stuttering rule is adopted so that the classic notion of acceptance for finite runs (thus words) would be included as a special case in Büchi automata. To apply the rule, we extend the alphabet Σ with a fixed predefined null-label ϵ, representing a no-op operation that is always executable and has no effect. For a Büchi automaton B, the stutter extension of finite run ρ with final state q_n is the ω-run ρ such that $(q_n, \epsilon, q_n)^\omega$ is the suffix of ρ. The final state of the run can be thought to repeat null action ϵ infinitely. It follows that such a run would satisfy the rules for Büchi acceptance if and only if the original final location q_n is in the set F of accepting locations. This means that it indeed generalizes the classical definition of the finite acceptance. In what follows, we denote Büchi automata with the stutter extension, simply as stutter-Büchi automata (SBA for short).

4 Extended Regular Expression

Corresponding to the stutter-Büchi automata, we define a kind of extended ω-regular expression (ERE) which is capable of defining both finite and infinite strings. Let $\Upsilon = \{r_1, ..., r_n\}$ be a finite set of symbols, namely alphabet. The extended ω-regular expressions are defined as follows,

$$\mathsf{ERE} \qquad R ::= \emptyset \mid \epsilon \mid r \mid R + R \mid R \bullet R \mid R^\omega \mid R^*$$

where $r \in \Upsilon$, ϵ denotes an empty string; $+$, $\bullet$ and $*$ are union, concatenation and Kleene (star) closure respectively; R^ω means infinitely many R are concatenated. In what follows, we use ERE to denote the set of extended regular expressions.

Before defining the language expressed by the extended regular expressions, we first introduce strings and operations on strings. A string is a finite or infinite sequence of symbols, $a_0 a_1 ... a_i ...$, where each a_i is chosen from the alphabet Υ. The length of a finite string w, denoted by $|w|$, is the number of the symbols in

w while the length of an infinite string is ω. For two strings w and w', $w \bullet w'$, w^* and w^ω are defined as follows,

$$w \bullet w' = \begin{cases} w, \text{ if } |w| = \omega \\ a_0...a_i a_{i+1}..., \text{ if } w = a_0...a_i, \text{ and } w' = a_{i+1}... \end{cases}$$

$$w^\omega = \begin{cases} w, \text{ if } |w| = \omega \\ \underbrace{a_0...a_i...a_0...a_i...}_{\omega \text{ times}}, \text{ if } w \text{ is finite and } w = a_0...a_i \end{cases}$$

$$w^* = \begin{cases} w, \text{ if } |w| = \omega \\ \exists n \in N_\omega, \underbrace{a_0...a_i...a_0...a_i}_{n \text{ times}}, \text{ if } w \text{ is finite and } w = a_0...a_i \end{cases}$$

Further, if W and W' are two sets of strings. Then $W \bullet W'$, W^ω and W^* are defined as follows,

$$\begin{aligned} W \bullet W' &= \{w \bullet w' \mid w \in W \text{ and } w' \in W'\} \\ W^\omega &= \{w^\omega \mid w \in W\} \\ W^* &= \{w^* \mid w \in W\} \end{aligned}$$

Accordingly, the language $L(R)$ expressed by extended regular expression R is given by,

Lr_1	$L(\emptyset) = \emptyset$	Lr_2	$L(r) = \{r\}$
Lr_3	$L(\epsilon) = \{\epsilon\}$	Lr_4	$L(R+R) = L(R) \cup L(R)$
Lr_5	$L(R \bullet R) = L(R) \bullet L(R)$	Lr_6	$L(R^\omega) = L(R)^\omega$
Lr_7	$L(R^*) = L(R)^*$		

For a string w, if $w \in L(R)$, w is called a word of expression R.

For convenience, we use PPTL* to denote the set of all PPTL with star formulas, SBA the set of all Stutter Bϋchi Automata, and ERE the set of all Extended ω-Regular Expressions. Further, the language classes determined by PPTL*, SBA and ERE are represented by L(PPTL*), L(SBA) and L(ERE) respectively. That is,

$$\begin{aligned} L(PPTL^*) &= \{L(P) | P \in PPTL^*\} \\ L(SBA) &= \{L(B) | B \in SBA\} \\ L(ERE) &= \{L(R) | R \in ERE\} \end{aligned}$$

5 Relationship between PPTL*, ERE and SBA

Even though the extended ω-regular expressions, PPTL with star formulas, and stutter-Büchi automata describe languages fundamentally in different ways, however, it turns out that they represent exactly the same class of languages, named the "full regular languages" as concluded in Theorem 1.

Theorem 1. PPTL with star formulas, extended regular expressions and stutter Bϋchi automata have the same expressiveness. □

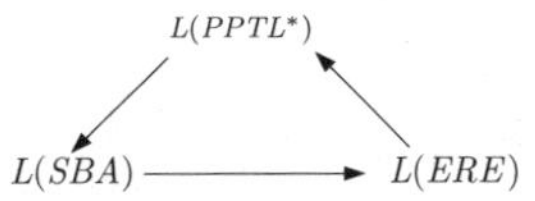

Fig. 1. The relationship between three language classes

In order to prove this theorem, we will show the following facts: (1) For any P, there is an SBA B such that $L(B) = L(P)$; (2) For any SBA B, there is an ERE R such that $L(R) = L(B)$; (3) For any ERE R, there is a PPTL with star formula P such that $L(P) = L(R)$. The relationship is depicted in Fig.1, where an arc from language class X to Y means that each language in X can also be defined in Y. This convinces us that three language classes are equivalence.

For the purpose of transformations between PPTL*, SBA and ERE, we define the set Q_p of atomic propositions appearing in PPTL with star formula Q, $|Q_p| = l$, and further need define alphabets Σ and Υ for SBA and ERE. To do so, We first define sets A_i, $1 \leq i \leq l$, as follows,

$$A_i = \{\{\dot{q_{j_1}}, ..., \dot{q_{j_i}}\} \mid q_{j_k} \in Q_p, \dot{q_{j_k}} \text{ denotes } q_{j_k} \text{ or } \neg q_{j_k}, 1 \leq k \leq i\}$$

Then, $\Sigma = \bigcup_{i=1}^{l} A_i \cup \{true\} \cup \{\epsilon\}$, and $\Upsilon = \bigcup_{i=1}^{l} A_i \cup \{true\}$. For instance, if $Q_p = \{p_1, p_2, p_3\}$, it is obtained that, $A_1 = \{\{\dot{p_1}\}, \{\dot{p_2}\}, \{\dot{p_3}\}\}$, $A_2 = \{\{\dot{p_1}, \dot{p_2}\}, \{\dot{p_1}, \dot{p_3}\}, \{\dot{p_2}, \dot{p_3}\}\}$, $A_3 = \{\{\dot{p_1}, \dot{p_2}, \dot{p_3}\}\}$. So, we have,$\Sigma = \bigcup_{i=1}^{3} A_i \cup \{true\} \cup \{\epsilon\} = \{\{\dot{p_1}\}, \{\dot{p_2}\}, \{\dot{p_3}\}, \{\dot{p_1}, \dot{p_2}\}, \{\dot{p_1}, \dot{p_3}\}, \{\dot{p_2}, \dot{p_3}\}, \{\dot{p_1}, \dot{p_2}, \dot{p_3}\}, true, \epsilon\}$, $\Upsilon = \bigcup_{i=1}^{3} A_i \cup \{true\} = \{\{\dot{p_1}\}, \{\dot{p_2}\}, \{\dot{p_3}\}, \{\dot{p_1}, \dot{p_2}\}, \{\dot{p_1}, \dot{p_3}\}, \{\dot{p_2}, \dot{p_3}\}, \{\dot{p_1}, \dot{p_2}, \dot{p_3}\}, true\}$. Obviously, for each $r \in \Upsilon$, r is a set of atomic propositions or their negations, denoted by $true$ or $\{\dot{q_i}, ..., \dot{q_j}\}$, where $1 \leq i \leq j \leq l$.

5.1 From PPTL with Star Formulas to SBAs

For PPTL with star formulas, their normal forms are the same as for PPTL formulas [6,7]. In [22], an algorithm was given for transforming a PPTL with star formula to its normal form. Further, based on the normal form, labeled normal form graph (LNFG) for PPTL with star formulas are constructed to precisely characterize the models of PPTL with star formulas. Also an algorithm was given to construct the LNFG of a PPTL with star formula [22]. The details about normal forms and LNFGs can be found in [7,22]. Here we focus on how to transform an LNFG G to an SBA B (corresponding SBA of G). For the clear presentation of the transformation, LNFGs are briefly introduced first.

The LNFG of formula P can be expressed as a graph, $G = (V, E, v_0, V_f)$, where V denotes the set of nodes in the LNFG, E is the set of directed edges among V, $v_0 \in V$ is the initial (or root) node, and $V_f \subseteq V$ denotes the set of nodes with finite label F. V and E in G can inductively be constructed by algorithm *LNFG* composed in [7]. Actually, in an LNFG, a node $v \in V$ denotes a PPTL with star formula and initial node is P itself while an edge from node v_i to v_j is a tuple (v_i, Q_e, v_j) where v_i and v_j are PPTL with star formulas and $Q_e \equiv \bigwedge_i \dot{q_i}$, q_i is an atomic proposition, $\dot{q_i}$ denotes q_i or $\neg q_i$. The following is an example of LNFG.

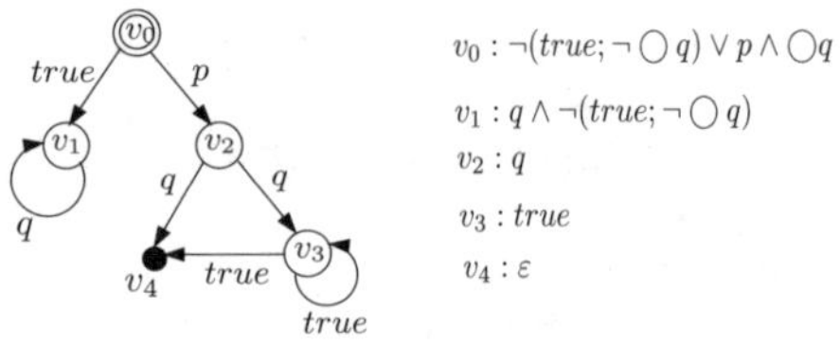

Fig. 2. LNFG of formula $\neg(true; \neg \bigcirc q) \vee p \wedge \bigcirc q$

Example 1. The LNFG of formula $\neg(true; \neg \bigcirc q) \vee p \wedge \bigcirc q$.
As shown in Fig.2, the LNFG of formula $\neg(true; \neg \bigcirc q) \vee p \wedge \bigcirc q$ is $G = \{V, E, v_0, V_f\}$, where $V = \{v_0, v_1, v_2, v_3, v_4\}$; $E = \{(v_0, true, v_1), (v_0, p, v_2), (v_1, q, v_1), (v_2, q, v_4), (v_2, q, v_3), (v_3, true, v_3), (v_3, true, v_4)\}$; the root node is v_0; and $V_f = \emptyset$. □

Factually, an LNFG contains all the information of the corresponding SBA. The set of nodes is in fact the set of locations in the corresponding SBA; each edge (v_i, Q_e, v_j) forms a transition; there exists only one initial location, the root node; the set of accepting locations consists of ε node and the nodes which can appear in infinite paths for infinitely many times. Given an LNFG $G = (V, E, v_0, V_f)$ of formula P, an SBA of formula P, $B = (Q, \Sigma, I, \delta, F)$, over an alphabet Σ can be constructed as follows.

• Sets of the locations Q and the initial locations I: $Q = V$, and $I = \{v_0\}$.
• Transition δ: Let $\dot{q_k}$ be an atomic proposition or its negation. We define a function $atom(\bigwedge_{k=1}^{m_0} \dot{q_k})$ for picking up atomic propositions or their negations appearing in $\bigwedge_{k=1}^{m_0} \dot{q_k}$ as follows,

$$\begin{aligned} atom(true) &= true \\ atom(\dot{q_k}) &= \begin{cases} \{q_k\}, & \text{if } \dot{q_k} \equiv q_k \ 1 \le k \le l \\ \{\neg q_k\}, & \text{otherwise} \end{cases} \\ atom(\bigwedge_{k=1}^{m_0} \dot{q_k}) &= atom(\dot{q_1}) \cup atom(\bigwedge_{k=2}^{m_0} \dot{q_k}) \end{aligned}$$

For each $e_i = (v_i, Q_e, v_{i+1}) \in E$, there exists $v_{i+1} \in \delta(v_i, atom(Q_e))$. For node ε, $\delta(\varepsilon, \epsilon) = \{\varepsilon\}$.
• Accepting locations F: We have proved in [6,7] that infinite paths in an LNFG precisely characterize the infinite models of the corresponding formula. In fact, there exists an infinite path if and only if there are some nodes appearing in the path for infinitely many times. Therefore, the nodes appearing in infinite paths for infinitely many times are defined as the accepting locations in the SBA. In addition, by employing the stutter extension rule, ε node is also an accepting location.

Lemma 1. For any PPTL with star formula P, there is an SBA B such that $L(B) = L(P)$. □

Formally, algorithm LNFG-SBA is useful for transforming an LNFG to an SBA. Also Example 2 is given to show how the algorithm works.

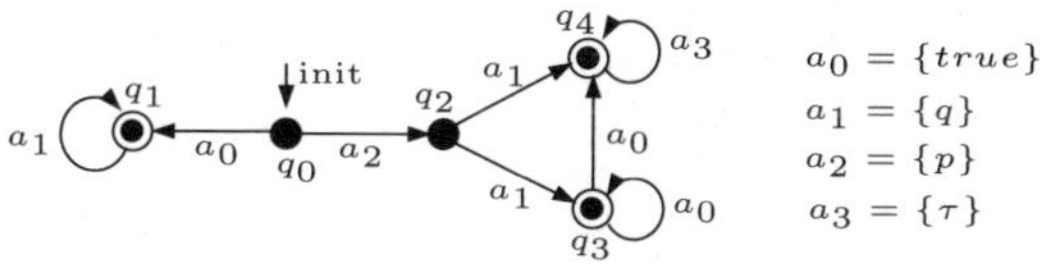

Fig. 3. Stutter-Büchi automaton of formula $P \equiv \neg(true; \neg \bigcirc q) \vee p \wedge \bigcirc q$

Example 2. Constructing the SBA, $B=(Q, \Sigma, I, \delta, F)$, from the LNFG in Example 1.

As depicted in Fig.3, the set of locations, $Q=\{q_0, q_1, q_2, q_3, q_4\}$, comes from V directly. The set of initial locations $I=\{q_0\}$ is root node v_0 in G. The set of the accepting locations $F=\{q_1, q_3, q_4\}$ consists of nodes v_1, v_3 appearing in loops and ε node in V. The transitions, $\delta(q_0, a_0)=\{q_1\}$, $\delta(q_0, a_2)=\{q_2\}$, $\delta(q_1, a_1)=\{q_1\}$, $\delta(q_2, a_1)=\{q_3, q_4\}$, $\delta(q_3, a_0)=\{q_3, q_4\}$, $\delta(q_4, a_3)=\{q_4\}$ are formalized according to the edges in E. □

```
Function LNFG-SBA(G)
/* precondition: G = (V, E, v0, Vf) is the LNFG of PPTL formula P*/
/* postcondition: LNFG-SBA(G) computes an SBA B = (Q, Σ, I, δ, F) from G*/
begin function
    Q = ∅; F = φ; I = ∅;
    for each node vi ∈ V,
        add a state qi to Q, Q = Q ∪ {qi};
        if vi is ε, F = F ∪ {qi}; δ(qi, ε) = {qi};
        else if vi appears in loops and vi ∉ Vf , F = F ∪ {qi};
    end for
    if q0 ∈ Q, I = I ∪ {q0};
    for each edge e = (vi, Pe, vj) ∈ E,
        qj ∈ δ(qi, atom(Pe));
    end for
   return B = (Q, Σ, I, δ, F)
end function
```

5.2 From SBAs to EREs

For the proof of the language $L(A)$ of any finite state automaton A being regular [24], Arden's rule [11] plays an important role.

Lemma 2. (Arden's Rule) For any sets of strings S and T, the equation $X = S \bullet X + T$ has $X = S^* \bullet T$ as a solution. Moreover, this solution is unique if $\epsilon \notin S$. □

From now on we shall often drop the concatenation symbol $\bullet$, writing SX for $S \bullet X$ etc. In the following, we show how Arden's rule is used to equivalently transform an SBA to an ERE.

Given a stutter-Büchi automaton B with $Q = \{q_0, ..., q_n\}$ and the starting location q_0. For $1 \leq i \leq n$, let X_i denote the ERE where $L(X_i)$ equals to the set of strings accepted by the sub-automaton of B starting at location q_i; thus $L(B) = L(X_0)$. We can write an equation for each X_i in terms of the languages defined by its successor locations. For example, for the stutter-Büchi automaton B in Example 2, we have,

$$(0)\ X_0 = a_0X_1 + a_2X_2 \quad (1)\ X_1 = a_1X_1 + a_1^\omega$$
$$(2)\ X_2 = a_1X_4 + a_1X_3 \quad (3)\ X_3 = a_0X_3 + a_0X_4 + a_0^\omega$$
$$(4)\ X_4 = a_3X_4 + a_3^\omega$$

Note that X_1, X_3 and X_4 contains a_1^ω, a_0^ω and a_3^ω respectively because q_1, q_3 and q_4 are accepting states with self-loops[2]. Now we use Arden's rule to solve the equations. First, for (4), since a_3 is ϵ,

$$X_4 = a_3X_4 + a_3^\omega = a_3^*a_3^\omega = a_3^\omega = \epsilon$$

Replacing X_4 in (3),

$$X_3 = a_0X_3 + a_0X_4 + a_0^\omega = a_0X_3 + a_0 + a_0^\omega = a_0^*a_0 + a_0^*a_0^\omega = a_0^*a_0 = true^*true$$

Replacing X_3 and X_4 in (2),

$$X_2 = a_1X_4 + a_1X_3 = \{q\} + \{q\}true^*true$$

For (1),

$$X_1 = a_1X_1 + a_1^\omega = a_1^*a_1^\omega = a_1^\omega = \{q\}^\omega$$

Finally, replacing X_1 and X_2 in (0), we have,

$$X_0 = a_0X_1 + a_2X_2 = a_0\{q\}^\omega + a_2(\{q\} + \{q\}true^*true)$$
$$= true\{q\}^\omega + \{p\}\{q\} + \{p\}\{q\}true^*true$$

Lemma 3. For any SBA B, there is an ERE R such that $L(R) = L(B)$. □

5.3 From EREs to PPTL with Star Formulas

Let Γ be the set of all models of PPTL with star. For any extended regular expression $R \in$ ERE, we can construct a PPTL with star formula F_R such that, (1) for any model $\sigma \in \Gamma$, if $\sigma \models F_R$, then $\Omega(\sigma) \in L(R)$; and (2) for any word $w \in L(R)$, there exists $\sigma \in \Gamma$, $\sigma \models F_R$ and $\Omega(\sigma) = w$. The mapping function $\Omega : \Gamma \to \Upsilon^*$ from models of PPTL formulas to words of extended regular expression is defined as follows,

$$\Omega(\sigma) = \begin{cases} \epsilon, & \text{if } |\sigma| = 0 \\ A(s_0)...A(s_{j-1}) & \text{if } \sigma \text{ is finite and } \sigma =< s_0, ..., s_j >, j \geq 1 \\ A(s_0)...A(s_j)... & \text{if } \sigma \text{ is infinite and } \sigma =< s_0, ..., s_j, ... > \end{cases}$$

where $A(s_i)$ denotes *true*, or the set of propositions and their negations holding at state s_i. It is not difficult to prove that $\Omega(\sigma_1 \circ \sigma_2) = \Omega(\sigma_1) \bullet \Omega(\sigma_2)$, $\Omega(\sigma^{\circ\omega}) = \Omega(\sigma)^\omega$ and $\Omega(\sigma^{\circ *}) = \Omega(\sigma)^*$. F_R is constructed inductively on the structure of R.

$$F_\emptyset \stackrel{\text{def}}{=} false$$
$$F_\epsilon \stackrel{\text{def}}{=} empty$$
$$F_r \stackrel{\text{def}}{=} \begin{cases} \dot{p_i} \wedge ... \wedge \dot{p_j} \wedge skip, & \text{if } r = \{\dot{p_i}, ..., \dot{p_j}\}, 1 \leq i \leq j \leq l \\ true \wedge skip, & \text{if } r = true \end{cases}$$

where $r \in \Upsilon$. Inductively, if R_1 and R_2 are extended regular expressions, then

$$F_{R_1+R_2} \stackrel{\text{def}}{=} F_{R_1} \vee F_{R_2}$$
$$F_{R_1 \bullet R_2} \stackrel{\text{def}}{=} F_{R_1}; F_{R_2}$$
$$F_{R^\omega} \stackrel{\text{def}}{=} F_R^* \wedge \Box more$$
$$F_{R_1^*} \stackrel{\text{def}}{=} F_{R_1}^*$$

[2] For finite state automata, X_i contains ϵ if q_i is accepted.

Now we need to prove that, for any $R \in$ ERE and $\sigma \in \Gamma$, if $\sigma \models F_R$, then $\Omega(\sigma) \in L(R)$; for any w, if $w \in L(R)$, then there exists $\sigma \in \Gamma$ such that $\Omega(\sigma) = w$ and $\sigma \models F_R$.

Lemma 4. For any ERE R, there is a PPTL with star formula P such that $L(P) = L(R)$. □

Example 3. Constructing PPTL with star formula from the extended regular expression, $true\{q\}^\omega + \{p\}\{q\} + \{p\}\{q\}true^*true$ obtained in Example 2.

$$
\begin{aligned}
& F_{true\{q\}^\omega+\{p\}\{q\}+\{p\}\{q\}true^*true} \\
\equiv\ & F_{true\{q\}^\omega} \vee F_{\{p\}\{q\}} \vee F_{\{p\}\{q\}true^*true} \\
\equiv\ & F_{true}; F_{\{q\}^\omega} \vee F_{\{p\}}; F_{\{q\}} \vee F_{\{p\}}; F_{\{q\}}; F_{true^*}; F_{true} \\
\equiv\ & true \wedge skip; F_{\{q\}^*} \wedge \Box more \vee p \wedge skip; q \wedge skip \vee p \wedge skip; q \wedge skip; \\
& (true \wedge skip)^*; true \wedge skip \\
\equiv\ & skip; (q \wedge skip)^* \wedge \Box more \vee p \wedge skip; q \wedge skip \vee p \wedge skip; q \wedge skip; \\
& skip^*; skip
\end{aligned}
$$
□

6 Conclusions

Further, it is readily to prove the following useful conclusions concerning characters of fragments of PPTL with star. To avoid abuse of notations, we use an expression like L(next, chop) to refer to the specific fragment of PPTL with star with temporal operators *next*, *chop* and the basic connections in the typical propositional logic.

1 L(chop) has the same expressiveness as star-free regular expressions without ϵ.
2 L(next, chop) has the same expressiveness as star-free regular expressions.
3 L(next, prj) has the same expressiveness as regular expressions without ω.
4 L(next, chop, chop*) has the same expressiveness as full regular expressions.

In this paper, we have proved that the expressiveness of PPTL with star is the same as the full regular expressions. Also, the proof itself provides approaches to translate a PPTL with star formula to an equivalent Buchi automaton, a Buchi automaton to an equivalent extended ω-regular expression, and an extended ω-regular expression to a PPTL with star formula. This enables us to specify and verify concurrent systems by compositional approach with PPTL with star. Further, we have developed a model checker based on SPIN for PPTL with star. Therefore, any systems with regular properties can be automatically verified within SPIN using PPTL with star.

References

1. Pnueli, A.: The temporal logic of programs. In: Proceedings of the 18th IEEE Symposium on Foundations of Computer Science, pp. 46–67.2 IEEE, New York (1977)
2. Kripke, S.A.: Semantical analysis of modal logic I: normal propositional calculi. Z. Math. Logik Grund. Math. 9, 67–96 (1963)
3. Rosner, R., Pnueli, A.: A choppy logic. In: First Annual IEEE Symposium on Logic In Computer Science. LICS, pp. 306–314 (1986)

4. Moszkowski, B.: Reasoning about digital circuits. Ph.D Thesis, Department of Computer Science, Stanford University. TRSTAN-CS-83-970 (1983)
5. Duan, Z.: An Extended Interval Temporal Logic and A Framing Technique for Temporal Logic Programming. PhD thesis, University of Newcastle Upon Tyne (May 1996)
6. Duan, Z., Tian, C.: Decidability of Propositional Projection Temporal Logic with Infinite Models. In: Cai, J.-Y., Cooper, S.B., Zhu, H. (eds.) TAMC 2007. LNCS, vol. 4484, pp. 521–532. Springer, Heidelberg (2007)
7. Duan, Z., Tian, C., Zhang, L.: A Decision Procedure for Propositional Projection Temporal Logic with Infinite Models. Acta Informatica 45, 43–78 (2008)
8. Tian, C., Duan, Z.: Model Checking Propositional Projection Temporal Logic Based on SPIN. In: Butler, M., Hinchey, M.G., Larrondo-Petrie, M.M. (eds.) ICFEM 2007. LNCS, vol. 4789, pp. 246–265. Springer, Heidelberg (2007)
9. Wolper, P.L.: Temporal logic can be more expressive. Information and Control 56, 72–99 (1983)
10. Holzmann, G.J.: The Model Checker Spin. IEEE Trans. on Software Engineering 23(5), 279–295 (1997)
11. Arden, D.: Delayed-logic and finite-state machines. In: Theory of Computing Machine Design, Univ. of Michigan Press, pp. 1–35 (1960)
12. Gabbay, D., Pnueli, A., Shelah, S., Stavi, J.: On the temporal analysis of fairness. In: POPL 1980: Proceedings of the 7th ACM SIGPLAN-SIGACT symposium on Principles of programming languages, pp. 163–173. ACM Press, New York (1980)
13. Sistla, A.P.: Theoretical issues in the design and verification of distributed systems. PhD thesis, Harvard University (1983)
14. Vardi, M.Y., Wolper, P.: Yet another process logic. In: Clarke, E., Kozen, D. (eds.) Logic of Programs 1983. LNCS, vol. 164, pp. 501–512. Springer, Heidelberg (1984)
15. Vardi, M.Y.: A temporal fixpoint calculus. In: POPL 1988, pp. 250–259 (1988)
16. McNaughton, R., Papert, S.A.: Counter-Free Automata (M.I.T research monograph no.65). The MIT Press, Cambridge (1971)
17. Barringer, H., Kuiper, R., Pnueli, A.: Now You May Compose Temporal Logic Specifications. In: Proc. 16th STOC, pp. 51–63 (1984)
18. Barringer, H., Kuiper, R., Pnueli, A.: The Compositional Temporal Approach to CSP-like Language. In: Proc.IFIP Conference, The Role of Abstract Models in Information Processing (January 1985)
19. Nguyen, V., Demers, A., Gries, D., Owicki, S.: Logic of Programs 1985. LNCS, vol. 193, pp. 237–254. Springer, Heidelberg (1985)
20. Nguyen, V., Gries, D., Owicki, S.: A Model and Temporal Proof System for Networks of Processes. In: Proc. 12th POPL, pp. 121–131 (1985)
21. Harel, D., Peleg, D.: Process Logic with Regular Formulas. Theoretical Computer Science 38, 307–322 (1985)
22. Tian, C., Duan, Z.: Complexity of Propositional Projection Temporal Logic with Star. Technical Report No.25, Institute of computing Theory and Technology, Xidian University, Xian P.R.China (2007)
23. Arden, D.: Delayed-logic and finite-state machines. In: Theory of Computing Machine Design, pp. 1–35. Univ. of Michigan Press (1960)
24. Milner, R.: Communicating and Mobile System: The -Calculus. Cambridge University Press, Cambridge (1999)

Genome Rearrangement Algorithms for Unsigned Permutations with $O(\log n)$ Singletons*

Xiaowen Lou and Daming Zhu

School of Computer Science and Technology,
Shandong University
xwlou@mail.sdu.edu.cn, dmzhu@sdu.edu.cn

Abstract. Reversal, transposition and transreversal are common events in genome rearrangement. The genome rearrangement sorting problem is to transform one genome into another using the minimum number of rearrangement operations. Hannenhalli and Pevzner discovered that singleton is the major obstacle for unsigned reversal sorting. They also gave a polynomial algorithm for reversal sorting on those unsigned permutations with $O(\log n)$ singletons. This paper involves two aspects. (1) We describe one case for which Hannenhalli and Pevzner's algorithm may fail, and propose a corrected algorithm for unsigned reversal sorting. (2) We propose a $(1+\varepsilon)$-approximation algorithm for the weighted sorting problem on unsigned permutations with $O(\log n)$ singletons. The weighted sorting means: sorting a permutation by weighted reversals, transpositions and transreversals, where reversal is assigned weight 1 and transposition(including transreversal) is assigned weight 2.

1 Introduction

The genome rearrangement problem is to find a shortest sequence of evolutionary events (such as reversals, transpositions, and transreversals) that transform one genome into another. Genomes can be represented as signed or unsigned permutations, depending on the data got from experiments. It is well known that this problem is equivalent to that of sorting a permutation by the same set of operations into the identity permutation.

Sorting unsigned permutations by reversals is proved to be NP-hard by Caprara [4]. This problem can be approximated to 1.375 [2] as we know. However, sorting signed permutations by reversals can be solved exactly in polynomial time [6]. Hannenhalli and Pevzner discovered that singleton is the major obstacle for sorting unsigned permutations by reversals. They designed a polynomial algorithm for unsigned permutations with $O(\log n)$ singletons [7]. However, there is one case for which Hannenhalli and Pevzner's algorithm may fail. We describe the case and propose a corrected algorithm.

Bafna and Pevzner suggested the sorting problem that considers reversals and transpositions simultaneously helps to understand the genome rearrangements related to the evolution of mammalian and viral [3]. Eriksen observed that an algorithm looking for the minimal number of such operations will produce a solution heavily biased towards transposition. Instead, he studied the weighted reversal/ transposition/ transreversal sorting

* Supported by (1) National Nature Science Foundation of China, 60573024. (2) Chinese National 973 Plan, previous special, 2005cca04500.

M. Agrawal et al. (Eds.): TAMC 2008, LNCS 4978, pp. 59–69, 2008.

problem, which assigns 1 to reversal and 2 to transposition or transreversal. The task is to sort a permutation by reversals or transpositions or transreversals with the minimum value of *rev*+2*trp*, where *rev* and *trp* is the number of reversals and transpositions/ transreversals respectively. Eriksen designed a (1+ε)-approximation algorithm for the weighted sorting of signed permutations [5].

In this paper, we focus on the same weighted sorting problem for unsigned permutations. We present a (1+ε)-approximation algorithm for the sorting of unsigned permutations by weighted reversal/ transposition/ transreversals, where the permutation is restricted to containing $O(\log n)$ singletons.

2 Preliminaries

Let $\pi=[g_1,\dots,g_n]$ be a permutation of $\{1,\dots,n\}$. It is also known as an unsigned permutation because each element of π has no signs. If each element of π has been added a sign of '+' or '−' to represent gene's direction, then it is called a *signed permutation*. Let $\pi=[g_1,\dots,g_n]$ be an unsigned permutation and $\pi'=[g'_1,\dots,g'_n]$ be a signed permutation. We say π' is a *spin* of π if either $g'_i=+g_i$ or $g'_i=-g_i$ for $1\le i\le n$. A *segment* of π is a sequence of consecutive elements in π. For example, segment $[g_i,\dots,g_j]$ $(i\le j)$ of π contains all the elements from g_i to g_j. We consider three kinds of rearrangement operations: reversal, transposition and transreversal. A *reversal* $r(i,j)$ $(i<j)$ cuts the segment $[g_{i+1},\dots,g_j]$ out of π, reverses it, and then pastes it back to exactly where it is cut; hence the permutation becomes $\sigma=[g_1,\dots,g_i,g_j,\dots,g_{i+1},g_{j+1},\dots,g_n]$. A *transposition* $t(i,j,k)$ $(i<j<k)$ exchanges two consecutive segments $[g_{i+1},\dots,g_j]$ and $[g_{j+1},\dots,g_k]$ of π, and the permutation becomes $\sigma=[g_1,\dots,g_i,g_{j+1},\dots,g_k,g_{i+1},\dots,g_j,g_{k+1},\dots,g_n]$. A *transreversal* $t_r(i,j,k)$ $(i<j<k)$ reverses the segment $[g_{i+1},\dots,g_j]$ while exchanging the two consecutive segments $[g_{i+1},\dots,g_j]$ and $[g_{j+1}\dots g_k]$, and the permutation becomes $\sigma=[g_1,\dots,g_i,g_{j+1},\dots,g_k,g_j,\dots,g_{i+1},g_{k+1},\dots,g_n]$. Reversing a segment of π' changes the direction of each element in the segment. Thus for a signed permutation π', each element in the reversed segment also has its sign flipped.

The *reversal sorting* problem asks to transform π into the identity permutation ι by the minimum number of reversals, where the signed identity permutation is $\iota=[+1,+2,\dots,+n]$, the unsigned identity permutation is $\iota=[1,2,\dots,n]$. And this number is called the *reversal distance* of π, denoted by $d_r(\pi)$. A *reversal scenario* is a sequence of reversal operations that transform π into ι. The *weighted sorting* problem asks to transform π into ι by the minimum weighted sum number of reversals, transpositions, and transreversals, where one reversal counts for once and one transposition or transreversal counts for twice. A *weighted scenario* is a sequence of such rearrangement operations that transform π into ι. Formally, the *weighted distance* of π is defined as: $d_{r+t}(\pi)=min_{\phi\in\Phi}\{rev(\phi)+2trp(\phi)\}$, where Φ is the set of weighted scenarios sorting π, $rev(\phi)$ and $trp(\phi)$ is the number of reversals and transposition/transreversals in a certain weighted scenario ϕ respectively. Among all the 2^n spins of an unsigned permutation π, if there is a spin π' satisfying $d_r(\pi')=d_r(\pi)$, call π' an *optimal r-spin*; if there is a spin π' satisfying $d_{r+t}(\pi')=d_{r+t}(\pi)$, call π' an *optimal rt-spin*.

Write $i\sim j$ if $|i-j|=1$. For an unsigned permutation π, a pair of consecutive elements (g_i,g_{i+1}) form an *adjacency* if $g_i\sim g_{i+1}$; otherwise a *breakpoint*. A segment $[g_i,\dots,g_j]$

of π is called a *strip* if each (g_k, g_{k+1}) is an adjacency for $i \le k < j$, but both (g_{i-1}, g_i) and (g_j, g_{j+1}) are breakpoints. A strip of one element is a *singleton*; a strip of two elements is called a *2-strip*; and otherwise a *long strip*. If π has no singletons, call it a *singleton-free permutation*. A strip $s=[g_i, \ldots, g_j]$ is *increasing (decreasing)* if $g_i<g_j$ $(g_i>g_j)$. Let π' be a spin of π, an increasing (decreasing) strip s of π is *canonical* in π' if all elements of s are positive (negative) in π'. An increasing (decreasing) strip s of π is *anti-canonical* in π' if all elements of s are negative (positive) in π'. An anti-canonical 2-strip is called an *anti-strip*. If every strip in π' is canonical, then call π' a *canonical spin* of π. Two spins π'_1 and π'_2 are *twins*(with respect to segment $s=[g_i, \ldots, g_j]$) if they differ only in the signs of elements in s.

For a signed permutation π', there exists a polynomial algorithm computing the reversal distance. Here are some details. First, transform a signed permutation $\pi' = [g'_1, \ldots, g'_n]$ of order n to a permutation of order $2(n+1)$ as follows: replace the positive element $+x$ by $(2x-1, 2x)$, and the negative element $-x$ by $(2x, 2x-1)$. Then add 0 at the beginning and $2n+1$ at the end. Such a permutation is called the *extended permutation* of π', denoted as π''. The reversal operation on π'' never touches 0 and $2n+1$, and never breaks the pair $(2x-1, 2x)$ or $(2x, 2x-1)$. Therefore, it has the same effect of operating on π'. The *breakpoint graph* $G(\pi'')$ of π'' is defined as follows: set a vertex of $G(\pi'')$ for each element of π''; draw a *gray edge* between vertices $2x$ and $2x+1$ if their positions are not consecutive in π''; draw a *black edge* between vertices g_{2i} and g_{2i+1} if they form a breakpoint. Note that every vertex has degree 2 in $G(\pi'')$, so it can be uniquely decomposed into *alternating* cycles, i.e., cycle with every two consecutive edges having distinct colors. We use l-cycle to denote an alternating cycle with l black edges. Note that the length of a cycle is at least 2. A cycle is *oriented* if, when we travel along it, at least one black edge is traveled from left to right and at least one black edge is traveled from right to left; otherwise, an *unoriented cycle*. Two gray edges $e_1=(g_{i_1}, g_{i_2})$ and $e_2=(g_{j_1}, g_{j_2})$ are *crossing* if $i_1<j_1<i_2<j_2$ or $j_1<i_1<j_2<i_2$. Cycles C_1 and C_2 are *crossing* if there exist crossing gray edges $e_1 \in C_1$ and $e_2 \in C_2$. If we take each alternating cycle of $G(\pi'')$ as a vertex, and draw an edge between two vertices if the cycles they present are crossing, we get another graph G_π. Those alternating cycles of $G(\pi'')$ denoted by vertices belonging to the same connected component of G_π together are referred to as a *component* of $G(\pi'')$. For every component of $G(\pi'')$, if it contains at least one oriented cycle, call it an *oriented component*; otherwise, an *unoriented component*. Imagining we bend a permutation into a circle of counterclockwise order and make $(0, 2n+1)$ an adjacency. If there is an interval on the circle which contains one and only one unoriented component, then this component is called a *hurdle*. If an unoriented component is not a hurdle, call it a *non-hurdle*. Note that a non-hurdle always stretches over an interval that contains a hurdle. A *hurdle* is called a *super hurdle* if removing it will turn a non-hurdle into a hurdle [5][6]. If $G(\pi'')$ has an odd number of hurdles and all these hurdles are super hurdles , then the permutation π' is referred to as a *fortress*. Let $b(\pi')$, $c(\pi')$, $h(\pi')$ denote the number of breakpoints, cycles and hurdles in $G(\pi'')$, respectively. If π' is a fortress, let $f(\pi')=1$; otherwise, $f(\pi')=0$. It is proved in [6] that:

$$d_r(\pi') = b(\pi') - c(\pi') + h(\pi') + f(\pi') \quad (1)$$

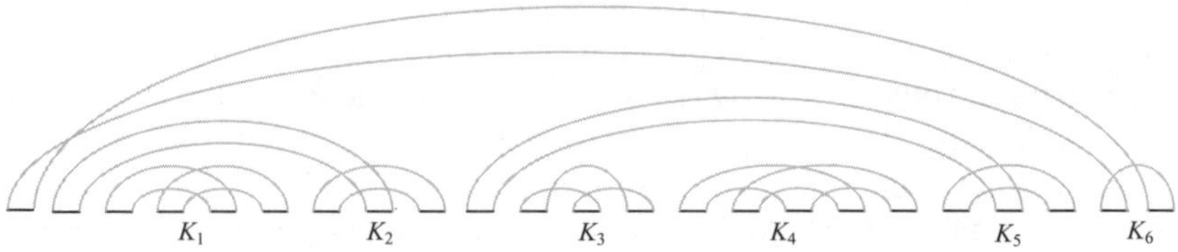

Fig. 1. A breakpoint graph with six unoriented components

Fig. 1 gives an example of a breakpoint graph with six components $K_1, \ldots, K_6$. All these components are unoriented components, where K_3, K_4 and K_6 are simple hurdles, K_1 is a super hurdle, K_2 and K_5 are non-hurdles.

For a signed permutation π', there exists a (1+ε)-approximation algorithm to sort π' by weighted reversals, transpositions, and transreversals. First construct $G(\pi'')$ from π' as aforementioned. For a component K of the breakpoint graph, we use $b(K)$ and $c(K)$ to denote the number of breakpoints and cycles in K respectively. If K is oriented, we can use $b(K)-c(K)$ times of reversals to transform K into $b(K)$ disjoint adjacencies, which is just an optimal weighted scenario sorting K. However, for an unoriented component, we cannot always have a method to sort it optimally by the weighted operations. Thus if $d_{r+t}(K)>b(K)-c(K)$, call K a *strongly unoriented component (SUC)*. If an unoriented component is not a SUC, call it a *non-SUC*. Again imagine bending a permutation into a circle, if there is an interval on the circle which contains one and only one SUC, then the SUC is a *strong hurdle*. A strong hurdle is a *super-strong-hurdle* if removing it will make another SUC become a strong hurdle. If a strong hurdle is not a super-strong-hurdle, call it a *simple-strong-hurdle*. If $G(\pi'')$ has an odd number of strong hurdles and all these strong hurdles are super-strong-hurdles, the permutation π' is called a *strong fortress*. Let $b(\pi')$, $c(\pi')$ and $h_t(\pi')$ denote the number of breakpoints, cycles and strong hurdles in $G(\pi'')$, respectively. If π' is a *strong fortress*, let $f_t(\pi')=1$; otherwise $f_t(\pi')=0$. It is proved in [5] that:

$$d_{r+t}(\pi') = b(\pi') - c(\pi') + h_t(\pi') + f_t(\pi') \tag{2}$$

Since $h_t(\pi')$ and $f_t(\pi')$ cannot be computed exactly, only an approximation algorithm is available [5]. As an example, in Fig. 1, K_1, K_2, K_4 and K_5 are all SUC's, K_3 and K_6 are non-SUC's, since each of them can be removed by one transposition, thus $d(K_3)=2=b(K_3)-c(K_3)$. And both K_1 and K_4 are super-strong-hurdles.

3 Corrected Algorithm for Unsigned Reversal Sorting

Using formula (1), Hannenhalli and Pevzner proved:

Lemma 1. [7] *For any unsigned permutation π, there exists an optimal r-spin π' of π such that all the long strips of π are canonical in π'; every 2-strip of π is either canonical or anti-canonical in π' .*

Lemma 2. [7] *For any unsigned permutation π, there exists an optimal r-spin π' of π such that (I) an unoriented component in π' does not contain any anti-strip, and (II) an oriented component in π' contains at most one anti-strip.*

Definition 1. super spin [7]: *Suppose π' is a canonical spin of π, for every unoriented component K of $G(\pi'')$, if K contains 2-strips, arbitrarily select any one of them and transform it into an anti-strip, all the other 2-strips remain canonical. Such a spin of π is called a super spin.*

Hannenhalli and Pevzner proved that every super spin of a singleton-free permutation is an optimal r-spin. However, such a super spin is not always optimal. Here is an example: Suppose π =[5, 6, 3, 4, 1, 2], the canonical spin of π is π'_1=[+5, +6, +3, +4, +1, +2] and consists of one unoriented component. According to their algorithm, suppose we select 2-strip [+5,+6] to be turned into anti-strip and get a super spin π'_2=[−5, −6, +3, +4, +1, +2]. From formula (1), $d_r(\pi'_1)$=4−2+1=3, $d_r(\pi'_2)$=5−1+0=4 > $d_r(\pi'_1)$; thus π'_2 can not be an optimal r-spin. Moreover, if we select the canonical 2-strip [+3,+4] or [+1,+2] to be turned into an anti-strip, the resulted super spin cannot be optimal either. The breakpoint graphs of $G(\pi''_1)$ and $G(\pi''_2)$ are shown in Fig. 2.

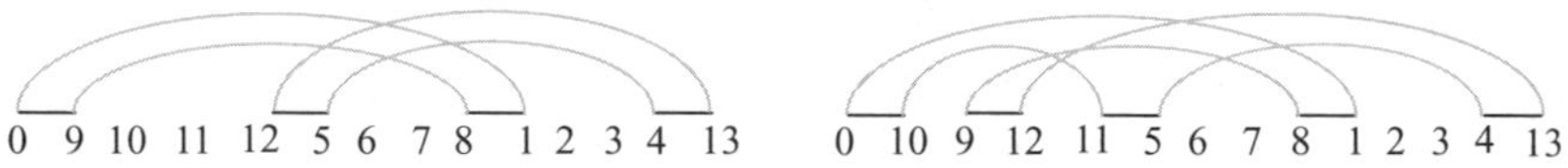

Fig. 2. Breakpoint graph $G(\pi''_1)$ and $G(\pi''_2)$

In fact, from lemma 1 and 2, to get an optimal r-spin from the canonical spin, it only needs to decide which canonical 2-strips of the canonical spin have to be turned into anti-strips. If one choose a canonical 2-strip which belongs to two cycles of the same unoriented component to be turned into anti-strip, the reversal distance will be increased. And this case is not considered in [7].

The following will consider rectifying Hannenhalli and Pevzner's algorithm. Without lose of generality, let s=[g_i, g_{i+1}] (g_i<g_{i+1}) be a 2-strip of an unsigned permutation π, its canonical form in a canonical spin π' is s=[$+g_i$, $+g_{i+1}$], which is transformed to [$2g_i-1, 2g_i, 2g_{i+1}-1, 2g_{i+1}$] in π''s extended permutation π''. If $2g_i-1$ and $2g_{i+1}$ belong to the same cycle of $G(\pi'')$, call s a *satisfied 2-strip*. Using this definition, we correct the definition of super spin as follows:

Definition 2. super-r-spin: *Suppose π' is a canonical spin of π, for every unoriented component K of $G(\pi'')$, if K contains satisfied 2-strips, arbitrarily select only one of them and transform it into an anti-strip; otherwise, all the strips of K remain canonical. We call such a spin a super-r-spin.*

The difference between super spin and super-r-spin is that only those satisfied 2-strips can be turned into anti-strips. In order to prove the super-r-spin's optimality, we present some simple lemmas firstly. In what follows, we use $flip(s)$ to denote flipping the sign of every element in s.

Lemma 3. *Let s be a canonical 2-strip, flip(s) will increase the number of breakpoints by 1.*

Proof. Let $s=[g_i, g_{i+1}]$ $(0<g_i<g_{i+1})$, its extended form is $[2g_i-1, 2g_i, 2g_{i+1}-1, 2g_{i+1}]$. We use $b_1=(x, 2g_i-1)$ and $b_2=(2g_{i+1}, y)$ to denote the two black edges incident to s. The anti-strip s' of s is $[-g_i, -g_{i+1}]$, its extended form is $[2g_i, 2g_i-1, 2g_{i+1}, 2g_{i+1}-1]$, where $(2g_i-1, 2g_{i+1})$ forms a new breakpoint, $(x, 2g_i)$ and $(2g_{i+1}-1, y)$ replace the old black edges incident to s. So the overall number of breakpoints will increase by 1. □

Lemma 4. *Let s be a canonical 2-strip of an unoriented component K, flip(s) will turn K into one oriented component K′.*

Proof. Let $s=[2g_i-1, 2g_i, 2g_{i+1}-1, 2g_{i+1}]$ in the extended form and b_1 and b_2 (n_1 and n_2) denote the two black edges (gray edges) incident to $2g_i-1$ and $2g_{i+1}$ respectively. From lemma 3, $flip(s)$ always adds a new black edge $m_1=(2g_i-1, 2g_{i+1})$ and a new gray edge $m_2=(2g_i, 2g_{i+1}-1)$ to the breakpoint graph. After $flip(s)$, black edge m_1 connects n_1 with n_2, and gray edge m_2 connects the altered b_1 with b_2.

Since K is unoriented, we can suppose that all the black edges of K are traveled from left to right, see Fig. 3. For convenience, we add an arrow to each edge, the head and tail of the arrow are called the *head* and *tail* of the edge. If we retain the direction of b_1, b_2, n_1, n_2, then in K', no matter s is a satisfied 2-strip or not, the newly created black edge m_1 joins n_2's head with n_1's tail, and the new gray edge m_2 joins b_1's head with b_2's tail. To see it clearly, we stretch out the alternating cycle into a circle, Fig. 4(a) shows

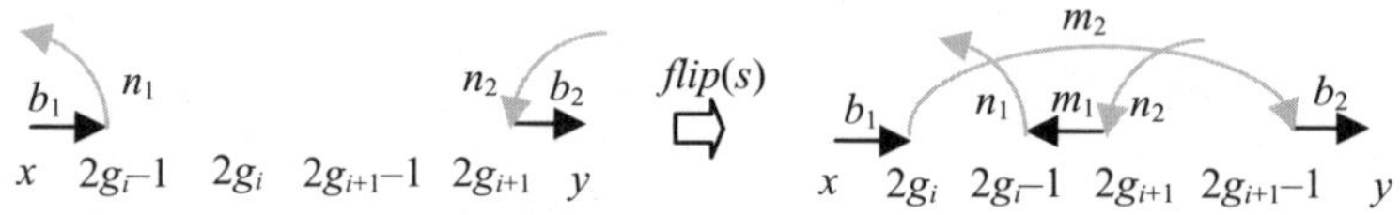

Fig. 3. Transform a 2-strip into an anti-strip, b_1 has the same direction with b_2

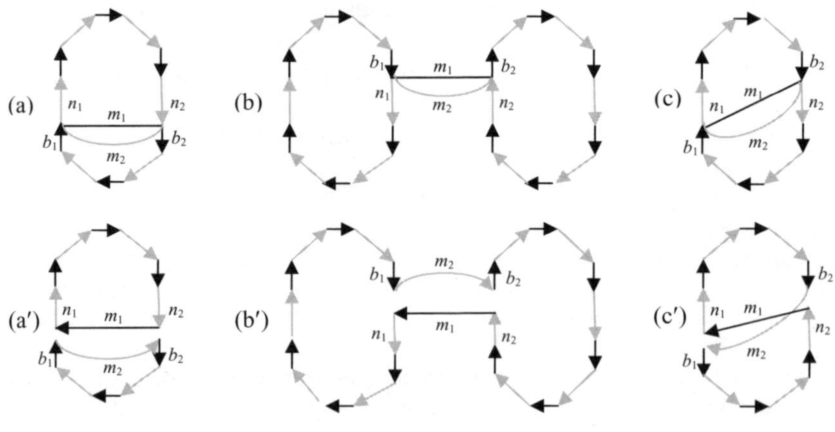

Fig. 4. Stretched circle

the case when s is a satisfied 2-strip, and Fig. 4(b) shows the case when s belongs to two cycles. Note that the newly created edges m_1 and m_2 can not form a 1-cycle, and all the cycles in K' must be alternating cycles. So the configuration of stretched cycles in K' must be the one shown in Fig. 4(a′) and Fig. 4(b′) correspondingly. It is clear that in each case, m_1 is directed from n_2's head to n_1's tail and the directions of all the other edges remain the same. What's more, in K', m_2 crosses with both n_1 and n_2, which makes K' remain one component. From Fig. 3, we know that in both cases, K' is an oriented component. □

Lemma 5. *Let s be a canonical 2-strip of π, the two black edges incident to it be b_1 and b_2. If s belongs to two cycles, $flip(s)$ will decrease $c(\pi)$ by 1; if s is a satisfied 2-strip, $flip(s)$ will increase $c(\pi)$ by 1 if b_1 has the same direction with b_2, or retain $c(\pi)$ if b_1 has the opposite direction with b_2.*

Proof. Let $s=[2g_i-1, 2g_i, 2g_{i+1}-1, 2g_{i+1}]$, its two incident black edges be $b_1=(x, 2g_i-1)$ and $b_2=(2g_{i+1}, y)$. If s belongs to two cycles, i.e., gray edge n_1 joins black edge b_1 through one more edges and gray edge n_2 joins b_2 through one more edges. After $flip(s)$, although b_1 and b_2 are altered, n_1 still joins x through one more edges and n_2 joins y through one more edges. On the other hand, in K', n_1 and n_2 are joined by the newly created black edge $m_1=(2g_i-1, 2g_{i+1})$, the altered b_1 and b_2 are joined by the newly created gray edge $m_2=(2g_i, 2g_{i+1}-1)$, so the two cycles are merged into one. Fig. 4(b) and (b′) shows the stretched cycle transformation when b_1 has the same direction with b_2.

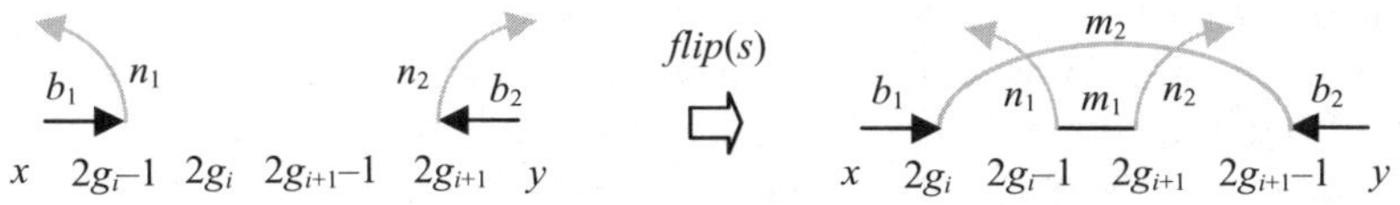

Fig. 5. Transform a 2-strip into an anti-strip, b_1 has the opposite direction with b_2

If s belongs to the same cycle, the change of $c(\pi)$ depends on the relative direction of b_1 and b_2. Case 1: b_1 and b_2 have the same direction, as shown in lemma 4 and Fig. 4(a′), in K', the new black edge m_1 is directed from n_2's head to n_1's tail, and the new gray edge m_2 is directed from b_1's head to b_2's tail, each in a closed cycle, thus increasing $c(\pi)$ by 1. Case 2: b_1 and b_2 have the opposite direction, see Fig. 5. If we retain the direction of b_1, b_2, n_1, n_2, then in K', m_1 joins two tails of gray edge n_1 and n_2, m_2 joins two heads of black edge b_1 and b_2. So the direction of some edges must change in K'. The corresponding stretched cycle is shown in Fig. 4(c). Since the newly created edges m_1 and m_2 can not form a 1-cycle, the cycle configuration of K' must be the one shown in Fig. 4(c′). It is clear that $c(\pi)$ remains the same. □

Theorem 1. *For a singleton-free unsigned permutation π, every super-r-spin π' of π is an optimal r-spin of π.*

Proof. Let π'_1 be an optimal r-spin of π containing the minimum number of anti-strips among all the optimal r-spins satisfying the conditions (I) and (II) of lemma 2. Let s

be an anti-strip of π'_1, and let π'_2 be the twin of π'_1 with canonical 2-strip s, i.e., $\pi'_1 = \pi'_2 \cdot flip(s)$. Then π'_2 also satisfies the conditions (I) and (II) of lemma 2. From lemma 3, $b(\pi'_1)=b(\pi'_2)+1$. Note that a 2-strip s can belong to two cycles of π'_2, or belong to the same cycle of π'_2. The following will discuss them respectively.

Case 1: If s belongs to two cycles of π'_2. From lemma 5, $c(\pi'_1)=c(\pi'_2)-1$. Since $flip(s)$ can affect at most two hurdles of π'_2, and when $flip(s)$ removes two hurdles, they must both be simple hurdles. From the definition of fortress, we know $(h+f)(\pi'_1) \geq (h+f)(\pi'_2)-2$. [1] From formula (1), $d_r(\pi'_2) \leq d_r(\pi'_1)$. This implies that π'_2 is an optimal r-spin of π satisfying conditions (I) and (II) of lemma 2, and that π'_2 contains fewer anti-strips than π'_1, a contradiction.

Case 2: If s belongs to the same cycle of π'_2. From lemma 5, the change of cycle number due to $flip(s)$ on π'_2 depends on the relative direction of black edges b_1 and b_2 which are incident to s.

Case 2.1 If b_1 and b_2 have the opposite direction, then $c(\pi'_1)=c(\pi'_2)$. Since s is in an oriented component of π'_2, $(h+f)(\pi'_1) \geq (h+f)(\pi'_2)$. From formula (1), $d_r(\pi'_1) > d_r(\pi'_2)$, a contradiction.

Case 2.2 If b_1 and b_2 have the same direction, then $c(\pi'_1)=c(\pi'_2)+1$. If b_1 and b_2 are in an oriented component of $G(\pi''_2)$, then $(h+f)(\pi'_1) \geq (h+f)(\pi'_2)$. From formula (1), $d_r(\pi'_1) \geq d_r(\pi'_2)$. This implies that π'_2 is an optimal r-spin of π satisfying conditions (I) and (II) of lemma 2, and that π'_2 contains fewer anti-strips than π'_1, a contradiction. So b_1 and b_2 must belong to one cycle of an unoriented component K of $G(\pi''_2)$, i.e., s is a satisfied 2-strip of K. In this case, $flip(s)$ will never increase $(h+f)(\pi'_2)$, and can at most decrease $(h+f)(\pi'_2)$ by 1. If $(h+f)(\pi'_1)=(h+f)(\pi'_2)$, then $d_r(\pi'_1)=d_r(\pi'_2)$, a contradiction. Therefore $(h+f)(\pi'_1)=(h+f)(\pi'_2)-1$ and $d_r(\pi'_1)=d_r(\pi'_2)-1$.

The above analysis shows that if π'_1 is an optimal r-spin of π containing the minimum number of anti-strips among all the optimal r-spins satisfying the conditions (I) and (II) of lemma 2, then its twin π'_2 with canonical s must have s in one cycle of an unoriented component of $G(\pi''_2)$ and $(h+f)(\pi'_1)=(h+f)(\pi'_2)-1$ holds.

Suppose that π'_1 contains t anti-strips $s'_1, s'_2, \ldots, s'_t$. Applying $flip(s)$ to all of them will turn them into canonical 2-strips $s_1, s_2, \ldots, s_t$, thus turning π'_1 into a canonical spin, denoted as δ. Note that $d_r(\delta)=d_r(\pi'_1)+t$ must hold, and each s_i ($1 \leq i \leq t$) must be a satisfied 2-strip of δ. Let $K_1, K_2, \ldots, K_t$ be the t unoriented components of δ containing satisfied 2-strips $s_1, s_2, \ldots, s_t$. Let π'_3 be a spin of π obtained from δ by arbitrarily choosing a satisfied 2-strip of K_i (for $1 \leq i \leq t$), and transforming it into an anti-strip. From lemma 3, $b(\pi'_3)=b(\delta)+t=b(\pi'_1)$, from lemma 5, $c(\pi'_3)=c(\delta)+t=c(\pi'_1)$. Moreover, $(h+f)(\pi'_1)=(h+f)(\pi'_3)$. This is because for an unoriented component K_i, if it contains more than one satisfied 2-strips, turning any one of them into an anti-strip has the same effect of turning K_i into an oriented component, thus having the same effect on $h+f$. Therefore, $d_r(\pi'_3)=d_r(\pi'_1)$.

If π'_3 does not have additional unoriented components containing satisfied 2-strips, π'_3 is a super-r-spin of π. Otherwise, suppose π'_3 has unoriented components K_{t+1},..., K_x, each of them contains satisfied 2-strips. Then for every K_j ($t+1 \leq j \leq x$), arbitrarily select a satisfied 2-strip $s_j \in K_j$, performing $flip(s_j)$ will transform K_j into an oriented component without increasing the value of $(b-c)(\pi'_3)$ and $(h+f)(\pi'_3)$, which means

[1] $(h+f)(\pi'_1)$ is a simplified notation of $h(\pi'_1)+f(\pi'_1)$.

$d_r(\pi'_3 \cdot flip(s_j)) \le d_r(\pi'_3)$. Let $\pi'_4 = \pi'_3 \cdot flip(s_{t+1}) \cdot \ldots \cdot flip(s_x)$. Then π'_4 is a super-r-spin of π, and $d_r(\pi'_4) \le d_r(\pi'_3) = d_r(\pi'_1)$, which implies that π'_4 is an optimal r-spin. This completes the proof of the theorem. □

The corrected algorithm is given as *Reversal_ Sorting*(π). For an unsigned permutation with k singletons, there are 2^k possible canonical spins. *Reversal_Sorting*(π) first gets the super-r-spin of each canonical spin, then computes the reversal distance of each super-r-spin, and finally takes the super-r-spin which has the minimum distance as the optimal r-spin. To decide whether a 2-strip is satisfied takes $O(n)$ time. Step 10 can be completed in $O(n)$ time [1], step 16 can be completed in $O(n\sqrt{nlogn})$ time [8]. Thus the time complexity of *Reversal_Sorting*(π) can reach $O(2^k{\cdot}n{+}n\sqrt{nlogn})$. If the number of singletons, k, is $O(logn)$, the computation can be completed in polynomial time.

Algorithm *Reversal_ Sorting*(π)
1 $d_r(\pi) \leftarrow n$;
2 for every canonical spin π' of π{
3 construct $G(\pi'')$;
4 for every unoriented component K_i of $G(\pi'')${
5 if K_i has satisfied 2-strips{
6 arbitrarily select one of them, denoted as s;
7 $\pi' \leftarrow \pi' \cdot flip(s)$;
8 }//endif
9 }//endfor
10 compute $d_r(\pi')$ according to formula (1);
11 if $d_r(\pi')<d_r(\pi)$ {
12 $d_r(\pi) \leftarrow d_r(\pi')$;
13 $\pi^* \leftarrow \pi'$;
14 }//endif
15 }//endfor
16 sort π^*;
17 sorting of π^* mimics optimal sorting of π by $d_r(\pi)$ reversals.

4 Approximation Algorithm for Unsigned Weighted Sorting

4.1 Algorithm for Singleton-Free Permutations

In this section, we consider sorting a singleton-free unsigned permutation π by weighted reversals, transpositions and transreversals. The weighted distance is defined as $d_{r+t}(\pi) = min_{\phi\in\Phi}\{rev(\phi) + 2trp(\phi)\}$. Eriksen studied the signed version of this problem, and gave a (1+ε)-approximation algorithm [5].

For a signed permutation π', a component K of $G(\pi'')$ is either an oriented component or an unoriented component, and an unoriented component is either a SUC or a non-SUC. If K is an oriented component, one can eliminate it using $b(K)-c(K)$ reversals. If K is a non-SUC, one can eliminate it using $((b(K)-c(K))/2$ transpositions, which contributes $b(K)-c(K)$ to the weighted distance, while it costs more than $b(K)-c(K)$ to eliminate a non-SUC by reversals. If K is a SUC, it takes more than

$((b(K)-c(K))/2$ transpositions to eliminate, while just $b(K)-c(K)+1$ reversals will do the job. However, it is difficult to distinguish between a SUC and a non-SUC, only approximation algorithm is available. The reason we do not mention transreversals is that in both signed case and unsigned case, each transreversal can be replaced by two reversals, without affecting the objective function [5]. So we only consider reversals and transpositions. Details about formula (2) see [5].

For an arbitrary unsigned permutation π of order n, let Π be the set of all 2^n spins of π. Using the same idea as reversal sorting, we can get the following lemmas and theorem. The proofs are similar to the those proved for reversal sorting and are given in the full version of this paper.

Lemma 6. *For any unsigned permutation* π, $d_{r+t}(\pi)=min_{\pi'\in\Pi}d_{r+t}(\pi')$.

Lemma 7. *Let* π'_1 *be a spin of* π *with canonical 3-element strip* $s=[g_i, g_{i+1}, g_{i+2}]$, *and let* π'_2 *be a twin of* π'_1 *with respect to s. Then* $d_{r+t}(\pi'_1)\leq d_{r+t}(\pi'_2)$

Lemma 8. *For any unsigned permutation* π, *there exists an optimal rt-spin* π' *of* π *such that all the long strips of* π *are canonical in* π'.

Lemma 9. *For any unsigned permutation* π, *there exists an optimal rt-spin* π' *of* π *such that every 2-strip of* π *is either canonical or anti-canonical in* π' .

Lemma 10. *For any unsigned permutation* π, *there exists an optimal rt-spin* π' *of* π *such that (I) an unoriented component in* π' *does not contain any anti-strip, and (II) an oriented component in* π' *contains at most one anti-strip.*

Theorem 2. *For a singleton-free unsigned permutation* π, *every super-r-spin* π' *of* π *is an optimal rt-spin of* π.

According to theorem 2, by running Eriksen's algorithm on the super-*r*-spin of π, we can also get a (1+ε)-approximation solution for sorting π by weighted reversal/ transposition/ transreversals. The algorithm is given as *Weighted_ Sorting*(π).

```
Algorithm Weighted_ Sorting(π)
construct the canonical spin π' of π and G(π");
for every unoriented component K_i of G(π") {
  if K_i has satisfied 2-strips
    arbitrarily select one of them and turn it into an anti-strip;
} // endfor
run the (1+ε)-approximation algorithm (Eriksen [5]) on the resulting π'.
```

4.2 Algorithm for Permutations with *O*(log *n*) Singletons

We show that algorithm *Weighted_ Sorting* can be applied to permutations with $O(\log n)$ singletons simply by enumerating of all the singleton's signs and guarantee the (1+ε)-approximation ratio.

Theorem 3. *For any unsigned permutation* π *with* $O(\log n)$ *singletons, a* (1+ε)*-approximation solution for sorting* π *by weighted reversal/ transposition/ transreversals can be computed in polynomial time.*

Proof. First we can enumerate all the singleton's signs to get the set of all the canonical spins of π. Let $P=\{\pi'_1, \pi'_2, \ldots, \pi'_m\}$ denote the set of canonical spins by the enumeration, δ_i denote the super-r-spin of π'_i, and $A_{r+t}(\delta_i)$ denote the distance of δ_i computed by running Eriksen's algorithm on δ_i. By running *Weighted_ Sorting* on every canonical spin in P and taking the minimum value, we can get the approximation distance for sorting π, i.e., $A_{r+t}(\pi)=min\{A_{r+t}(\delta_1), A_{r+t}(\delta_2), \ldots, A_{r+t}(\delta_m)\}=A_{r+t}(\delta_x)(1\leq x\leq m)$. On the other hand, from theorem 2, the optimal weighted distance for sorting π is $d_{r+t}(\pi) = min\{d_{r+t}(\delta_1), d_{r+t}(\delta_2), \ldots, d_{r+t}(\delta_m)\}$. Let $d_{r+t}(\pi)=d_{r+t}(\delta_y)(1\leq y\leq m)$. Since $\frac{A_{r+t}(\delta_i)}{d_{r+t}(\delta_i)}\leq 1+\varepsilon$ for $1\leq i \leq m$, we have:

$$\frac{A_{r+t}(\pi)}{d_{r+t}(\pi)} = \frac{min\{A_{r+t}(\delta_1), A_{r+t}(\delta_2), \ldots, A_{r+t}(\delta_m)\}}{min\{d_{r+t}(\delta_1), d_{r+t}(\delta_2), \ldots, d_{r+t}(\delta_m)\}} = \frac{A_{r+t}(\delta_x)}{d_{r+t}(\delta_y)} \leq \frac{A_{r+t}(\delta_y)}{d_{r+t}(\delta_y)} \leq 1+\varepsilon$$

When there are $O(\log n)$ singletons in the permutation, the size of set P is $O(n^k)$, where k is a fixed constant. Similarly to section 3, we can get a polynomial time algorithm on such permutations and guarantee the $(1+\varepsilon)$-approximation ratio. □

Acknowledgments. We would like to thank the reviewers for their helpful suggestions.

References

1. Bader, D.A., Moret, B.M.E., Yan, M.: A linear-time algorithm for computing inversion distance between signed permutations with an experimental study. J. Comput. Biol. 8, 483–491 (2001)
2. Berman, P., Hannenhalli, S., Karpinki, M.: 1.375-approximation algorithm for sorting by reversals. In: Möhring, R.H., Raman, R. (eds.) ESA 2002. LNCS, vol. 2461, pp. 200–210. Springer, Heidelberg (2002)
3. Bafna, V., Pevzner, P.A.: Sorting by transpositions. SIAM J. Discrete Math. 11, 272–289 (1998)
4. Caprara, A.: Sorting by reversals is difficult. In: Proc. 1st Annu. Int. Conf. Res. Comput. Mol. Biol. (RECOMB 1997), pp. 75–83 (1997)
5. Eriksen, N.: (1+ε)-approximation of sorting by reversals and transpositions. Theor. Comput. Sci. 289, 517–529 (2002)
6. Hannenhalli, S., Pevzner, P.A.: Transforming cabbage into turnip (polynomial algorithm for sorting signed permutations by reversals). In: Proc. 27th Annu. ACM Symp. Theory Computing (STOC 1995), pp. 178–189 (1995)
7. Hannenhalli, S., Pevzner, P.A.: To cut … or not to cut (applications of comparative physical maps in molecular evolution). In: Proc. 7th Annu. ACM-SIAM Symp. Discrete Algorithms (SODA 1996), pp. 304–313 (1996)
8. Tannier, E., Sagot, M.-F.: Sorting by Reversals in Subquadratic Time. In: Sahinalp, S.C., Muthukrishnan, S.M., Dogrusoz, U. (eds.) CPM 2004. LNCS, vol. 3109, pp. 1–13. Springer, Heidelberg (2004)

On the Complexity of the Hidden Subgroup Problem

Stephen Fenner[1,⋆] and Yong Zhang[2,⋆⋆]

[1] Department of Computer Science & Engineering, University of South Carolina
Columbia, SC 29208 USA
fenner@cse.sc.edu
[2] Department of Mathematical Sciences, Eastern Mennonite University
Harrisonburg, VA 22802, USA
yong.zhang@emu.edu

Abstract. We show that several problems that figure prominently in quantum computing, including Hidden Coset, Hidden Shift, and Orbit Coset, are equivalent or reducible to Hidden Subgroup. We also show that, over permutation groups, the decision version and search version of Hidden Subgroup are polynomial-time equivalent. For Hidden Subgroup over dihedral groups, such an equivalence can be obtained if the order of the group is smooth. Finally, we give nonadaptive program checkers for Hidden Subgroup and its decision version.

1 Introduction

The Hidden Subgroup problem generalizes many interesting problems that have efficient quantum algorithms but whose known classical algorithms are inefficient. While we can solve Hidden Subgroup over abelian groups satisfactorily on quantum computers, the nonabelian case is more challenging. Although there are many families of groups besides abelian ones for which Hidden Subgroup is known to be solvable in quantum polynomial time, the overall successes are considered limited. People are particularly interested in solving Hidden Subgroup over two families of nonabelian groups, permutation groups and dihedral groups, since solving them will immediately give solutions to Graph Isomorphism [Joz00] and Shortest Lattice Vector [Reg04], respectively.

To explore more fully the power of quantum computers, researchers have also introduced and studied several related problems. Van Dam, Hallgren, and Ip [vDHI03] introduced the Hidden Shift problem and gave efficient quantum algorithms for some instances. Their results provide evidence that quantum computers can help to recover shift structure as well as subgroup structure. They also introduced the Hidden Coset problem to generalize Hidden Shift and Hidden Subgroup. Recently, Childs and van Dam [CvD07] introduced the Generalized Hidden Shift problem, extending Hidden Shift from a

⋆ Partially supported by NSF grant CCF-0515269.
⋆⋆ Partially supported by an EMU Summer Research Grant 2006.

M. Agrawal et al. (Eds.): TAMC 2008, LNCS 4978, pp. 70–81, 2008.

different angle. In an attempt to attack HIDDEN SUBGROUP using a divide-and-conquer approach over subgroup chains, Friedl et al. [FIM+03] introduced the ORBIT COSET problem, which they claimed to be an even more general problem including HIDDEN SUBGROUP and HIDDEN SHIFT[1] as special instances. They called ORBIT COSET a *quantum* generalization of HIDDEN SUBGROUP and HIDDEN SHIFT, since the definition of ORBIT COSET involves *quantum functions*.

In Section 3, we show that all these related problems are equivalent or reducible to HIDDEN SUBGROUP with different underlying groups. In particular,

1. HIDDEN COSET is polynomial-time equivalent to HIDDEN SUBGROUP,
2. ORBIT COSET is equivalent to HIDDEN SUBGROUP if we allow functions in the latter to be quantum functions, and
3. HIDDEN SHIFT and GENERALIZED HIDDEN SHIFT reduce to instances of HIDDEN SUBGROUP over a family of wreath product groups.[2]

There are a few results in the literature about the complexity of HIDDEN SUBGROUP. HIDDEN SUBGROUP over abelian groups is in class **BQP** [Kit95, Mos99]. Ettinger, Hoyer, and Knill [EHK04] showed that HIDDEN SUBGROUP (over arbitrary finite groups) has polynomial quantum query complexity. Arvind and Kurur [AK02] showed that HIDDEN SUBGROUP over permutation groups is in the class $\mathbf{FP}^{\mathbf{SPP}}$ and is thus low for the counting complexity class **PP**. In Section 4 we study the relationship between the decision and search versions of HIDDEN SUBGROUP, denoted as HIDDEN SUBGROUP$_D$ and HIDDEN SUBGROUP$_S$, respectively. It is well known that **NP**-complete sets such as SAT are self-reducible, which implies that the decision and search versions of **NP**-complete problems are polynomial-time equivalent. We show this is also the case for HIDDEN SUBGROUP and HIDDEN SHIFT over permutation groups. There are evidences in the literature showing that HIDDEN SUBGROUP$_D$ over permutation groups is difficult [HRTS03, GSVV04, MRS05, KS05, HMR+06, MRS07]. In particular, Kempe and Shalev [KS05] showed that under general conditions, various forms of the Quantum Fourier Sampling method are of no help (over classical exhaustive search) in solving HIDDEN SUBGROUP$_D$ over permutation groups. Our results yield evidence of a different sort that this problem is difficult—namely, it is just as hard as the search version. For HIDDEN SUBGROUP over dihedral groups, our results are more modest. We show the search-decision equivalence for dihedral groups of smooth order, i.e., where the largest prime dividing the order of the group is small.

Combining our results in Sections 3 and 4, we obtain nonadaptive program checkers for HIDDEN SUBGROUP and HIDDEN SUBGROUP$_D$ over permutation groups. We give the details in Section 5.

---

[1] They actually called it the Hidden Translation problem.

[2] Friedl et al. [FIM+03] gave a reduction from HIDDEN SHIFT to instances of HIDDEN SUBGROUP over semi-direct product groups. However, their reduction only works when the group G is abelian.

2 Preliminaries

2.1 Group Theory

Background on general group theory and quantum computation can be found in textbooks [Sco87] and [NC00]. A special case of the wreath product groups plays an important role in several proof.

Definition 1. For any finite group G, the *wreath product* $G \wr \mathbb{Z}_n$ of G and $\mathbb{Z}_n = \{0, 1, \ldots, n-1\}$ is the set $\{(g_1, g_2, \ldots, g_n, \tau) \mid g_1, g_2, \ldots, g_n \in G, \ \tau \in \mathbb{Z}_n\}$ equipped with the group operation $\circ$ such that

$$(g_1, g_2, \ldots, g_n, \tau) \circ (g'_1, g'_2, \ldots, g'_n, \tau') = (g_{\tau'(1)}g'_1, g_{\tau'(2)}g'_2, \ldots, g_{\tau'(n)}g'_n, \tau\tau').$$

We abuse notation here by identifying τ and τ' with permutations over the set $\{1, \ldots, n\}$ sending x to $x + \tau \bmod n$ and to $x + \tau' \bmod n$, respectively.

Let Z be a set and S_Z be the *symmetric group* of permutations of Z. We define the composition order to be from left to right, i.e., for $g_1, g_2 \in S_Z$, g_1g_2 is the permutation obtained by applying g_1 first and then g_2. For $n \geq 1$, we abbreviate $S_{\{1,2,\ldots,n\}}$ by S_n. Subgroups of S_n are the *permutation groups* of degree n. For a permutation group $G \leq S_n$ and an element $i \in \{1, \ldots, n\}$, let $G^{(i)}$ denote the stabilizer subgroup of G that fixes the set $\{1, \ldots, i\}$ pointwise. The chain of the stabilizer subgroups of G is $\{id\} = G^{(n)} \leq G^{(n-1)} \leq \cdots \leq G^{(1)} \leq G^{(0)} = G$. Let C_i be a complete set of right coset representatives of $G^{(i)}$ in $G^{(i-1)}$, $1 \leq i \leq n$. Then the cardinality of C_i is at most $n-i$ and $\cup_{i=1}^{n} C_i$ forms a *strong generator set* for G [Sim70]. Any element $g \in G$ can be written uniquely as $g = g_n g_{n-1} \cdots g_1$ with $g_i \in C_i$. Furst, Hopcroft, and Luks [FHL80] showed that given any generator set for G, a strong generator set can be computed in polynomial time. For $X \subseteq Z$ and $G \leq S_Z$, we use G_X to denote the subgroup of G that stablizes X setwise. It is evident that G_X is the direct sum of S_X and $S_{Z \setminus X}$. We are particularly interested in the case when G is S_n. In this case, a generating set for G_X can be easily computed.

Let G be a finite group. Let Γ be a set of mutually orthogonal quantum states. Let $\alpha : G \times \Gamma \to \Gamma$ be a group action of G on Γ, i.e., for every $x \in G$ the function $\alpha_x : \Gamma \to \Gamma$ mapping $|\phi\rangle$ to $|\alpha(x, |\phi\rangle)\rangle$ is a permutation over Γ, and the map h from G to the symmetric group over Γ defined by $h(x) = \alpha_x$ is a homomorphism. We use the notation $|x \cdot \phi\rangle$ instead of $|\alpha(x, |\phi\rangle)\rangle$, when α is clear from the context. We let $G(|\phi\rangle)$ denote the orbit of $|\phi\rangle$ with respect to α, i.e., the set $\{|x \cdot \phi\rangle : x \in G\}$, and we let $G_{|\phi\rangle}$ denote the stabilizer subgroup of $|\phi\rangle$ in G, i.e., $\{x \in G : |x \cdot \phi\rangle = |\phi\rangle\}$. Given any positive integer t, let α^t denote the group action of G on $\Gamma^t = \{|\phi\rangle^{\otimes t} : |\phi\rangle \in \Gamma\}$ defined by $\alpha^t(x, |\phi\rangle^{\otimes t}) = |x \cdot \phi\rangle^{\otimes t}$. We need α^t because the input superpositions cannot be cloned in general.

Definition 2. Let G be a finite group.

- Given a generating set for G and a function f that maps G to some finite set S, where the values of f are constant on a subgroup H of G and distinct on each left (right) coset of H. The HIDDEN SUBGROUP problem is to find a generating set for H. The decision version of HIDDEN SUBGROUP, denoted as HIDDEN SUBGROUP$_D$, is to determine whether H is trivial. The search version, denoted as HIDDEN SUBGROUP$_S$, is to find a nontrivial element, if there exists one, in H.

- Given a generating set for G and n injective functions $f_1, f_2, \ldots, f_n$ defined on G, with the promise that there is a "shift" $u \in G$ such that for all $g \in G$, $f_1(g) = f_2(gu), f_2(g) = f_3(gu), \ldots, f_{n-1}(g) = f_n(gu)$, the GENERALIZED HIDDEN SHIFT problem is to find u. If $n = 2$, this problem is called the HIDDEN SHIFT problem.
- Given a generating set for G and two functions f_1 and f_2 defined on G such that for some shift $u \in G$, $f_1(g) = f_2(gu)$ for all g in G, the HIDDEN COSET problem is to find the set of all such u.
- Given a generating set for G and two quantum states $|\phi_0\rangle, |\phi_1\rangle \in \Gamma$, the ORBIT COSET problem is to either reject the input if $G(|\phi_0\rangle) \cap G(|\phi_1\rangle) = \emptyset$, or else output both a $u \in G$ such that $|u \cdot \phi_1\rangle = |\phi_0\rangle$ and also a generating set for $G_{|\phi_1\rangle}$.

2.2 Program Checkers

Let π be a computational decision or search problem. Let x be an input to π and $\pi(x)$ be the output of π. Let P be a deterministic program (supposedly) for π that halts on all inputs. We are interested in whether P has any bug, i.e., whether there is some x such that $P(x) \neq \pi(x)$. A efficient *program checker* C for P is a probabilistic expected-polynomial-time oracle Turing machine that uses P as an oracle and takes x and a positive integer k (presented in unary) as inputs. The running time of C does not include the time it takes for the oracle P to do its computations. C will output CORRECT with probability $\geq 1 - 1/2^k$ if P is correct on all inputs (no bugs), and output BUGGY with probability $\geq 1 - 1/2^k$ if $P(x) \neq \pi(x)$. This probability is over the sample space of all finite sequences of coin flips C could have tossed. However, if P has bugs but $P(x) = \pi(x)$, we allow C to behave arbitrarily. If C only queries the oracle nonadaptively, then we say C is a *nonadaptive checker*. See Blum and Kannan [BK95] for more details.

3 Several Reductions

The HIDDEN COSET problem is to find the set of all shifts of the two functions f_1 and f_2 defined on the group G. In fact, the set of all shifts is a coset of a subgroup H of G and f_1 is constant on H (see [vDHI03] Lemma 6.1). If we let f_1 and f_2 be the same function, this is exactly HIDDEN SUBGROUP. On the other hand, if f_1 and f_2 are injective functions, this is HIDDEN SHIFT.

Theorem 1. HIDDEN COSET *is polynomial-time equivalent to* HIDDEN SUBGROUP.

Proof. Let G and f_1, f_2 be the input of HIDDEN COSET. Let the set of shifts be Hu, where H is a subgroup of G and u is a coset representative. Define a function f that maps $G \wr \mathbb{Z}_2$ to $S \times S$ as follows: for any $(g_1, g_2, \tau) \in G \wr \mathbb{Z}_2$,

$$f(g_1, g_2, \tau) = \begin{cases} (f_1(g_1), f_2(g_2)) \text{ if } \tau = 0, \\ (f_2(g_2), f_1(g_1)) \text{ if } \tau = 1. \end{cases}$$

The values of f are constant on the set $K = (H \times u^{-1}Hu \times \{0\}) \cup (u^{-1}H \times Hu \times \{1\})$, which is a subgroup of $G \wr \mathbb{Z}_2$. Furthermore, the values of f are distinct on all left cosets of K. Given a generating set of K, there is at least one generator of the form $(k_1, k_2, 1)$. Pick k_2 to be the coset representative u of H. Form a generating set S of H as follows. S is initially empty. For each generator of K, if it is of the form $(k_1, k_2, 0)$, then add k_1 and uk_2u^{-1} to S; if it is of the form $(k_1, k_2, 1)$, then add uk_1 and k_2u^{-1} to S.

Corollary 1. HIDDEN COSET *has polynomial quantum query complexity.*

It was mentioned in Friedl et al. [FIM$^+$03] that HIDDEN COSET in general is of exponential (classical) query complexity.

Recently Childs and van Dam [CvD07] proposed GENERALIZED HIDDEN SHIFT where there are n injective functions (encoded in a single function). Using a similar approach, we show GENERALIZED HIDDEN SHIFT essentially addresses HIDDEN SUBGROUP over a different family of groups.

Proposition 1. GENERALIZED HIDDEN SHIFT *reduces to* HIDDEN SUBGROUP *in time polynomial in n.*

Proof. The input for GENERALIZED HIDDEN SHIFT is a group G and n injective functions $f_1, f_2, \ldots, f_n$ defined on a group G such that for all $g \in G$, $f_1(g) = f_2(gu), \ldots, f_{n-1}(g) = f_n(gu)$. Consider the group $G \wr \mathbb{Z}_n$. Define a function f such that for any element in $(g_1, \ldots, g_n, \tau) \in G \wr \mathbb{Z}_n$, $f((g_1, \ldots, g_n, \tau)) = (f_{\tau(0)}(g_0), \ldots, f_{\tau(n)}(g_n))$. The function values of f are constant and distinct for left cosets of the n-element cyclic subgroup generated by $(u^{-(n-1)}, u, u, \cdots, u, 1)$.

Van Dam, Hallgren, and Ip [vDHI03] introduced the Shifted Legendre Symbol problem as a natural instance of HIDDEN SHIFT. They claimed that assuming a conjecture this problem can also be reduced to an instance of HIDDEN SUBGROUP over dihedral groups. By Proposition 1, this problem can be reduced to HIDDEN SUBGROUP over wreath product groups without any conjecture.

Next we show ORBIT COSET is not a more general problem either, if we allow the function in HIDDEN SUBGROUP to be a quantum function. We need this generalization since the definition of ORBIT COSET involves quantum functions, i.e., the ranges of the functions are sets of orthogonal quantum states. In HIDDEN SUBGROUP, the function is implicitly considered by most researchers to be a classical function, mapping group elements to a classical set. For the purposes of quantum computation, however, this generalization to quantum functions is natural and does not affect any existing quantum algorithms for HIDDEN SUBGROUP.

Proposition 2. ORBIT COSET *is quantum polynomial-time equivalent to* HIDDEN SUBGROUP.

Proof. Let G and two orthogonal quantum states $|\phi_0\rangle, |\phi_1\rangle \in \Gamma$ be the inputs of ORBIT COSET. Define the function $f : G \wr \mathbb{Z}_2 \to \Gamma \otimes \Gamma$ as follows:

$$f(g_1, g_2, \tau) = \begin{cases} |g_1 \cdot \phi_0\rangle \otimes |g_2 \cdot \phi_1\rangle \text{ if } \tau = 0, \\ |g_2 \cdot \phi_1\rangle \otimes |g_1 \cdot \phi_0\rangle \text{ if } \tau = 1. \end{cases}$$

The values of the function f are identical and orthogonal on each left coset of the following subgroup H of $G \wr \mathbb{Z}_2$: If there is no $u \in G$ such that $|u \cdot \phi_1\rangle = |\phi_0\rangle$, then $H = G_{|\phi_0\rangle} \times G_{|\phi_1\rangle} \times \{0\}$. If there is such a u, then $H = (G_{|\phi_0\rangle} \times G_{|\phi_1\rangle} \times \{0\}) \cup (G_{|\phi_1\rangle} u^{-1} \times u G_{|\phi_1\rangle} \times \{1\})$. For $i, j \in \{0, 1\}$, let $g_i \in G$ be the i'th coset representative of $G_{|\phi_0\rangle}$ (i.e., $|g_i \cdot \phi_0\rangle = |\phi_i\rangle$), and let $g_j \in G$ be the j'th coset representative of $G_{|\phi_1\rangle}$ (i.e., $|g_j \cdot \phi_1\rangle = |\phi_j\rangle$). Then elements of the left coset of H represented by $(g_i, g_j, 0)$ will all map to the same value $|\phi_i\rangle \otimes |\phi_j\rangle$ via f.

4 Decision Versus Search

For **NP**-complete problems, the decision and search version are polynomial-time equivalent. This equivalence was also obtained for problems which are not known to be NP-complete, such as GRAPH ISOMORPHISM [Mat79] and GROUP INTERSECTION [AT01]. Since both problems reduce to instances of HIDDEN SUBGROUP, we ask the question whether HIDDEN SUBGROUP has such property. In this section we show that over permutation groups S_n, this decision-search equivalence can actually be obtained for HIDDEN SUBGROUP and also HIDDEN SHIFT. On the other hand, over dihedral groups D_n, this equivalence can only be obtained for HIDDEN SUBGROUP when n has small prime factors.

Lemma 1. *Given (generating sets for) a group $G \leq S_n$, a function $f : G \to S$ that hides a subgroup $H \leq G$, and a sequence of subgroups $G_1, \ldots, G_k \leq S_n$, an instance of* HIDDEN SUBGROUP *can be constructed to hide the group $D = \{(g, g, \ldots, g) \mid g \in H \cap G_1 \cap \cdots \cap G_k\}$ inside $G \times G_1 \times \cdots \times G_k$.*

Proof. Define a function f' over the direct product group $G \times G_1 \times \cdots \times G_k$ so that for any element $(g, g_1, \ldots, g_k)$, $f'(g, g_1, \ldots, g_k) = (f(g), gg_1^{-1}, \ldots, gg_k^{-1})$. The values of f' are constant and distinct over left cosets of D.

In the following, we will use the tuple $\langle G, f \rangle$ to represent a standard HIDDEN SUBGROUP input instance, and $\langle G, f, G_1, \ldots, G_k \rangle$ to represent a HIDDEN SUBGROUP input instance constructed as in Lemma 1.

We define a natural isomorphism that identifies $S_n \wr \mathbb{Z}_2$ with a subgroup of S_Γ, where $\Gamma = \{(i, j) \mid i \in \{1, \ldots, n\},\ j \in \{1, 2\}\}$. This isomorphism can be viewed as a group action, where the group element (g_1, g_2, τ) maps (i, j) to $(g_j(i), \tau(j))$. Note that this isomorphism can be efficiently computed in both directions.

Theorem 2. *Over permutation groups,* HIDDEN SUBGROUP$_S$ *is truth-table reducible to* HIDDEN SUBGROUP$_D$ *in polynomial time.*

Proof. Suppose f hides a nontrivial subgroup H of G, first we compute a strong generating set for G, corresponding to the chain $\{id\} = G^{(n)} \leq G^{(n-1)} \leq \cdots \leq G^{(1)} \leq G^{(0)} = G$. Define f' over $G \wr \mathbb{Z}_2$ such that f' maps (g_1, g_2, τ) to $(f(g_1), f(g_2))$ if τ is 0, and $(f(g_2), f(g_1))$ otherwise. It is easy to check that for the group $G^{(i)} \wr \mathbb{Z}_2$, $f'|_{G^{(i)} \wr \mathbb{Z}_2}$ hides the subgroup $H^{(i)} \wr \mathbb{Z}_2$.

Query the HIDDEN SUBGROUP$_D$ oracle with inputs

$$\left\langle G^{(i)} \wr \mathbb{Z}_2, f'|_{G^{(i)} \wr \mathbb{Z}_2}, (S_\Gamma)_{\{(i,1),(j,2)\}}, (S_\Gamma)_{\{(i,2),(j',1)\}}, (S_\Gamma)_{\{(k,1),(\ell,2)\}} \right\rangle$$

for all $1 \leq i \leq n$, all $j, j' \in \{i+1, \ldots, n\}$, and all $k, \ell \in \{i, \ldots, n\}$.

Claim. Let i be such that $H^{(i)} = \{id\}$ and $H^{(i-1)} \neq \{id\}$. For all $i < j, j' \leq n$ and all $i \leq k, l \leq n$, there is a (necessarily unique) permutation $h \in H^{(i-1)}$ such that $h(i) = j$, $h(j') = i$ and $h(k) = \ell$ if and only if the query

$$\left\langle G^{(i-1)} \wr \mathbb{Z}_2, f'|_{G^{(i-1)} \wr \mathbb{Z}_2}, (S_\Gamma)_{\{(i,1),(j,2)\}}, (S_\Gamma)_{\{(i,2),(j',1)\}}, (S_\Gamma)_{\{(k,1),(\ell,2)\}} \right\rangle$$

to the HIDDEN SUBGROUP$_D$ oracle answers "nontrivial."

Proof of Claim. For any $j > i$, there is at most one permutation in $H^{(i-1)}$ that maps i to j. To see this, suppose there are two distinct $h, h' \in H^{(i-1)}$ both of which map i to j. Then $h'h^{-1} \in H^{(i)}$ is a nontrivial permutation, contradicting the assumption $H^{(i)} = \{id\}$. Let $h \in H^{(i-1)}$ be a permutation such that $h(i) = j$, $h(j') = i$, and $h(k) = \ell$. Then $(h, h^{-1}, 1)$ is a nontrivial element in the group $H^{(i-1)} \wr \mathbb{Z}_2 \cap (S_\Gamma)_{\{(i,1),(j,2)\}} \cap (S_\Gamma)_{\{(i,2),(j',1)\}} \cap (S_\Gamma)_{\{(k,1),(\ell,2)\}}$, and thus the oracle answers "nontrivial."

Conversely, if the oracle answers "nontrivial," then the nontrivial element must be of the form $(h, h', 1)$ where $h, h' \in H^{(i-1)}$, since the other form $(h, h', 0)$ will imply that h and h' both fix i and thus are in $H^{(i)} = \{id\}$. Therefore, h will be a nontrivial element of $H^{(i-1)}$ with $h(i) = j$, $h(j') = i$, and $h(k) = \ell$. This proves the Claim.

Find the largest i such that the query answers "nontrivial" for some $j, j' > i$ and some $k, \ell \geq i$. Clearly this is the smallest i such that $H^{(i)} = \{id\}$. A nontrivial permutation in $H^{(i-1)}$ can be constructed by looking at the query results that involve $G^{(i-1)} \wr \mathbb{Z}_2$.

Corollary 2. HIDDEN SUBGROUP$_D$ *and* HIDDEN SUBGROUP$_S$ *are polynomial-time equivalent over permutation groups,.*

Next we show that the search version of HIDDEN SHIFT, as a special case of HIDDEN SUBGROUP, also reduces to the corresponding decision problem.

Definition 3. Given a generating set for a group G and two injective functions f_1, f_2 defined on G, the problem HIDDEN SHIFT$_D$ is to determine whether there is a shift $u \in G$ such that $f_1(g) = f_2(gu)$ for all $g \in G$.

Theorem 3. *Over permutation groups,* HIDDEN SHIFT$_D$ *and* HIDDEN SHIFT$_S$ *are polynomial-time equivalent.*

Proof. We show that if there is a translation u for the two injective functions defined on G, we can find u with the help of an oracle that solves HIDDEN SHIFT$_D$. First compute the strong generator set $\cup_{i=1}^{n} C_i$ of G using the procedure in [FHL80]. Note that $\cup_{i=k}^{n} C_i$ generates $G^{(k-1)}$ for $1 \leq k \leq n$. We will proceed in steps along the stabilizer subgroup chain $G = G^{(0)} \geq G^{(1)} \geq \cdots \geq G^{(n)} = \{id\}$.

Claim. With the help of the HIDDEN SHIFT$_D$ oracle, finding the translation u_i for input $(G^{(i)}, f_1, f_2)$ reduces to finding another translation u_{i+1} for input $(G^{(i+1)}, f_1', f_2')$. In particular, we have $u_i = u_{i+1}\sigma_i$.

Proof of Claim. Ask the oracle whether there is a translation for the input instance $(G^{(i+1)}, f_1|_{G^{(i+1)}}, f_2|_{G^{(i+1)}})$. If the answer is yes, then we know $u_i \in G^{(i+1)}$ and therefore set $\sigma_i = id$ and $u_i = u_{i+1}\sigma_i$.

If the answer is no, then we know that u is in some right coset of $G^{(i+1)}$ in $G^{(i)}$. For every $\tau \in C_{i+1}$, define a function f_τ such that $f_\tau(x) = f_2(x\tau)$ for all $x \in G^{(i+1)}$. Ask the oracle whether there is a translation for input $(G^{(i+1)}, f_1|_{G^{(i+1)}}, f_\tau)$. The oracle will answer yes if and only if u and τ are in the same right coset of $G^{(i+1)}$ in $G^{(i)}$, since

$$\begin{aligned}
&u \text{ and } \tau \text{ are in the same right coset of } G^{(i+1)} \text{ in } G^{(i)} \\
\iff& u = u'\tau \text{ for some } \tau' \in G^{(i+1)} \\
\iff& f_1(x) = f_2(xu) = f_2(xu'\tau) = f_\tau(xu') \text{ for all } x \in G^{(i)} \\
\iff& u' \text{ is the translation for } (G^{(i+1)}, f_1|_{G^{(i+1)}}, f_\tau).
\end{aligned}$$

Then we set $\sigma_i = \tau$.

We apply the above procedure $n-1$ times until we reach the trivial subgroup $G^{(n)}$. The translation u will be equal to $\sigma_n\sigma_{n-1}\cdots\sigma_1$. Since the size of each C_i is at most $n-i$, the total reduction is in classical polynomial time.

For HIDDEN SUBGROUP over dihedral groups D_n, we can efficiently reduce search to decision when n has small prime factors. For a fixed integer B, we say an integer n is B-smooth if all the prime factors of n are less than or equal to B. For such an n, the prime factorization can be obtained in time polynomial in $B + \log n$. Without loss of generality, we assume that the hidden subgroup is an order-two subgroup of D_n [EH00].

Theorem 4. *Let n be a B-smooth number,* HIDDEN SUBGROUP *over the dihedral group D_n reduces to* HIDDEN SUBGROUP$_D$ *over dihedral groups in time polynomial in $B + \log n$.*

Proof. Without loss of generality, we assume the generator set for D_n is $\{r, \sigma\}$, where the order of r and σ are n and 2, respectively. Let $p_1^{e_1}p_2^{e_2}\cdots p_k^{e_k}$ be the prime factorization of n. Since n is B-smooth, $p_i \leq B$ for all $1 \leq i \leq k$. Let the hidden subgroup H be $\{id, r^a\sigma\}$ for some $a < n$.

First we find $a \bmod p_1^{e_1}$ as follows. Query the HIDDEN SUBGROUP$_D$ oracle with input groups (we will always use the original input function f) $\langle r^{p_1}, \sigma\rangle$, $\langle r^{p_1}, r\sigma\rangle, \ldots, \langle r^{p_1}, r^{p_1-1}\sigma\rangle$. It is not hard to see that the HIDDEN SUBGROUP$_D$ oracle will answer "nontrivial" only for the input group $\langle r^{p_1}, r^{m_1}\sigma\rangle$ where $m_1 = a \bmod p_1$. The next set of input groups to the HIDDEN SUBGROUP$_D$ oracle are $\langle r^{p_1^2}, r^{m_1}\sigma\rangle, \langle r^{p_1^2}, r^{p_1+m_1}\sigma\rangle, \ldots, \langle r^{p_1^2}, r^{(p_1-1)p_1+m_1}\sigma\rangle$. From the oracle answers we obtain $m_2 = a \bmod p_1^2$. Repeat the above procedure until we find $a \bmod p_1^{e_1}$.

Similarly, we can find $a \bmod p_2^{e_2}, \ldots, a \bmod p_k^{e_k}$. A simple usage of the Chinese Remainder Theorem will then recover a. The total number of queries is $e_1p_1 + e_2p_2 + \cdots + e_kp_k$, which is polynomial in $\log n + B$.

5 Nonadaptive Checkers

An important concept closely related to self-reducibility is that of a *program checker*, which was first introduced by Blum and Kannan [BK95]. They gave program checkers for some group-theoretic problems and selected problems in **P**. They also characterized the class of problems having polynomial-time checkers. Arvind and Torán [AT01] presented a nonadaptive **NC** checker for GROUP INTERSECTION over permutation groups. In this section we show that HIDDEN SUBGROUP$_D$ and HIDDEN SUBGROUP over permutation groups have nonadaptive checkers.

For the sake of clarity, we give the checker for HIDDEN SUBGROUP$_D$ first. Let P be a program that solves HIDDEN SUBGROUP$_D$ over permutation groups. The input for P is a permutation group G given by its generating set and a function f that is defined over G and hides a subgroup H of G. If P is a correct program, then $P(G, f)$ outputs TRIVIAL if H is the trivial subgroup of G, and NONTRIVIAL otherwise. The checker $C^P(G, f, 0^k)$ checks the program P on the input G and f as follows:

Begin
Compute $P(G, f)$.
if $P(G, f) =$ NONTRIVIAL, **then**
Use Theorem 2 and P (as if it were bug-free) to search for a nontrivial element h of H.
if $f(h) = f(id)$, **then**
return CORRECT
else
return BUGGY
if $P(G, f) =$ TRIVIAL, **then**
Do k times (in parallel):
generate a random permutation $u \in G$.
define f' over G such that $f(g) = f'(gu)$ for all $g \in G$, use (G, f, f') to be an input instance of HIDDEN SHIFT
use Theorem 1 to convert (G, f, f') to an input instance $(G \wr \mathbb{Z}_2, f'')$ of HIDDEN SUBGROUP[3]

[3] Using the natural isomorphism we define in Section 4, the group $G \wr \mathbb{Z}_2$ is still considered as a permutation group.

use Theorem 2 and P to search for a nontrivial element h of the subgroup of $G \wr \mathbb{Z}_2$ that f'' hides.

if $h \neq (u^{-1}, u, 1)$, **then return** BUGGY

End-do

return CORRECT

End

Theorem 5. *If P is a correct program for* HIDDEN SUBGROUP$_D$, *$C^P(G, f, 0^k)$ always outputs CORRECT. If $P(G, f)$ is incorrect, the probability of $C^P(G, f, 0^k)$ outputting CORRECT is $\leq 2^{-k}$. Moreover, $C^P(G, f, 0^k)$ runs in polynomial time and queries P nonadaptively.*

Proof. If P is a correct program and $P(G, f)$ outputs NONTRIVIAL, then $C^P((G, f, 0^k)$ will find a nontrivial element of H and outputs CORRECT. If P is a correct program and $P(G, f)$ outputs TRIVIAL, the function f' constructed by $C^P(G, f, 0^k)$ will hide the two-element subgroup $\{(id, id, 0), (u, u^{-1}, 1)\}$. Therefore, $C^P(G, f, 0^k)$ will always recover the random permutation u correctly, and output CORRECT.

On the other hand, if $P(G, f)$ outputs NONTRIVIAL while H is actually trivial, then $C^P(G, f, 0^k)$ will fail to find a nontrivial element of H and thus output BUGGY. If $P(G, f)$ outputs TRIVIAL while H is actually nontrivial, then the function f'' constructed by $C^P(G, f, 0^k)$ will hide the subgroup $(H \times u^{-1}Hu \times \{0\}) \cup (u^{-1}H \times Hu \times \{1\})$. P correctly distinguishes u and other elements in the coset Hu only by chance. Since the order of H is at least 2, the probability that $C^P(G, f, 0^k)$ outputs CORRECT is at most 2^{-k}.

Clearly, $C^P(G, f, 0^k)$ runs in polynomial time. The nonadaptiveness follows from Theorem 2.

Similarly, we can construct a nonadaptive checker $C^P(G, f, 0^k)$ for a program $P(G, f)$ that solves HIDDEN SUBGROUP over permutation groups. The checker makes k nonadaptive queries.

Begin

Run $P(G, f)$, which outputs a generating sets S.

Verify that elements of S are indeed in H.

Do k times (in parallel):

generate a random element $u \in G$.

define f' over G such that $f(g) = f'(gu)$ for all $g \in G$, use (G, f, f') to be an input instance of HIDDEN COSET

use Theorem 1 to convert (G, f, f') to an input instance $(G \wr \mathbb{Z}_2, f'')$ of HIDDEN SUBGROUP

$P(G \wr \mathbb{Z}_2, f'')$ will output a set S' of generators and a coset representative u'

if S and S' don't generate the same group **or** u and u' are not in the same coset of S, **then**

return BUGGY

End-do

return CORRECT

End

The proof of correctness for the above checker is similar to the proof of Theorem 5.

6 Discussion

The possibility of achieving exponential quantum savings is closely related to the underlying algebraic structure of a problem. Some researchers argued that generalizing from abelian groups to nonabelian groups may be too much for quantum computers [qua], since nonabelian groups exhibit radically different characteristics comparing with abelian groups. The results in this paper seem to support this point of view. Given the rich set of nonabelian group families, such as wreath product groups, HIDDEN SUBGROUP is indeed a "robust" and difficult problem. Recently Childs, Schulman, and Vazirani [CSV07] suggested an alternative generalization of abelian HIDDEN SUBGROUP, namely to the problems of finding nonlinear structures over finite fields. They gave two such problems, HIDDEN RADIUS and HIDDEN FLAT OF CENTERS, which exhibit exponential quantum speedup. An interesting open problem is whether these two problems are instances of HIDDEN SUBGROUP over some nonabelian group families.

Acknowledgments

We thank Andrew Childs and Wim van Dam for valuable comments on a preliminary version of this paper. We also thank anonymous referees for helpful suggestions.

References

[AK02] Arvind, V., Kurur, P.P.: Graph Isomorphism is in SPP. In: Proceedings of the 43rd IEEE Symposium on Foundations of Computer Science, IEEE, New York (2002)

[AT01] Arvind, V., Torán, J.: A nonadaptive NC checker for permutation group intersection. Theoretical Computer Science 259, 597–611 (2001)

[BK95] Blum, M., Kannan, S.: Designing programs that check their work. Journal of the ACM 42(1), 269–291 (1995)

[CSV07] Childs, A.M., Schulman, L.J., Vazirani, U.V.: Quantum algorithms for hidden nonlinear structures. In: FOCS 2007: Proceedings of the 48th Annual IEEE Symposium on Foundations of Computer Science (FOCS 2007), pp. 395–404. IEEE Computer Society, Washington (2007)

[CvD07] Childs, A.M., van Dam, W.: Quantum algorithm for a generalized hidden shift problem. In: SODA 2007: Proceedings of the eighteenth annual ACM-SIAM symposium on Discrete algorithms, pp. 1225–1232 (2007)

[EH00] Ettinger, M., Høyer, P.: On quantum algorithms for noncommutative hidden subgroups. Advances in Applied Mathematics 25, 239–251 (2000)

[EHK04] Ettinger, M., Høyer, P., Knill, E.: The quantum query complexity of the hidden subgroup problem is polynomial. Information Processing Letters 91(1), 43–48 (2004)

[FHL80] Furst, M.L., Hopcroft, J.E., Luks, E.M.: Polynomial-time algorithms for permutation groups. In: Proceedings of the 21st IEEE Symposium on Foundations of Computer Science, pp. 36–41 (1980)

[FIM+03] Friedl, K., Ivanyos, G., Magniez, F., Santha, M., Sen, P.: Hidden translation and orbit coset in quantum computing. In: Proceedings of the 35th ACM Symposium on the Theory of Computing, pp. 1–9 (2003)

[GSVV04] Grigni, M., Schulman, J., Vazirani, M., Vazirani, U.: Quantum mechanical algorithms for the nonabelian hidden subgroup problem. Combinatorica 24(1), 137–154 (2004)

[HMR+06] Hallgren, S., Moore, C., Rötteler, M., Russell, A., Sen, P.: Limitations of quantum coset states for graph isomorphism. In: STOC 2006: Proceedings of the thirty-eighth annual ACM symposium on Theory of computing, pp. 604–617 (2006)

[HRTS03] Hallgren, S., Russell, A., Ta-Shma, A.: The hidden subgroup problem and quantum computation using group representations. SIAM Journal on Computing 32(4), 916–934 (2003)

[Joz00] Jozsa, R.: Quantum factoring, discrete algorithm and the hidden subgroup problem (manuscript, 2000)

[Kit95] Kitaev, A.Y.: Quantum measurements and the Abelian Stabilizer problem, quant-ph/9511026 (1995)

[KS05] Kempe, J., Shalev, A.: The hidden subgroup problem and permutation group theory. In: Proceedings of the sixteenth annual ACM-SIAM symposium on Discrete algorithms, Vancouver, British Columbia, January 2005, pp. 1118–1125 (2005)

[Mat79] Mathon, R.: A note on the graph isomorphism counting problem. Information Processing Letters 8, 131–132 (1979)

[Mos99] Mosca, M.: Quantum Computer Algorithms. PhD thesis, University of Oxford (1999)

[MRS05] Moore, C., Russell, A., Schulman, L.J.: The symmetric group defies strong fourier sampling. In: FOCS 2005: Proceedings of the 46th Annual IEEE Symposium on Foundations of Computer Science, pp. 479–490 (2005)

[MRS07] Moore, C., Russell, A., Sniady, P.: On the impossibility of a quantum sieve algorithm for graph isomorphism. In: STOC 2007: Proceedings of the thirty-ninth annual ACM symposium on Theory of computing, pp. 536–545 (2007)

[NC00] Nielsen, M.A., Chuang, I.L.: Quantum Computation and Quantum Information. Cambridge University Press, Cambridge (2000)

[qua] Quantum pontiff, `http://dabacon.org/pontiff/?p=899`

[Reg04] Regev, O.: Quantum computation and lattice problems. SIAM Journal on Computing 33(3), 738–760 (2004)

[Sco87] Scott, W.R.: Group Theory. Dover Publications, Inc. (1987)

[Sim70] Sims, C.C.: Computational methods in the study of permutation groups. In: Computational problems in abstract algebra, pp. 169–183 (1970)

[vDHI03] van Dam, W., Hallgren, S., Ip, L.: Quantum algorithms for some hidden shift problems. In: Proceedings of the 14th annual ACM-SIAM symposium on Discrete algorithms, pp. 489–498 (2003)

An $O^*(3.52^{3k})$ Parameterized Algorithm for 3-Set Packing*

Jianxin Wang and Qilong Feng

School of Information Science and Engineering, Central South University
jxwang@mail.csu.edu.cn

Abstract. Packing problems have formed an important class of NP-hard problems. In this paper, we provide further theoretical study on the structure of the problems, and design improved algorithm for packing problems. For the 3-Set Packing problem, we present a deterministic algorithm of time $O^*(3.52^{3k})$, which significantly improves the previous best algorithm of time $O^*(4.61^{3k})$.

1 Introduction

In the complexity theory, packing problem forms an important class of NP-hard problems, which are used widely in scheduling and code optimization fields. We first give some related definitions [1].

Assume all the elements used in this paper are from U.

Set Packing: Given a pair (S, k), where S is a collection of n sets and k is an integer, find a largest subset S' such that no two sets in S' have the common elements.

(Parameterized) 3-Set Packing: Given a pair (S, k), where S is a collection of n sets and k is an integer, each set contains 3elements, either construct a k-packing or report that no such packing exists.

3-Set Packing Augmentation: Given a pair (S, P_k), where S is a collection of n sets and P_k is a k-packing in S, either construct $(k+1)$-packing or report that no such packing exists.

Recently, Downey and Fellows [2] proved that the 3-D Matching problem is Fixed Parameter Tractable (FPT), and gave an algorithm with time complexity $O^*((3k)!(3k)^{9k+1})$, which can be applied to solve 3-Set Packing problem. Jia, Zhang and Chen [3] reduced the time complexity to $O^*((5.7k)^k)$ using greedy localization method. Koutis [4] proposed a randomized algorithm with time

* This work is in part supported by the National Natural Science Foundation of China under Grant No. 60773111 and No. 60433020, Provincial Natural Science Foundation of Hunan (06JJ10009), the Program for New Century Excellent Talents in University No. NCET-05-0683 and the Program for Changjiang Scholars and Innovative Research Team in University No. IRT0661.

M. Agrawal et al. (Eds.): TAMC 2008, LNCS 4978, pp. 82–93, 2008.

complexity $O^*(10.88^{3k})$ and a deterministic algorithm with time complexity at least $O^*(32000^{3k})$. Fellows et al [5] gave an algorithm with time complexity at least $O^*(12.67^{3k}T(k))$ for the 3-D Matching problem, (based on current technology, $T(k)$ is at least $O^*(10.4^{3k})$). Kneis, Moelle, S.R and Rossmanith [6] presented a deterministic algorithm of time $O^*(16^{3k})$ using randomized divide and conquer. Chen, Lu, S.H.S and Zhang [7] gave a randomized algorithm with time complexity $O^*(2.52^{3k})$ based on the divide and conquer method, whose deterministic algorithm is of time complexity $O^*(12.8^{3k})$. Liu, Lu, Chen and H Sze [1] gave a deterministic algorithm of time $O^*(4.61^{3k})$ based on greedy localization and color coding, which is currently the best result in the world.

In this paper, we will discuss how to construct a $(k+1)$-packing from a k-packing, so as to solve the 3-Set Packing problem. After further analyzing the structure of the problem, we can get the following property: if the 3-Set Packing Augmentation problem can not be solved in polynomial time, then each set in P_{k+1} should contain at least one element of P_k. Based on the above property, we can get a randomized algorithm of time $O^*(3.52^{3k})$ using randomized divide and conquer. According to the structure analysis and the derandomization method given in [8], we can get a deterministic algorithm with the same time complexity $O^*(3.52^{3k})$, which greatly improves the current best result $O^*(4.61^{3k})$.

2 Related Terminology and Lemmas

We first introduce two important lemmas [1].

Lemma 1. *For any constant $c > 1$, the 3-Set Packing Augmentation problem can be solved in time $O^*(c^k)$ if and only if the 3-Set Packing problem can be solved in time $O^*(c^k)$.*

Lemma 2. *Let (S, P_k) be an instance of 3-Set Packing Augmentation, where P_k is a k-packing in S. If S also has $(k+1)$-packings, then there exists a $(k+1)$-packing P_{k+1} in S such that every set in P_k contains at least two elements in P_{k+1}.*

By lemma 1, reducing the time complexity of 3-Set Packing Augmentation problem is our objective.

For the convenience of analyzing the structure of the 3-Set Packing Augmentation problem, we give the following definitions.

Definition 1. Let (S, P_k) be an instance of 3-Set Packing Augmentation problem, where P_k is a k-packing in S. Assume there exists P_{k+1}, for a certain $(k+1)$-packing P and a set ρ_i in P_k, if only one element of ρ_i is contained in P, ρ_i is called *1-Set*; if no element of ρ_i is contained in P, it is called *0-Set*. The collection of all the 1-Set and 0-Set in P_k is called $(1,0)$-*Collection* of the $(k+1)$-packing P. If the $(1,0)$-Collection of P is null, then P is called a $(0,1)$-*free packing*.

We need to point out that: different $(k+1)$-packing may have different $(1,0)$-Collection.

It is easy to get the following lemma from relation between P_k and P_{k+1}.

Lemma 3. *Let (S, P_k) be an instance of 3-Set Packing Augmentation problem, where P_k is k-packing in S. Assume there exists P_{k+1}, any $(k+1)$-packing can be transformed into a $(0,1)$-free packing in polynomial time.*

Proof. For an instance of 3-Set Packing Augmentation problem (S, P_k). Assume there exists P_{k+1}, for any $(k+1)$-packing P, do the following process.

Find out all the 1-Set and 0-Set in P_k, denoted by W. For each set ρ_i in W, discuss it in the following two cases.

Case 1: ρ_i is a 1-Set. Assume one element a in ρ_i is contained in P and the set ρ'_j in P contains the element a. Use ρ_i to replace ρ'_j in P such that the number of 1-Set in P_k is reduced by one. Because of the replacement, there may produce new 1-Set or 0-Set in P_k. Find out all the new 1-Set and 0-Set in P_k. If a new 1-Set or 0-Set has not existed in W, add this set into W.

Case 2: ρ_i is a 0-Set. Since P_k has k sets and P_{k+1} has $k+1$ sets, there must exists one set ρ'_j in P that is not contained in P_k. Use ρ_i to replace ρ'_j in P such that the number of 0-Set in P_k is reduced by one. Because of the replacement, there may produce new 1-Set or 0-Set in P_k. Find out all the new 1-Set and 0-Set in P_k. If a new 1-Set or 0-Set has not existed in W, add this set into W.

After processing all the sets in W, there are no 1-Set and 0-Set in P_k. Therefore, the $(k+1)$-packing P is converted into a $(0,1)$-free packing.

Now, we prove that the above process can be done in polynomial time.

In order to find out all the 1-Set and 0-Set, for each set ρ_i in P_k, we need to consider all the sets in P. Obviously, the time complexity of this process is bounded by $O(k^2)$. When a set in P is replaced by 1-Set or 0-Set,it needs to redetermine the 1-Set and 0-Set in P_k. The whole time complexity of this process is bounded by $O(k^3)$. This completes the proof of the lemma. □

Based on the lemma 3, we can get following lemma.

Lemma 4. *Let (S, P_k) be an instance of 3-Set Packing Augmentation problem, where P_k is a k-packing in S. Assume there exists P_{k+1}, for any $(k+1)$-packing P, if P is a $(0,1)$-free packing, then each set in P_k should have at least 2 elements be contained in P.*

Proof. Assume there exists P_{k+1} in S, for any $(k+1)$-packing P, if P is a $(0,1)$-free packing, the $(1,0)$-Collection of P is null, that is, there is no 1-Set or 0-Set in P_k. Therefore, each set in P_k should have at least 2 elements be contained in P. □

Combing lemma 4 with the structure analysis of P_{k+1}, we can get the following lemma.

Lemma 5. *Given an instance of 3-Set Packing Augmentation problem (S, P_k), where P_k is k-packing in S. For any $(0,1)$-free packing P, assume there are $2k+x$ $(0 \le x \le k)$ elements from P_k contained in P, if the 3-Set Packing Augmentation problem can not be solved in polynomial time, each set in P should contain at least one of those $2k+x$ elements.*

Proof. By the lemma 4, each $(0, 1)$-free packing contains at least $2k$ elements of P_k.

We use contradiction method to prove. Assume that: although the 3-Set Packing Augmentation problem can not be solved in polynomial time, one or more sets in P contain none of those $2k + x$ elements.

Assume that there is a set α in P containing none of those $2k + x$ elements. Except those $2k + x$ elements, other elements in P_k are definitely not in P. Therefore, α and all sets in P_k have no common elements.

According to the relation between elements and sets, construct the bipartite graph $G = (V_1 \cup V_2, E)$, where the vertices in V_1 correspond to the elements in U, and the vertices in V_2 denote the sets in S. If an element is contained in a set, then the corresponding vertices will be connected by an edge. In graph G, we can find out all the sets having no common elements with P_k, thus, α is definitely in those sets. A $(k + 1)$-packing can be constructed by the k sets in P_k and α, which can be done in polynomial time. This contradicts with the assumption. This completes the proof of the lemma. □

3 The Randomized Algorithm

In this part, randomized divide and conquer will be used efficiently to solve the 3-Set Packing Augmentation problem. Based on the lemma 3, we can assume that all the $(k + 1)$-packing used in the following are $(0, 1)$-free packing.

By lemma 5, we can get the following lemma.

Lemma 6. *Given an instance of* 3*-Set Packing Augmentation problem* (S, P_k)*, where* P_k *is a* k*-packing in* S*. Assume there exists* P_{k+1}*, for a* $(k+1)$*-packing* P*, if* P *can not be found in polynomial time,* P *is composed of the following there parts.*

(1) P *has* r *sets, each of which contains only one element of* P_k*,* $0 \le r \le \lceil \frac{k+3}{2} \rceil$.

(2) P *has* s *sets, each of which contains only two elements of* P_k*,* $0 \le s \le k+1$.

(3) P *has* t *sets, each of which contains three elements of* P_k*,* $0 \le t \le k - 1$.

where $r + s + t = k + 1$.

Proof. By lemma 5, if there exists P_{k+1} and could not find a $(k + 1)$-packing in polynomial time, each set in P_{k+1} should contains at least one element of P_k. Therefore, for the $(k+1)$-packing P, each set in P may contain 1, 2 or 3 elements of P_k, which is one of the three types given in the lemma.

Now we will prove that there are at most $\lceil \frac{k+3}{2} \rceil$ sets in P, each of which contains only one element of P_k. Assume that P contains $2k + x$ $(0 \le x \le k)$ elements of P_k. If $s = 0$ and each set in P has already contained one of those $2k + x$ elements. When the remaining $2k + x - (k + 1) = k + x - 1$ elements are used to form the sets containing three elements of P_k, r gets the maximum value: $k + 1 - \lfloor \frac{k+x-1}{2} \rfloor \le \lceil \frac{k+3}{2} \rceil$.

When each set in P contains only two elements of P_k, s gets the maximum value $k + 1$, thus, $s \le k + 1$.

Because each set in P contains at least one element of P_k, the maximum number of sets in P containing three elements of P_k is $k - 1$, thus, $t \le k - 1$. □

Let C_i $(1 \leq i \leq 3)$ denote all the sets having i common elements with P_k, which can be found in polynomial time based on the relation of elements and sets. Assume U_{P_k} denotes the $3k$ elements in P_k and U_{S-P_k} denotes the elements in $S - P_k$. Therefore, each set in C_2 contains 2 elements from U_{P_k}, and each set in C_3 contains 3 elements from U_{P_k}. Let $U_{C_2-P_k}$ denote the elements in C_2 but not in U_{P_k}, then $U_{C_2-P_k} \subseteq U_{S-P_k}$.

By lemma 6, if P_{k+1} exists, there are r sets in P_{k+1} such that each of which contains only one element of P_k, which are obviously included in C_1. To find the r elements from U_{P_k}, there are $\binom{3k}{r}$ enumerations. Let H be the collection of sets in C_1 containing one of those r elements.

Assume $U_{P_{k+1}-P_k}$ denotes all the elements in P_{k+1} but not in P_k, and the size of the $U_{P_{k+1}-P_k}$ is denoted by $y = |U_{P_{k+1}-P_k}|$. By lemma 5, P_{k+1} contains at least $2k$ elements of U_{P_k}, thus, $U_{P_{k+1}-P_k}$ contains at most $k+3$ elements of U_{S-P_k}, that is, $y \leq k+3$. It can be seen that the elements in $U_{P_{k+1}-P_k}$ are either in H or in $C_2 \cup C_3$. Assume that $U'_{P_{k+1}-P_k}$ denotes the elements of $U_{P_{k+1}-P_k}$ belonging to H. When the elements in $U_{P_{k+1}-P_k}$ are partitioned, the probability that the elements in $U'_{P_{k+1}-P_k}$ are exactly partitioned into H is $\frac{1}{2^y}$.

The general ideal of our randomized algorithm is as follows:

Divide P_{k+1} into two parts to handle, one of which is in H and the other in $C_2 \cup C_3$. For the part contained in $C_2 \cup C_3$, we use dynamic programming to find a $(k+1-r)$-packing; For the part in H, we use randomized divide and conquer to handle.

3.1 Use Dynamic Programming to find a $(k+1-r)$-Packing in $C_2 \cup C_3$

For the convenience of describing the algorithm, we first give the concept of symbol pair. For each set $\rho_i \in C_2$, the elements from U_{P_k} in set ρ_i is called a symbol pair.

The algorithm of finding a $k' = k+1-r$ $(k' \leq k+1)$ packing in $C_2 \cup C_3$ is given in figure 1.

Theorem 1. *If there exists k'-packing in $C_2 \cup C_3$, algorithm SP will definitely return a collection of symbol pairs and 3-sets with size k', and the time complexity is bounded by $O^*(2^{3k})$.*

Proof. If there exists k'-packing in $C_2 \cup C_3$, assume that the number of sets in C_2 contained in the k'-packing is k'', $0 \leq k'' \leq k'$. Therefore, the k'' sets of k'-packing in C_2 can form a k''-packing. We need to prove the following two parts.

(1) After the execution of the for-loop in step 3, Q_1 must contain a collection of symbol pairs with size k''.

(2) After the execution of the for-loop in step 5, Q_1 must contain a collection of symbol pairs and 3-sets with size k'.

The proof of the first part is as follows.

It can be seen from step 3.1-3.4 that the C' added into Q_1 in the step 3.7 is a collection of symbol pairs from the right packing. We get a induction for the i in

Algorithm SP

Input: C_2, C_3, k', U_{P_k}

Output: if there exists k'-packing in $C_2 \cup C_3$, return a collection of symbol pairs and 3-sets with size k'

1. assume the elements in $U_{C_2 - P_k}$ are $x_1, x_2, \ldots x_m$;
2. Q_1=$\{\phi\}$; Q_{new}=$\{\phi\}$;
3. **for** $i = 1$ **to** m **do**

3.1 **for** each collection C in Q_1 **do**

3.2 **for** each 3-set ρ in C_2 having element x_i **do**

3.3 **if** C has no common element with ρ **then**

3.4 $C' = C \cup$ {elements in ρ belonging to U_{P_k}};

3.5 **if** C' is not larger than k' and no collection in Q_{new} has used exactly the same elements as that used in C' **then**

3.6 add C' into Q_{new};

3.7 $Q_1 = Q_{new}$;

4. assume the 3-sets in C_3 are $z_1, z_2, \ldots z_l$;
5. **for** $h = 1$ **to** l **do**

5.1 **for** each collection C in Q_1 **do**

5.2 **if** C does not have common elements with z_h **then**$C' = C \cup \{z_h\}$;

5.4 **if** C' is not larger than k' and no collection in Q_{new} has used exactly the same elements as that used in C' **then** add C' into Q_{new};

5.6 $Q_1 = Q_{new}$;

6. **if** there is a collection of symbol pairs and 3-sets with size k' in Q_1 **then** return the collection.

Fig. 1. Use dynamic programming to find a $(k+1-r)$-packing in $C_2 \cup C_3$

the step 3 so as to prove that: if there exists k''-packing in C_2, Q_1 must contain a collection of symbol pairs with size k''.

There are m different elements in $U_{C_2 - P_k}$: $x_1, x_2, \ldots x_m$. For any arbitrary i $(1 \le i \le m)$, assume that X_i denotes all the sets containing the element in $\{x_1, x_2, \ldots x_i\}$. Therefore, we only need to prove the following claim.

Claim 1. If there exists a j-packing symbol pairs collection P_j in X_i, then after i-thexecution of the for-loop in step 3, Q_1 contains a j-packing symbol pairs collection P'_j, which uses the same $2j$ elements with P_j.

In the step 2, Q_1=$\{\phi\}$. Thus, if X_i has a 0-packing, the claim is true.

When $i \ge 1$, assume there exists a j-packing symbol pair collection $P_j = \{\varphi_{l_1}, \varphi_{l_2}, \ldots, \varphi_{l_j}\}$, where $1 \le l_1 < l_2 < \cdots < l_j \le i$, then there must exists a $(j-1)$-packing symbol pair collection $P_{j-1} = \{\varphi_{l_1}, \varphi_{l_2}, \ldots, \varphi_{l_j - 1}\}$ in $X_{l_j - 1}$. By the induction assumption, after the $(l_j - 1)$-th execution of for-loop in step 3, Q_1 contains a $(j-1)$-packing symbol pairs collection P'_{j-1}, which use the same $2(j-1)$ elements of U_{P_k} with P_{j-1}. By the assumption, X_i contains a j-packing symbol pair collection P_j. Therefore, when the set containing the φ_{l_j} is considered in step 3.2, the elements belonging to U_{P_k} in P'_{j-1} are totally different from the elements in φ_{l_j}. As a result, if there is no collection of symbol pairs in Q_1 containing the same $2j$ elements with $P'_{j-1} \cup \{\varphi_{l_j}\}$, the j-packing symbol pairs $P'_{j-1} \cup \{\varphi_{l_j}\}$ will be added into Q_1. Because all the collection of symbol pairs in Q_1 are not removed from Q_1 and $l_j \leq i$, after the i-th execution of the for-loop in step 3, Q_1 must contains a j-packing symbol pairs collection using the same $2j$ elements with P_j. When $i = m$, if there exists k''-packing $P_{k''}$ in C_2, Q_1 must contain a collection of symbol pairs using the same $2k''$ elements of U_{P_k} with $P_{k''}$.

The proof of the second part is similar to the first part, which is neglected here.

At last, we analyze the time complexity of algorithm SP. If considering C_2 only, for each j ($0 \leq j \leq k'$) and any subset of U_{P_k} containing $2j$ elements, Q_1 record at most one collection symbol pairs using those $2j$ elements, thus, Q_1 contains at most $\sum_{j=0}^{k'+1} \binom{3k}{2j}$ collections. If considering C_3 only, for each j ($0 \leq j \leq k'$) and any subset of containing $3j$ elements, Q_1 record at most one collection of 3-sets using those $3j$ elements, thus, Q_1 contains at most $\sum_{j=0}^{k'-1} \binom{3k}{3j}$ collections. Therefore, the time complexity of algorithm SP is bounded by $\max\{O^*(\sum_{j=0}^{k'+1} \binom{3k}{2j}), O^*(\sum_{j=0}^{k'-1} \binom{3k}{3j})\} = O^*(2^{3k})$. □

If there exists k'-packing in $C_2 \cup C_3$, the collection returned by algorithm SP may contain symbol pairs, which can be converted into 3-sets using the bipartite maximum matching.

3.2 Use Randomized Divide and Conquer to find a r-Packing in H

Assume that we have already picked r ($0 \leq r \leq \lceil \frac{k+3}{2} \rceil$) elements from U_{P_k}, and let H be the collection of sets in C_1 containing one of those r elements. The algorithm of finding a r-packing in H is given in figure 2.

Theorem 2. *If H has r-packing, algorithm RSP will return a collection D containing the r-packing with probability larger than 0.75, and the time complexity is bounded by $O^*(4^{2r})$.*

Proof. Algorithm RSP divides H into two parts to handle: H_1, H_2. If there exists r-packing P_r, P_r has $2r$ elements of U_{S-P_k}, denoted by $U_{P_r-P_k}$. Assume that $U'_{P_r-P_k}$ denotes the elements belonging to $U_{P_r-P_k}$ in H_1, thus, the elements belonging to $U_{P_r-P_k}$ in H_2 can be denoted by $U_{P_r-P_k} - U'_{P_r-P_k}$. Mark all the elements from U_{S-P_k} in H_1 and H_2 with red and blue colors. When elements in $U'_{P_r-P_k}$ are exactly marked with red and elements in $U_{P_r-P_k} - U'_{P_r-P_k}$ are marked with blue, it is called that elements in H_1 and H_2 are rightly marked, which occurs with probability $\frac{1}{2^{2r}}$. Therefore, the probability that the elements in H_1 and H_2 are not rightly marked is $1 - \frac{1}{2^{2r}}$.

If there exists r-packing in H, let δ_r be the probability that algorithm RSP can not find the r-packing. In the step 3, H is divided into two parts: H_1, H_2. Therefore, the probability that RSP(H_1) and RSP(H_2) can not find the corresponding packing respectively is $\delta_{r/2}$. After $3 \cdot 2^{2r}$ iterations, the probability that algorithm RSP could not find the P_{k+1} is $(1 - \frac{1}{2^{2r}} + \frac{1}{2^{2r-1}} \cdot \delta_{r/2})^{3 \cdot 2^{2r}}$. We need

Algorithm RSP
Input: H, r
Output: return a collection D of packings

1. **if** $r = 0$ **then** return ϕ;
2. **if** $r = 1$ **then** return H;
3. randomly pick $\lceil \frac{r}{2} \rceil$ elements from the r elements, and let H_1 denote all the sets containing one of those $\lceil \frac{r}{2} \rceil$ elements;
4. $H_2 = H - H_1$; $D = \phi$;
5. **for** $3 \cdot 2^{2r}$ times **do**

5.1 mark all the elements from U_{S-P_k} in H_1 and H_2 with red and blue;
for each set ρ in H_1, if the colors of the elements belonging to U_{S-P_k} are not both red, delete the set ρ;
for each set ρ' in H_2, if the colors of the elements belonging to U_{S-P_k} are not both blue, delete the set ρ';

5.2 D_1=RSP(H_1, $\lceil \frac{r}{2} \rceil$);

5.3 D_2=RSP(H_2, $\lfloor \frac{r}{2} \rfloor$);

5.4 **for** each packing α in D_1 **do**
  for each packing β in D_2 **do**
    if there does not exist $\alpha \cup \beta$ in D, add $\alpha \cup \beta$ into D;

6. return D;

Fig. 2. Use randomized divide and conquer to find a r-packing in H

to prove that: for any r, $\delta_r \leq 1/4$. It is obvious that $\delta_1 = 0$. Assume $\delta_{r/2} \leq 1/4$. By the induction assumption, we can get that: $\delta_r = (1 - \frac{1}{2^{2r}} + \frac{1}{2^{2r-1}} \cdot \delta_{r/2})^{3 \cdot 2^{2r}} \leq (1 - \frac{1}{2^{2r}} + \frac{1}{2^{2r-1}} \cdot 1/4)^{3 \cdot 2^{2r}} = (1 - \frac{1}{2^{2r+1}})^{\frac{3}{2} \cdot 2^{2r+1}} \leq e^{-3/2} < 1/4$.

Let T_r denote the number of recursive calls in algorithm RSP, then $T_r \leq 3 \cdot 2^{2r} \cdot (T_{\lceil \frac{r}{2} \rceil} + T_{\lfloor \frac{r}{2} \rfloor}) \leq 3 \cdot 2^{2r+1} \cdot T_{\lceil \frac{r}{2} \rceil} = O(3^{\log 2r} 4^{2r}) = O((2r)^{\log 3} 4^{2r}) = O^*(4^{2r})$. □

3.3 The General Algorithm for 3-Set Packing Augmentation

Based on the above two algorithm, the general algorithm for 3-Set Packing Augmentation problem is given in figure 3.

Theorem 3. *If S has $(k+1)$-packing, algorithm GSP will return the $(k+1)$-packing with probability larger than 0.75, and the time complexity is bounded by $O^*(3.52^{3k})$.*

Proof. In the above algorithm, we need to consider all the enumeration of r. If S has a $(k+1)$-packing, there must exist a r and an enumeration satisfying the condition. The algorithm divides P_{k+1} into two parts to handle, one of which is in H and the other in $C_2 \cup C_3$. When elements in $U'_{P_r - P_k}$ are exactly marked with white and elements in $U_{P_r - P_k} - U'_{P_r - P_k}$ are marked with black, it is called that elements in U_{S-P_k} are rightly marked, which occurs with probability $\frac{1}{2^y}$.

Algorithm GSP
Input: S, k
Output: whether there is a $(k+1)$-packing in S

1. **for** $r = 0$ **to** $\lceil \frac{k+3}{2} \rceil$ **do**
1.1 enumerate r elements from U_{P_k}, and get $\binom{3k}{r}$ enumerations;
1.2 **for** each enumeration **do**
  let H be the collection of sets in C_1 containing one of those r elements;
1.3 **for** $24 \cdot 2^k$ times **do**
  $C_2' = C_2$; $C_3' = C_3$;
  for each element a in U_{P_k}, if a belongs to H, then delete all the sets in $C_2' \cup C_3'$ containing a;
  use colors black and white to mark all the elements in U_{S-P_k};
  for each set ρ in C_2', if the color of the element belonging to U_{S-P_k} is not black, delete the set ρ;
  for each set ρ' in H, if the colors of the elements belonging to U_{S-P_k} are not both white, delete the set ρ';
  $Q_2 = \text{SP}(C_2', C_3', k+1-r, U_{P_k})$;
  $Q_3 = \text{RSP}(H, r)$;
  use the bipartite maximum matching algorithm to convert the symbol pairs in Q_2 into 3-sets;
  if Q_2 is a $(k+r-1)$-packing and Q_3 has a r-packing
  then return the $(k+1)$-packing; stop;
2. return no $(k+1)$-packing in S;

Fig. 3. The general algorithm for 3-Set Packing Augmentation

Therefore, the probability that the elements in U_{S-P_k} are not rightly marked is $1 - \frac{1}{2^y}$. If S has P_{k+1}, let δ_k denote the probability that algorithm GSP can not find the P_{k+1}. If H has r-packing, let δ_r denote the probability that algorithm RSP can not find the r-packing. By theorem 3, we know that $\delta_r \leq 1/4$. If P_{k+1} exists, for a certain r and an enumeration, after $24 \cdot 2^k$ iterations of step 1.3, the probability that algorithm GSP could not find the P_{k+1} is $(1 - \frac{1}{2^y} + \frac{1}{2^{y-1}} \cdot \delta_r)^{24 \cdot 2^k}$, that is,

$\delta_k = (1 - \frac{1}{2^y} + \frac{1}{2^{y-1}} \cdot \delta_r)^{24 \cdot 2^k} \leq (1 - \frac{1}{2^y} + \frac{1}{2^{y-1}} \cdot \frac{1}{4})^{24 \cdot 2^k} = (1 - \frac{1}{2^{y+1}})^{24 \cdot 2^k} \leq e^{-3/2} < 1/4.$

By theorem 1, If there exists $(k+1-r)$-packing in $C_2 \cup C_3$, algorithm SP will definitely return a collection of symbol pairs and 3-sets with size $k+1-r$. By theorem 2, if H has r-packing, algorithm RSP will return the r-packing with probability larger than 0.75. Therefore, if S has $(k+1)$-packing, algorithm GSP will return the $(k+1)$-packing with probability larger than 0.75.

Now we analyze time complexity of the above algorithm. For each r, there are $\binom{3k}{r}$ ways to enumerate r elements from U_{P_k}. By theorem 1, the time complexity of algorithm SP is bounded by $O^*(2^{3k})$. By theorem 2, the time complexity of algorithm RSP is $O^*(4^{2r})$. Because of $0 \leq r \leq \lceil \frac{k+3}{2} \rceil$, the running time of algorithm RSP is bounded by $O^*(4^k)$. Using bipartite maximum matching algorithm to covert symbol pairs in Q_2 can be done in polynomial time. Therefore, the total time complexity of algorithm GSP is $\sum_{r=0}^{\lceil \frac{k+3}{2} \rceil} \binom{3k}{r}(2^k(2^{3k-r}+4^k)) = O^*(3.52^{3k})$. □

4 Derandomization

When there exists $(k+1)$-packing, in order to make failure impossible, we need to derandomize the above algorithm. We first point out that: partitioning a set means dividing the set into two parts.

Because the size of the $U_{P_{k+1}-P_k}$ is y, there are 2^y ways to partition $U_{P_{k+1}-P_k}$. Therefore, there must exist a partition satisfying the following property: elements in $U'_{P_{k+1}-P_k}$ are exactly partitioned into H, and elements in $U_{P_{k+1}-P_k} - U'_{P_{k+1}-P_k}$ are exactly in $C_2 \cup C_3$. However, the problem is that: $U_{P_{k+1}-P_k}$ is unknown.

Naor, Schulman and Srinivasan [9] gave the solution for the above problem. Moreover, Chen and Lu [8] presented a more detailed description of that method.

Now we introduce a very important lemma in [10].

Lemma 7. *Let n, k be two integers such that $0 < k \leq n$. There is an (n,k)-universal set P of size bounded by $O(n2^{k+12\log^2 k+12\log k+6})$, which can be constructed in time $O(n2^{k+12\log^2 k+12\log k+6})$.*

The (n,k)-universal set in the above lemma is a set F of splitting functions, such that for every k-subset W of $\{0,1,\cdots,n-1\}$ and any partition (W_1, W_2) of W, there is a splitting function f in F that implements (W_1, W_2).

In the construction of the above lemma, Chen and Lu [8] constructed a fuction $h(x)$=((ix mod p)mod k^2) from $\{0,1,\cdots,n-1\}$ to $\{0,1,\cdots,k^2-1\}$, and used the fact that there are at most $2n$ $h(x)$ to get the above lemma. However, the bound $2n$ is not tight. Now, we introduce an important lemma in [9].

Lemma 8. *There is an explicit (n,k,k^2)-splitter $A(n,k)$ of size $O(k^6 \log k \log n)$.*

In the above lemma, the (n,k,k^2)-splitter $A(n,k)$ denotes the function from $\{0,1,\cdots,n-1\}$ to $\{0,1,\cdots,k^2-1\}$. Thus, the number of functions from $\{0,1,\cdots,n-1\}$ to $\{0,1,\cdots,k^2-1\}$ are bounded by $O(k^6 \log k \log n)$.

Based on lemma 7, lemma 8, we can get the following lemma.

Lemma 9. *Let n, k be two integers such that $0 < k \leq n$. There is an (n,k)-universal set P of size bounded by $O(\log n 2^{k+12\log^2 k+18\log k})$, which can be constructed in time $O(\log n 2^{k+12\log^2 k+18\log k})$.*

By the lemma 9, we can get the following theorem.

Theorem 4. *3-Set Packing Augmentation problem can be solved deterministically in time $O^*(3.52^{3k})$.*

Proof. Given an instance of 3-Set Packing Augmentation problem (S, P_k), if there exists P_{k+1}, by lemma 5, P_{k+1} contains at least $2k$ elements of U_{P_k}, thus, $U_{P_{k+1}-P_k}$ contains at most $k+3$ elements of U_{S-P_k}. After picking r elements from U_{P_k}, P_{k+1} is divided into two parts to handle in the randomized algorithm, one of which is in H and the other in $C_2 \cup C_3$. By lemma 9, we can construct the $(S-P_k, k+3)$-universal set, whose size is bounded by $O(\log n 2^{k+3+12\log^2(k+3)+18\log(k+3)})$.

For each partition to U_{S-P_k} in the $(S-P_k, k+3)$-universal set, let U_{H-P_k} denote the elements partitioned into H. If H has a r-packing P_r, there are $2r$ elements of U_{S-P_k} in P_r, denoted by $U_{P_r-P_k}$. Assume $U'_{P_r-P_k}$ denotes the elements of $U_{P_r-P_k}$ in H_1. In order to find P_r, $U'_{P_r-P_k}$ should be partitioned into H_1, and $U_{P_r-P_k} - U'_{P_r-P_k}$ should be in H_2. By lemma 9, we can construct $(U_{H-P_k}, 2r)$-universal set, whose size is bounded by $O(\log n 2^{k+3+12\log^2(k+3)+18\log(k+3)})$.

In the derandomization of algorithm RSP, the time complexity is:

$$\begin{aligned} T_r &\le \log n 2^{k+3+12\log^2(k+3)+18\log(k+3)}(T_{\lceil \frac{r}{2}\rceil} + T_{\lfloor \frac{r}{2}\rfloor}) \\ &\le \log n 2^{k+3+12\log^2(k+3)+18\log(k+3)+1}T_{\lceil \frac{r}{2}\rceil} \\ &= O((k+3)^{\log\log n}2^{2(k+3)+4\log^3(k+3)+15\log^2(k+3)+13\log(k+3)}). \end{aligned}$$

In the practical point of view, T_r is bounded by: $O(2^{2(k+3)+4\log^3(k+3)+15\log^2(k+3)+11\log(k+3)})$.

If there exists P_{k+1}, on the basis of $(S-P_k, k+3)$-universal set and the above result, we can get the P_{k+1} deterministically with time complexity:

$$\sum_{r=0}^{\lceil \frac{k+3}{2}\rceil}\binom{3k}{r}(\log n 2^{k+3+12\log^2(k+3)+18\log(k+3)}$$
$$(2^{3k-r} + 2^{2(k+3)+4\log^3(k+3)+15\log^2(k+3)+13\log(k+3)})) = O^*(3.52^{3k}).$$ □

By lemma 1 and theorem 4, we can get the following corollary.

Corollary 1. 3-Set Packing can be solved in $O^*(3.52^{3k})$.

5 Conclusions

For the 3-Set Packing problem, we construct a $(k+1)$-packing P_{k+1} from a k-packing P_k. After further analyzing the structure of the problem, we can get the following property: for any $(0,1)$-free packing P, if the 3-Set Packing Augmentation problem can not be solved in polynomial time, each set in P should contains at least one element of P_k. On the basis of the above property, we get a randomized algorithm of time $O^*(3.52^{3k})$. Based on the derandomization method given in [10], we can get a deterministic algorithm with the same time complexity, which greatly improves the current best result $O^*(4.61^{3k})$. Our results also imply improved algorithms for various triangle packing problems in graphs [10].

References

1. Liu, Y., Lu, S., Chen, J., Sze, S.H.: Greedy localization and color-coding:improved matching and packing algorithms. In: Bodlaender, H.L., Langston, M.A. (eds.) IWPEC 2006. LNCS, vol. 4169, pp. 84–95. Springer, Heidelberg (2006)
2. Downey., R., Fellows, M.: Parameterized complexity. Springer, New York (1999)
3. Jia, W., Zhang, C., Chen, J.: An efficient parameterized algorithm for m-SET PACKING. J. Algorithms 50, 106–117 (2004)
4. Koutis, I.: A faster parameterized algorithm for set packing. Information Processing Letters 94, 7–9 (2005)
5. Fellows, M., Knauer, C., Nishimura, N., Ragde, P., Rosamond, F., Stege, U., Thilikos, D., Whitesides, S.: Faster fixed-parameter tractable algorithms for matching and packing problems. In: Albers, S., Radzik, T. (eds.) ESA 2004. LNCS, vol. 3221, Springer, Heidelberg (2004)
6. Kneis, J., Moelle, D., Richter, S., Rossmanith, P.: Divide-and-Color. In: Fomin, F.V. (ed.) WG 2006. LNCS, vol. 4271, pp. 58–67. Springer, Heidelberg (2006)
7. Chen, J., Lu, S., Sze, S.-H., Zhang, F.: Improved algorithms for path, matching, and packing problems. In: Proc. 17th Annual ACM-SIAM Symposium on Discrete Algorithms (SODA 2007), pp. 298–307 (2007)
8. Chen, J., Lu, S.: Improved parameterized set splitting algorithms: A probabilistic approach. Algorithmica (to appear)
9. Naor, M., Schulman, L., Srinivasan, A.: Splitters and near-optimal derandomization. In: Proceedings of the 36th Annual Symposium on Foundations of Computer Science (FOCS 1995), pp. 182–190 (1995)
10. Mathieson, L., Prieto, E., Shaw, P.: Packing edge disjoint triangles: A parameterized view. In: Downey, R.G., Fellows, M.R., Dehne, F. (eds.) IWPEC 2004. LNCS, vol. 3162, pp. 127–137. Springer, Heidelberg (2004)

Indistinguishability and First-Order Logic

Skip Jordan and Thomas Zeugmann

Division of Computer Science
Hokkaido University, N-14, W-9, Sapporo 060-0814, Japan
{skip,thomas}@ist.hokudai.ac.jp

Abstract. The "richness" of properties that are indistinguishable from first-order properties is investigated. Indistinguishability is a concept of equivalence among properties of combinatorial structures that is appropriate in the context of testability. All formulas in a restricted class of second-order logic are shown to be indistinguishable from first-order formulas. Arbitrarily hard properties, including RE-complete properties, that are indistinguishable from first-order formulas are shown to exist. Implications on the search for a logical characterization of the testable properties are discussed.

Keywords: Property testing, logic, graph theory, descriptive complexity.

1 Introduction

In property testing we are interested in efficiently deciding whether a given structure possesses, or is far from possessing, a desired property. Although algorithmic efficiency is often defined as polynomial time, this is not always ideal. For example, we may not have an explicit representation of the input but instead only the ability to query an oracle for individual bits. These queries may be expensive and even a linear-time algorithm, which requires us to compute the entire input explicitly, may be unacceptable. If we are only concerned about distinguishing with high probability between the input *satisfying* and being *far from* satisfying a property, a sub-linear number of queries may be sufficient.

Testers are randomized algorithms that, given the size of the input, are restricted to a certain number of queries which *does not* depend on the input size. We shall give formal definitions below, but mention here that such algorithms are probabilistic approximation algorithms.

In a certain sense, the whole area can be traced back to Freivalds [14] who introduced a program result checker for matrix multiplication over a finite field. Subsequently, the study of property testing originally arose in the context of program verification (see Blum *et al.* [7] and Rubinfeld and Sudan [22]), and the first explicit definition appears to be in [22]. For surveys of the field see e.g., Fischer [12] and Ron [21].

Characterizing the testable properties has been called the most important problem in the area (see Alon and Shapira [5]). The class of regular languages was shown to be testable in Alon *et al.* [4], a result that was extended in Chockler and Kupferman [8]. However, there are context-free languages that are not

M. Agrawal et al. (Eds.): TAMC 2008, LNCS 4978, pp. 94–104, 2008.

testable [12]. It is then perhaps at first surprising that there are many natural, testable properties that are considerably harder. Testers for NP-complete graph properties including k-color were given in Goldreich *et al.* [16].

Restricting ourselves to characterizations of testable graph properties, the first step towards a logical characterization was obtained by Alon *et al.* [2], later extended by Fischer [13]. They show that all properties expressible in first-order logic (FO) with all quantifier alternations of type "$\exists\forall$" are testable, whereas there exists a property expressible with a single quantifier alternation of type "$\forall\exists$" that is not testable. It is useful to note the equivalence of first-order logic with arithmetic and uniform AC^0 (see Barrington *et al.* [6]). Later, a characterization of the graph properties testable with one-sided error by algorithms that are unaware of the input size was given in [5], a result that was extended to hypergraphs by Rödl and Schacht [20]. An exact combinatorial characterization of the graph properties testable with a constant number of queries has been obtained by Alon *et al.* [3].

In the present paper we focus on a question raised in [13]: the expressive power of first-order logic in the context of indistinguishability and testing. The concept of *indistinguishability* was introduced by Alon *et al.* [2] as a suitable form of equivalence in the context of testing. First-order logic with arithmetic is equivalent to uniform AC^0, and so it is strictly contained in NP (see Furst *et al.* [15]). Therefore, properties that are complete for NP under first-order reductions such as k-color cannot be expressed in FO (see Allender *et al.* [1]). However, it has been noted in Alon *et al.* [2] that there are first-order expressible properties that are *indistinguishable* from such properties, including k-color. In this sense, the descriptive power of first-order logic with indistinguishability is surprisingly rich. We examine the set of properties that are indistinguishable from FO-expressible properties and show that this set is larger than was previously known.

The paper is organized as follows. We begin by giving more formal definitions. In Section 3 we show that all graph properties expressible in a restriction of monadic second-order existential logic (MSO$\exists$) are indistinguishable from FO properties. We next prove that there are arbitrarily-hard properties, including RE-complete properties, that are indistinguishable from FO properties (cf. Section 4). In fact, we can construct arbitrarily-hard *testable* properties. Finally, we discuss the implications of the results obtained (cf. Section 5).

2 Preliminaries

We generally restrict our attention to graph properties, and so give the following definitions for graphs. We consider only finite, undirected graphs without loops. We use G and H to refer to graphs, $n = |G|$ as the number of vertices in a graph, and P, Q and R to refer to properties.

Let G and G' be two graphs having the same vertex set and let $\varepsilon > 0$. If G' can be constructed from G by adding and removing no more than εn^2 edges then we say that G and G' *differ in no more than* εn^2 *places*.

Definition 1 (Alon *et al.* [2]). *Let P be a property of graphs and let $\varepsilon > 0$.*

(1) *A graph G with n vertices is called ε-*far from satisfying *P if no graph G' with the same vertex set, which differs from G in no more than εn^2 places, satisfies P.*
(2) *An ε-*test *for P is a randomized algorithm which, given the quantity n and the ability to make queries whether or not a desired pair of vertices of an input graph G with n vertices are adjacent, distinguishes with probability at least $\frac{2}{3}$ between the case of G satisfying P and the case of G being ε-far from satisfying P.*

Note that in Definition 1 the choice of $\frac{2}{3}$ is of course traditional and arbitrary. Any probability strictly greater than $\frac{1}{2}$ can be chosen and the resulting test can be iterated a constant number of times to achieve any desired accuracy strictly less than one, see e.g., Hromkovič [19].

Definition 2 (Alon *et al.* [2]). *The property P is called* testable *if for every fixed $\varepsilon > 0$ there exists an ε-test for P whose total number of queries is bounded only by a function of ε, which is independent of the size of the input graph.*

We allow the tester to know the size of the input, and to make its queries in any computable fashion. In [17] this was shown to be equivalent to the "non-adaptive" model, where a tester uniformly chooses a set of vertices, receives the induced subgraph, and makes a decision based on whether that subgraph has some fixed property. We note that this definition of testability is not an $o(n)$ number of queries and that "ε-far" clearly depends on n.

In general we do not require a uniformity condition. However, it is very natural to require the ε-tests for P in the above definition to be computable given ε. We refer to properties satisfying this additional condition as *uniformly testable* where the uniform and non-uniform cases differ (Proposition 1).

We note that testers given in the literature are generally presented as a single algorithm that takes ε as a parameter and therefore uniform.

Definition 3 (Alon *et al.* [2]). *Two graph properties P and Q are called* indistinguishable *if for every $\varepsilon > 0$ there exists $N = N(\varepsilon)$ satisfying the following. For every graph G having $n > N$ vertices that satisfies P there exists a graph G' with the same vertex set, differing from G in no more than εn^2 places, which satisfies Q; and for every graph H with $n > N$ vertices satisfying Q there exists a graph H' with the same vertex set, differing from H in no more than εn^2 places, which satisfies P.*

As notation, we use ϕ, ψ and γ to refer to logical formulas. Whether these formulas are first- or second-order will be clear from context. We use Φ to denote a logical interpretation of variables. First-order variables are denoted by x_i, y_i and t_i while second-order variables are denoted by C_i. Members of the universe (nodes in the graph) are referred to as u_i when we wish to distinguish between variables and the nodes bound to them. We write $E(x, y)$ to denote the predicate "there is an edge between x and y." Since we consider only finite, undirected graphs without loops, it follows that $E(x, y)$ implies $E(y, x)$ for all x, y and that $E(x, x)$ is false for all x.

Logical structures such as graphs allow us to interpret predicate symbols. The combination of structures and interpretations, e.g., (G, Φ), then allows us to interpret formulas. We write $G \models \phi$, read G models ϕ, if formula ϕ holds when interpreting the edge predicate according to G. Inductively we use interpretations to interpret bound variables, and write $(G, \Phi) \models \phi$ if formula ϕ holds when interpreting bound variables according to Φ and the edge predicate (assuming it isn't bound by a second-order quantifier) according to G. For a more formal introduction to the logics used, see e.g., Enderton [9].

We assume that ordering and arithmetic are not present in the logics considered, however ordering and arithmetic can be defined in logics containing existential second-order. In proofs we treat the universal quantifier ($\forall$) as the dual of the existential quantifier ($\exists$). Although properties are often implicitly assumed to be computable, we shall see that there are implications to this assumption, and so explicitly state that we do not make this assumption.

Finally, we define DTIME($f(n)$) in the usual manner as the set of decision problems computable on a deterministic Turing machine in $f(n)$ steps.

3 Monadic Second-Order Existential Logic

In this section we show that all graph properties expressible in a restriction of monadic second-order existential logic (see Definition 4 below) are indistinguishable from FO properties.

Definition 4. *Let rMSO∃ denote the graph properties expressible in monadic second-order existential logic that satisfy the following. For all $P \in$rMSO∃ expressible with r second-order quantifiers, there exists an N such that in all graphs G satisfying P with $n > N$ vertices*

(1) *there exists a set of r vertices such that removing them does not affect P, and*
(2) *adding r disconnected vertices to the graph does not affect P.*

Note that rMSO∃ contains natural problems such as k-color. The first restriction is similar to but weaker than hereditariness. The proof of the following theorem is a generalization of a result regarding the indistinguishability of k-color from FO properties (see, e.g., Alon *et al.* [2]).

Theorem 1. *All properties expressible in rMSO∃ are indistinguishable from properties expressible in FO.*

Proof. Let P be the rMSO∃ property expressed by formula $\phi := \exists C_1 C_2 \ldots C_r \psi$, where ψ is first-order. We construct a first-order formula ϕ' expressing property Q such that P and Q are indistinguishable.

$$\phi' := \exists t_1 t_2 \ldots t_r \psi',$$

where the symbols t_i do not occur in ψ (if they do occur, simply rename them). Formula ψ' is derived from ψ with the following changes:

1. Replace all occurrences of $C_i(x)$ with $E(t_i, x)$.
2. Replace all quantifiers with restricted quantifiers:
   $\exists x : \gamma$ with $\exists x : (x \neq t_1 \wedge \ldots \wedge x \neq t_r) \wedge \gamma$ and
   $\forall x : \gamma$ with $\forall x : (x \neq t_1 \wedge \ldots \wedge x \neq t_r) \rightarrow \gamma$.

First, let $\varepsilon > 0$ be arbitrary and let G model ϕ. Assume $|G| = n > N$. Choose the set of r vertices guaranteed to exist by Restriction (1) of Definition 4, call them $u_1 \ldots u_r$ and remove them. Call this graph G^-. Take an interpretation Φ under which G^- models ϕ. Replace the u_i as disconnected vertices. Connect u_i and x iff $C_i(x)$ holds in Φ. Call the resulting graph G'. Note that we have changed at most $r(n-1)$ edges, which is less than εn^2 for sufficiently large n.

Claim 1. Graph G' models ϕ'.

Proof. We construct a satisfying interpretation Φ' from Φ. The only change is to bind t_i to u_i. Recall that the t_i appear only where we added them above and thus the u_i are only referred to in these contexts.

Note that:

- $(G^-, \Phi) \models x = y \iff (G', \Phi') \models x = y$,
  because the u_i cannot appear here and all other members are retained.
- $(G^-, \Phi) \models C_i(x) \iff (G', \Phi') \models E(t_i, x)$, by construction.
- Logical operators $\wedge$, $\neg$ are preserved inductively.
- $(G^-, \Phi) \models \exists x : \gamma \iff (G', \Phi') \models \exists x : (x \neq t_1 \wedge \ldots \wedge x \neq t_r) \wedge \gamma$,
  because G^- does not contain the vertices referred to by the t_i.

Therefore, G' models ϕ' and thus has the first-order property Q. □ Claim 1

Next, let $\varepsilon > 0$ be arbitrary and assume that graph H models ϕ'. Assume further that $|H| > N$. Let Φ' be an interpretation satisfying ψ'. Let the vertices bound to the t_i be called u_i. Recall that because of the restricted quantifiers in ϕ', the u_i are referred to only as t_i, and the t_i only occur where we explicitly added them.

Remove the u_i from H and call this graph H^-. Next, re-add the u_i as r isolated vertices and call this graph H'. We claim that both H^- and H' model ϕ, and construct a satisfying interpretation Φ from Φ'. Because of Restriction (2) in Definition 4, it is sufficient to prove that H^- models ϕ as adding r isolated vertices will not affect property P.

We set $C_i(x)$ to be true in Φ iff there is an edge between u_i and x in H.

Claim 2. $(H^-, \Phi) \models \phi$.

Proof. Note that:

- $(H^-, \Phi) \models x = y \iff (H, \Phi') \models x = y$,
  because x and y cannot be bound to u_i on the right.
- $(H^-, \Phi) \models C_i(x) \iff (H, \Phi') \models E(t_i, x)$, by construction.

- Logical operators $\wedge$, $\neg$ are preserved inductively.
- $(H^-, \Phi) \models \exists x : \gamma \iff (H, \Phi') \models \exists x : (x \neq t_1 \wedge \ldots \wedge x \neq t_r) \wedge \gamma$, because H^- does not contain the vertices referred to by the t_i.

□ Claim 2

Therefore, H^- has property P and by Restriction (2) of Definition 4, H' does too. We have changed at most $r(n-1)$ edges, which is less than εn^2 for sufficiently large n.

Properties P and Q are therefore indistinguishable. □

4 Hard Properties

In the previous section, we showed that every property expressible in a restriction of monadic second-order existential logic is indistinguishable from some FO-expressible property. All properties expressible in this logic are contained in NP by Fagin's [10] theorem. For graphs it is known that MSO∃ is strictly less expressive than SO∃: there are graph properties in NP that are not expressible in MSO∃ (see, Fagin *et al.* [11]). In this section we continue our study of the set of properties indistinguishable from FO-expressible properties, and show that it contains *much harder* properties. We show that this set contains uncomputable properties, and also, for every $f(n)$, computable (and testable) properties that are not computable in time $f(n)$. In this sense, the power of first-order logic in the context of indistinguishability is even larger than previously known. However, we also show that there exist computable properties that are *distinguishable* from every first-order expressible property.

In the next proof we use the concept of RE-completeness. A decision problem is in RE (recursively enumerable) iff there exists a Turing machine that halts and accepts all positive instances and does not accept negative instances. The machine is not required to halt on negative instances. A decision problem D is *RE-complete* iff it is in RE and all other problems in RE can be *decided* by machines given an oracle for D. The halting problem is the canonical RE-complete problem.

Theorem 2. *There exists an RE-complete graph property that is indistinguishable from a FO-expressible property.*

Proof. We define property P such that graph G satisfies it iff

(1) there is a vertex i such that all edges are incident to i, and
(2) taking the degree d of vertex i as the number of a Turing machine M_d in some canonical enumeration $(M_i)_{i \in \mathbb{N}}$ of Turing machines where every Turing machine appears at least once, provided M_d halts on the empty string.

If M_d does not halt on the empty string, then graph G does not have property P. For convenience, we require that machine M_0 in the enumeration halts on the empty string. We shall show that P is an RE-complete property that is indistinguishable from the FO-expressible property of being an empty graph.

Lemma 1. *P is RE-complete.*

Proof. Let HALT=$\{a \mid M_a$ halts on the empty string$\}$. We show that P is RE-complete by reducing HALT to it. On input a, representing Turing machine M_a in the enumeration, we output a graph on $a+1$ vertices. There is an edge between vertices i and j iff exactly one of them is zero. All edges are then incident to the zero vertex, and as the degree of vertex zero is $n-1 = a$, the graph has property P iff machine M_a halts on the empty string. □ Lemma 1

Lemma 2. *P is indistinguishable from the FO property of being an empty graph ($\forall x, y : \neg E(x, y)$).*

Proof. Assume graph G satisfies property P and let $\varepsilon > 0$ be arbitrary. Let i denote the vertex that is mentioned in Property (1) of the Theorem, and let d be its degree. Remove the $d \leq n-1$ edges incident to i. By assumption, all edges in G were incident to i, and so the resulting graph is empty. For n sufficiently large, $n - 1 < \varepsilon n^2$.

Now assume graph G is an empty graph. By assumption, M_0 halts. Choose an arbitrary vertex i and note that all zero edges are incident to it. Consequently, G has property P. We have not changed any edges, and therefore $0 < \varepsilon n^2$ for all non-zero n. □ Lemma 2

Lemma 1 and 2 directly yield the theorem. □

The following proposition is essentially obvious: an ε-tester is a probabilistic machine. It remains only to mention that we can, by choosing $\varepsilon > 0$ appropriately as a function of n, remove the "approximation" in "probabilistic approximation algorithm." Of course, we then make a number of queries that depends on the input size.

It is important to note that this proposition does not hold in the non-uniform case.

Proposition 1. *All uniformly testable graph properties can be decided by a probabilistic Turing machine with success probability at least $\frac{2}{3}$.*

Proof. Assume graph property P is testable. We construct a probabilistic machine deciding P. On input G of size n, choose ε such that $\varepsilon n^2 < 1$, for example, $\varepsilon = \frac{1}{n^2+1}$. Run the ε-test on G and output the result. Because $\varepsilon n^2 < 1$, being *at most* ε-far from G implies that we may not change *any* edges. We therefore distinguish with probability at least $\frac{2}{3}$ between the case of G satisfying P and G not satisfying P. □

Corollary 1. *All uniformly testable properties are decidable by probabilistic machines.*

Proof. Simply modify εn^2 in the proof to the appropriate definition of ε-far for the vocabulary in question. □

The following also follows immediately, as randomization does not allow us to compute uncomputable functions. We provide a proof sketch for completeness.

Corollary 2. *All uniformly testable properties are recursive.*

Proof. We convert the probabilistic machine into a deterministic machine using the following generic construction.

All probabilistic machines can be modified such that their randomness is taken from a special binary "random tape" that is randomly fixed when the machine is started, in which each digit is 0 or 1 with equal probability.

All halting probabilistic machines must eventually halt, regardless of the random choices made. We can then simulate the machine over all initial segments of increasing lengths, keeping track of "accepting," "rejecting" and "still running" states. Once any given segment has halted, all random strings beginning with that initial segment must also halt. Therefore, the percentage of halting paths is increasing, and we shall eventually reach a length such that at least 70% of the paths have halted. Our error probability is at most $\frac{1}{3}$, strictly less than half of 70% and so we can output the decision of the majority of the halting paths. □

Theorem 3 (Alon *et al.* [2]). *Every first-order property P of the form*

$$\exists x_1, \dots, x_t \forall y_1, \dots, y_s : A(x_1, \dots, x_t, y_1, \dots, y_s),$$

where A is quantifier-free, is testable.

Note that our RE-complete property P defined in the proof to Theorem 2, is indistinguishable from a first-order property of this form ($t = 0$).

Theorem 4 (Alon *et al.* [2]). *If P and Q are indistinguishable graph properties, then P is testable if and only if Q is testable.*

However, we now see a contradiction in the uniform case. Our RE-complete property P, defined in the proof to Theorem 2, is indistinguishable from a FO-property that, according to Theorem 3, is testable (which it is). Theorem 4 then implies that this RE-complete property P is also testable, which contradicts Corollary 2. Therefore, Theorem 4 is not strictly correct in the uniform case. The proof given in [2] assumes that if input G has strictly less than $N = N(\varepsilon)$ vertices, there exists a decision procedure that gives "accurate output according to whether it satisfies" the property in question. Of course, no such (uniform) procedure exists for RE-complete properties. Theorem 4 holds in the uniform case when restricted to recursive properties.

We can however use a similar construction to Theorem 2 to obtain the following, restricting ourselves to recursive properties. By the time hierarchy theorem, for every computable $f(n)$ there exist computable properties that cannot be decided in time $f(n)$, see Hartmanis and Stearns [18].

We define *arbitrarily-hard properties* to be any set of *computable* properties that "for each computable $f(n)$, contains properties that cannot be computed in DTIME($f(n)$)."

It is of course possible to use other complexity measures. We have restricted these sets to computable properties and so they are obviously infinite.

The reduction in the following proof increases the input length by an exponential factor. However, because we are interested in *arbitrarily-hard* properties and by the time hierarchy theorem, we can choose Q such that it is not computable in, e.g., DTIME($2^{f(n)}$). Then, after the input length is increased exponentially, we have a property that cannot be computed in DTIME($f(n)$). This can be done for all $f(n)$.

Theorem 5. *There are arbitrarily-hard testable properties.*

Proof. Let Q be an arbitrarily-hard property. We define property R such that Q is reducible to R. Let $q(x)$ be the characteristic function for Q with an appropriate encoding of the input. Similar to Theorem 2, we define R to hold in graph G of size n iff

1. there is a vertex i such that all edges are incident to i,
2. and either the degree of i is zero or $q(n) = 1$.

We can obviously reduce Q to R by computing the encoding x, and outputting a graph on x vertices with one edge. We can therefore construct arbitrarily hard properties R.

Using the same proof as that used for Theorem 2, we see that all such R are also indistinguishable from the empty graph. The property of being the empty graph is testable by Theorem 3, and so R is testable by Theorem 4, if it is decidable. We can therefore construct arbitrarily hard, testable properties. □

Computable graph properties that are distinguishable from all first-order properties do exist however. We show the following by a simple diagonalization argument. We define *distinguishable* as "not indistinguishable."

Theorem 6. *There exist computable graph properties that are distinguishable from all first-order properties.*

Proof. We let first-order formula ϕ_i denote the i'th formula in some enumeration of first-order formulas on graphs, in which all such formulas occur infinitely often. We define property Q such that G has property Q iff $\phi_{|G|}$ does not hold on any graph with $|G| = n$ vertices.

We show that Q is distinguishable from all first-order properties by contradiction. Assume that first-order ψ expresses a property that is indistinguishable from Q. Find the first $i > N$ such that $\psi = \phi_i$. There are two cases. First, assume that there is a graph G on i vertices such that G satisfies ψ. Then, there is no graph on i vertices with property Q, by construction. We therefore cannot obtain a graph G' on i vertices with property Q by changing at most εn^2 edges in G, because no such graph exists.

There must then be no such graph G on i vertices satisfying ψ. In this case, by definition all graphs on i vertices have property Q. Taking any of them, we see that we cannot obtain a graph on i vertices satisfying ψ by modifying at most εn^2 edges, because again no such graph exists. There must then be no first-order ψ expressing a property indistinguishable from Q. □

5 Discussion

The descriptive power of first-order logic with indistinguishability is surprisingly large. As we have seen, we can construct testable properties of arbitrary hardness. However, as seen in [2], there are problems in FO with quantifier alternations "$\forall\exists$" that are not testable. So, although the class of testable properties contains arbitrarily hard properties, it does not strictly contain uniform AC^0 or even the context-free languages. In this sense, it is a rather odd class.

A complete, logical characterization of the testable properties must then not contain FO entirely, but must contain arbitrarily hard properties. However, in the uniform case we have seen that all testable properties are recursive, and so a characterization of the uniformly testable properties must not be able to express e.g. RE-complete properties.

We also believe that the distinction between the *uniformly* testable and *non-uniformly* testable properties is important. In particular, given our motivation of searching for a class that is very efficiently computable, it seems undesirable to admit uncomputable properties. In this sense, the uniform case is preferable (Proposition 1). It would also be worthwhile to consider other possible definitions of uniformity.

Acknowledgements. We wish to thank Osamu Watanabe for an inspiring discussion regarding the importance of uniformity conditions.

References

[1] Allender, E., Balcázar, J.L., Immerman, N.: A first-order isomorphism theorem. SIAM J. Comput. 26(2), 539–556 (1997)
[2] Alon, N., Fischer, E., Krivelevich, M., Szegedy, M.: Efficient testing of large graphs. Combinatorica 20(4), 451–476 (2000)
[3] Alon, N., Fischer, E., Newman, I., Shapira, A.: A combinatorial characterization of the testable graph properties: It's all about regularity. In: STOC 2006: Proceedings of the 38th Annual ACM Symposium on Theory of Computing, pp. 251–260. ACM, New York, NY, USA (2006)
[4] Alon, N., Krivelevich, M., Newman, I., Szegedy, M.: Regular languages are testable with a constant number of queries. SIAM J. Comput. 30(6), 1842–1862 (2001)
[5] Alon, N., Shapira, A.: A characterization of the (natural) graph properties testable with one-sided error. In: Proceedings, 46th Annual IEEE Symposium on Foundations of Computer Science, FOCS 2005, pp. 429–438. IEEE Computer Society Press, Los Alamitos (2005)
[6] Barrington, D.A.M., Immerman, N., Straubing, H.: On uniformity within NC1. J. of Comput. Syst. Sci. 41(3), 274–306 (1990)
[7] Blum, M., Luby, M., Rubinfeld, R.: Self-testing/correcting with applications to numerical problems. J. of Comput. Syst. Sci. 47(3), 549–595 (1993)
[8] Chockler, H., Kupferman, O.: ω-regular languages are testable with a constant number of queries. Theoret. Comput. Sci. 329(1-3), 71–92 (2004)
[9] Enderton, H.B.: A Mathematical Introduction to Logic, 2nd edn. Academic Press, London (2000)

[10] Fagin, R.: Generalized first-order spectra and polynomial-time recognizable sets. In: Karp, R.M. (ed.) Complexity of Computation, SIAM-AMS Proceedings, vol. VII, Amer. Mathematical Soc., pp. 43–73 (1974)
[11] Fagin, R., Stockmeyer, L.J., Vardi, M.Y.: On monadic NP vs. monadic co-NP. Inform. Comput. 120(1), 78–92 (1995)
[12] Fischer, E.: The art of uninformed decisions. Bulletin of the European Association for Theoretical Computer Science 75(97) (October 2001); Columns: Computational Complexity
[13] Fischer, E.: Testing graphs for colorability properties. Random Struct. Algorithms 26(3), 289–309 (2005)
[14] Freivalds, R.: Fast probabilistic algorithms. In: Becvar, J. (ed.) MFCS 1979. LNCS, vol. 74, pp. 57–69. Springer, Heidelberg (1979)
[15] Furst, M.L., Saxe, J.B., Sipser, M.: Parity, circuits, and the polynomial-time hierarchy. Mathematical Systems Theory 17(1), 13–27 (1984)
[16] Goldreich, O., Goldwasser, S., Ron, D.: Property testing and its connection to learning and approximation. J. ACM 45(4), 653–750 (1998)
[17] Goldreich, O., Trevisan, L.: Three theorems regarding testing graph properties. Random Struct. Algorithms 23(1), 23–57 (2003)
[18] Hartmanis, J., Stearns, R.E.: On the computational complexity of algorithms. Transactions of the American Mathematical Society 117, 285–306 (1965)
[19] Hromkovič, J.: Design and Analysis of Randomized Algorithms: Introduction to Design Paradigms. Springer, Heidelberg (2005)
[20] Rödl, V., Schacht, M.: Property testing in hypergraphs and the removal lemma. In: STOC 2007: Proceedings of the 39th Annual ACM Symposium on Theory of Computing, pp. 488–495. ACM, New York, NY, USA (2007)
[21] Ron, D.: Property testing ch. 15. In: Rajasekaran, S., Pardalos, P.M., Reif, J.H., Rolim, J. (eds.) Handbook of Randomized Computing, vol. II, ch. 15, pp. 597–649. Kluwer Academic Publishers, Dordrecht (2001)
[22] Rubinfeld, R., Sudan, M.: Robust characterizations of polynomials with applications to program testing. SIAM J. Comput. 25(2), 252–271 (1996)

Derandomizing Graph Tests for Homomorphism

Angsheng Li[1,⋆] and Linqing Tang[1,2,⋆]

[1] State Key Lab. of Computer Science, Institute of Software, Chinese Academy of Sciences
[2] Graduate University of Chinese Academy of Sciences
{angsheng,linqing}@ios.ac.cn

Abstract. In this article, we study the randomness-efficient graph tests for homomorphism over arbitrary groups (which can be used in locally testing the Hadamard code and PCP construction). We try to optimize both the amortized-tradeoff (between number of queries and error probability) and the randomness complexity of the homomorphism test simultaneously. For an abelian group $G = \mathbb{Z}_p^m$, by using the λ-biased set S of G, we show that, on any given bipartite graph $H = (V_1, V_2; E)$, the graph test for linearity over G is a test with randomness complexity $|V_1| \log |G| + |V_2| O(\log |S|)$, query complexity $|V_1| + |V_2| + |E|$ and error probability at most $p^{-|E|} + (1 - p^{-|E|}) \cdot \delta$ for any f which is $1 - p^{-1}(1 + \frac{\sqrt{\delta^2 - \lambda}}{2})$-far from being affine linear. For a non-abelian group G, we introduce a random walk of some length, ℓ say, on expander graphs to design a probabilistic homomorphism test over G with randomness complexity $\log |G| + O(\log \log |G|)$, query complexity $2\ell + 1$ and error probability at most $1 - \frac{\delta^2 \ell^2}{(\delta\ell + \delta^2\ell^2 - \delta^2\ell) + 2\psi(\lambda, \ell)}$ for any f which is $2\delta/(1-\lambda)$-far from being affine homomorphism, here $\psi(\lambda, \ell) = \sum_{0 \le i < j \le \ell - 1} \lambda^{j-i-1}$.

1 Introduction

For any two finite groups G and Γ, a *homomorphism* is a function $f : G \to \Gamma$ such that for every $g_1, g_2 \in G$ we have $f(g_1 \cdot g_2) = f(g_1) \cdot f(g_2)$. If G and Γ are abelian groups, we may also call f as *linear* function. An *affine homomorphism* is a function f such that $f(0)^{-1} \cdot f$ is a homomorphism, here 0 is the unit element of group G. For two functions f and g, define the *distance* $d(f, g) = Pr_{x \in G}[f(x) \neq g(x)]$.

A homomorphism test, which tests the proximity of a function to a family of homomorphism functions, was first raised by Blum, Luby and Rubinfeld [2]. In the last decade, some interesting results about the test focused on two different objects have been achieved: one is the graph test of Samorodnitsky and Trevisan [4] which tried to optimize the tradeoff between the number of queries and the

⋆ The authors are partially supported by NSFC grant number 60325206, and NSFC Grand International Joint Project No. 60310213.

M. Agrawal et al. (Eds.): TAMC 2008, LNCS 4978, pp. 105–115, 2008.

error probability of the test. Another is the randomness-efficient version of linearity test of Ben-Sasson et al [11], which tried to save the randomness used by the test.

1.1 The Affine Homomorphism Testing

A (δ, ϵ, q, r)-*test* for (affine) homomorphisms from G to Γ is a test which queries q values of f, uses r random bits, always accepts homomorphism f and rejects any f which is ϵ-far from being a (affine) homomorphism with probability at least δ.

Blum, Luby and Rubinfeld [2] first gives a $(\delta, 9\delta/2, 3, 2\log|G|)$-test (conditioned on $\delta < 2/9$) for functions over abelian groups, so we always call the basic linearity test BLR-test. The BLR-test just picks two random elements x, y from the abelian group G and checks whether $f(x)f(y) = f(xy)$. Ben-Or et al [1] extended the result and showed that the test works for functions over any general group, they showed that:

Theorem 1. *For any $\delta < 2/9$, there is a $(\delta, \tau, 3, 2\log|G|)$-test for homomorphism over any group G, where τ is the smaller root of $3x - 6x^2 = \delta$.*

Note that $\tau = \delta/3 + O(\delta^2)$ and the analysis is nearly optimal since the best possible τ is at least $\delta/3$.

1.2 The Graph Test for Linearity

In order to reuse the queried bits and get better tradeoff between the number of queries and the error probability of the test, Samorodnitsky and Trevisan [4] introduced a notion of *graph test* and obtained very strong results for linearity test (for PCP construction too). Given a graph $H = (V, E)$, the graph test for linearity chooses a random element from the group $G = \mathbb{Z}_p^m$ for each vertex $v \in V$, implements a basic BLR-test on each edge $(x, y) = e \in E$ using the values of f at x, y and xy. Samorodnitsky and Trevisan [4] used Fourier analysis and proved (the analysis was simplified by Håstad and Wigderson [6] later):

Theorem 2. *For a graph $H = (V, E)$, if the graph test for linearity over abelian group $G = \mathbb{Z}_p^m$ accepts $f : G \to \mu_p$ with probability $p^{-|E|} + \epsilon$, then $|\hat{f}_\alpha| \geq \epsilon$ for some $\alpha \in \mathbb{Z}_p^m$. In particular, f has agreement $\geq p^{-1}(1 + \epsilon/2)$-with some affine linear function. Moreover, the test queries $|V| + |E|$ values from f and uses $|V| log |G|$ random bits.*

1.3 The Randomness-Efficient Linearity Test

Reducing the number of random bits required by the homomorphism test is also a very intriguing problem. Several authors have supplied methods for this purpose. Ben-Sasson et al [11] used ϵ-biased sets to derandomize the BLR-test for linearity over abelian group $G = \mathbb{Z}_p^m$, which uses only $(1 + o(1)) \log |G|$ random bits whereas the original test uses $2 \log |G|$ random bits, while just lost a quite small quantity in soundness.

Theorem 3. *For group $G = \mathbb{Z}_p^m$ and $\lambda > 0$, $f : G \to \mu_p$, let S be a λ-biased set in G, the test randomly chooses $x \in G$, $y \in S$ and checks whether $f(x)f(y) = f(xy)$. If $Pr($ test accepts $) \geq \frac{1}{p} + (1 - \frac{1}{p}) \cdot \delta$, then $|\hat{f^a_\alpha}| \geq \sqrt{\delta^2 - \lambda}$ for some $\alpha \in \mathbb{Z}_p^m$ and $1 \leq a \leq p-1$, here $f^a(x) = (f(x))^a$. In particular, f has agreement $\geq p^{-1}(1 + \frac{\sqrt{\delta^2-\lambda}}{2})$ with some affine linear function. Moreover, the test queries 3 values of f and uses $log|G| + log|S|$ random bits.*

The main technique for analysis of Ben-Sasson et al [11] is Fourier analysis, which seems hard to be generalized to non-abelian groups. Recently Shpilka and Wigderson [3] extended the result of Ben-Sasson et al [11] to general groups by using combinatorial arguments and properties of expanding Cayley graphs. They showed:

Theorem 4. *For any group G, Γ and any $\lambda > 0$, let $S \subseteq G$ be an expanding generating set with $\lambda(Cay(G;S)) < \lambda$, there is a $\big(\delta, 4\delta/(1-\lambda), 3, log|G|+log|S|\big)$-test for affine homomorphism from G to Γ, given that $12\delta/(1-\lambda) < 1$.*

1.4 Our Results

For the linearity test over abelian groups, Samorodnitsky and Trevisan [4] and Ben-Sasson et al [11] tried to optimize different parameter but fail to consider the other one. A natural question asks how about trying to optimize both the amortized-tradeoff (between number of queries and error probability) and the randomness complexity simultaneously. In this paper we study this problem, we extend the result of [4] to a derandomized version. We show the following

Theorem 5. *For any $\delta > 0$, group $G = \mathbb{Z}_p^m$ with a λ-biased set S and bipartite graph $H = (V_1, V_2; E)$, there is a derandomized graph test for $f : G \to \mu_p$ on graph H, if the test accepts f with probability $p^{-|E|} + (1 - p^{-|E|}) \cdot \delta$, then f has agreement $\geq p^{-1}(1+\frac{\sqrt{\delta^2-\lambda}}{2})$ with some affine linear function. Moreover, the test queries $|V_1| + |V_2| + |E|$ values from f and uses $|V_1| \log|G| + |V_2| \log|S|$ random bits.*

Note that for group $G = \mathbb{Z}_p^m$, λ-biased set of size $O(\log|G|)$ can be efficiently constructed (see [13], [12]). Specifically, when graph H is a star graph with k leaves, we only use $\log|G| + k \log\log|G|$ random bits compared to the original graph test uses $(k+1)\log|G|$ random bits. We are able to extend this result to graph test on graphs which is the union of a bipartite graph $H = (V_1, V_2; E')$ and the clique graph over V_1 easily.

We also try to extend the derandomized graph test for homomorphism to general groups. We partially realized this goal. The main technique we use is the random walk of length ℓ over an expanding Cayley graph, and for each edge on the walk, we do a basic homomorphism test. ℓ is a parameter to be fixed later. It can be thought that the test is a graph test on a path.

Theorem 6. *For groups G, Γ, $\delta > 0$ and a symmetric subset S of G. Let X_0 be chosen uniformly at random and $X_0, ..., X_\ell$ be a random walk on the expanding Cayley graph $Cay(G; S)$ starting at X_0, the test checks whether $f(x_i)f(x_i^{-1}x_{i+1}) = f(x_{i+1})$ for each edge (X_i, X_{i+1}). Then if*

$$Pr(\text{ test accepts } f) \geq 1 - \frac{\delta^2\ell^2}{(\delta\ell + \delta^2\ell^2 - \delta^2\ell) + 2\psi(\lambda, \ell)}$$

then f is $\frac{2\delta}{1-\lambda}$-close to some affine homomorphism, here $\psi(\lambda, \ell) = \sum_{0\leq i<j\leq \ell-1} \lambda^{j-i-1}$. Moreover, the test queries $2\ell + 1$ values from f and uses $\log |G| + \ell O(\log\log |G|)$ random bits.

In order to keep the randomness complexity of the test to be $(1 + o(1)) \log |G|$, we may let $\ell = O(\log\log |G|)$. Note that the test of Shpilka and Wigderson [3] is just the case of $\ell = 1$.

We remark that the error probability bound of derandomized homomorphism test of Shpilka and Wigderson [3] can be improved slightly. We have (compare to Theorem 4):

Theorem 7. *For any groups G, H ,$\delta > 0$ and let $S \subseteq G$ be an expanding generating set with $\lambda(Cay(G; S) < \lambda$, the test that picks uniformly an edge $e = (x, y)$ in $Cay(G; S)$ and checks whether $f(x)f(x^{-1}y) = f(y)$ is a $(\delta, 2\delta/(1 - \lambda)), 3, \log |G| + O(\log\log |G|))$-test for homomorphism given $6\delta/(1 - \lambda) < 1$.*

By using a little more random bits (but still $\log |G| + O(\log\log |G|)$), we can decrease the part corresponding to λ in the error probability exponentially. It's easy to have

Theorem 8. *For any groups G, H ,$\delta > 0$ and let $S \subseteq G$ be an expanding generating set with $\lambda(Cay(G; S) < \lambda$, the test that picks uniformly a random walk of lengh ℓ on $Cay(G; S)$ (with starting and ending points x and y respectively) and checks whether $f(x)f(x^{-1}y) = f(y)$ is a $(\delta, 2\delta/(1 - \lambda^\ell)), 3, \log |G| + \ell O(\log\log |G|))$-test for homomorphism given $6\delta/(1 - \lambda^\ell) < 1$.*

2 Preliminaries

In this section, we give some necessary background.

2.1 Fourier Transformation over Abelian Groups, λ-Biased Sets

Let G be an Abelian group with $|G| = k$. Let

$$\mathcal{F} = \{f \mid G \to \mathbf{C}^*\}$$

where $\mathbf{C}^*$ is the multiplicative group of complex numbers. We define the standard inner product over $\mathcal{F}$:

$$\langle f, g\rangle = |G|^{-1} \sum_{x\in G} f(x)\overline{g(x)}$$

where $\overline{a}$ is the conjugation of a in C^*.

A *character* of G is a homomorphism

$$\chi : G \to \mathbf{C}^* \ ,$$

it is easy to verify that the set of characters forms an orthonormal basis for the vector space $\mathcal{F}$, so for any $f \in \mathcal{F}$, we can represent it as

$$f = \sum_{\chi} \hat{f}_\chi \chi$$

where $\hat{f}_\chi = \langle f, \chi \rangle$ is the Fourier coefficient corresponding to χ.

Let G be a group, a set $S \subseteq G$ is called *symmetric* if it is closed under inverses, that is $s \in S$ iff $-s \in S$.

Definition 1. *For G a finite abelian group, let $S \subseteq G$ be a symmetric multiset. We call S λ-*biased *if for all nontrivial characters χ, we have*

$$|S|^{-1} |\sum_{x \in S} \chi(x)| \leq \lambda$$

We note that G is 0-biased set. We can define inner product $\langle f, g \rangle_S = |S|^{-1} \sum_{x \in S} f(x)\overline{g(x)}$ and ℓ_2-norm $\|f\|_S = \sqrt{\langle f, f \rangle_S}$ on vector space $\mathbf{C}^S$ similar to that in G.

Naor and Naor [13] first defined and gave a construction of λ-biased sets. Since then many other constructions appeared for various groups (see [13], [12]).

2.2 Random Walk on Expanding Cayley Graphs

Let $H = (V, E)$ be a graph on n vertices. Let A_H be its adjacency matrix. For two sets $A, B \in V$ denote $E(A, B) = \{(u, v) | u \in A, v \in B\}$. Let $e(A, B) = |E(A, B)|$. Denote with $\lambda_1 \geq ... \geq \lambda_n$ the normalized eigenvalues of A_H. If G is a d-regular graph, then $\lambda_1 = 1$. Let

$$\lambda(H) = max(\lambda_2, |\lambda_n|)$$

We denote $\lambda = \lambda(H)$ sometimes when H is obviously known.

Definition 2. *An (n, d, α)-expanding graph is an n-vertice, d-regular graph with* $\max(\lambda_2, |\lambda_n|) \leq \alpha$.

Definition 3. *For G a finite group, let $S \subseteq G$ be a symmetric generating set for G. Define the expanding Cayley graph $Cay(G; S) = (V, E)$ as follows: let the vertices be the elements of G and the edges be the pairs (g, gs) for any $g \in G$ and $s \in S$.*

It is easy to see that if $(g_1, g_2) \in E$, then $(xg_1, xg_2) \in E$ for any $x \in G$.

Lemma 1. *[WX [12]] Given a group G of size n and $\lambda < 1$, there exists an algorithm running in time poly(n, λ), constructs a symmetric set $S \subseteq G$ of size $O(\log n)$ such that $\lambda(Cayley(G; S)) \leq \lambda$.*

For the relation between the expansion of H and $\lambda(H)$, Alon and Chung [9] proved the following:

Lemma 2. *[expander mixing lemma] Given an (n, d, α)-expanding graph $H = (V, E)$, for any two sets $A, B \subseteq V$ we have*

$$\left| e(A, B) - \frac{d(|A| \cdot |B|)}{n} \right| \leq \lambda \cdot d \cdot \sqrt{|A| \cdot |B|}$$

Specifically, Shpilka and Wigderson showed that (as Corollary 2 in [3])

Lemma 3. *Let $H = (V, E)$ be an (n, d, α)-expanding graph, then:*

1. *For set $A \subseteq V$ we have*

$$\min(|A|, n - |A|) \leq \frac{2}{1 - \lambda} \cdot \frac{e(A, A^c)}{d}$$

2. *There exists a connected component of size at least $(1 - \frac{4\delta}{1-\lambda}) \cdot n$ in G if removing only $2\delta dn < \frac{1-\lambda}{6} \cdot dn$ edges from it (here conditioned on $\frac{12\delta}{1-\lambda} < 1$).*

If we regard the expanding Cayley graph $Cay(G; S) = (V, E)$ as a directed graph, i.e., there is an directed edge (y, ys) from y to ys for any $y \in G$ and $s \in S$, then each vertex in $Cay(G; S)$, since S is symmetric, has both out-degree and in-degree to be d. As usual, we say a directed graph is *weakly connected* if it is possible to reach any node starting from any other node by traversing edges in some direction (i.e., not necessarily in the direction they point).

We remark that there is a similar result as Lemma 3 for the size of weakly connected component in $Cay(G; S)$.

Lemma 4. *There exists a weakly connected component of size at least $\dfrac{2\delta}{1 - \lambda} \cdot n$ in $Cay(G; S)$ if removing only $2\delta dn$ directed edges from it (here conditioned on $\dfrac{6\delta}{1 - \lambda} < 1$).*

We delay the proof of Lemma 4 to the full version.

A *random walk* on an expanding Cayley graph $H = (V, E)$ is a stochastic process $(X_0, X_1, ...)$: chooses a vertex X_0 from some initial distribution on V and X_{i+1} uniformly from the neighbors of X_i.

Recently, Dinur [8] proved:

Proposition 1. *Let $H = (V, E)$ be a d-regular graph with $\lambda(H) = \lambda$. Let $F \subseteq E$ be a set of edges, and let K be the distribution on vertices induced by selecting a random edge in F, and then a random endpoint. The probability p that a random walk that starts with distribution K takes the $i + 1$st step in F, is bounded by $\frac{|F|}{|E|} + \lambda^i$.*

Corollary 1. *Let $H = (V, E)$ be a (n, d, λ)-graph and $F \subset E$ be of size $\mu|E|$, $X_0, ...X_\ell$ is a random walk on G with X_0 uniformly chosen from V, then for $i < j$, we have*

$$Pr(\text{ edge } (X_j, X_{j+1}) \text{ is in } F \mid \text{edge } (X_i, X_{i+1}) \text{ is in } F\ ) \leq \mu + \lambda^{j-i-1} \quad .$$

3 Derandomized Graph Test for Linearity in Abelian Groups

In this section, we prove Theorem 5. Let $G = \mathbb{Z}_p^m$ be an abelian group, S be a λ-biased set of G. $f : G \to \mu_p$ is the function needs to be tested, where μ_p is the group of p'th roots of unity.

Given a graph $H = (V_1, V_2; E)$ with $|V_1| = k_1$ and $|V_2| = k_2$, the graph test (**Test 1**) works as follows:

1. For each vertex $u \in V_1$, chooses a random element x_u from G uniformly.
2. For each vertex $v \in V_2$, chooses a random element y_v from S uniformly.
3. For each edge $e = (u, v) \in V_1 \times V_2$, checks whether $f(x_u)f(y_v) = f(x_u y_v)$. The test accepts iff it's true for all edges in E.

Now we analyze the accept probability of the above test. It is straightforward that the test always accept a linear function. Moreover, we have the following:

Lemma 5. *If **Test 1** accepts f with probability at least $p^{-|E|} + (1 - p^{-|E|})\delta$, then there exists some character χ of G and some $1 \leq a \leq p - 1$ such that $\hat{f}_\chi^a \geq \sqrt{\delta^2 - \lambda}$, here $\hat{f}_\chi^a = \langle f^a, \chi \rangle$ and $f^a(x) = (f(x))^a$.*

We delay the proof of Lemma 5 to the appendix. Note that Ben-Sasson et al [11] have showed that (lemma 3.3 in [11])

Lemma 6. *If $|\hat{f}_\chi^a| \geq \sqrt{\delta^2 - \lambda}$, then f has agreement $\geq p^{-1}(1 + \frac{\sqrt{\delta^2-\lambda}}{2})$ with some affine function.*

Theorem 5 follows from the combining of the above two lemmas.

4 Derandomized Homomorphism Test by Random Walk on Expander Graphs

4.1 Proof Theorem 7

Proof. We divided the proof into three claims. Define function $\phi(x) = Plurality_{y \in G} f(xy)f(y)^{-1}$.

Claim 1. For any $x \in G$, $Pr_{y \in G}[f(xy)f(y)^{-1} = \phi(x)] \geq 1 - 2\delta/(1 - \lambda)$.

Proof. For a fixed $x \in G$, we know

$$Pr_{y \in G, s \in S}[f(y)f(s) \neq f(ys)] = \delta$$
$$Pr_{y \in G, s \in S}[f(xy)f(s) \neq f(xys)] = \delta.$$

Construct a subgraph of $Cay(G, S)$ as follows: delete edge $y \to ys$ from the $Cay(G; S)$ (take it as directed graph) if $f(y)f(s) \neq f(ys)$ or $f(xy)f(s) \neq f(xys)$. By the above two equations, we delete at most $2\delta dn$ directed edges, by lemma 4 we know that the remaining graph H_x contains a weakly connected component C_x of size at least $\big(1 - 2\delta/(1-\lambda)\big)n$. We claim that for every two elements $u, v \in C_x$, we have $f(xu)f(u)^{-1} = f(xv)f(v)^{-1}$, then since $|C_x| > |G|/2$, claim 1 follows. For the last claimed equation, we may assume $v = v_1, ..., v_t = u$ be a path (do not consider the direction of edges) between v and u in C_x, since C_x is weakly connected, such path always exists. W.l.o.g., we may assume the direction is $v_i \to v_{i+1}$ for edge $\langle v_i, v_{i+1}\rangle$, i.e., $v_{i+1} = v_i \cdot s_i$ for some s_i. Since the edge $\langle v_i, v_{i+1}\rangle$ is in C_x, we know that $f(v_i)f(s_i) = f(v_i s_i) = f(v_{i+1})$ and $f(xv_i)f(s_i) = f(xv_i s_i) = f(xv_{i+1})$, then $f(v_i)^{-1}f(v_{i+1}) = f(s_i) = f(xv_i)^{-1}f(xv_{i+1})$ and $f(xv_i)f(v_i)^{-1} = f(xv_{i+1})f(v_{i+1})^{-1}$. So $f(xv)f(v)^{-1} = f(xu)f(u)^{-1}$ follows.

Claim 2. ϕ is a homomorphism.

Claim 3. There exists $\gamma \in \Gamma$ such that $Pr_{x\in G}[\phi(x) = f(x)\gamma] \geq 1 - 2\delta/(1-\lambda)$.

The proof of claim 2 and claim 3 is same as that in Shpilka and Wigderson [3]. We omit it here.

We finish our proof by combining the three claims.

4.2 Proof Theorem 8

We do the following test: Choose a random vertex x in G, then independently choose ℓ elements $s_1, ..., s_\ell$ from S, check whether $f(x)f(s_1 \cdot ... \cdot s_\ell) = f(x \cdot s_1 \cdot ... \cdot s_\ell)$, the test accepts if it is true. The above test may be thought as taking a random walk of length ℓ on Cayley graph $Cay(G; S)$, then use the values of f at the starting point and the terminal point of the random walk to check whether the homomorphism condition is satisfied. We construct a new graph $H' = (V', E')$ as follows: The vertex set V' is still all the group elements. For each path of length ℓ from u to v in $Cay(G; S)$, construct a corresponding edge (u, v) in E' (parallel edges are permitted).

Since graph $Cay(G; S)$ is d-regular graph with $\lambda(Cay(G; S)) = \lambda$, it is easy to see that graph H' is an $(n, d^\ell, \lambda^\ell)$-expander. The new test we defined is just taking a random edge from H', and tests whether it satisfies the homomorphism property. So we just need similar analysis to that in proving Theorem7, to prove that the test is a $(\delta, 2\delta/(1-\lambda^\ell), 3, log|G| + O(log|S|)$-test for affine homomorphism, which proves Theorem 8.

5 Derandomized Graph Test for Homomorphism in General Groups

5.1 Proof of Theorem 6

Ben-or et al [1] proved that the BLR-test works over non-abelian groups too. Later Shpilka and Wigderson [3] used the Expanding Cayley graphs to derandomize the homomorphism testing in general groups. However, derandomized graph test for homomorphism over non-abelian groups seems untouched till now. We will try to go further in this area.

However, in order to maintain the quantity of random bits used by the test to be $\log|G| + O(\log\log|G|)$, we are only able to analyze the case that the test for homomorphism uses a random walk of length ℓ, queries $\ell+1$ values of f, and runs ℓ basic tests.

For any group G, Γ, and subset S of G. Let X_0 be chosen uniformly at random and $X_0, ..., X_\ell$ be a random walk on the expanding Cayley graph $Cay(G;S)$ starting at X_0. We define a homomorphism test (**Test 2**) as follows: checks whether $f(x_i)f(x_i^{-1}x_{i+1}) = f(x_{i+1})$ for each edge (X_i, X_{i+1}) met on the random walk. It may be viewed as a graph test for homomorphism on a path of length ℓ.

First we introduce a useful lemma for our proof.

Lemma 7. *[MU [7]] Let X_i be a 0-1 random variable and $X = \sum_{i=0}^{\ell-1} X_i$, then*

$$Pr[X > 0] \geq \sum_{i=0}^{\ell-1} \frac{Pr(X_i = 1)}{E[X|X_i = 1]} \ .$$

Now we proof the following key lemma for Theorem 6.

Lemma 8. *If $Pr($ **Test 2** accepts $f\,) \geq 1 - A(\delta,\lambda)$, then f is $2\delta/(1-\lambda)$-close to some affine homomorphism, where $A(\delta,\lambda) = \dfrac{\delta^2\ell^2}{(\delta\ell + \delta^2\ell^2 - \delta^2\ell) + 2\delta\psi(\lambda,\ell)}$ and $\psi(\lambda,\ell) = \sum_{0\leq i<j\leq \ell-1} \lambda^{j-i-1} = \sum_{t=0}^{\ell-1} t\cdot\lambda^{\ell-1-t}$.*

Proof. Define random variables X_i for $i = 0, ..., \ell-1$ such that:

$$X_i = \begin{cases} 1 & \text{if } f(x_i)f(x_i^{-1}x_{i+1}) \neq f(x_{i+1}) \\ 0 & \text{otherwise} \end{cases}$$

and $X = \sum_{i=0}^{\ell-1} X_i$.

Define set $F = \{(x, xs) \in Cay(G;S) | f(x)f(s) \neq f(xs)\}$ and let $\mu = \frac{|F|}{|E(Cay(G;S))|}$, then we know

$$E[X_i] = Pr(X_i = 1) = \mu$$

$$E[X] = \sum_{i=0}^{\ell-1} E(X_i) = \mu\ell$$

$$\begin{aligned} & Pr[\text{ test rejects f }] = Pr[X > 0] \\ & \geq \sum_{i=0}^{\ell-1} \frac{Pr(X_i = 1)}{E[X|X_i = 1]} = \sum_{i=0}^{\ell-1} \frac{\mu}{E[X|X_i = 1]} \\ & \geq \frac{\ell^2\mu}{\sum_{i=0}^{\ell-1} E[X|X_i = 1]} \end{aligned}$$

where the first inequality is from lemma 7 and the last inequality is from Jensen's inequality. And

$$\begin{aligned}\sum_{i=0}^{\ell-1} E[X|X_i=1] &= \sum_{i,j=0}^{\ell-1} E[X_{j=1}|X_i=1]\\ &= \sum_{i,j=0}^{\ell-1} Pr(X_{j=1}|X_i=1) = \mu^{-1}\sum_{i,j=0}^{\ell-1} Pr(X_{j=1},X_i=1)\\ &= \mu^{-1}E[X^2].\end{aligned}$$

Moreover

$$E[X^2] = \sum_{i,j=0}^{\ell-1} E[X_iX_j] = \sum_{i=0}^{\ell-1} E[X_i^2] + 2\sum_{0\le i<j\le \ell-1} E[X_iX_j]$$

Since each X_i is 0-1 variable, so $E[X_i^2] = E[X_i] = \mu$ and for $i<j$

$$E[X_iX_j] = Pr(X_i=1)\cdot Pr(X_j=1|X_i=1) \le \mu(\mu+\lambda^{j-i-1})$$

where the last inequality is from corollary 1. It implies

$$E[X^2] \le \mu\ell + 2\sum_{0\le i<j\le \ell-1} \mu(\mu+\lambda^{j-i-1}) = (\mu\ell+\mu^2\ell^2-\mu^2\ell) + 2\mu\sum_{0\le i<j\le \ell-1} \lambda^{j-i-1}$$

Define $\psi(\lambda,\ell) = \sum_{0\le i<j\le\ell-1}\lambda^{j-i-1} = \sum_{t=0}^{\ell-1} t\cdot\lambda^{\ell-1-t}$ then

$$E[X^2] \le (\mu\ell+\mu^2\ell^2-\mu^2\ell) + 2\mu\psi(\lambda,\ell)$$

and

$$\begin{aligned}Pr[\text{ test rejects f }] = Pr[X>0] &\ge \sum_{i=0}^{\ell-1}\frac{Pr(X_i=1)}{E[X|X_i=1]}\\ &\ge \frac{\mu^2\ell^2}{(\mu\ell+\mu^2\ell^2-\mu^2\ell)+2\mu\psi(\lambda,\ell)} = A(\mu,\lambda)\ .\end{aligned}$$

So if

$$Pr(\textbf{ Test 2 } \text{accepts f }) \ge 1 - A(\delta,\lambda)$$

then

$$Pr(\textbf{ Test 2 } \text{rejects f }) \le A(\delta,\lambda)$$

which implies

$$Pr(\text{ test of SW [3] rejects } f\ ) = Pr_{y\in G,s\in S}(f(y)f(s)\ne f(ys)) \le \delta$$

By the analysis of Theorem 7, we know that in this case f is $2\delta/(1-\lambda)$-close to some affine homomorphism.

Remark 1: When δ is relatively large, say $\delta > \frac{1-\alpha}{2}$, we are able to prove a better bound on the error probability, say $Pr($ **Test 3** accepts $) \leq (1 - \delta + \frac{1-\alpha}{2})^{\ell-1}$.

Remark 2: Be different from the abelian group case, there exists some kind of graph test may not work over general groups.

References

1. Ben-Or, M., Coppersmith, D., Luby, M., Rubinfeld, R.: Non-Abelian Homomorphism Testing, and Distributions Close to their Self-Convolutions. In: Jansen, K., Khanna, S., Rolim, J.D.P., Ron, D. (eds.) RANDOM 2004. LNCS, vol. 3122, Springer, Heidelberg (2004)
2. Blum, M., Luby, M., Rubinfeld, R.: Self-Testing/Correcting with Applications to Numerical Problems. Journal of Computer and System Sciences 47(3), 549–595 (1993)
3. Shpilka, A., Wigderson, A.: Derandomizing Homomorphism Testing in General Groups. In: Proceedings of the 36th Annual ACM Symposium on Theory of Computing (STOC), pp. 427–435 (2004)
4. Samorodnitsky, A., Trevisan, L.: A PCP characterization of NP with optimal amortized query complexity. In: Proceedings of the Thirty-Second Annual ACM Symposium on Theory of Computing, Portland, OR, USA, May 21-23, 2000, pp. 191–199 (2000)
5. Ajtai, M., Komlós, J., Szemerédi, E.: Deterministic simulation in LOGSPACE. In: Proceedings of the 19th Annual ACM Symposium on Theory of Computing, pp. 132–140 (1987)
6. Håstad, J., Wigderson, A.: Simple Analysis of Graph Tests for Linearity and PCP. Random Structures and Algorithms 22(2), 139–160 (2003)
7. Mitzwnmacher, M., Upfal, E.: Probability and Computing: Randomized Algoriths and Probabilistic Analysis. Cambridge University Press, Cambridge (2005)
8. Dinur, I.: The PCP Theorem by Gap Amplification. In: Proceedings of the 38th Annual ACM Symposium on Theory of Computing, pp. 241–250 (2006)
9. Alon, N., Chung, F.R.K.: Explicit construction of linear sized tolerant networks. Discrete Mathematics 72, 15–19 (1988)
10. Bellare, M., Coppersmith, D., Hastad, J., Kiwi, M., Sudan, M.: Linearity testing in characteristic two. In: The 36th Annual Symposium on Foundations of Computer Science, p. 432 (1995)
11. Ben-Sasson, E., Sudan, M., Vadhan, S., Wigderson, A.: Randomness-efficient Low Degree Tests and Short PCPs via Epsilon-Biased Sets. In: Proceedings of the Thirty-fifth Annual ACM Symposium on Theory of Computing, San Diego, CA, USA, June 9-11, pp. 612–621 (2003)
12. Wigderson, A., Xiao, D.: Derandomizing the AW matrix-valued Chernoff bound using pessimistic estimators and applications. In: ECCC TR06-105 (2006)
13. Naor, J., Naor, M.: Small Bias Probability Spaces: Efficient Constructions and Applications. SIAM Journal on Computing 22(4), 838–856 (1993)

Definable Filters in the Structure of Bounded Turing Reductions

Angsheng Li[1,⋆], Weilin Li[1,2,⋆], Yicheng Pan[1,2,⋆], and Linqing Tang[1,2,⋆]

[1] State Key Lab. of Computer Science, Institute of Software, Chinese Academy of Sciences
[2] Graduate University of Chinese Academy of Sciences
{angsheng,weilin,yicheng,linqing}@ios.ac.cn

Abstract. In this article, we show that there exist c.e. bounded Turing degrees $\mathbf{a}$, $\mathbf{b}$ such that $\mathbf{0} < \mathbf{a} < \mathbf{0}'$, and that for any c.e. bounded Turing degree $\mathbf{x}$, $\mathbf{b} \vee \mathbf{x} = \mathbf{0}'$ if and only if $\mathbf{x} \geq \mathbf{a}$. The result gives an unexpected definability theorem in the structure of bounded Turing reducibilities.

1 Introduction

A set $A \subseteq \omega$ is called *computably enumerable* (c.e., for short), if there is an algorithm to enumerate the elements of it. Given sets $A, B \subseteq \omega$, we say that A is Turing *reducible to* B, if there is an oracle Turing machine, Φ say, such that $A = \Phi^B$ (denoted by $A \leq_{\mathrm{T}} B$), and furthermore, if the bits of oracle queries are bounded by a computable function, we say that A is *bounded Turing reducible* to B (written $A \leq_{\mathrm{bT}} B$). A Turing and a bounded Turing (or bT, for short) degree is the equivalence class of a set under the Turing reductions and the bounded Turing reductions respectively. A degree is called *computably enumerable* (c.e.), if it contains a c.e. set.

Let $\mathcal{E}$ and $\mathcal{E}_{\mathrm{bT}}$ be the structures of the c.e. degrees under the Turing reductions and the bounded Turing reductions respectively. During the past decades, the studies of both structures focused on that of the algebraic properties, leading to major achievements such as the decidability results of the Σ_1-theory of $\mathcal{E}$, and the Σ_2-theory of $\mathcal{E}_{\mathrm{bT}}$ (Ambos-Spies, P. Fejer, S. Lempp and M. Lerman [1996]), and the undecidability results of the Σ_3-theory of $\mathcal{E}$ (Lempp, Nies, and Slaman [1998]), and of the Σ_4-theory of $\mathcal{E}_{\mathrm{bT}}$ (Nies and Lempp [1995]). This progress brings the decidability problems of the Σ_2-theory of $\mathcal{E}$, and the Σ_3-theory of $\mathcal{E}_{\mathrm{bT}}$ into sharper focus, for which new ingredients are welcome.

In recent years, the study of the computably enumerable degrees has focused on Turing definability in the structure $\mathcal{E}$. For instance, Slaman asked in 1985 if there are any c.e. degrees that are incomplete and nonzero which are definable in the c.e. degrees $\mathcal{E}$. A natural approach to this problem is to find some definable substructures of $\mathcal{E}$ that have nontrivial minimal/maximal and/or least/greatest

⋆ The authors are partially supported by NSFC Grant No. 60325206, and No. 60310213.

M. Agrawal et al. (Eds.): TAMC 2008, LNCS 4978, pp. 116–124, 2008.

members. This resumes interests in topics such as the continuity of the c.e. degrees, started by Lachlan early in 1967.

Harrington and Soare [1992] showed that there is no maximal minimal pair in the c.e. Turing degrees, and Seetapun [1991] proved a stronger result that for any c.e. degree $\mathbf{b} \neq \mathbf{0}, \mathbf{0}'$, there is a c.e. Turing degree $\mathbf{a} > \mathbf{b}$ such that for any c.e. Turing degree $\mathbf{x}$, $\mathbf{a} \wedge \mathbf{x} = \mathbf{0}$ if and only if $\mathbf{b} \wedge \mathbf{x} = \mathbf{0}$.

In the dual case, Ambos-Spies, Lachlan and Soare [1993] showed that for any c.e. Turing degrees $\mathbf{x}, \mathbf{y}$, if $\mathbf{x}$ and $\mathbf{y}$ are nontrivial splitting of $\mathbf{0}'$, then there exists a c.e. Turing degree $\mathbf{a} < \mathbf{x}$ such that $\mathbf{a} \vee \mathbf{y} = \mathbf{0}'$. Remarkably, Cooper and Li [ta] (the major subdegree theorem) showed that for any c.e. Turing degree $\mathbf{b} \neq \mathbf{0}, \mathbf{0}'$, there exists a c.e. Turing degree $\mathbf{a} < \mathbf{b}$ such that for any c.e. Turing degree $\mathbf{x}$, $\mathbf{b} \vee \mathbf{x} = \mathbf{0}'$ if and only if $\mathbf{a} \vee \mathbf{x} = \mathbf{0}'$ (.

However, the study of the continuity and the definability in the bounded Turing structures is started just recently. Paul Brodhead, Angsheng Li and Weilin Li [ta] (BLL) showed that the Seetapun's result holds in the c.e. bounded Turing degrees, that is, for any c.e. bounded Turing (bT) degree $\mathbf{b} \neq \mathbf{0}, \mathbf{0}'$, there exists a c.e. bT-degree $\mathbf{a} > \mathbf{b}$ such that for any c.e. bT-degree $\mathbf{x}$, $\mathbf{a} \wedge \mathbf{x} = \mathbf{0}$ if and only if $\mathbf{b} \wedge \mathbf{x} = \mathbf{0}$. In the present paper, we consider the dual case of this result, the BLL result. It is very unexpected that the dual case fails badly. In fact we are able to prove:

Theorem 1. *There exist c.e. bounded Turing degrees* $\boldsymbol{a}, \boldsymbol{b}$ *with the following properties:*

(1) $\boldsymbol{0} < \boldsymbol{a} < \boldsymbol{0}'$,

(2) For any c.e. bounded Turing degree $\boldsymbol{x}$, $\boldsymbol{b} \vee \boldsymbol{x} = \boldsymbol{0}'$ *if and only if* $\boldsymbol{x} \geq \boldsymbol{a}$.

An immediate corollary of the theorem is that there exist c.e. bounded Turing degrees $\mathbf{a}$, $\mathbf{b}$ such that $\mathbf{a} \vee \mathbf{b} = \mathbf{0}'$ but for no c.e. bT-degree $\mathbf{x} < \mathbf{a}$ with $\mathbf{x} \vee \mathbf{b} = \mathbf{0}'$, and that the Cooper-Li major subdegree theorem fails badly in the c.e. bT-degrees.

However the result gives a very nice theorem in the Turing definability of the c.e. bounded Turing degrees. That is, there is a principal filter $[\mathbf{a}, \mathbf{0}']$ which is definable by equation $\mathbf{x} \vee \mathbf{b} = \mathbf{0}'$ for some nonzero and incomplete c.e. bT-degree $\mathbf{b}$. The result may provide ingredients to the decidability/undecidability problem of the Σ_3-theory of the c.e. bounded Turing degrees.

The rest of this paper is devoted to proving theorem 1.1, our main result. In section 2, we formulate the conditions of the theorem by requirements, and for each requirement, we give corresponding strategy to satisfy it; in section 3, we arrange all strategies to satisfy the requirements on nodes of a tree, or more precisely, the *priority tree* T. In section 4, we use the priority tree to describe a stage-by-stage construction of the objects we need. Finally, in section 5 we verify that the construction in section 4 satisfies all of the requirements, finishing the proof of the theorem.

Our notation and terminology are standard and generally follow Soare [1987]. During the course of a construction, notations such as A, Φ are used to denote the current approximations to these objects, and if we want to specify the values immediately at the end of stage s, then we denote them by A_s, $\Phi[s]$ etc. For a *partial computable* (p.c., or for simplicity, also a Turing) functional, Φ say, the use function is denoted by the corresponding lower case letter ϕ. The value of the use function of a converging computation is the greatest number which is actually used in the computation. For a Turing functional, if a computation is not defined, then we define its use function $= -1$. During the course of a construction, whenever we define a parameter, p say, as *fresh*, we mean that p is defined to be the least natural number which is greater than any number mentioned so far. In particular, if p is defined as fresh at stage s, then $p > s$. The notion of bounded Turing reducibility is taken from Soare's new book: Computability Theory and Applications [ta] (Definition 3.4.1).

2 Requirements and Strategies: Theorem 1.1

2.1 The Requirements

We will build c.e. sets A, B, C and D to satisfy the following requirements:

$$\begin{array}{ll} \mathcal{T}: & K \leq_{\mathrm{bT}} A \oplus B \\ \mathcal{P}_e: & A \neq \theta_e \\ \mathcal{S}_e: & C \neq \Psi_e(A) \\ \mathcal{R}_e: & D = \Phi_e(X_e, B) \longrightarrow A \leq_{\mathrm{bT}} X_e \end{array}$$

where $e \in \omega$, $\{(\theta_e, \Phi_e, \Psi_e, X_e) : e \in \omega\}$ is an effective enumeration of all quadruples (θ, Φ, Ψ, X) of all computable partial functions θ, of all bounded Turing (bT, for short) reductions Φ, and Ψ, and of all c.e. sets X; and K is a fixed creative set.

Let $\mathbf{a}$, $\mathbf{b}$ and $\mathbf{x}$ be the bT-degrees of A, B and X respectively. By the $\mathcal{P}$-requirements and the $\mathcal{S}$-requirements, we have $\mathbf{0} < \mathbf{a} < \mathbf{0}'$. By the $\mathcal{R}$-requirements, we know that for any c.e. bT degree $\mathbf{x}$, if $\mathbf{x} \vee \mathbf{b} = \mathbf{0}'$, then $\mathbf{a} \leq \mathbf{x}$. By the $\mathcal{T}$-requirement, for any c.e. bounded Turing degree $\mathbf{x}$, if $\mathbf{a} \leq \mathbf{x}$, then $\mathbf{0}' \leq \mathbf{a} \vee \mathbf{b} \leq \mathbf{x} \vee \mathbf{b}$. Therefore satisfying all the requirements is sufficient to prove the theorem.

We introduce some conventions of the bounded Turing reductions for describing the strategies. We will assume that for any given bounded Turing reduction Φ or Ψ, the use functions ϕ and ψ will be increasing in arguments. Now we are ready to describe the strategies.

2.2 The $\mathcal{T}$-Strategy

To satisfy the $\mathcal{T}$-requirement, $K \leq_{\mathrm{bT}} A \oplus B$, we need to construct a bounded Turing reduction Γ such that $K = \Gamma(A, B)$.

We construct the Γ by coding K into A and B as follows:

For any k, if $k \in K$, then either $2k \in A$ or $2k \in B$.

Therefore, $k \in K$ if and only if either $A(2k) = 1$ or $B(2k) = 1$. K is bounded Turing reducible to $A \oplus B$, the $\mathcal{T}$-requirement is satisfied.

2.3 A $\mathcal{P}$-Strategy

A $\mathcal{P}$-strategy satisfying a $\mathcal{P}$-requirement, $A \neq \theta$ say, is a Friedberg-Muchnik procedure, and proceeds as follows:

1. Define a witness a as a fresh odd number.
2. Wait for a stage, v say, at which $\theta(a) \downarrow = 0 = A(a)$, then
   — enumerate a into A and terminate.

By the strategy, we know if step 2 of the procedure occurs, then $\theta(a) \downarrow = 0 \neq 1 = A(a)$ is created, otherwise $\theta(a) \neq 0 = A(a)$ occurs at each stage. In either case, the $\mathcal{P}$-requirement is satisfied. We use a node on a tree (the priority tree as we'll see later), γ say, to denote a $\mathcal{P}$-strategy.

2.4 An $\mathcal{S}$-Strategy

Suppose that we want to satisfy an $\mathcal{S}$-requirement, $C \neq \Psi(A)$ say. It will be satisfied by a Friedberg-Muchnik procedure as follows:

1. Define a witness c as a fresh odd number.
2. Wait for a stage, v say, at which $\Psi(A;c)[v] \downarrow = 0 = C_v(c)$, then
   — enumerate c into C;
   — define an A-restraint $r^A = \psi(c)$.

The key point to the satisfaction of $\mathcal{S}$ is that the A-restraint $r^A = \psi(c)$ will be preserved once step 2 of the strategy occurs. In this case, we have that if step 2 occurs, then $\Psi(A;c) \downarrow = 0 \neq 1 = C(c)$ is created and preserved, else if step 2 never occurs, then $\Psi(A;c) \neq 0 = C(c)$. In either case, the $\mathcal{S}$-requirement is satisfied. We use a node β, say, on the priority tree to denote the $\mathcal{S}$-strategy.

2.5 An $\mathcal{R}$-Strategy

Suppose we want to satisfy an $\mathcal{R}$ requirement, $\mathcal{R}$: $D = \Phi(X, B) \longrightarrow A \leq_{\mathrm{bT}} X$, say. We will build a bounded Turing reduction Δ such that if $D = \Phi(X, B)$ then $\Delta(X) = A$. The bounded Turing reduction Δ will be built by an ω-sequence of cycles n, cycle n will be responsible for defining $\Delta(X;n)$.

The $\mathcal{R}$-strategy will proceed as follows:

1. Let n be the least x, such that $\Delta(X;x) \uparrow$. Define a *witness block* U_n such that $min\{y \in U_n\}$ is fresh, and that $|U_n| = n+1$.
2. Wait for a stage, v say, at which the following conditions occur: For all $y \in U_{n'}$ for some $n' \leq n$, we have
   $\Phi(X, B; y) \downarrow = D(y)$
   then:
   — define $\Delta(X;n) \downarrow = A(n)$ with $\delta(n) = max\{\phi(y) | y \in U_{n'}, n' \leq n\}$.
3. If there is an x such that $\Delta(X;x) \downarrow \neq A(x)$, then let n be the least such x, and go on to step 4.

4. If for all $y \in U_n$, $\Phi(X, B; y) \downarrow= D(y)$, then
   — let x be the least $y \in U_n \backslash D$;
   — enumerate x into D;
   — define a *conditional restraint* $\overrightarrow{r} = (n, \delta(n)]$.
5. Suppose that a conditional restraint $\overrightarrow{r} = (n, \delta(n)]$ is kept, then there is no element b with $n < b \leq \delta(n)$ that can be enumerated into B.

Note that if X changes below $\delta(n)$, then $\Delta(X; n)$ becomes undefined, and simultaneously the conditional restraint $\overrightarrow{r} = (n, \delta(n)]$ drops.

The new idea of this proof is the notion of the witness block, which seems the first time it becomes available, and the notion of conditional restraints. The later notion was largely the first author's idea that has already been used several times in the literature.

Now we analyze the correctness of the $\mathcal{R}$-strategy. We proceed the arguments by cases:

Case 1: $\Delta(X)$ is built infinitely often.

In this case we will prove that $\Delta(X)$ is a total function (i.e., $\Delta(X; x)$ is defined for every x). Note that for every n, $\Delta(X; n)$ is redefined only finitely many times. So there exists some s , such that for any stage s' with $s' > s$, $\Delta(X; n)[s']$ does not change any more.

Assume by contradiction, there exists an x such that $\Delta(X; x) \uparrow$ eventually, let n be the least such x. Assume after stage v, the $\Delta(X; n)$ would never been defined any more. By step 2 of the $\mathcal{R}$-strategy, we know that for each stage s with $s \geq v$, there exists some $y \in U_{n'}$ for some $n' \leq n$ such that $\Phi(X, B; y) \downarrow\neq D(y)$ (otherwise $\Delta(X; n)$ will be defined again by step 2 of the $\mathcal{R}$-strategy), then for any undefined $\Delta(X; m)$ with $m > n$, it would never be defined after stage s with $s \geq v$. So $\Delta(X)$ is finitely built, a contradiction. That means that $\Delta(X)$ is a total function.

Now if $\Delta(X) \neq A$, i.e., there exists an x such that $\Delta(X; x) \downarrow\neq A(x)$, we prove that $\Phi(X, B) \neq D$ under this assumption.

Let n be the least such x, we prove that there is an $x \in U_n$ such that $\Phi(X, B; x) \downarrow= 0 \neq 1 = D(x)$. Let s_n be the stage at which the permanent computation $\Delta(X; n)$ was created, and let $t_n > s_n$ be the stage at which n is enumerated into A. By step 4 of the $\mathcal{R}$-strategy, for any $s > t_n$, if $\Phi(X, B; y) \downarrow= D(y)$ holds for all $y \in U_n$, then we enumerate an element $x \in U_n$ into D, and create a conditional restraint $(n, \delta(n)]$. By the choice of s_n, the conditional restraint will be kept forever. By the conditional restraint, no number x with $n < x \leq \delta(n)$ could be enumerated into B, therefore $r_n \doteq \{\Phi(X, B; y) | y \in U_n\}$ can be injured only by numbers $x \leq n$. Once r_n is injured, we may waste a witness $x \in U_n$. However, r_n can be injured at most n many times and $|U_n| = n + 1$. Therefore after up to n many times injury, there is a witness $y \in U_n$ which can be used to create a permanent inequality between $\Phi(X, B)$ and D. Let $v_n > t_n$ be a minimal stage after which B will never change below n. Therefore if at a stage $s > t_n$ we have that

(1) $\Delta(X;n)\downarrow\neq A(n)$, and
(2) for any $y\in U_n$, $\Phi(X,B;y)[s]\downarrow=D_s(y)$,

then we enumerate the least $x\in U_n\backslash D$, x_0 say, into D. By the choice of n, v_n, and by the conditional restraint $\overrightarrow{r}=(n,\delta(n)]$, $\Phi(X,B;x_0)\downarrow=0\neq 1=D(x_0)$ holds permanently, which means $\Phi(X,B)\downarrow\neq D$.

Case 2: Δ is finitely built.

In this case, assume w.l.o.g. that $\Delta(X;0)\downarrow,\cdots,\Delta(X;n-1)\downarrow$, and $\Delta(X;n)\uparrow$ eventually, we can assume after s_n, $\Delta(X;n)$ doesn't have chance to be defined any more (since if it has infinitely many chances to be defined, then it must be defined). By step 2 of the $\mathcal{R}$-strategy, for any stage s with $s>s_n$, there exists a $y\in U_{n'}$ for some $n'\leq n$ such that $\Phi(X,B;y)\neq D(y)$ (otherwise $\Delta(X;n)$ will be defined again by step 2 of the $\mathcal{R}$-strategy). Since $\bigcup_{n'\leq n}U_{n'}$ is finite, so there exists some n' with some $y_0\in U_{n'}$ such that $\Phi(X,B;y_0)[s]\neq D_s(y_0)$ for infinitely many stages s, that means $\Phi(X,B)\neq D$.

In either case, the $\mathcal{R}$-requirement is satisfied. We use a node on the priority tree, α say, to denote the $\mathcal{R}$-strategy.

Note in the case $\Delta(X)=A$, there is no permanent conditional restraint $\overrightarrow{r}$. However, there may be infinitely many stages at which we create $\overrightarrow{r[s]}=(a[s],b[s]]$. The key to the proof is that this unbounded conditional restraints $\overrightarrow{r[s]}$ are harmless, because:

(1) A conditional restraint $\overrightarrow{r[s]}=(a[s],b[s]]$ controls elements x only if $a[s]<x\leq b[s]$.

(2) The conditional restraints ensure that the $\mathcal{R}$-strategy has no influence on any number no larger than $a[s]$. We say it's not important because in this case, $a[s]$ will be unbounded. This guarantees that the $\mathcal{R}$-strategy would not injure a lower priority $\mathcal{S}$-strategy after some fixed stage. That is to say, a lower priority $\mathcal{S}$-strategy is injured by the $\mathcal{R}$-strategy only finitely many times. This is the reason why we can only use conditional restraints here, instead of the typical restraints, which is of course one of the main contributions of this paper.

We thus define the *possible outcomes of* the $\mathcal{R}$-strategy α by

$$0<_{\mathrm{L}}1$$

to denote infinite and finite actions respectively.

3 The Priority Tree

In this section, we will build a *priority tree of strategies* $T\subset\Lambda^{<\omega}$, with $\Lambda=\{0,1\}$. Let $\mathcal{P}<\mathcal{R}$ denote that the priority ranking of $\mathcal{P}$ is higher than that of $\mathcal{R}$. Also let $<_L$ be a *left-to-right* ordering of the nodes on the priority tree, as given below.

Definition 1. Define the *priority ranking of the requirements* such that $\mathcal{T}$ has the highest priority, and $\forall e \in \omega$: $\mathcal{R}_e < \mathcal{S}_e < \mathcal{P}_e < \mathcal{R}_{e+1}$.

Note that both a $\mathcal{P}$-strategy and an $\mathcal{S}$-strategy have only one possible outcome, we denote it by 1. However, as we mentioned before, an $\mathcal{R}$-strategy has two outcomes which we denote by $0 <_L 1$.

We now define the priority tree T inductively as follows.

Definition 2 (The Priority Tree).

(i) Define the root node $\emptyset$ of the tree to be an $\mathcal{R}_0$-strategy.
(ii) The immediate successors of a node are the possible outcomes of the corresponding strategy. We say a requirement $\mathcal{X}$ is satisfied at some node ξ, if there is a node $\xi' \subseteq \xi$ working on the requirement $\mathcal{X}$.
(iii) The immediate successors of a node, ξ say, will work on the highest priority ranking requirement which is not satisfied on path ξ.

Definition 3. The *index $I(\xi)$ of a node ξ* is the index of the requirement on which the node acts. For example, if ξ is an $\mathcal{R}_e$- or a $\mathcal{P}_e$- or an $\mathcal{S}_e$-strategy, we define $I(\xi) = e$.

4 The Construction

Our construction will perform different actions at even and odd stages. Suppose that K is enumerated at odd stages only, and that there is exactly one element that enters K at each odd stage. At even stages, strategies on the tree will act to satisfy the requirements.

During the course of the construction, we may initialize a node, ξ say, which means that all the actions taken by ξ previously, are canceled, or set to be totally undefined.

Now we give the construction stage-by-stage.

Definition 4. (The Construction) The construction is defined as follows:

Stage $s = 0$. Set $A = B = C = D = \emptyset$, and initialize every node of the priority tree T.

Stage $s = 2n + 1$. For $k_s \in K_s \backslash K_{s-1}$. Let $x = 2k_s$, we need to decide either $x \in A$ or $x \in B$:

Case 1: Find the $<$-least ξ such that (a) or (b) below occurs:

(a) $\xi = \alpha$ is an $\mathcal{R}$-strategy such that α has conditional restraint $\overrightarrow{r[s]} \downarrow = (a[s], b[s]]$ and $a[s] < x \leq b[s]$;

(b) $\xi = \beta$ is an $\mathcal{S}$-strategy such that $r^A(\beta) \downarrow \geq x$.

then:

Subcase 1: $\xi = \alpha$, i.e., (a) occurs

— enumerate x into A;

— initialize all nodes ξ to the right of $\alpha\hat{\ }\langle 0\rangle$.

Subcase 2: $\xi = \beta$, i.e., (b) occurs

— enumerate x into B;
— initialize all nodes ξ with $\xi \not\leq \beta$.
Case 2: Otherwise, we can't find node ξ such that (a) or (b) occurs for ξ:
— enumerate x into A.

Stage $s = 2n + 2$. We specify certain strategies to be eligible to act. First we allow the root node to be eligible to act at substage $t = 0$. At each substage t, we let the strategy be eligible to act run its program, and then, either close the current stage or specify a node to be eligible to act next, i.e., at the next substage of stage s.

Substage t. Suppose that node ξ is eligible to act at substage t of stage s. If $t = s$, then initialize all nodes $\xi' \not\leq \xi$, and close the current stage. Otherwise, corresponding to different types of the strategy, there are three cases:

Case 1. $\xi = \beta$ is an $\mathcal{S}$-strategy. Then run the following:

Program β:

1. If $c(\beta)\downarrow$, and $\Psi_\beta(A; c(\beta))\downarrow = 0 = C(c(\beta))$, then
   (a) If for any α with $\alpha\hat{}\langle 0\rangle \subseteq \beta$ we have $dom(\alpha) > \psi_\beta(c)$, then
   — enumerate $c(\beta)$ into C, and
   — define $r^A(\beta) = \psi(c(\beta))$,
   where $dom(\alpha)$ is the size of the domain of Δ_α.
   (b) Otherwise, then do nothing,
   in either case, initialize all $\xi' \not\leq \beta$ and go to stage $s + 1$.
2. If $c(\beta)\uparrow$, then
   — define $c(\beta)$ as fresh;
   — initialize all $\xi' \not\leq \beta$ and go to stage $s + 1$.
3. Otherwise, let $\beta\hat{}\langle 1\rangle$ be eligible to act next.

Case 2. $\xi = \gamma$ is a $\mathcal{P}$-strategy. Run the following

Program γ:

1. If $a(\gamma)\downarrow$, and $\theta_\gamma(a(\gamma))\downarrow = 0 = A(a(\gamma))$, then
   — enumerate $a(\gamma)$ into A,
   – initialize all nodes ξ with $\xi \not\leq \gamma$, and go to stage $s + 1$.
2. If $a(\gamma)\uparrow$, then
   — define $a(\gamma)$ as fresh;
   — initialize all $\xi' \not\leq \gamma$ and go to stage $s + 1$.
3. Otherwise, let $\gamma\hat{}\langle 1\rangle$ be eligible to act next.

Case 3. $\xi = \alpha$ is an $\mathcal{R}$-strategy. In this case, we perform the following

Program α:

1. If there is a $k > b(\alpha)$, such that $\Delta_\alpha(X_\alpha; k)\downarrow \neq A(k)$, then:
   (a) If $\overrightarrow{r}(\alpha)\downarrow$, then:
   — Let $\alpha\hat{}\langle 1\rangle$ be eligible to act next.

(b) Otherwise and $\forall y \in U_k^\alpha$, $\Phi_\alpha(X_\alpha, B; y)[s] \downarrow = D(y)$, then:
— let i be the least k such that $\Delta_\alpha(X_\alpha; k) \downarrow \neq A(k)$,
— let x be the least $y \in U_i^\alpha \backslash D$, enumerate x into D;
— define a conditional restraint $\overrightarrow{r}(\alpha) = (i, \delta(i)]$;
— initialize all nodes ξ to the right of $\alpha\hat{}\langle 0\rangle$, and go to stage $s+1$.
(c) Otherwise:
- let $\alpha\hat{}\langle 1\rangle$ be eligible to act next.

2. Otherwise and for $k = \min\{x > b(\alpha) | \Delta_\alpha(X_\alpha; x) \uparrow\}$ we have:
(a) $U_k^\alpha \downarrow$;
(b) $\forall y \in U_{k'}^\alpha$ for some $k' \leq k$, $\Phi_\alpha(X_\alpha, B; y)[s] \downarrow = D(y)$.
Then:
— define $\Delta_\alpha(X_\alpha; k) \downarrow = A(k)$ with $\delta_\alpha(k) = \max\{\phi_\alpha(y) | y \in U_{k'}^\alpha, k' \leq k\}$;
— let $\alpha\hat{}\langle 0\rangle$ be eligible to act next.
3. Otherwise and for $k = \min\{x > b(\alpha) | \Delta_\alpha(X_\alpha; x) \uparrow\}$ we have $U_k^\alpha \uparrow$, then:
— define $U_k^\alpha = (l, l + k + 1)$, where l is a fresh number;
— initialize all nodes ξ to the right of $\alpha\hat{}\langle 0\rangle$, and close the current stage.
4. Otherwise and $b(\alpha) \uparrow$, then:
— define $b(\alpha)$ as fresh;
— initialize all $\xi' \not\leq \alpha$ and go to stage $s+1$.
5. Otherwise, then let $\alpha\hat{}\langle 1\rangle$ be eligible to act next.

This completes the description of the construction. The verification of this construction will be given in the full version of the paper.

References

1. Ambos-Spies, K., Fejer, P.A., Lempp, S., Lerman, M.: Decidability of the two-quantifier theory of the recursively enumerable weak truthtable degrees and other distributive upper semi-lattices. Journal of Symbolic Logic 61, 880–905 (1996)
2. Ambos-Spies, K., Lachlan, A.H., Soare, R.I.: The continuity of cupping to 0'. Annals of Pure and Applied Logic 64, 195–209 (1993)
3. Brodhead, P., Li, A., Li, W.: Continuity of capping in $\mathcal{E}_{dT}$ (to appear)
4. Cooper, S.B., Li, A.: On Lachlan's major subdegree problem (to appear)
5. Harrington, L., Soare, R.I.: Games in recursion theory and continuity properties of capping degrees. In: Judah, H., Just, W., Woodin, W.H. (eds.) Set Theory and the Continuum, Proceedings of Workshop on Set Theory and the Continuum, MSRI, Berkeley, CA, October, 1989, pp. 39–62. Springer, Heidelberg (1992)
6. Nies, A., Lempp, S.: The undecidability of the Π_4-theory for the r.e. wtt- and Turing degrees. Journal of Symbolic Logic 60, 1118–1136 (1995)
7. Nies, A., Lempp, S., Slaman, T.A.: The Π_3-theory of the enumerable Turing degrees is undecidable. Transactions of the American Mathematical Society 350, 2719–2736 (1998)
8. Seetapun, D.: Contributions to Recursion Theory, Ph. D thesis, Trinity College (1991)
9. Soare, R.: Recursively Enumerable Sets and Degrees. Springer, Heidelberg (1987)
10. Soare, R.I.: Computability Theory and Applications. Springer, Heidelberg (to appear)

Distance Constrained Labelings of Trees

Jiří Fiala[1], Petr A. Golovach[2,*], and Jan Kratochvíl[1]

[1,**]Institute for Theoretical Computer Science
and Department of Applied Mathematics,
Charles University, Prague, Czech Republic
{fiala,honza}@kam.mff.cuni.cz
[2] Institutt for informatikk,
Universitetet i Bergen, Norway
petr.golovach@ii.uib.no

Abstract. An $H(p,q)$-labeling of a graph G is a vertex mapping $f : V_G \to V_H$ such that the distance between $f(u)$ and $f(v)$ (measured in the graph H) is at least p if the vertices u and v are adjacent in G, and the distance is at least q if u and v are at distance two in G. This notion generalizes the notions of $L(p,q)$- and $C(p,q)$-labelings of graphs studied as particular graph models of the Frequency Assignment Problem. We study the computational complexity of the problem of deciding the existence of such a labeling when the graphs G and H come from restricted graph classes. In this way we extend known results for linear and cyclic labelings of trees, with a hope that our results would help to open a new angle of view on the still open problem of $L(p,q)$-labeling of trees for fixed $p > q > 1$ (i.e., when G is a tree and H is a path).

We present a polynomial time algorithm for $H(p,1)$-labeling of trees for arbitrary H. We show that the $H(p,q)$-labeling problem is NP-complete when the graph G is a star. As the main result we prove NP-completeness for $H(p,q)$-labeling of trees when H is a symmetric q-caterpillar.

1 Introduction

Motivated by models of wireless communication, the notion of so called distance constrained graph labelings has received a lot of interest in Discrete Mathematics and Theoretical Computer Science in recent years.

In the simplest case of constraints at distance two, the typical task is, given a graph G and parameters p and q, to assign integer labels to vertices of G such that labels of adjacent vertices differ by at least p, while vertices that share a common neighbor have labels at least q apart. The aim is to minimize the span, i.e., the difference between the smallest and the largest labels used.

The notion of proper graph coloring is a special case of this labeling notion — when $(p,q) = (1,0)$. Thus it is generally NP-hard to decide whether such

* Supported by Norwegian Research Council.
** Supported by the Ministry of Education of the Czech Republic as project 1M0021620808.

M. Agrawal et al. (Eds.): TAMC 2008, LNCS 4978, pp. 125–135, 2008.

a labeling exists. On the other hand, in some special cases polynomial-time algorithms exist. For example, when G is a tree and $p \geq q = 1$, the algorithm of Chang and Kuo based on dynamic programming finds a labeling of the minimum span [3,6] in superlinear time $O(n\Delta^5)$, where Δ is the maximum degree of the tree.

Distance constrained labelings can be generalized in several ways. For example, constraints on longer distances can be involved [18,14], or the constraints on the difference of labels between close vertices can be directly implemented by edge weights — the latter case is referred to as the Channel Assignment Problem [22].

Alternatively, the cyclic metric or even more complex metrics on the label set are considered. In particular, if the metric is given as the distance between vertices in some graph H, we get the following mapping:

Definition 1. *Given two positive integers p and q, we say that f is an $H(p,q)$-labeling of a graph G if f maps the vertices of G onto vertices of H such that the following hold:*

- *if u and v are adjacent in G, then $dist_H(f(u), f(v)) \geq p$,*
- *if u and v are nonadjacent but share a common neighbor, then $dist_H(f(u), f(v)) \geq q$.*

Observe that if the graph H is a path, the ordinary linear metric is obtained. This has been introduced by Roberts and studied in a number of papers — see, e.g., recent surveys [23,1]. The cyclic metric (corresponding to the case when H is a cycle) was studied in [17,20]. The general approach was suggested in [8] and several (both P-time and NP-hardness) results for various fixed graphs H were presented in [7,10].

Several computational problems can be defined by restricting the graph classes of the input graphs, and/or by fixing some values as parameters of the most general problem which we refer to as follows:

Distance Labeling	DL
Instance: G, H, p and q	
Question: Does G allow an $H(p,q)$-labeling?	

As it was already mentioned, the linear metric is often considered, i.e. when $H = P_{\lambda+1}$ is a path of length λ. In this case we use the traditional notation "$L(p,q)$-labeling of span λ" for "$P_{\lambda+1}(p,q)$-labeling" and also define the problem explicitly:

$L(p,q)$-Distance Labeling	$L(p,q)$-DL
Parameters: p, q	
Instance: G and λ	
Question: Does G allow an $L(p,q)$-labeling of span λ?	

We focus our attention on various parameterized versions of the DL problem. Griggs and Yeh [15] showed NP-hardness of the $L(2,1)$-DL of a general graph, which means that DL is NP-complete for fixed $(p, q) = (2, 1)$ and H being a path. Later Fiala et al. [9] showed that the $L(2, 1)$-DL problem remains NP-complete for every fixed $\lambda \geq 4$. Similarly, the labeling problem with the cyclic metric is NP-complete for a fixed span [8], i.e., when $(p, q) = (2, 1)$ and $H = C_\lambda$ for an arbitrary $\lambda \geq 6$.

From the very beginning it was noticed that the distance constrained labeling problem is in certain sense more difficult than ordinary coloring. The first polynomial time algorithm for $L(2, 1)$-DL for trees came as a little surprise [3]. Attention was then paid to input graphs that are trees and their relatives, either paths and caterpillars as special trees, or graphs of bounded treewidth. Table 1 briefly summarizes the known results on the complexity of the DL problem on these graph classes and the new results presented in this paper. Notice in particular that $L(2, 1)$-DL belongs to a handful of problems that are solvable in polynomial time on graphs of treewidth one and NP-complete on treewidth two [5].

Table 1. Summary of the complexity of the DL problem

Class of G	Class of H	p	q	Complexity
bounded tw	finite class	on input	on input	P [*]
$tw \leq 2$	paths	2	1	NP-c [5]
$tw \leq 2$	cycles	2	1	NP-c [5]
$pw \leq 2$	all graphs	2	1	NP-c [5]
stars	all graphs	fixed	fixed, ≥ 2	NP-c [*]
trees, $\mathcal{L}$	paths	fixed	fixed, ≥ 2	NP-c [11]
trees	all graphs	on input	1	P [3,2, *]
trees	cycles	on input $\geq q$	on input ≥ 1	P [20,19]
trees	q-caterpillars	fixed, $\geq 2q + 1$	fixed, ≥ 2	NP-c [*]
trees	**paths**	**fixed**	**fixed, ≥ 2**	**Open**

The symbol tw means treewidth; pw is pathwidth; $\mathcal{L}$ indicates the list version of the DL problem where every vertex has prescribed set (list) of possible labels; P are problems solvable in polynomial time; NP-c are NP-complete problems; the reference [*] indicates results of this paper.

We start with two observations. For the seventh line, the algorithm of Chang and Kuo [3] can be easily modified to work for arbitrary p and H on the input. It only suffices to modify all tests whether labels of adjacent vertices have difference at least two into tests whether they are mapped onto vertices at distance at least p in H.

The first line of the table is a corollary of a strong theorem of Courcelle [4], who proved that properties expressible in Monadic Second Order Logic (MSOL) can be recognized in polynomial time on any class of graphs of bounded treewidth. If H belongs to a finite graph class, then the graph property "to allow an $H(p, q)$-labeling" straightforwardly belongs to MSOL. A special case arises, of course, if a single graph H is a fixed parameter of the problem.

In particular, if λ is fixed, then $L(p,q)$-DL is polynomially solvable for trees. On the other hand, without this assumption on λ the computational complexity of $L(p,q)$-DL on trees remains open so far. It is regarded as one of the most interesting open problems in the area.

We focus on the following variant when both p and q are fixed:

(p,q)-DISTANCE LABELING	(p,q)-DL
Parameters: p and q	
Instance: G and H	
Question: Does G allow an $H(p,q)$-labeling?	

Our main contribution is given by Theorems 1 and 2 showing two NP-hardness results for the (p,q)-DL problem: in the case when G is a star (with no restriction on H), and in the case when G is a tree and H is a specific symmetric q-caterpillar. The latter choice of H is motivated by the similarity of a caterpillar with a path, which might provide a useful step in a proof of NP-hardness of the $L(p,q)$-labeling problem.

2 Preliminaries

Throughout the paper the symbol $[a,b]$ means the interval of integers between a and b (including the bounds), i.e., $[a,b] := \{a, a+1, \ldots, b\}$. We also define $[a] := [1,a]$. The relation $i \equiv_q j$ means that i and j are congruent modulo q, i.e., q divides $i-j$. For $i \equiv_q j$ we define $[i,j]_{\equiv q} := \{i, i+q, i+2q, \ldots, j-q, j\}$. A set M of integers is *t-sparse* if the difference between any two elements of M is at least t. We say that a set of integers M is *λ-symmetric* if for every $x \in M$, it holds that $\lambda - x \in M$.

All graphs are assumed to be finite, undirected, and simple, i.e., without loops or multiple edges. Throughout the paper V_G stands for the set of vertices, and E_G for the set of edges, of a graph G.

We use standard terminology: a *path* P_n is a sequence of n consecutively adjacent vertices (its length being $n-1$); a *cycle* is a path where the first and the last vertex are adjacent as well; a graph is *connected* if each pair of vertices is joined by a path; the *distance* between two vertices is the length of a shortest path that connects them; a *tree* is a connected graph without a cycle; and a *leaf* is a vertex of degree one. For precise definitions of these terms see, e.g., the textbooks [21,16]. A *q-caterpillar* is a tree that can be constructed from a path, called the *backbone*, by adding new disjoint paths of length q, called *legs*, and merging one end of each leg with some backbone vertex. We say that a caterpillar is *symmetric*, if it allows an automorphism that reverses the backbone. Obviously, a path can be viewed as a symmetric caterpillar.

As a technical tool for proving NP-hardness results we use the following problem of finding distant representatives:

SYSTEM OF q-DISTANT REPRESENTATIVES	Sq-DR
Parameter: q	
Instance: A collection of sets $M_i, i \in [m]$ of integers.	
Question: Is there a collection of elements $u_i \in M_i$, $i \in [m]$ that pairwise differ by at least q?	

It is known that the S1-DR problem allows a polynomial time algorithm (by finding a maximum matching in a bipartite graph), while for all $q \geq 2$ the Sq-DR problem is NP-complete, even if each set M has at most three elements [12].

We conclude this section with several observations specific to the $H(p,q)$-labelings when H is a symmetric caterpillar. If f is such a labeling then the "reversed" mapping f' defined as f composed with the backbone reversing automorphism is a valid $H(p,q)$-labeling as well. Hence, if we look for a specific graph construction where only a fixed label is allowed on a certain vertex, we can not avoid symmetry of labelings f and f'. For that purpose we need a stronger concept of systems of q-distant representatives.

Lemma 1. *For any $q \geq 2$ and $t \geq q$, the* Sq-DR *problem remains* NP*-complete even when restricted to instances whose sets are of size at most 6, t-sparse and λ-symmetric for some λ.*

Proof. We extend the construction from [12] where an instance of the well known NP-complete problem 3-SATISFIABILITY (3-SAT) [13, problem L02] was transformed into an instance of the Sq-DR problem as follows:

Assume that the given Boolean formula is in the conjunctive normal form and consists of m clauses, each of size at most three, over n variables, each with one positive and two negative occurrences.

- The three literals for a variable x_i are represented by a triple $a_i, b_i, a_i + q$ such that $a_i < b_i < a_i + q$. The number b_i represents the positive literal, while $a_i, a_i + q$ the negative ones.
- Triples representing different variables are at least t apart (e.g., the elements a_i form an arithmetic progression of step $q + t$).
- The sets $M'_i, i \in [m]$ represent clauses and are composed from at most three numbers, each uniquely representing one literal of the clause.

The equivalence between the existence of a satisfying assignment and the existence of a set of q-distant representatives is straightforward (for details see [12]).

Without loss of generality assume that there are positive integers α and β such that $M'_i \subset [\alpha, \beta]$ for every $i \in [m]$ (we may assume that these bounds are arbitrarily high, but sufficiently apart).

We set $\lambda := 2\beta + t$ and construct the family of sets $M_i, i \in [m+n]$ as follows:

- for $i \in [m] : M_i := \{a, \lambda - a : a \in M'_i\}$,
- for $i \in [n] : M_{m+i} := \{b_i, \lambda - b_i\}$.

If we choose the representatives for the sets $M_{m+i}, i \in [n]$ arbitrarily, and exclude infeasible numbers from the remaining sets, then the remaining task is equivalent to the original instance $M'_i, i \in [m]$ of the Sq-DR problem.

3 Distance Labeling of Stars

We first prove NP-hardness of the (p,q)-DL problem when G belongs to a very simple class of graphs, namely to the class of stars.

Theorem 1. *For any $p \geq 0$ and $q \geq 2$, the (p,q)-DL problem is NP-complete even when the graph G is required to be a star.*

Proof. We reduce the INDEPENDENT SET (IS) problem, which for a given graph G and an integer k asks whether G has k pairwise nonadjacent vertices. The IS problem is well known to be NP-complete [13, problem GT20].

Let H_0 and k be an instance of IS. We construct a graph H such that the star $K_{1,k}$ has an $H(p,q)$-labeling if and only if H_0 has k independent vertices.

The construction of H goes in three steps:

Firstly, if $q = 2$, then we simply let $H_1 := H_0$ and $M := V_{H_0}$. Otherwise, i.e., for $q \geq 2$, we replace each edge of H_0 by a path of length $q-1$ to obtain H_1. Now let M be the set of the middle points of the replacement paths (i.e., M is of size $|E_{H_0}|$ when q is odd; otherwise M is twice bigger).

For the second step we first prepare a path of length $\max\{0, \lceil \frac{2p-q-1}{2} \rceil\}$ (observe that this path consists of a single vertex when $2p \leq q+1$). We make one its its ends adjacent to all vertices in the set M. We denote the other end of the path by w.

Finally, if q is odd, we insert new edges to make all vertices in M pairwise adjacent, i.e., the set M now induces a clique. This concludes the construction of the graph H.

The properties of H can be summarized as follows:

- If two vertices were adjacent in H_0, then they have distance $q-1$ in H. Analogously, they have distance q if they were non-adjacent.
- Every original vertex is at distance at least p from w, and this bound is attained whenever $2p \geq q-1$.
- If $p \leq q$ then every newly added vertex except w is at distance less than q from any other vertex of H.
- If $p \geq q$ then every newly added vertex is at distance less than p from w.

Straightforwardly, if H_0 has an independent set S of size k, then we map the center of $K_{1,k}$ onto w and the leaves of $K_{1,k}$ bijectively onto S. This yields a valid $H(p,q)$-labeling of $K_{1,k}$.

For the opposite implication assume that $K_{1,k}$ has an $H(p,q)$-labeling f. We distinguish three cases:

- When $p < q$, then the images of the leaves of $K_{1,k}$ are pairwise at distance q in H. Hence, by the properties of H, these are k original vertices that form an independent set in H_0.
- If $p = q$, then the q-distant vertices of H are some nonadjacent original vertices together with the vertex w. As the image of the center of $K_{1,k}$ belongs into this set as well, H_0 has at least k independent vertices.
- If $p > q$, then w is the image of the center of $K_{1,k}$ (unless $k = 1$, but then the problem is trivial). Analogously as in the previous cases, images of the leaves of $K_{1,k}$ form an independent set of H_0.

4 Distance Labeling between Two Trees

In this section we show that the DL problem is NP-complete for any $q \geq 2$ and $p \geq 2q + 1$ even when both graphs G and H are required to be trees. Before we state the theorem, we describe the target graph H and explore its properties.

Let p and q be given, such that $q \geq 2$ and $p \geq 2q+1$. Assume that $l > 2(p-q)$ and for $i \in [l]$ define $m_i := l^3 + il$. For convenience we also let $m_i := m_{2l-i} = l^3 + (2l - i)l$ for $i \in [l + 1, 2l - 1]$.

We construct a graph H_l as follows: We start with a path of length $2l - 2$ on vertices $v_1, \ldots, v_{2l-1}$, called the backbone vertices. For each vertex $v_i, i \in [2l-1]$, we prepare m_i paths of length q and identify one end of each of these m_i paths with the vertex v_i. By symmetry, every vertex v_{2l-i} has the same number of pending q-paths as the vertex v_i. Observe that the resulting graph depicted in Fig. 1 is a q-caterpillar.

For $i \in [2l - 1]$ and $j \in [m_i]$, let $u_{i,j}$ denote the final vertex of the j-th path hanging from the vertex v_i.

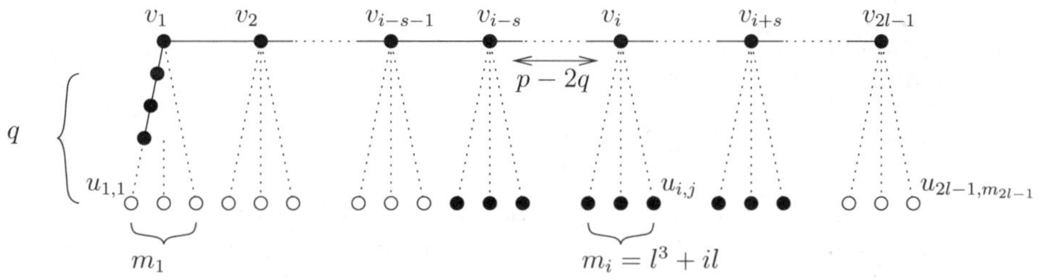

Fig. 1. Construction of the target tree H_l. White vertices define n_i as well as a lower bound on $r(u_{i,j})$.

Observe that the total number of leaves in H_l is $2l^4$.

We define $s := p - 2q$ to shorten some expressions.

For $i \in [2l - 1]$, let n_i be the number of leaves of H_l at distance at least $p - q$ from v_i, i.e.,

$$n_i := \sum_{j \in S_i} m_i = \begin{cases} 2l^4 - (s+i)l^3 - (s+2i)il + \frac{ls(s+1)}{2} & \text{if } i \in [s-1], \\ 2l^4 - (2s+1)l^3 - (2s+1)il & \text{if } i \in [s, l-s], \text{ and} \\ 2l^4 - (2s+1)l^3 - (2s+1)il + \\ \quad + (s+i-l)(s+i-l+1)l & \text{if } i \in [l-s+1, l], \end{cases}$$

where $S_i := [2l+1] \setminus [i-s, i+s]$. By symmetry, $n_i := n_{2l-i}$ for $i \in [l+1, 2l-1]$. Observe that the sequence $n_1, \ldots, n_{l-s}$ is decreasing.

For a vertex $u \in V_{H_l}$, we further define $r(u)$ to be the maximum size of a set of vertices of H_l that are pairwise at least q apart, and that are also at distance at least p from u. In other words, $r(u)$ is an upper bound on the degree of a vertex which is mapped onto u in an $H_l(p, q)$-labeling.

We claim that for the leaves $u_{i,j}$ with $i \in [l], j \in [m_i]$ the $r(u_{i,j})$ can be bounded by

$$n_i \leq r(u_{i,j}) \leq n_i + \frac{2l-p}{q},$$

since the desired set can be composed from the n_i leaves that are at distance at least $p-q$ from v_i together with suitable backbone vertices. Consequently, $r(u_{i,j}) < n_{i-1}$ for $i \in [2, l-s]$.

Finally, observe that if u is a non-leaf vertex of H, then $r(u) < n_{l-s}$, since with every step away from a leaf the size of the set of p distant vertices decreases by the factor of $\Omega(l^3)$.

By the properties of the values n_i and $r(u)$ we get that:

Lemma 2. *For given p, q and l such that $p \geq 2q+1$ and $l > 2(p-q)+1$, let T be a tree of three levels such that for every $i \in [l-s]$ and $j \in [m_i]$, the root y of T has two children $x_{i,j}, x_{2l-i,j}$, both of degree n_i.*

Every $H_l(p,q)$-labeling f of T satisfies that $f(y) \in \{u_{l,j} \mid j \in [m_l]\}$, and

$$\forall i \in [l-s] : \{f(x_{i,j}), f(x_{2l-i,j}) \mid j \in [m_i]\} = \{u_{i,j}, u_{2l-i,j} \mid j \in [m_i]\}.$$

In addition, T has an $H_l(p,q)$-labeling such that the leaves of T are mapped onto the leaves of H_l.

Proof. Assume by induction that all vertices $x_{k,j}$ and $x_{2l-k,j}$ with $k < i$ are mapped onto the set $W = \{u_{k,j}, u_{2l-k,j} \mid k < i, j \in [m_k]\}$.

Then vertices $x_{i,j}$, $x_{2l-i,j}$ with $j \in [m_k]$ must be mapped onto vertices that are at least q apart from W, i.e., on the backbone vertices or inside a path under some $v_{i'}$ with $i' \in [i, 2l-i]$. Among those vertices only leaves $u_{i,j}$ and $u_{2l-i,j}$ satisfy $r(u_{i,j}) = r(u_{2l-i,j}) \geq n_i$ and can be used as images for $x_{i,j}$, $x_{2l-i,j}$.

When the labels of all $x_{i,j}$ are fixed, the root must be mapped onto a vertex that is at distance at least p from all $u_{i,j}$ with $i \leq l-p+2q$ or $i \geq l+p-2q$. The only such vertices are $u_{l,j}$ with $j \in [m_l]$.

If the first two levels of T are partially labeled as described above, then the children of $x_{i,j}$ can be labeled by vertices $u_{k,j}$ with $|k-i| > s$, only the label of the root y must be avoided. This provides a valid $H_l(p,q)$-labeling of T.

For $i \in [l-s]$, let T_i denote the tree T rearranged such that its root is one of the children of $x_{i,1}$.

We are ready to prove the main theorem of this section.

Theorem 2. *For any $q \geq 2$ and $p \geq 2q+1$, the (p,q)-DL problem is NP-complete, even if G is a tree and H is a symmetric q-caterpillar.*

Proof. We reduce the 3-SAT problem and extend the reduction to the Sq-DR problem exposed in Lemma 1.

For a formula with n variables, we set $\alpha := p-q+1$, $\beta := 2p-3q$, $t := p$, $l := \alpha + (n-1)p + nq + \beta$, and $\lambda := 2l$. According to these parameters we build the graph H_l and transform the given formula into a collection of t-sparse sets $M_i, i \in [m+n]$.

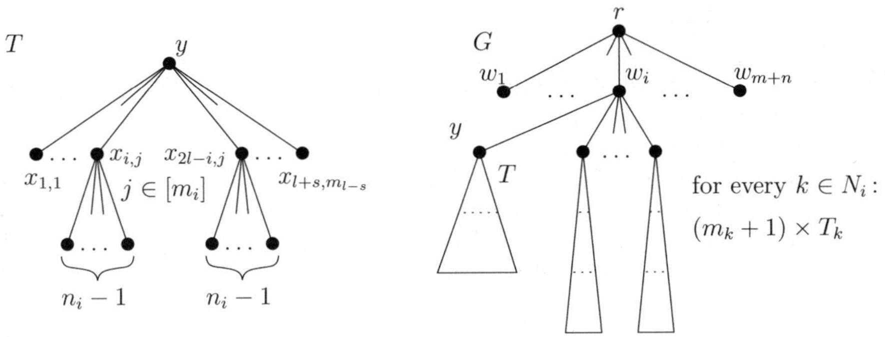

Fig. 2. Construction of trees T and G

We construct the tree G of six levels as follows: the root r has children $w_i, i \in [m+n]$, representing sets M_i.

For each $i \in [m+n]$, let N_i be the set containing all numbers of $[l-s]$ that are at least $p-q$ apart from any number of M_i. Formally, $N_i := [l-s] \setminus \bigcup_{j \in M_i} [j-p+q+1, j+p-q-1]$.

For each $k \in N_i$, we take $m_k + 1$ copies of the tree T_k and add $m_k + 1$ edges between the roots of these trees and the vertex w_i. Finally, we insert a copy of the tree T and insert a new edge so that the root of this T is also a child of w_i.

We repeat the above construction for all $i \in [m+n]$ to obtain the desired graph G. (See Fig. 2.)

We claim that if f is an arbitrary $H_l(p,q)$-labeling of G then every vertex w_i is mapped on some vertex v_j with $j \in M_i$.

The child of w_i, which is the root y, maps on some $u_{l,j}$. Also the children of y map onto all leaves of form $u_{i,j}, i \in [l-p+q], j \in [m_i]$. Hence, the image of w_i is one of the backbone vertices v_i with $i \in [l-p+q] \cup [l+p-q, 2l-1]$.

On the other hand, for any $k \in N_i$ the image of w_i is at least p apart from some $u_{k,l}$ as well as from some $u_{2l-k,l'}$ with $l, l' \in [m_k]$. This follows from the fact that w_i has in T_k more children than there are the leaves under v_k (or under v_{2l-k}), so both k and $l-k$ appear as the first index of the leaf which is the image of a child of w_i. This proves the claim.

Therefore, the existence of such mapping f yields a valid solution of the original 3-SAT and Sq-DR problems.

In the opposite direction observe that any valid solution of the Sq-DR problem transforms naturally to the mapping on vertices $w_i, i \in [m+n]$. We extend this partial mapping onto the remaining vertices of G such that the root r is mapped onto $u_{1,1}$, all vertices y onto $u_{l,1}$. The copies of trees T are labeled as described in Lemma 2.

For every w_i, we map its children in copies of T_k onto distinct vertices of the set $\{u_{k,j}, u_{2l-k,j} \mid j \in [m_k]\} \setminus \{u_{1,1}, u_{l,1}\}$. Then we extend the labeling onto the entire copy of each T_k like in Lemma 2 without causing any conflicts with other labels. In particular, every child of w_i in T_k is of degree n_k+1 and its children are labeled by leaves of H_l, while its parent (the vertex w_i) by a backbone vertex.

5 Conclusion

In this paper we have studied the computational complexity of the $H(p,q)$-labeling problem when both the input graph G and the label space graph H are trees. This could hopefully pave the way to the solution of the $L(p,q)$-labeling of trees (with both p and q fixed), which is the most interesting open problem in the area of computational complexity of distance constrained labeling problems. Another persistent open problem is the complexity of the $L(2,1)$-labeling problem for graphs of bounded pathwidth.

References

1. Calamoneri, T.: The $L(h,k)$-labeling problem: A survey and annotated bibliography. Computer Journal 49(5), 585–608 (2006)
2. Chang, G.J., Ke, W.-T., Liu, D.D.-F., Yeh, R.K.: On $l(d,1)$-labellings of graphs. Discrete Mathematics 3(1), 57–66 (2000)
3. Chang, G.J., Kuo, D.: The $L(2,1)$-labeling problem on graphs. SIAM Journal of Discrete Mathematics 9(2), 309–316 (1996)
4. Courcelle, B.: The monadic second-order logic of graphs. I: Recognizable sets of finite graphs. Inf. Comput. 85(1), 12–75 (1990)
5. Fiala, J., Golovach, P.A., Kratochvíl, J.: Distance Constrained Labelings of Graphs of Bounded Treewidth. In: Caires, L., Italiano, G.F., Monteiro, L., Palamidessi, C., Yung, M. (eds.) ICALP 2005. LNCS, vol. 3580, pp. 360–372. Springer, Heidelberg (2005)
6. Fiala, J., Kloks, T., Kratochvíl, J.: Fixed-Parameter Complexity of λ-Labelings. In: Widmayer, P., Neyer, G., Eidenbenz, S. (eds.) WG 1999. LNCS, vol. 1665, Springer, Heidelberg (1999)
7. Fiala, J., Kratochvíl, J.: Complexity of Partial Covers of Graphs. In: Eades, P., Takaoka, T. (eds.) ISAAC 2001. LNCS, vol. 2223, pp. 537–549. Springer, Heidelberg (2001)
8. Fiala, J., Kratochvíl, J.: Partial covers of graphs. Discussiones Mathematicae Graph Theory 22, 89–99 (2002)
9. Fiala, J., Kratochvíl, J., Kloks, T.: Fixed-parameter complexity of λ-labelings. Discrete Applied Mathematics 113(1), 59–72 (2001)
10. Fiala, J., Kratochvíl, J., Pór, A.: On the computational complexity of partial covers of theta graphs. Electronic Notes in Discrete Mathematics 19, 79–85 (2005)
11. Fiala, J., Kratochvíl, J., Proskurowski, A.: Distance Constrained Labeling of Precolored Trees. In: Restivo, A., Ronchi Della Rocca, S., Roversi, L. (eds.) ICTCS 2001. LNCS, vol. 2202, pp. 285–292. Springer, Heidelberg (2001)
12. Fiala, J., Kratochvíl, J., Proskurowski, A.: Systems of distant representatives. Discrete Applied Mathematics 145(2), 306–316 (2005)
13. Garey, M.R., Johnson, D.S.: Computers and Intractability. W. H. Freeman and Co., New York (1979)
14. Golovach, P.A.: Systems of pair of q-distant representatives and graph colorings (in Russian). Zap. nau. sem. POMI 293, 5–25 (2002)
15. Griggs, J.R., Yeh, R.K.: Labelling graphs with a condition at distance 2. SIAM Journal of Discrete Mathematics 5(4), 586–595 (1992)
16. Harary, F.: Graph theory. Addison-Wesley Series in Mathematics IX (1969)

17. Leese, R.A.: A fresh look at channel assignment in uniform networks. In: EMC 1997 Symposium, Zurich, pp. 127–130 (1997)
18. Leese, R.A.: Radio spectrum: a raw material for the telecommunications industry. In: 10th Conference of the European Consortium for Mathematics in Industry, Goteborg (1998)
19. Leese, R.A., Noble, S.D.: Cyclic labellings with constraints at two distances. Electr. J. Comb. 11(1) (2004)
20. Liu, D.D.-F., Zhu, X.: Circulant distant two labeling and circular chromatic number. Ars Combinatoria 69, 177–183 (2003)
21. Matoušek, J., Nešetřil, J.: Invitation to Discrete Mathematics. Oxford University Press, Oxford (1998)
22. McDiarmid, C.: Discrete mathematics and radio channel assignment. In: Recent advances in algorithms and combinatorics. CMS Books Math./Ouvrages Math. SMC, vol. 11, pp. 27–63. Springer, Heidelberg (2003)
23. Yeh, R.K.: A survey on labeling graphs with a condition at distance two. Discrete Mathematics 306(12), 1217–1231 (2006)

A Characterization of NC^k by First Order Functional Programs

Jean-Yves Marion and Romain Péchoux

Loria-INPL, École Nationale Supérieure des Mines de Nancy, B.P. 239, 54506
Vandœuvre-lès-Nancy Cedex, France
Jean-Yves.Marion@loria.fr, Romain.Pechoux@loria.fr

Abstract. This paper is part of a research on static analysis in order to predict program resources and belongs to the implicit computational complexity line of research. It presents intrinsic characterizations of the classes of functions, which are computable in NC^k, that is by a uniform, poly-logarithmic depth and polynomial size family of circuits, using first order functional programs. Our characterizations are new in terms of first order functional programming language and extend the characterization of NC^1 in [9]. These characterizations are obtained using a complexity measure, the sup-interpretation, which gives upper bounds on the size of computed values and captures a lot of program schemas.

1 Introduction

Our work is related to machine independent characterizations of functional complexity classes initiated by Cobham's work [14] and studied by the *Implicit computational complexity (ICC)* community, including safe recursion of Bellantoni and Cook [4], data tiering of Leivant [23], linear type disciplines by Girard et al. [17,18], Lafont [22], Baillot-Mogbil [2], Gaboardi-Ronchi Della Rocca [16] and Hofmann [19] and studies on the complexity of imperative programs using matrix algebra by Kristiansen-Jones [21] and Niggl-Wunderlich [28]. Traditional results of the ICC focus on capturing all functions of a complexity class and we should call this approach extensional whereas our approach, which tries to characterize a class of programs, which represents functions in some complexity classes, as large as possible, is rather intensional. In other words, we try to delineate a broad class of programs using a certain amount of resources.

Our approach relies on methods combining term rewriting systems and interpretation methods for proving complexity upper bounds by static analysis. It consists in assigning a function from real numbers to real numbers to some symbols of a program. Such an assignment is called a sup-interpretation if it satisfies some specific semantics properties introduced in [27]. Basically, a sup-interpretation provides upper bounds on the size of computed values. Sup-interpretation is a generalization of the notion of quasi-interpretation of [10]. The problem of finding a quasi-interpretation or sup-interpretation of a given program, called synthesis problem, is crucial for potential applications of the

M. Agrawal et al. (Eds.): TAMC 2008, LNCS 4978, pp. 136–147, 2008.

method. It consists in automatically finding an interpretation of a program in order to determine an upper bound on its complexity. It was demonstrated in [1,8] that the synthesis problem is decidable in exponential time for small classes of polynomials. Quasi-interpretations and sup-interpretations have already been used to capture the sets of functions computable in polynomial time and space [26,7] and capture a broad class of algorithms, including greedy algorithms and dynamic programming algorithms. Consequently, it is a challenge to study whether this approach can be adapted to characterize small parallel complexity classes. Parallel algorithms are difficult to design. Employing the sup-interpretation method leads to delineate efficient parallel programs amenable to circuit computing. Designing parallel implementations of first order functional programs with interpretation methods for proving complexity bounds, might be thus viable in the near future.

A circuit C_n is a directed acyclic graph built up from Boolean gates *And*, *Or* and *Not*. Each gate has an in-degree less or equal to two and an out-degree equal to one. A circuit has n input nodes and $g(n)$ output nodes, where $g(n) = O(n^c)$, for some constant $c \geq 1$. Thus, a circuit C_n computes a function $f_n : \{0,1\}^n \rightarrow \{0,1\}^{g(n)}$. A family of circuits is a sequence of circuits $C = (C_n)_n$, which computes a family of finite functions $(f_n)_n$ over $\{0,1\}^*$. A function f is computed by a family of circuits $(C_n)_n$ if the restriction of f to inputs of size n is computed by C_n. A uniformity condition ensures that there is a procedure which, given n, produces a description of the circuit C_n. Such a condition is introduced to ensure that a family of circuits computes a reasonable function. All along, we shall consider U_{E^*}-uniform family of circuits defined in [29]. The complexity of a circuit depends on its depth (the longest path from an input to an output gate) and its size (the number of gates). The class NC^k is the class of functions computable by a U_{E^*}-uniform family of circuits of size bounded by $O(n^d)$, for some constant d, and depth bounded by $O(\log^k(n))$. Intuitively, it corresponds to the class of functions computed in poly-logarithmic time with a polynomial number of processors. Following [3], the main motivation in the introduction of such classes was the search for separation results: "NC^1 is the at the frontier where we obtain interesting separation results". NC^1 contains binary addition, substraction, prefix sum of associative operators. Buss [11] has demonstrated that the evaluation of boolean formulas is a complete problem for NC^1. A lot of natural algorithms belong to the distinct levels of the NC^k hierarchy. In particular, the reachability problem in a graph or the search for a minimum covering tree in a graph are two problems in NC^2.

In this paper, we define a restricted class of first order functional programs, called *fraternal and arboreal* programs, using the notion of sup-interpretation of [27]. We demonstrate that functions, which are computable by these programs at some rank k, are exactly the functions computed in NC^k. This result generalizes the characterization of NC^1 established in [9]. To our knowledge, these are the first results, which connect small parallel complexity classes and first order functional programs.

2 First Order Functional Programs

2.1 Syntax of Programs

We define a generic first order functional programming language. The vocabulary $\Sigma = \langle \mathit{Var}, \mathit{Cns}, \mathit{Op}, \mathit{Fct}\rangle$ is composed of four disjoint sets of symbols. The arity of a symbol is the number n of arguments that it takes. A program $\mathbf{p}$ consists in a vocabulary and a set of rules $\mathcal{R}$ defined by the following grammar:

$$
\begin{array}{llll}
\texttt{(Values)} & \mathcal{T}(\mathit{Cns}) \ni \mathbf{v} & ::= & \mathbf{c} \mid \mathbf{c}(\mathbf{v}_1, \cdots, \mathbf{v}_n) \\
\texttt{(Patterns)} & \mathcal{T}(\mathit{Var}, \mathit{Cns}) \ni p & ::= & \mathbf{c} \mid x \mid \mathbf{c}(p_1, \cdots, p_n) \\
\texttt{(Expressions)} & \mathcal{T}(\mathit{Var}, \mathit{Cns}, \mathit{Op}, \mathit{Fct}) \ni e & ::= & \mathbf{c} \mid x \mid \mathbf{c}(e_1, \cdots, e_n) \\
 & & & \mid \mathbf{op}(e_1, \cdots, e_n) \mid \mathtt{f}(e_1, \cdots, e_n) \\
\texttt{(Rules)} & \mathcal{R} \ni r & ::= & \mathtt{f}(p_1, \cdots, p_n) \to e
\end{array}
$$

where $x \in \mathit{Var}$ is a variable, $\mathbf{c} \in \mathit{Cns}$ is a constructor symbol, $\mathbf{op} \in \mathit{Op}$ is an operator, $\mathtt{f} \in \mathit{Fct}$ is a function symbol, $p_1, \cdots, p_n \in \mathcal{T}(\mathit{Var}, \mathit{Cns})$ are patterns and $e_1, \cdots, e_n \in \mathcal{T}(\mathit{Var}, \mathit{Cns}, \mathit{Op}, \mathit{Fct})$ are expressions. The program's main function symbol is the first function symbol in the program's list of rules.

Throughout the paper, we only consider orthogonal programs having disjoint and linear rule patterns. Consequently, each program is confluent [20]. We will use the notation $\overline{e}$ to represent a sequence of expressions, that is $\overline{e} = e_1, \ldots, e_n$.

2.2 Semantics

The domain of computation of a program $\mathbf{p}$ is the constructor algebra $\texttt{Values} = \mathcal{T}(\mathit{Cns})$. Set $\texttt{Values}^* = \texttt{Values} \cup \{\mathbf{Err}\}$, where $\mathbf{Err}$ is a special symbol associated to runtime errors. An operator $\mathbf{op}$ of arity n is interpreted by a function $[\![\mathbf{op}]\!]$ from $\texttt{Values}^n$ to $\texttt{Values}^*$. Operators are essentially basic partial functions like destructors or characteristic functions of predicates like $=$.

Set $\texttt{Values}^{\#} = \texttt{Values} \cup \{\mathbf{Err}, \bot\}$, where $\bot$ means that a program is non-terminating. Given a program $\mathbf{p}$ of vocabulary $\langle \mathit{Var}, \mathit{Cns}, \mathit{Op}, \mathit{Fct}\rangle$ and an expression $e \in \mathcal{T}(\mathit{Cns}, \mathit{Op}, \mathit{Fct})$, the computation of e, noted $[\![e]\!]$, is defined by $[\![e]\!] = \mathbf{w}$ iff $e \stackrel{*}{\to} \mathbf{w}$ and $\mathbf{w} \in \texttt{Values}^*$, otherwise $[\![e]\!] = \bot$, where $\stackrel{*}{\to}$ is the reflexive and transitive closure of the rewriting relation $\to$ induced by the rules of $\mathcal{R}$. By definition, if no rule is applicable, then an error occurs and $[\![e]\!] = \mathbf{Err}$. A program of main function symbol $\mathtt{f}$ computes a partial function $\phi : \texttt{Values}^n \to \texttt{Values}^*$ defined by $\forall \mathbf{u}_1, \cdots, \mathbf{u}_n \in \texttt{Values}$, $\phi(\mathbf{u}_1, \cdots, \mathbf{u}_n) = w$ iff $[\![\mathtt{f}(\mathbf{u}_1, \cdots, \mathbf{u}_n)]\!] = w$.

Definition 1 (Size). *The size of an expression e is defined by $|e| = 0$, if e is a 0-arity symbol, and $|b(e_1, \cdots, e_n)| = \sum_{i \in \{1,\ldots,n\}} |e_i| + 1$, if $e = b(e_1, \cdots, e_n)$.*

Example 1. Consider the following program which computes the logarithm function over binary numbers using the constructor symbols $\{\mathbf{0}, \mathbf{1}, \epsilon\}$:

$$
\begin{aligned}
\mathtt{f}(x) &\to \mathbf{rev}(\log(x)) \\
\log(\mathbf{i}(x)) &\to \mathbf{if}(\mathbf{Msp}(\mathbf{Fh}(\mathbf{i}(x)), \mathbf{Sh}(\mathbf{i}(x))), \mathbf{0}(\log(\mathbf{Fh}(\mathbf{i}(x)))), \mathbf{1}(\log(\mathbf{Fh}(\mathbf{i}(x))))) \\
\log(\epsilon) &\to \epsilon
\end{aligned}
$$

where $\mathbf{if}(u, v, w)$ is an operator, which outputs v or w depending on whether u is equal to $\mathbf{1}(\epsilon)$ or $\mathbf{0}(\epsilon)$, **rev** is an operator, which reverses a binary value given as input, **Msp** is an operator, which returns $\mathbf{1}(\epsilon)$ if the leftmost $|x| \ominus |y|$ bits of x (where $\forall x, y \in \mathbb{N}, x \ominus y = 0$ if $y > x$ and $x - y$ otherwise) are equal to the empty word ϵ and returns $\mathbf{0}(\epsilon)$ otherwise, **Fh** is an operator, which outputs the leftmost $\lfloor |x|/2 \rfloor$ bits of x, and **Sh** is an operator, which outputs the rightmost $\lceil |x|/2 \rceil$ bits of x.

The algorithm tests whether the number $\lfloor |x|/2 \rfloor$ of leftmost bits is equal to the number $\lceil |x|/2 \rceil$ of rightmost bits in the input x using the operator **Msp**. In this case, the last digit of the logarithm is a 0, otherwise it is a 1. Finally, the computation is performed by applying a recursive call over half of the input digits and the result is obtained by reversing the output, using the operator **rev**. For any binary value **v**, we have $[\![\log(\mathbf{v})]\!] = \mathbf{u}$, where **u** is the value representing the binary logarithm of the input value **v**. For example, $[\![\log(\mathbf{1}(\mathbf{0}(\mathbf{0}(\mathbf{0}(\epsilon)))))]\!] = \mathbf{1}(\mathbf{0}(\mathbf{0}(\epsilon)))$.

2.3 Call-Tree

We now describe the notion of call-tree which is a representation of a program state transition sequences induced by the rewrite relation $\rightarrow$ using a call-by-value strategy. In this paper, the notion of call-tree allows to control the successive function calls corresponding to a recursive rule. First, we define the notions of context and substitution. A *context* is an expression $\mathsf{C}[\diamond_1, \cdots, \diamond_r]$ containing one occurrence of each $\diamond_i$, with $\diamond_i$ new variables which do not appear in Σ. A substitution is a finite mapping from variables to $\mathcal{T}(Var, Cns, Op, Fct)$. The substitution of each $\diamond_i$ by an expression d_i in the context $\mathsf{C}[\diamond_1, \cdots, \diamond_r]$ is noted $\mathsf{C}[d_1, \cdots, d_r]$. A ground substitution σ is a mapping from variables to `Values`. Throughout the paper, we use the symbol σ and the word "substitution" to denote a ground substitution. The application of a substitution σ to an expression (or a sequence of expressions) e is noted $e\sigma$.

Definition 2. *Suppose that we have a program* $\mathbf{p}$. *A* state $\langle \mathtt{f}, \boldsymbol{u}_1, \cdots, \boldsymbol{u}_n \rangle$ *of* $\mathbf{p}$ *is a tuple where* $\mathtt{f}$ *is a function symbol of arity* n *and* $\boldsymbol{u}_1, \cdots, \boldsymbol{u}_n$ *are values of* `Values`*.

There is state transition, *noted* $\eta_1 \rightsquigarrow \eta_2$, *between two states* $\eta_1 = \langle \mathtt{f}, \boldsymbol{u}_1, \ldots, \boldsymbol{u}_n \rangle$ *and* $\eta_2 = \langle \mathtt{g}, \boldsymbol{v}_1, \cdots, \boldsymbol{v}_m \rangle$ *if there are a rule* $\mathtt{f}(p_1, \cdots, p_n) \rightarrow e$ *of* $\mathbf{p}$, *a substitution* σ, *a context* $\mathsf{C}[-]$ *and expressions* $e_1, \cdots, e_m$ *such that* $\forall i \in \{1, n\}$ $p_i\sigma = \boldsymbol{u}_i$, $\forall j \in \{1, m\}$, $[\![e_j\sigma]\!] = \boldsymbol{v}_j$ *and* $e = \mathsf{C}[\mathsf{g}(e_1, \cdots, e_m)]$. *We write* $\overset{*}{\rightsquigarrow}$ *to denote the reflexive and transitive closure of* $\rightsquigarrow$. *A call-tree of* $\mathbf{p}$ *of root* $\langle \mathtt{f}, \boldsymbol{u}_1, \cdots, \boldsymbol{u}_n \rangle$ *is the following tree:*

- *the root is the node labeled by the state* $\langle \mathtt{f}, \boldsymbol{u}_1, \cdots, \boldsymbol{u}_n \rangle$.
- *the nodes are labeled by states of* $\{\eta \mid \langle \mathtt{f}, \boldsymbol{u}_1, \cdots, \boldsymbol{u}_n \rangle \overset{*}{\rightsquigarrow} \eta\}$,
- *there is an edge between two nodes* η_1 *and* η_2 *if there is a transition between both states which label the nodes (i.e.* $\eta_1 \rightsquigarrow \eta_2$*).*

A branch of the call-tree is a sequence of states of the call-tree $\eta_1, \cdots, \eta_k$ *such that* $\eta_1 \rightsquigarrow \eta_2 \ldots \rightsquigarrow \eta_{k-1} \rightsquigarrow \eta_k$. *Given a branch B of a call-tree, the depth of the branch* $\mathtt{depth}(B)$ *is the number of states in the branch, i.e. if* $B = \eta_1, \ldots, \eta_{i-1}, \eta_i$, *then* $\mathtt{depth}(B) = i$.

Notice that a call-tree may be infinite if it corresponds a non-terminating program.

2.4 Fraternity

The fraternity is the main syntactic notion, we use in order to restrict the computational power of considered programs.

Definition 3. *Given a program* $\mathbf{p}$, *the* precedence $\geq_{Fct}$ *is defined on function symbols by* $\mathtt{f} \geq_{Fct} \mathtt{g}$ *if there is a rule* $\mathtt{f}(\overline{p}) \to \mathsf{C}[\mathtt{g}(\overline{e})]$ *in* $\mathbf{p}$. *The reflexive and transitive closure of* $\geq_{Fct}$ *is also noted* $\geq_{Fct}$. *Define* $\approx_{Fct}$ *by* $\mathtt{f} \approx_{Fct} \mathtt{g}$ *iff* $\mathtt{f} \geq_{Fct} \mathtt{g}$ *and* $\mathtt{g} \geq_{Fct} \mathtt{f}$ *and define* $>_{Fct}$ *by* $\mathtt{f} >_{Fct} \mathtt{g}$ *iff* $\mathtt{f} \geq_{Fct} \mathtt{g}$ *and not* $\mathtt{g} \geq_{Fct} \mathtt{f}$. *We extend the precedence to operators and constructor symbols by* $\forall \mathtt{f} \in Fct$, $\forall b \in Cns \cup Op$, $\mathtt{f} >_{Fct} b$.

Definition 4. *Given a program* $\mathbf{p}$, *an expression* $\mathsf{C}[\mathtt{g}_1(\overline{e_1}), \ldots, \mathtt{g}_r(\overline{e_r})]$ *is a* fraternity *activated by* $\mathtt{f}(p_1, \cdots, p_n)$ *if:*

1. $\mathtt{f}(p_1, \cdots, p_n) \to \mathsf{C}[\mathtt{g}_1(\overline{e_1}), \ldots, \mathtt{g}_r(\overline{e_r})]$ *is a rule of* $\mathbf{p}$,
2. *For each* $i \in \{1, r\}$, $\mathtt{g}_i \approx_{Fct} \mathtt{f}$,
3. *For every symbol* b *in the context* $\mathsf{C}[\diamond_1, \cdots, \diamond_r]$, $\mathtt{f} >_{Fct} b$.

Notice that a fraternity corresponds to a recursive rule.

Example 2. $\mathbf{if}(\mathbf{Msp}(\mathbf{Fh}(\mathbf{i}(x)), \mathbf{Sh}(\mathbf{i}(x))), \mathbf{0}(\log(\mathbf{Fh}(\mathbf{i}(x)))), \mathbf{1}(\log(\mathbf{Fh}(\mathbf{i}(x)))))$ is the only fraternity in the program of example 1. It is activated by $\log(\mathbf{i}(x))$ by taking $\mathsf{C}[\diamond_1, \diamond_2] = \mathbf{if}(\mathbf{Msp}(\mathbf{Fh}(\mathbf{i}(x)), \mathbf{Sh}(\mathbf{i}(x))), \mathbf{0}(\diamond_1), \mathbf{1}(\diamond_2))$ since $\log \approx_{Fct} \log$.

3 Sup-interpretations

3.1 Monotonic, Polynomial and Additive Partial Assignments

Definition 5. *A partial assignment* $\mathcal{I}$ *is a partial mapping from the vocabulary* Σ *such that, for each symbol* b *of arity* n *in the domain of* $\mathcal{I}$, *it yields a partial function* $\mathcal{I}(b) : (\mathbb{R}^+)^n \longmapsto \mathbb{R}^+$, *where* $\mathbb{R}^+$ *is the set of non-negative real numbers. The domain of a partial assignment* $\mathcal{I}$ *is noted* $dom(\mathcal{I})$ *and satisfies* $Cns \cup Op \subseteq dom(\mathcal{I})$.

A partial assignment $\mathcal{I}$ *is monotonic if for each symbol* $b \in dom(\mathcal{I})$, *we have* $\forall i \in \{1, \ldots, n\}$, $\forall X_i, Y_i \in \mathbb{R}^+$, $X_i \geq Y_i \Rightarrow \mathcal{I}(b)(X_1, \cdots, X_n) \geq \mathcal{I}(b)(Y_1, \cdots, Y_n)$.

A partial assignment $\mathcal{I}$ *is* polynomial *if for each symbol* b *of* $dom(\mathcal{I})$, $\mathcal{I}(b)$ *is a max-polynomial function ranging over* $\mathbb{R}^+$. *That is,* $\mathcal{I}(b) = \max(P_1, \ldots, P_k)$, *with* P_j *polynomials.*

A partial assignment is additive *if the assignment of each constructor symbol* $\mathbf{c}$ *of arity* n *is of the shape* $\mathcal{I}(\mathbf{c})(X_1, \cdots, X_n) = \sum_{i=1}^{n} X_i + \alpha_{\mathbf{c}}$, *with* $\alpha_{\mathbf{c}} \geq 1$, *whether* $n > 0$, *and* $\mathcal{I}(\mathbf{c}) = 0$ *otherwise.*

If each function symbol of a given expression e having m variables $x_1, \cdots, x_m$ belongs to dom($\mathcal{I}$) then, given m fresh variables $X_1, \cdots, X_m$ ranging over $\mathbb{R}^+$, we define the homomorphic extension $\mathcal{I}^*(e)$ of the assignment $\mathcal{I}$ inductively by:

1. If x_i is a variable of *Var*, then $\mathcal{I}^*(x_i) = X_i$
2. If b is a symbol of Σ of arity 0, then $\mathcal{I}^*(b) = \mathcal{I}(b)$.
3. If b is a symbol of arity $n > 0$ and $e_1, \cdots, e_n$ are expressions, then

$$\mathcal{I}^*(b(e_1, \cdots, e_n)) = \mathcal{I}(b)(\mathcal{I}^*(e_1), \ldots, \mathcal{I}^*(e_n))$$

Given a sequence $\overline{e} = e_1, \cdots, e_n$, we will sometimes use the notation $\mathcal{I}^*(\overline{e})$ to denote $\mathcal{I}^*(e_1), \ldots, \mathcal{I}^*(e_n)$.

3.2 Sup-interpretations

Definition 6. *A sup-interpretation is a partial assignment θ which satisfies the three conditions below:*

1. *θ is a monotonic assignment.*
2. *For each $v \in$ `Values`, $\theta^*(v) \geq |v|$*
3. *For each symbol $b \in dom(\theta)$ of arity n and for each value $v_1, \ldots, v_n$ of `Values`, if $[\![b(v_1, \ldots, v_n)]\!] \in$ `Values`, then*

$$\theta^*(b(v_1, \ldots, v_n)) \geq \theta^*([\![b(v_1, \ldots, v_n)]\!])$$

A sup-interpretation is additive (respectively polynomial) if it is an additive (respectively polynomial) assignment.

Notice that if a sup-interpretation is an additive assignment then the second condition of the above definition is always satisfied. Intuitively, the sup-interpretation is a special program interpretation which bounds from above the output size of the function denoted by the program, as demonstrated in the following lemma:

Lemma 1 ([27]). *Given a sup-interpretation θ and an expression e defined over $dom(\theta)$, if $[\![e]\!] \in$ `Values` then we have $|[\![e]\!]| \leq \theta^*([\![e]\!]) \leq \theta^*(e)$*

Example 3. Consider the program of example 1. Define the additive assignment θ by $\theta(\mathbf{1})(X) = \theta(\mathbf{0})(X) = X + 1$, $\theta(\epsilon) = 0$, $\theta(\mathbf{Msp})(X, Y) = X \ominus Y = \max(X - Y, 0)$, $\theta(\mathbf{Fh})(X) = X/2$ and $\theta(\mathbf{Sh})(X) = X/2$. We claim that θ is an additive and polynomial sup-interpretation. Indeed, all these functions are monotonic. Moreover, for every binary value $\mathbf{v}$, we have $\theta^*(\mathbf{v}) = |\mathbf{v}|$ since the sup-interpretation of a value is equal to its size. Finally, such an assignment satisfies the third condition of the above definition. In particular, we check that, for every binary values $\mathbf{u}$ and $\mathbf{v}$, if $\mathbf{Msp}(\mathbf{u}, \mathbf{v}) \in$ `Values` then we have:

$$\begin{aligned}
\theta^*(\mathbf{Msp}(\mathbf{u}, \mathbf{v})) &= \theta(\mathbf{Msp})(\theta^*(\mathbf{u}), \theta^*(\mathbf{v})) && \text{By definition of assignments} \\
&= \theta^*(\mathbf{u}) \ominus \theta^*(\mathbf{v}) && \text{By definition of } \theta(\mathbf{Msp}) \\
&= |\mathbf{u}| \ominus |\mathbf{v}| && \text{Since } \forall \mathbf{w} \in \texttt{Values},\ \theta^*(\mathbf{w}) = |\mathbf{w}| \\
&= |[\![\mathbf{Msp}(\mathbf{u}, \mathbf{v})]\!]| && \text{By definition of } \mathbf{Msp} \\
&= \theta^*([\![\mathbf{Msp}(\mathbf{u}, \mathbf{v})]\!]) && \text{Since } \forall \mathbf{w} \in \texttt{Values},\ \theta^*(\mathbf{w}) = |\mathbf{w}|
\end{aligned}$$

Notice that θ is a partial assignment since it is not defined on the symbols **if** and log. However, we can extend θ by $\theta(\mathbf{if})(X, Y, Z) = \max(Y, Z)$ and $\theta(\log)(X) = X$ in order to obtain a total, additive and polynomial sup-interpretation.

4 Arboreal and Fraternal Programs

In this section, we give two restrictions on programs (i) the arboreal condition which ensures that the number of successive recursive calls corresponding to a function symbol (in the depth of a call-tree) is bounded logarithmically by the input size (ii) and the fraternal condition which ensures that the size of each computed value is polynomially bounded by the input size.

4.1 Arboreal Programs

An arboreal program is a program whose recursion depth (number of successive recursive calls) is bounded logarithmically by the input size. This logarithmic upper bound is obtained by ensuring that some complexity measure, corresponding to the combination of sup-interpretations and motonoinc and polynomial partial assignments, is divided by a fixed constant $K > 1$ at each recursive call.

Definition 7. *A program* $\mathbf{p}$ *is arboreal iff there are a polynomial and additive sup-interpretation* θ*, a monotonic and polynomial partial assignment* ω *and a constant* $K > 1$ *such that for every fraternity* $\mathsf{C}[\mathtt{g}_1(\overline{e_1}), \ldots, \mathtt{g}_r(\overline{e_r})]$ *activated by* $\mathtt{f}(\overline{p})$*, the following conditions are satisfied:*

- *For any substitution* σ, $\omega(\mathtt{f})(\theta^*(\overline{p}\sigma)) \geq 1$
- *For any substitution* σ *and* $\forall i \in \{1, \ldots, r\}$, $\omega(\mathtt{f})(\theta^*(\overline{p}\sigma)) \geq K \times \omega(\mathtt{g}_i)(\theta^*(\overline{e_i}\sigma))$
- *There is no function symbol* $\mathtt{h} \in \overline{e_i}$ *such that* $\mathtt{h} \approx_{Fct} \mathtt{f}$.

Example 4. In the program of example 1, there is one fraternity:

$$\mathbf{if}(\mathbf{Msp}(\mathbf{Fh}(\mathbf{i}(x)), \mathbf{Sh}(\mathbf{i}(x))), \mathbf{0}(\log(\mathbf{Fh}(\mathbf{i}(x)))), \mathbf{1}(\log(\mathbf{Fh}(\mathbf{i}(x)))))$$

activated by $\log(\mathbf{i}(x))$. Taking the additive and polynomial sup-interpretation θ of example 3, the polynomial partial assignment $\omega(\log)(X) = X$ and the constant $K = 2$, we check that the program is arboreal:

$$\omega(\log)(\theta^*(\mathbf{i}(x))) = X + 1 \geq 2 \times (X+1)/2 = K \times \omega(\log)(\theta^*(\mathbf{Fh}(\mathbf{i}(x)))) \geq 1$$

Lemma 2. *Assume that* $\mathbf{p}$ *is an arboreal program. Then* $\mathbf{p}$ *is terminating. That is, for every function symbol* $\mathtt{f}$ *and for any values* $\overline{\mathbf{u}}$ *in* `Values`, $[\![\mathtt{f}(\overline{\mathbf{u}})]\!]$ *is in* `Values`*.

Moreover, for each branch $B = \langle \mathtt{f}, \boldsymbol{u}_1, \cdots, \boldsymbol{u}_n \rangle, \ldots, \langle \mathtt{g}, \boldsymbol{v}_1, \cdots, \boldsymbol{v}_m \rangle$ *of a call-tree corresponding to one execution of* $\mathbf{p}$ *and such that* $\mathtt{f} \approx_{Fct} \mathtt{g}$*, we have:*

$$\mathtt{depth}(B) \leq \alpha \times \log(\omega(\mathtt{f})(\theta^*(\boldsymbol{u}_1), \ldots, \theta^*(\boldsymbol{u}_n))), \text{ for some constant } \alpha$$

Proof. Consider an arboreal program $\mathbf{p}$, a call-tree of $\mathbf{p}$ and one of its branch of the shape $\langle \mathtt{f}, \mathbf{u}_1, \cdots, \mathbf{u}_n \rangle \rightsquigarrow \langle \mathtt{g}, \mathbf{v}_1, \cdots, \mathbf{v}_m \rangle$, with $\mathtt{f} \approx_{Fct} \mathtt{g}$. We know, by definition of a call-tree, that there are a rule of the shape $\mathtt{f}(p_1, \cdots, p_n) \to \mathsf{C}[\mathtt{g}(e_1, \cdots, e_m)]$ and a substitution σ such that $p_i\sigma = \mathbf{u}_i$ and $[\![e_j\sigma]\!] = \mathbf{v}_j$. We obtain:

$$\begin{aligned}
\omega(\mathtt{f})(\theta^*(\mathbf{u}_1), \ldots, \theta^*(\mathbf{u}_n)) &= \omega(\mathtt{f})(\theta^*(p_1\sigma), \ldots, \theta^*(p_n\sigma)) \\
&\geq K \times \omega(\mathtt{g}_i)(\theta^*(e_1\sigma), \ldots, \theta^*(e_m\sigma)) \quad \text{By definition 7} \\
&\geq K \times \omega(\mathtt{g}_i)(\theta^*(\mathbf{v}_1), \ldots, \theta^*(\mathbf{v}_m)) \quad \text{By lemma 1}
\end{aligned}$$

Applying the same reasoning, we demonstrate, by induction on the depth of a branch, that for each branch $B = \langle \mathtt{f}, \mathbf{u}_1, \cdots, \mathbf{u}_n \rangle \rightsquigarrow \ldots \rightsquigarrow \langle \mathtt{g}, \mathbf{v}_1, \cdots, \mathbf{v}_m \rangle$, with $\mathtt{f} \approx_{Fct} \mathtt{g}$ and $\mathtt{depth}(B) = i$:

$$\omega(\mathtt{f})(\theta^*(\mathbf{u}_1), \ldots, \theta^*(\mathbf{u}_n)) \geq K^i \times \omega(\mathtt{g})(\theta^*(\mathbf{v}_1), \ldots, \theta^*(\mathbf{v}_m))$$

Consequently, the depth is bounded by $\log_K(\omega(\mathtt{f})(\theta^*(\mathbf{u}_1), \ldots, \theta^*(\mathbf{u}_n)))$, because of the first condition of definition 7, whenever $\langle \mathtt{f}, \mathbf{u}_1, \cdots, \mathbf{u}_n \rangle$ is the first state of the considered branch. It remains to combine this result with the equality $\log(x) = \frac{\log_K(x)}{\log_K(2)}$ in order to obtain the required result. Since every branch corresponding to a recursive call has a bounded depth and we are considering confluent programs, the program is terminating. □

4.2 Fraternal Programs

Definition 8. *A program $\mathbf{p}$ is fraternal if there is a polynomial and additive sup-interpretation θ such that for each fraternity $\mathsf{C}[\mathtt{g}_1(\overline{e_1}), \ldots, \mathtt{g}_r(\overline{e_r})]$ activated by $\mathtt{f}(p_1, \cdots, p_n)$ and for each symbol b of arity m appearing in C or in $\overline{e_j}$, there are constants $\alpha^b_{i,j}, \beta^b_j \in \mathbb{R}^+$ satisfying:*

$$\theta(b)(X_1, \cdots, X_m) = \max_{j \in J}\left(\sum_{i=1}^{m} \alpha^b_{i,j} \times X_i + \beta^b_j\right)$$

where J is a finite set of indices.

In other words, a program is fraternal if every symbol in a context or in an argument of a fraternity admits an affinely bounded sup-interpretation.

Example 5. $\mathsf{C}[\log(\mathbf{Fh}(\mathbf{i}(x))), \log(\mathbf{Fh}(\mathbf{i}(x)))]$ is the only fraternity in the program of example 1, where $\mathsf{C}[\diamond_1, \diamond_2] = \mathbf{if}(\mathbf{Msp}(\mathbf{Fh}(\mathbf{i}(x)), \mathbf{Sh}(\mathbf{i}(x))), \mathbf{0}(\diamond_1), \mathbf{1}(\diamond_2))$. Consequently, we have to check that the symbols **if**, **Msp**, **Fh**, **Sh**, **0** and **1** admit affine sup-interpretations. This is the case by taking the polynomial sup-interpretation of example 3 and, consequently, the program is fraternal.

Lemma 3. *Given a sup-interpretation θ and a monotonic and polynomial partial assignment ω for which the program $\mathbf{p}$ is arboreal and fraternal, there is a polynomial P such that for each sequence of values $\mathbf{u}_1, \cdots, \mathbf{u}_n$ and for each function symbol $\mathtt{f}$ of arity n, we have: $|[\![\mathtt{f}(\mathbf{u}_1, \cdots, \mathbf{u}_n)]\!]| \leq P(\max_{i=1..n}(|\mathbf{u}_i|))$.*

Proof (Sketch). Lemma 2 states that the number of successive recursive calls occurring in the depth of the call-tree is logarithmically bounded by the input size. By definition 8, the contexts and arguments of a recursive call (fraternity) of a fraternal program are affinely bounded by the input size. Consequently, a logarithmic composition of affinely bounded functions remains polynomially bounded. □

5 Characterizations of NC^k and NC

Similarly to Buss' encoding [11], we represent constructors and destructors (implicitly used in pattern matching definitions) by U_{E^*}-uniform circuits of constant depth and polynomial size. Given such an encoding *code* and a function $\phi : \texttt{Values}^n \to \texttt{Values}^*$ computed by some program $\mathbf{p}$, we define a function $\tilde{\phi} : \{0,1\}^* \to \{0,1\}^*$ by $\forall \overline{\mathtt{u}} \in \texttt{Values}^n\ \tilde{\phi}(code(\overline{\mathtt{u}})) = code(\phi(\overline{\mathtt{u}}))$. A function ϕ of `Values` is computable in NC^k relatively to the encoding *code* if and only if $\tilde{\phi}$ is computable by a U_{E^*}-uniform family of circuits in NC^k.

Now, we define a notion of rank in order to make a distinction between the levels of the NC^k hierarchy:

Definition 9. *Given a program $\mathbf{p}$ composed by a vocabulary Σ, a set of rules $\mathcal{R}$ and an encoding code, the rank of a symbol b, $\mathrm{rk}(b)$, and the rank of a symbol b relatively to an expression e, $\nabla(b,e)$, are partial functions ranging over $\mathbb{N}$ and defined by induction over the precedence $\geq_{Fct}$:*

- *If b is a variable or a constructor symbol then* $\mathrm{rk}(b) = 0$.
- *If b is an operator, which can be computed by a U_{E^*}-uniform family of circuits of polynomial size and depth bounded by* $\log^k$ *relatively to the encoding code, then* $\mathrm{rk}(b) = k$.
- *If b is a function symbol we define its rank relatively to an expression e by:*
  - *If $\forall b' \in e$, $b >_{Fct} b'$ then* $\nabla(b,e) = \max_{b' \in e}(\mathrm{rk}(b'))$
  - *Otherwise $\exists b'' \in e$ such that $b \approx_{Fct} b''$ and $e = b'(e_1, \cdots, e_n)$:*
    - *If $b >_{Fct} b'$ then* $\nabla(b,e) = \max(\mathrm{rk}(b') + 1, \nabla(b,e_1), \ldots, \nabla(b,e_n))$
    - *Otherwise $b \approx_{Fct} b'$ then* $\nabla(b,e) = \max(\nabla(b,e_1), \ldots, \nabla(b,e_n)) + 1$
- *Finally, we define the rank of a function symbol b by:*
  - $\mathrm{rk}(b) = \max_{b(\overline{p}) \to e \in \mathcal{R}}(\nabla(b,e))$

where $b \in e$ means that the symbol b appears in the expression e. The rank of a program is defined to be the highest rank of a symbol of a program.

Example 6. In the program of example 1, the operators **if**, **Fh**, **Sh**, **rev** and **Msp** are of rank 0 since they all belong to NC^0 (Cf.[5] for **Fh**, **Sh** and **Msp**) using an encoding *code* which returns the binary value in $\mathcal{T}(\{\mathit{0}, \mathit{1}, \epsilon\})$ corresponding to a binary string in $\{0,1\}^*$. Consequently, we obtain that:

$$\begin{aligned}
\mathrm{rk}(\log) &= \nabla(\log, \mathbf{if}(\mathbf{Msp}(\mathbf{Fh}(\mathbf{i}(x)), \mathbf{Sh}(\mathbf{i}(x))), \mathbf{0}(\log(\mathbf{Fh}(\mathbf{i}(x)))), \mathbf{1}(\log(\mathbf{Fh}(\mathbf{i}(x)))))) \\
&= \max(\mathrm{rk}(\mathbf{if}) + 1, \nabla(\log, \mathbf{0}(\log(\mathbf{Fh}(\mathbf{i}(x))))), \nabla(\log, \mathbf{1}(\log(\mathbf{Fh}(\mathbf{i}(x)))))) \\
&= \max(1, \mathrm{rk}(\mathbf{0}) + 1, \mathrm{rk}(\mathbf{1}) + 1, \nabla(\log, \log(\mathbf{Fh}(\mathbf{i}(x))))) \\
&= \max(1, \nabla(\log, \mathbf{Fh}(\mathbf{i}(x))) + 1) = 1
\end{aligned}$$

Theorem 1. *A function ϕ from* `Values`^{n} *to* `Values`* *is computed by a fraternal and arboreal program of rank $k \geq 1$ (resp. $k \in \mathbb{N}$) if and only if ϕ is computable in NC^k (resp. NC).*

Proof (Sketch). We can show, by induction on the rank and the precedence $\geq_{Fct}$, that an arboreal and fraternal program of rank k can be simulated by a U_{E^*}-uniform family of circuits of polynomial size and $\log^k$ depth, using a discriminating circuit of constant depth which, given some inputs, picks the right rule to apply. The $\log^k$ depth relies on a logarithmic number of $\log^{k-1}$ depth circuits compositions by lemma 2. The polynomial size is a consequence of lemma 3. Conversely, we use the characterization of Clote [12] to show the completeness. Clote's algebra is based on two recursion schemas called Concatenation Recursion on Notation (CRN) and Weak bounded Recursion on Notation (WBRN) defined over a function algebra from natural numbers to natural numbers when considering the following initial functions $\texttt{zero}(x) = 0$, $s_0(x) = 2 \times x$, $s_1(x) = 2 \times x + 1$, $\pi_k^n(x_1, \cdots, x_n) = x_k$, $|x| = \lceil \log_2(x+1) \rceil$, $x\#y = 2^{|y| \times |x|}$, $\texttt{bit}(x, i) = \lfloor x/2^i \rfloor \ \texttt{mod}\ 2$ and a function `tree`, which computes alternations of bitwise conjunctions and bitwise disjunctions. All Clote's initial functions can be simulated by operators of rank 1 since they are in AC^0 ([12]). We show that a function of rank k in Clote's algebra can be simulated by an arboreal and fraternal program of rank $k+1$, using divide-and-conquer algorithms (This requirement is needed because of the arboreal property). A difficulty to stress here is that Clote's rank differs from the notion of rank we use because of the function `tree` and the CRN schema. □

6 Comparison with Previous Works

This work extends the characterization of NC^1 in [9] to the NC^k hierarchy. For that purpose, we have substituted the more general notion of fraternal program to the notion of explicitly fraternal of [9] and we have introduced a notion of rank. In the literature, there are many characterizations of NC^1 using several computational models, like ATM or functions algebra. Compton and Laflamme [15] gave a characterization of NC^1 based on finite functions. Bloch [5] used ramified recurrence schema relying on a divide and conquer strategy to characterize NC^1. Leivant and Marion [25] have established another characterization using ramified recurrence over a specific data structure, well balanced binary trees. We can show that our characterization strictly generalizes the ones of Bloch and Leivant-Marion since they both rely on divide-and-conquer strategies. The characterizations of the NC^k classes by Clote [13], using a bounded recurrence schema à la Cobham, are distinct from our characterizations since they do not rely on a divide and conquer strategy. However, like Cobhams' work, Clote's WBRN schema requires an upper bound on the computed function which is removed from our characterizations. Other characterizations of NC are also provided in [24,6]. All these purely syntactic characterizations capture a few algorithmic patterns. On the contrary, our work tries to delineate a broad class of algorithms using the semantics notion of sup-interpretation.

References

1. Amadio, R.: Synthesis of max-plus quasi-interpretations. Fundamenta Informaticae 65(1–2) (2005)
2. Baillot, P., Mogbil, V.: Soft lambda-calculus: A language for polynomial time computation. In: Walukiewicz, I. (ed.) FOSSACS 2004. LNCS, vol. 2987, pp. 27–41. Springer, Heidelberg (2004)
3. Barrington, D., Immerman, N., Straubing, H.: On uniformity within NC. J. of Computer System Science 41(3), 274–306 (1990)
4. Bellantoni, S., Cook, S.: A new recursion-theoretic characterization of the poly-time functions. Computational Complexity 2, 97–110 (1992)
5. Bloch, S.: Function-algebraic characterizations of log and polylog parallel time. Computational complexity 4(2), 175–205 (1994)
6. Kahle, R., Marion, J.-Y., Bonfante, G., Oitavem, I.: Towards an implicit characterization of NC^k. In: Ésik, Z. (ed.) CSL 2006. LNCS, vol. 4207, pp. 212–224. Springer, Heidelberg (2006)
7. Bonfante, G., Marion, J.-Y., Moyen, J.-Y.: On lexicographic termination ordering with space bound certifications. In: Bjørner, D., Broy, M., Zamulin, A.V. (eds.) PSI 2001. LNCS, vol. 2244, Springer, Heidelberg (2001)
8. Bonfante, G., Marion, J.-Y., Moyen, J.-Y., Péchoux, R.: Synthesis of quasi-interpretations. In: LCC2005, LICS affiliated Workshop (2005), `http://hal.inria.fr`
9. Bonfante, G., Marion, J.-Y., Péchoux, R.: A characterization of alternating log time by first order functional programs. In: Hermann, M., Voronkov, A. (eds.) LPAR 2006. LNCS (LNAI), vol. 4246, pp. 90–104. Springer, Heidelberg (2006)
10. Bonfante, G., Marion, J.Y., Moyen, J.Y.: Quasi-interpretations, a way to control resources. TCS (2007)
11. Buss, S.: The boolean formula value problem is in ALOGTIME. In: STOC, pp. 123–131 (1987)
12. Clote, P.: Sequential, machine-independent characterizations of the parallel complexity classes ALOGTIME, AC^k, NC^k and NC. In: Buss, R., Scott, P. (eds.) Workshop on Feasible Math, pp. 49–69. Birkhäuser, Basel (1989)
13. Clote, P.: Computational models and function algebras. In: Leivant, D. (ed.) LCC 1994. LNCS, vol. 960, pp. 98–130. Springer, Heidelberg (1995)
14. Cobham, A.: The intrinsic computational difficulty of functions. In: Conf. on Logic, Methodology, and Philosophy of Science, pp. 24–30. North-Holland, Amsterdam (1962)
15. Compton, K.J., Laflamme, C.: An algebra and a logic for NC. Inf. Comput. 87(1/2), 240–262 (1990)
16. Gaboardi, M., Rocca, S.R.D.: A soft type assignment system for λ-calculus. In: Duparc, J., Henzinger, T.A. (eds.) CSL 2007. LNCS, vol. 4646, pp. 253–267. Springer, Heidelberg (2007)
17. Girard, J.-Y.: Light linear logic. Inf. and Comp. 143(2), 175–204 (1998)
18. Girard, J.Y., Scedrov, A., Scott, P.: Bounded linear logic. Theoretical Computer Science 97(1), 1–66 (1992)
19. Hofmann, M.: Programming languages capturing complexity classes. SIGACT News Logic Column 9 (2000)
20. Huet, G.: Confluent reductions: Abstract properties and applications to term rewriting systems. Journal of the ACM 27(4), 797–821 (1980)

21. Kristiansen, L., Jones, N.D.: The flow of data and the complexity of algorithms. In: Cooper, S.B., Löwe, B., Torenvliet, L. (eds.) CiE 2005. LNCS, vol. 3526, pp. 263–274. Springer, Heidelberg (2005)
22. Lafont, Y.: Soft linear logic and polynomial time. Theoretical Computer Science 318(1-2), 163–180 (2004)
23. Leivant, D.: Predicative recurrence and computational complexity I: Word recurrence and poly-time. In: Feasible Mathematics II, pp. 320–343. Birkhäuser, Basel (1994)
24. Leivant, D.: A characterization of NC by tree recurrence. In: FOCS 1998, pp. 716–724 (1998)
25. Leivant, D., Marion, J.-Y.: A characterization of alternating log time by ramified recurrence. TCS 236(1-2), 192–208 (2000)
26. Marion, J.-Y., Moyen, J.-Y.: Efficient first order functional program interpreter with time bound certifications. In: Parigot, M., Voronkov, A. (eds.) LPAR 2000. LNCS (LNAI), vol. 1955, pp. 25–42. Springer, Heidelberg (2000)
27. Marion, J.-Y., Péchoux, R.: Resource analysis by sup-interpretation. In: Hagiya, M., Wadler, P. (eds.) FLOPS 2006. LNCS, vol. 3945, pp. 163–176. Springer, Heidelberg (2006)
28. Niggl, K.-H., Wunderlich, H.: Certifying polynomial time and linear/polynomial space for imperative programs. SIAM J. on Computing (to appear)
29. Ruzzo, W.: On uniform circuit complexity. J. of Computer System Science 22(3), 365–383 (1981)

The Structure of Detour Degrees

Lars Kristiansen[1] and Paul J. Voda[2]

[1] Department of Mathematics, University of Oslo
larsk@math.uio.no
http://www.iu.hio.no/~larskri
[2] Institute of Informatics, Comenius University Bratislava
voda@fmph.uniba.sk
http://www.fmph.uniba.sk/~voda

Abstract. We introduce a new subrecursive degree structure: *the structure of detour degrees.* We provide definitions and basic properties of the detour degrees, including that they form a lattice and admit a jump operator. Our degree structure sheds light upon the open problems involving the small Grzegorcyk classes. There are also connections to complexity theory and complexity classes defined by resource-bounded Turing machines.

1 Introduction

Two main categories of subrecursive degrees are known from the literature: (i) degrees of honest functions and (ii) degrees of computable sets.

An honest function is a total computable and monotone increasing function with a simple graph, i.e., the characteristic function for the relation $f(x) = y$ is of degree $\mathbf{0}$. (E.g., if we are working with elementary degrees, the graph should be elementary; if we are working with primitive recursive degrees, the graph should be primitive recursive.) Typical reducibility relations between honest functions will be "being (Kalmar) elementary in" and "being primitive recursive in". Strong resource bounded reducibilities like "being polynomial time computable in" do not work when we are dealing with honest functions, and hence, computer scientists and complexity theoreticians have never been very attracted to honest degree theory. Honest subrecursive degree structures were studied in the early seventies by e.g. Basu [2], Meyer & Ritchie [20], Machtey [16][17][18]; and more recently by the first author, see [7][8][9][10]. The structure of honest elementary degrees is a distributive lattice with strong density properties. Kristiansen studies a jump operator on the honest elementary degrees, see e.g. [7] [10].

The study of degrees of computable sets started off in the mid seventies with Ladner's seminal paper [15]. In contrast to honest functions, computable sets admit degree structures induced by strong resource bounded reducibility relations, like e.g. "polynomial time computable in" and "logarithmic space computable in". The resulting degree structures, which of course are highly interesting from a complexity-theoretic point of view, are all upper semi lattices with strong density properties. See Merkle [19] for very general embedding results. Ladner [15]

M. Agrawal et al. (Eds.): TAMC 2008, LNCS 4978, pp. 148–159, 2008.

proved that neither the structure of polynomial time m-degrees nor the structure of polynomial time T-degrees are lattices, and his proof methods and results generalise to wide variety of reducibilities. See Ambos-Spies [1] for more on degrees of computable sets.

In this paper we introduce a new subrecursive degree structure that does not fall into either of the two categories discussed above. The theory we develop has a flavour of honest degree theory, but our reducibility relation is sufficiently strong to make the theory interesting from a complexity-theoretic point of view. We prove that our degree structure is a distributive lattice, and we introduce a jump operator.

2 Preliminaries

We will use some notation and terminology from Clote [3]. An *operator*, here also called *(definition) scheme*, is a mapping from functions to functions. Let $\mathcal{X}$ be a set of functions (possibly given in a slightly informal notation), and let OP be a collection of operators. The *function algebra* $[\mathcal{X}; \text{OP}]$ is the smallest set of functions containing $\mathcal{X}$ and closed under the operations of OP. Let COMP denote the scheme of *composition*, i.e. the scheme $f(\vec{x}) = h(g_1(\vec{x}), \ldots, g_m(\vec{x}))$ where $m \geq 0$, and let PR denotes the scheme of *primitive recursion*, that is, the scheme

$$f(\vec{x}, 0) = g(\vec{x}) \qquad\qquad f(\vec{x}, y+1) = h(\vec{x}, y, f(\vec{x}, y)) \ .$$

Further, *bounded minimalisation* is the scheme $f(\vec{x}) = (\mu i \leq y)[R(\vec{x}, i)]$ where

$$(\mu i \leq y)[R(\vec{x}, i)] = \begin{cases} \text{the least } i \leq y \text{ such that the relation } R(\vec{x}, i) \text{ holds} \\ 0 \text{ if no such } i \text{ exists.} \end{cases}$$

Any function $f : \mathbb{N}^n \to \mathbb{N}$ will be identified with a relation $R_f \subseteq \mathbb{N}^n$. The relation $R_f(\vec{n})$ holds iff $f(\vec{x}) = 0$, and the relation R_f belongs to a class of functions $\mathcal{F}$ iff f belongs to $\mathcal{F}$. A *problem* is a subset of $\mathbb{N}$, or equivalently, a unary relation. We will use $A, B, C, \ldots$ to denote problems, and a function f *decides* a problem A when $f(x) = 0$ iff $x \in A$. For any set $\mathcal{F}$ of number-theoretic functions, $\mathcal{F}_*$ denotes the set of problem decided by the functions in $\mathcal{F}$.[1]

Let $I_i^n : \mathbb{N}^n \to \mathbb{N}$ denote the projection function, i.e. $I_i^n(x_1, \ldots, x_n) = x_i$ where $\vec{x} = x_1, \ldots, x_n$ and $1 \leq i \leq n$. Let I denote the set of all such projection functions. Let C_k denote the constant function yielding the natural number k. The *small Grzegorczyk classes* are defined by $\mathcal{E}^0 := [I, C_0, S, \max; \text{COMP}, \text{BR}]$, $\mathcal{E}^1 := [I, C_0, S, +; \text{COMP}, \text{BR}]$ and $\mathcal{E}^2 := [I, C_0, S, +, x^2 + 2; \text{COMP}, \text{BR}]$ where S denotes the successor function and BR denotes the scheme of bounded primitive recursion, that is, the scheme

$$f(\vec{x}, 0) = g(\vec{x}) \qquad f(\vec{x}, y+1) = h(\vec{x}, y, f(\vec{x}, y)) \qquad f(\vec{x}, y) \leq j(\vec{x}, y) \ .$$

[1] Our use of the subscript $*$ differs slightly from the standard in the literature. Normally, $\mathcal{F}_*$ denotes the 0-1 valued functions in $\mathcal{F}$ whereas we use $\mathcal{F}_*$ to denote the set of problem decided by the functions in $\mathcal{F}$. This is a matter of convenience, and the deviation has no essential mathematical implications.

3 Detour Functions and $\mathcal{P}^-$-Computability

Definition. $\mathcal{P}^- := [I, C_1; \text{COMP}, \text{PR}]$. □

Roughly speaking, $\mathcal{P}^-$ contains the primitive recursive functions which can be defined without the successor function, and it is easy to see that we have $f(\vec{x}) \leq \max(\vec{x}, 1)$ for any $f \in \mathcal{P}^-$. Even though $\mathcal{P}^-$ contains only non-increasing functions, surprisingly powerful relations and predicates belong to the class. Indeed, it is an open problem if $\mathcal{P}^-_* = \text{LINSPACE}$; we have $\mathcal{P}^-_* = \mathcal{E}^0_* \subseteq \mathcal{E}^1_* \subseteq \mathcal{E}^2_* = \text{LINSPACE}$, and it is not known if any of the inclusions are strict.[2] For more on $\mathcal{P}^-$ and similar non-increasing function algebras, see e.g. Kristiansen [5] [6] and Kristiansen & Voda [13] [14].

Lemma 1 (Basic functions). *The following number-theoretic functions belong to $\mathcal{P}^-$: (i) 0,1 (constant functions); (ii) $P(x)$ (predecessor); (iii) $x \dot{-} y$ (modified subtraction); (iv) c where $c(x, y_1, y_2) = y_1$ if $x = 0$ and $c(x, y_1, y_2) = y_2$ if $x \neq 0$; (v) $\max(x, y)$ and $\min(x, y)$; (vi) f where $f(x, m) = x + 1 \pmod{m + 1}$ for $x \leq m$; (vii) f where $f(x, y, m) = x + y \pmod{m + 1}$ for $x, y \leq m$; (viii) f where $f(x, y, m) = x \times y \pmod{m+1}$ for $x, y \leq m$; (ix) f where $f(x, y, m) = x^y \pmod{m + 1}$ for $x, y \leq m$; (x) for each $k \in \mathbb{N}$, the almost everywhere constant function ν_k where $\nu_k(x) = \min(x, k)$. Moreover, the class $\mathcal{P}^-$ is closed under bounded minimalisation.*

Proof. To define the constant function 0 is slightly nontrivial. Define g by primitive recursion such that $g(x, 0) = x$ and $g(x, y+1) = y$. Then we can define the predecessor P from g since $P(x) = g(x, x)$. Further, we can define the constant function 0 by $0 = P(1)$. This proves that (i) and (ii) hold. (iii) holds since we have $x \dot{-} 0 = x$ and $x \dot{-} (y + 1) = P(x \dot{-} y)$. (iv) holds since $c(0, y_1, y_2) = I^2_1(y_1, y_2)$ and $c(x + 1, y_1, y_2) = I^4_2(y_1, y_2, x, c(x, y_1, y_2))$. Furthermore, (v) holds since $\max(x, y) = c(1 \dot{-} (x \dot{-} y), x, y)$ and $\min(x, y) = c(1 \dot{-} (x \dot{-} y), y, x)$. (vi) holds since $c(m \dot{-} x, 0, m \dot{-} ((m \dot{-} x) \dot{-} 1)) = x + 1 \pmod{m + 1}$ for $x \leq m$. We omit the rest of the proof. □

Lemma 2 (Basic relations). *(i) The relations $x \leq y$ and $x = y$ belong to $\mathcal{P}^-$. (ii) The relations of $\mathcal{P}^-$ are closed under the propositional operations (and, or, not) and under bounded existential and bounded universal quantification, i.e., $\exists x \leq y[\ldots]$ and $\forall x \leq y[\ldots]$.*

Proof. Apply the functions given by Lemma 1. We have $y \dot{-} x = 0$ iff $x \leq y$; we have $c(f(\vec{x}), c(g(\vec{x}), 0, 1), 1) = 0$ iff $f(\vec{x}) = 0$ and $g(\vec{x}) = 0$; we have $c(f(\vec{x}), 1, 0) = 0$ iff $f(\vec{x}) \neq 0$; and so on. Use the bounded minimalisation operator to define the bounded quantifiers. □

---

[2] The notorious problem $\mathcal{E}^0_* \overset{?}{\subseteq} \mathcal{E}^1_* \overset{?}{\subseteq} \mathcal{E}^2_*$ is more than 50 years old and was posed in Grzegorczyk [4]. It is trivial that $\mathcal{P}^-_* \subseteq \mathcal{E}^0_*$, and it follows from Theorem 5 in this paper that $\mathcal{P}^-_* = \mathcal{E}^0_*$. (The problem $\mathcal{P}^-_* \overset{?}{\subseteq} \mathcal{E}^0_*$ was posed as open in Kristiansen & Barra [11].) Ritchie [22] proved that $\mathcal{E}^2_* = \text{LINSPACE}$.

Definition. A *detour function* $f : \mathbb{N} \rightarrow \mathbb{N}$ satisfies the following requirements: (1) $x \leq f(x)$; (2) $f(x) \leq f(x+1)$; and (3) the graph of f, i.e. the relation $f(x) = y$, is in $\mathcal{P}^-$. Henceforth we will use Latin letters f, g, h, etc., to denote detour functions. and Greek letters ϕ, ψ, ξ, etc., to denote arbitrary functions. A function ψ is *$\mathcal{P}^-$-computable in the detour function* f if there exists $\phi \in \mathcal{P}^-$ such that $\psi(\vec{x}) = \phi(\vec{x}, z)$ for all $z \geq f(\max(\vec{x}))$. Let $\mathcal{P}^-(f)$ denote the set of functions $\mathcal{P}^-$-computable in f. □

What motivates our definitions? Well, we are mainly interested in $\mathcal{P}^-(f)_*$, that is, the set of problems $\mathcal{P}^-$-computable in f. A problem A will be $\mathcal{P}^-$-computable in a detour function f if f provides enough resources for a $\mathcal{P}^-$-computation to solve the problem. Any $\mathcal{P}^-$-function is non-increasing, and thus, a $\mathcal{P}^-$-computation solving a problem on input x in a detour function f, can make *detours* via any number less than $f(x)$, but no further detours are possible. Our definitions assure that if g provides more resources (longer detours) than f, that is if $f(x) \leq g(x)$, then any problem $\mathcal{P}^-$-computable in f will also be $\mathcal{P}^-$-computable in g.

It is essential that we require the graph of a detour function to be in $\mathcal{P}^-$. This prevents us from coding a set A of high computational complexity into a detour function f, e.g. by letting $f(x) = 2x$ when $x \in A$ and $f(x) = 2x+1$ when $x \notin A$, and thus, the requirement ensures that a detour function provides nothing but resources to a $\mathcal{P}^-$-computation. (Without the requirement our degree structure would simply degenerate to a degree structure of computable sets.)

Definition. For any set $A \subseteq \mathbb{N}$, we define the *characteristic function* χ_A by $\chi_A(x) = 0$ if $x \in A$ and $\chi_A(x) = 1$ if $x \notin A$. We define the function f^{-1} by $f^{-1}(x) = (\mu i \leq x)[f(i) \geq x]$, and when f is a detour function, we will say that f^{-1} is the *inverse function* of f. For any $f : \mathbb{N} \rightarrow \mathbb{N}$ and $g : \mathbb{N} \rightarrow \mathbb{N}$, we define the functions $\max[f, g]$ and $\min[f, g]$ by respectively $\max[f, g](x) = \max(f(x), g(x))$ and $\min[f, g](x) = \min(f(x), g(x))$ □

The two following lemmas are collections of very basic facts. The lemmas will be applied frequently throughout the paper, but we will only occasionally refer explicitly to the lemmas.

Lemma 3. *Let f be any detour function. (i) There exists $\xi \in \mathcal{P}^-$ such that $\phi(\vec{x}) = \xi(\vec{x}, f(\max(\vec{x})))$ if, and only if, there exist $\psi \in \mathcal{P}^-$ such that $\phi(\vec{x}) = \psi(\vec{x}, z)$ for any $z \geq f(\max(\vec{x}))$; (ii) $A \in \mathcal{P}^-(f)_*$ iff $\chi_A \in \mathcal{P}^-(f)$; (iii) $\mathcal{P}^-(f)_*$ is closed under unions, intersections and complements; (iv) $f^{-1} \in \mathcal{P}^-$ and $f^{-1}(f(x)) \leq x$, besides, if f is strictly monotone, we also have $f^{-1}(f(x)) = x$.*

Proof. We prove the only-if-direction of (i). (The if-direction is trivial.) Assume there exists $\xi \in \mathcal{P}^-$ such that $\phi(\vec{x}) = \xi(\vec{x}, f(\max(\vec{x})))$. Let $\xi_0(\vec{x}, y) = (\mu i \leq y)[f(\max(\vec{x})) = i]$. The graph of f belongs to $\mathcal{P}^-$, and $\mathcal{P}^-$ is closed under bounded minimalisation, and thus, we have $\xi_0 \in \mathcal{P}^-$. Let $\xi_1(\vec{x}, z) = \xi(\vec{x}, \xi_0(\vec{x}, z))$. Now we have $\xi_1 \in \mathcal{P}^-$ and $\phi(\vec{x}) = \xi_1(\vec{x}, z)$ for any $z \geq f(\max(\vec{x}))$. Next we prove (iv). By the definition of f^{-1} we have $f^{-1}(f(x)) = (\mu i \leq f(x))[f(i) \geq f(x)]$. Since f is monotone, it cannot be the case that $f^{-1}(f(x)) > x$, and if f is strictly monotone, it cannot be the case that $f^{-1}(f(x)) < x$. Thus, $f^{-1}(f(x)) = x$ when f is strictly monotone. The graph of f belongs to $\mathcal{P}^-$, and $\mathcal{P}^-$ is closed under bounded minimalisation. Thus, $f^{-1} \in \mathcal{P}^-$. We omit the rest of the proof. □

Lemma 4. *(i) Any non-constant polynomial $p(x)$ (generated from x and the constants of $\mathbb{N}$ by $+$ and $\times$) is a detour function. (ii) The function 2^x is a detour function. (iii) The detour functions are closed under compositions. (iv)* $\max[f,g]$ *and* $\min[f,g]$ *are detour functions if f and g are detour functions.*

Proof. (i) and (ii) follow from Lemma 1. Let f and g be detour functions. Obviously, we have $x \leq f(g(x))$ and $f(g(x)) \leq f(g(x+1))$, and thus, (iii) follows from Lemma 2 since $f(g(x)) = y \;\Leftrightarrow\; \exists z \leq y[g(x) = z \;\wedge\; f(z) = y]$. Further, (iv) follows from Lemma 2 since $\max[f,g](x) = y \;\Leftrightarrow$

$$(f(x) = y \;\wedge\; \exists z \leq y[g(x) = z]) \;\vee\; (g(x) = y \;\wedge\; \exists z \leq y[f(x) = z])$$

and $\min[f,g](x) = y \;\Leftrightarrow$

$$(f(x) = y \;\wedge\; \neg\exists z < y[g(x) = z]) \;\vee\; (g(x) = y \;\wedge\; \neg\exists z < y[f(x) = z])\,.$$ □

4 The Lattice of Detour Degrees

Definition. We define $f \preceq g :\Leftrightarrow \mathcal{P}^-(f)_* \subseteq \mathcal{P}^-(g)_*$; $f \equiv g :\Leftrightarrow f \preceq g \wedge g \preceq f$; $f \prec g :\Leftrightarrow f \preceq g \wedge g \not\preceq f$. Now, $\equiv$ is an equivalence relation, and the *detour degrees* are the $\equiv$-equivalence classes of the detour functions. We denote the equivalence class of f by $\mathrm{dg}(f)$, that is, $\mathrm{dg}(f) = \{g \mid g \equiv f\}$, and following the tradition of classical computability theory, we use boldface lowercase Latin letters $\mathbf{a}, \mathbf{b}, \mathbf{c}, \ldots$ to denote our degrees. Furthermore, we use $<$ and $\leq$ to denote the relations induced on the degrees by respectively $\prec$ and $\preceq$, and we use $\mathcal{D}$ to denote the set of detour degrees. □

Lemma 5. *Let f, g, h be detour functions. (i) If $h \preceq f$ and $h \preceq g$, then $h \preceq$* $\min[f,g]$. *(ii) If $f \preceq h$ and $g \preceq h$, then* $\max[f,g] \preceq h$.

Proof. The following simple claim is the key to the proof.

> **(Claim)** Let f and g be detour functions, and let $B = \{x | f(x) \leq g(x)\}$. We have $B \in \mathcal{P}^-(\min[f,g])_*$ (and thus $B \in \mathcal{P}^-(f)_*$ and $B \in \mathcal{P}^-(g)_*$).

Let

$$\xi(x) \;=\; (\mu i \leq \min[f,g](x))\,[f(x) = i \vee g(x) = i]$$

and let $\chi_B(x) = c(f(x) = \xi(x), 0, 1)$. Now, χ_B is the characteristic function of B, and $\chi_B \in \mathcal{P}^-(\min[f,g])$ by the lemmas in Section 3. This proves the claim.

We prove (i). Assume $h \preceq f$ and $h \preceq g$ (*). (We will prove $h \preceq \min[f,g]$.) Assume $A \in \mathcal{P}^-(h)_*$. (We will prove $A \in \mathcal{P}^-(\min[f,g])_*$.) By (*), we have $A \in \mathcal{P}^-(f)_*$ and $A \in \mathcal{P}^-(g)_*$, and hence, we have $\phi_1, \phi_2 \in \mathcal{P}^-$ such that $\chi_A(x) = \phi_1(x, f(x))$ and $\chi_A(x) = \phi_2(x, g(x))$. By (Claim) we have $\eta \in \mathcal{P}^-$ such that

$\eta(x, \min[f,g](x)) = 0$ iff $f(x) \leq g(x)$. Let $\psi(x,z) = c(\eta(x,z), \phi_1(x,z), \phi_2(x,z))$. Then, $\psi \in \mathcal{P}^-$ and $\chi_A(x) = \psi(x, \min[f,g](x))$, and thus, $A \in \mathcal{P}^-(\max[f,g])_*$. This completes the proof of (i).

In order to prove (ii), we assume $f \preceq h$ and $g \preceq h$ (*) and prove $\max[f,g] \preceq h$. To prove $\max[f,g] \preceq h$, we assume $A \in \mathcal{P}^-(\max[f,g])_*$ and prove $A \in \mathcal{P}^-(h)_*$. Since $A \in \mathcal{P}^-(\max[f,g])_*$, we have $\phi \in \mathcal{P}^-$ such that $\chi_A(x) = \phi(x, \max[f,g](x))$. Let $B = \{x | f(x) \leq g(x)\}$, and let

$$x \in A_1 \Leftrightarrow \phi(x, f(x)) = 0 \wedge x \notin B \quad \text{and} \quad x \in A_2 \Leftrightarrow \phi(x, g(x)) = 0 \wedge x \in B\,.$$

By (Claim), we have $A_1 \in \mathcal{P}^-(f)_*$ and $A_2 \in \mathcal{P}^-(g)_*$, and by (*), we have $A_1, A_2 \in \mathcal{P}^-(h)_*$. By Lemma 3, we have $A_1 \cup A_2 \in \mathcal{P}^-(h)_*$. This completes the proof because $A = A_1 \cup A_2$. □

Lemma 6. *For any detour functions f, f_1, g, g_1, we have (i) if $f \equiv f_1$ and $g \equiv g_1$, then* $\max[f,g] \equiv \max[f_1,g_1]$*; (ii) if $f \equiv f_1$ and $g \equiv g_1$, then* $\min[f,g] \equiv \min[f_1,g_1]$.

Proof. We prove $f \preceq f_1 \wedge g \preceq g_1 \Rightarrow \max[f,g] \preceq \max[f_1,g_1]$. It follows that (i) holds. Assume $f \preceq f_1$ and $g \preceq g_1$. Since $f_1 \preceq \max[f_1,g_1]$ and $g_1 \preceq \max[f_1,g_1]$, we have $f \preceq \max[f_1,g_1]$ and $g \preceq \max[f_1,g_1]$. By Lemma 5 (ii), we have $\max[f,g] \preceq \max[f_1,g_1]$. This completes the proof of (i). The proof of (ii) is symmetric; apply Lemma 5 (i) in place of Lemma 5 (ii). □

Lemma 4 (iv) and Lemma 6 show that the minimum and the maximum operators on the detour functions induce operators on the detour degrees, and thus, the next definition make sense.

Definition. Let $f \in \mathbf{a}$ and $g \in \mathbf{b}$ be detour functions. We define $\mathbf{a} \cup \mathbf{b}$ and $\mathbf{a} \cap \mathbf{b}$ by respectively $\mathrm{dg}(\max[f,g])$ and $\mathrm{dg}(\min[f,g])$. □

Theorem 1 (The Lattice of Detour Degrees). *The structure $\langle \mathcal{D}, \leq, \cup, \cap \rangle$ is a distributive lattice, that is, for any* $\mathbf{a}, \mathbf{b}, \mathbf{c} \in \mathcal{D}$ *we have (i)* $\mathbf{a} \cap \mathbf{b}$ *is the greatest lower bound (glb) of* $\mathbf{a}$ *and* $\mathbf{b}$ *under the ordering $\leq$; (ii)* $\mathbf{a} \cup \mathbf{b}$ *is the least upper bound (lub) of* $\mathbf{a}$ *and* $\mathbf{b}$ *under the ordering $\leq$; (iii)* $\mathbf{a} \cup (\mathbf{b} \cap \mathbf{c}) = (\mathbf{a} \cup \mathbf{b}) \cap (\mathbf{a} \cup \mathbf{c})$ *and* $\mathbf{a} \cap (\mathbf{b} \cup \mathbf{c}) = (\mathbf{a} \cap \mathbf{b}) \cup (\mathbf{a} \cap \mathbf{c})$.

Proof. We prove (i). Since we have $\min[f,g] \preceq f$ and $\min[f,g] \preceq g$, it is obvious that $\mathbf{a} \cap \mathbf{b}$ is a lower bound of $\mathbf{a}$ and $\mathbf{b}$. To see that $\mathbf{a} \cap \mathbf{b}$ indeed is the greatest lower bound of $\mathbf{a}$ and $\mathbf{b}$, pick any degree $\mathbf{c}$ that lies below both $\mathbf{a}$ and $\mathbf{b}$. By Lemma 5 (i), we have that $\mathbf{c}$ also lies below $\mathbf{a} \cap \mathbf{b}$. Thus, (i) holds. The proof of (ii) is symmetric (use Lemma 5 (ii) in place of Lemma 5 (i)). Finally, (iii) holds since the max operator distributes over the min operator and vice versa. □

5 The Jump Operator

Lemma 7. *Let f, g, h be detour functions, and let h be strictly monotone. If $f \preceq g$, then $f \circ h \preceq g \circ h$.*

Proof. For any set A, let $\#_A(x) = \sum_{i\leq x} 1 \dot{-} \chi_A(i)$. The function $\#_A(x)$ gives the cardinality of the set $\{y \mid y \in A \wedge y \leq x\}$, and when $f(x) \geq x+1$, we have

$$A \in \mathcal{P}^-(f)_* \;\Leftrightarrow\; \#_A \in \mathcal{P}^-(f) \qquad \text{(Claim)}$$

We skip the proof of the claim and assume $f \preceq g$. (We will prove $f \circ h \preceq g \circ h$.)

To prove $f \circ h \preceq g \circ h$, we have to prove that $A \in \mathcal{P}^-(f \circ h)_*$ entails $A \in \mathcal{P}^-(g \circ h)_*$. We assume $A \in \mathcal{P}^-(f \circ h)_*$ (and prove $A \in \mathcal{P}^-(g \circ h)_*$.)

By (Claim), we have $\#_A \in \mathcal{P}^-(f \circ h)$, and thus, there exists $\phi \in \mathcal{P}^-$ such that $\#_A(x) = \phi(x, fh(x))$. Let B be the set given by $\#_B(x) = \phi(h^{-1}(x), f(x))$. Then, $B \in \mathcal{P}^-(f)_*$ and

$$\#_B(h(x)) \;=\; \phi(h^{-1}h(x), fh(x)) \;=\; \phi(x, fh(x)) \;=\; \#_A(x)\,.$$

Now, since $B \in \mathcal{P}^-(f)_*$ and $f \preceq g$, we have $B \in \mathcal{P}^-(g)_*$, and then, by (Claim), there exists $\psi \in \mathcal{P}^-$ such that $\#_B(x) = \psi(x, g(x))$. Thus,

$$\#_A(x) \;=\; \#_B(h(x)) \;=\; \psi(h(x), gh(x))\,. \qquad (*)$$

Let $h'(x,y) = (\mu z \leq y)[h(x) = z]$ and $\xi(x,y) = \psi(h'(x,y), y)$. Then, $\xi \in \mathcal{P}^-$ and

$$\xi(x, gh(x)) \;=\; \psi(h'(x, gh(x)), gh(x)) \;=\; \psi(h(x), gh(x)) \;\overset{(*)}{=}\; \#_A(x)\,.$$

Thus, we have $\#_A \in \mathcal{P}^-(g \circ h)$, and by (Claim), we also have $A \in \mathcal{P}^-(g \circ h)_*$. □

It follows from Lemma 7 that $f \equiv g \Rightarrow f \circ h \equiv g \circ h$ for any detour functions f, g and any strictly monotone detour function h. Now, 2^x is a strictly monotone detour function, and furthermore, the detour functions are closed under composition. Hence, the next definition makes sense.

Definition. We define the *jump operator* $\cdot'$ on the detour functions by $f'(x) = f(2^x)$. We define the *jump operator* $\cdot'$ on the detour degrees by $\mathbf{a}' = \mathrm{dg}(f')$ where f is some detour function such that $\mathbf{a} = \mathrm{dg}(f)$. Let $\mathbf{0}$ denote the degree $\mathrm{dg}(I_1^1)$. □

It is easy to see that we have $\mathbf{0} \leq \mathbf{a}$ for any degree $\mathbf{a}$. (We have $\mathbf{a} = \mathrm{dg}(f)$ for some f such that $f(x) \geq I_1^1(x) = x$, and thus $\mathcal{P}^-(I_1^1)_* \subseteq \mathcal{P}^-(f)_*$.) To prove that $\mathbf{0} < \mathbf{0}' < \mathbf{0}'' < \mathbf{0}''' < \ldots$, we have to resort to a machine model of computation. We have chosen the Turing machines.

Lemma 8. *There exists $\phi \in \mathcal{P}^-$ such that for any $A \in \mathcal{P}^-(f)_*$ we have a fixed $k \in \mathbb{N}$ and fixed $e_0, e_1, e_2, \ldots \in \mathbb{N}$ such that*

$$\chi_A(x) = \phi(e_i, x, z) \quad \textit{for any } z \geq f(x)^k \textit{ and any } i \in \mathbb{N}.$$

Proof. We will work with Turing machines accepting pairs of natural numbers. The two inputs numbers are given to the machines in dyadic notation, and $|z|$ denotes the number of bits required to represent the number z. We assume some computable enumeration $M_0, M_1, M_2, \ldots$ of such machines, furthermore, we can w.l.o.g. assume that each machine occurs infinitely many times in the enumeration. We define the function Φ: Let $\Phi(e,x,y,z) = 0$ if M_e accepts the input x, y in space $|z|$ and time z; let $\Phi(e,x,y,z) = 1$ otherwise. We skip the proof of the following claim.

(Claim 1) $\Phi \in \mathcal{P}^-$.

Obviously, we have $\mathcal{P}^- \subseteq \mathcal{E}^2$, and it is well known that any function in $\mathcal{E}^2$ can be computed by a Turing machine working in linear space in the length of the input (Ritchie [22]). Thus, if ϕ is a binary function in $\mathcal{P}^-$, then $\phi(x,y)$ is computable by a Turing machine M running in space $k_1|\max(x,y)|$ for some fixed $k_1 \in \mathbb{N}$. Now, if M runs in space $k_1|\max(x,y)|$, there also exists a fixed $k \in \mathbb{N}$ such that M runs in space $|\max(x,y)^k|$ and time $\max(x,y)^k$. Hence, the next claim holds.

(Claim 2) Let ψ be a binary 0-1 valued function in $\mathcal{P}^-$. There exists a fixed $k \in \mathbb{N}$ and infinitely many $e \in \mathbb{N}$ such that $\psi(x,y) = \Phi(e,x,y,\max(x,y)^k)$.

Let $\langle\cdot,\cdot\rangle$ be a standard bijection from $\mathbb{N}\times\mathbb{N}$ into $\mathbb{N}$, e.g. $\langle x,y\rangle = (\sum_{i\leq x+y} i) + y$, and let π_1,π_2 be the corresponding decoding functions, i.e. $\pi_1(\langle x,y\rangle) = x$ and $\pi_2(\langle x,y\rangle) = y$. The pairing function is not in $\mathcal{P}^-$, but the decoding functions π_1,π_2 are. Furthermore, the binary function $\lfloor x^{\frac{1}{y}}\rfloor$ belongs to $\mathcal{P}^-$. We define the function ϕ by

$$\phi(w,x,y) \;=\; \Phi(\pi_1(w),x,\lfloor y^{\frac{1}{\pi_2(w)}}\rfloor,y)\;.$$

It follows from our discussion and (Claim 1) that we have $\phi \in \mathcal{P}^-$. To complete our proof, we will prove that for any $A \in \mathcal{P}^-(f)_*$ there exist $k \in \mathbb{N}$ and infinitely many $e \in \mathbb{N}$ such that $\chi_A(x) = \phi(e,x,z)$ holds for any $z \geq f(x)^k$.

By the definition of ϕ we have

$$\phi(\langle e,k\rangle,x,z^k) \;=\; \Phi(e,x,z,z^k)\;. \qquad (*)$$

Let A be any set in $\mathcal{P}^-(f)_*$. Then we have $\psi \in \mathcal{P}^-$ such that $\chi_A(x) = \psi(x,z)$ for any $z \geq f(x)$. Furthermore, for some e,k and and any $z \geq f(x)$, we have

$$\begin{aligned}\chi_A(x) &= \psi(x,z) & \\ &= \Phi(e,x,z,\max(x,z)^k) & \text{(Claim 2)} \\ &= \Phi(e,x,z,z^k) & \text{since } x \leq f(x) \leq z \\ &= \phi(\langle e,k\rangle,x,z^k)\;. & (*)\end{aligned}$$

and indeed, by (Claim 2) there are infinitely many numbers e such that these equalities hold. Let $d_0,d_1,d_2,\ldots$ denote these numbers, and let $e_i = \langle d_i,k\rangle$. Now we have $\chi_A(x) = \phi(e_i,x,z^k)$ for any $i \in \mathbb{N}$ and any $z \geq f(x)$, and thus, $\chi_A(x) = \phi(e_i,x,z)$ for any $i \in \mathbb{N}$ and any $z \geq f(x)^k$. □

Theorem 2. *Let f be any detour function such that for any $k \in \mathbb{N}$ we have $f(x)^k \leq f(2^x)$ for all but finitely many x. Then, we have $f' \not\preceq f$.*

Proof. Let B be the set given by

$$\chi_B(x) = \begin{cases} 0 \text{ if } \phi(x,x,f'(x)) = 1 \\ 1 \text{ otherwise.}\end{cases} \qquad (*)$$

where ϕ is the function given by Lemma 8. Obviously, we have $B \in \mathcal{P}^-(f')$. We prove that $B \notin \mathcal{P}^-(f)$.

Assume for the sake of a contradiction that $B \in \mathcal{P}^-(f)$. Then, by Lemma 8 and (*), we have e_i and k such that

$$\chi_B(e_i) \;=\; \phi(e_i, e_i, f(e_i)^k) \;=\; \phi(e_i, e_i, f'(e_i)) \;\neq\; \chi_B(e_i)\,.$$

(Since we have infinitely many e_i's, it is possible to pick e_i such that $f(e_i)^k \leq f(2^{e_i}) = f'(e_i)$.) It is a contradiction that $\chi_B(e_i) \neq \chi_B(e_i)$. □

Theorem 3. *Let* $\mathbf{0}^0 = \mathbf{0}$ *and* $\mathbf{0}^{n+1} = (\mathbf{0}^n)'$. *We have* $\mathbf{0}^n < \mathbf{0}^{n+1}$ *for any* $n \in \mathbb{N}$.

Proof. It is trivial that $\mathbf{0}^n \leq \mathbf{0}^{n+1}$. Let $g_0(x) = x$ and $g_{i+1}(x) = 2^{g_i(x)}$. Then we have $\mathrm{dg}(g_i) = \mathbf{0}^i$. Further, for any fixed k, we have $g_n(x)^k < g_n(2^x) = g_{n+1}(x)$ for all sufficiently large x. Hence, we have $\mathbf{0}^{n+1} \not\leq \mathbf{0}^n$ by Theorem 2. □

Theorem 4 (Jump Inversion). *Let* $\mathbf{a}$ *be any detour degree above* $\mathbf{0}'$, *i.e.* $\mathbf{0}' \leq \mathbf{a}$. *There exists a detour degree* $\mathbf{b}$ *such that* $\mathbf{b}' = \mathbf{a}$.

Proof. Let $\mathbf{a} = \mathrm{dg}(f)$. We can w.l.o.g. assume that $f(x) \geq 2^x$. Let $\lceil \log_2 x \rceil$ be the inverse function of 2^x, and let $g(x) = f(\lceil \log_2 x \rceil)$. Now, g is a detour function, and obviously we have $g'(x) = f(\lceil \log_2 2^x \rceil) = f(x)$, and hence, $\mathrm{dg}(g)' = \mathbf{a}$. □

6 A Few Theorems and Some Comments

The implication $f(x) \leq g(x) \Rightarrow \mathrm{dg}(f) \leq \mathrm{dg}(g)$ follows straightforwardly from our definitions. The reverse implication is not true. (So, our degree structure is not trivial.)

Theorem 5. $\mathcal{P}^-(x+1)_* \subseteq \mathcal{P}^-(x)_*$.

By Lemma 7 and Theorem 5, we have $\mathrm{dg}(f(x)+a) = \mathrm{dg}(f(x)+b)$ for any $a, b \in \mathbb{N}$ and any detour function f. The long and technical proof of the Theorem 5 is available in a preprint posted on the Internet [12]. By extending the ideas of this proof we expect to be able to prove that we have $\mathrm{dg}(x^n) = \mathrm{dg}(x^n + p(x))$ for any $n \in \mathbb{N}$ and any detour polynomial p of degree strictly less than n.

Theorem 6. *For any decidable problem (computable set) A there exists a detour function f such that* $A \in \mathcal{P}^-(f)_*$.

Proof. Let A be any computable set. By Kleene's Normal Form Theorem (see e.g. Odifreddi [21] p. 179) there exists a primitive recursive function $\mathcal{U}$ and a primitive recursive relation $\mathcal{T}$ such that $\chi_A(x) = \mathcal{U}((\mu i)[\mathcal{T}(x,i)])$. It can be proved that $\mathcal{U}, \mathcal{T} \in \mathcal{P}^-$. Let $g(x) = (\mu i)[\mathcal{T}(x,i)]$ and $f(x) = \sum_{i \leq x} g(i)$. Now, f is a detour function and $A \in \mathcal{P}^-(f)_*$. □

The previous theorem shows that the detour degrees span all the computable sets. The next theorems show that the degree structure is sufficiently fine-grained to shed light upon (open) problems in subrecursion theory and complexity theory. We omit the proofs of these theorems due to space limitations. (Recall that any unary number-theoretic function generated by composition from constants, $+$, $\times$ and 2^x, is a detour function. The small Grzegorczyk classes are defined in Section 2.)

Theorem 7 (Turing Machine Space Classes). *For $i \in \mathbb{N}$, let* SPACE 2_i^{LIN} *be the set of problems decidable by a deterministic Turing machine working in space $2_i^{c|x|}$ for some $c \in \mathbb{N}$ (where $2_0^x = x$ and $2_{i+1}^x = 2^{2_i^x}$). Then, $A \in$* SPACE 2_i^{LIN} *iff $A \in \mathcal{P}^-(2_i^{p(x)})_*$ for some polynomial p. Moreover, for any $i \in \mathbb{N}$ and any polynomial p, the function $2_i^{p(x)}$ is a detour function*

Theorem 8 (Small Grzegorczyk Classes). *(i) $A \in \mathcal{E}_*^0$ iff $A \in \mathcal{P}^-(x+k)_*$ for some k. (ii) $A \in \mathcal{E}_*^1$ iff $A \in \mathcal{P}^-(kx)_*$ for some k. (iii) $A \in \mathcal{E}_*^2$ iff $A \in \mathcal{P}^-(p)_*$ for some polynomial p.*

Now, it follows from Theorem 5 that $\mathcal{P}_*^- = \mathcal{E}_*^0$, and it is an open problem whether any of the inclusions $\mathcal{E}_*^0 \subseteq \mathcal{E}_*^1 \subseteq \mathcal{E}_*^2$ are strict. Thus, it will be very hard to prove that $\mathbf{0} \neq \text{dg}(p)$ for some polynomial p. It might even not be true. However, in our degree structure the open problem $\mathcal{E}_*^0 \overset{?}{\subseteq} \mathcal{E}_*^1 \overset{?}{\subseteq} \mathcal{E}_*^2$ manifests itself as a special case of a more general open problem: Do there exist $n \in \mathbb{N}$ and a detour polynomial p such that $\mathbf{0}^n < \text{dg}(2_n^{p(x)})$?

Finally, we will argue that there exist incomparable degrees below $\mathbf{0}'$, i.e., we have $\mathbf{a}, \mathbf{b} < \mathbf{0}'$ such that $\mathbf{a} \not\leq \mathbf{b}$ and $\mathbf{b} \not\leq \mathbf{a}$. Let $d_0 = 0$ and $d_{i+1} = 2^{d_i}$, and for $j = 0, 1, 2$, let

$$f_j(x) = \begin{cases} 2^x & \text{if } d_{3i+j} \leq x < d_{3i+j+1} \\ \max(f_j(x \dot{-} 1), x) & \text{otherwise.} \end{cases}$$

Now, f_0, f_1, f_2 are detour functions such that $\max(f_0(x), f_1(x), f_2(x)) = 2^x$ and $\min(f_0(x), f_1(x), f_2(x)) = x$. Let $\mathbf{a} = \text{dg}(f_0)$, $\mathbf{b} = \text{dg}(f_1)$ and $\mathbf{c} = \text{dg}(f_2)$. Then, we have $\mathbf{a} \cup \mathbf{b} \cup \mathbf{c} = \mathbf{0}'$ and $\mathbf{a} \cap \mathbf{b} \cap \mathbf{c} = \mathbf{0}$. Now, since $\mathbf{0} \neq \mathbf{0}'$, any two of the three degrees will probably be incomparable, but we are not sure how we best can prove this. (Since the three degrees are uniformly constructed it is extremely unlikely that one of them should equal $\mathbf{0}$ and one of them should equal $\mathbf{0}'$.) The construction can be generalised to yield incomparable degrees between any two degrees $\mathbf{a}, \mathbf{b}$ where $\mathbf{a} < \mathbf{b}$.

In general, we expect our degree structure to be interesting and challenging from a degree-theoretic point of view. The methods required to investigate the structure will probably be a mixture of the number-theoretic techniques developed by Kristiansen for honest degrees and the techniques developed for degrees of computable sets, e.g. delayed diagonalisation.

7 Stronger Reducibility Relations

Our reducibility relation is slightly weak from a complexity-theoretic point of view as the number of steps in a $\mathcal{P}^-$-computation is not bounded by a polynomial in the *length* of the input. (The number of steps in a $\mathcal{P}^-$-computation is bounded by a polynomial in the input x and thus by a number-theoretic expression of the form $2^{k|x|}$ where $|x|$ denotes the length of the input.) However, we believe that we can develop the structure of detour degrees for stronger reducibilities, e.g. by resorting to the function algebra $\mathcal{P}_b^-$. This function algebra is defined as $\mathcal{P}^-$ but the scheme of primitive recursion is replaced by the scheme of (primitive) recursion on notation, i.e., the scheme

$$\phi(\vec{x},0) = \xi(\vec{x}) \quad \phi(\vec{x},2y) = \psi_0(\vec{x},y,\phi(\vec{x},y)) \quad \phi(\vec{x},2y+1) = \psi_1(\vec{x},y,\phi(\vec{x},y)) \,.$$

The number of steps in a $\mathcal{P}_b^-$-computation is bounded by a polynomial in the length of the input, and $\mathcal{P}_{b*}^-$ is included in LOGSPACE.

References

1. Ambos-Spies, K.: Polynomial time reducibilities and degrees. In: Griffor, E. (ed.) Handbook of Computability Theory, Elsevier, Amsterdam (1999)
2. Basu, S.: On the structure of subrecursive degrees. J. Comput. System Sci. 4, 452–464 (1970)
3. Clote, P.: Computation models and function algebra. In: Griffor, E. (ed.) Handbook of Computability Theory, Elsevier, Amsterdam (1999)
4. Grzegorczyk, A.: Some classes of recursive functions. Rozprawy Matematyczne IV, Warszawa (1953)
5. Kristiansen, L.: Neat function algebraic characterizations of Logspace and Linspace. Computational Complexity 14, 72–88 (2005)
6. Kristiansen, L.: Complexity-theoretic hierarchies induced by fragments of Gödel's *T*. In: Theory of Computing Systems, Springer, Heidelberg (July 2007)
7. Kristiansen, L.: A jump operator on honest subrecursive degrees. Archive for Mathematical Logic 37, 105–125 (1998)
8. Kristiansen, L.: Subrecursive degrees and fragments of Peano arithmetic. Archive for Mathematical Logic 40, 365–397 (2001)
9. Kristiansen, L.: Information content and computational complexity of recursive sets. In: Gödel 1996. Logical Foundations of Mathematics, Computer Science and Physics – Kurt Gödel Legacy. Lecture Notes in Logic., vol. 6, pp. 235–246. Springer, Heidelberg (1996)
10. Kristiansen, L.: Low_n, $high_n$, and intermediate subrecursive degrees. In: Combinatorics, computation and logic. Proceedings of DMTCS 1999 and CATS 1999, Auckland, New Zealand. Australian Computer Science Communications, vol. 21, pp. 235–246. Springer, Heidelberg (1996)
11. Kristiansen, L., Barra, G.: The small Grzegorczyk classes and the typed λ-calculus. In: Cooper, S.B., Löwe, B., Torenvliet, L. (eds.) CiE 2005. LNCS, vol. 3526, pp. 252–262. Springer, Heidelberg (2005)
12. Kristiansen, L., Voda, P.: Constant detours do not matter and so $\mathcal{P}_*^- = \mathcal{E}_*^0$, `http://www.ii.fmph.uniba.sk/~voda/E0.ps`
13. Kristiansen, L., Voda, P.: Programming languages capturing complexity classes. Nordic Journal of Computing 12, 1–27 (2005) (special issue for NWPT 2004)
14. Kristiansen, L., Voda, P.: The trade-off theorem and fragments of Gödel's *T*. In: Cai, J.-Y., Cooper, S.B., Li, A. (eds.) TAMC 2006. LNCS, vol. 3959, pp. 654–674. Springer, Heidelberg (2006)

15. Ladner, R.: On the structure of polynomial time reducibility. J. Assoc. Comput. Mach. 22, 155–171 (1975)
16. Machtey, M.: On the density of honest subrecursive classes. J. Comput. System Sci. 10, 183–199 (1975)
17. Machtey, M.: The honest subrecursive classes are a lattice. Information and Control 24, 247–263 (1974)
18. Machtey, M.: Augmented loop languages and classes of computable functions. J. Comput. System Sci. 6, 603–624 (1972)
19. Merkle, W.: Lattice embeddings for abstract bounded reducibilities. SIAM Journal on Computing 31, 1119–1155 (2002)
20. Meyer, A., Ritchie, D.: A classification of the recursive functions. Z. Math. Logik Grundlagen Math. 18, 71–82 (1972)
21. Odifreddi, P.: Classical recursion theory. In: Studies in Logic and the Foundations of Mathematics, vol. 125, (Paperback edition 1992 edn.) North-Holland Publishing Co., Amsterdam (1989)
22. Ritchie, R.W.: Classes of predictably computable functions. Transactions of the American Mathematical Society 106, 139–173 (1963)

Hamiltonicity of Matching Composition Networks with Conditional Edge Faults

Sun-Yuan Hsieh* and Chia-Wei Lee

Department of Computer Science and Information Engineering,
National Cheng Kung University
hsiehsy@mail.ncku.edu.tw, cwlee@csie.ncku.edu.tw

Abstract. In this paper, we sketch structure characterization of a class of networks, called Matching Composition Networks (MCNs), to establish necessary conditions for determining the conditional fault hamiltonicity. We then apply our result to n-dimensional restricted hypercube-like networks, including n-dimensional crossed cubes, and n-dimensional locally twisted cubes, to show that there exists a fault-free Hamiltonian cycle if there are at most $2n-5$ faulty edges in which each node is incident to at least two fault-free edges. We also demonstrate that our result is worst-case optimal with respect to the number of faulty edges tolerated.

Keywords: Algorithmica aspect of network problems, conditional edge faults, fault-tolerance, graph theory, Hamiltonian cycles, Hamiltonicity, matching composition networks, multiprocessor systems, restricted hypercube-like networks.

1 Introduction

Cycles (rings), the most fundamental class of network topologies for parallel and distributed computing, are suitable for designing simple algorithms with low communication costs. Numerous efficient algorithms based on cycles for solving various algebraic problems and graph problems can be found in [1,10]. These algorithms can also be used as control/data flow structures for distributed computing in arbitrary networks.

Usually, when the Hamiltonicity of a graph G is an issue, investigations center in whether G is Hamiltonian or Hamiltonian-connected. A *Hamiltonian cycle* (*Hamiltonian path*) in a graph is a cycle (path) goes through every node of G exactly once. G is called *Hamiltonian* if there is a Hamiltonian cycle in G, and *Hamiltonian-connected* if there is a Hamiltonian path between every two distinct nodes of G. Examples of applying Hamiltonian paths and cycles to practical problems include on-line optimization of a complex flexible manufacturing system [2] and wormhole routing [16,17]. These applications motivated us to embed Hamiltonian cycles in networks.

* Corresponding author.

M. Agrawal et al. (Eds.): TAMC 2008, LNCS 4978, pp. 160–169, 2008.

Since link (edge) faults may occur when a network is put into use, it is important to consider faulty network. Investigating the Hamiltonicity for various faulty networks has received a great deal of attention in recent years[4,5,6,7,9,12,13,14]. Among the proposed works, there are two assumptions for faulty edges: One is the *standard fault-assumption* in which there is no restriction on the distribution of faulty edges. This assumption is also adapted in [4,7,12,14]. The other is the *conditional fault-assumption* in which each node is incident to at least two fault-free edges. This assumption is adapted in [5,6,9,13].

In this paper, we investigate the hamiltonicity of a wide class of interconnection networks, called the Matching Composition Networks (MCNs), under the conditional fault-assumption. Basically, each MCN is constructed from two graphs G_0 and G_1 with the same number of nodes by adding a perfect matching between the nodes of G_0 and G_1. We first establish necessary conditions for determining the hamiltonicity of an MCN with faulty edges. We then apply our result to a subclass of MCNs, called n-dimensional restricted hypercube-like networks, which include n-dimensional crossed cubes, and n-dimensional locally twisted cubes. We show that each mentioned cube contains a fault-free Hamiltonian cycle if there are at most $2n-5$ faulty edges in which each node is incident to at least two fault-free edges. Two of these particular applications are previously known results [6,9] using rather lengthy proofs. Our approach unifies these special cases and our proof is much simpler. We also demonstrate that our result is optimal in the worst case, with respect to the number of faulty edges tolerated.

The remainder of this paper is organized as follows: Section 2 introduces some definitions, notations, and properties. In Sections 3 and 4, we investigate the respective Hamiltonicity of MCNs and restricted hypercube-like networks, under the conditional fault-assumption. In Section 5, we apply our result to two multiprocessor systems. Finally, some concluding remarks are given in Section 6.

2 Preliminaries

A *graph* $G = (V, E)$ is comprised of a *node set* V and an *edge set* E, where V is a finite set and E is a subset of $\{(u, v)|\ (u, v)$ is an unordered pair of $V\}$. We also use $V(G)$ and $E(G)$ to denote the node set and edge set of G, respectively. Two nodes u and v are *adjacent* if (u, v) is an edge in G. For a node v, we call the nodes adjacent to it the *neighbors* of v, denoted by $N_G(v)$. The *degree* of node v, denoted by $d_G(v)$, is the number of edges incident to it, i.e., $d_G(v) = |N_G(v)|$. Let $\delta(G) = min\{d_G(v)|\ v \in V(G)\}$. A *path* $P[v_0, v_t] = \langle v_0, v_1, ..., v_t\rangle$, is a sequence of distinct nodes such that any two consecutive nodes are adjacent. A path may contain other subpaths, denoted by $\langle v_0, v_1, ..., v_i, P[v_i, v_j], v_j, v_{j+1}, ..., v_t\rangle$, where $P[v_i, v_j] = \langle v_i, v_{i+1}, ..., v_{j-1}, v_j\rangle$. A *cycle* $\langle v_0, v_1, \ldots, v_k, v_0\rangle$ for $k \geq 2$ is a sequence of nodes in which any two consecutive nodes are adjacent, where $v_0, v_1, \ldots, v_k$ are all distinct.

A cycle that traverses each node of G exactly once is a *Hamiltonian cycle* and a graph is said to be *Hamiltonian* if there exists a Hamiltonian cycle. A *Hamiltonian path* is a path that contains every node of G and a graph is *Hamiltonian-connected* if there exists a Hamiltonian path between any two distinct nodes of G. A graph G is called *k-fault Hamiltonian* (respectively, *k-fault Hamiltonian-connected*) if there exists a Hamiltonian cycle (respectively, a Hamiltonian path between each pair of distinct nodes) in $G - F$ for any set F of faulty edges with $|F| \leq k$. A graph G is *k-mixed-fault Hamiltonian-connected* if there exists a Hamiltonian path between each pair of distinct nodes in $G - F$ for any set F of faulty elements (nodes and/or edges) with $|F| \leq k$. The term *k-mixed-fault Hamiltonian* can be defined similarly. A graph G is k-fault edge-Hamiltonian-connected if there exists a Hamiltonian path $P[x, y]$ between arbitrary two adjacent nodes x and y in G with $|F| \leq k$. Furthermore, a graph G is said to be *conditional k-fault Hamiltonian* if there exists a Hamiltonian cycle in $G - F$ for any set F of faulty edges with $|F| \leq k$, under the conditional fault-assumption.

Lemma 1. [12] *If a graph G is k-mixed-fault Hamiltonian-connected and $(k+1)$-mixed-fault Hamiltonian, then $k \leq \delta(G) - 3$ and thus $k + 4 \leq |V(G)|$.*

A *matching* in a graph G is a set of non-loop edges with no shared end-nodes. The nodes incident to the edges of a matching M are *saturated* by M; the others are *unsaturated.* A *perfect matching* in a graph is a matching that saturates every node. The class of *Matching Composition Networks* (MCNs) is defined as follows.

Definition 1. Let G_0 and G_1 be two graphs with the same number of nodes. Let PM be an arbitrary perfect matching between $V(G_0)$ and $V(G_1)$, i.e., PM is a set of edges connecting $V(G_0)$ and $V(G_1)$ in a one to one fashion; the resulting composition graph, denoted by $G_0 \bigoplus G_1$, is called a *Matching Composition Network* (MCN), where the symbol "$\bigoplus$" represents an operation that connects G_0 and G_1 by adding a perfect matching between $V(G_0)$ and $V(G_1)$. For convenience, G_0 and G_1 are called the *components* of an MCN. Note that $V(G_0 \bigoplus G_1) = V(G_0) \bigcup V(G_1)$ and $E(G_0 \bigoplus G_1) = E(G_0) \bigcup E(G_1) \bigcup PM$.

For two end-nodes of an edge in PM, one is said to be the *crossing neighbor* of the other. For convenience, we use $\bar{x}$ to denote the crossing neighbor of x.

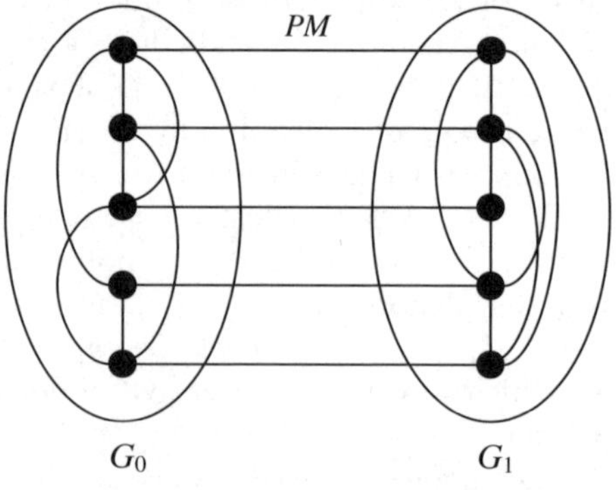

Fig. 1. An example of an MCN

Vaidya *et al.*[15] introduced a class of hypercube-like networks, called *HL-graphs*, which can be defined recursively as follows: (1) $HL_0 = \{K_1\}$, where K_1 is a single node, and (2) for $n \geq 1$, $HL_n = \{G_0 \bigoplus G_1 | \ G_0, G_1 \in HL_{n-1}\}$. A graph which belongs to HL_n is called an *n-dimensional HL-graph.* Clearly, HL_n belongs to MCNs. Note that each graph in HL_n has 2^n nodes and is regular with the common degree n. Furthermore, non-bipartite HL-graphs are said to be *restricted HL-graphs* (*RHL*), which can be recursively defined as follows [12]: $RHL_0 = \{K_1\}, RHL_1 = \{K_2\}, RHL_2 = \{C_4\}, RHL_3 = \{G(8,4)\}$, and $RHL_n = \{G_0 \bigoplus G_1 | \ G_0, G_1 \in RHL_{n-1}\}$ for $n \geq 4$, where C_4 is a cycle graph with 4 nodes and $G(8,4)$ is a recursive circulant defined in [11]. A graph which belongs to RHL_n is called an *n-dimensional restricted HL-graph.*

Definition 2. A graph G is said to be *2-disjoint-path-coverable* (*2-DPC* for short) if given four distinct nodes x, y, u, and v in G, there exists two paths, $P[x,y]$ and $P[u,v]$, such that $V(P[x,y]) \bigcap V(P[u,v]) = \emptyset$ and $V(P[x,y]) \bigcup V(P[u,v]) = V(G)$.

Hereafter, let F be the set of faulty edges in $G = G_0 \bigoplus G_1$ and let F_0 and F_1 be the sets of faulty edges in G_0 and G_1, respectively. Also, let F_{PM} be the set of faulty edges in PM. Note that $F = F_0 \bigcup F_1 \bigcup F_{PM}$. For convenience, let $f_0 = |F_0|$, $f_1 = |F_1|$, and $f_{PM} = |F_{PM}|$.

Definition 3. A graph G is said to be *k-fault super-Hamiltonian* if the following properties hold: (1) G is k-mixed-fault Hamiltonian-connected, (2) G is $(k+1)$-fault Hamiltonian, (3) G is $(k+1)$-fault edge-Hamiltonian-connected, and (4) G is conditional $(2k+1)$-fault Hamiltonian.

Definition 4. An MCN $G = G_0 \bigoplus G_1$ has the *edge-property* if for a Hamiltonian cycle C_i in G_i, $i = 0, 1$, there exist two edges (x,y) and (u,v) in C_i such that $(\bar{x}, \bar{y})$ and $(\bar{u}, \bar{v})$ are two edges in G_{1-i}, where x, y, u, v are all distinct.

3 Conditional Fault Hamiltonicity of MCNs

Given the set of faulty edges F in a graph G, the notation $\delta(G - F) \geq 2$ means that each node is incident to at least two fault-free edges in $G - F$. The following result determines the conditional fault hamiltonicity of MCNs.

Theorem 1. *Let $G = G_0 \bigoplus G_1$ be an MCN and let k_i be a positive integer such that $|V(G_i)| \geq 4k_i + 7$ for $i = 0, 1$. Then, G is conditional $(2k+3)$-fault Hamiltonian, where $k = \min\{k_0, k_1\}$, if the following conditions hold: (1) each G_i is 2-DPC, (2) G_i is k_i-fault super-Hamiltonian, and (3) G has the edge-property.*

Proof. Without loss of generality, we assume that $f = 2k+3$ and $f_0 \geq f_1$. Hence, $f_1 \leq \lfloor (2k+3)/2 \rfloor = k+1$. We have the following three cases.

Case 1: $f_0 \leq 2k+1$. Since G_0 is k_0-mixed-fault Hamiltonian-connected and (k_0+1)-fault Hamiltonian, then by Lemma 1, $\delta(G_0) \geq k_0 + 3$. Hence, there is at most one node incident to only one fault-free edge in G_0; otherwise,

$f_0 \geq (\delta(G_0) - 1) + (\delta(G_0) - 1) - 1 = 2\delta(G_0) - 3 \geq 2k_0 + 3 \geq 2k + 3$, which contradicts to the fact that $f_0 \leq 2k + 1$, which leads to the following subcases.

Case 1.1: $\delta(G_0 - F_0) \geq 2$. Since $f_0 \leq 2k+1 \leq 2k_0+1$ and G_0 is conditional $(2k_0+1)$-fault Hamiltonian, there exists a Hamiltonian cycle C_0 in $G_0 - F_0$. We have the following scenarios.

Case 1.1.1: $f_1 = k + 1$; hence, $f_{PM} \leq 2k + 3 - 2(k + 1) = 1$. Since G has the edge-property, there are two edges (x, y) and (u, v) in C_0 such that $(\bar{x}, \bar{y})$ and $(\bar{u}, \bar{v})$ are also two edges in G_1, where x, y, u, v are all distinct. Since $f_{PM} \leq 1$, $|\{(x, \bar{x}), (y, \bar{y}), (u, \bar{u}), (v, \bar{v})\} \bigcap F_{PM}| \leq 1$ and thus there are at least three fault-free edges in $\{(x, \bar{x}), (y, \bar{y}), (u, \bar{u}), (v, \bar{v})\}$. Without loss of generality, we assume that $(x, \bar{x})$ and $(y, \bar{y})$ are both fault-free. By deleting (x, y) from C_0, we obtain a path $P_0[x, y]$. Moreover, since $f_1 = k + 1 \leq k_1 + 1$ and G_1 is $(k_1 + 1)$-fault edge-Hamiltonian-connected, there exists a Hamiltonian path $P_1[\bar{y}, \bar{x}]$ in

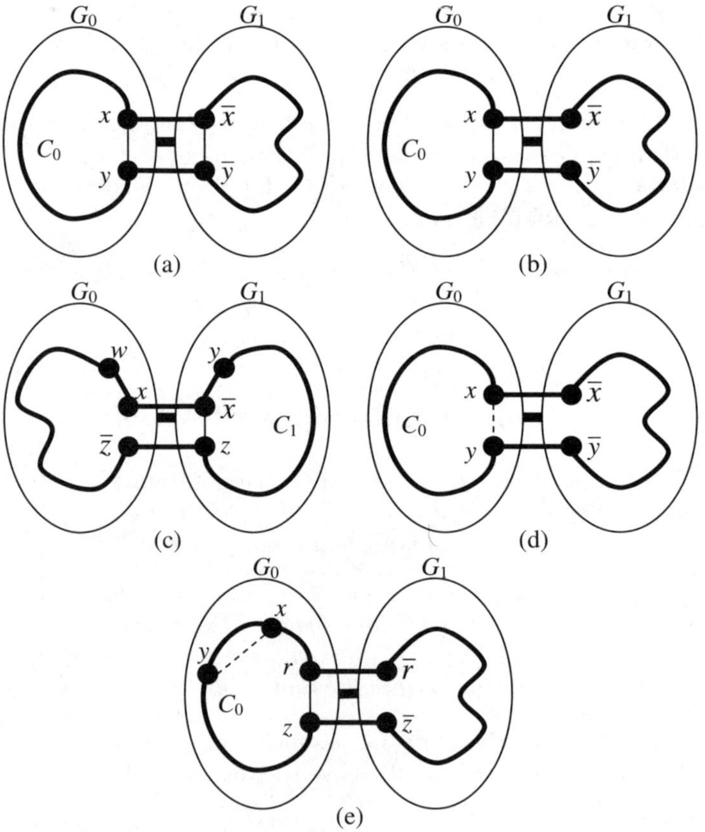

Fig. 2. Illustration of the proof of Theorem 1, where dotted lines represent faulty edges

$G_1 - F_1$. Therefore, a fault-free Hamiltonian cycle can be constructed (see Figure 2(a)).

Case 1.1.2: $f_1 \leq k$. Since $|V(G_0)| \geq 4k_0 + 7 \geq 4k + 7$, there exists an edge (x, y) in C_0 such that $(x, \bar{x})$ and $(y, \bar{y})$ are both fault-free[1]. By deleting (x, y) from C_0, we obtain a path $P_0[x, y]$. Moreover, since $f_1 \leq k = \min\{k_0, k_1\}$ and G_1 is k_1-mixed-fault Hamiltonian-connected, there exists a Hamiltonian path $P_1[\bar{y}, \bar{x}]$ in $G_1 - F_1$. A fault-free Hamiltonian cycle can be constructed (see Figure 2(b)).

Case 1.2: $\delta(G_0 - F_0) = 1$. Let x be a unique node incident to only one fault-free edge in G_0. Note that $f_0 \geq \delta(G_0) - 1 \geq k_0 + 2 \geq k + 2$, $f_1 + f_{PM} \leq (2k + 3) - (k + 2) = k + 1$, and $(x, \bar{x})$ is fault-free. We have the following scenarios.

Case 1.2.1: $f_1 = k + 1$; hence, $f_{PM} = 0$ and $f_0 = k + 2$. Since $f_1 = k + 1 \leq 2k+1 \leq 2k_1+1$ and G_1 is conditional $(2k_1+1)$-fault hamiltonian, there is a Hamiltonian cycle C_1 in $G_1 - F_1$. Let (w, x) be the unique fault-free edge incident to x in G_0. Since $f_0 = k+2$ and the number of faulty edges incident to x in G_0 is at least $\delta(G_0) - 1 \geq k_0 + 2 \geq k + 2$, the number of faulty edges in G_0 are all incident to x. Let y and z be two neighbors of $\bar{x}$ in C_1. Clearly, $\bar{y} \neq w$ or $\bar{z} \neq w$. Without loss of generality, we assume that $\bar{z} \neq w$. Moreover, since G_0 is k_0-mixed-fault Hamiltonian connected, there is a Hamiltonian path in $G_0 - x$, where $G_0 - x$ is the graph obtained by deleting x from G_0. Therefore, a fault-free Hamiltonian cycle can be constructed (see Figure 2(c)).

Case 1.2.2: $f_1 \leq k$. Since $f_{PM} \leq k+1$ and there are at least $k+2$ faulty edges incident to x, there is an edge $(x, y) \in F_0$ such that $(y, \bar{y}) \notin F_{PM}$. Since $f_1 \leq k \leq k_1$ and G_1 is k_1-mixed-fault Hamiltonian-connected, there exists a Hamiltonian path $P_1[\bar{y}, \bar{x}]$ in $G_1 - F_1$. Since $\delta(G_0 - (F_0 - \{(x, y)\})) = 2$, there is a Hamiltonian cycle C_0 in $G_0 - (F_0 - \{(x, y)\})$ such that C_0 contains (x, y). Therefore, a fault-free Hamiltonian cycle can be constructed (see Figure 2(d)).

Case 2: $f_0 = 2k + 2$; hence, $f_1 + f_{PM} \leq 1$. Note that there is at most one node that is incident to only one fault-free edge in G_0. There are the following two subcases.

Case 2.1: $\delta(G_0 - F_0) \geq 2$. We can select an edge $(x, y) \in F_0$ such that $(x, \bar{x})$ and $(y, \bar{y})$ are both fault-free. Since $|F_0 - \{(x, y)\}| = 2k + 1 \leq 2k_0 + 1$ and G_0 is conditional $(2k_0+1)$-fault Hamiltonian, there is a Hamiltonian cycle C_0 in $G_0 - (F_0 - \{(x, y)\})$. If C_0 contains (x, y), then a fault-free Hamiltonian cycle can be constructed similar as Case 1.2.2. However, if C_0 does not contain (x, y), we select an arbitrary edge (r, z) in C_0. By replacing x with r, y with z, $\bar{x}$ with $\bar{r}$, and $\bar{y}$ with $\bar{z}$, a fault-free Hamiltonian cycle can also be constructed (see Figure 2(e)).

Case 2.2: $\delta(G_0 - F_0) = 1$. Let x be a unique node incident to only one fault-free edge in G_0. Since $f_{PM} \leq 1$, we can select an edge $(x, y) \in F_0$ such

[1] Since $|V(G_0)| \geq 4k + 7$, C_0 contains at least $4k + 7$ edges and contributes total $4k + 7$ choices. If such an edge does not exist, then $f_{PM} \geq \lfloor \frac{4k+7}{2} \rfloor > 2k + 3 \geq f$, which is a contradiction.

that $(y, \bar{y}) \notin F_{PM}$. Since $|F_0 - \{(x, y)\}| = 2k+1$, $\delta(G_1 - (F_0 - \{(x, y)\})) = 2$, and G_0 is conditional $(2k_0 + 1)$-fault Hamiltonian, there is a Hamiltonian cycle C_0 in $G_0 - (F_0 - \{(x, y)\})$ such that C_0 contains (x, y). Therefore, a fault-free Hamiltonian cycle can be constructed similar as Case 1.2.2.

Case 3: $f_0 = 2k + 3$; hence, $f_1 = f_{PM} = 0$. In this case, there are at most two nodes in which each node is incident to only one fault-free edge in G_0. We can construct a fault-free Hamiltonian cycle similar to that used in above cases.

By combining above cases, we complete the proof. □

4 Conditional Fault Hamiltonicity of RHL_n

In this session, we will apply Theorem 1 to n-dimensional restricted hypercube-like networks RHL_n. We will demonstrate that the conditions described in Theorem 1 can be further simplified because RHL_n is recursively constructed.

Lemma 2. [12] *Every RHL_n for $n \geq 3$ is $(n-3)$-mixed-fault Hamiltonian-connected and $(n-2)$-mixed-fault Hamiltonian.*

The following lemmas show that RHL_n is 2-DPC for $n \geq 4$, and $(n-2)$-fault edge-Hamiltonian-connected for $n \geq 6$. Due to the space limitation, the proofs are omitted.

Lemma 3. *For any four distinct nodes x, y, u, and v in a graph $G \in RHL_n$ for $n \geq 4$, there exist two paths, $P[x, y]$ and $P[u, v]$, such that $P[x, y] \bigcap P[u, v] = \emptyset$ and $P[x, y] \bigcup P[u, v] = V(G)$.*

Lemma 4. *Let $G = G_0 \bigoplus G_1$ be a graph in RHL_n for $n \geq 6$. If G has the edge property and $G_0, G_1 \in RHL_{n-1}$ are $(n-3)$-fault edge-Hamiltonian-connected, then G is $(n-2)$-fault edge-Hamiltonian-connected.*

Theorem 2. *Let $G = G_0 \bigoplus G_1$ be a graph in RHL_n with $|V(G)| \geq 8n - 18$. Then, G is conditional $(2n-5)$-fault Hamiltonian if the following conditions hold: (1) each G_i is $(n-3)$-fault edge-Hamiltonian-connected, (2) each G_i is conditional $(2n-7)$-fault Hamiltonian, and (3) G has the edge-property.*

Proof. Note that $G_i \in RHL_{n-1}$. We prove this theorem by checking whether all the conditions of Theorem 1 hold. First, by Lemma 3, each G_i is 2-DPC. Secondly, since $|V(G)| \geq 8n - 18$, $|V(G_i)| \geq \frac{8n-18}{2} = 4(n-4) + 7$. By Lemma 2 and Conditions 2–3, each G_i is $(n-4)$-fault super-Hamiltonian. Moreover, the edge-property hold by Condition 3. Therefore, all the conditions of Theorem 1 are satisfied and the result follows. □

5 Application to Multiprocessor Systems

In this section, we apply Theorem 2 to two popular multiprocessor systems: n-dimensional crossed cubes, and n-dimensional locally twisted cubes, all of which belong to RHL_n. It is of course possible to apply them to many other potentially useful systems. Interested readers can refer to [3,18] for the definitions of the above networks.

To prove Theorem 3, we first show two useful lemmas in Lemma 5 and Lemma 6. Due to the space limitation, the proofs are omitted.

Lemma 5. *If C is a Hamiltonian cycle in CQ^i_{n-1} for $i = 0, 1$, where $n \geq 3$, then there exist two edges (x, y) and (u, v) in C such that $(\bar{x}, \bar{y})$ and $(\bar{u}, \bar{v})$ are also two edges in CQ^{1-i}_{n-1}, where x, y, u and v are all distinct.*

Lemma 6. *CQ_n for $n \geq 5$ is $(n-2)$-fault edge-Hamiltonian-connected.*

We can prove the following theorem by induction on n. The basis case for $n = 3, 4, 5$ can be shown using a computer program. For $n \geq 6$, we show the result by Theorem 2, Lemmas 5 and 6. Then, we have the following theorem.

Theorem 3. *CQ_n for $n \geq 3$ is conditional $(2n-5)$-fault Hamiltonian.*

The following lemmas demonstrate that LTQ_n has edge-property, and LTQ_n is $(n-2)$-fault-Hamiltonian-connected. The proofs of lemmas are omitted.

Lemma 7. *If C is a Hamiltonian cycle in LTQ^i_{n-1} for $i = 0, 1$, where $n \geq 3$, then there exist two edges (x, y) and (u, v) in C such that $(\bar{x}, \bar{y})$ and $(\bar{u}, \bar{v})$ are also two edges in LTQ^{1-i}_{n-1}, where nodes x, y, u and v are distinct.*

Lemma 8. *LTQ_n for $n \geq 5$ is $(n-2)$-fault edge-Hamiltonian-connected.*

We can prove the following theorem by induction on n. The basis case for $n = 3, 4, 5$ can be shown using a computer program. For $n \geq 6$, we show the result by Theorem 2, Lemmas 7 and 8. Thus, we have Theorem 4.

Theorem 4. *LTQ_n for $n \geq 3$ is conditional $(2n-5)$-fault Hamiltonian.*

6 Concluding Remarks

In this paper, we have focused on investigating conditional fault hamiltonicty of matching composition networks. By applying the established theorem to restricted hypercube-like networks, we have successfully determined the conditional fault-tolerant hamiltonicities of two popular multiprocessor systems, including crossed cubes, and locally twisted cubes.

Since an n-dimensional restricted hypercube-like network G, where $n \geq 3$, studied in this paper possesses the edge-property, the girth (the length of the shortest cycle) equals four. This leads to a worst-case scenario: Let $\langle u, t, v, s, u \rangle$ forms a cycle of length 4. Also, assume that $(u, s), (u, t), (v, s)$ and (v, t) are

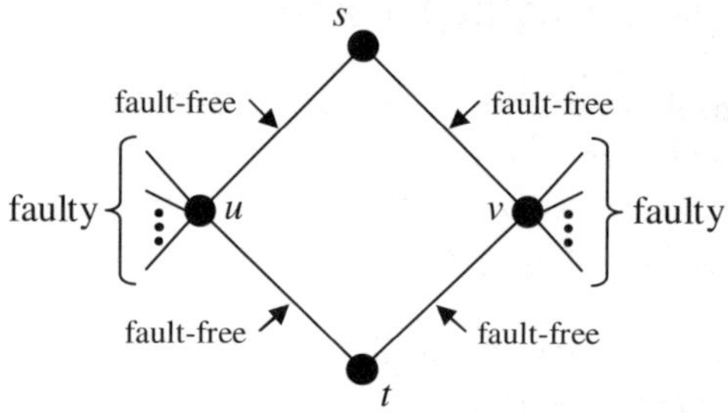

Fig. 3. A worst case scenario

fault-free. Clearly, any fault-free cycle containing nodes u and v must contain $(u, s), (u, t), (v, s)$ and (v, t). Since G is regular of the common degree n, the above worst case contributed total $2n - 4$ faulty edges; moreover, it is impossible to generate a fault-free Hamiltonian cycle containing nodes u, t, v, and s. Therefore, $2n-5$ is worst-case optimal with respect to the number of faulty edges tolerated.

References

1. Akl, S.G.: Parallel Computation: Models and Methods. Prentice Hall, Englewood Cliffs (1997)
2. Ascheuer, N.: Hamiltonian Path Problems in the On-Line Optimization of Flexible Manufacturing Systems. PhD thesis, University of Technology, Berlin, Germany (1995)
3. Efe, K.: The crossed cube architecture for parallel computation. IEEE Transactions on Parallel and Distributed Systems 3(5), 513–524 (1992)
4. Fu, J.-S.: Fault-tolerant cycle embedding in the hypercube. Parallel Comptuing 29(6), 821–832 (2003)
5. Fu, J.-S.: Conditional fault-tolerant hamiltonicity of star graphs. In: 7th Parallel and Distributed Computing, Applications and Technologies (PDCAT 2006), pp. 11–16. IEEE Computer Society, Los Alamitos (2006)
6. Fu, J.-S.: Conditional fault-tolerant hamiltonicity of twisted cubes. In: 7th Parallel and Distributed Computing, Applications and Technologies (PDCAT 2006), pp. 5–10. IEEE Computer Society, Los Alamitos (2006)
7. Hsieh, S.-Y., Ho, C.-W., Chen, G.-H.: Fault-free hamiltoninan cycles in faulty arrangement graphs. IEEE Transactions on Parallel and Distributed Systems 10(3), 223–237 (1999)
8. Hsieh, S.-Y., Chen, G.-H., Ho, C.-W.: Longest fault-free paths in star graphs with edge faults. IEEE Transactions on Computers 50(9), 960–971 (2001)
9. Hung, H.-S., Chen, G.-H., Fu, J.-S.: Conditional fault-tolerant cycle-embedding of crossed graphs. In: 7th Parallel and Distributed Computing, Applications and Technologies (PDCAT 2006), pp. 90–95. IEEE Computer Society, Los Alamitos (2006)
10. Leighton, F.T.: Introduction to parallel algorithms and architecture: arrays· trees· hypercubes. Morgan Kaufmann, San Mateo, CA (1992)
11. Park, J.-H., Chwa, K.-Y.: Recursive circulant: A new topology for multicomputer networks. In: Internation Symposium Parallel Architectures, Algorithms and Networks (ISPAN 1994), pp. 73–80. IEEE press, New York (1994)

12. Park, J.-H., Kim, H.-C., Lim, H.-S.: Fault-hamiltonicity of hypercube-like interconnection networks. In: 19th IEEE International Parallel and Distributed Processing Symposium (IPDPS 2005), IEEE Computer Society, Los Alamitos (2005)
13. Tsai, C.-H.: Linear array and ring embeddings in conditional faulty hypercubes. Theoretical Computer Science 314, 431–443 (2004)
14. Tseng, Y.-C., Chang, S.-H., Sheu, J.-P.: Fault-tolerant ring embedding in a star graph with both link and node failures. IEEE Transactions on Parallel and Distributed Systems 8(12), 1185–1195 (1997)
15. Vaidya, A.S., Rao, P.S.N., Shankar, S.R.: A class of hypercube-like networks. In: 5th IEEE Symposium on Parallel and Distributed Proceeding (SPDP 1993), pp. 800–803. IEEE Press, Los Alamitos (1993)
16. Wang, N.-C., Chu, C.-P., Chen, T.-S.: A dual-hamiltonian-path-based multicasting strategy for wormhole-routed star graph interconnection networks. Journal of Parallel and Distributed Computing 62(12), 1747–1762 (2002)
17. Wang, N.-C., Yan, C.-P., Chu, C.-P.: Multicast communication in wormhole-routed symmetric networks with hamiltonian cycle model. Journal of Systems Architecture 51(3), 165–183 (2005)
18. Yang, X., Evans, D.J., Megson, G.M.: The locally twisted cubes. International Journal of Computer Mathematics 82(4), 401–413 (2005)

Local 7-Coloring for Planar Subgraphs of Unit Disk Graphs

J. Czyzowicz[1], S. Dobrev[2], H. González-Aguilar[3], R. Kralovic[4], E. Kranakis[5], J. Opatrny[6], L. Stacho[7], and J. Urrutia[8]

[1] Département d'informatique, Université du Québec en Outaouais, Gatineau, Québec J8X 3X7, Canada. Research supported in part by NSERC.
[2] Slovak Academy of Sciences, Bratislava, Slovakia. Research partially funded by grant APVV 0433-06.
[3] Centro de Investigacion en Matematicas, Guanajuato, Gto., C.P. 36000, Mexico.
[4] Department of Computer Science, Faculty of Mathematics, Physics and Informatics Comenius University, Bratislava, Slovakia. Research partially funded by grant APVV 0433-06.
[5] School of Computer Science, Carleton University, 1125 Colonel By Drive, Ottawa, Ontario, Canada K1S 5B6. Research supported in part by NSERC and MITACS.
[6] Department of Computer Science, Concordia University, 1455 de Maisonneuve Blvd West, Montréal, Qubec, Canada, H3G 1M8. Research supported in part by NSERC.
[7] Department of Mathematics, Simon Fraser University, 8888 University Drive, Burnaby, British Columbia, Canada, V5A 1S6. Research supported in part by NSERC.
[8] Instituto de Matemáticas, Universidad Nacional Autónoma de México, Área de la investigación cientifica, Circuito Exterior, Ciudad Universitaria, Coyoacán 04510, México, D.F. México.

Abstract. The problem of computing locally a coloring of an arbitrary planar subgraph of a unit disk graph is studied. Each vertex knows its coordinates in the plane, can directly communicate with all its neighbors within unit distance. Using this setting, first a simple algorithm is given whereby each vertex can compute its color in a 9-coloring of the planar graph using only information on the subgraph located within at most 9 hops away from it in the original unit disk graph. A more complicated algorithm is then presented whereby each vertex can compute its color in a 7-coloring of the planar graph using only information on the subgraph located within a constant number of hops away from it.

1 Introduction

We are interested in the graph vertex coloring as applicable to wireless ad-hoc networks. The wireless ad-hoc networks of interest to us are geometrically embedded in the plane and consist of a number of *location aware* nodes, say n, whereby two nodes are adjacent if and only if they are within the transmission range of each other. If all the nodes have the same transmission range then these networks are known as *unit disk graphs*. For such graphs there have been several papers in the literature addressing the coloring problem. Among these it is

M. Agrawal et al. (Eds.): TAMC 2008, LNCS 4978, pp. 170–181, 2008.

worth mentioning the work by Marathe et al. [10] (presenting an on-line coloring heuristic which achieves a competitive ratio of 6 for unit disk graphs: the heuristic does not need a geometric representation of unit disk graphs which is used only in establishing the performance guarantees of the heuristics), Graf et al. [7] (which improves on a result of Clark, Colbourn and Johnson (1990) and shows that the coloring problem for unit disk graphs remains NP-complete for any fixed number of colors $k \geq 3$), Caragiannis et al. [2] (which proves an improved upper bound on the competitiveness of the on-line coloring algorithm First-Fit in disk graphs which are graphs representing overlaps of disks on the plane) and Miyamoto et al. [11] (which constructs multi-colorings of unit disk graphs represented on triangular lattice points).

There are also several papers on coloring restricted to planar graphs of which we note Ghosh et al. [6] because it is concerned with a self-stabilizing algorithm for coloring such graphs. Their algorithm achieves a 6 coloring by transforming the planar graph into a DAG of out-degree at most five. However, this algorithm needs the full knowledge of the topology of the graph. The specificity of the problem for ad-hoc networks requires a different approach. An ad-hoc network can be a very large dynamic system, and in some cases a node can join or leave a network at any time. Thus, the full knowledge of the topology of an ad-hoc network might not be available, or possible for each node of the network at all times. Thus algorithms that can make computations in a fully distributed manner, using in each node only information about the network within a fixed distance neighborhood of the node, are of particular interest in ad hoc networks. Examples of algorithms of this type are the Gabriel test [5] for constructing a planar spanner, face routing [8],[1] or an approximation of the minimum spanning tree [14], [3].

To reduce network complexity, the unit disk graph G is sometimes reduced to a much smaller subset P of its edges called a spanner. A good spanner must have some properties so that certain parameters of communication within P are preserved. To ensure all to all communication P must be connected. An important property is having a constant stretch factor s, guaranteeing that the length of a path joining two nodes in G is at most s times shorter that the shortest path joining these nodes in P. A desired property of P is planarity, which, on one hand, permits an efficient routing scheme based on *face routing* and, on the other hand, ensures linear complexity of P with respect to its number of nodes. Planar graphs also have low chromatic number, hence a small set of frequencies is sufficient to realize radio communication.

In this paper, we are interested in *local* distributed coloring algorithms whereby messages emanating from any node can propagate for only a constant number of hops. This model was first introduced in the seminal paper of Linial [9]. One of the advantages of this model is that it aims to obtain algorithms that could cope with a dynamically changing infrastructure in a network. In this approach, each node may communicate with nodes at a bounded distance from it and thus a local change in the network only needs a local adjustment of a solution. In Linial's model of communication, locality results in a constant-time distributed algorithm.

1.1 Network Model and Results of the Paper

We are given a set S of n points in the plane and a planar subgraph G of the unit disk graph induced by S. We assume that G is connected and all the nodes either know their exact (x, y) coordinates (which could be achieved for example by having the nodes equipped with GPS devices) or have a consistent relative coordinate system. If G is not connected, all reasoning can be applied to each connected component of G independently, in which case the coloring constructed applies to each connected component. We propose two local coloring algorithms. The first simple algorithm computes a 9-coloring of the planar graph. We assume that each node knows its 9-neighborhood in the original unit disk graph (i.e. all nodes at distance at least 9 hops away from it), it can communicate with each node of its neighborhood and it is aware which of these nodes belong to the planar subgraph. We then present a more complicated algorithm whereby each vertex can compute its color in a 7-coloring of the planer graph using only information on the subgraph located within at most a constant number $h = 201$ of hops away from it. The constant h is quite large though in practice nodes at much smaller distance will need to communicate. The algorithm does not determine locally either what the different connected components are or even what are the local parts of a component connected somewhere far away. Moreover, the correctness of the algorithm is independent of the connectivity of the planar subgraph.

2 Simple Local Coloring Algorithm

The basic idea of the coloring algorithms in this paper is to partition the plane containing G into fixed sized areas, compute a coloring of the subgraph of G within each such area independently and possibly adjust colors of some vertices that are on the border of an area and thus are adjacent to nodes in another area. This is possible to do consistently and without any pre-processing because the nodes know their coordinates and thus can determine the area in which they belong. Since each area is of fixed size, a subgraph of the given unit disk graph belonging to this area is of a bounded diameter. Hence a constant number of hops is needed for a node to communicate with each other node of the same area.

2.1 Coloring with Regular Hexagonal Tilings

The simplest partitioning we consider is obtained by tiling the plane with regular hexagons having sides of size 1. We suppose that two edges of the hexagons are horizontal and one of the hexagons is centered in coordinates $[0, 0]$. To assure the disjointness of the hexagon areas we assume that only the upper part of the boundary and the leftmost vertex belongs to each hexagon area while the rightmost vertex does not belong to it. Under such conditions, two vertices of G can be connected by an edge only if they are in the same or adjacent hexagon areas.

As each vertex knows its own coordinates, it can calculate which hexagon it belongs to. In the first step each vertex communicates with vertices within its hexagon and learns the part of the subgraph located in its hexagon. By Lemma 1, communication inside a hexagon may be done using at most nine hops.

Lemma 1. *Any connected component of the subgraph of a unit disk graph induced by its nodes belonging to a regular hexagon with sides of size* 1 *has a diameter smaller or equal to nine. Moreover, there exist configurations of nodes inside the hexagon of diameter equal to nine.*

We could apply the standard 4-coloring algorithm for each connected component of the graph induced by G on points within each hexagon. Since the tiling of the plane by hexagons can be 3-colored, three disjoint sets of four colors can be used, one set of colors for vertices in each hexagon area of the same color. This

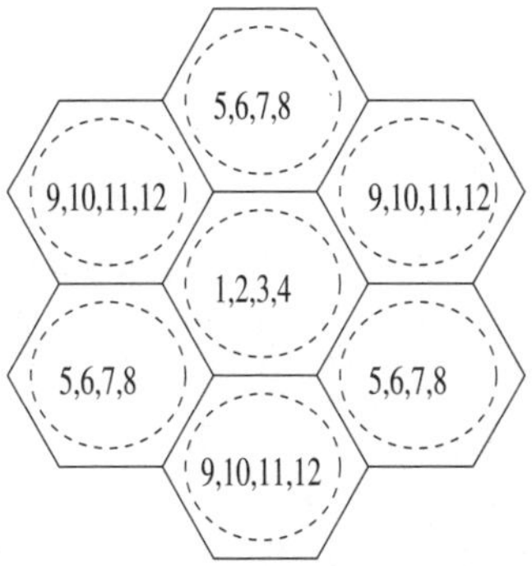

Fig. 1. Coloring of a hexagonal tiling of the plane using 12 colors

would lead to a 12 coloring of G (see the coloring scheme of the hexagonal tiling depicted in Figure 1).

The number of colors can be reduced by coloring the outer face of each component in a hexagon with prescribed colors, using the result of the following theorem.

Theorem 1 (3 + 2 Coloring, Thomassen [12]). *Given a planar graph* G*,* 3 *prescribed colors and an outer face* F *of* G*, graph* G *can be* 5*-colored while the vertices of* F *use only the prescribed three colors.*

The precise result in Thomassen's paper [12] (see also Dörre [4]) states that every planar graph is L-list colorable for every list assignment with lists of size 5. The proof of this result however implies the following statement taken from Tuza et al. [13] that for a planar graph G with outer face F, every pre-coloring of two adjacent vertices v_1, v_2 of F can be extended to a list coloring of G for every list assignment with $|L(v)| = 3$ if $v \in V(F) - \{v_1, v_2\}$ and $|L(v)| = 5$ if $v \notin V(F)$, where $V(F)$ denotes the vertex set of F.

Using Theorem 1 the number of colors can be reduced to 9 as follows. The idea is that the vertices of G in a hexagon can be adjacent only to the outer face vertices of the graphs induced in the neighboring hexagons. Three disjoint

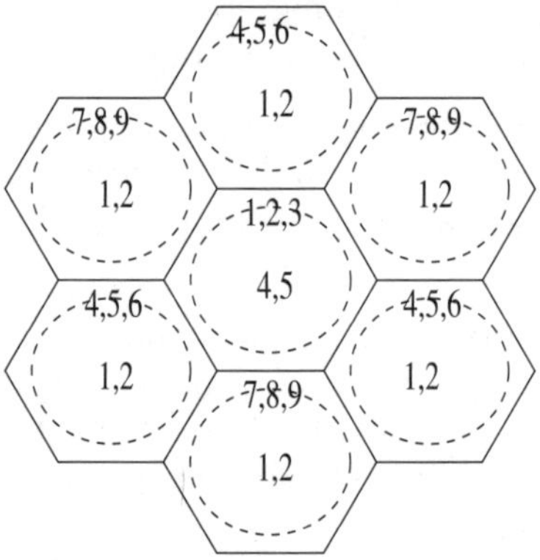

Fig. 2. Colorings of a hexagonal tiling of the plane using 9-colors

sets of colors of size three are used as the prescribed colors of the vertices on the outer faces, the inner vertices of G in a hexagon can employ, in addition to the three colors used on the outer face of its hexagon, two additional colors of the outer faces of other hexagons (see Figure 2).

Theorem 2. *Using the partitioning of a plane into regular hexagons with sides of size 1, a vertex can compute its color in a 9-coloring of the planar graph using only information on the subgraph located within at most 9 hops away from it.*

3 The 7-Coloring Algorithm

In this section we give a 7 coloring algorithm and prove its correctness. The trade-off is the larger area of the network a node needs to examine in this algorithm and a more complex partitioning of the plane.

3.1 Reducing the Number of Colors Using Mixed Tilings

We will employ the tiling using octagons and squares shown in Figure 3. Each square is of size $5 + \epsilon$ while the slanted part of an octagon border is of length $3 + \epsilon$, meaning that these sides can be chosen arbitrarily close to but greater than 5 and 3, respectively. We shall assume that one of the octagons is centered in coordinates $[0, 0]$. The reasons for choosing the sizes of tiles this way is to isolate the meeting places of octagons (the slanted border part) from each other so that each of them can be dealt with independently and locally, and to ensure that there is no edge between two vertices in different squares or between two vertices of the same squares that could be recolored for different crossings in the algorithm. In handling a meeting place, only vertices at a distance at most 2 from it will be recolored, therefore it is impossible to have two neighboring vertices recolored due to different crossing.

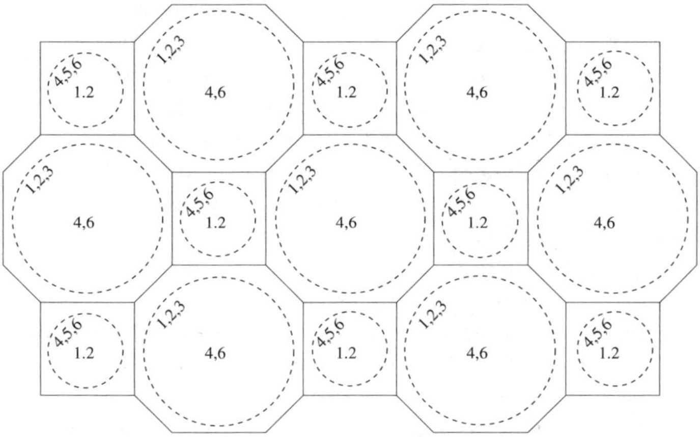

Fig. 3. A coloring of the set of points based on the octagon/square tiling of the plane

Similarly as in the previous coloring, the subgraph induced by the vertices in a tile is colored using 5 colors, three of them are used on the outer face and these three colors plus the additional two colors are used on vertices not on the outer face using Theorem 1. Since each node knows its location, it can determine which tile it belongs to. By Lemma 2 the communication inside an octagon may be done using at most 201 hops.

Lemma 2. *Any subgraph of a unit disk graph induced by its vertices belonging to an octagon used in the algorithm has a diameter smaller or equal to* 201*.*

Despite the fact that the simple, surface comparing argument leaves some room for improvement (the packing density is at most $\pi\sqrt{3}/6 = 0.907$), it is possible to construct configurations of nodes, centered inside the octagon, inducing a graph of diameter at least 183.

Since the square tile admits a smaller hop diameter, any node can determine the subgraph induced by the vertices in its tile by examining nodes at hop distance at most 201.

The color sets used in the tiles are as specified in Figure 3. The resulting coloring is using only 6 colors. Due to the chosen sizes of the tiles and the chosen coloring scheme, an edge of G crossing from a square to an octagon is between vertices of different colors. After this initial color assignment, any edge of G whose endpoints are of the same color is necessarily an edge crossing the slanted part of the border of two adjacent octagons. The following construction shows that using one additional color and with careful attention to detail near the common border of adjacent octagons, some of the vertices can be recolored in order to achieve a 7-coloring of G. Details follow in Subsection 3.2.

3.2 Adjusting the Coloring

As seen from Figure 3, the centers of the tiles form an infinite regular mesh. We shall denote the hexagon tile that is centered at coordinates $[0, 0]$ as $S_{0,0}$. For a tile denoted $S_{i,j}$, its horizontal left, horizontal right, vertical down, vertical up neighboring tile is denoted $S_{i-1,j}$, $S_{i+1,j}$, $S_{i,j+1}$, $S_{i,j-1}$, respectively. Let $G_{i,j}$ denote the subgraph of G induced by the vertices located within $S_{i,j}$ and suppose that $F_{i,j}$ denotes the outer face of $G_{i,j}$. In case that $G_{i,j}$ is not connected, we consider each connected component of $G_{i,j}$ separately. (Notice that we only need to consider those components of $G_{i,j}$ that contain vertices adjacent to more than one octagon, for otherwise the coloring of such a component could be added to the coloring of $G_{i+1,j+1}$.)

Let $S_{i,j}$ be one of the hexagonal tiles, i.e., $i + j$ is even. Consider the place where $S_{i,j}$ and $S_{i+1,j+1}$ meet (we call it a $CR_{i,j}$ crossing). Consider the sequence of vertices obtained in a counterclockwise cyclical traversal of the outer face $F_{i,j}$ of $G_{i,j}$ and let $L = \{u_1, u_2, \ldots, u_k\}$ be the *shortest subsegment* of this traversal containing all the vertices connected to $G_{i+1,j+1}$, i.e., as in Figure 4.

Notice that L could contain some vertex more than once if the outer face is not a simple curve. Define $M = \{v_1, v_2, \ldots, v_l\}$ analogously for $G_{i+1,j+1}$, using clockwise traversal.

We say that crossing $CR_{i,j}$ is *simple* if no inside vertex of L different from u_1 and u_k is connected both to a vertex of M and to a vertex in $G_{i,j+1}$ or $G_{i+1,j}$, and the same analogous condition holds for the inside vertices of M. If the crossing is simple, the problem of having some edges between vertices of L or M having both endpoints of the same color can be resolved using the following lemma.

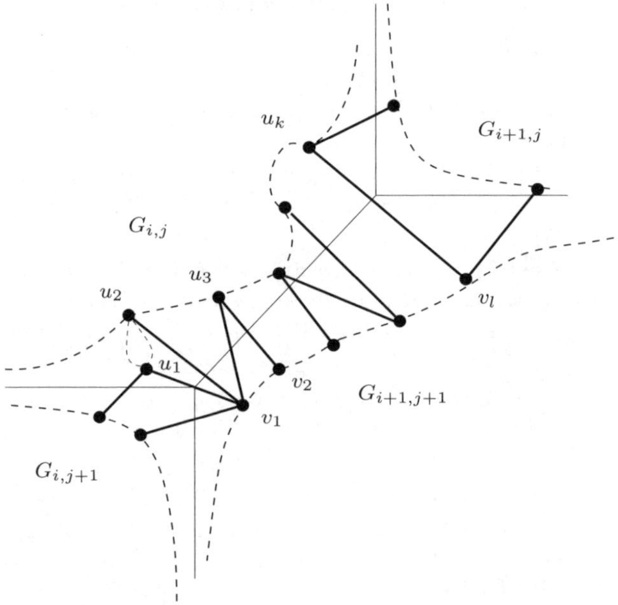

Fig. 4. A typical simple crossing

Lemma 3. *Let $CR_{i,j}$ be a simple crossing (see Figure 4). Then the vertices of L and M and some of the neighbors of u_1, v_1, u_k and v_k can be recolored, possibly with the help of color 7, in such a way that no edge incident to L or M or a neighbor of either of L or M has both endpoints of the same color.*

Proof. After the initial coloring, the only edges which might have endpoints of the same color are the edges connecting vertices of L to vertices of M. Without loss of generality we may assume that the vertices of L and M use colors $1, 2, 3$, the inside of $G_{i,j}$ uses in addition colors $4, 5$, while the inside of $G_{i+1,j+1}$ uses colors $4, 6$ in addition to $1, 2, 3$.

The recoloring is done in two steps, where in the first step the conflict vertices of L and M (i.e. the ones with a neighbor of the same color) are recolored as indicated in Table 1.

Table 1. First step of recoloring

Conflict vertex of	old color	new color
L	1	6
M	2	5
L	3	7

This ensures that no edge incident to an *inner vertex* of L or M, i.e, different from $\{u_1, v_1, u_k, v_l\}$, has both endpoints of the same color since:

- if there was an edge from L to M connecting two vertices of the same color, one endpoint of this edge has been recolored.
- As no color 6 was used in $G_{i,j}$, the vertices of L recolored to 6 have no neighbors of color 6 in $G_{i,j}$ (and they cannot be neighbors, as both had color 1 in the coloring of $G_{i,j}$). They also do not have neighbors of color 6 in $G_{i+1,j+1}$ as all their neighbors in $G_{i+1,j+1}$ are on the outer face and thus of colors $1, 2, 3$ (and newly 5). Finally, from the second property of simple crossing it follows that the inner conflict vertices of L and M do not have neighbors in $G_{i,j+1}$ and $G_{i+1,j}$,
- Analogous argument applies for vertices recolored to 5 in M and vertices recolored to 7 in L.

It remains to consider edges incident to $\{u_1, v_1, u_k, v_l\}$ that might have endpoints of the same color; for example u_1 was recolored from 1 to 6 but it has a neighbor of color 6 in $S_{i,j+1}$ (note that there is no problem if the new color was 7). In such a case, these same-color neighbors in $S_{i,j+1}$ are recolored to color 7. We claim that this does not create same-color edges. First note that if u_1 recolored its neighbors of color 6 in $G_{i,j+1}$, then v_1 necessarily kept its original color, since v_1 might change its color only if it was originally 2. Hence, by the simplicity of the crossing, only the neighbors of v_1 of color 6 (or, by symmetry, only the neighbors of u_1 of color 5) need to be recolored to color 7. The cases for other extreme vertices of L and M are analogous. Since the width of the gap

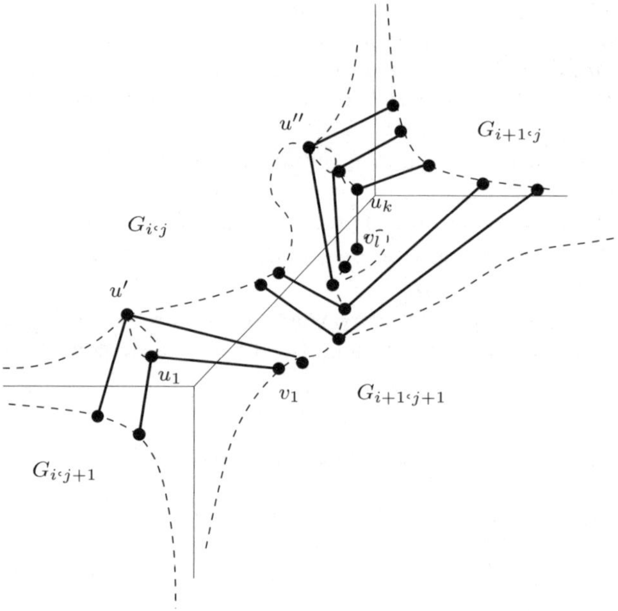

Fig. 5. A not simple crossing

between the squares is greater than 3, it ensures that the recoloring of vertices in $G_{i,j+1}$ and $G_{i+1,j}$ does not create any conflict in coloring.

Furthermore, any two vertices of $G_{i,j+1}$ which were recolored to 7 due to different crossings with octagons cannot be neighbors since the size of the squares is $5 + \epsilon$.

While the width of the gap between the squares ensures the first condition for a crossing to be simple is always satisfied, there can be a case when several different vertices of L are connected both to a vertex of M and to a vertex in $G_{i,j+1}$ or $G_{i+1,j}$, see Figure 5. Notice that this may happen when some inner vertices of L or M are cut vertices in $G_{i,j}$ or $G_{i+1,j+1}$.

We resolve the problem of a crossing that is not simple by a pre-processing phase in which some of the vertices in the octagons are assigned to the neighboring squares, with the goal to make the crossing simple.

Consider a crossing $CR_{i,j}$ and let L and M be defined as before. Let u' be the last (in L) occurrence of a node connected to both M and $G_{i,j+1}$ (if there is no such node, set $u' = u_1$). Similarly, let u'' be the first occurrence in L of a node connected to both M and $G_{i+1,j}$. Define v' and v'' analogously in M, using clockwise traversal, see Figure 5. Any vertex of L which is connected to both $G_{i,j+1}$ and $G_{i+1,j+1}$ must occur in the segment of L from u_1 to u', since the edges incident with u' connecting it to $G_{i,j+1}$ and $G_{i+1,j+1}$ act as separators in the planar graph. Similarly, any node of L which is connected to both $G_{i+1,j}$ and $G_{i+1,j+1}$ must occur in the segment of L from u'' to u_k (see Figure 6).

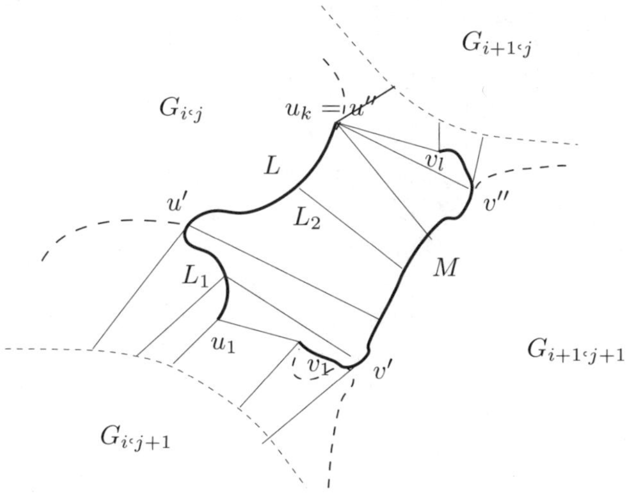

Fig. 6. L and M in a crossing

We now partition L into three parts: Let L_1 be the shortest initial segment of L from u_1 to to first occurrence of u' so that all vertices connected to both $G_{i,j+1}$ and $G_{i+1,j+1}$ are contained in L_1, let L_3 be the shortest final segment of L starting with an occurrence of u'' so that all vertices connected to both $G_{i+1,j}$ and $G_{i+1,j+1}$ are contained in L_3, and L_2 be the remaining part of L. We define M_1, M_2, and M_3 analogously as segments of M using v' and v''.

To make the crossing simple, we assign the components of $G_{i,j}$ separated by u' and encountered in the traversal of L_1 to $G_{i,j+1}$, and the components of $G_{i,j}$ separated by u'' and encountered in the traversal of L_3 are assigned to are $G_{i+1,j}$. The same is applied to the segments of $G_{i+1,j+1}$ separated by v' and v'' and encountered in the traversal of M_1 and M_3, (see Figure 7). All the components that are assigned to $G_{i,j+1}$ are inside the area bordered by the edges connecting u' or v' to $G_{i+1,j+1}$ and $G_{i,j+1}$ or $G_{i,j}$ and $G_{i,j+1}$. Similarly all the components that are assigned to $G_{i+1,j}$ are inside the area bordered by the edges connecting u'' or v'' to $G_{i+1,j+1}$ and $G_{i+1,j}$ or $G_{i,j}$ and $G_{i+1,j}$. Since the length of the crossing is more than 3, there cannot be any edge between vertices assigned to $G_{i+1,j}$ and $G_{i,j+1}$. Furthermore, after this reassignment, u' is the only vertex in $G_{i,j}$ that can be connected to both $G_{i+1,j+1}$ and $G_{i,j+1}$ and u'' is the only vertex in $G_{i,j}$ that can be connected to both $G_{i+1,j+1}$ and $G_{i+1,j}$. The analogous statement can be made about v' and v'', Thus the crossing L' and M' between $G_{i,j}$ and $G_{i+1,j+1}$ is a subset of u', L_2, u'' and v', M_2, v'' and this modified crossing $CR_{i,j}$ satisfies both conditions of a simple crossing, see Figure 7, and we can thus proceed with the coloring as stated in Lemma 3.

The following lemma, together with the size of the squares being selected as $5 + \epsilon$, allows us to apply Lemma 3 to each crossing independently.

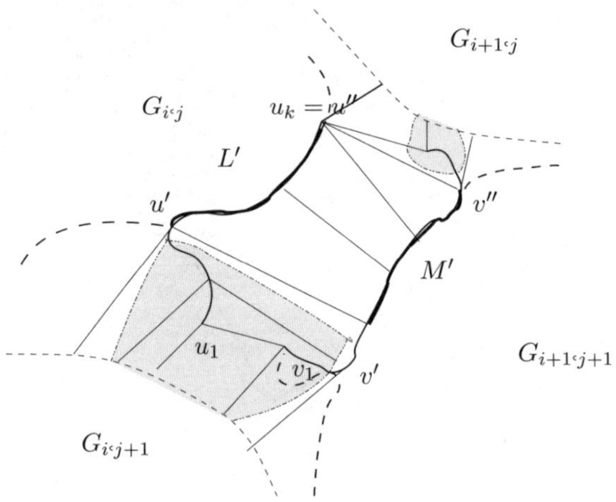

Fig. 7. L' and M' after the transformation. The shaded areas belong to $G_{i,j+1}$ and $G_{i+1,j}$.

Lemma 4. *Any vertex recolored due to resolving conflicts in crossing $CR_{i,j}$ is at a distance at most 2 from the line separating $S_{i,j}$ and $S_{i+1,j+1}$.*

3.3 Local 7-Coloring Algorithm

Putting the pieces together we have the following local, fully distributed algorithm that is executed at each vertex of the graph to obtain a valid 7-coloring of the graph.

Algorithm 1. The local 7-coloring algorithm for a vertex v

1: Learn your neighborhood up to distance 201 // *Note that all steps can be performed locally using the information learned in the first (communication) step, without incurring further communication.*
2: From your coordinates, identify the square/octagon $S_{i,j}$ you are located in, and calculate the connected component of $G_{i,j}$ you belong to.
// *The next step is for vertices near a crossing*
3: Calculate L and M, and then L' and M'. Determine whether you have been shifted to a neighboring square. Determine whether L' and M' are connected, if not but the squares are now connected, repeat the process until the final L^* and M^* are computed.
4: Apply the $3+2$ coloring algorithm from Theorem 1 for each $G_{i,j}$, as in Figure 3
5: Apply the recoloring from Lemma 3.

The results of this section can be summarized in the following theorem.

Theorem 3. *Given a planar subgraph of the unit disk graph whose vertices correspond to hosts that are each aware of its geometric location in the plane, Algorithm 1 computes locally a 7-coloring of this subgraph using only information on the subgraph located within a constant number of hops away from it.*

An interesting remaining open problem is whether the number of colors needed for a local coloring of a planar subgraphs of a unit disk graph can be decreased any further.

References

1. Bose, P., Morin, P., Stojmenovic, I., Urrutia, J.: Routing with guaranteed delivery in ad hoc wireless networks. wireless networks 7, 609–616 (2001)
2. Caragiannis, I., Fishkin, A.V., Kaklamanis, C., Papaioannou, E.: A tight bound for online coloring of disk graphs. In: Pelc, A., Raynal, M. (eds.) SIROCCO 2005. LNCS, vol. 3499, pp. 78–88. Springer, Heidelberg (2005)
3. Chavez, E., Dobrev, S., Kranakis, E., Opatrny, J., Stacho, L., Urrutia, J.: Local construction of planar spanners in unit disk graphs with irregular transmission ranges. In: Cardoso, J., Sheth, A.P. (eds.) SWSWPC 2004. LNCS, vol. 3387, pp. 286–297. Springer, Heidelberg (2005)
4. Dörre, P.: Every planar graph is 4-colourable and 5-choosable a joint proof. Fachhochschule Südwestfalen (University of Applied Sciences) (unpublished note)
5. Gabriel, K.R., Sokal, R.R.: A new statistical approach to geographic variation analysis. Systemic Zoology 18, 259–278 (1972)
6. Ghosh, S., Karaata, M.H.: A self-stabilizing algorithm for coloring planar graphs. Distributed Computing 7, 55–59 (1993)
7. Gräf, A., Stumpf, M., Weißenfels, G.: On coloring unit disk graphs. Algorithmica 20(3), 277–293 (1998)
8. Kranakis, E., Singh, H., Urrutia, J.: Compass routing on geometric networks. In: Proc. of 11th Canadian Conference on Computational Geometry, August 1999, pp. 51–54 (1999)
9. Linial, N.: Locality in distributed graph algorithms. SIAM J. COMP. 21(1), 193–201 (1992)
10. Marathe, M.V., Breu, H., Hunt III, H.B., Ravi, S.S., Rosenkrantz, D.J.: Simple heuristics for unit disk graphs. Networks 25(1), 59–68 (1995)
11. Miyamoto, Y., Matsui, T.: Multicoloring unit disk graphs on triangular lattice points. In: SODA, pp. 895–896. SIAM, Philadelphia (2005)
12. Thomassen, C.: Every planar graph is 5-choosable. Combinatorial Theory Series B 62(1), 180–181 (1994)
13. Tuza, Z., Voigt, M.: A note on planar 5-list colouring: non-extendability at distance 4. Discrete Mathematics 251(1), 169–172 (2002)
14. Wang, Y., Li, X.-Y.: Localized construction of bounded degree and planar spanner for wireless ad hoc networks. In: DialM: Proceedings of the Discrete Algorithms and Methods for Mobile Computing & Communications (2003)

Generalized Domination in Degenerate Graphs: A Complete Dichotomy of Computational Complexity

Petr Golovach[1,⋆] and Jan Kratochvíl[2,⋆⋆]

[1] Department of Informatics, University of Bergen, PB 7803, 5020 Bergen, Norway
petrg@ii.uib.no

[2] Department of Applied Mathematics, and Institute for Theoretical Computer Science, Charles University, Malostranské nám. 25, 118 00 Praha 1, Czech Republic
honza@kam.mff.cuni.cz

Abstract. The so called (σ, ρ)-domination, introduced by J.A. Telle, is a concept which provides a unifying generalization for many variants of domination in graphs. (A set S of vertices of a graph G is called (σ, ρ)-*dominating* if for every vertex $v \in S$, $|S \cap N(v)| \in \sigma$, and for every $v \notin S$, $|S \cap N(v)| \in \rho$, where σ and ρ are sets of nonnegative integers and $N(v)$ denotes the open neighborhood of the vertex v in G.) It is known that for any two nonempty finite sets σ and ρ (such that $0 \notin \rho$), the decision problem whether an input graph contains a (σ, ρ)-dominating set is NP-complete, but that when restricted to some graph classes, polynomial time solvable instances occur. We show that for every k, the problem performs a complete dichotomy when restricted to k-degenerate graphs, and we fully characterize the polynomial and NP-complete instances. It is further shown that the problem is polynomial time solvable if σ, ρ are such that every k-degenerate graph contains at most one (σ, ρ)-dominating set, and NP-complete otherwise. This relates to the concept of ambivalent graphs previously introduced for chordal graphs.

Subject: Computational complexity, graph algorithms.

1 Introduction and Overview of Results

We consider finite undirected graphs without loops or multiple edges. The vertex set of a graph G is denoted by $V(G)$ and its edge set by $E(G)$. The open neighborhood of a vertex is denoted by $N(u) = \{v\colon (u, v) \in E(G)\}$. The closed neighborhood of a vertex u is the set $N[u] = N(u) \cup \{u\}$. If $U \subset V(G)$, then $G[U]$ denotes the subgraph of G induced by U.

Let σ, ρ be a pair of nonempty sets of nonnegative integers. A set S of vertices of G is called (σ, ρ)-*dominating* if for every vertex $v \in S$, $|S \cap N(v)| \in \sigma$, and for every $v \notin S$, $|S \cap N(v)| \in \rho$. The concept of (σ, ρ)-domination was introduced by

⋆ Supported by Norwegian Research Council.

⋆⋆ Supported by the Czech Ministry of Education as Research Project No. 1M0545.

M. Agrawal et al. (Eds.): TAMC 2008, LNCS 4978, pp. 182–191, 2008.

J.A. Telle [4,5] (and further elaborated on in [6,2]) as a unifying generalization of many previously studied variants of the notion of dominating sets. In particular, ($\mathbb{N}_0$,$\mathbb{N}$)-dominating sets are ordinary dominating sets, ($\{0\}$,$\mathbb{N}_0$)-dominating sets are independent sets, ($\mathbb{N}_0$,$\{1\}$)-dominating sets are efficient dominating sets, ($\{0\}$,$\{1\}$)-dominating sets are 1-perfect codes (or independent efficient dominating sets), ($\{0\}$,$\{0, 1\}$)-dominating sets are strong stable sets, ($\{0\}$,$\mathbb{N}$)-dominating sets are independent dominating sets, ($\{1\}$,$\{1\}$)-dominating sets are total perfect dominating sets, or ($\{r\}$,$\mathbb{N}_0$)-dominating sets are induced r-regular subgraphs ($\mathbb{N}$ and $\mathbb{N}_0$ denote the sets of positive and nonnegative integers, respectively).

We are interested in the complexity of the problem of existence of a (σ, ρ)-dominating set in an input graph, and we denote this problem by $\exists(\sigma, \rho)$-DOMINATION. It can be easily seen that if $0 \in \rho$, then the $\exists(\sigma, \rho)$-DOMINATION problem has a trivial solution $S = \emptyset$. So throughout our paper we suppose that $0 \notin \rho$. We consider only finite sets σ and ρ and use the notation $p_{\min} = \min \sigma$, $p_{\max} = \max \sigma$, $q_{\min} = \min \rho$, and $q_{\max} = \max \rho$.

It is known that for any nontrivial combination of finite sets σ and ρ (considered as fixed parameters of the problem), $\exists(\sigma, \rho)$-DOMINATION is NP-complete [4]. It is then natural to pay attention to restricted graph classes for inputs of the problem. The problem is shown polynomial time solvable for interval graphs in [3], where also the study of its complexity for chordal graphs was initiated. A full dichotomy for chordal graphs was proved in [1], where a direct connection between the complexity of $\exists(\sigma, \rho)$-DOMINATION and the so called ambivalence of the parameter sets σ, ρ was noted. A pair (σ, ρ) is called *ambivalent* for a graph class $\mathcal{G}$ if there exists a graph in $\mathcal{G}$ containing at least two different (σ, ρ)-dominating sets (such a graph will be called (σ, ρ)-*ambivalent*), and the pair (σ, ρ) is called *non-ambivalent* otherwise.

It is shown in [1] that for finite sets σ, ρ, $\exists(\sigma, \rho)$-DOMINATION is polynomial time solvable for chordal graphs if the pair (σ, ρ) is non-ambivalent (for chordal graphs), and it is NP-complete otherwise. It should be noted that the characterization which is given in [1] is nonconstructive in the sense that the authors did not provide a structural description of ambivalent (or non-ambivalent) pairs σ, ρ (and there is indication that such a description will not be simple).

In this paper we consider the connection between ambivalence and computational complexity of $\exists(\sigma, \rho)$-DOMINATION for k-degenerate graphs. A graph G is called *k-degenerate* (with k being a positive integer) if every induced subgraph of G has a vertex of degree at most k. For example, trees are exactly connected 1-degenerate graphs, every outerplanar graph is 2-degenerate, and every planar graph are 5-degenerate. An ordering of vertices $v_1, v_2, \ldots, v_n$ is called a *k-degenerate ordering* if every vertex v_i has at most k neighbors among the vertices $v_1, v_2, \ldots, v_{i-1}$. It is well known that a graph is k-degenerate if and only if it allows a k-degenerate ordering of its vertices.

It is known [6] that for trees (and for graphs of bounded treewidth), $\exists(\sigma, \rho)$-DOMINATION can be solved in polynomial time. Thus we assume $k \geq 2$ throughout the paper. We prove that also in the case of k-degenerate graphs, ambivalence and NP-hardness of $\exists(\sigma, \rho)$-DOMINATION go hand in hand.

Theorem 1. *For finite sets σ, ρ, $\exists(\sigma,\rho)$-DOMINATION is polynomial (linear) time solvable for k-degenerate graphs if the pair (σ,ρ) is non-ambivalent for k-degenerate graphs (moreover, the problem can be solved by an algorithm which is polynomial not only in the size of the graph, but also in $p_{\max}$ and $q_{\max}$), and it is NP-complete otherwise.*

Unlike the case of chordal graphs, for k-degenerate graphs we are able to describe a complete and constructive classification of ambivalent and non-ambivalent pairs.

Theorem 2. *Let σ, ρ be finite sets, and $k \geq 2$. If $p_{\min} > k$, then no k-degenerate graph has a (σ,ρ)-dominating set. If $p_{\min} \leq k$, then the pair (σ,ρ) is non-ambivalent for k-degenerate graphs if and only if one of the following two conditions holds:*

1. $(\sigma \cup \rho) \cap \{0, 1, \ldots, k\} = \{0\}$,
2. *for every $p \in \sigma$ and every $q \in \rho$, $|p - q| > k$.*

The last section of the paper is devoted to planar graphs. These undoubtedly form one of the most interesting k-degenerate classes of graphs ($k = 5$ in this case). Here we end up with several open problems. We are able to prove the NP-hardness part of an analog of Theorem 1.

Theorem 3. *For finite sets σ, ρ, $\exists(\sigma,\rho)$-DOMINATION is NP-complete for planar graphs if the pair (σ,ρ) is ambivalent for planar graphs.*

However, we do not know if non-ambivalence implies polynomial time recognition algorithm in this case. We are able to classify ambivalent and non-ambivalent pairs for some special pairs of sets σ and ρ, e.g., one-element sets, but even in this case the proof of non-ambivalence is nonconstructive and does not yield an algorithm.

Theorem 4. *Let σ, ρ be one-element sets, $\sigma = \{p\}$, $\rho = \{q\}$, and $0 \neq q$. If $p > 5$, then no planar graph has a (σ,ρ)-dominating sets. And if $p \leq 5$, then the pair (σ,ρ) is non-ambivalent for planar graphs if and only if $q - p > 3$ or $p - q > 2$.*

2 Classification of Ambivalent and Non-ambivalent Pairs for *k*-Degenerate Graphs

In this section we present a structural characterization of ambivalent and non-ambivalent pairs of sets (σ,ρ) for k-degenerate graphs. We also describe an algorithm which (in case of a non-ambivalent pair (σ,ρ)) constructs the unique (σ,ρ)-dominating set (if it exists) in an input k-degenerate graph. We start with the following simple statement.

Lemma 1. *Let σ, ρ be finite sets, and let k be a positive integer. If $p_{\min} > k$, then no k-degenerate graph contains a (σ,ρ)-dominating set.*

Proof. Let $v_1, v_2, \ldots, v_n$ be a k-degenerate ordering of a k-degenerate graph G. Since $\deg v_n \le k < p_{\min}$, the vertex v_n does not belong to any (σ, ρ)-dominating set. But then every (σ, ρ)-dominating set of G is also a (σ, ρ)-dominating set of the subgraph of G induced by $\{v_1, v_2, \ldots, v_{n-1}\}$. By repeating this argument inductively, we conclude that only the empty set can be (σ, ρ)-dominating. And since $0 \notin \rho$, this is impossible. □

Now we assume that $p_{\min} \le k$ and we prove that the conditions given in Theorem 2 are sufficient for non-ambivalence of σ, ρ. Towards this end, we describe greedy algorithms which construct the unique candidate for a (σ, ρ)-dominating set.

Let G be a k-degenerate graph with n vertices, and suppose that $v_1, v_2, \ldots, v_n$ is a k-degenerate ordering of the vertices of G. We consider two cases, and in each if them a set S, which is a unique candidate for (σ, ρ)-dominating set in G, is constructed.

Case 1. $(\sigma \cup \rho) \cap \{0, 1, \ldots, k\} = \{0\}$.

```
Procedure Construct A;
U := V(G), S := ∅;
while U ≠ ∅ do
    i := max{j: v_j ∈ U};
    S := S ∪ {v_i}, U := U \ N[v_i];
Return S
```

Case 2. $(\sigma \cup \rho) \cap \{0, 1, \ldots, k\} \neq \{0\}$, and for every $p \in \sigma$ and every $q \in \rho$, $|p - q| > k$.

```
Procedure Construct B;
U := V(G), S := ∅;
for i := n downto 1 do
    r := |N(v_i) ∩ S|, s := |N(v_i) ∩ U|;
    if there is p ∈ σ such that r ≤ p ≤ r + s then
        S := S ∪ {v_i}, U := U \ {v_i}
    else
        if there is q ∈ ρ such that r ≤ q ≤ r + s then
            U := U \ {v_i}
        else
            Return There is no (σ,ρ)-dominating set, Halt;
Return S
```

Even if the procedures `Construct A` or `Construct B` construct a set S, it is still possible that this set is not (σ, ρ)-dominating. So we have to test for this property:

Procedure `Test`;
for $i := 1$ **to** n **do**
  if $(v_i \in S$ *and* $|N(v_i) \cap S| \notin \sigma)$ *or* $(v_i \notin S$ *and* $|N(v_i) \cap S| \notin \rho)$ **then**
    Return `There is no` (σ, ρ)`-dominating set`, **Halt;**
Return S

The properties of the algorithms are summarized in the next statement.

Lemma 2. *Let σ, ρ be finite sets. Suppose that k is a positive integer, $p_{\min} \leq k$ and either $(\sigma \cup \rho) \cap \{0, 1, \ldots, k\} = \{0\}$, or for every $p \in \sigma$ and every $q \in \rho$, $|p - q| > k$. Then the described algorithms correctly construct the (σ, ρ)-dominating set (if it exists) for any k-degenerate graph G, and this set is unique. The running time is $O((p_{\max} + q_{\max})(n + m))$, where n is the number of vertices of G, and m is the number of its edges.*

Proof. The correctness of the procedure `Construct A` is straightforward. The loop invariant of the procedure is that no vertex of U has a neighbor in S. Hence if $(\sigma \cup \rho) \cap \{0, 1, \ldots, k\} = \{0\}$, then every vertex $v \in U$ with degree no more than $k < q_{\min}$ must belong to every (σ, ρ)-dominating set, and vertices of $N(v)$ can not belong to any such set.

The correctness of the procedure `Construct B` follows from the following observation: If for every $p \in \sigma$ and every $q \in \rho$, $|p - q| > k$, then the set $\{r, r + 1, \ldots, s\}$ can not contain elements of both sets σ and ρ, since $s \leq k$. And since the number of S-neighbors of v_i (in the final (σ, ρ)-dominating set S) will end up in the interval $[r, r + s]$, the justification is clear.

It is known that a k-degenerate ordering can be constructed in time $O(n+m)$. Then the estimate of the running time immediately follows from the description of the algorithms. (Note here that we can assume that $p_{\max}, q_{\max} \leq n$ since otherwise we can truncate the sets σ and ρ.) □

To complete the proof of Theorem 2 we have to prove that the conditions given in the theorem are not only sufficient but also necessary. We do so by constructing graphs with at least two different (σ, ρ)-dominating sets. Let σ, ρ be finite sets, and let $k \geq 2$ be a positive integer. Suppose that $p_{\min} \leq k$, $(\sigma \cup \rho) \cap \{0, 1, \ldots, k\} \neq \{0\}$, and there are $p \in \sigma$ and $q \in \rho$ such that $|p - q| \leq k$. We consider 3 different cases.

Case 1. $\max(\sigma \cap \{0, 1, \ldots, k\}) = 0$. Since $(\sigma \cup \rho) \cap \{0, 1, \ldots, k\} \neq \{0\}$, there is a $q \in \rho$ such that $q \leq k$. Then each class of bipartition of the complete bipartite graph $K_{q,q}$ is a (σ, ρ)-dominating set in $K_{q,q}$ and $K_{q,q}$ (and consequently the pair (σ, ρ)) is ambivalent.

Case 2. $1 \in \sigma$. If $p < q$, then we start the construction with the complete bipartite graph $K_{q-p,q-p}$. Let the bipartition of its vertex set be $\{u_1, u_2, \ldots, u_{q-p}\}$ and $\{v_1, v_2, \ldots, v_{q-p}\}$. We further join every pair of vertices u_i and v_i by p different paths of length 2. Let us denote by X the set of the middle vertices of these paths. Since $k \geq 2$ and $q - p \leq k$, the graph constructed is k-degenerate. And it has two different (σ, ρ)-dominating sets: $\{u_1, u_2, \ldots, u_{q-p}\} \cup X$ and $\{v_1, v_2, \ldots, v_{q-p}\} \cup X$.

If $p \geq q$, then the construction starts with two copies of the complete graph K_{p-q+1}, with vertex sets $\{u_1, u_2, \ldots, u_{p-q+1}\}$ and $\{v_1, v_2, \ldots, v_{p-q+1}\}$. Again, we join every pair of vertices u_i and v_i by q different paths of length 2, and we let X denote the set of the middle vertices of these paths. Since $k \geq 2$ and $p-q \leq k$, the graph constructed in this way is k-degenerate. And it has two different (σ, ρ)-dominating sets: $\{u_1, u_2, \ldots, u_{p-q+1}\} \cup X$ and $\{v_1, v_2, \ldots, v_{p-q+1}\} \cup X$.

Case 3. $r \in \sigma$ for some $2 \leq r \leq k$. Let H denote the complete graph K_{r+1} with one edge deleted, and let u, v be the endvertices of this edge. We will further refer to these vertices as the *poles* of H. If $p < q$, then we start the construction with two copies of the complete bipartite graph $K_{q-p,q-p}$ with the bipartition of the vertex sets $\{u_1, u_2, \ldots, u_{q-p}\}$, $\{v_1, v_2, \ldots, v_{q-p}\}$ and $\{x_1, x_2, \ldots, x_{q-p}\}$, $\{y_1, y_2, \ldots, y_{q-p}\}$, respectively. Then for every $i \in \{1, 2, \ldots, q-p\}$, we introduce p copies of H and join one pole of each of them with u_i and v_i by edges, and the other pole with x_i and y_i. Let X be the union of the sets of vertices of all added graphs H. The resulting graph is k-degenerate (first the non-pole vertices of H's have degrees $r \leq k$, after their deletion the pole vertices have degrees $2 \leq k$, and after the deletion of these all the remaining vertices have degrees $p - q \leq k$), and it has two different (σ, ρ)-dominating sets: $\{u_1, u_2, \ldots, u_{q-p}\} \cup \{x_1, x_2, \ldots, x_{q-p}\} \cup X$ and $\{v_1, v_2, \ldots, v_{q-p}\} \cup \{y_1, y_2, \ldots, y_{q-p}\} \cup X$.

If $p \geq q$, then we start the construction with four copies of the complete graph K_{p-q+1}, with vertex sets $\{u_1, u_2, \ldots, u_{p-q+1}\}$, $\{v_1, v_2, \ldots, v_{p-q+1}\}$, $\{x_1, x_2, \ldots, x_{p-q+1}\}$, and $\{y_1, y_2, \ldots, y_{p-q+1}\}$. For every $i \in \{1, 2, \ldots, p-q+1\}$, we add q copies of the graph H and join one pole of each of them with u_i and v_i, and the other one with x_i and y_i. Again let X be the union of the sets of vertices of the added copies of H. The resulting graph is k-degenerate, and it has two different (σ, ρ)-dominating sets: $\{u_1, u_2, \ldots, u_{p-q+1}\} \cup \{x_1, x_2, \ldots, x_{p-q+1}\} \cup X$ and $\{v_1, v_2, \ldots, v_{p-q+1}\} \cup \{y_1, y_2, \ldots, y_{p-q+1}\} \cup X$.

Unifying the claims of Lemmas 1, 2 and these constructions we have completed the proof of Theorem 2. Also since we presented polynomial time algorithms which construct unique (σ, ρ)-dominating sets (if they exist) for the non-ambivalent pairs (σ, ρ), the polynomial part of Theorem 1 is proved.

To conclude this section, let us point out a property of the constructed graphs which will be used in the next section.

Lemma 3. *For every ambivalent pair (σ, ρ), there is a k-degenerate graph G with at least two different (σ, ρ)-dominating sets, which has a k-degenerate ordering $v_1, v_2, \ldots, v_n$ such that for some ℓ, the first ℓ vertices $v_1, \ldots, v_\ell$ belong to one and are not included to the other (σ, ρ)-dominating set.*

3 NP-Completeness of $\exists(\sigma, \rho)$-DOMINATION for Ambivalent Pairs

It this section we outline the proofs of the NP-hardness part of Theorem 1 and of Theorem 3.

We use a reduction from a special covering problem. Let r be a positive integer. An instance of the COVER BY NO MORE THAN r SETS is a pair (X, M), where X is a nonempty finite set and M is a collection of sets of elements of X. We ask about the existence of a collection $M' \subset M$ of sets such that every element of X belongs to at least one and to at most r sets of M'. The graph $G(X, M)$ of an instance (X, M) is the bipartite graph with the vertex set $X \cup M$ and edge set $\{xm | x \in m \in M\}$. The proof of the following lemma will appear in the full version of the paper.

Lemma 4. *For every fixed $r \geq 1$, the* COVER BY NO MORE THAN r SETS *problem is NP-complete even for instances (X, M) for which the graph $G(X, M)$ is 2-degenerate. It also stays NP-complete if the graph $G(X, M)$ is planar.*

The main technical part of the NP-hardness proof is the construction of a gadget which "enforces" on a given vertex the property of "not belonging to any (σ, ρ)-dominating set", and which guarantees that this vertex has a given number of neighbors in any (σ, ρ)-dominating set in the gadget:

Lemma 5. *Assume that $k \geq 2$ and $p_{\min} \leq k$. Let r be a nonnegative integer. Then there is a rooted graph F with the root u such that:*

1. *there is set $S \subset V(F) \setminus \{u\}$ such that for every $x \in S$, $|N(x) \cap S| \in \sigma$, and for every $x \notin S$, $x \neq u$, $|N(x) \cap S| \in \rho$;*
2. *for every such set S, $|N(u) \cap S| = r$;*
3. *for every set $S \subset V(F)$ such that $u \in S$, either there is $x \in S$, $x \neq u$, for which $|N(x) \cap S| \notin \sigma$, or there is $x \notin S$ for which $|N(x) \cap S| \notin \rho$;*
4. *F has a k-degenerate ordering with u as the first vertex.*

The construction of F (which will appear in the full version of the paper) is technical and requires a lengthy case analysis. A specific variant of the gadget F' for planar graphs is also constructed, and the construction will also appear in the full version of the paper.

Now we complete the proof of Theorem 1. Suppose that σ, ρ are finite sets of integers, $k \geq 2$, and the pair (σ, ρ) is ambivalent for k-degenerate graphs. Let $r = \max\{i \in \mathbb{N}_0 : i \notin \rho, i+1 \in \rho\}$. Since $0 \notin \rho$, r is correctly defined. We are going reduce COVER BY NO MORE THAN t SETS for $t = q_{\max} - r$.

Suppose that (X, M) is an instance of COVER BY NO MORE THAN t SETS such that the graph $G(X, M)$ is 2-degenerate, $X = \{x_1, x_2, \ldots, x_n\}$ and $M = \{s_1, s_2, \ldots, s_m\}$. Let H be a k-degenerate ambivalent rooted graph with root u, such that u belongs to some (σ, ρ)-dominating set and u is not included in some other (σ, ρ)-dominating set, and H has a k-degenerate ordering for which the root is the first vertex. The existence of such a graph was proved in Lemma 3. For every vertex s_i of the graph $G(X, M)$, we take a copy of the graph H and identify its root with s_i. For every vertex x_j of our graph, a copy of the graph F (cf. Lemma 5) is constructed and its root is identified with x_j. Denote the graph obtained in this way by G. Clearly, G is k-degenerate.

We claim that the graph G has a (σ, ρ)-dominating set if and only if (X, M) allows a cover by no more than t sets. Since the graphs H and F depend only on σ and ρ, G has $O(n+m)$ vertices, our reduction is polynomial and the proof will be concluded.

Suppose first that G has a (σ,ρ)-dominating set S. Let $M' = \{s_i \in M : s_i \in S\}$. It follows from the properties of the forcing gadget F that every vertex x_j has exactly r neighbors in the gadget with root x_j and $x_j \notin S$. Then x_j has at least one S-neighbor in the set $\{s_1, s_2, \ldots, s_m\}$, but no more than $t = q_{\max} - r$ such neighbors. So, M' is a cover of X by no more than t sets.

Suppose now that $M' \subseteq M$ is a cover of X by no more than t sets. For every $i = 1, 2, \ldots, m$, we choose a (σ,ρ)-dominating set S_i in the copy of H with the root s_i such that $s_i \in S_i$ if and only if $s_i \in M'$. Let $S'_1, S'_2, \ldots, S'_n$ be (σ,ρ)-dominating sets in the copies of F. Since $\{t+1, t+2, \ldots, q_{\max}\} \subseteq \rho$, $S = S_1 \cup S_2 \cup \cdots \cup S_m \cup S'_1 \cup S'_2 \cup \ldots S'_n$ is a (σ,ρ)-dominating set in G.

The proof of Theorem 3 follows along the same lines. Suppose that (X, M) is an instance of Cover by no more than t sets such that the graph $G(X, M)$ is planar, $X = \{x_1, x_2, \ldots, x_n\}$ and $M = \{s_1, s_2, \ldots, s_m\}$. Let H' be a planar ambivalent rooted graph with root u, such that u belongs to some (σ,ρ)-dominating set and u is not included in some other (σ,ρ)-dominating set. For every vertex s_i of the graph $G(X, M)$, we construct a copy of the graph H' and unify its root with s_i. For every vertex x_j of our graph, a copy of the forcing gadget F' is constructed and the root of F' is identified with x_j. Let G be the resulting graph. Obviously, G is planar and G has a (σ,ρ)-dominating set if and only if (X, M) allows a cover by no more than t sets.

4 Ambivalence and Non-ambivalence for Planar Graphs

Since planar graphs are 5-degenerate, Theorem 2 gives sufficient conditions for non-ambivalence, but these conditions are not necessary for planar graphs. In this section we give some new sufficient conditions for non-ambivalence for planar graphs for certain cases of sets σ and ρ, and prove that in some cases these conditions are also necessary. We start with the case $q_{\min} > p_{\max}$.

Lemma 6. *Let σ, ρ be finite sets, and $p_{\min} \le 5$. If $q_{\min} - p_{\max} > 3$, then the pair (σ,ρ) is non-ambivalent for planar graphs.*

Proof. Assume that $q_{\min} - p_{\max} > 3$ and let G be a planar graph with two different (σ,ρ)-dominating sets S_1 and S_2. Let $X = S_1 \cap S_2$, $Y_1 = S_1 \setminus S_2$ and $Y_2 = S_2 \setminus S_1$. If $x \in Y_1$, then since $x \in S_1$, $|N(x) \cap X| \le p_{\max}$, and since $x \notin S_2$, $|N(x) \cap S_2| \ge q_{\min}$. So, x has at least 4 neighbors in Y_2. Similarly for $y \in Y_2$. Hence $G[Y_1 \cup Y_2]$ is a planar bipartite graph such that every vertex has degree at least 4, but this is impossible, since planar bipartite graphs are 3-degenerate. □

Now we consider the case $q_{\max} < p_{\min}$.

Lemma 7. *Let $\sigma = \{p\}$ for some $p \le 5$ and $0 \notin \rho$. If $p - q_{\max} > 2$, then the pair (σ,ρ) is non-ambivalent for planar graphs.*

Proof. Suppose that $p - q_{\max} > 2$ and let G be a planar graph with two different (σ, ρ)-dominating sets S_1 and S_2. Let $X = S_1 \cap S_2$, $Y_1 = S_1 \setminus S_2$ and $Y_2 = S_2 \setminus S_1$. If $x \in Y_1$, then since $x \notin S_2$, $|N(x) \cap X| \le q_{\max}$, and since $x \in S_1$, $|N(x) \cap S_1| = p$. So x has at least 3 neighbors in Y_1, i.e., $G[Y_1]$ and $G[Y_2]$ are planar graphs with all vertices of degree at least 3. Assume that a vertex $x \in Y_1$ is not adjacent to any vertex of Y_2. Then x has some neighbor $y \in X$. The vertex y must be adjacent to some vertex $z \in Y_2$ because σ contains exactly one element. Hence every vertex of Y_1 is either adjacent to some vertex of Y_2, or is connected with some vertex of Y_2 by a path of length two with the middle vertex from X.

Consider a plane embedding of G. It induces plane embeddings of $G[Y_1]$ and $G[Y_2]$. Let $x \in Y_1$. It is joined by an edge or by a path of length two to some vertex $y \in Y_2$, which belongs to some component H of $G[Y_2]$. This graph H lies completely in one face of $G[Y_1]$. Since all vertices of H have degrees at least 3, the graph H is not outerplanar. Then there is a vertex $z \in V(H)$ which does not belong to the boundary of the external face of H. By repeating the same arguments for z instead of x, we conclude that some component of $G[Y_1]$ lies completely in some internal face of H, and so on. Since the number of components of $G[Y_1]$ and $G[Y_2]$ is finite, this immediately gives a contradiction. □

The conditions given in Lemmas 6 and 7 are not only sufficient but also necessary for one-element sets σ and ρ, and this completes the proof of Theorem 4. The proof of this claim is provided by examples which are omitted here and will be given the full version of the paper. We conclude the section and the paper by explicitly stating some questions left open for planar graphs.

Problem 1. Is $\exists(\sigma, \rho)$-DOMINATION polynomial (NP-complete) when restricted to planar graphs if and only if the pair σ, ρ is non-ambivalent (ambivalent, respectively) for planar graphs? We believe that it would be interesting to solve this problem even for one-element sets σ and ρ.

Problem 2. Complete the characterization of ambivalent pairs σ, ρ for planar graphs.

References

1. Golovach, P., Kratochvil, J.: Computational complexity of generalized domination: A complete dichotomy for chordal graphs. In: Brandstaedt, A., Kratsch, D., Müller, H. (eds.) Graph-theoretic Concepts in Computer Science, Proceedings WG 2007. LNCS, vol. 4769, pp. 1–11. Springer, Heidelberg (2007)
2. Heggernes, P., Telle, J.A.: Partitioning graphs into generalized dominating sets. Nordic J. Comput. 5, 173–195 (1998)
3. Kratochví, J., Manuel, P., Miller, M.: Generalized domination in chordal graphs. Nordic Journal of Computing 2, 41–50 (1995)

4. Telle, J.A.: Complexity of domination-type problems in graphs. Nordic Journal of Computing 1, 157–171 (1994)
5. Telle, J.A.: Vertex partitioning problems: characterization, complexity and algorithms on partial k-trees, Ph. D tesisis, Department of Computer Science, Universiy of Oregon, Eugene (1994)
6. Telle, J.A., Proskurowski, A.: Algorithms for vertex partitioning problems on partial k-trees. SIAM J. Discrete Math. 10, 529–550 (1997)

More on Weak Bisimilarity of Normed Basic Parallel Processes*

Haiyan Chen[1,2]

[1] State Key Laboratory Computer Science of Institute of Software, Chinese Academy of Sciences
[2] Graduate School of the Chinese Academy of Sciences
chy@ios.ac.cn

Abstract. Deciding strong and weak bisimilarity of BPP are challenging because of the infinite nature of the state space of such processes. Deciding weak bisimilarity is harder since the usual decomposition property which holds for strong bisimilarity fails. Hirshfeld proposed the notion of bisimulation tree to prove that weak bisimulation is decidable for totally normed BPA and BPP processes. In this paper, we present a tableau method to decide weak bisimilarity of totally normed BPP. Compared with Hirshfeld's bisimulation tree method, our method is more intuitive and more direct. Moreover from the decidability proof we can derive a complete axiomatisation for the weak bisimulation of totally normed BPP.

1 Introduction

A lot of attention has been devoted to the study of decidability and complexity of verification problems for infinite-state systems [1,15,16]. In [2], Baeten, Bergstra, and Klop proved the remarkable result that bisimulation equivalence was decidable for irredundant context-free grammars (without the empty product). Subsequently, many algorithms in this domain were proposed. In [7], Hans Hüttel and Colin Stirling proved the decidability of normed BPA by using a tableau method, which can also be used as a decision procedure. Decidability of strong bisimilarity for BPP processes has been established in [13]. Furthermore, [14] proved that deciding strong bisimilarity of BPP is PSPACE-complete.

For weak bisimilarity, much less is known. Semidecidability of weak bisimilarity for BPP has been shown in [5]. In [6] it is shown that weak bisimilarity is decidable for those BPA and BPP processes which are "totally normed". P.Jančar conjectured that the method in [14] might be used to show the decidability of weak bisimilarity for general BPP. However, the problem of decidability of weak bisimilarity for general BPP is open.

Our work is inspired by Hirshfeld's idea. In [6] Hirshfeld proposed the notion of bisimulation tree to prove the decidability of weak bisimulation of totally normed

* Supported by the National Natural Science Foundation of China under Grant Nos. 60673045, 60496321.

M. Agrawal et al. (Eds.): TAMC 2008, LNCS 4978, pp. 192–203, 2008.

BPP. Based on the idea, we show that weak bisimulation for totally normed BPP is decidable by a tableau method. In [13], S. Christensen, Y. Hirshfield and F. Moller proposed a tableau decision procedure for deciding strong bisimilarity of normed BPP. The key for tableau method to work is a nice decomposition property which holds for strong bisimulation, but fails for weak bisimulation. In our work, instead of using decomposition property, we apply Hirshfeld's idea to control the size of the tableaux to make the tableau method work correctly. This approach not only provides us a more direct decision method, but also has the advantage of providing a completeness proof of an equational theory for weak bisimulation of totally normed BPP processes, similar to the tableau method of [13] provides such a completeness proof for strong bisimulation of normed BPP processes. Moreover, the termination proof for tableau is greatly simplified.

The paper is organized as follows. Section 2 introduces the notion of BPP processes and weak bisimulation and describes weak bisimulation equivalence. Section 3 gives the tableau decision method and presents the soundness and completeness results. In Section 4 we prove the completeness of the equational theory. Finally, Section 5 sums up conclusions and gives suggestions for further work.

2 BPP Processes and Weak Bisimulation Equivalence

Assuming a set of variables $\mathcal{V}$, $\mathcal{V}=\{X,Y,Z,\cdots\}$ and a set of actions Act_τ, $Act_\tau = \{\tau,a,b,c,\cdots\}$ which contains a special element τ, we consider the set of BPP expressions $\mathcal{E}$ given by the following syntax; we shall use $E,F,\ldots$ as metavariables over $\mathcal{E}$.

$$
\begin{array}{lll}
E ::= & 0 & \text{(inaction)} \\
 & \mid X & \text{(variables, } X \in \mathcal{V}) \\
 & \mid E_1 + E_2 & \text{(summation)} \\
 & \mid \mu E & (\mu \in Act_\tau) \\
 & \mid E_1|E_2 & \text{(merge)}
\end{array}
$$

A BPP process is defined by a finite family of recursive process equations

$$\Delta = \{X_i \stackrel{def}{=} E_i | 1 \leq i \leq n\}$$

where the $X_i \in \mathcal{V}$ are distinct variables and each E_i is BPP expressions, and free variables in each E_i range over set $\{X_1,\ldots,X_n\}$. In this paper, we concentrate on guarded BPP systems.

Definition 1. *A BPP expression E is guarded if each occurrence of variable is within the scope of an atomic action, and a BPP system is guarded if each E_i is guarded for $1 \leq i \leq n$.*

Definition 2. *The operational semantics of a guarded BPP system can be simply given by a labeled transition system $(\mathcal{S}, Act_\tau, \longrightarrow)$ where the transition relation $\longrightarrow$ is generated by the rules in Table 1.*

Table 1. Transition rules

act	$aE \xrightarrow{a} E$	rec	$\dfrac{E \xrightarrow{a} E'}{X \xrightarrow{a} E'}$ $(X = E \in \Delta)$
sum1	$\dfrac{E \xrightarrow{a} \alpha}{E + F \xrightarrow{a} \alpha}$	sum2	$\dfrac{F \xrightarrow{a} \beta}{E + F \xrightarrow{a} \beta}$
par1	$\dfrac{E \xrightarrow{a} \alpha}{E\|F \xrightarrow{a} \alpha\|F}$	par2	$\dfrac{F \xrightarrow{a} \beta}{E\|F \xrightarrow{a} E\|\beta}$

where the state space $\mathcal{S}$ consists of finite parallel of BPP processes, and the transition relation $\longrightarrow \subseteq \mathcal{S} \times Act_\tau \times \mathcal{S}$ is generated by the rules in Table 1, in which (as also later) we use Greek letters $\alpha, \beta, \cdots$ as meta variables ranging over elements of $\mathcal{S}$, Each such α denotes a BPP process by forming the product of the elements of α, i.e. by combining the elements of α in parallel using the merge operator. We write ϵ for empty sequence. We shall write X^n to represent the term $X|\cdots|X$ consisting of n copies of X combined in parallel. By length(α) we denote the cardinality of α.

It is shown in [3] that any guarded system can be effectively transformed into a 3-GNF normal form

$$\{X_i \stackrel{def}{=} \Sigma_{j=1}^{n_i} a_{ij}\alpha_{ij} | 1 \leq i \leq m\}$$

where for all i, j such that $1 \leq i \leq m$, $1 \leq j \leq n_i$, $length(\alpha_{ij}) < 3$. So we only considered BPP processes given in 3-GNF in this paper.

Moreover, we write $\alpha \stackrel{\epsilon}{\Longrightarrow} \beta$ for $\alpha(\stackrel{\tau}{\longrightarrow})^*\beta$, and write $\alpha \stackrel{a}{\Longrightarrow} \beta$ for $\alpha \stackrel{\epsilon}{\Longrightarrow} \stackrel{a}{\longrightarrow} \stackrel{\epsilon}{\Longrightarrow} \beta$. Let $\hat{}: Act_\tau \rightarrow \{Act_\tau - \tau\} \cup \epsilon$ be the function such that $\hat{a} = a$ when $a \neq \tau$ and $\hat{\tau} = \epsilon$, then the following general definition of weak bisimulation on $\mathcal{S}$ is standard.

Definition 3. *A binary relation $R \subseteq \mathcal{S} \times \mathcal{S}$ is a weak bisimulation if for all $(\alpha, \beta) \in R$ the following conditions hold:*

1. *whenever $\alpha \xrightarrow{a} \alpha'$, then $\beta \stackrel{\hat{a}}{\Longrightarrow} \beta'$ for some β' with $(\alpha', \beta') \in R$;*
2. *whenever $\beta \xrightarrow{a} \beta'$, then $\alpha \stackrel{\hat{a}}{\Longrightarrow} \alpha'$ for some α' with $(\alpha', \beta') \in R$.*

Two states α and β are said to be weak bisimulation equivalent, written $\alpha \approx \beta$, if there is a weak bisimulation R such that $(\alpha, \beta) \in R$.

It is standard to prove that $\approx$ is an equivalence relation between processes. Moreover it is a congruence with respect to composition on $\mathcal{S}$:

Proposition 1. *If $\alpha_1 \approx \beta_1$ and $\alpha_2 \approx \beta_2$ then $\alpha_1|\alpha_2 \approx \beta_1|\beta_2$.*

Proposition 2. *$\alpha|\beta \approx \beta|\alpha$.*

Proposition 3. $(\alpha|\beta)|\gamma \approx \alpha|(\beta|\gamma)$.

Definition 4. *A process α is said to be* normed *if there exists a finite sequence $\alpha \longrightarrow \ldots \longrightarrow \alpha_n \longrightarrow \epsilon$ transitions from α to ϵ, and un-normed otherwise. The weak norm of a normed α is the length of the shortest transition sequence of the form $\alpha \stackrel{a_1}{\Longrightarrow} \ldots \stackrel{a_n}{\Longrightarrow} \epsilon$, where each $a_i \neq \tau$ and $\stackrel{a_i}{\Longrightarrow}$ is counted as 1. We denote by $||\alpha||$ the weak norm of α. Also, for unnormed α, we follow the convention that $||\alpha|| = \infty$ and $\infty > n$ for any number n. A BPP system Δ is totally normed if for every variable X appears Δ, $0 < \|X\| < \infty$.*

With this definition, it is obvious that weak norm is additive: for normed $\alpha, \beta \in \mathcal{S}$, $||\alpha|\beta|| = ||\alpha|| + ||\beta||$. Moreover, the following proposition says that weak norm is respected by $\approx$.

Proposition 4. *If $\alpha \approx \beta$, then either both α, β are un-normed, or both are normed and $||\alpha|| = ||\beta||$.*

3 The Tableau Decision Method

From now on, we restrict our attention to the totally normed BPP processes in 3-GNF, i.e. processes of a parallel labeled rewrite system $\langle \mathcal{S}, Act_\tau, \longrightarrow \rangle$ where $\infty > ||X|| > 0$ for all $X \in \mathcal{V}$. And throughout the rest of the paper, we assume that all the processes considered are totally normed unless stated otherwise.

With the preparation of the previous section, in this section we can devise a tableau decision method. The rules of the tableau system are built around equations $\alpha = \beta$, where $\alpha, \beta \in \mathcal{S}$. Each rule of the tableau system has the form

$$\mathsf{name} \quad \frac{\alpha = \beta}{\alpha_1 = \beta_1 \ldots \alpha_n = \beta_n} \quad \text{side condition.}$$

The premise of a rule represents the goal to be achieved while the consequents are the subgoals. There are three rules altogether. One for unfolding. Two rules for substituting the states. We now explain the three rules in turn.

Table 2. Tableau rules

$$\mathsf{subl} \quad \frac{\alpha_1|\beta_1 = \alpha_2|\beta_2}{\alpha_1|\beta_1 = \alpha_1|\beta_2} \quad (if\ there\ is\ \alpha_1 \prec \alpha_2\ and\ a\ dominated\ node\ labeled\ \alpha_1 = \alpha_2\ or\ \alpha_2 = \alpha_1\ )$$

$$\mathsf{subr} \quad \frac{\alpha_1|\beta_1 = \alpha_2|\beta_2}{\alpha_2|\beta_1 = \alpha_2|\beta_2} \quad (if\ there\ is\ \alpha_2 \prec \alpha_1\ and\ a\ dominated\ node\ labeled\ \alpha_1 = \alpha_2\ or\ \alpha_2 = \alpha_1\ )$$

$$\mathsf{unfold} \quad \frac{\alpha = \beta}{\{\alpha' = \beta' \mid (\alpha', \beta') \in M\}} \quad M\ is\ a\ match\ for\ (\alpha, \beta)$$

3.1 Substituting the States

The next two rules can be used to substitute the expressions in the goal. The rules are based on the following observation.

Definition 5. (dominate and improve)[6]

1. *The pair* $(\alpha_1|\alpha_2, \beta_1|\beta_2)$ *dominates the pair* (α_1, β_1).
2. $X_1^{k_1}|\cdots|X_{i_0}^{k_{i_0}}|\cdots|X_n^{k_n}$ *improves* $X_1^{m_1}|\cdots|X_{i_0}^{m_{i_0}}|\cdots|X_n^{m_n}$ *iff there is some* i_0 *such that for* $i < i_0$ *the (total) number of occurrences of* X_i *is equal in both pairs, i.e.* $k_i = m_i$ *while the number of occurrences of* X_{i_0} *is smaller in* $(X_1^{k_1}|\cdots|X_{i_0}^{k_{i_0}}|\cdots|X_n^{k_n})$ *than in* $(X_1^{m_1}|\cdots|X_{i_0}^{m_{i_0}}|\cdots|X_n^{m_n})$ *i.e.* $k_{i_0} < m_{i_0}$.

Proposition 5. *Every sequence of pairs in which every pair improves the previous one is finite.*

Proposition 6. *Every sequence of pairs in which no pair dominates a previous one is finite.*

Definition 6. *By* $\prec$ *we denote the well-founded ordering on* $\mathcal{S}$ *given as follows:* $X_1^{k_1}|\cdots|X_n^{k_n} \prec X_1^{l_1}|\cdots|X_n^{l_n}$ *iff there exists* j *such that* $k_j < l_j$ *and for all* $i < j$ *we have* $k_i = l_i$.

It is easy to show that $\prec$ is well-founded. Moreover, We shall rely on the fact that $\prec$ is total in the sense that for any $\alpha, \beta \in \mathcal{S}$ such that $\alpha \neq \beta$ we have $\alpha \prec \beta$ or $\beta \prec \alpha$. Also we shall rely on the fact that $\alpha \prec \beta$ implies $\alpha|\alpha' \prec \beta|\alpha'$ as well as $\alpha \prec \beta|\alpha'$ for any $\alpha' \in \mathcal{S}$. All these properties are easily seen to hold for $\prec$.

When building tableaux basic nodes might dominate other basic nodes; we say a basic node $n : \alpha_1|\beta_1 = \alpha_2|\beta_2$ or $\alpha_2|\beta_2 = \alpha_1|\beta_1$ dominates any node $n' : \alpha_1 = \alpha_2$ or $n' : \alpha_2 = \alpha_1$ which appears above n in the tableau. There $n' : \alpha_1 = \alpha_2$ or $n' : \alpha_2 = \alpha_1$ is called the dominated node.

Definition 7. *We define a weight function* ω, *s.t. for* $\alpha = X_1^{k_1}|\cdots|X_n^{k_n}, \beta = Y_1^{m_1}|\cdots|Y_n^{m_n}$, $\omega(\alpha, \beta) = 1\times k_1 + 1\times k_2 + \cdots + 1\times k_n + 1\times m_1 + 1\times m_2 + \cdots + 1\times m_n$.

Proposition 7. *For every* α_1, β_1, *if* $\alpha \approx \beta$, *then* $\alpha|\alpha_1 \approx \beta|\beta_1$ *iff* $\alpha|\alpha_1 \approx \alpha|\beta_1$ *iff* $\beta|\alpha_1 \approx \beta|\beta_1$.

Proof. For the only if direction, suppose $\alpha|\alpha_1 \approx \beta|\beta_1$, since $\alpha \approx \beta$ and $\beta_1 \approx \beta_1$, then $\alpha|\beta_1 \approx \beta|\beta_1$ by Proposition 1, by $\alpha|\alpha_1 \approx \beta|\beta_1$, so $\alpha|\alpha_1 \approx \alpha|\beta_1$ since $\approx$ is an equivalence. For the if direction, suppose $\alpha|\alpha_1 \approx \alpha|\beta_1$, since $\alpha \approx \beta$ and $\beta_1 \approx \beta_1$, $\alpha|\beta_1 \approx \beta|\beta_1$ by Proposition 1, by $\alpha|\alpha_1 \approx \alpha|\beta_1$, so $\alpha|\alpha_1 \approx \beta|\beta_1$ since $\approx$ is an equivalence. For β it is similar to previous proof. □

Proposition 8. *One of the pairs* $(\alpha|\alpha_1, \alpha|\beta_1)$ *or* $(\beta|\alpha_1, \beta|\beta_1)$ *is an improvement of* $(\alpha|\alpha_1, \beta|\beta_1)$ *where* $\alpha \neq \beta$.

This proposition guarantees the soundness and backwards soundness of subl, subr rules.

In fact in section 2, from Proposition 5 we know that every sequence of pairs in which every pair improves the previous one is finite. So this means that there are only finitely many different ways to apply the rules.

3.2 Unfolding by Matching the Transitions

Definition 8. *Let $(\alpha, \beta) \in \mathcal{S} \times \mathcal{S}$. A binary relation $M \subseteq \mathcal{S} \times \mathcal{S}$ is a match for (α, β) if the following hold:*

1. *whenever $\alpha \xrightarrow{a} \alpha'$ then $\beta \xRightarrow{\hat{a}} \beta'$ for some $(\alpha', \beta') \in M$;*
2. *whenever $\beta \xrightarrow{a} \beta'$ then $\alpha \xRightarrow{\hat{a}} \alpha'$ for some $(\alpha', \beta') \in M$;*
3. *whenever $(\alpha', \beta') \in M$ then $||\alpha'|| = ||\beta'||$ and either $\alpha \xrightarrow{a} \alpha'$ or $\beta \xrightarrow{a} \beta'$ for some $a \in Act_\tau$.*

It is easy to see that for a given $(\alpha, \beta) \in \mathcal{S} \times \mathcal{S}$, there are finitely many possible $M \subseteq \mathcal{S} \times \mathcal{S}$ which satisfies 3. Above and moreover each of them must be finite. And for such M it is not difficult to see that it is decidable whether M is a match for (α, β).

The rule can be used to obtain subgoals by matching transitions, and it is based on the following observation.

Proposition 9. *Let $\alpha, \beta \in \mathcal{S}$. Then $\alpha \approx \beta$ if and only if there exists a match M for (α, β) such that $\alpha' \approx \beta'$ for all $(\alpha', \beta') \in M$.*

This proposition guarantees the soundness and backwards soundness of unfold rule.

As pointed out above there are finitely many matches for a given (α, β), so there are finitely many ways to apply this rule on (α, β).

3.3 Constructing Tableau

We determine whether $\alpha \approx \beta$ by constructing a tableau with root $\alpha = \beta$ using the three rules introduced above. A tableau is a finite tree with nodes labeled by equations of the form $\alpha = \beta$, where $\alpha, \beta \in \mathcal{S}$.

Moreover if $\alpha = \beta$ labels a non-leaf node, then the following are satisfied:

1. $||\alpha|| = ||\beta||$;
2. its sons are labeled by $\alpha_1 = \beta_1 \ldots \alpha_n = \beta_n$ obtained by applying rule subl, subr or unfold in Table 2 to $\alpha = \beta$, in that priority order;
3. no other non-leaf node is labelled by $\alpha = \beta$.

A tableau is a successful tableau if the labels of all its leaves have the forms:

1. $\alpha = \beta$ where there is a non-leaf node is also labeled $\alpha = \beta$;
2. $\alpha \equiv \beta$

3.4 Decidability, Soundness, and Completeness

Lemma 1. *Every tableau with root $\alpha = \beta$ is finite, Furthermore, there is only a finite number of tableaux with root $\alpha = \beta$.*

Theorem 1. *If $\alpha \approx \beta$ then there exists a successful tableau with root labeled $\alpha = \beta$.*

Proof. Suppose $\alpha \approx \beta$. If we can construct a tableau $T(\alpha = \beta)$ for $\alpha = \beta$ with the property that any node $n : \alpha = \beta$ of $T(\alpha = \beta)$ satisfies $\alpha \approx \beta$, then by Lemma 1 that construction must terminate and each terminal will be successful. Thus the tableau itself will be successful.

We can construct such a $T(\alpha = \beta)$ if we verify that each rule of the tableau system is forward sound in the sense that if the antecedent relates bisimilar processes then it is possible to find a set of consequents relating bisimilar processes. For the rule subl or subr we know from Proposition 7. For the rest of the tableau rules it is easily verified that they are forward sound in the above sense. □

Finally we must show soundness of the tableau system, namely that the existence of a successful tableau for $\alpha = \beta$ indicates that $\alpha \approx \beta$. This follows from the fact that the tableau system tries to construct a family of binary relations which are bisimilar.

Definition 9. *A sound tableau is a tableau such that if* $\alpha = \beta$ *is a label in it then* $\alpha \approx \beta$.

Theorem 2. *A successful tableau is a sound tableau.*

Proof. Let T be a successful tableau. We define $W = \{B \subseteq \mathcal{S} \times \mathcal{S}\}$ to be the smallest binary relations satisfies the following:

1. if $\alpha \equiv \beta$ labels a node in T then $(\alpha, \beta) \in W$;
2. if there is a node in T labeled $\alpha = \beta$ and on which rule unfold is applied then $(\alpha, \beta) \in W$;
3. if $(\alpha_1, \alpha_2) \in W$, $(\alpha_1|\beta_1, \alpha_1|\beta_2) \in W$ where $\alpha_1 \prec \alpha_2$ then $(\alpha_1|\beta_1, \alpha_2|\beta_2) \in W$;
4. if $(\alpha_1, \alpha_2) \in W$, $(\alpha_2|\beta_1, \alpha_2|\beta_2) \in W$ where $\alpha_2 \prec \alpha_1$then $(\alpha_1|\beta_1, \alpha_2|\beta_2) \in W$.

We will prove the following properties about W:

A. If $\alpha = \beta$ labels a node in T then $(\alpha, \beta) \in W$.
B. If $(\alpha, \beta) \in W$, then the following hold:
   (a) if $\alpha \xrightarrow{a} \alpha'$ then $\beta \overset{\hat{a}}{\Longrightarrow} \beta'$ for some β' such that $(\alpha', \beta') \in W$;
   (b) if $\beta \xrightarrow{a} \beta'$ then $\alpha \overset{\hat{a}}{\Longrightarrow} \alpha'$ for some α' such that $(\alpha', \beta') \in W$.

Clearly property B. implies that

$$B = \{(\alpha, \beta) \mid (\alpha, \beta) \in W\}$$

is a weak bisimulation. Then together with property A. it implies that T is a sound tableau.

We prove A. by induction on weight $\omega' = \omega(\alpha, \beta)$. If $\alpha = \beta$ is a label of an non-leaf node, there are three cases according to which rule is applied on this node. If unfold is applied, then by rule 2. of the construction of W clearly $(\alpha, \beta) \in W$. If subl is applied, in this case $\alpha = \beta$ is of the form $\alpha_1|\beta_1 = \alpha_2|\beta_2$, and the node has sons labeled by $\alpha_1|\beta_1 = \alpha_1|\beta_2$. Clearly $\omega(\alpha_1|\beta_1, \alpha_1|\beta_2) < \omega(\alpha_1|\beta_1, \alpha_2|\beta_2)$, then by the induction hypothesis $(\alpha_1|\beta_1, \alpha_1|\beta_2) \in W$. Then by rule 3. in the construction of W, $(\alpha_1|\beta_1, \alpha_2|\beta_2) \in W$. If subr is applied, it is similar to subl proof. If $\alpha = \beta$ is a label of a leaf node, then since T is a successful tableau either there is a non-leaf node also labeled by $\alpha = \beta$ and in this case we have proved that $(\alpha, \beta) \in W$, or $\alpha \equiv \beta$ must hold and in this case by rule 1. in the construction of W we also have $(\alpha, \beta) \in W$.

We prove B. by induction on the four rules define W. Suppose $(\alpha, \beta) \in W$, there are the following cases.

Case of rule 1. i.e. $\alpha \equiv \beta$. It is obvious B. holds.

Case of rule 2. i.e. there exists M which is a match for (α, β) such that $\alpha' = \beta'$ is a label of T for all $(\alpha', \beta') \in M$. Then by A. it holds that $(\alpha', \beta') \in W$ for all $(\alpha', \beta') \in M$, then by definition of a match, clearly B. holds.

Case of rule 3. i.e. there exist $(\alpha_1, \alpha_2) \in W$, $(\alpha_1|\beta_1, \alpha_1|\beta_2) \in W$ where $\alpha_1 \prec \alpha_2$ and $\alpha = \alpha_1|\beta_1$. If $\alpha_1|\beta_1 \xrightarrow{a} \alpha''$, we have to match this by looking for a β'' such that $\alpha_2|\beta_2 \stackrel{\hat{a}}{\Longrightarrow} \beta''$ and $(\alpha'', \beta'') \in W$. By transition rule for α'' has two cases: the first case $\alpha_1 \xrightarrow{a} \alpha_1'$ then $\alpha'' = \alpha_1'|\beta_1$. Now $(\alpha_1, \alpha_2) \in W$, by the induction hypothesis there exists $\alpha_2' \in \alpha$ such that $\alpha_2 \stackrel{\hat{a}}{\Longrightarrow} \alpha_2'$ and $(\alpha_1', \alpha_2') \in W$. since $(\alpha_1|\beta_1, \alpha_1|\beta_2) \in W$, by the induction hypothesis there exists $\alpha_1'|\beta_2 \in \alpha$ such that $\alpha_1|\beta_2 \stackrel{\hat{a}}{\Longrightarrow} \alpha_1'|\beta_2$ and $(\alpha_1'|\beta_1, \alpha_1'|\beta_2) \in W$. By rule 3 we have $(\alpha_1'|\beta_1, \alpha_2'|\beta_2) \in W$. The another direction can be proved in a similar way; the second case $\beta_1 \xrightarrow{a} \beta_1'$ then $\alpha'' = \alpha_1|\beta_1'$. since $(\alpha_1|\beta_1, \alpha_1|\beta_2) \in W$, by the induction hypothesis there exists $\alpha_1|\beta_1' \in \alpha$ such that $\alpha_1|\beta_2 \stackrel{\hat{a}}{\Longrightarrow} \alpha_1|\beta_2'$ and $(\alpha_1|\beta_1', \alpha_1|\beta_2') \in W$. Now $(\alpha_1, \alpha_2) \in W$, by rule 3 we have $(\alpha_1|\beta_1', \alpha_2|\beta_2') \in W$. The another direction can be proved in a similar way.

Case of rule 4. it is similar rule 3. proof. □

Theorem 3. *Let $\alpha, \beta \in \mathcal{S}$ be totally normed. Then $\alpha \approx \beta$ if and only if there exists a successful tableau with root $\alpha = \beta$.*

4 The Equational Theory

We will develop the equational theory proposed by Sϕren Chirstensen, Yoram Hirshfeld, Faron Moller in [13] for strong bisimulation on normed BPP processes given in 3-GNF. We now describe a sound and complete axiomatisation for totally normed BPP processes. We pay attention to BPP processes in 3-GNF. The axiomatisation shall be parameterised by Δ and consists of axioms and inference rules that enable one to derive the root of successful tableaux.

The axiomatisation is built around sequences of the form $\Gamma \vdash_\Delta E = F$ where Γ is a finite set of assumptions of the form $\alpha = \beta$ and E, F are BPP expressions. Let Δ be a finite family of BPP processes in 3-GNF. A sequent is interpreted as follows:

Definition 10. *We write $\Gamma \models_\Delta E = F$ when it is the case that if the relation $\{(\alpha, \beta) | \alpha = \beta \in \Gamma\} \cup \{(X_i, E_i) | X_i \stackrel{def}{=} E_i \in \Delta\}$ is part of a bisimulation then $E \approx F$.*

Thus, the special case $\emptyset \models_\Delta E = F$ states that $E \approx F$ (relative to the system of process equations Δ).

For the presentation of rule unfold we introduce notation $unf(\alpha)$ to mean the unfolding of α given as follows (assuming that $\alpha \equiv Y_1|Y_2|\cdots|Y_m$):

$$unf(\alpha) = \sum_{i=1}^{m}\{a_i\gamma_i : a_i\alpha_i \in Y_i\},$$

where $\gamma_i = \alpha_i|(\prod_{j=1,j\neq i}^{m} Y_j)$ and the notation $a\alpha \in Y$ means that $a\alpha$ is a summand of the defining equation for Y.

The proof system is presented in Table 3. Equivalence and congruence rules are R1-6. In [13] the rule R5 of the axiomatisation for normed BPP processes can not directly apply in our rules, since we know that weak bisimulation is not preserved by summation, i.e. if $E_1 \approx E_2$ and $F_1 \approx F_2$, but we can't get $E_1 + F_1 \approx E_2 + F_2$. So we increase two rules R5-6 to achieve summation. The rules R7-15 correspond to the BPP laws; notably we have associativity and commutativity for merge. Finally, we have two rules characteristic for this axiomatisation; R16 is an assumption introduction rule underpinning the role of the assumption list Γ and R17 is an assumption elimination rule and also a version of fixed point induction. The special form of R17 has been dictated by the rule unfold of the tableau system presented in Table 2.

Definition 11. *A proof of $\Gamma \vdash_\Delta E = F$ is a proof tree with root labeled $\Gamma \vdash_\Delta E = F$, instances of the axioms R1 and R7-R16 as leaves and where the father of a set of nodes is determined by an application of one of the inference rules R2-R6 or R17.*

Definition 12. *The relations $\approx_o$ for ordinals o are defined inductively as follows, where we assume that l is a limit ordinal*

$E \approx_0 F$ for all E, F

$E \approx_{o+1} F$ iff for $a \in (Act_\tau \cup \{\epsilon\})$

$E \stackrel{a}{\Longrightarrow} E'$, *then* $\exists F'.F \stackrel{\hat{a}}{\Longrightarrow} F'$ *and* $E' \approx_o F'$

$F \stackrel{a}{\Longrightarrow} F'$, *then* $\exists E'.E \stackrel{\hat{a}}{\Longrightarrow} E'$ *and* $E' \approx_o F'$.

$E \approx_l F$ iff $\forall o < l. E \approx_o F$

So we can get a fact that is $\approx = \bigcap_{n=0}^{\infty} \approx_n$.

Theorem 4. (Soundness) *If $\Gamma \vdash_\Delta E = F$ then we have $\Gamma \models_\Delta E = F$. In particular if $\vdash_\Delta E = F$ then $E \approx F$.*

The similar proof for soundness can be found in [13].

Lemma 2. *If $\Gamma \vdash_\Delta E = F$ then $\Gamma, \Gamma' \vdash_\Delta E = F$ for any Γ'.*

The completeness proof rests on a number of lemmas and definitions which tell us how to determine our sets of hypotheses throughout a proof of $E \approx F$ from a successful tableau for $E \approx F$. We prove completeness from [13] idea.

Table 3. The axiomatisation

Equivalence		Congruence	
R1	$\Gamma \vdash_\Delta E = E$	R4	$\dfrac{\Gamma \vdash_\Delta E_1 = F_1 \ \Gamma \vdash_\Delta E_2 = F_2}{\Gamma \vdash_\Delta E_1 \vert E_2 = F_1 \vert F_2}$
R2	$\dfrac{\Gamma \vdash_\Delta E = F}{\Gamma \vdash_\Delta F = E}$	R5	$\dfrac{\Gamma \vdash_\Delta E = F}{\Gamma \vdash_\Delta E = F + \tau E}$
R3	$\dfrac{\Gamma \vdash_\Delta E = F \ \Gamma \vdash_\Delta F = G}{\Gamma \vdash_\Delta E = G}$	R6	$\dfrac{\Gamma \vdash_\Delta E = F}{\Gamma \vdash_\Delta aE + R = aF + R}$
Axioms			
R7	$\Gamma \vdash_\Delta E + (F + G) = (E + F) + G$	R12	$\Gamma \vdash_\Delta E \vert F = F \vert E$
R8	$\Gamma \vdash_\Delta E + F = F + E$	R13	$\Gamma \vdash_\Delta E \vert 0 = E$
R9	$\Gamma \vdash_\Delta E + E = E$	R14	$\Gamma \vdash_\Delta \tau E = E$
R10	$\Gamma \vdash_\Delta E + 0 = E$	R15	$\Gamma \vdash_\Delta a(E + \tau F) + aF = a(E + \tau F)$
R11	$\Gamma \vdash_\Delta E \vert (F \vert G) = (E \vert F) \vert G$		
Recursion			
R16	$\Gamma, \alpha = \beta \vdash_\Delta \alpha = \beta$	R17	$\dfrac{\Gamma, \alpha = \beta \vdash_\Delta unf(\alpha) = unf(\beta)}{\Gamma \vdash_\Delta \alpha = \beta}$

Definition 13. *For any node n of a tableau, $Rn(n)$ denotes the set of labels of the nodes above n to which the rule* unfold *is applied. In particular, Rn(r)=$\emptyset$ where r is the root of the tableau.*

Theorem 5. (Completeness)*If $\alpha \approx \beta$ then $\Gamma \vdash_\Delta \alpha = \beta$*

Proof. If $\alpha \approx \beta$, then there exists a finite successful tableau with root labeled $\alpha = \beta$. Let $T(\alpha = \beta)$ be such a tableau. We shall prove that for any node $n : E = F$ of $T(\alpha = \beta)$ we have $Rn(n) \vdash_\Delta E = F$. In particular, for the root $r : \alpha = \beta$, this reduces to $\vdash_\Delta \alpha = \beta$, so we shall have our result.

We prove $Rn(n) \vdash_\Delta E = F$ by induction on the depth of the subtableau rooted at n. As the tableau is built modulo associativity and commutativity of merge and by removing 0 components sitting in parallel or in sum we shall assume that the axioms R12-R14 are used whenever required to accomplish the proof.

Firstly, if $n : E = F$ is a terminal node then either E and F are identical terms α, so $Rn(n) \vdash E = F$ follows from R1.

Hence assume that $n : E = F$ is a internal nodes. We proceed to apply to n according to the tableau rule.

(i) Suppose unfold is applied. Then n is the label $E = F$ and the son n' of n is labeled $E_i = F_i (i \in \{1 \cdots n\}$, $E_i = F_i$ is match of $E = F)$, by induction hypothesis $Rn(E_i = F_i) \vdash E_i = F_i$, $Rn(E_i = F_i) - \{E = F\}, E = F \vdash E_i = F_i$, $Rn(E_i = F_i) - \{E = F\} \vdash E = F$ by R17, we know $Rn(E = F) = Rn(E_i = F_i) - \{E = F\}$, so $Rn(E = F) \vdash E = F$.

(ii) Suppose sub is applied wlog that is subl. Then the label $E = F$ is of the form $E_1 | F_1 = E_2 | F_2$ with the corresponding node n'' labeled $E_1 = E_2$ and the son n' of n is labeled $E_1|F_1 = E_1|F_2$, by induction hypothesis $Rn(E_1|F_1 = E_1|F_2) \vdash E_1|F_1 = E_1|F_2$ since $Rn(E_1|F_1 = E_1|F_2) = Rn(E_1|F_1 = E_2|F_2)$, and $E_1 = E_2 \in Rn(E_1|F_1 = E_2|F_2)$, so $Rn(E_1|F_1 = E_2|F_2) \vdash E_1 = E_2$ by R16. Hence from R1, R4, R3 we have $Rn(E_1|F_1 = E_2|F_2) \vdash E_1|F_1 = E_2|F_2$, last $Rn(E = F) \vdash E = F$ is required.

This completes the proof. □

5 Conclusions and Directions for Further Work

In this paper we proposed a tableau method to decide whether a pair of totally normed BPP processes is a weak bisimilar relation. The whole procedure is direct and easy to understand, while the termination proof is also very simple. This tableau method also helps us to show the completeness of Sϕren Chirstensen, Yoram Hirshfeld, Faron Moller's equational theory on totally normed BPP systems. Recent results by Richard Mayr show that weak bisimulation of Basic Parallel Processes is $\prod_2^P$-hard[18].

The study of bisimulation decision problems in the fields of BPA and BPP processes has been already rather sophisticated. All the results were recorded and updated by J.Srba[17], as well as open problems in this field. About algorithms, the things left should be concerned with lowing complexity and improving the efficiency. As the equational theory depends on assumptions, it is somewhat different from Milner's equational theory for regular processes[9]. One direction of interest is the construction of equational theory of $\approx$ since many decision results for weak bisimulation are already given.

References

1. Moller, F.: Infinite results. In: Sassone, V., Montanari, U. (eds.) CONCUR 1996. LNCS, vol. 1119, pp. 195–216. Springer, Heidelberg (1996)
2. Baeten, J.C.M., Bergstra, J.A., Klop, J.W.: Decidability of bisimulation equivalence for processes generating context-free languages. Journal of the Association for Computing Machinery 93(40), 653–682 (1993)
3. Christensen, S.: Forthcoming Ph.D. thesis. University of Edinburgh
4. Hirshfeld, Y., Jerrum, M., Moller, F.: Bisimulation equivalence is decidable for normed process algebra. In: Wiedermann, J., Van Emde Boas, P., Nielsen, M. (eds.) ICALP 1999. LNCS, vol. 1644, pp. 412–421. Springer, Heidelberg (1999)
5. Esparza, J.: Petri nets, commutative context-free grammars, and basic parallel processes. In: Reichel, H. (ed.) FCT 1995. LNCS, vol. 965, pp. 221–232. Springer, Heidelberg (1995)
6. Hirshfeld, Y.: Bisimulation trees and the decidability of weak bisimulations. In: INFINITY 1996. Proceedings of the 1st International Workshop on Verification of Infinite State Systems, Germany, vol. 5 (1996)
7. Hüttel, H., Stirling, C.: Actions speak louder than words. Proving bisimilarity for context-free processes. In: LICS 1991. Proceedings of 6th Annual symposium on Logic in Computer Science, Amsterdam, pp. 376–386 (1991)
8. R., Milner: A Calculus of Communicating Systems. LNCS, vol. 92. Springer, Heidelberg (1980)

9. Milner, R.: Communication and Concurrency. Prentice-Hall International, Englewood Cliffs (1989)
10. Bergstra, J.A., Klop, J.W.: Process theory based on bisimulation semantics. In: Prodeedings of REX Workshop, pp. 50–122. The Netherlands, Amsterdam (1988)
11. Caucal, D.: Graphes canoniques de graphes algébriques. Informatique théorique et Applications(RAIRO) 90-24(4), 339–352 (1990)
12. Dixon, L.E.: Finiteness of the odd perfct and primitive abundant numbers with distinct factors. American Journal of Mathematics 35, 413–422 (1913)
13. Christensen, S., Hirshfield, Y., Moller, F.: Decomposability, Decidability and Axiomatisability for Bisimulation Equivalence on Basic Parallel Processes. In: LICS 1993. Proceedings of the Eighth Annual Symposium on Logic in Computer Science, Montreal, Canada, pp. 386–396 (1993)
14. Jančar, P.: Strong bisimilarity on basic parallel processes is PSPACE-complete. In: LICS 2003. Proceedings of the 18th Annual IEEE Symposium on Logic in Computer Science, Ottawa, Canada, pp. 218-227 (2003)
15. Esparza, J.: Decidability of model checking for infinite-state concurrent systems. Acta Informatica 97-34, 85–107 (1997)
16. Burkart, O., Esparza, J.: More infinite results. Electronic Notes in Theoretical computer Science 5 (1997)
17. Srba, J.: Roadmap of Infinite Results. Bulletin of the European Association for Theoretical Computer Science 78, 163–175 (2002), columns:Concurrency
18. Mayr, R.: On the Complexity of Bisimulation Problems for Basic Parallel Processes. In: Welzl, E., Montanari, U., Rolim, J.D.P. (eds.) ICALP 2000. LNCS, vol. 1853, pp. 329–341. Springer, Heidelberg (2000)
19. Mayr, R.: Weak Bisimulation and Model Checking for Basic Parallel Processes. In: Chandru, V., Vinay, V. (eds.) FSTTCS 1996. LNCS, vol. 1180, pp. 88–99. Springer, Heidelberg (1996)

Extensions of Embeddings in the Computably Enumerable Degrees

Jitai Zhao[1,2,⋆]

[1] State Key Lab. of Computer Science, Institute of Software, Chinese Academy of Sciences
[2] Graduate University of Chinese Academy of Sciences
jitai@ios.ac.cn

Abstract. In this paper, we study the extensions of embeddings in the computably enumerable Turing degrees. We show that for any c.e. degrees $\mathbf{x} \not\leq \mathbf{y}$, if either $\mathbf{y}$ is low or $\mathbf{x}$ is high, then there is a c.e. degree $\mathbf{a}$ such that both $\mathbf{0} < \mathbf{a} \leq \mathbf{x}$ and $\mathbf{x} \not\leq \mathbf{y} \cup \mathbf{a}$ hold.

1 Introduction

A set $A \subseteq \omega$ is called *computably enumerable* (c.e.) if and only if either $A = \emptyset$ or A is the range of a computable function.

A set A is *simple* if A is c.e. and $\bar{A}$, the complement of A, is infinite but contains no infinite c.e. set. Clearly if A is simple then A is not computable because $\bar{A}$ can not be c.e., since otherwise it contains an infinite c.e. set, i.e., $\bar{A}$.

Given sets $A, B \subseteq \omega$, we say that A is *Turing reducible* to B, if there is an oracle Turing machine Φ such that $A = \Phi(B)$ (denoted by $A \leq_{\mathrm{T}} B$). $A \equiv_{\mathrm{T}} B$ if $A \leq_{\mathrm{T}} B$ and $B \leq_{\mathrm{T}} A$. The *Turing degree* of A is defined to be $\mathbf{a} = deg(A) = \{B : B \equiv_{\mathrm{T}} A\}$.

A degree $\mathbf{a} \leq \mathbf{0}'$ is *low*, if $\mathbf{a}' = \mathbf{0}'$, and *high* if $\mathbf{a}' = \mathbf{0}''$. A set $A \leq_{\mathrm{T}} \emptyset'$ is *low* (*high*), if $deg(A)$ is low (high).

The degrees $\mathcal{D}$ form a partially ordered set under the relation $deg(A) \leq deg(B)$ if and only if $A \leq_{\mathrm{T}} B$. We write $deg(A) < deg(B)$ if $A \leq_{\mathrm{T}} B$ and $B \not\leq_{\mathrm{T}} A$.

A degree is called *computably enumerable* (c.e.), if it contains a c.e. set. Let $\mathcal{E}$ denote the class of c.e. degrees with the some ordering as that for $\mathcal{D}$. As we know $(\mathcal{D}; \leq, \cup)$ and $(\mathcal{E}; \leq, \cup)$ will form upper semi-lattices.

Given r.e. degrees $\mathbf{0} < \mathbf{b} < \mathbf{a}$, we say that $\mathbf{b}$ *cups to* $\mathbf{a}$ if there exists an r.e. degree $\mathbf{c} < \mathbf{a}$ such that $\mathbf{b} \cup \mathbf{c} = \mathbf{a}$; if no such $\mathbf{c}$ exists then $\mathbf{b}$ is an *anti-cupping witness* for $\mathbf{a}$. The r.e. degree $\mathbf{a}$ has *the anti-cupping (a.c.) property* if it has an anti-cupping witness.

Extensions and non-extensions of embeddings in the computably enumerable Turing degrees have been an extensively studied phenomena in the past decades since Shoenfield [1965] published his conjecture. The conjecture was soon proved false by the minimal pair theorem of Lachlan [1966]. However the characterization of the structure satisfying Shoenfield's conjecture has become a successful

⋆ The author is partially supported by NSFC Grant No. 60325206, and No. 60310213.

M. Agrawal et al. (Eds.): TAMC 2008, LNCS 4978, pp. 204–211, 2008.

research in the local Turing degrees, leading to the final resolution of Slaman and Soare [1995] of the full characterization of the problem.

The interests in the strong extensions of embeddings in the computably enumerable degrees come from the close relationship between the problem and the decidability/undecidability of the Σ_2-fragment of the c.e. degrees. For instance, Slaman asked in [1983] the following question:

For any two c.e. degrees $\mathbf{x}$ and $\mathbf{y}$, with $\mathbf{x} \not\leq \mathbf{y}$, does there exist a c.e. degree $\mathbf{a}$, satisfying:

1. $\mathbf{0} < \mathbf{a} \leq \mathbf{x}$,
2. $\mathbf{x} \not\leq \mathbf{y} \cup \mathbf{a}$?

Intuitively, the problem wants to construct a c.e. degree $\mathbf{a}$ which should "code more information than" $\mathbf{0}$, and which cannot compute $\mathbf{x}$ even if it joins with $\mathbf{y}$.

In fact it is not always possible to find such a c.e. degree $\mathbf{a}$, which is stated in Slaman and Soare [2001]. This problem has recently been resolved negatively in progress by Barmpalias, Cooper, Li, Xia, and Yao.

In this article, we consider possible partial results on the positive side of the problem above. We consider two special cases, i.e., when $\mathbf{y}$ is low and $\mathbf{x}$ is high.

We will show the following two theorems:

Theorem 1. *For any two c.e. degrees $\boldsymbol{x}$ and $\boldsymbol{l}$, with $\boldsymbol{x} \not\leq \boldsymbol{l}$ and $\boldsymbol{l}$ low, there is a c.e. degree $\boldsymbol{a}$, satisfying:*

1. *$\boldsymbol{0} < \boldsymbol{a} \leq \boldsymbol{x}$, and*
2. *$\boldsymbol{x} \not\leq \boldsymbol{l} \cup \boldsymbol{a}$.*

Theorem 2. *For any two c.e. degrees $\boldsymbol{h}$ and $\boldsymbol{y}$, with $\boldsymbol{h} \not\leq \boldsymbol{y}$ and $\boldsymbol{h}$ high, there is a c.e. degree $\boldsymbol{a}$, satisfying:*

1. *$\boldsymbol{0} < \boldsymbol{a} \leq \boldsymbol{h}$, and*
2. *$\boldsymbol{h} \not\leq \boldsymbol{y} \cup \boldsymbol{a}$.*

Note that Theorem 2 can be directly deduced from the following theorem which appears in Miller [1981].

Theorem 3. *Every high r.e. degree $\mathbf{h}$ has the a.c. property via a high r.e. witness $\boldsymbol{a}$.*

We briefly describe how to show Theorem 2 by Theorem 3: for a given high r.e. degree $\mathbf{h}$, there exists a high r.e. degree $\mathbf{a}$, $\mathbf{0} < \mathbf{a} < \mathbf{h}$, and for a given r.e. $\mathbf{y}$, $\mathbf{h} \leq \mathbf{y} \cup \mathbf{a}$ implies $\mathbf{h} \leq \mathbf{y}$, i.e., $\mathbf{h} \not\leq \mathbf{y}$ implies $\mathbf{h} \not\leq \mathbf{y} \cup \mathbf{a}$. Therefore $\mathbf{a}$ is a desired r.e. degree in Theorem 2.

The rest of the paper is devoted to proving Theorem 1, our main result.

Our notations and terminology are standard and generally follow Soare [1987] and Cooper [2003]. During the course of a construction, notations such as A, Φ are used to denote the current approximations to these objects, and if we want to specify the values immediately at the end of stage s, then we denote them by A_s, $\Phi[s]$ etc. For a *computable partial functional* (c.p., or for simplicity, also a Turing functional), Φ say, the use function is denoted by the corresponding lower case letter ϕ. The value of the use function of a converging computation is the greatest number which is actually used in the computation. For a Turing functional, if a computation is not defined, then we define its use function equal to -1.

2 Proof of Theorem 1

2.1 Requirements and Strategies

Given c.e. sets $X \in \mathbf{x}$, and $L \in \mathbf{l}$, with $X \leq_T L$ and L low, we will build a c.e. set A to satisfy the following requirements:

$$\mathcal{T}: A \leq_T X$$
$$\mathcal{P}_e: W_e \ infinite \Rightarrow W_e \cap A \neq \emptyset$$
$$\mathcal{N}_e: X \neq \Phi_e(L \oplus A)$$

where $e \in \omega$, $\{\Phi_e : e \in \omega\}$ is an effective enumeration of all Turing reductions Φ, and W_e is the e-th c.e. set.

Let $\mathbf{a}$ be the Turing degree of A. By the $\mathcal{T}$-requirement, $\mathbf{a} \leq \mathbf{x}$, by the $\mathcal{P}$-requirements, A is simple, so it is not computable, i.e., $\mathbf{a} > \mathbf{0}$, and by the $\mathcal{N}$-requirements, $\mathbf{x} \not\leq \mathbf{l} \cup \mathbf{a} = deg(L) \cup deg(A) = deg(L \oplus A)$. Therefore the requirements are sufficient to prove the theorem.

Let $\{X_s\}_{s\in\omega}, \{L_s\}_{s\in\omega}$ be computable enumerations of X, L respectively.

During the construction, the requirements may be divided into the *positive* requirements $\mathcal{P}_e$, which attempt to put elements *into* A, and the *negative* requirements $\mathcal{N}_e$, which attempt to keep elements *out of* A, i.e., impose an A-restraint function with priority $\mathcal{N}_e$. The priority rank of the requirements is $\mathcal{N}_e < \mathcal{P}_e < \mathcal{N}_{e+1}$, for all $e \in \omega$.

First we introduce an easy method in Yates [1965] for constructing a c.e. set A which is computable in a given non-computable c.e. set B by enumerating an element x in A at some stage s only when B permits x in the sense that some element $y < x$ appears in B at the same stage s.

Proposition 1. *If $\{A_s\}_{s\in\omega}$ and $\{B_s\}_{s\in\omega}$ are computable enumerations of c.e. sets A and B respectively, such that $x \in A_{s+1} - A_s$ implies $(\exists y < x)[y \in B - B_s]$, then $A \leq_T B$.*

Proof. To B-recursively compute whether $x \in A$, find a stage s such that $B_s \upharpoonright x = B \upharpoonright x$. Now $x \in A$ if and only if $x \in A_s$. □

The strategy for meeting the $\mathcal{T}$-requirement is attached onto the positive requirements. When an element x is enumerated into A, it must satisfy that $X_{s+1} \upharpoonright x \neq X_s \upharpoonright x$ so that $A \leq_T X$ holds according to Proposition 1 (Soare [1987]). Note that this kind of x can always be found because X is not computable in L.

The strategy for meeting a single requirement $\mathcal{P}_e$ is the same as for Post's simple set. Intuitively, enumerate W_e until the first element $> 2e$ appears in W_e simultaneously satisfying other conditions and put it into A.

We now give a property of a low c.e. set, as found in Soare [1987].

Proposition 2. *If L is a low set then*

$$C = \{j : (\exists n \in W_j)[D_n \subseteq \bar{L}]\} \leq_{\mathrm{T}} \emptyset'. \tag{1}$$

where W_j is the j-th c.e. set, and D_n is a finite set with canonical index n.

Proof. Clearly, C is $\sum_1^L$, so $C \leq_{\mathrm{T}} L'$. If L is low then $L' \leq_{\mathrm{T}} \emptyset'$, so $C \leq_{\mathrm{T}} \emptyset'$. □

According to the Limit Lemma, let $g(e, s)$ be a computable function such that $\lim_s g(e, s)$ is the characteristic function of C.

Now we state the basic strategy for meeting a requirement $\mathcal{N}_e$, without loss of generality, let $A \subseteq 2\omega$, the even numbers, and $L \subseteq 2\omega + 1$, the odd numbers. Note that $A \oplus L \equiv_{\mathrm{T}} A \cup L$, so we use the latter from now on, i.e.,

$$\mathcal{N}_e : X \neq \Phi_e(L \cup A).$$

We follow some basic idea in Robinson [1971] of proving the Robinson Low Splitting Theorem which also can be found in Soare [1987].

Fix e and x. Intuitively, we use the lowness of L to help to "L-certify" a computation $\Phi_e((L \cup A) \restriction u; x)[s]$ where $u = \phi_e(L \cup A; x)[s]$ is the use function of this computation as follows:

Let $D_n = \bar{L}_s \restriction u$. Enumerate n into a c.e. V that we shall build during the construction. By the Recursion Theorem we may assume that we have in advance an index j such that $V = W_j$. Find the least $t \geq s$ such that either $D_n \cap L_t \neq \emptyset$, in which case the computation is obvious disturbed, or $g(j, t) = 1$, in which case we "L-certify" the computation and guess that it is L-correct. It may happen that we were wrong and $L \restriction u \neq L_s \restriction u$, but this happens at most finitely often by Proposition 2 and the Limit Lemma. Since we are really using g as an oracle to inquire whether $D_n \subseteq \bar{L}$ for the current $D_n = \bar{L}_s \restriction u$, it is very important that there are no previous $m \in V_s$ unless $D_m \cap L_s \neq \emptyset$. Thus, whenever an L-certified computation first becomes A-invalid by $A_t \restriction u \neq A_s \restriction u$, we abandon the old c.e. set V and start with a new version of V and hence a new index j such that $W_j = V$.

This L-certification process is best formalized by transforming the function $\Phi_e(L \cup A; x)$ to a computable function $\hat{\Phi}_e(L \cup A; x)$. When we have fixed e, for notational convenience, we let

$$\hat{\Phi}_s(x) \;\leftrightarrow\; \hat{\Phi}_e(L \cup A; x)[s],$$

and

$$u_x^s \;\leftrightarrow\; \phi_e(L \cup A; x)[s].$$

If $\hat{\Phi}_{s-1}(x) \downarrow$ but $\hat{\Phi}_s(x) \uparrow$ we say that the (e, x)-computation $\hat{\Phi}_{s-1}(x) \downarrow$ becomes *A-invalid* if

$$(\exists z < u_x^{s-1})[z \in A_s - A_{s-1}]$$

and otherwise becomes *L-invalid.*

Fix e, x and s, we define $\hat{\Phi}_s(x)$ as follows: given A_t and L_t, $t \leq s$ and assume that

$$\Phi_e(L \cup A; x)[s] \downarrow = y,$$

and

$$\neg(\exists z < u_x^{s-1})[z \in (A_s \cup L_s) - (A_{s-1} \cup L_{s-1})].$$

Let $D_n = \bar{L}_s \upharpoonright u_x^s$. Enumerate n into $V_s^{e,x}$. Let v be the greatest stage less than s at which an (e, x)-computation becomes A-invalid, and $v = 0$ if no such stage exists. By the Recursion Theorem, choose j such that $W_j = \bigcup\{V_t^{e,x} : t > v\}$. Find the least $t \geq s$ such that either

$$D_n \cap L_t \neq \emptyset, \tag{2}$$

or

$$g(j, t) = 1. \tag{3}$$

If the latter holds, define $\hat{\Phi}_s(x) \downarrow = y$. Otherwise, $\hat{\Phi}_s(x) \uparrow$.

We use a strategy which is similar to the Sacks' *preserving agreement* strategy to meet a negative requirement. Here we want to construct a c.e A to meet a requirement of the form $X \neq \Phi_e(L \cup A)$. During the construction we preserve agreement between X and $\Phi_e(L \cup A)$. Sufficient preservation will guarantee that if $X = \Phi_e(L \cup A)$, then in fact $X \leq_{\mathrm{T}} L$, contrary to hypothesis.

As usual, we define the computable functions:

$$(length\ function)\ \hat{l}(e, s) = max\{x : (\forall y < x)[X_s(y) = \hat{\Phi}_s(y)]\},$$

$$(restraint\ fucntion)\ \hat{r}(e, s) = max\{u_x^s : x \leq \hat{l}(e, s)\ \&\ \hat{\Phi}_s(x) \downarrow\}.$$

We say that x *injures* $\mathcal{N}_e$ at stage $s + 1$ if $x \in A_{s+1} - A_s$ and $x \leq \hat{r}(e, s)$. Define the *injury set* for $\mathcal{N}_e$,

$$(injury\ set)\ \hat{I}(e) = \{x : (\exists s)[x \in A_{s+1} - A_s\ \&\ x \leq \hat{r}(e, s)]\}.$$

The positive requirements of course are never injured.

2.2 Construction and Verification

Proof of Theorem 1.

Construction of A.

Stage $s = 0$. Set $A_0 = \emptyset$.

Stage $s + 1$. Since A_s has already been defined, we can define, for all e, the length function $\hat{l}(e, s)$ and restraint function $\hat{r}(e, s)$.

We say $\mathcal{P}_e$ *requires attention* at stage $s + 1$ if

$$W_{e,s} \cap A_s = \emptyset,$$

Then find if $\exists x$,

$$x \in W_{e,s},$$

$$x > 2e,$$

$$X_{s+1} \upharpoonright x \neq X_s \upharpoonright x,$$

and

$$(\forall i \leq e)[\hat{r}(i, s) < x].$$

Choose the least $i \leq s$ such that $\mathcal{P}_i$ requires attention, and then enumerate the least such x into A_{s+1}, and we say that $\mathcal{P}_i$ *receives attention*. Hence $W_{i,s} \cap A_{s+1} \neq \emptyset$ and $(\exists x \in A_{s+1})[X_{s+1} \upharpoonright x \neq X_s \upharpoonright x]$, so $\mathcal{P}_i$ is satisfied and never again requires attention.

If i does not exist, do nothing.

Let $A = \bigcup A_s$. This ends the construction.

To verify that the construction succeeds we must prove the following lemmas.

Lemma 1. $(\forall e)$ $[\hat{I}(e)$ *is finite*$]$. *($\mathcal{N}_e$ is injured at most finitely often.)*

Proof. Note that once $\mathcal{P}_i$ receives attention, it will become satisfied and remain satisfied forever. Hence each $\mathcal{P}_i$ contributes at most one element to A, and $\mathcal{N}_e$ can be injured by $\mathcal{P}_i$ only if $i < e$. So $|\hat{I}(e)| \leq e$. □

Lemma 2. $(\forall e)$ $[X \neq \Phi_e(A \cup L)]$. *($\mathcal{N}_e$ is met.)*

Proof. Assume for a contradiction that $X = \Phi_e(A \cup L)$. Then $\lim_s \hat{l}(e, s) = \infty$. By Lemma 1, choose s_1 such that $\mathcal{N}_e$ is never injured after stage s_1. We shall show that $X \leq_T L$ contrary to hypothesis. To L-recursively compute $X(p)$ for $p \in \omega$, find some stage $s > s_1$ such that $\hat{l}(e, s) > p$ and each computation $\hat{\Phi}_s(x)$, $x \leq p$, is L-correct, namely, $L_s \upharpoonright u_x^s = L \upharpoonright u_x^s$. It follows by induction on $t \geq s$ that

$$(\forall t \geq s)[\hat{l}(e, t) > p \ \& \ \hat{r}(e, t) \geq max\{u_x^s : x \leq p\}], \quad (4)$$

and hence that for all $t \geq s$,

$$\hat{\Phi}_t(p) = \Phi_e(A \cup L; p) = X(p).$$

So X is computable in L.

To prove (4), when $t = s$, clearly it is true. Assume that it holds for t. Then by the definition of $\hat{r}(e, t)$ and $s > s_1$, for any $x \leq p$, it ensures that $(A_{t+1} \cup L_{t+1}) \upharpoonright z = (A_t \cup L_t) \upharpoonright z$ for all numbers z used in a computation $\hat{\Phi}_t(x) \downarrow = y$. Hence,

$$\hat{\Phi}_{t+1}(x) \downarrow = \hat{\Phi}_t(x) \downarrow = X_t(x).$$

So $\hat{l}(e, t+1) > p$ unless $X_{t+1}(x) \neq X_t(x)$. But if $X_t(x) \neq X_s(x)$ for some $t \geq s$, since X is c.e., the *disagreement* $\hat{\Phi}_t(x) \downarrow \neq X_t(x)$ is preserved forever, so $X(x) = X_t(x) \neq \hat{\Phi}_t(x) \downarrow = \Phi_e(A \cup L; x)$, contrary to the hypothesis $X = \Phi_e(A \cup L)$. □

Lemma 3. $(\forall e)[\lim_s \hat{r}(e, s)$ *exists and is finite*$]$.

Proof. By Lemma 1, choose s_1 such that $\mathcal{N}_e$ is never injured after stage s_1. By Lemma 2, choose $p = (\mu x)[X(x) \neq \Phi_e(A \cup L; x)]$. Choose $s_2 \geq s_1$ sufficiently large so that for all $s \geq s_2$,

$$(\forall x < p)[\hat{\Phi}_s(x) \downarrow = \Phi_e(A \cup L; x)],$$

and

$$(\forall x \leq p)[X_s(x) = X(x)].$$

Case 1. $\Phi_e(A \cup L; p) \downarrow \neq X(p)$. Choose $s_3 \geq s_2$ such that for all $s \geq s_3$, $\hat{\Phi}_s(p) \downarrow = q \neq X(p)$. Hence, for all $s \geq s_3$, $\hat{l}(e,s) = \hat{l}(e,s_3)$ and $\hat{r}(e,s) = \hat{r}(e,s_3)$.

Case 2. $\Phi_e(A \cup L; p) \uparrow$. We shall find a stage v such that for all $s \geq v$, $\hat{\Phi}_s(p) \uparrow$. Hence, for all $s \geq v$, $\hat{r}(e,s) = \hat{r}(e,v)$.

Note that if $\hat{\Phi}_s(p) \downarrow$ for any $s \geq s_2$ then $\hat{r}(e,s) \geq u_p^s$, so by induction on $t \geq s$, the computation $\hat{\Phi}_t(p) = \hat{\Phi}_s(p)$ holds as long as it remains L-valid. Let s' be the least t such that no (e,p)-computation becomes A-invalid at any stage $\geq t$. By the Recursion Theorem, choose j such that $W_j = \bigcup\{V_s^{e,p} : s \geq s'\}$. Since $\Phi_e(A \cup L; p) \uparrow$, any computation $\hat{\Phi}_s(p)$, $s \geq s'$, becomes L-invalid at some stage $t > s$, at which time $D_m \bigcap C_t \neq \emptyset$ for every $m \in V_t^{e,p}$. Hence, $\lim_s g(j,s) = 0$ by (1). Choose $v > s_2$ such that $\hat{\Phi}_v(p) \uparrow$ and $g(j,s) = 0$ for all $s \geq v$. We claim that $\hat{\Phi}_s(p) \uparrow$ for all $s \geq v$. Suppose $s > v$, $\hat{\Phi}_{s-1}(p) \uparrow$ and $\hat{\Phi}_s(p) \downarrow$. Then we enumerate $n \in V_s^{e,p}$, where $D_n = \bar{L}_s \upharpoonright u_p^s$, and we choose the least $t \geq s$ satisfying (2) or (3). But (3) could not occur by the choice of v, so (2) occurs and $\hat{\Phi}_s(p) \uparrow$. □

Lemma 4. $(\forall e)$ $[W_e$ *infinite* $\Rightarrow W_e \cap A \neq \emptyset]$. *($\mathcal{P}_e$ is met, simultaneously, $\mathcal{T}$ is met.)*

Proof. By the above lemmas, for all $i \leq e$, let

$$\hat{r}(i) = \lim_s \hat{r}(i,s)$$

and

$$\hat{R}(e) = max\{\hat{r}(i) : i \leq e\}.$$

Choose s_0 such that

$$(\forall t \geq s_0)(\forall i \leq e)[\hat{r}(e,t) = \hat{r}(e)],$$

and no $\mathcal{P}_i$, $i < e$, receives attention after stage s_0.

Now choose $s \geq s_0$, if $\exists x$,

$$x \in W_{e,s},$$

$$x > 2e,$$

$$X_{s+1} \upharpoonright x \neq X_s \upharpoonright x,$$

and

$$\hat{R}(e) < x.$$

Now either $W_{e,s} \cap A_s \neq \emptyset$ or else $\mathcal{P}_e$ receives attention at stage $s+1$, then in either case $W_{e,s} \cap A_{s+1} \neq \emptyset$, so $\mathcal{P}_e$ is met by the end of stage $s+1$. And by Proposition 1, $A \leq_{\mathrm{T}} X$ is obviously met.

It is remarkable that searching for an x with the condition $X_{s+1} \upharpoonright x \neq X_s \upharpoonright x$ does not impact the requirement $\mathcal{P}_e$. Suppose for the sake of contradiction that if W_e infinite, while $A \cap W_e = \emptyset$, then $A \subseteq \overline{W}_e$, choose an increasing c.e. sequence of elements $x_1 < x_2 < \cdots$ in W_e such that $x_1 > \hat{R}(e)$ for all stage $s \geq s_0$. Choose s_k minimal such that $s_k > s_0$ and $x_k \in W_{e,s_k}$. Now $X_{s_k} \upharpoonright x_k = X \upharpoonright x_k$ so X is computable, contrary to $X \not\leq_{\mathrm{T}} L$. □

Note that $\bar{A}$ is infinite by the clause "$x > 2e$". To see this, note that A contains at most e elements in $\{0, 1, \ldots, 2e\}$, hence $card(\bar{A} \upharpoonright (2e+1)) \geq 2e+1-e = e+1$. A is simple.

This ends the proof of Theorem 1. ■

References

1. Lachlan, A.H.: Lower bounds for pairs of recursively enumerable degrees. Proc. London Math. Soc. 16, 537–569 (1966)
2. Barmpalias, Cooper, Li, Xia, Yao, Super minimal pairs (in progress)
3. Yates, C.E.M.: Three theorems on the degrees of recursively enumerable sets. Duke Math. J.32, 461–468 (1965)
4. Miller, D.: High recursively enumerable degrees and the anti-cupping property. In: Lerman, Schmerl, and Soare, pp. 230–245 (1981)
5. Shoenfield, J.R.: Application of model theory to degrees of unsolvability. In: Addison, Henkin, and Tarski, pp. 359–363 (1965)
6. Soare, R.I.: Recursively Enumerable Sets and Degrees. Springer, Heidelberg (1987)
7. Robinson, R.W.: Interpolation and embedding in the recursively enumerable degrees. Ann. of Math. 2(93), 285–314 (1971)
8. Cooper, S.B.: Computability Theory. Chapman Hall/CRC Mathematics Series, vol. 26 (2003)
9. Slaman, T.A.: The recursively enumerable degrees as a substructure of the Δ_2^0 degrees (1983) (unpublished notes)
10. Slaman, T.A., Soare, R.I.: Algebraic aspects of the computably enumerable degrees. Proceedings of the National Academy of Science, USA 92, 617–621 (1995)
11. Slaman, T.A., Soare, R.I.: Extension of embeddings in the computably enumerable degrees. Ann. of Math (2) 154(1), 1–43 (2001)

An Improved Parameterized Algorithm for a Generalized Matching Problem*

Jianxin Wang, Dan Ning, Qilong Feng, and Jianer Chen

School of Information Science and Engineering, Central South University
jxwang@mail.csu.edu.cn

Abstract. We study the parameterized complexity of a generalized matching problem, the P_2-packing problem. The problem is NP-hard and has been studied by a number of researchers. In this paper, we provide further study of the structures of the P_2-packing problem, and propose a new kernelization algorithm that produces a kernel of size $7k$ for the problem, improving the previous best kernel size $15k$. The new kernelization leads to an improved algorithm for the problem with running time $O^*(2^{4.142k})$, improving the previous best algorithm of time $O^*(2^{5.301k})$.

1 Introduction

Packing problem has formed an important class of NP-hard problems. In particular, as one of the graph packing problem, the H-packing problem has gained more attention, which arises in applications such as scheduling, wireless sensor tracking, wiring-board design and code optimization, etc. The problem is defined as follows [1].

Definition 1. Given a graph $G = (V, E)$ and a fixed graph H. An H-packing of G is a set of vertex disjoint subgraphs of G, each of which is isomorphic to H.

From the optimization point of view, the problem of MAXIMUN H-packing is to find the maximum number of vertex disjoint copies of H in G. If the H is the complete graph K_2, the MAXIMUN H-packing becomes the familiar maximum matching problem in bipartite graph, which can be solved in polynomial time. When the graph H is a connected graph with at least three vertices, D. G. Kirkpatrick and P. Hell [2] gave that the problem is NP-complete. From the approximation point of view, V. Kann [3] proved that the MAXIMUN H-packing problem is MAX-SNP-complete. C. Hurkens and A. Schrijver [4] presented an approximation algorithm with ratio $|V_H|/2 + \varepsilon$ for any $\varepsilon > 0$.

* This work is in part supported by the National Natural Science Foundation of China under Grant No. 60773111 and No. 60433020, Provincial Natural Science Foundation of Hunan (06JJ10009), the Program for New Century Excellent Talents in University No. NCET-05-0683 and the Program for Changjiang Scholars and Innovative Research Team in University No. IRT0661.

M. Agrawal et al. (Eds.): TAMC 2008, LNCS 4978, pp. 212–222, 2008.

Recently, parameterized complexity theory has been used to design efficient algorithms for H-packing problem. M. Fellows et. al. [5] proposed a parameterized algorithm with time complexity of $O(2^{O(|H|k \log k + k|H| \log |H|)})$ for any arbitrary graph H. For the edge disjoint triangle packing problem, L. Mathieson, E. Prieto and P. Shaw [6] proved that the problem has a $4k$ kernel and gave a parameterized algorithm of running time $O(2^{\frac{9k}{2} \log k + \frac{9k}{2}})$ based on the kernel.

When H belongs to the restricted family of graphs $K_{1,s}$, a star with s leaves, we can get the $K_{1,s}$-packing problem, which is defined as follows:

Definition 2. Parameterized $K_{1,s}$-PACKING(k-$K_{1,s}$-PACKING): Given a graph $G = (V, E)$ and a positive integer k, whether there are at least k vertex disjoint $K_{1,s}$ in G?

M. Fellows, E. Prieto and C. Sloper [7] gave that the parameterized $K_{1,s}$-packing problem is fixed-parameter tractable and got a $O(k^3)$ kernel. M. Fellows [7] et.al. also studied the P_2-packing problem, where P_2 is a path of three vertices (one center vertex and two endpoints) and two edges, which is defined as follows:

Definition 3. Parameterized P_2-PACKING (k-P_2-PACKING): Given a graph $G = (V, E)$ and a positive integer k, whether there are at least k vertex disjoint P_2 in G?

In [7], for the P_2-packing problem, M. Fellows et.al. gave a kernel of size at most $15k$ and proposed an algorithm with time complexity $O^*(2^{5.301k})$.

In this paper, we mainly focus on the kernelization of the k-P_2-packing problem and give a kernel of size at most $7k$. Based on the kernel, we present a parameterized algorithm with time complexity $O^*(2^{4.142k})$, which greatly improves the current best result $O^*(2^{5.301k})$.

This paper is organized as follows. In section 2, we introduce some related definitions and lemmas. In section 3, we present all the steps of the kernelization algorithm, and prove that the k-P_2-packing problem has a size of $7k$ kernel. In section 4, we give the general algorithm solving the k-P_2-packing. In section 5, we draw some final conclusions.

2 Related Definitions and Lemmas

We first give some concepts and terminology about graph [8].

Assume $G = (V, E)$ denotes a simple, undirected, connected graph, where $|V| = n$. The neighbors of a vertex v are denoted as $N(v)$. The induced subgraph of $S \subseteq V$ is denoted $G[S]$. For an arbitrary subgraph H of G, let $N(H)$ denote the vertices that are not in H but connect with at least one vertex in H. We use the simpler $G \backslash v$ to denote $G[V \backslash v]$ for a vertex v and $G \backslash e$ to denote $G = (V, E \backslash e)$ for an edge e. Likewise, $G \backslash V'$ denotes $G[V \backslash V']$ and $G \backslash E'$ denotes $G = (V, E \backslash E')$ where V' is a set of vertices and E' is a set of edges.

For the convenience of description, we firstly introduce the definitions of 'double crown' decomposition and 'fat crown' decomposition [7].

Definition 4. A double crown decomposition (H, C, R) in a graph $G = (V, E)$ is a partitioning of the vertices of the graph into three sets H, C and R that have the following properties:

(1) H (the head) is a separator in G such that there are no edges in G between vertices belonging to C and vertices belonging to R.

(2) $C = C_u \cup C_m \cup C_{m2}$ (the crown) is an independent set in G.

(3) $|C_m| = |H|$, $|C_{m2}| = |H|$ and there is a perfect matching between C_m and H, and a perfect matching between C_{m2} and H.

Definition 5. A fat crown decomposition (H, C, R) in a graph $G = (V, E)$ is a partitioning of the vertices of the graph into three sets H, C and R that have the following properties:

(1) H (the head) is a separator in G such that there are no edges in G between vertices belonging to C and vertices belonging to R.

(2) $G[C]$ is a forest where each component is isomorphic to K_2.

(3) $|C| \geq |H|$, and there is a perfect matching M between H and a subset of cardinality $|H|$ in C, where one endpoint of each edge in M is in H, and the other is the endpoint of K_2 in C.

We introduce the following lemmas [7] about the 'double crown' decomposition and 'fat crown' decomposition that will be used in our algorithm.

Lemma 1. *A graph $G = (V, E)$ that admits a 'double crown'-decomposition (H, C, R) has a k-P_2-packing if and only if $G \backslash (H \cup C)$ has a $(k - |H|)$-P_2-packing.*

Lemma 2. *A graph $G = (V, E)$ that admits a 'fat crown'-decomposition (H, C, R) has a k-P_2-packing if and only if $G \backslash (H \cup C)$ has a $(k - |H|)$-P_2-packing.*

Lemma 3. *A graph G with an independent set I, where $|I| \geq 2|N(I)|$, has a double crown decomposition (H, C, R), $H \subseteq N(I)$, which can be constructed in linear time.*

Lemma 4. *A graph G with a collection J of independent K_2s, where $|J| \geq |N(J)|$, has a fat crown decomposition (H, C, R), $H \subseteq N(J)$, which can be constructed in linear time.*

3 Kernelization Algorithm for the k-P_2-Packing Problem

In this section we propose a kernelizaiton algorithm that can get a kernel of size at most $7k$ for the parameterized version of P_2-packing problem.

Assume W denotes a maximal P_2-packing and the vertices in W are denoted by $V(W)$. Let W be $\{L_1, ..., L_t\}$, $t \leq k - 1$, where each of $L_i (1 \leq i \leq t)$ is a subgraph in G that is isomorphic to P_2. Let L_i be (e_1, c, e_2), $1 \leq i \leq t$, where e_1 and e_2 are two endpoints of L_i, and c is the center vertex of L_i. Therefore, each connected component of the graph induced by $Q = V \backslash V(W)$ is either a single vertex or a single edge [7]. Let Q_0 be the set of all vertices such that each vertex in Q_0 makes a connected component of the graph induced by Q, and each vertex in Q_0 will be called a Q_0-vertex. Let Q_1 be the set of all edges such that each edge in Q_1 makes a connected component of the graph induced by Q. Each edge in Q_1 will be called a Q_1-edge and each vertex in Q_1 will be called a Q_1-vertex.

3.1 RPLW Algorithm

Based on the kernelization algorithm given in [7], the kernelization process we propose is to apply the algorithm RPLW repeatedly. By using the kernelization algorithm in [7], we can get a graph G, which consists of a maximal packing W and $Q = V \backslash V(W)$. The algorithm RPLW is to further reduce the vertices in W and Q to get a better kernel, whose general idea is given in the following:

Algorithm RPLW deals with the Q_0-vertices and Q_1-edges in Q. When the size of W is not changed, the algorithm aims at reducing the number of Q_0-vertices in Q. When the number of Q_1-edges in Q is reduced, the size of W becomes larger (the number of disjoint P_2 in W is increased) and the algorithm returns the larger W. Then we call the algorithm for the larger W. If the 'double crown' decomposition or the 'fat crown' decomposition is found, the parameter k becomes smaller and the algorithm returns the smaller parameter. Then we call the algorithm for the smaller parameter.

For the convenience of analyzing the RPLW algorithm, we first discuss the following two structures as shown in Fig.1 and Fig.2, which use solid circles and thick lines for vertices and edges in the maximal P_2-packing W, and use hollow circles and thin lines for vertices and edges not in W. (In particular, thin lines that connect two hollow circles are Q_1-edge.)

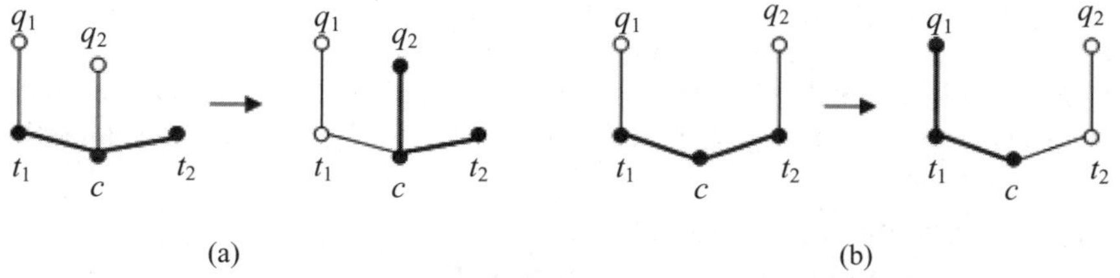

Fig. 1. Reduce the number of Q_0-vertex

The general idea of Fig.1 is that: In order to decrease the number of Q_0-vertices in Q, replace the L_i in W. The specific process is as follows.

In Fig.1(a), assume the P_2 is the L_i in W whose center vertex is c and two endpoints are t_1, t_2. Q_0-vertex q_2 is adjacent to c, and Q_0-vertex q_1 is adjacent to t_1. Vertices q_2, c and t_2 can form a new P_2. Let the new P_2 be L'_i. If L_i is replaced by L'_i in W, the number of Q_0-vertices in Q is just reduced by 2 (q_1 and t_1 form a Q_1-edge in Q).

In Fig.1(b), assume the P_2 is the L_i in W whose center vertex is c and two endpoints are t_1, t_2. Q_0-vertex q_2 is adjacent to t_2, and Q_0-vertex q_1 is adjacent to t_1. Vertices $q_1 t_1$ and c can form a new P_2. Let the new P_2 be L'_i. If L_i is replaced by L'_i in W, the number of Q_0-vertices in Q is just reduced by 2 (q_2 and t_2 form a Q_1-edge in Q).

The general idea of Fig.2 is that: In order to decrease the number of Q_1-edges in Q and increase the number of disjoint P_2 in W, replace L_i in W . The specific process is as follows.

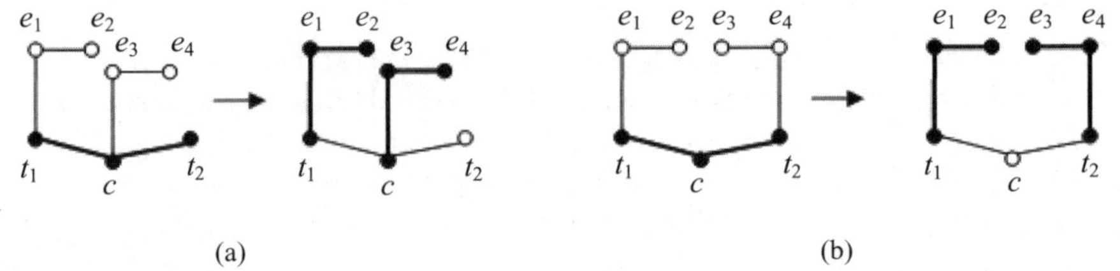

Fig. 2. Reduce the number of Q_1-edge

In Fig.2(a), assume the P_2 is the L_i in W whose center vertex is c and two endpoints are t_1, t_2. Q_1-edge (e_1, e_2) is adjacent to t_1, and Q_1-edge (e_3, e_4) is adjacent to c. Vertices e_1, e_2 and t_1 can form a new P_2, which can be denoted as L_i'. Vertices e_3, e_4 and c can also form a new P_2, which is denoted as L_i''. If L_i is replaced by L_i'' and L_i' in W, the number of Q_1-edges in Q is just reduced by 1 and the number of P_2 in W is increased by 1.

In Fig.2(b), assume the P_2 is the L_i in W whose center vertex is c and two endpoints are t_1, t_2. Q_1-edge (e_1, e_2) is adjacent to t_1, and Q_1-edge (e_3, e_4) is adjacent to t_2. Vertices e_1, e_2 and t_1 can form a new P_2 which is denoted as L_i'. Vertices e_3, e_4 and t_2 can also form a new P_2 which is denoted as L_i''. If L_i is replaced by L_i'' and L_i' in W , the number of Q_1-edges in Q is just reduced by 1and the number of P_2 in W are increased by 1.

Fig.1 and Fig.2 vividly illustrate how to replace a L_i in W to change the number of Q_0-vertices and Q_1-edges. According to Fig.1 and Fig.2, we can obtain the following rules.

Rule1. If a L_i in W has two vertices that each is adjacent to a different Q_0-vertex, then apply the processes described in Fig.1 to decrease the number of Q_0-vertices by 2 (and increase the number of Q_1-edges by 1).

Rule2. If a L_i in W has two vertices that each is adjacent to a different Q_1-edge, then apply the processes described in Fig.2 to decrease the number of Q_1-edges by 1(and increase the size of the maximal P_2-packing by 1).

The RPLW algorithm tries to reduce the number of Q_0-vertices and the number of Q_1-edges by applying Rule1 and Rule2 consecutively. Note that these rules cannot be applied forever. As shown in Fig.3, the while-loop in step1 of the algorithm tries to reduce the number of Q_0-vertices in Q. Because the number of vertices in input graph G is limited (at most $15k$ [7]) and each applications of Rule1 reduces the number of Q_0-vertices by 2, the number of consecutive applications of Rule1 is bonded by $7.5k$. During the applications of these rules, the resulting W may becomes non-maximal. In this cases, we simply first make W maximal again, using any proper greedy algorithm in step2 of the algorithm before we further apply the rules. Thus, the P_2 founded in Q can be put into W to make W larger. Assume the larger packing is W', then call the algorithm for W'.

During the process of the replacement of Q_1-edges in step3 of the algorithm, since each application of Rule2 increases the number of P_2 in W by 1, the

Algorithm RPLW
Input: G, W, k
Output: a maximal P_2-packing W and $|W'| > |W|$, or a smaller parameter k' and $|k'| < |k|$, or a reduced graph G'

1. **while** W is a maximal P_2-packing and a P_2 in W has two vertices that each is adjacent to two different Q_0-vertices **do**
   apply Rule1 to replace W by a packing of the same size with reduced Q_0-vertices;
2. **if** W is not maximal **then**
   use greedy algorithm to construct a larger P_2-packing W';
   return (G, W', k).
3. **if** two Q_1-edges are adjacent to two different vertices on a P_2 in W **then**
   apply Rule2 to obtain a larger P_2-packing W';
   return (G, W', k).
4. **if** $|Q_0| \geq 2|W|$ **then**
   construct a double crown decompositon (H, C, R), then $k' = k - |H|$;
   return (G, W, k').
5. **if** $|Q_1| \geq |W|$ **then**
   construct a fat crown decompositon (H, C, R), then $k' = k - |H|$;
   return (G, W, k').
6. Assume the reduced graph is G', return (G', W, k).

Fig. 3. RPLW algorithm

total number of applications of Rule2 is bounded by k. Step4 and step5 of the algorithm aim at finding 'double crown' and 'fat crown' in G induced by the replacement of Q_0-vertices and Q_1-edges. Once 'double crown' or 'fat crown' is found, the parameter k must be reduced ($k' = k - |H|$).

For completeness, we verify the algorithm's correctness, and analyze its precise complexity.

Lemma 5. *Repeatedly calling the algorithm RPLW will either find a k-P_2-packing or reduce the size of G, and those can be done in $O(k^3)$.*

Proof. From step2 and step3 of the RPLW algorithm, it can be seen that the number of disjoint P_2 in W is increased by the replacement of Q_0-vertices and Q_1-edges. By calling the algorithm repeatedly, when the number of disjoint P_2 in W is k, a k-P_2-packing is found in graph G. Because of the replacements in step4 and step5 of the algorithm, 'double crown'-decomposition or 'fat crown'-decomposition will be found in G. Therefore, the parameter k is decreased and the number of disjoint P_2 needed to be found is also decreased. By calling the algorithm repeatedly, when the parameter k is reduced to 0, a k-P_2-packing is found in graph G. On the other hand, because the replacement in step1-3 of the algorithm limits the number of Q_0-vertices and Q_1-edges, and some vertices are removed by the 'double crown'-decomposition or 'fat crown'-decomposition in step4-5 of the algorithm, the size of G will be reduced. Thus, if a k-P_2-packing is not found in the algorithm, the algorithm returns a reduced G'.

At last, we analyze the time complexity of those whole process. Calling the RPLW algorithm repeatedly is to apply Rule1 and Rule2 consecutively, which can be finished in polynomial time. The number of consecutive applications of Rule1 is bonded by $7.5k$, and the total number of applications of Rule2 is bounded by k. When 'double crown'-decomposition or 'fat crown'-decomposition is applicable, the parameter k is reduced accordingly. The 'double crown'-decomposition or 'fat crown'-decomposition can be founded in $O(k^2)$ [7]. The algorithm must return when the number of disjoint P_2 in W is increased or the parameter k is reduced, and the algorithm is called again for the larger packing W' or the smaller parameter k'. If a k-P_2-packing is not found in the algorithm, the algorithm returns a reduced G'. Therefore, the algorithm is called at most $9.5k$ times. In consequence, the whole process can be computed in $O(k^3)$ time. □

Our kernelization process is to apply the RPLW algorithm repeatedly, i.e, to apply Rule1 and Rule2 repeatedly by starting with a maximal P_2-packing. The whole process can finish in polynomial time. The kernelization will either find a k-P_2-packing or reduce the size of G until Rule1 and Rule2 are not applicable. The reduced G can be considered as a kernel of the k-P_2-packing problem. Note that when Rule1 and Rule2 are not applicable, the maximal P_2-packing W (for each L_i in W) has the following properties:

Property 1. If more than one Q_0-vertices are adjacent to L_i, then all these Q_0-vertices must be adjacent to the same (and unique) vertex in L_i.

Property 2. If more than one vertex in L_i are adjacent to Q_0-vertices, then all these vertices in L_i must be adjacent to the same (and unique) Q_0-vertex.

Property 3. If more than one Q_1-edges are adjacent to L_i, then all these Q_1-edges must be adjacent to the same (and unique) vertex in L_i.

Property 4. If more than one vertex in L_i are adjacent to Q_1-edges, then all these vertices in L_i must be adjacent to the same (and unique) Q_1-edge.

3.2 A Smaller Kernel

In the following, we first analyze the number of Q_0-vertices and Q_1-edges in Q after the kernelization. Then we will present how the kernel of size at most $7k$ is obtained for k-P_2-packing problem.

We first analyze the number of Q_0-vertices in Q.

Theorem 1. *The number of Q_0-vertices is bounded by $2(k-1)$, that is, $|Q_0$-vertex$| \leq 2(k-1)$, or else we can find a double crown decomposition in polynomial time.*

Proof. When Rule1 and Rule2 are not applicable, let W be the maximal packing $W = L_1, \cdots, L_t$, $t \leq k-1$, which is a collection of disjoint P_2. We partition the disjoint P_2 in W into two groups: $\{L_1, \cdots, L_d\}$, $\{L_{d+1}, \cdots, L_t\}$, which satisfy the following property: for each L_i, $1 \leq i \leq d$, each Q_0-vertex adjacent to L_i can be adjacent to more than one vertex in L_i, and we denote these Q_0-vertex as Q_{0i} $(1 \leq i \leq d)$; for each L_j, $j > d$, each Q_0-vertex adjacent to L_j is at most adjacent to one vertex in L_j.

Consider the vertex set Q_0'-vertex=Q_0-vertex$-\{Q_{01}, \cdots, Q_{0d}\}$, and let $W' = \{v_1, \cdots, v_s\}$ be the set of vertices in $L_{d+1} \cup \cdots \cup L_t$ such that each vertex in W' has neighbor in Q_0-vertex. By the above partition property, each $L_j (j > d)$ has at most one vertex in W'. Thus, $s \leq t - d$. Moreover, by Property2, no vertex in Q_0'-vertex is adjacent to any L_i. Therefore, each vertex in Q_0'-vertex has all its neighbors in W', that is, $W' = N(Q_0'\text{-vertex})$.

Assume the total number of vertices in Q_0' -vertex is p. If $p > 2s$, there is a 'double crown'-decomposition in the input graph (note that the set of Q_0'-vertex is an independent set). We can call the RPLW algorithm again, which contradicts that Rule1 and Rule2 are not applicable. On the other hand, if $p \leq 2s$, the total number of Q_0-vertices in the graph is that: $|Q_0\text{-vertex}| = |Q_0'\text{-vertex}| + d = p + d \leq 2s + d \leq 2(s + d) \leq 2t \leq 2(k - 1)$. This completes the proof. □

In the following, we analyze the number of Q_1-vertices in Q.

Theorem 2. *The number of Q_1-vertices is bounded by $2(k - 1)$, that is, $|Q_1\text{-edge}| \leq k - 1$, or else we can find a double crown decomposition in polynomial time.*

Proof. When Rule1 and Rule2 are not applicable, let W be the maximal packing $W = L_1, \cdots, L_t$, $t \leq k - 1$, which is a collection of disjoint P_2. We partition the disjoint P_2 in W into two groups: $\{L_1, \cdots, L_d\}$, $\{L_{d+1}, \cdots, L_t\}$, which satisfy the following property: for each L_i, $1 \leq i \leq d$, each Q_1-edge adjacent to L_i can be adjacent to more than one vertex in L_i, and we denote these Q_1-edges as Q_{1i} $(1 \leq i \leq d)$; for each L_j, $j > d$, each Q_1-edge adjacent to L_j is at most adjacent to one vertex in L_j.

Consider the vertex set Q_1'-edge=Q_1-edge$-\{Q_{11}, \cdots, Q_{1d}\}$, and let $W' = \{v_1, \cdots, v_s\}$ be the set of vertices in $L_{d+1} \cup \cdots \cup L_t$ such that each vertex in W' has neighbors in Q_1-vertex. By the above partition property, each $L_j (j > d)$ has at most one vertex in W'. Thus, $s \leq t - d$. Moreover, by Property4, no vertex in Q_1'-edge is adjacent to any L_i, $1 \leq i \leq d$. Therefore, each vertex in Q_1'-edge has all its neighbors in W', that is, $W' = N(Q_1'\text{-edge})$.

Assume the total number of edges in Q_1'-edge is p. If $p > s$, there is a 'fat crown'-decomposition in the input graph (note that the set of Q_1'-edge is an independent set of K_2). We can call the RPLW algorithm again, which contradicts that Rule1 and Rule2 are not applicable. On the other hand, if $p \leq s$, the total number of Q_1-edges in the graph is that: $|Q_1\text{-edge}| = |Q_1'\text{-edge}| + d = p + d \leq s + d \leq t \leq k - 1$. Each Q_1-edge has two Q_1-vertices, therefore, the number of Q_1-vertices is bounded by $|Q_1\text{-vertex}| \leq 2(k - 1)$. This completes the proof. □

Based on theorem 1 and theorem 2, we can get the following theorem.

Theorem 3. *The k-P_2-packing problem has a kernel of size at most $7k - 7$.*

Proof. By applying the RPLW algorithm repeatedly until the two rules are not applicable. The vertices in G consist of the vertices in W and Q. Assume $V(G)$ denotes the vertices in G. The vertices in Q contains only Q_0-vertices and Q_1-vertices. By Theorem1, we can get that $|Q_0\text{-vertex}| \le 2(k-1)$. By Theorem2, we can get that $|Q_1\text{-vertex}| \le 2(k-1)$. Since $|V(W)| \le 3(k-1)$, thus, $|V(G)| = |V(W)| + |Q_0| + |Q_1| \le 3(k-1) + 2(k-1) + 2(k-1) = 7k - 7$. Therefore, the k-P_2-packing problem has a kernel of size at most $7k - 7$. □

4 The Improved Parameterized Algorithm

For the k-P_2-packing problem, we proposed an improved parameterized algorithm based on the $7k$ kernel. We first apply the kernelizaiton algorithm to obtain a kernel for the problem. Since each P_2 has a center vertex, in order to find k vertex disjoint P_2, we just need to find k center vertices in brute force manner on the $7k$ kernel. The specific algorithm is given in figure 4.

Algorithm KPPW
Input: $G = (V, E)$
Output: a k-P_2-packing in G, or can not find a k-P_2-packing in G

1. compute a maximal P_2-packing W with a greedy algorithm;
2. apply the RPLW(G, W, k) until rule1 and rule2 are not applicable;
3. **if** $|V(G)| > 7k$ **then**
   report "there exists a k-P_2-packing" and stop;
4. find all possible subsets C of size k in reduced G;
5. **for** each C **do**
6. for each vertex v in C, produce a copy vertex v';
7. Construct a bipartite graph $G' = (V_1 \cup V_2, E)$ in the following way: the edges connecting to v are also connected to v'. The k vertices and its copy vertices are put into V_1, and the neighbors of the k vertices in C are put into V_2;
8. use Maximum bipartite matching algorithm to find the k center vertices;
9. **if** all the vertices on V_1 are matched **then**
   report "there exists a k-P_2-packing in G" and stop;
10. report "there is no a k-P_2-packing in G" and stop;

Fig. 4. KPPW agorithm

Theorem 4. *If there exists a k-P_2-packing, the KPPW algorithm will find the k-P_2-packing in time $O^*(2^{4.142k})$.*

Proof. It can be seen from the algorithm, the step2 is the whole kernelization process applying the RPLW algorithm repeatedly. As a result, we can obtain a kernel of size at most $7k$. We check the number of vertices in reduced G in Step3. If the number of vertices is more than $7k$, there must be a k-P_2-packing in G, and the KPPW algorithm does not need to run. We apply a straightforward brute-force method on the kernel to find the optimal solution from Step 4 to Step 8. The general idea is as follows:

We find all possible subsets C of size k in reduced graph G. For each C, we will construct a bipartite graph $G' = (V_1 \cup V_2, E)$ in the following way: first, for each vertex v in C, we produce a copy vertex v' with the property that the edges connecting to v are also connected to v'. The k vertices and its copy vertices are put into V_1, and the neighbors of the k vertices in C are put into V_2. If all the vertices on the V_1 are matched by a maximum bipartite matching, the k vertices in C must be the center vertices of a k-P_2-packing, therefore, report "there exists a k-P_2-packing in G" , and the algorithm stops. If for all C, we cannot find k center vertices, report "there does not exist a k-P_2-packing in G", and the algorithm stops.

In the following, we analyze the time complexity of algorithm KPPW.

Step1: Using greedy algorithm to find a maximal packing can be done in time $O(|E|)$.

Step2: The kernelization process given in [7] can be done in $O(n^3)$ time, and the whole process that call the algorithm RPLW repeatedly until Rule1 and Rule2 are not applicable runs in $O(k^3)$ time, therefore, the time complexity of Step2 is $O(n^3 + k^3)$.

Step3: Obviously, the running time of Step 3 is linear in the size of V.

Step4-Step8: We find the center vertices of the P_2-packing in a brute force manner, which has $\binom{7k}{k}$ enumerations. By Stirling's formula, this is bounded by $2^{4.142k}$. We construct a bipartite graph $G' = (V_1 \cup V_2, E)$ with k vertices in C and its copy vertices in V_1, and the neighbors of k vertices in V_2. Thus, the original question is transformed to find the maximum matching problem in bipartite G' which can be solved in time $O(\sqrt{|V_1 + V_2|}|E|) = O(k^{2.5})$. Therefore, the total running time of Step4-Step8 is $O(2^{4.142k}k^{2.5})$.

As a result, the total running time of algorithm KPPW is bounded by $O(|E| + |n^3 + k^3 + |V| + k + 2^{4.142k}k^{2.5}) = O^*(2^{4.142k})$. □

5 Conclusions

In this paper, we mainly focus on the kernelization for the k-P_2-packing problem. We give further structure analysis of the problem, and propose a kernelization algorithm obtaining a kernel of size at most $7k$. Comparing with the kerneliztion given in [7], our algorithm makes further optimization on the vertices of any P_2 in W and their Q_0-vertex neighbors and Q_1-edge neighbors, which reduces the number of Q_0-vertices and Q_1-edges in Q. Based on the $7k$ kernel, we also present an improved parameterized algorithm with time complexity $O^*(2^{4.142k})$, which greatly improves the current best result $O^*(2^{5.301k})$.

References

1. Hell, P.: Graph Packings. Electronic Notes in Discrete Mathematics 5 (2000)
2. Kirkpatrick, D.G., Hell, P.: On the complexity of general graph factor problems. SIAM J. Comput. 12, 601–609 (1983)
3. Kann, V.: Maximum bounded H-matching is MAX-SNP-complete. J. nform. Process. Lett. 49, 309–318 (1994)
4. Hurkens, C., Schrijver, A.: On the size of systems of sets every t of which have an SDR, with application to worst case ratio of Heuristics for packing problems. ISIAM J. Discrete Math. 2, 68–72 (1989)
5. Fellows, M., Heggernes, P., Rosamond, F., Sloper, C., Telle, J.A.: Exact algorithms for finding k disjoint triangles in an arbitrary graph. In: Hromkovič, J., Nagl, M., Westfechtel, B. (eds.) WG 2004. LNCS, vol. 3353, pp. 235–244. Springer, Heidelberg (2004)
6. Mathieson, L., Prieto, E., Shaw, P.: Packing edge disjoint triangles: a parameterized view. In: Downey, R.G., Fellows, M.R., Dehne, F. (eds.) IWPEC 2004. LNCS, vol. 3162, pp. 127–137. Springer, Heidelberg (2004)
7. Prieto, E., Sloper, C.: Look at the stars. Theoretical Computer Science 351, 437–445 (2006)
8. Chen, J.: Parameterized computation and complexity: a new approach dealing with NP-hardness. Journal of Computer Science and Technology 20, 18–37 (2005)

Deterministic Hot-Potato Permutation Routing on the Mesh and the Torus

Andre Osterloh

FernUniversität in Hagen
Germany
Andre.Osterloh@FernUni-Hagen.de

Abstract. In this paper we consider deterministic hot-potato routing algorithms on $n \times n$ meshes and tori. We present algorithms for the permutation routing problem on these networks and achieve new upper bounds. The basic ideas used in the presented algorithms are sorting, packet concentration and fast algorithms for one-dimensional submeshes. Using this ideas we solve the permutation routing problem in $3.25n+o(n)$ steps on an $n \times n$ mesh and in $2.75n + o(n)$ steps on an $n \times n$ torus.

1 Introduction

This paper studies routing in a synchronous network with bidirectional links in which at most one packet is able to traverse any link in each time step and direction. We consider two of the most studied network with a fixed interconnection, the two-dimensional $n \times n$ mesh and torus.

The problem of routing packets through a network is fundamental to the study of parallel computation. One of the best studied routing problems is the problem where each processor is source and destination of one packet, the *permutation routing problem.* In the literature many different approaches to solve permutation routing problems have been studied, e.g. adaptive, oblivious, cut-through, wormhole, fault-tolerant, local, etc. [8]. Here we consider a routing strategy known as hot-potato or deflection routing. In hot-potato routing no buffers are used for storing packets. In this routing strategy in each step and in each processor all incoming packets, unless they have reached their destination, have to leave the processor in the following step (see Figure 1). Hot-potato algorithms are attractive for practical applications because they tend to be simple and due to the lack of buffers they have efficient hardware realizations. They have been observed to work well in practice and have been used successfully on several parallel machines or optical networks [10, 18, 19, 1, 9].

Although hot-potato routing algorithms are rather simple they are very hard to analyze. Compared to other routing strategies the gap between the known upper and lower bounds for hot-potato routing algorithms on meshes (and tori) is large. To give an example, for greedy[1] hot-potato routing no asymptotically

[1] In greedy hot-potato routing a packet has to use an outgoing link in the direction of its destination whenever it is possible.

M. Agrawal et al. (Eds.): TAMC 2008, LNCS 4978, pp. 223–233, 2008.

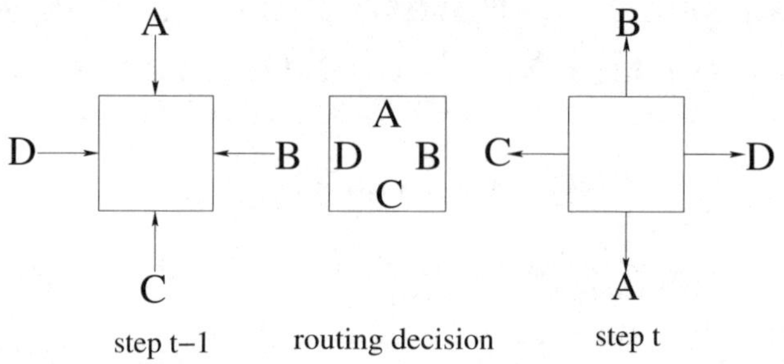

Fig. 1. One step in a processor in hot-potato routing

optimal routing algorithm is known. Nearly all fast routing algorithms for the mesh and torus use sorting to achieve their bounds, e.g. see [8, 11, 17]. In the case of hot-potato routing it is surprising that sorting can be used. Newman and Schuster in [16] introduced it for hot-potato routing and we use a variant of their technique here. Another well known technique to achieve fast routing algorithms is to concentrate packets in a smaller subnetwork and to solve the problem in this subnetwork. In combination with hot-potato routing the concentration technique was not used before. Since in hot-potato routing no buffers are allowed to store packets the possibilities to concentrate packets are very limited. In the algorithm we concentrate the packets in a subnetwork with the half number of nodes, so there are two packets on each node in a step. This can be done without buffers. In routing algorithms for two-dimensional meshes or tori, one-dimensional routing is often used as a subroutine. We use one-dimensional routing where each processor is source and destination of at most two packets at several points of algorithm. Hence we analyze one-dimensional routing and achieve new upper bounds for two-dimensional hot-potato routing on meshes and tori.

The rest of the paper is organized as follows. In Section 2 we present related work. In Section 3 we give some necessary definitions. In Section 4 we analyze one-dimensional routing with the farthest destination first strategy and give some simple results. In Section 5 we present and analyze our $3.25n+o(n)$ $(2.75n+o(n))$ step permutation routing algorithms for the mesh (torus). In Section 6 we give a short conclusion.

2 Related Work

Several randomized or deterministic hot-potato routing algorithms have been designed for the $n \times n$-mesh or torus [2, 3, 4, 5, 6, 7, 11, 14, 16]. In [2] Ben-Aroya *et al.* provide a deterministic greedy hot-potato routing algorithm for the $n \times n$ mesh that delivers a batch of l packets in $2(l-1) + d_{\max}$ steps, where $d_{\max}$ is the maximal initial soure to destination distance of a packet. In their batch problems a processor is source of at most four and destination of at most l

packets. Since $l = n^2$ in permutation routing this results an $O(n^2)$ bound for the permutation routing problem. In [4] Ben-Dor *et al.* achieve a $O(n\sqrt{l})$ bound for the batch problem in an $n \times n$ mesh with the help of potential function analysis. For the two-dimensional mesh, Kunde in [14] presents the first deterministic greedy hot-potato routing algorithm with a $o(n^2)$ bound for permutation routing. Kunde achieves a bound of $O(n\sqrt{n}\log n)$ steps. In [5] Busch *et al.* present a randomized greedy hot-potato routing algorithm for the $n \times n$ mesh that solves any permutation routing problem in $O(n \ln n)$ and batch routing problems in $O(m \ln n)$ steps with high probability, where $m = \min\{m_r, m_c\} \in \Theta(n)$ and m_r (m_c) is the maximum number of packets targeted to a single row (column). In [6] the same authors study batch routing where each processor is source of at most one and destination of at most $4n^2$ packets. Their algorithm needs $O(LB \cdot \log^3 n)$ steps with high probability, where $LB \in \Omega(n)$ is a lower bound based on $d_{\max}$ and the maximum congestion of an instance.

For non-greedy hot-potato routing, Feige and Raghavan [7] present a randomized algorithm that routes any random destination problem in $2n + O(\ln n)$ steps with high probability. They also solve any permutation routing problem in $9n$ steps with high probability. Newman and Schuster [16] give a deterministic non-greedy hot-potato routing algorithm for permutation routing that needs $7n + o(n)$ steps on the mesh and $4n + o(n)$ steps on the torus. The algorithms for the mesh and torus in [16] use sorting to solve the routing problem. In [11] Kaufmann *et al.* improve this result to $3.5n + o(n)$. Hence $3.5n + o(n)$ was the so far best known bound for deterministic hot-potato routing on an $n \times n$ mesh or torus.

For non hot-potato routing several optimal $2n - 2$ step algorithms for the permutation routing problem are known [8].

3 Basic Definitions

In an $n \times n$ mesh each processor (node) is given as a pair (r, c), $1 \le r, c \le n$. A processor (r, c) lies in row r and column c and is connected to at most four adjacent nodes by bidirectional links. In a mesh a processor (r, c) is connected with (r', c') iff $|r - r'| + |c - c'| = 1$, $1 \le r, r', c, c' \le n$. The diameter of a mesh is $2n - 2$ and the bisection width is n. We call a node (r, c) where $r + c$ is even a *black* node and a node where $r + c$ is odd a *white* node. Packets that are initially on a black (white) node are called black (white) packets. In hot-potato routing white and black packets never meet in any node. We denote the (at most) four incoming and outgoing links of a processor by 1, 2, 3 or 4 (see Figure 2).

A torus is a mesh with *wrap-around links.* Node $(i, 1)$ is connected with (i, n) and node $(1, j)$ is connected with (n, j), $1 \le i, j \le n$. The advantage of a torus is that all nodes can be treated in the same way since no border or corner nodes exists. The diameter of a torus is half the diameter of the corresponding mesh and its bisection width is twice as large.

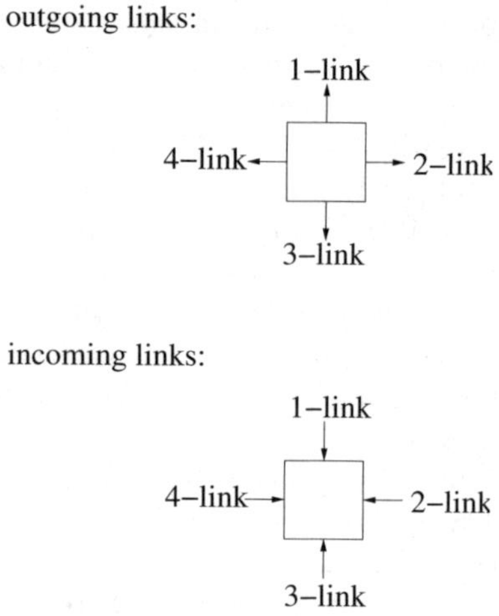

Fig. 2. A node (r, c) in an $n \times n$ mesh or torus. In the case of the mesh $1 < r, c < n$

4 Basic Problems

First we consider the following problem in row i, $1 < i \leq n$, of the mesh[2]. We call the problem **2-2 basic routing**. We want to route up to n packets in row i using hot-potato routing. Initially there are at most two packets on black (white) nodes and no packets on white (black) nodes. Each node is destination of at most two packets. During the routing the packets are allowed to use the 2-links and 4-links in row i and the 1-links and 3-links between row i and row $i - 1$. We assume that a packet is taken out of the mesh when it arrives at its destination.

For a problem instance R of 2-2 basic routing we define $r_R(k, l)$ ($l_R(l, k)$) as the number of packets that have to pass from left to right (right to left) node (i, k) and node (i, l).

We route the packets greedily to their destination using the farthest destination first strategy. In the case that during routing two packets in a node want to use the same 2-link or 4-link, the packet with lower priority uses the 1-link and go to row $i - 1$. In the next step it uses the 3-link to come back into the node in row i. If a packet uses neighboring nodes to wait, we say that the packet *vibrates.*

The following holds (for a similar result for (non hot-potato) routing see Lemma 3.1 in [12]):

Lemma 1. *Solving a 2-2 basic routing problem R in row i, using the farthest destination first strategy, takes at most* $\max_{k,l \in X}\{2(\max\{r_R(k, l), l_R(l, k)\} - 1) + l - k\}$ *steps, where* $X = \{(k, l) \mid k < l \wedge (r_R(k, l) \neq 0 \vee l_R(l, k) \neq 0)\}$.

Proof. In this proof we restrict our attention to packets traveling from left to right. For the other packets the bound could be proved analogously. We show the bound by proving

[2] Row i consists of processors (i, x), $1 \leq x \leq n$.

$$\forall n > 0.\, n = \max_{k<l, r_R(k,l)\neq 0}\{2(r_R(k,l) - 1) + l - k\}$$
$$\Longrightarrow$$
It takes at most n steps to solve R.

with induction on n. For $n = 1$ this is obviously true. In the following R is an instance of 2-2 basic routing for which the bound is $n + 1$. We perform two routing steps on R and get an instance R' of 2-2 basic routing. For R' we prove that $\max_{k<l, r_{R'}(k,l)\neq 0} 2(r_{R'}(k,l) - 1) + l - k$ is at most $n - 1$. Hence we are able to apply the induction hypothesis to R'.

For $1 \leq a < b \leq n$ we define $f_R(a,b) := 2(r_R(a,b) - 1) + b - a$ and $M_R(a,b)$ as the set of packets in R with a source in columns $s \leq a$ and a destination in column $d \geq b$. Obviously $f_{R'}(a,b) \leq f_R(a,b)$ for all a, b. Now choose k_{max}, l_{max} such that $n+1 = f_R(k_{max}, l_{max})$ and set $M := M(k_{max}, l_{max})$. Since $f_R(k_{max}, l_{max})$ is a maximum over all a, b, there exists a packet in M with source column k_{max} (otherwise $f_R(k, l_{max}) > f_R(k_{max}, l_{max})$ for a $k < k_{max}$). If there are one or two packets from M in k_{max}, we get $f_{R'}(k_{max}, l_{max}) = n - 1$, since $r_{R'}(k_{max}, l_{max}) = r_R(k_{max}, l_{max}) - 1$. If there are two packets in k_{max} from which only one is in M, the packet in M has higher priority because the destination of the other packet is smaller than l_{max}. So $f_{R'}(k_{max}, l_{max}) = n - 1$ also for this case.

Now assume that there are a, b such that $f_{R'}(a,b) = n$. No packet of $M_R(a,b)$ has passed column a in the two steps otherwise $f_R(a,b) > n + 1$. So there are two possibilities: (1) There is no packets in a in instance R or (2) all packets in a have a destination column $< b$. In both cases we get $M_R(a-1,b) = M_R(a,b)$ and $f_R(a-1,b) \geq f_R(a,b) + 1 = n + 1$. So we get the maximum of $n+1$ for columns $a - 1$ and b. For a maximum we know that a packet of $M_R(a-1,b)$ leaves $a - 1$ in the first step and hence leaves a in the second step. A contradiction to (1) and (2). •

The result of Lemma 1 holds also for row 1. In row 1 the packets vibrate using row 2. Since the vibration of packets in different rows and the vibration of black and white packets to do not interfere we get:

Lemma 2. *The 2-2 basic routing problem can be solved simultaneously for all n rows and black and white packets in at most* $\max_{k<l} 2(\max\{\hat{r}(k,l), \hat{l}(l,k)\} - 1)$ $+ l - k$ *steps. Here $\hat{r}(k,l)$ ($\hat{l}(k,l)$) is the maximum of $r_R(k,l)$ ($l_R(k,l)$) over all rows.*

With the help of a variant of Odd-Even-Transposition Sort the following is easy to see:

Lemma 3. *In row i, $1 \leq i \leq n$, of the $n \times n$ mesh let n packets be located in the black (white) nodes such that initially there are two packets in each node. The packets can be sorted in n steps using a variant of Odd-Even-Transposition Sort such that finally there are two packets in each black (white) node.*

As in the case for 2-2 basic routing the sorting can be done simultaneously for all rows, black and white packets. Packets in row 1 use row 2 to vibrate, packets in a row $i > 1$ use row $i - 1$. Now we want to note that it is possible to sort an $n \times n$ mesh in $O(n)$ steps. To do so a variant of the sorting algorithm presented in [15] or [17] can be used.

Lemma 4. *Let each black (white) node of the $n \times n$ mesh have two packets. There is a hot-potato algorithm that sorts simultaneously black and white packets in snake-like column-major order in $O(n)$ time. (Finally the black (white) packets are in black (white) nodes.)*

Of course the result of Lemma 4 also holds for column-major, row-major snake-like row-major and similar indexing schemes. For the algorithm presented in the next section it is sufficient to sort an $n^{3/4} \times n^{3/4}$ mesh in $o(n)$ steps. For this purpose also algorithms like Shearsort which are not asymptotically optimal can be used. The benefit of Shearsort is that it is simple and easy to implement.

5 Permutation Routing on the Mesh and Torus

We begin with a rough version of the algorithm and give a more detailed version of step 2 and step 3 later. The mesh is devided in four submeshes of $n \times n/4$ nodes. We call these submeshes A, B, C, and D (see Figure 3). In step 1 the packets are concentrated in the middle $n/2$ columns of the mesh (in BC). In step 2 the algorithm uses a kind of 2-2 sorting to solve the routing problem. The packets are sorted according to their destination column. To save time we do not totally sort the packets and stop after $3n/2 + o(n)$ steps. In step 3 the packets are transported to their final destination.

Algorithm Permutation Routing

(1) Move the packets $n/4$ steps horizontally from submesh A to submesh B and from submesh D to submesh C. ($n/4$ steps)
(2) In the $n/2 \times n$ mesh BC partially sort the packets according to the destination column of a packet. ($3n/2 + o(n)$ steps)
(3) Route packets to their destination:
  (3a) Horizontally to their destination column. ($n/2 + o(n)$ steps)
  (3b) Vertically within the destination column. ($n + o(n)$ steps)

Lemma 5. *Algorithm Permutation Routing on a $n \times n$ mesh needs* $3.25n + o(n)$ *steps.*

Since black and white packets never interact we can restrict our attention to black packets. All the following results and proofs are given only for black packets. They also hold for white packets.

Details of Step 2. Sorting in BC.
In the following we use column-major and snake-like column-major in the usual sense and assume that $n/4$, $n^{1/4}/4$ are integer[3]. The algorithm is based on Schnorr/Shamir's algorithm from [17]. We divide BC into blocks of size $2n^{3/4} \times$

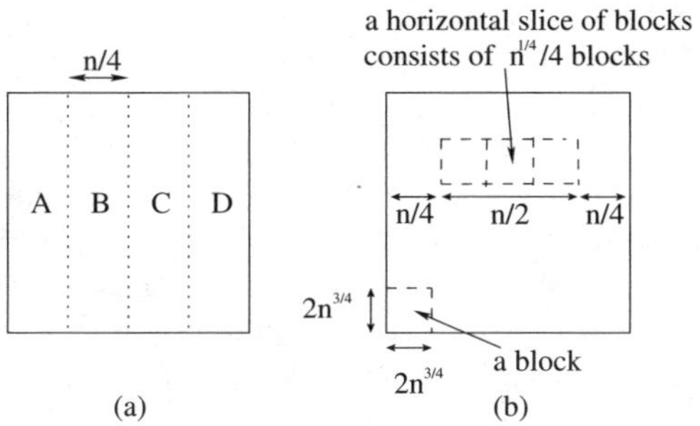

Fig. 3. Division of the Mesh for Permutation Routing

$2n^{3/4}$. We call a row of $n^{1/4}/4$ blocks a horizontal slice and a column of $n^{1/4}/2$ blocks a vertical slice. BC consists of $n^{1/4}/2$ horizontal and $n^{1/4}/4$ vertical slices.

Algorithm Partially Sorting

(1) Sort each block in snake-like column-major order. ($o(n)$ steps)
(2) Perform an $n^{1/4}/2$-way unshuffle of the rows.
(The packets move within the vertical slices). (n steps)
(3) Sort each block into snake-like column-major order. ($o(n)$ steps)
(4) Sort each row in linear order. ($n/2$ steps)
(5) Collectively sort blocks 1 and 2, blocks 3 and 4, blocks 5 and 6, etc. of each horizontal slice into column-major order. ($o(n)$ steps)
(6) Collectively sort blocks 2 and 3, blocks 4 and 5, blocks 6 and 7, etc. of each horizontal slice into column-major order. ($o(n)$ steps)

In the following proofs we use the 0-1 principle([13]) several times. For k, $1 \leq k \leq n$, we mark all packets with a destination column $< k$ as 0s, all packets with destination column k with 1s and all packets with destination column $> k$ with 2s.

Lemma 6. *After step 6 of algorithm Partially Sorting the following holds:*

(0) In any horizontal slice there are at most $n^{3/4} + n^{1/2}/4$ 1s.
(1) There are at most 3 neighboring columns in BC that contain black 1s.
(2) There are at most 2 black packets in each row with the same column destination.

Proof. In step 2 of algorithm Partially Sorting the number of 0s (1s, 2s) a block sends to any two blocks of a vertical slice (vs for short) differs at most by two. Since there are $n^{1/4}/2$ blocks in a vs, the number of 0s (1s, 2s) in any two blocks in a vs differs at most by $n^{1/4}$ after step 2. There are $n^{1/4}/4$ blocks in a horizontal slice (hs for short). Therefore, the number of 0s (1s, 2s) in any two hs differs at most by $n^{1/2}/4$ after step 2. Steps (3) to (6) do not change the number of 0s (1s, 2s) in a hs. Hence after step 6 the number of 0s (1s, 2s) in any two hs differ at most by $n^{1/2}/4$. Furthermore, steps 3 to 6 sort the horitzontal slices in

[3] Since $n^{1/4}/4$ is integer we have $n \geq 4^4$.

column-major order. Let $l_{1s,\min}$ be the minimal number of 1s in a hs, $l_{1s,\max}$ the maximal number of 1s in a hs and t_{1s} be the total number of 1s in BC. Then

$$2t_{1s}/n^{1/4} - n^{1/2}/4 < l_{1s,\min} \leq l_{1s,\max} < 2t_{1s}/n^{1/4} + n^{1/2}/4.$$

So (0) holds. Since $(2t_{1s}/n^{1/4} + 2n^{1/2}/4)/(2n^{3/4}) < 2$ (the additional $(n^{1/2}/4)/(2n^{3/4})$ stems from the difference of 0s within a hs) (1) holds. Although the 1s could be in three neighboring columns, the cut of these three columns and any row consists of at most two neighboring nodes. This is due to the sorting in the blocks. Since from any two neighboring nodes in a row, one node is a black node and one node is a white node (2) is fulfilled. •

Note that BC is not sorted in column-major but the horizontal slices are. Furthermore, for each destination column i, $1 \leq i \leq n$, the black packets with destination column i can be found in at most three neighboring columns of the mesh. Additionally there are at most two black packets in a row with destination column i. Finally we consider the running time of the sorting step:

Lemma 7. *Algorithm Partially Sorting needs* $1.5n + O(n^{3/4})$ *steps.*

Proof. The running time of steps (1), (3), (5), and (6) follows from Lemma 4. The running time of step (2) follows from Lemma 2 and the running time for step (4) follows from Lemma 3. •

Details of Step 3. Greedy Routing to the destination.
In step (3a) we have to solve an instance of 2-2 basic routing. The destination of a packet in this problem is its destination column. This, together with the fact that the packets are partially sorted according to their destination column, ensures that there is no overtaking of packets during the horizontal routing, i.e. if two packets π and π' in a row are initially in column i and i', $i < i'$, and their destinations are j and j', then $j \leq j'$. A packet that has reached its destination column in step (3a) vibrates until step (3b) starts. We have to make sure that the packets are able to vibrate before the beginning of step (3b). Packets in row i with destination column j use nodes $(i-1, j)$ and $(i+1, j)$ to vibrate. So columns 1 and n could be a problem. Since there is no overtaking, node $(i, 1)$ is able to vibrate the two packets to node $(i, 2)$ and node $(i+1, 1)$ if $i < n$. Analogously node (i, n) can vibrate packets to node $(i, n-1)$ and node $(i+1, n)$. For the nodes $(n, 1)$ and (n, n) a special treatment is necessary. One packet of the possibly two packets can vibrate using node $(n, 2)$ ($(n, n-1)$). The other packet have to vibrate from node $(n, 2)$ to node $(n-1, 2)$ (node $(n, n-1)$ and $(n-1, n-1)$). In step (3b) this special treatment of the two corner nodes only results in $O(1)$ extra steps.

The time bound of $n/2 + O(1)$ for step (3a) can be seen with the help of Lemma 2.

Lemma 8. *Step (3a) can be done in* $n/2 + O(1)$ *steps.*

Proof. (Sketch) We only consider packets traveling from left to right. The bounds for packets traveling from right to left are the same. To avoid notational clutter, we omit constants that stem from the fact that packets with a destination column k, $1 \leq k \leq n$, are located in three neighboring columns. Instead we assume that BC is sorted. This reduces the running time only by a constant. We have to give an upper bound for $\max_{(i,j)\in X}\{2(\max\{r(i,j), l(i,j)\} - 1) + j - i\}$. To do so we construct a worst case, i.e. we place the packets such that $\max_{(i,j)\in X}\{2(\max\{r(i,j), l(i,j)\} - 1) + j - i\}$ is maximal. We have to maximize $2(r(i,j) - 1) + j - i$. Such a worst case occurs when all black packets have a destination in the right half of the mesh. In the following $n/4 \leq i < j \leq 3n/4$, $l := j - i$, and $k := 3n/4 - j$. Note that n black packets fit in one column of BC and that a column of the mesh is destination of n packets. Packets located in columns j to $3n/4$ in the beginning of step 3a have to go to columns $n - k$ to n. Hence at most $(n - k - j) \cdot n$ free destinatons exist in columns t where $t \geq j$. Since $n - k - j = n/4$ we have at most $n^2/4$ free destinations. Packets between columns i and j have a destination to the right of column i. Hence at most $n^2/4 - ln$ free destinations to the right of column i exist. Since $n^2/4 - ln$ packets fit in $n/4 - l$ columns, at most $n/4 - l$ black packets in a row have to pass i and j from left to right. So we get $j - i + 2(n/4 - l) = n/2 - j + i \leq n/2$. •

Simulations of the algorithm have shown that the average time the packets need in step (3a) is approximately $n/4$.

In the following we describe how packets are routed in step (3b). We restrict our attention to packets with destination column k, $1 < k < n$, and begin with the description of the initial situation. We come back to the description of the situation for packet with destination column 1 or n later. Initially the packets are in columns $k - 1$, k, and $k + 1$ such that in column k there are at most two packets on any black node and no packets on any white node. In columns $k - 1$ and $k + 1$ there is at most one packet on any black node and no packet on any white node. The packets are routed to their destination in the following way. Packets in column $k - 1$ and $k + 1$ are routed horizontally to column k. Packets in column k are routed vertically towards their destination. If two packets in column k compete for a vertical link (1-link or 3-link) the packet with the farthest destination wins. The other packet is routed horizontally to column $k - 1$ or k. For the time of step (3b) we get

Lemma 9. *For packets with destination column k, $1 < k < n$, step (3b) can be done in $n + o(n)$ steps.*

Proof. (sketch) Let $u(i,j)$ be the number of packets that have to pass row i and j downwards and $d(i,j)$ be the number of packets that have to pass row i and j upwards. ($u(i,j)$ and $d(i,j)$ are defined as $r(i,j)$ and $l(i,j)$ for 2-2 basic routing). Then a bound of $\max_{(k,l)\in X}\{2(\max\{u(k,l), d(l,k)\} - 1) + l - k\} + O(1)$ steps can be proved analogously to the proof of Lemma 2. The additional $O(1)$ steps result from the fact that initially some packets are on a node in column $k - 1$ or $k + 1$. So, to get a bound for the running time of step (3b), we have to find an upper bound for the number of packets that pass node i and node j. We restrict our attention to black packets that have to travel downwards. From Lemma 6.(0) we know that in an horizontal slice at most $2n^{3/4} + O(n^{1/2})$ packets with destination column k exist. So in r horizontal slices there are at most $r(2n^{3/4} + O(n^{1/2}))$ packets with destination column k. Since we consider permutation routing, each node is destination of exactly one packet. So at most

$$\min\{\lceil i/(2n^{3/4})\rceil(2n^{3/4} + O(n^{1/2}), n - j\} = min\{i + o(n), n - j\}$$

packets pass i and j. Since $2(d(i,j) - 1) + j - i \leq 2min\{i + o(n), n - j\} + j - i = n + o(n)$ we have proved the desired bound. •

Finally we have to consider destination columns 1 and n. We restrict our attention to column n. Column 1 can be treated in the same way. In a horizontal slice there are at most $2n^{3/4} + O(n^{1/2})$ black packets with destination column n. Nearly all (up to $O(n^{1/2})$) of these packets are in column $3n/4$ after step 2 of algorithm Partially Sorting. So only from row $i \geq n - O(n^{1/2})$ each row has two packets. In rows $i < n - O(n^{1/2})$ only every second row has two packets. Hence we have a situation very similar to a 2-2 basic routing problem that can be solved in $n + o(n)$ steps. We omit a proof here. We get

Theorem 1. *There is a deterministic hot-potato routing algorithm that solves Permutaion Routing on an $n \times n$ mesh in $3.25n + o(n)$ steps.*

Algorithm Permutation Routing also works on a torus but does not benefit from the connections between row 1 (column 1) and row n (column n). On a torus no special treatment for columns (rows) 1 or n is necessary and step (2) of algorithm Partially Sorting can be done within $n/2$ steps. At the moment we do not know whether step (3b) can be performed faster than in n steps on a torus. We assume that it can be done in $3n/4 + o(n)$ steps. To get a better bound for step (3b) a deeper analysis of hot-potato algorithms for 2-2 basic routing problems on rings is necessary. So we have:

Theorem 2. *There is a deterministic hot-potato routing algorithm that solves Permutaion Routing on an $n \times n$ torus in $2.75n + o(n)$ steps.*

Under the assumption that step (3b) can be done in $3n/4 + o(n)$ steps a bound of $2.5n + o(n)$ could be achieved.

6 Conclusion

In this paper we have shown new upper bounds for deterministic hot-potato routing on the mesh and torus. For permutation routing we reduced the gap between the known upper and lower bound. To achieve the new bound we analyzed routing problems on one-dimensional meshes. The results achieved there could be useful for the design of fast hot-potato routing algorithms for higher dimensional meshes or similar networks. To design fast (permutation routing) algorithms for the torus it would be helpful to have a similar result for rings. For local routing problems we achieved asymtotically optimal deterministic hot-potato algorithms. Our algorithms use sorting to solve the routing problem. As far as we know no asymptotically optimal deterministic hot-potato routing algorithm for the Permutation Routing problem exists that does not use sorting.

References

[1] Acampora, A.S., Shah, S.I.A.: Multihop Lightwave Networks: Acomparison of Store-and-Forward and Hot-Potato Routing. In: Proceedings of IEEE INFOCOM, pp. 10–19 (1991)

[2] Ben-Aroya, I., Eilam, T., Schuster, A.: Greedy Hot-Potato Routing on the Two-Dimensional Mesh. Distributed Computing 9(1), 3–19 (1995)

[3] Borodin, A., Rabani, Y., Schieber, B.: Beterministic Many-To-Many Hot Potato Routing. Theory of Computer Systems 31(1), 41–61 (1998)

[4] Ben-Or, A., Halevi, S., Schuster, A.: Randomized Single-Target Hot-Potato Routing. Journal of Algorithms 23(1), 101–120 (1997)

[5] Busch, C., Herlihy, M., Wattenhofer, R.: Randomized Greedy Hot-Potato Routing. In: Proceedings of the Eleventh Annual ACM-SIAM Symposium on Discrete Algorithms (SODA 2000), San Francisco, Calefornia, USA, January 2000, pp. 458–466 (2000)

[6] Busch, C., Herlihy, M., Wattenhofer, R.: Hard-potato Routing. In: Proceedings of the Thirty-second Annual ACM Symposium on Theory of Computing (STOC 2000), Portland, Oregon, USA, May 2000, pp. 278–285 (2000)

[7] Feige, U., Raghavan, P.: Exact Analysis of Hot-Potato Routing. In: IEEE Proceedings of the 33rd Annual Symposium on Foundations of Computer Science (FOCS 1992), Pittsburgh, Pennsylvania, USA, October 1992, pp. 553–562 (1992)

[8] Grammatikakis, M.D., Hsu, D.F., Sibeyn, J.F.: Packet routing in fixed-connection networks: A survey. Journal of Parallel and Distributed Computing 54(2), 77–132 (1998)

[9] Greenber, A.G., Goodmann, J.: Sharp Approximate Models of Deflection Routing in Mesh Networks. IEEE Transactions on Communications 41(1), 210–223 (1993)

[10] Hillis, W.D.: The Connection Machine. MIT Press, Cambridge (1985)

[11] Kaufmann, M., Lauer, H., Schröder, H.: Fast Deterministic Hot-Potato Routing on Meshes. In: Du, D.-Z., Zhang, X.-S. (eds.) ISAAC 1994. LNCS, vol. 834, pp. 273–282. Springer, Heidelberg (1994)

[12] Kaufmann, M., Rajasekaran, S., Sibeyn, J.F.: Matching the Bisection Bound for Routing and Sorting on the Mesh. In: Proceedings of the 4th Symposium on Parallel Algorithms and Architectures (SPAA 1992), pp. 31–40 (1992)

[13] Knuth, D.: The Art of Computer Programming, Sorting and Searching, vol. III. Addison-Wesley, Reading (1973)

[14] Kunde, M.: A New Bound for Pure Greedy Hot Potato Routing. In: Thomas, W., Weil, P. (eds.) STACS 2007. LNCS, vol. 4393, pp. 49–60. Springer, Heidelberg (2007)

[15] Leighton, T.: Introduction to Parallel Algorithms and Architectures: Arrays-Trees-Hypercubes. Morgan-Kaufmann Publishers, San Francisco (1992)

[16] Newman, I., Schuster, A.: Hot-Potato Algorithms for Permutation Routing. IEEE Transactions on Parallel and Distributed Systems 6(11), 1168–1176 (1995)

[17] Schnorr, C.P., Schamir, A.: An Optimal Sorting Algorithm for Mesh Connected Computers. In: Prooceedings of the 18th Symposium on Theory of Computing (STOC 1986), pp. 255–263 (1986)

[18] Seitz, C.L.: Mosaic C: An Experimental, Fine-Grain Multicomputer. In: Bensoussan, A., Verjus, J.-P. (eds.) INRIA 1992. LNCS, vol. 653, pp. 69–85. Springer, Heidelberg (1992)

[19] Smith, B.J.: Architecture and Applications of the HEP multiprocessor computer. Soc. Photocopti. Instrum. Eng. 298, 241–248

Efficient Algorithms for Model-Based Motif Discovery from Multiple Sequences

Bin Fu[1], Ming-Yang Kao[2], and Lusheng Wang[3]

[1] Dept. of Computer Science, University of Texas - Pan American
TX 78539, USA
binfu@cs.panam.edu

[2] Department of Electrical Engineering and Computer Science,
Northwestern University, Evanston, IL 60208, USA
kao@northwestern.edu

[3] Department of Computer Science, The City University of Hong Kong,
Kowloon, Hong Kong
lwang@cs.cityu.edu.hk

Abstract. We study a natural probabilistic model for motif discovery that has been used to experimentally test the quality of motif discovery programs. In this model, there are k background sequences, and each character in a background sequence is a random character from an alphabet Σ. A motif $G = g_1 g_2 \ldots g_m$ is a string of m characters. Each background sequence is implanted a randomly generated approximate copy of G. For a randomly generated approximate copy $b_1 b_2 \ldots b_m$ of G, every character is randomly generated such that the probability for $b_i \neq g_i$ is at most α. In this paper, we give the first analytical proof that multiple background sequences do help for finding subtle and faint motifs.

1 Introduction

Motif discovery is an important problem in computational biology and computer science. For instance, it has applications to coding theory [3,4], locating binding sites and conserved regions in unaligned sequences [18,10,6,17], genetic drug target identification [9], designing genetic probes [9], and universal PCR primer design [13,2,16,9].

This paper focuses on the application of motif discovery to finding conserved regions in a set of given DNA, RNA, or protein sequences. Such conserved regions may represent common biological functions or structures. Many performance measures have been proposed for motif discovery. Let C be a subset of 0-1 sequences of length n. The covering radius of C is the smallest integer r such that each vector in $\{0,1\}^n$ is at a distance at most r from a set of 0-1 sequence of length n. The decision problem associated with the covering radius for a set of binary sequences is NP-complete [3]. Another similar problem called closest string problem was also proved to be NP-hard [3,9]. Some approximation algorithms have also been proposed. Li et al. [12] gave an approximation scheme for the closest string and substring problems. The related consensus patterns problem is that give n sequences $s_1, \cdots, s_n$, it asks for a region of length L in each

M. Agrawal et al. (Eds.): TAMC 2008, LNCS 4978, pp. 234–245, 2008.

s_i, and a median string s of length L so that the total Hamming distance from s to these regions is minimized. Approximation algorithms for the consensus patterns problem were also reported in [11]. Furthermore, a number of heuristics and programs have been developed [15,7,8,19,1].

In many applications, motifs are faint and may not be apparent when two sequences alone are compared but may become clearer when more sequences are together [5]. For this reason, it has been conjectured that comparing more sequences together can help identifying faint motifs. In this paper, we give the first analytical proof for this conjecture.

In this paper, we study a natural probabilistic model for motif discovery. In this model, there are k background sequences and each character in the background sequence is a random character from an alphabet Σ. A motif $G = g_1 g_2 \ldots g_m$ is a string of m characters. Each background sequence is implanted a randomly generated approximate copy of G. For a randomly generated approximate copy $b_1 b_2 \ldots b_m$ of G, every character is randomly generated such that the probability for $b_i \neq g_i$ is at most α. This model was first proposed in [15] and has been widely used in experimentally testing motif discovery programs [7,8,19,1].

We design an algorithm that for a reasonably large k can discover the implanted motif with high probability. Specifically, we prove that for $\alpha < 0.1771$ and any constant $x \geq 8$, there exist constants $t_0, \delta_0, \delta_1 > 0$ such that if the length of the motif is at least $\delta_0 \log n$, the alphabet has at least t_0 characters, and there are at least $\delta_1 \log n_0$ input sequences, then in $O(n^3)$ time the algorithm finds the motif with probability at least $1 - \frac{1}{2^x}$, where n is the longest length of any input sequence and $n_0 \leq n$ is an upper bound for the length of the motif. When x is considered as a parameter of order $O(\log n)$, the parameters $t_0, \delta_0, \delta_1 > 0$ do not depend on x. We also show some lower bounds that imply our conditions for the length of the motif and the number of input sequences are tight to within a constant multiplicative factor. This algorithm's time complexity depends on the length of input sequences and is independent of the number of the input sequences. This is because that for a fixed x, $\Theta(\log n)$ sequences are sufficient to guarantee the probability of at least $1 - \frac{1}{2^x}$ to discover the motif. In contrast to the NP-hardness of other variants of the common substring problem, motif discovery is solvable in $O(n^3)$ time in this probabilistic model.

Our algorithm is an exact algorithm that has provable high probability to return the motif. The algorithm employs novel methods that extract similar consecutive regions among multiple sequences while tolerating noises. The algorithm needs the motif to be long enough, but does not need to have the length of the motif as an input. The algorithm allows the motif to appear any position at each sequence, and each mutation in a motif to be arbitrary (a mutation lets a character to be changed to an arbitrary character without any probabilistic condition). We also derive lower bounds that indicate the upper bounds are almost optimal.

We give a brief description about the algorithm as section 3. Before giving the algorithm, we set up a few parameters and constants that will affect the algorithm at section 4.1. Then entire Algorithm Find-Noisy-Motif is described.

We give the analysis and proof about Algorithm Find-Noisy-Motif and state it in our main theorem (Theorem 1). Two lower bounds are presented at section 5.

2 Notations

For a set A, $|A|$ denotes the number of elements in A. Σ is an alphabet with $|\Sigma| = t \geq 2$. For an integer $n \geq 0$, Σ^n is the set of sequences of length n with characters from Σ. For a sequence $S = a_1 a_2 \cdots a_n$, $S[i]$ denotes the character a_i, and $S[i,j]$ denotes the substring $a_i \cdots a_j$ for $1 \leq i \leq j \leq n$. $|S|$ denotes the length of the sequence S. We use $\emptyset$ to represent the empty sequence, which has length 0.

Let $G = g_1 g_2 \cdots g_m$ be a fixed sequence of m characters. G is the motif to be discovered by our algorithm. A $\Theta_\alpha(n, G)$-sequence has the form $S = a_1 \cdots a_{n_1} b_1 \cdots b_m a_{n_1+1} \cdots a_{n_2}$, where $n_2 + m \leq n$, each a_i has probability $\frac{1}{t}$ to be equal to π for each $\pi \in \Sigma$, and b_i has probability at most α not equal to g_i for $1 \leq i \leq m$, where $m = |G|$. $\aleph(S)$ denotes the motif region $b_1 \cdots b_m$ of S. The motif region $b_1 \cdots b_m$ of S may start at an arbitrary or worst-case position in S. Also, a mutation may convert a character g_i in the motif into an arbitrary or worst-case different character b_i only subject to the restriction that g_i will mutate with probability at most α.

A mutation converts a character g_i in the motif into an arbitrary different character b_i without probability restriction. This allows a character g_i in the motif to change into any character b_i in $\Sigma - \{g_i\}$ with even different probability.

For two sequences $S_1 = a_1 \cdots a_m$ and $S_2 = b_1 \cdots b_m$ of the same length, let $\text{diff}(S_1, S_2) = \frac{|\{i | a_i \neq b_i \text{ for } i=1,\cdots,m\}|}{m}$, i.e., the ratio of difference between the two sequences.

Definition 1. *Assume that $S = a_1 a_2 \cdots a_n$ is a sequence. For its substring $S' = S[i_1, j_1]$ and $S'' = S[i_2, j_2]$, define* $\text{shift}_S(S', S'') = \min(|i_1 - i_2|, |j_1 - j_2|)$.

The analysis of our algorithm employs the well known Chernoff bound [14].

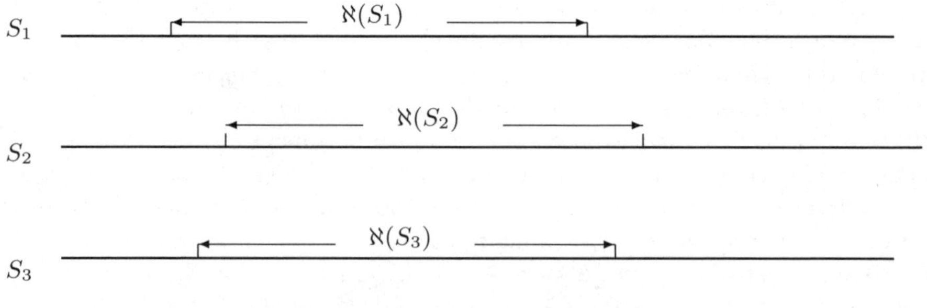

Fig. 1. The motif regions of S_1, S_2 and S_3 are not aligned

3 A Sketch of the Algorithm Find-Noisy-Motif

Our Algorithm Find-Noisy-Motif has two phases. The first phase exploits the fact that with high probability, the motif area in some sequences conserves the first and last characters. Furthermore, the middle area of the motif changes with a small ratio. We will select enough pairs of $\Theta_\alpha(n, G)$-sequences S', S'' and find their substrings G' and G'' of S' and S'', respectively such that G' and G'' match in their left and right most characters. Furthermore, G' and G'' only have a relatively small difference in the middle area. For each such pair S' and S'', the substring G'' of S'' is extracted.

During the second phase, a new set of $\Theta_\alpha(n, G)$-sequences $S_1, S_2, \cdots, S_{k_2}$ will be used. For each G'' extracted from a pair of sequences in the first phase, it is used to match a substring G_i of S_i for $i = 1, 2, \cdots, k_2$. Assume that $G_1, \cdots, G_{k_2}$ are derived from matching G'' to all sequences $S_1, S_2, \cdots, S_{k_2}$. Some G_i may be an empty sequence if G'' can not match well to any substring of S_i. If G'' has the same length as that of motif G and is very similar to G, then the number of non-empty sequences among $G_1, \cdots, G_{k_2}$ is much larger than $\frac{k_2}{2}$ and the i-th character $G[i]$ of G can be recovered from voting among $G_1[i], \cdots, G_{k_2}[i]$. In other words, $G[i]$ is the character that appears more than $\frac{k_2}{2}$ times in $G_1[i], \cdots, G_{k_2}[i]$. We prove that with high probability, such a G'' exists. The conversion from figure 1 to figure 2 shows how we recover the motif via voting.

On the other hand, if $|G''| > |G|$ or G'' does not match G well, we can prove that the number of non-empty sequences among $G_1, \cdots, G_{k_2}$ is less than $\frac{k_2}{2}$. Our algorithm's time complexity depends on the length of the input sequences and is independent of the number of the input sequences. This is because that for a fixed x, $\Theta(\log n)$ sequences are sufficient to guarantee the probability of at least $1 - \frac{1}{2^x}$ the motif will be discovered. Additional sequences can improve the probability but are not needed for the high probability guarantee.

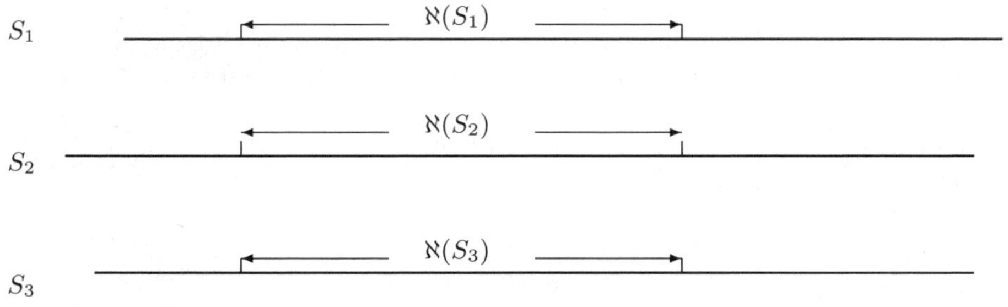

Fig. 2. S_1, S_2 and S_3 with their motif in the same column region

4 Algorithm Find-Noisy-Motif

In this section, we give an algorithm that any motif G can be discovered in $O(n^3)$ time. It requires that the size of alphabet is larger than a fixed constant.

Some parameters and constants will be used in Algorithm Find-Noisy-Motif . In section 4.1, we give a list of assignments for some parameters and constants that are used in the algorithm. The description of Algorithm Find-Noisy-Motif is given at section 4.2. The analysis of the algorithm is given at section 4.3.

4.1 Parameters

As multiple parameters affect the performance of Algorithm Find-Noisy-Motif, we list the parameters and discuss some useful inequalities here.

- Let x be any constant at least 8. We will prove that Algorithm Find-Noisy-Motif has probability at least $1 - \frac{1}{2^x}$ to output the motif G. Let α be any constant with $\alpha < 0.1771$. Note that $(1-\alpha)^2 - \alpha > \frac{1}{2}$. Let $\eta = \frac{1}{6}$ and let $\rho_0 = \frac{1}{24}$. Let $\epsilon > 0$ be any constant such that $(1-\alpha)^2 - \alpha - 3\epsilon > \frac{1}{2}$.
- Select any constant $r_0 > 0$ such that

$$(1-\alpha)^2 - \alpha - 3\epsilon - 2r_0 > \frac{1}{2}. \tag{1}$$

- Let v be the least integer that satisfies the inequalities below:

$$1 \le v, \tag{2}$$

$$(1-\alpha)^2 - \frac{2c^v}{1-c} - \alpha - 3\epsilon - 2r_0 > \frac{1}{2}, \tag{3}$$

$$\frac{2c_2 v^3 c^v}{1-c} < \rho_0, \tag{4}$$

$$\frac{2c^v}{1-c} < \frac{r_0}{2}, \tag{5}$$

$$\frac{2c^v}{1-c} < \rho_0, \tag{6}$$

where $c = e^{-\frac{\epsilon^2}{3}}$. Note that the existence of v for (3) follows from (1).
- We define the following Q_0. It will be first used in Lemma 1. Let $Q_0 = (1-\alpha)^2 - \frac{2c^v}{1-c}$.
- Let c_2 be a constant to be specified in Lemma 6.
- Let t_0 be any constant such that

$$\frac{2(v-1)}{t_0} \le \frac{r_0}{2}, \tag{7}$$

$$\frac{c_2 v^3}{t_0} \le \rho_0, \tag{8}$$

$$\frac{t_0 - 1}{t_0} - \beta > \epsilon, \tag{9}$$

where R is to be defined in Lemma 8. In the remainder of this paper, we always assume the parameter $t \ge t_0$. Combining (3), (5), (7) and the definition of R, we have $Q_0 - \alpha - 3\epsilon - 2R > \frac{1}{2}$.

- Let $\beta = 2\alpha + 2\epsilon$.
- The constant z is selected so that $z \geq v$, and $\frac{4e^{-\frac{\epsilon^2}{3}z}}{1-e^{-\frac{\epsilon^2}{3}}} \leq \rho_0$.
- The number k_1 is selected such that

$$(1 - Q_1)^{k_1} \leq \frac{\eta}{2^x}, \tag{10}$$

where Q_1 is defined in Lemma 6, and is at least $\frac{1}{12}$. Note that $k_1 = O(1)$ is a constant independent of the length of the input sequences.
- Select a constant $\delta_0 > 0$ and let $d = \delta_0 \log n$ such that $n^2 e^{-d} \leq \frac{\eta}{2^x}$, $n^2 e^{-\frac{\epsilon^2}{3}d} \leq \rho_0$.
- We require that the length of the motif G is at least d. Let n be the largest length of an input $\Theta_\alpha(n, G)$-sequence. Let parameter $n_0 \in [d, n]$ be a given upper bound on the length of the motif G that will be discovered by Algorithm Find-Noisy-Motif .
- Select a constant $\delta_1 > 0$ and let $k_2 = \delta_1 \log n_0 - 2k_1$ so that $n_0 k_2 e^{-\frac{\epsilon^2}{3}k_2} \leq \frac{\eta}{2^x}$, and $k_1 e^{-\frac{\epsilon^2}{3}k_2} \leq \frac{\eta}{2^x}$.

The motif G is a pattern unknown to Algorithm Find-Noisy-Motif , and Algorithm Find-Noisy-Motif will attempt to recover G from a series of $\Theta_\alpha(n, G)$-sequences generated by the probabilistic model, which is controlled by the parameters α, n, and G. The source of randomness comes entirely from the input sequence.

Let's imagine how a sequence S is generated in this model. 1). Generate a sequence S' with $n - |G|$ characters, in which each character is a random character Σ. 2). Generate G' such that with probability at most α, $G'[i] \neq G[i]$. For $G'[i] \neq G[i]$, it represents a mutation. Note that there is no restriction about how a character will change to in a mutation. 3). Insert G', which servers the motif region $\aleph(S)$ of S, into any position of S'.

Let Z_0 be a set of k_1 pairs of random $\Theta_\alpha(n, G)$-sequences $(S_1', S_1''), \cdots, (S_{k_1}', S_{k_1}'')$. Let Z_1 be the set $\Theta_\alpha(n, G)$-sequences $\{S_1', S_1'', \cdots, S_{k_1}', S_{k_1}''\}$ in the k_1 pairs of sequences in Z_0, where k_1 is defined by inequality (10). Let Z_2 be a set of k_2 sequences used in the second phase of Algorithm Find-Noisy-Motif . Let $k = 2k_1 + k_2$ be the total number of $\Theta_\alpha(n, G)$-sequences that are used as the input to Algorithm Find-Noisy-Motif . In the remainder of this paper, we assume that the alphabet has $t \geq t_0$ characters.

4.2 Description of Algorithm Find-Noisy-Motif

Algorithm Find-Noisy-Motif has two phases. The input to Phase 1 is k_1 pairs of $\Theta_\alpha(n, G)$-sequences in the set Z_0. The input to Phase 2 is k_2 $\Theta_\alpha(n, G)$-sequences in the set Z_2 and the output result from Phase 1. All the $\Theta_\alpha(n, G)$-sequences are independent random $\Theta_\alpha(n, G)$-sequences. Note that k_1 is constant, $k_2 = O(\log n_0)$, and $n_0 (\leq n)$ is an upper bound for the length of the motif G according to the setting in Section 4.1. Algorithm Find-Noisy-Motif is a deterministic algorithm, which is based on the randomness of those sequences in both Z_0 and Z_2 and the independence in selecting them. Algorithm Find-Noisy-Motif is deterministic, but its input is generated by a probabilistic model. The following steps generate data sequenced for Algorithm Find-Noisy-Motif for Z_0 and Z_2.

Step 1. Randomly select $2k_1$ $\Theta_\alpha(n,G)$-sequences $S'_1, S''_1, S'_2, S''_2, \cdots, S'_{k_1}, S''_{k_1}$ and let $Z_0 = \{(S'_1, S''_1), (S'_2, S''_2), \cdots, (S'_{k_1}, S''_{k_1})\}$.

Step 2. Randomly select k_2 $\Theta_\alpha(n,G)$-sequences $S_1, \cdots, S_{k_2}$ and let $Z_2 = \{S_1, \cdots, S_{k_2}\}$.

Definition 2. – *Two sequences X_1 and X_2 are left matched if (1) $|X_1| = |X_2|$, (2) $X_1[1] = X_2[1]$, and (3) $\mathrm{diff}(X_1[1,i], X_2[1,i]) \le \beta$ for all integers i, $v \le i \le |X_1|$.*
- *Two sequences X_1 and X_2 are right matched if X_1^R and X_2^R are left matched, where $X^R = a_n \cdots a_1$ is the inverse sequence of $X = a_1 \cdots a_n$.*
- *Two sequences X_1 and X_2 are matched if X_1 and X_2 are both left and right matched.*

The function Extract(S_1, S_2) below extracts the longest similar region between two sequences S_1 and S_2.

Function Extract(S_1, S_2)
Input: a pair of $\Theta_\alpha(n,G)$-sequences S_1 and S_2
Output: a subsequence of S_2 which is similar to a subsequence of S_1.
Steps:
for $h = \min(|S_1|, |S_2|)$ to d (recall from Section 4.1 that $|G| \ge d$)
  for $i = 1$ to $|S_1|$
    for $j = 1$ to $|S_2|$
      let $i' = i + h - 1$ and $j' = j + h - 1$;
      if $S_1[i, i']$ and $S_2[j, j']$ are both left and right matched (see Definition 2)
      then return $S_2[j, j']$;
return $\emptyset$ (the empty sequence);
End of Extract

The following are the steps of Phase 1 of Algorithm Find-Noisy-Motif :
Phase 1:
Input: $Z_0 = \{(S'_1, S''_1), (S'_2, S''_2), \cdots, (S'_{k_1}, S''_{k_1})\}$, a set of pairs of sequences generated at Step 1 in the initial stage of the algorithm.
Output: a set W that contains a similar region of each pair in Z_0.
Steps:
let $W = \emptyset$ (empty set);
for each pair of sequence $(S, S') \in Z_0$
  let $G' = \mathrm{Extract}(S, S')$ and put G' into W;
return W, which will be used in Phase 2;
End of Phase 1

After a set of motif candidates W is produced from Phase 1 of Algorithm Find-Noisy-Motif, we use this set to match with another set of sequences to recover the hidden motif via voting.

Function Match(G', S_i)

Input: a motif candidate G', which is returned from the function Extract(), and a sequence S from the group Z_2;

Output: either a subsequence G_i of S_i of the same length as G' or an empty sequence. G_i will be considered the motif region $\aleph(S_i)$ of S_i if it is not empty, and the empty sequence means the failure in extracting the motif region $\aleph(S_i)$ of S_i.

Steps:

find a substring G_i of S_i with $|G| = |G_i|$ such that
  G' and G_i are matched (see Definition1)
if such a G_i does not exist, let $G_i = \emptyset$ (empty string).
Output G_i;

End of Match

The function Vote$(G_1, G_2, \cdots, G_{k'})$ is to generate another sequence G' via voting, where $G'[i]$ is the most frequent character among $G_1[i], G_2[i], \cdots, G_{k'}[i]$.

Function Vote$(G_1, G_2, \cdots, G_{k'})$

Input: sequences $G_1, G_2, \cdots, G_{k'}$ of the same length with $k' \le k_2$;

Output: a sequence G', which is derived from voting at every position of the input sequences.

Steps:

let $m = |G_1|$;
for each $j = 1, \cdots, m$
  if strictly more than $\frac{k_2}{2}$ characters from $G_1[j], \cdots, G_{k'}[j]$ are equal
    to some character a
  then let $a_j = a$
  else return "failure";
return $G' = a_1 \cdots a_m$;

End of Vote

The following are the steps of Phase 2 of Algorithm Find-Noisy-Motif . It uses the candidates of motif derived in the Phase 1 to extract the motif regions of another set Z_2 of sequences, and recover the motif via voting.

Phase 2:

let $Z_2 = \{S_1, \cdots, S_{k_2}\}$ as defined in the begining of Section 4.2.
for each $G' \in W$, let $G_i = \text{Match}(G', S_i)$ for $i = 1, \cdots, k_2$.
  let $G'_1, \cdots, G'_{k'_2}$ be the list of all non-empty sequences in the list $G_1, \cdots, G_{k_2}$ (Note: For every non-empty sequence that appears multiple times in the second list, it also appears the same number of times in the first list.)
  If $k'_2 \ge (Q_0 - 2R - 2\epsilon)k_2$
    then output Vote$(G'_1, G'_2, \cdots, G'_{k'_2})$ (which will be proven to be identical to G with probability at least $1 - \frac{1}{2^x}$).

End of Phase 2

4.3 Analysis of Phase 1 of Algorithm Find-Noisy-Motif

We present Lemma 1 that shows that with high probability, the initial part and last part of motif region in a $\Theta_\alpha(n, G)$-sequence do not change much.

Lemma 1. *With probability at least* $Q_0 = (1-\alpha)^2 - \frac{2c^v}{1-c}$, *a* $\Theta_\alpha(n, G)$*-sequence* S *contains* $G' = \aleph(S)$ *satisfying the following conditions: (1)* $G'[1] = G[1]$*; (2)* $G'[m] = G[m]$*; (3)* $\text{diff}(G'[1,h], G[1,h]) \leq \frac{\beta}{2}$ *for all* $h = v, v+1, \cdots, m$*; (4)* $\text{diff}(G'[m-h, m], G[m-h, m]) \leq \frac{\beta}{2}$ *for* $h = v-1, v+1, \cdots, m-1$*, where* $c = e^{-\frac{\epsilon^2}{3}}$ *and* $m = |G|$ *as defined in Sections 4.1 and 2, respectively.*

Lemma 2 shows that with small probability, a sequence can match a random sequence. It will be used to prove that when two subsequences in two different $\Theta_\alpha(n, G)$-sequences are similar, they are unlikely to stay away the motif regions in the two $\Theta_\alpha(n, G)$-sequences, respectively.

Lemma 2. *Assume that* X_1 *and* X_2 *are two independent sequences of the same length and that every character of* X_2 *is a random character from* Σ*. Then*

1. *if* $1 \leq |X_1| = |X_2| < v$*, then the probability that* X_1 *and* X_2 *are matched is* $\leq \frac{1}{t}$*; and*
2. *if* $v \leq |X_1| = |X_2|$*, then the probability for* $\text{diff}(X_1, X_2) \leq \beta$ *is at most* $e^{-\frac{\epsilon^2|X_1|}{3}}$.

Function Extract(S_1, S_2) returns a subsequence of S_2. We expect that Extract($S_1, S_2)$ is the motif region $\aleph(S_2)$ in S_2. Lemma 3 shows that with small probability, the region for Extract(S_1, S_2) in S_2 has no overlap with the motif region $\aleph(S_2)$ of S_2.

Lemma 3. *With probability at most* ρ_0, Extract(S_1, S_2) *and* $\aleph(S_2)$ *are not overlaping substrings of* S_2*. In other words, with probability is at most* ρ_0, Extract$(S_1, S_2) = S_2[j, j']$, $\aleph(S_2) = S_2[t, t']$, *and the two intervals* $[j, j']$ *and* $[f, f']$ *have no overlap* ($[j, j'] \cap [f, f'] = \emptyset$)*.*

In order to show that Extract(S_1, S_2) is efficient to find a motif region in S_2, we give Lemma 4 show that with small probability, the region to fetch Extract(S_1, S_2) in S_2 shift much from the motif region $\aleph(S_2)$ of S_2.

Lemma 4. *For every* $z > 0$*, the probability is at most* $H_1 = 2\rho_0$ *that for a pair of sequences* (S_1, S_2) *from* Z_0, $\text{shift}_{S_2}(M, \aleph(S_2)) \geq z$ *and* $|M| \geq |G|$*, where* $M = \text{Extract}(S_1, S_2)$.

We need the Lemma 5, which will be useful to give the upper bound of probability analysis. It is derived by the standard methods in calculus.

Lemma 5. *Let* a *be a real constant in interval* $(0, 1)$ *and* j *be an integer* ≥ 1*. Then, 1.* $\sum_{i=j}^{\infty} ia^i = \frac{ja^j - (j-1)a^{j+1}}{(1-a)^2} < \frac{ja^j}{(1-a)^2}$*; and*

2. $\sum_{i=j}^{\infty} i^2 a^i = a^j\left(\frac{(j^2-(j-1)(j+1)a)(1-a)-(j-(j-1)a)2(-a)}{(1-a)^3}\right) < \frac{2j^2 a^j}{(1-a)^3}$.

Lemma 6 gives a lower bound for the probability that Extract(S_1, S_2) returns the motif region $\aleph(S_2)$ of S_2. Furthermore, the motif region $\aleph(S_2)$ of S_2 does not have much difference with the original motif G.

Lemma 6. *Given two independent $\Theta_\alpha(n, G)$-sequences S_1 and S_2, it has the probability at least $Q_1 = Q_0^2 - H_2 - H_1 \geq Q_0^2 - 4\rho_0$ that $G' = \text{Extract}(S_1, S_2)$ is $\aleph(S_2)$, and $\aleph(S_2)$ satisfies the conditions of G' Lemma 1, where H_1 is defined in Lemma 4, $H_2 = c_2 v^3(\frac{1}{t} + c^v)$ and $c_2 = O(1)$ is a constant.*

By (3), we have $Q_0 \geq \frac{1}{2}$. By Lemma 6 and $\rho_0 = \frac{1}{24}$ defined in Section 4.1, we have $Q_1 \geq Q_0^2 - 4\rho_0 \geq \frac{1}{12}$. Since the number k_1 is selected to be large enough that $(1 - Q_1)^{k_1} \leq \frac{\eta}{2^x}$ (see (10)), the probability is at least $1 - (1 - Q_1)^{k_1} \geq 1 - \frac{\eta}{2^x}$ (by Lemma 6) that there is $G_0 = \text{Extract}(S_1, S_2) = \aleph(S_2)$, where S_1 and S_2 satisfy the conditions of Lemma 1. We now assume there is such a G_0 that satisfies the conditions described above.

4.4 Analysis of Phase 2 of Algorithm Find-Noisy-Motif

Lemma 7 shows that with small probability, Z_1 generated in the initial stage (step 2) of Algorithm Find-Noisy-Motif has a sequence whose motif region has many mutations.

Lemma 7. *With probability at most $2k_1 e^{-\frac{\epsilon^2}{3}d}$, there is a sequence S in Z_1 that changes more than $\frac{\beta}{2}|G|$ characters in its motif region $\aleph(S)$.*

Lemma 8 shows that with high probability, phase 2 of Algorithm Find-Noisy-Motif extracts motif regions from the sequences in Z_1.

Lemma 8. *1. Assume that $G'' = \text{Extract}(S_i', S_i'')$ with $|G| \leq |G''|$. Let S be a $\Theta_\alpha(n, G)$-sequence with $M = \text{Match}(G'', S)$ and let w_0 be the number of characters of M that are not in the region of $\aleph(S)$. Then the probability is at most $R = 2(\frac{v-1}{t} + \frac{c^v}{1-c})$ that $w_0 \geq 1$.*
2. The probability is at least $Q_0 - R$ that given a random $\Theta_\alpha(n, G)$-sequence S, $\aleph(S) = \text{Match}(G_0, S)$.

Lemma 9 shows that we can use G' to extract most of the motif regions for the sequences in Z_2 if $G' = G_0$ (recall that G_0 is close to the original motif G and G_0 is defined right after Lemma 6).

Lemma 9. *Assume that $|G'| \geq |G|$ and $G_i = \text{Match}(G', S_i)$ for $S_i \in Z_2 = \{S_1, \cdots, S_{k_2}\}$ and $i = 1, \cdots, k_2$ (Recall that each sequence G_i is either an empty sequence or a sequence of the length $|G'|$).*

1. *If $G' = G_0$, then the probability is at least $1 - e^{-\frac{\epsilon^2 k_2}{3}}$ that there are more than $(Q_0 - R - \epsilon)k_2$ sequences G_i with $G_i = \aleph(S_i)$.*
2. *The probability is at least $1 - e^{-\frac{\epsilon^2 k_2}{3}}$ that for every G', $|\{i | G_i \neq \aleph(S_i)(i = 1, \cdots, k_2)\}| \leq (R + \epsilon)k_2$.*

Theorem 1 (Main). *Assume that α is a constant less than* 0.1771. *There exist constants t_0, δ_0, and δ_1 such that if the size t of the alphabet Σ is at least t_0 and the length of the motif G is at least $\delta_0 \log n$, then given k independent $\Theta_\alpha(n, G)$-sequences with $k \geq \delta_1 \log n_0$, Algorithm Find-Noisy-Motif outputs G with probability $\geq 1 - \frac{1}{2^x}$ and runs in $O(n^3)$ time, where n is the longest length of any input sequences and $n_0 \leq n$ is a given upper bound for the length of G.*

5 Lower Bounds on the Parameters

In this section, we show some lower bounds for the length of the motif and the number of input sequences that are needed to recover the motif with high probability.

Theorem 2 shows that when the motif is short, it is impossible to recover it with a small number $O(\log n)$ of sequences. Thus, the upper bounds of Algorithm Find-Noisy-Motif and the lower bounds here have constant factor multiplicative.

Theorem 2. *Assume that constant $\epsilon > 0$ and the alphabet has constant number t characters. There is a constant $\delta > 0$ such that with probability at least $1 - o(1)$ that given $n^{1-\epsilon}$ independent random $\Theta_\alpha(n, G)$-sequences $S_1, \cdots, S_{n^{1-\epsilon}}$, every sequence of length $m_0 = \lceil \delta \log n \rceil$ is a substrings of each S_i for $i = 1, 2, \cdots, n^{1-\epsilon}$.*

We consider the lower bound for the number of sequences needed for recovering the motif. Theorem 3 shows that if the number of sequences is $o(\log n)$, it is impossible to recover the motif correctly.

Theorem 3. *There exists a constant δ such that no algorithm can recover the motif G with at most $\delta \log n$ $\Theta_\alpha(n, G)$-sequences.*

Open Problems: An interesting open problem is whether there exists an algorithm to recover all the motifs for the alphabet with four characters.

Acknowledgements. We thank Miklós Csürös and Manan Sanghi for helpful discussions. Ming-Yang Kao is supported in part by National Science Foundation Grant CNS-0627751. Lusheng Wang is fully supported by a grant from the Research Grants Council of the Hong Kong Special Administrative Region, China (Project No. CityU 1196/03E).

References

1. Chin, F., Leung, H.: Voting algorithms for discovering long motifs. In: Proceedings of the 3rd Asia-Pacific Bioinformatics Conference, pp. 261–272 (2005)
2. Dopazo, J., Rodríguez, A., Sáiz, J.C., Sobrino, F.: Design of primers for PCR amplification of highly variable genomes. Computer Applications in the Biosciences 9, 123–125 (1993)
3. Frances, M., Litman, A.: On covering problems of codes. Theoretical Computer Science 30, 113–119 (1997)

4. Gąsieniec, L., Jansson, J., Lingas, A.: Efficient approximation algorithms for the Hamming center problem. In: Proceedings of the Tenth Annual ACM-SIAM Symposium on Discrete Algorithms, pp. S905–S906 (1999)
5. Gusfield, D.: Algorithms on Strings, Trees, and Sequences. Cambridge University Press, Cambridge (1997)
6. Hertz, G., Stormo, G.: Identification of consensus patterns in unaligned DNA and protein sequences: a large-deviation statistical basis for penalizing gaps. In: Proceedings of the 3rd International Conference on Bioinformatics and Genome Research, pp. 201–216 (1995)
7. Keich, U., Pevzner, P.: Finding motifs in the twilight zone. Bioinformatics 18, 1374–1381 (2002)
8. Keich, U., Pevzner, P.: Subtle motifs: defining the limits of motif finding algorithms. Bioinformatics 18, 1382–1390 (2002)
9. Lanctot, J.K., Li, M., Ma, B., Wang, L., Zhang, L.: Distinguishing string selection problems. In: Proceedings of the 10th Annual ACM-SIAM Symposium on Discrete Algorithms, pp. 633–642 (1999)
10. Lawrence, C., Reilly, A.: An expectation maximization (EM) algorithm for the identification and characterization of common sites in unaligned biopolymer sequences. Proteins 7, 41–51 (1990)
11. Li, M., Ma, B., Wang, L.: Finding similar regions in many strings. In: Proceedings of the Thirty-first Annual ACM Symposium on Theory of Computing, pp. 473–482 (1999)
12. Li, M., Ma, B., Wang, L.: On the closest string and substring problems. Journal of the ACM 49(2), 157–171 (2002)
13. Lucas, K., Busch, M., Mossinger, S., Thompson, J.: An improved microcomputer program for finding gene- or gene family-specific oligonucleotides suitable as primers for polymerase chain reactions or as probes. Computer Applications in the Biosciences 7, 525–529 (1991)
14. Motwani, R., Raghavan, P.: Randomized Algorithms. Cambridge University Press, Cambridge (2000)
15. Pevzner, P., Sze, S.: Combinatorial approaches to finding subtle signals in DNA sequences. In: Proceedings of the 8th International Conference on Intelligent Systems for Molecular Biology, pp. 269–278 (2000)
16. Proutski, V., Holme, E.C.: Primer master: a new program for the design and analysis of PCR primers. Computer Applications in the Biosciences 12, 253–255 (1996)
17. Stormo, G.: Consensus patterns in DNA. In: Doolitle, R.F. (ed.) Molecular evolution: computer analysis of protein and nucleic acid sequences. Methods in Enzymolog, 183, 211–221 (1990)
18. Stormo, G., Hartzell III, G.: Identifying protein-binding sites from unaligned DNA fragments. Proceedings of the National Academy of Sciences of the United States of America 88, 5699–5703 (1991)
19. Wang, L., Dong, L.: Randomized algorithms for motif detection. Journal of Bioinformatics and Computational Biology 3(5), 1039–1052 (2005)

Ratio Based Stable In-Place Merging

Pok-Son Kim[1] and Arne Kutzner[2]

[1] Kookmin University, Department of Mathematics, Seoul 136-702, Rep. of Korea
pskim@kookmin.ac.kr
[2] Seokyeong University, Department of Computer Science, Seoul 136-704, Rep. of Korea
kutzner@skuniv.ac.kr

Abstract. We investigate the problem of stable in-place merging from a ratio $k = \frac{n}{m}$ based point of view where m, n are the sizes of the input sequences with $m \leq n$. We introduce a novel algorithm for this problem that is asymptotically optimal regarding the number of assignments as well as comparisons. Our algorithm uses knowledge about the ratio of the input sizes to gain optimality and does not stay in the tradition of Mannila and Ukkonen's work [8] in contrast to all other stable in-place merging algorithms proposed so far. It has a simple modular structure and does not demand the additional extraction of a movement imitation buffer as needed by its competitors. For its core components we give concrete implementations in form of Pseudo Code. Using benchmarking we prove that our algorithm performs almost always better than its direct competitor proposed in [6].

As additional sub-result we show that stable in-place merging is a quite simple problem for every ratio $k \geq \sqrt{m}$ by proving that there exists a primitive algorithm that is asymptotically optimal for such ratios.

1 Introduction

Merging denotes the operation of rearranging the elements of two adjacent sorted sequences of size m and n, so that the result forms one sorted sequence of $m+n$ elements. An algorithm merges two sequences *in place* when it relies on a fixed amount of extra space. It is regarded as *stable*, if it preserves the initial ordering of elements with equal value.

There are two significant lower bounds for merging. The lower bound for the number of assignments is $m + n$ because every element of the input sequences can change its position in the sorted output. As shown e.g. in Knuth [7] the lower bound for the number of comparisons is $\Omega(m \log(\frac{n}{m} + 1))$, where $m \leq n$. A merging algorithm is called *asymptotically fully optimal* if it is asymptotically optimal regarding the number of comparisons as well as assignments.

We will inspect the merging problem on the foundation of a ratio based approach. In the following k will always denote the ratio $k = \frac{n}{m}$ of the sizes of the input sequences. The lower bounds for merging can be expressed on the foundation of such a ratio as well. We get $\Omega(m \log(k+1))$ as lower bound for the number of comparisons and $m \cdot (k + 1)$ as lower bound for the number of assignments.

M. Agrawal et al. (Eds.): TAMC 2008, LNCS 4978, pp. 246–257, 2008.

In the first part of this paper we will show that there is a simple asymptotically fully optimal stable in-place merging algorithm for every ratio $k \geq \sqrt{m}$. Afterward we will introduce a novel stable in-place merging algorithm that is asymptotically fully optimal for any ratio k. The new algorithm has a modular structure and does not rely on the techniques described by Mannila and Ukkonen [8] in contrast to all other works ([10,4,2,6]) known to us. Instead it exploits knowledge about the ratio of the input sizes to achieve optimality. In its core our algorithm consists of two separated operations named "Block rearrangement" and "Local merges". The separation allowed the omitting of the extraction of an additional movement imitation buffer as e.g. necessary in [6]. For core parts of the new algorithm we will give an implementation in Pseudo-Code. Some benchmarks will show that it performs better than its competitor proposed in [6] for a wide range of inputs.

A first conceptual description of a stable asymptotically fully optimal in-place merging algorithm can be found in the work of Symvonis [10]. Further work was done by Geffert et al. [4] and Chen [2] where Chen presented a simplified variant of Geffert et al's algorithm. All three publications delivered neither an implementation in Pseudo-Code nor benchmarks. Recently Kim and Kutzner [6] published a further algorithm together with benchmarks. These benchmarks proved that stable asymptotically fully optimal in-place merging algorithms are competitive and don't have to be viewed as theoretical models merely.

2 A Simple Asymptotically Optimal Algorithm for $k \geq \sqrt{m}$

We now introduce some notations that will be used throughout the paper. Let u and v be two ascending sorted sequences. We define $u \leq v$ ($u < v$) iff $x \leq y$ ($x < y$) for all elements $x \in u$ and for all elements $y \in v$. $|u|$ denotes the size of the sequence u. Unless stated otherwise, m and n ($m \leq n$) are the sizes of two input sequences u and v respectively. δ always denotes some block-size with $\delta \leq m$.

Tab. 1 contains the complexity regarding comparisons and assignments for six elementary algorithms that we will use throughout this paper. Brief descriptions

Table 1. Complexity of the Toolbox-Algorithms

Algorithm	Arguments	Comparisons	Assignments
Hwang and Lin	u, v with $\lvert u\rvert \leq \lvert v\rvert$ let $m = \lvert u\rvert, n = \lvert v\rvert$	$m(t+1) + n/2^t$ where $t = \lfloor \log(n/m) \rfloor$	
(1) - ext. buffer			$2m + n$
(2) - m rotat.			$n + m^2 + m$
Block Swapping	u, v with $\lvert u\rvert = \lvert v\rvert$	-	$3\lvert u\rvert$
Block Rotation	u, v	-	$\lvert u\rvert + \lvert v\rvert + \gcd(\lvert u\rvert, \lvert v\rvert) \leq 2(\lvert u\rvert + \lvert v\rvert)$
Binary Search	u, x (searched element)	$\lfloor \log \lvert u\rvert \rfloor + 1$	-
Minimum Search	u	$\lvert u\rvert - 1$	-
Insertion Sort	u, let $m = \lvert u\rvert$	$\frac{m(m-1)}{2} + (m-1)$	$\frac{m(m+1)}{2} - 1$

of these algorithms except for "Minimum Search" can be found in [6]. In the case of "Minimum Search" we assume that u is unsorted, therefore a linear search is necessary.

First we will now show that there is a simple stable merging algorithm called Block-Rotation-Merge that is asymptotically fully optimal for any ratio $k \geq \sqrt{m}$. Afterward we will prove that there is a relation between the number of different elements in the shorter input sequence u and the number of assignments performed by the rotation based variant of Hwang and Lin's algorithm [5].

Algorithm 1: Block-Rotation-Merge (u, v, δ)

1. We split the sequence u into blocks $u_1 u_2 \ldots u_{\lceil \frac{m}{\delta} \rceil}$ so that all sections u_2 to $u_{\lceil \frac{m}{\delta} \rceil}$ are of equal size δ and u_1 is of size $m \,\mathbf{mod}\, \delta$. Let x_i be the last element of u_i $(i = 1, \cdots, \lceil \frac{m}{\delta} \rceil)$. Using binary searches we compute a splitting of v into sections $v_1 v_2 \ldots v_{\lceil \frac{m}{\delta} \rceil}$ so that $v_i < x_i \leq v_{i+1} (i = 1, \cdots, \lceil \frac{m}{\delta} \rceil - 1)$.
2. $u_1 u_2 \ldots u_{\lceil \frac{m}{\delta} \rceil} v_1 v_2 \ldots v_{\lceil \frac{m}{\delta} \rceil}$ is reorganized to $u_1 v_1 u_2 v_2 \ldots u_{\lceil \frac{m}{\delta} \rceil} v_{\lceil \frac{m}{\delta} \rceil}$ using $\lceil \frac{m}{\delta} \rceil - 1$ many rotations.
3. We locally merge all pairs $u_i v_i$ using $\lceil \frac{m}{\delta} \rceil$ calls of the rotation based variant of Hwang and Lin's algorithm ([5]).

The steps 2 and 3 are interlaced as follows: After creating a new pair $u_i v_i$ $(i = 1, \cdots, \lceil \frac{m}{\delta} \rceil)$ as part of the second step we immediately locally merge this pair as described in step 3.

Lemma 1. Block-Rotation-Merge *performs* $\frac{m^2}{2 \cdot \delta} + 2n + 6m + m \cdot \delta$ *many assignments at most if we use the optimal algorithm from Dudzinski and Dydek [3] for all block-rotations.*

Proof. For the first rotation from $u_1 u_2 \cdots u_{\lceil \frac{m}{\delta} \rceil} v_1$ to $u_1 v_1 u_2 \cdots u_{\lceil \frac{m}{\delta} \rceil}$ the algorithm performs $|u_2| + \cdots + |u_{\lceil \frac{m}{\delta} \rceil}| + |v_1| + \gcd(|u_2| + \cdots + |u_{\lceil \frac{m}{\delta} \rceil}|, |v_1|)$ assignments. The second rotation from $u_2 u_3 \cdots u_{\lceil \frac{m}{\delta} \rceil} v_2$ to $u_2 v_2 u_3 \cdots u_{\lceil \frac{m}{\delta} \rceil}$ requires $|u_3| + \cdots + |u_{\lceil \frac{m}{\delta} \rceil}| + |v_2| + \gcd(|u_3| + \cdots + |u_{\lceil \frac{m}{\delta} \rceil}|, |v_2|)$ assignments, and so on. For the last rotation from $u_{\lceil \frac{m}{\delta} \rceil - 1} u_{\lceil \frac{m}{\delta} \rceil} v_{\lceil \frac{m}{\delta} \rceil - 1} v_{\lceil \frac{m}{\delta} \rceil}$ to $u_{\lceil \frac{m}{\delta} \rceil - 1} v_{\lceil \frac{m}{\delta} \rceil - 1} u_{\lceil \frac{m}{\delta} \rceil} v_{\lceil \frac{m}{\delta} \rceil}$ the algorithm requires $|u_{\lceil \frac{m}{\delta} \rceil}| + |v_{\lceil \frac{m}{\delta} \rceil - 1}| + \gcd(|u_{\lceil \frac{m}{\delta} \rceil}|, |v_{\lceil \frac{m}{\delta} \rceil - 1}|)$ assignments. Additionally $\frac{m}{\delta}(3\delta + 3\delta + \delta^2) = 6m + m \cdot \delta$ assignments are required for the local merges. Altogether the algorithm performs $\delta \cdot ((\frac{m}{\delta} - 1) + (\frac{m}{\delta} - 2) + \cdots + 1) + n + n + 6m + m \cdot \delta = \frac{m^2}{2 \cdot \delta} - \frac{m}{2} + 2 \cdot n + 6m + m \cdot \delta \leq \frac{m^2}{2 \cdot \delta} + 2 \cdot n + 6m + m \cdot \delta$ assignments at most. □

Lemma 2. Block-Rotation-Merge *is asymptotically optimal regarding the number of comparisons.*

Corollary 1. *If we assume a block-size of* $\lfloor \sqrt{m} \rfloor$ *then* Block-Rotation-Merge *is asymptotically fully optimal for all* $k \geq \sqrt{m}$.

So, for $k \geq \sqrt{m}$ there is a quite primitive asymptotically fully optimal stable in-place merging algorithm. In the context of complexity deliberations in the next section we will rely on the following Lemma.

Lemma 3. *Let λ be the number of different elements in u. Then the number of assignments performed by the rotation based variant of Hwang and Lin's algorithm is $O(\lambda \cdot m + n) = O((\lambda + k) \cdot m)$.*

Proof. Let $u = u_1 u_2 \dots u_\lambda$, where every $u_i (i = 1, \cdots, \lambda)$ is a maximally sized section of equal elements. We split v into sections $v_1 v_2 \dots v_\lambda v_{\lambda+1}$ so that we get $v_i < u_i \leq v_{i+1}$ $(i = 1, \cdots, \lambda)$. (Some v_i can be empty.) We assume that Hwang and Lin's algorithm already merged a couple of section and comes to the first elements of the section $u_i (i = 1, \cdots, \lambda)$. The algorithm now computes the section v_i and moves it in front of u_i using one rotation of the form $\cdots u_i \dots u_\lambda v_i \cdots$ to $\cdots v_i u_i \dots u_\lambda \cdots$. This requires $|u_i| + \cdots + |u_\lambda| + |v_i| + \gcd(|u_i| + \cdots + |u_\lambda|, |v_i|) \leq 2(m + |v_i|)$ many assignments. Afterward the algorithm continues with the second element in u_i. Obviously there is nothing to move at this stage because all elements in u_i are equal and the smaller elements from v were already moved in the step before. Because we have only λ different sections we proved our conjecture. □

Corollary 2. *Hwang and Lin's algorithm is fully asymptotically optimal if we have either $k \geq m$ or $k \geq \lambda$ where λ is the number of different elements in the shorter input sequence u .*

3 Novel Asymptotically Optimal Stable In-Place Merging Algorithm

We will now propose a novel stable in-place merging algorithm called STABLE-OPTIMAL-BLOCK-MERGE that is fully asymptotically optimal for any ratio. Notable properties of our algorithm are: It does not rely on the block management techniques described in Mannila and Ukonnen's work [8] in contrast to all other such algorithms proposed so far. It degenerates to the simple BLOCK-ROTATION-MERGE algorithm for roughly $k \geq \sqrt{m}/2$. The internal buffer for local merges and the movement imitation buffer share a common buffer area. The two operations "block rearrangement" and "local merges" stay separated and communicate using a common block distribution storage. There is no lower bound regarding the size of the shorter input sequence.

Algorithm 2: STABLE-OPTIMAL-BLOCK-MERGE

Step 1: Block distribution storage assignment
Let $\delta = \lfloor \sqrt{m} \rfloor$ be our block-size. We split the input sequence u into $u = s_1 t s_2 u'$ so that s_1 and s_2 are two sequences of size $\lfloor m/\delta \rfloor + \lfloor n/\delta \rfloor$ and t is a sequence of maximal size with elements equal to the last element of s_1. We assume that there are enough elements to get a nonempty u' and call s_1 together with s_2 our *block distribution storage* (in the following shortened to bd-storage).

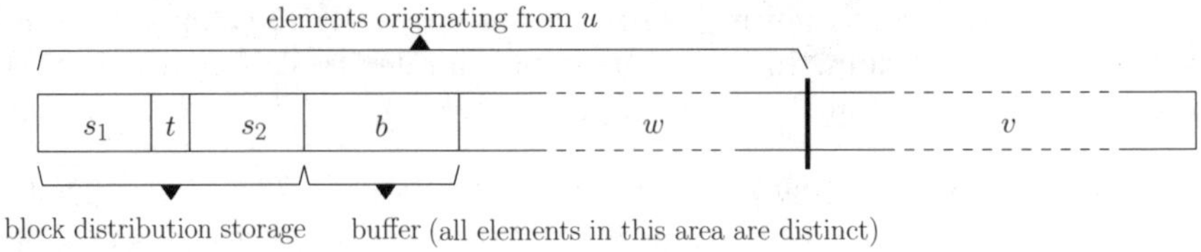

Fig. 1. Segmentation after the buffer extraction

Step 2: Buffer extraction
In front of the remaining sequence u' we extract an ascending sorted buffer b of size δ so that all pairs of elements inside b are distinct. Simple techniques how to do so are proposed e.g. in [9] or [6]. Once more we assume that there are enough elements to do so. Now let w be the remaining right part of u' after the buffer extraction.

The segmentation of our input sequences after the buffer extraction is shown in Fig. 1.

Step 3: Block rearrangement
We logically split the sequence wv into blocks of equal size δ as shown in Fig. 2 (a). The two blocks w_1 and $v_{\lfloor \frac{n}{\delta} \rfloor +1}$ are undersized and can even be empty. In the following we call every block originating from w a w-block and every block originating from v a v-block. The minimal w-block of a sequence of w-blocks is always the w-block with the lowest order (smallest elements) regarding the original order of these blocks.

We rearrange all blocks except of the two undersized blocks w_1 and $v_{\lfloor \frac{n}{\delta} \rfloor +1}$, so that the following 3 properties hold:
(1) If a v-block is followed by a w-block, then the the last element of the v-block must be smaller than the first element of the w block (Fig. 2(b)).

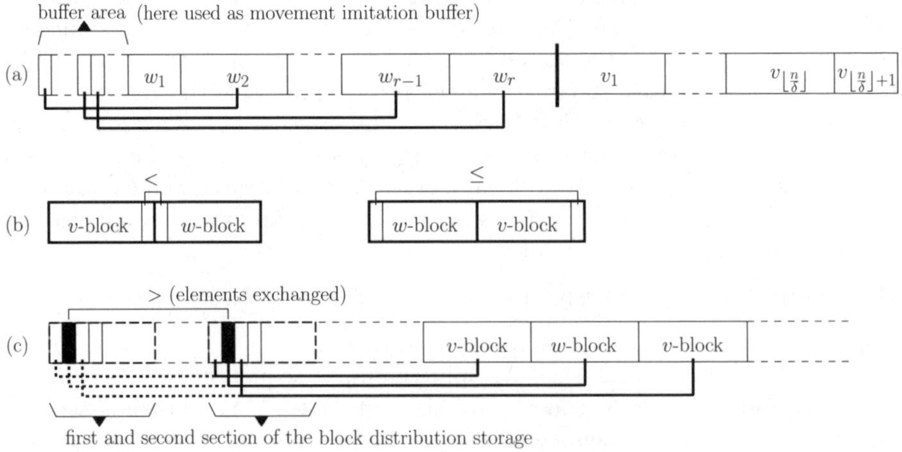

Fig. 2. Graphical remarks to the block rearrangement process

(2) If a w-block is followed by a v-block, then the first element of the w-block must be smaller or equal to the last element of the v-block (Fig. 2(b)).
(3) The relative order of the v-blocks as well as w-blocks stays unchanged.

This rearrangement can be easily realized by "rolling" the w-blocks through the v-blocks and "drop" minimal w-blocks so that the above properties are fulfilled. During this rolling the w-blocks stay together as group but they can be moved out of order. So, due to the need for stability, we have to track their positions. For this reason we mirror all block replacements in the buffer area using a technique called movement imitation (The technique of movement imitation is described e.g. in [10] and [6]). Every time when a minimal w-block was dropped we can find the position of the next minimal block using this buffer area.

Later we will have to find the positions of w-blocks in the block-sequence created as output of the rearrangement process. For this purpose we store the positions of w-blocks in the block distribution storage as follows:

The block distribution storage consist of two sections of size $\lfloor m/\delta \rfloor + \lfloor n/\delta \rfloor$ and the i-th element of the first section together with the i-th element of the second section belong to the i-th block in the result of the rearrangement process. Please note that, due to the technique used for constructing the bd-storage, such pairs of elements are always different with the first one smaller than the second one. If the i-th block originates from w we exchange the corresponding elements in the bd-storage otherwise we leave them untouched. Fig. 2(c) shows this graphically.

Step 4: Local merges
We visit every w-block and proceed as follows:

Let p be the w-block to be merged and let q be the sequence of all v-originating elements immediately to the right of p that are still unmerged. Further let x be the first element of p.

(1) Using a binary search we split q into $q = q_1 q_2$ so that we get $q_1 < x \leq q_2$. It holds $|q_1| < \delta$ due to the block rearrangement applied before. (2) We rotate $p q_1 q_2$ to $q_1 p q_2$. (3) We locally merge p and q_2 by Hwang and Lin's algorithm, where we use the buffer area as internal buffer.

This visiting process starts with the rightmost w-block and moves sequentially w-block by w-block to the left. The positions of the w-blocks are detected using the information hold in the bd-storage. Every time when we locate the position of a w-block in the bd-storage we bring the corresponding bd-storage elements back to their original order. So, after finishing all local merges both sections of the bd-storage are restored to their original form.

Step 5: Final sweeping up
On the left there is a still unmerged sub-sequence $s_1 t s_2 b w_1 v'$ where v' is the subsection of v that consists of the remaining unmerged elements. We proceed as follows: (1) We split v' into $v' = v'_1 v'_2$ so that $v'_1 < x \leq v'_2$ where x is the last element of s_2. Afterward we rotate $b w_1 v'_1 v'_2$ to $v'_1 b w_1 v'_2$ and locally merge w_1 and v'_2 using Hwang and Lin's algorithm with the internal buffer. (2) In the same way we split v'_1 into $v' = v'_{1,1} v'_{1,2}$ so that we get $v'_{1,1} < y \leq v'_{1,2}$ where y is the last element of s_1. We rotate $s_1 t s_2 v_{1,1} v_{1,2}$ to $s_1 v_{1,1} t s_2 v_{1,2}$ and locally merge s_1

with $v'_{1,1}$ and s_2 with $v_{1,2}$' using the Block-Rotation-Merge algorithm with a block-size of $\lfloor\sqrt{m}\rfloor$. (3) We sort the buffer area using Insertion-Sort and merge it with all elements right of it using the rotation based variant of Hwang and Lin's algorithm.

Lack of Space in Step 1
The inputs are so asymmetric that u' becomes empty. Using a binary search we split v into $v = v_1 v_2$ so that we get $v_1 < t \leq v_2$ and rotate $s_1 t s_2 v_1 v_2$ to $s_1 v_1 t s_2 v_2$. Using the Block-Rotation-Merge algorithm with a block-size $\lfloor\sqrt{m}\rfloor$ we locally merge s_1 with v_1 and s_2 with v_2. If s_2 is empty we ignore it and directly merge s_1 with v in the same style.

Extracted buffer smaller than $\lfloor\sqrt{m}\rfloor$ in Step 2
We assume that we could extract a buffer of size λ with $\lambda < \lfloor\sqrt{m}\rfloor$. We change our block-size δ to $\lfloor|u|/\lambda\rfloor$ and apply the algorithm as described but with the modification that we use the rotation based variant of Hwang and Lin's algorithm for all local merges.

Corollary 3. Stable-Optimal-Block-Merge *is stable.*

Theorem 1. *The* Stable-Optimal-Block-Merge *algorithm requires* $O(m+n) = O(m \cdot (k+1))$ *assignments..*

Proof. It is enough to prove that every step is performed with $O(m+n)$ assignments. In the first step no assignments occur at all. The buffer extraction in step 2 requires $O(m)$ assignments, as shown in [6]. In step 3 the "rolling" of the w-blocks through the v-blocks together with the "dropping" of the minimal w-blocks requires $3\sqrt{m}\cdot(\sqrt{m}+\frac{n}{\sqrt{m}}) = O(m+n)$ assignments. The rotations for the integrated "movement imitation" contribute $O(\sqrt{m}\cdot(\sqrt{m}+\frac{n}{\sqrt{m}})) = O(m+n)$ assignments. The marking of the positions of the w-blocks in the bd-storage needs $O(\sqrt{m})$ assignments. So, altogether step 3 requires $O(m+n)$ assignments. In step 4 each w-block rotation requires $\sqrt{m}+\sqrt{m}+\gcd(\sqrt{m},\sqrt{m}) = 3\sqrt{m}$ assignments at most. So all w-block rotations need $3\sqrt{m}\cdot\sqrt{m} = O(m)$ assignments. The local mergings using Hwang and Lin's algorithm consume $2m+n$ assignments altogether. The reconstruction of the original order of the exchanged elements in the bd-storage contributes $O(\sqrt{m})$ assignments. In step 5 the first rotation requires $4\sqrt{m}$ assignments at most and the local merging of w_1 and v'_2 needs $3\sqrt{m}$ assignments at most. The second rotation requires $3\sqrt{m} + \frac{n}{\sqrt{m}}$ assignments at most. The success in step 1 implies that roughly $k \leq \sqrt{m}/2$, so we get $k \cdot \sqrt{m} \leq m$. Further we have $\lfloor m/\delta\rfloor + \lfloor n/\delta\rfloor$ is roughly equal to $(k+1)\cdot\sqrt{m} = \frac{m+n}{\sqrt{m}}$. So, according to Lemma 1 each local merging with Block-Rotation-Merge needs $\frac{(k\sqrt{m})\cdot k\sqrt{m}}{2\cdot\sqrt{m}} + 2\cdot n + 6k\sqrt{m} + k(\sqrt{m}\sqrt{m}) \leq \frac{k\cdot m}{2} + 2n + 6m + k\cdot m$ assignments at most. The buffer sorting using insertion sort contributes $O(m)$ assignments and the final call of Hwang and Lin's algorithm requires $n+m+\sqrt{m}$ assignments. So, step 5 needs altogether $O(m+n)$ assignments as well.

In the first exceptional case "Lack of Space in Step 1" we have roughly $k \geq \sqrt{m}/2$ and directly switch to Block-Rotation-Merge. According to Corollary 1 Block-Rotation-Merge is fully asymptotically optimal for such k.

In the second exceptional case "Extracted buffer smaller than $\lfloor\sqrt{m}\rfloor$" we change the block-size to $\lfloor |u| / \lambda \rfloor$ with $\lambda < \sqrt{m}$ and use the rotation based variant of Hwang and Lin's algorithm for local merges. A recalculation of the steps 3 to 5, were we use Lemma 3 in the context of all local merges, proves that the number of assignments is still $O(m+n)$. □

Lemma 4. *If $k = \sum_{i=1}^{n} k_i$ for any $k_i > 0$ and integer $n > 0$, then $\sum_{i=1}^{n} \log k_i \leq n \log(k/n)$.*

Proof. It holds because the function $\log x$ is concave. □

Theorem 2. *The* STABLE-OPTIMAL-BLOCK-MERGE *algorithm requires $O(m \log(\frac{n}{m}+1)) = O(m \log(k+1))$ comparisons.*

Proof. As in the case of the assignments it is enough to show that every step keeps the asymptotic optimality. Step 1 contains one binary search over m merely. The buffer extraction in step 2 requires m comparisons at most, as shown in [6]. The rearrangement of all blocks except of the two undersized blocks w_1 and $v_{\lfloor \frac{n}{\delta} \rfloor + 1}$ in step 3 requires $2\sqrt{m} + \frac{n}{\sqrt{m}}$ comparisons at most. The detection of the minimal element in the movement imitation buffer demands $\sqrt{m} \cdot \sqrt{m}$ many comparisons at most. In step 4 the binary searches for splitting the q-sequences cost $\sqrt{m} \cdot \log \sqrt{m}$ comparisons at most. Now let $(m_1, n_1), (m_2, n_2), \cdots, (m_r, n_r)$ be the sizes of all r-groups that are locally merged by Hwang and Lin's algorithm. According to Lemma 4, Table 1 and since $r < \sqrt{m}$ this task requires $\sum_{i=1}^{r}(m_i(\log(\frac{n_i}{m_i}) + 1) + m_i) = \sum_{i=1}^{r}(m_i \log(\frac{n_i}{m_i}) + 2m_i) \leq \sum_{i=1}^{r} m_i \log(\frac{n_i}{m_i}) + 2m = \sum_{i=1}^{r}(m_i \log n_i - m_i \log m_i) + 2m = \sqrt{m} \sum_{i=1}^{r}(\log n_i - \log m_i) + 2m \leq \sqrt{m}(\sqrt{m} \log \frac{n}{r} - \sqrt{m} \log \frac{m}{r}) + 2m \leq m(\log(\frac{n}{m}+1)) + 2m = O(m \log(\frac{n}{m}+1))$ comparisons. The asymptotic optimality in step 5 as well as in the exceptional case "Lack of Space in Step 1" is obvious due to Lemma 2. The change of the block-size in the second exceptional case "Extracted buffer smaller than $\lfloor\sqrt{m}\rfloor$" triggers a simple recalculation of step 3 and step 4, where we leave the details to the reader. □

Corollary 4. STABLE-OPTIMAL-BLOCK-MERGE *is an asymptotically fully optimal stable in-place merging algorithm.*

Table 2. Pseudo-code Definitions of the Toolbox Algorithms

Pseudo-code Definition	Description of the Arguments
HWANG-AND-LIN($A, first1, first2, last$)	u is in $A[first1 : first2 - 1]$, v is in $A[first2 : last - 1]$
BINARY-SEARCH($A, first, last, x$)	delivers the position of the **first** occurrence of x in $A[first : last - 1]$
MINIMUM($A, pos1, pos2$)	delivers the index of the minimal element in $A[pos1 : pos2 - 1]$
BLOCK-SWAP($A, pos1, pos2, len$)	u is in $A[pos1 : pos1 + len - 1]$, v is in $A[pos2 : pos2 + len - 1]$
BLOCK-ROTATE($A, first1, first2, last$)	u, v as in HWANG-AND-LIN

EXCHANGE($A, pos1, pos2$) is equal to BLOCK-SWAP($A, pos1, pos2, 1$).

Algorithm 1. Pseudo-code of the procedure for the block rearrangement

```
REARRANGE-BLOCKS(A, first1, first2, last, buf, bds1, bds2, blockSize)
1  ▷ w_2 ... w_x is in A[first1 : first2 − 1], v_1 ... v_{y−1} is in A[first2 : last − 1]
2  ▷ buffer b is in A[buf : buf + ⌊√m⌋ − 1]
3  ▷ bd-storage s_{1|2} is in A[bds{1|2} : bds{1|2} + ⌊√m⌋ + ⌊n/√m⌋ − 1]
4
5  bufEnd ← buf + (first2 − first1) / blockSize
6  minBlock ← first1
7  while first1 < first2
8        do if first2+blockSize < last and A[first2+blockSize−1] < A[minBlock]
9              then BLOCK-SWAP(A, first1, first2, blockSize)
10                  BLOCK-ROTATION(A, buf, buf + 1, bufEnd)
11                  if minBlock = first1
12                     then minBlock ← first2
13                  first2 ← first2 + blockSize
14             else BLOCK-SWAP(A, minBlock, first1, blockSize)
15                  EXCHANGE(A, buf, buf + (minBlock − first1) / blockSize)
16                  EXCHANGE(A, bds1, bds2)
17                  buf ← buf + 1
18                  if buf < end
19                     then minIndex ← MINIMUM(A, buf, bufEnd)
20                          minBlock ← first1 + (minIndex − buf) ∗ blockSize
21           bds1 ← bds1 + 1; bds2 ← bds2 + 1
22           first1 = first1 + blockSize
```

Pseudo-code implementations for the core operations "block rearrangement" and "local merges" are given in Alg. 1 and Alg. 2, respectively. Both code segments contain calls of the toolbox algorithms mentioned in section 2. The Pseudo-code definitions for these toolbox algorithms are summarized in Tab. 2.

3.1 Optimizations

We now report about several optimizations that help improving the performance of the algorithm without any impact on its asymptotic properties. The immediate mirroring of all w-block movements in the movement imitation buffer (occurs in Step 3) triggers a rotation (line 10 in Alg. 1) every time when a v-block is moved into front of the group of w-blocks. The number of necessary rotations can be reduced by first counting the number of v-blocks moved into front of the w-blocks. This counting follows a single update of the movement imitation buffer if the placement of a minimal w-block happens. In the context of the movement of v-blocks into front of w-blocks (Step 3) the floating hole technique (for a description see [4] or [6]) can be applied for reducing the number of assignments. Similarly the floating hole technique can also be applied during the local merges (Step 4) by combining the block swap to the internal buffer with the rotation

Algorithm 2. Pseudo-code of the function for local merges

```
LOCAL-MERGES(A, first, last, buf, bds1, bds2, blockSize, numBlocks)
▷ A[first : last − 1] contains all blocks in distributed form

index ← ((last − first) / blockSize) − 1
while numBlocks > 0
      do while A[bsd1 + index] < A[bsd2 + index]
              do index ← index − 1
         first2 ← first + ((index + 1) ∗ blockSize)
         if first2 < last
           then b ← BINARY-SEARCH(first2, last, A[first2 − blockSize])
                BLOCK-ROTATION(A, first2 − blockSize, first2, b)
                HWANG-LIN(A, b − blockSize, b, last, buf)
                last ← b − blockSize
         EXCHANGE(A, bds1 + index, bds2 + index)
         numBlocks ← numBlocks − 1; index ← index − 1
return last
```

that moves smaller v-originating elements to the front of the w-block. In the special case "Extracted buffer smaller then $\lfloor\sqrt{m}\rfloor$" the sorting of the buffer b in Step 5 is unnecessary because the buffer is already sorted after Step 3 and stays unchanged during Step 4. Insertion-Sort can be replace by some more efficient sorting algorithm. Please note that there is no need for stability in the context of the buffer sorting because all buffer elements are distinct.

4 Experimental Work

We did some experimental work with our algorithm in order to get an impression of its performance. We compared it with the stable fully asymptotically optimal algorithm presented in [6] as well as the simple standard algorithm that relies

Table 3. Practical comparison of various merge algorithms

n	m	STABLE-O.-B.-MERGE			SOFSEM 2006 Alg.			Linear Standard Alg.		
		#comp	*#assign*	t_e	*#comp*	*#assign*	t_e	*#comp*	*#assign*	t_e
2^{21}	2^{21}	5843212	37551852	227	5961524	49666369	335	4194239	8388608	121
2^{21}	2^{18}	1500433	15866835	100	1505766	17182008	122	2359288	4718592	71
2^{21}	2^{15}	280611	17350896	87	280412	12681115	68	2129890	4259840	64
2^{21}	2^{12}	43611	4422493	35	47330	10512479	53	2100804	4202496	63
2^{23}	2^{9}	8057	16350956	133	8589	38150052	202	8373039	16778240	251
2^{23}	2^{6}	1200	15459824	131	1271	30749720	161	8234508	16777344	254
2^{23}	2^{3}	172	11322991	119	170	7535160	68	7572307	16777232	301
2^{23}	2^{0}	23	4163489	55	24	4163489	55	4225121	16777218	261

t_e : Execution time in ms, *#comp* : Number of comp., m, n : Lengths of inp. seq.

on external space of size m. The results of our experimental work summarizes Tab. 3 where every line shows average values for 50 runs with different data. We took a standard desktop computer with 2GHz processor as hardware platform. All coding happened in the C programming language. For the measurement of the number of assignments we applied the optimal block rotation algorithm presented in [3]. Although this algorithm is optimal regarding the number of assignments it is quite slow in practice due to its high computational demands. Therefore for the time measurements we applied a block-swap based algorithm presented e.g. in [1] using identical data.

Regarding the buffer extraction (Step 2) there are several alternatives. The extraction process can be started from the left end as well as from the right end of the input and we can choose between a binary search and linear search for the determination of the next element. All 4 possible combinations keep the asymptotic optimality. However, there is no clear "best choice" among them because the most advantageous combination can vary depending on the structure of the input. In the context of the Stable-Optimal-Block-Merge algorithm we decided for the variant "starting from the left combined with binary search", the SOFSEM 2006 algorithm already originally chose "starting from the right combined with linear search".

Except for two combinations of input sizes our new algorithm is always faster than its predecessor. The bad performance in the case $(2^{21}, 2^{15})$ reflects the lack of the implementation of the floating hole technique as mentioned in the section about optimizations. The application of Block-Rotation-Merge triggers unnecessary rotations in the case $(2^{23}, 2^{3})$. This can be fixed by introduction of a check whether $k \geq m$ and a direct switch to the rotation based variant of Hwang and Lin's algorithm if true.

5 Conclusion

We investigated the problem of stable in-place merging from a ratio based point of view by introducing a ratio $k = \frac{n}{m}$, where m,n are the sizes of the input sequences with $m \leq n$. We could show that there is a simple asymptotically fully optimal (optimal regarding the number of comparisons as well as assignments) stable in-place merging algorithm for any ratio $k \geq \sqrt{m}$.

In the second part of this paper we introduced a novel asymptotically fully optimal stable in-place merging algorithm which is constructed on the foundation of deliberations regarding the ratio of the input sizes. Highlights of this algorithm are: It has a modular structure and does not rely on techniques described by Mannila and Ukkonen [8] in contrast to all its known competitors ([10,4,6]). The tasks "block-distribution" and "local block mergings" are modular separated. As side effect they can share a common buffer area and the extraction of a separated movement imitation buffer is not necessary. The algorithm demands no lower bound for the size of the shorter input sequence (32 elements in case of the alg. in [4] and 10 elements for the alg. in [6]).

Our algorithm performs for a wide range of inputs remarkably better than its direct competitor presented in [6]. There is a superiority in particular for symmetrically sized inputs, a fact that is of importance in the context of the Merge-sort algorithm.

The number of comparisons and assignments are good measurements for the efficiency of merging algorithms. However, the impact of other operations as e.g. numerical calculations and index comparisons deserves investigation as well. As motivation we would like to refer to a well known effect with the optimal block-rotation algorithm introduced by Dudzinski and Dydek in [3]. Their algorithm is optimal regarding the number of assignments but has a bad performance due to a included computation of a greatest common divisor. For our further work we plan to include deliberations regarding such so far uncounted operations.

References

1. Bentley, J.: Programming Pearls, 2nd edn. Addison-Wesley, Reading (2000)
2. Chen, J.: Optimizing stable in-place merging. Theoretical Computer Science 302(1/3), 191–210 (2003)
3. Dudzinski, K., Dydek, A.: On a stable storage merging algorithm. Information Processing Letters 12(1), 5–8 (1981)
4. Geffert, V., Katajainen, J., Pasanen, T.: Asymptotically efficient in-place merging. Theoretical Computer Science 237(1/2), 159–181 (2000)
5. Hwang, F.K., Lin, S.: A simple algorithm for merging two disjoint linearly ordered sets. SIAM J. Comput. 1(1), 31–39 (1972)
6. Kim, P.-S., Kutzner, A.: On optimal and efficient in place merging. In: Wiedermann, J., Tel, G., Pokorný, J., Bieliková, M., Štuller, J. (eds.) SOFSEM 2006. LNCS, vol. 3831, pp. 350–359. Springer, Heidelberg (2006)
7. Knuth, D.E.: The Art of Computer Programming. Sorting and Searching, vol. 3. Addison-Wesley, Reading (1973)
8. Mannila, H.: A simple linear-time algorithm for in situ merging. Information Processing Letters 18, 203–208 (1984)
9. Pardo, L.T.: Stable sorting and merging with optimal space and time bounds. SIAM Journal on Computing 6(2), 351–372 (1977)
10. Symvonis, A.: Optimal stable merging. Computer Journal 38, 681–690 (1995)

A Characterisation of the Relations Definable in Presburger Arithmetic

Mathias Barra

Dept. of Mathematics, University of Oslo, P.B. 1053, Blindern, 0316 Oslo, Norway
georgba@math.uio.no
http://folk.uio.no/georgba

Abstract. Four sub-recursive classes of functions, $\mathcal{B}$, $\mathcal{D}$, $\mathcal{BD}$ and $\mathcal{BDD}$ are defined, and compared to the classes G^0, G^1 and G^2, originally defined by Grzegorczyk, based on *bounded minimalisation*, and characterised by Harrow in [5]. $\mathcal{B}$ is essentially G^0 with *predecessor* substituted for *successor*; $\mathcal{BD}$ is G^1 with *(truncated) difference* substituted for *addition*. We prove that the induced relational classes are preserved ($G^0_\star = \mathcal{B}_\star$ and $G^1_\star = \mathcal{BD}_\star$). We also obtain $\mathcal{D}_\star = \mathfrak{PrA}^{\mathrm{qf}}_\star$ (the quantifier free fragment of Presburger Arithmetic), and $\mathcal{BD}_\star = \mathfrak{PrA}_\star$, and $\mathcal{BDD}_\star = G^2_\star$, where $\mathcal{BDD}$ is G^2 with *integer division* and *remainder* substituted for *multiplication*, and where $G^2_\star$ is known to be equal to the predicates definable by a bounded formula in *Peano Arithmetic*.

1 Introduction and Notation

Introduction: This paper emerges from some investigations into (very) small sub-recursive classes. The original motivation has been to discard all increasing functions, and to compare the resulting classes to otherwise similar classes. This approach—banning all growth—has proved successful in the past, and has repeatedly yielded surprising and enlightening results, see e.g. Jones [7,8]; Kristiansen and Voda in [12,13] (with functionals of higher types, and imperative programming languages respectively); Kristiansen and Barra [11] (with function algebras and λ-calculus); and Kristiansen [9,10].

The author has been interested in further investigations into the consequences of omitting increasing functions from the set of initial functions of various *inductively defined classes* (see definitions below).

The seminal paper by A. Grzegorczyk *Some classes of recursive functions* [4] from 1953 was the source of great inspiration to many researchers during the decades to follow. One significant contribution emerged with Harrow's Ph.D. dissertation *Sub-elementary classes of functions and relations* [5], and his findings were later summarised and enhanced in *Small Grzegorczyk classes and limited minimum* [6]. He there answered several questions (in the negative), originally posed by Grzegorczyk, and with regard to the interchangeability of the schemata *bounded primitive recursion* and *bounded minimalisation*. Another result from [5] is that $G^1_\star$ (this and other classes mentioned below are defined in the sequel)

M. Agrawal et al. (Eds.): TAMC 2008, LNCS 4978, pp. 258–269, 2008.

is identical to the set of predicates $\mathfrak{PrA}_\star$, i.e. those subsets of $\mathbb{N}^k$ definable by a formula in the language of *Presburger Arithmetic.*

More precisely, for $i = 0, 1, 2$, we study the three classes G^i which are defined analogously to Grzegorczyk's $\mathcal{E}^i$, but where the schema of *bounded minimalisation* is substituted for *bounded primitive recursion.* These classes contain the increasing functions *successor* (in G^0 and $\mathcal{E}^0$), *addition* (in G^1 and $\mathcal{E}^1$) and *multiplication* (in G^2 and $\mathcal{E}^2$).

We will show that the classes G^i contain redundancies, in the sense that the *increasing* functions 'S', '+' and '×' can be substituted with their nonincreasing inverses *predecessor*, *(truncated) difference*, and *integer division* and *remainder*, without affecting the induced relational classes $G^i_\star$. That is, the growth provided by e.g. addition, in the restricted framework of composition and bounded minimalisation, does not contribute to the number of computable predicates. In fact, we show that the quantifier free fragment of Presburger Arithmetic may be captured in a much weaker system: essentially only truncated difference and composition is necessary.

Notation: Unless otherwise specified, a *function* means a function $f : \mathbb{N}^k \to \mathbb{N}$ The *arity* of f, denoted $\mathrm{ar}(f)$, is then k. When $\mathcal{F}$ is a set of functions, $\mathcal{F}^k$ denotes the k-ary functions in $\mathcal{F}$.

A function is *nonincreasing* if, for some $c_f \in \mathbb{N}$ we have[1] $f(\vec{x}) \leq \mathsf{max}(\vec{x}, c_f)$ for all $\vec{x} \in \mathbb{N}^k$.

We say that f has *top index* i if $f(\vec{x}) \leq \mathsf{max}(x_i, c_f)$. If $c_f = 0$, we say that f is *strictly nonincreasing*, and that i is a *strict top index.*

The *bounded* f, denoted $\hat{f}$, is the $(\mathrm{ar}(f)+1)$-ary function $\hat{f}(\vec{x}, b) \stackrel{\text{def}}{=} \mathsf{min}(f(\vec{x}), b)$. These bounded versions, in particular the bounded versions of increasing functions like $\hat{\mathsf{S}}$, the *bounded successor function*, will be of major importance for the ensuing developments. The *predecessor function*, denoted P, is defined by $\mathsf{P}(x) \stackrel{\text{def}}{=} \mathsf{max}(x - 1, 0)$. The *case function*, denoted C, and the *(truncated) difference function*, denoted $\dot{-}$, are defined by

$$\mathsf{C}(x, y, z) \stackrel{\text{def}}{=} \begin{cases} x & \text{, if } z = 0 \\ y & \text{, else} \end{cases} \quad \text{and} \quad x \dot{-} y \stackrel{\text{def}}{=} \mathsf{max}(x - y, 0) \stackrel{\text{def}}{=} \begin{cases} 0 & \text{, if } x \leq y \\ x - y & \text{, if } x > y \end{cases}$$

When we use the symbol '$-$' without the dot in an expression or formula, we mean the usual minus on $\mathbb{Z}$.

Let $\phi(x, y, n, r) \stackrel{\text{def}}{\Leftrightarrow} 0 \leq r < y \land x = ny + r$. The *remainder function* and *integer division function*, denoted rem and $\left\lfloor \frac{\cdot}{\cdot} \right\rfloor$ respectively, are defined by

$$\left\lfloor \frac{x}{y} \right\rfloor \stackrel{\text{def}}{=} \begin{cases} x & \text{, if } y = 0 \\ n & \text{, if } \phi(x, y, n, r) \end{cases} \quad \text{and} \quad \mathsf{rem}(x, y) \stackrel{\text{def}}{=} \begin{cases} x & \text{, if } y = 0 \\ r & \text{, if } \phi(x, y, n, r) \end{cases},$$

respectively. We define e.g. $\left\lfloor \frac{x}{0} \right\rfloor = x$ in order to make the functions total on $\mathbb{N}^2$. Note that we have $\left\lfloor \frac{x}{y} \right\rfloor y + \mathsf{rem}(x, y) = x$ for all x and y.

---

[1] The bound which holds for f in G^0 and $\mathcal{E}^0$ is $f(\vec{x}) \leq \mathsf{max}(\vec{x}) + c_f$; note the distinction. The latter bound is sometimes referred to as 0-*boundedness*, while $f(\vec{x}) \leq c_f \mathsf{max}(\vec{x})$ as 1-*boundedness*.

$\mathcal{I}$ is the set of all *projections* $\mathsf{I}_i^k(\vec{x}) = x_i$, and $\mathcal{N}$ is the set of all *constant functions* $\mathsf{c}(x) = c$ for all $c \in \mathbb{N}$.

An *inductively defined class of functions* (*idc.*), is generated from a set $\mathcal{X}$ called the *initial*, *primitive* or *basic* functions, as the least class containing $\mathcal{X}$ and closed under the *schemata*, *functionals* or *operations* of some set OP of functionals. We write $[\mathcal{X}; \text{OP}]$ for this set[2]. The schemata *composition*, denoted COMP, and *bounded minimalisation*[3], denoted BMIN, will be used extensively. We write $h \circ \vec{g}$ for the function $f(\vec{x}) = h(g_1(\vec{x}), \ldots, g_m(\vec{x}))$, and $\mu_{z \leq y}[g_1 = g_2]$ for the function $f(\vec{x}, y)$ which equals the least $z \leq y$ satisfying the equation $g_1(\vec{x}, z) = g_2(\vec{x}, z)$, if such exists, and 0 else.

A *relation* is a subset R of $\mathbb{N}^k$ for some k. Relations are interchangeably called *predicates*. Sets of predicates are usually subscripted with a $\star$. For a set $\mathcal{F}_\star$ of relations, we say that $\mathcal{F}_\star$ *is boolean*, when $\mathcal{F}_\star$ is closed under finite intersections and complements. When $\mathcal{F}_\star$ is some set of relations of varying arity, $\mathcal{F}_\star^k$ denotes the k-ary relations of $\mathcal{F}_\star$.

When $R = f^{-1}(0)$, the function f is referred to as *a characteristic function of* R, and is denoted χ_R. Let $\mathcal{F}$ be a set of functions. $\mathcal{F}_\star$ denotes the set of *relations of* $\mathcal{F}$, viz. those subsets $R \subseteq \mathbb{N}^k$ with $\chi_R \in \mathcal{F}$. Formally

$$\mathcal{F}_\star^k = \{f^{-1}(0) \subseteq \mathbb{N}^k | f \in \mathcal{F}^k\} \quad \text{and} \quad \mathcal{F}_\star = \bigcup_{k \in \mathbb{N}} \mathcal{F}_\star^k \ .$$

The *graph of* f, denoted Γ_f, is the relation $\{(\vec{x}, y) \in \mathbb{N}^{\mathrm{ar}(f)+1} | f(\vec{x}) = y\}$. We overload Γ_f to also denote its characteristic function.

Whenever a symbol occurs under an arrow, e.g. $\vec{x}$, we usually do not point out the length of the list—by convention it shall be k unless otherwise specified.

2 The Classes $\mathcal{B}$ and G^0

Let the class $\mathcal{B}$ be defined by $\mathcal{B} \stackrel{\text{def}}{=} [\mathcal{I} \cup \mathcal{N} \cup \{\mathsf{P}, \hat{\mathsf{C}}\}; \text{COMP}, \text{BMIN}]$. Note that the case function $\hat{\mathsf{C}}$ takes four arguments: $\hat{\mathsf{C}}(x, y, z, b) \stackrel{\text{def}}{=} \min(\mathsf{C}(x, y, z), b)$. G^0 is defined by (in our notation) $[\mathcal{I} \cup \{0, \mathsf{S}, \mathsf{P}\}; \text{COMP}, \text{BMIN}]$. The essential difference between the two classes is that G^0 includes the increasing initial function S, whereas $\mathcal{B}$ only includes nonincreasing initial functions. It is clear that $\mathcal{N} \subseteq G^0$, and it is easy to prove that $\hat{\mathsf{C}} \in G^0$. Hence $\mathcal{B} \subseteq G^0$. The inclusion is strict, since all functions in $\mathcal{B}$ satisfy

Lemma 1 (top index). *Let $f \in \mathcal{B}^k$. Then f has a top index. Furthermore, if $f(\mathbb{N}^k)$ is infinite, the top index is strict.* □

We skip the proof, which is easy and by induction on $f \in \mathcal{B}$. Observe next that $\hat{\mathsf{C}}(1, 0, f, 1) = 1 \dot{-} f$ and $\hat{\mathsf{C}}(0, 1, f_1, f_2) = 0 \Leftrightarrow f_1 = 0 \vee f_2 = 0$. Since $\mathcal{B}$ is closed under BMIN by definition, we obtain

[2] This notation is adopted from Clote [2], where an i.d.c is called a *function algebra.*

[3] a.k.a. *bounded search* a.k.a. *limited minimum.*

Proposition 1. *$\mathcal{B}_\star$ is boolean, and is closed under $\exists_{y\leq x}$ type quantifiers.* □

This explains the inclusion of $\hat{\mathsf{C}}$ amongst $\mathcal{B}$'s initial functions[4]. As the main result of this section will show, the only essential use of S in [6] is to produce functions like e.g. $1 \dot{-} x$ via the successor and bounded minimalisation.

Harrow proves in [6] that $G^0 = \mathcal{E}^0 \cap \mathcal{PL}$, where $\mathcal{PL}$ is the set of *piecewise linear* functions. This class is defined as the functions which may be put on the form $f(\vec{x}) = L_i(\vec{x}) \Leftrightarrow P_i(\vec{x})$, where each L_i is either constant or of the form $x_j \dot{-} c$ or $x_j + c$, and where the P_i are conjunctions of predicates of the form $L_r \leq L_s$. Note that there should be a finite number of clauses[5].

Note that I^1_1 is a characteristic function for the relation 'equals zero'. Also, $x < y \Leftrightarrow y \neq 0 \wedge (x = 0 \vee \mu_{z\leq y}[\mathsf{P}(z) = x] \neq 0)$, and $x = y \Leftrightarrow \neg(x < y \vee y < x)$. Importantly, for $f \overset{\text{def}}{=} \mu_{z\leq b}[\mathsf{P}(z) = x]$, we have $\hat{\mathsf{S}}(x, b) = \hat{\mathsf{C}}(b, f, f, b)$, and so $x + 1 = y \Leftrightarrow x < y \wedge \hat{\mathsf{S}}(x, y) = y$. Clearly this generalizes to relations $x + c = y$ for $c \in \mathbb{N}$.

Combining the above, it is not hard to see that we may decide in $\mathcal{B}_\star$ whether $P(\vec{x})$ for any relation P of the form specified in the definition of $\mathcal{PL}$, and—for ditto L—whether $L(\vec{x}) = y$.

But now, for $f \in \mathcal{PL}$, clearly $f(\vec{x}) = y \Leftrightarrow \bigvee_j (P_j(\vec{x}) \wedge L_j(\vec{x}) = y)$, with the latter relation in $\mathcal{B}_\star$, whence

Theorem 1. $\mathcal{B}_\star = G^0_\star = \mathcal{PL}_\star$.

Proof. Since $\mathcal{B} \subseteq G^0$, inclusion to the right is trivial. If $R \in G^0_\star$, by definition $R = f^{-1}(0)$ for some $f \in G^0 \subseteq \mathcal{PL}$. Hence $\vec{x} \in R \Leftrightarrow f(\vec{x}) = 0 \in \mathcal{B}_\star$. □

Hence, only nonincreasing functions are necessary to carry out the analysis. In SECTION 4 we show the analogous result holds also of Harrow's G^1. We first dedicate some time to the truncated difference function. As the results of the next section show, it is surprisingly strong from a computational point of view.

3 The Class $\mathcal{D}$

Define $\mathcal{D} \overset{\text{def}}{=} [\mathcal{I} \cup \mathcal{N} \cup \{\dot{-}\}; \text{COMP}]$. The lemma below is the starting point for showing that the class $\mathcal{D}$ is surprisingly powerful. The reader is asked to note that composition is the *sole* closure operation of $\mathcal{D}$. A top-index lemma also holds for $\mathcal{D}$.

Lemma 2 (top index). *Let $f \in \mathcal{D}^k$. Then f has a top index. Furthermore, if $f(\mathbb{N}^k)$ is infinite, the top index is strict.* □

---

[4] We also conjecture that $\hat{\mathsf{C}}$ cannot be omitted, e.g. PROPOSITION 1 would not hold. However, we can prove that with an alternative version of BMIN, which returns 'y' rather than '0' upon a failed search, $\mathcal{I} \cup \mathcal{N}$ suffice as initial functions. The proofs of this claim are straightforward, but involved.

[5] E.g. $f(x_1, x_2) = \begin{cases} 5 & \text{, if } x_1 \leq x_2 \wedge 3 \leq x_1 \\ x_2 + 4 & \text{, else} \end{cases}$ is a member of $\mathcal{PL}$, since the 'else clause' can be split into the two clauses $x_2 + 1 \leq x_1$ and $x_1 \leq 2$.

Lemma 3 (bounded addition). *The function* $\min(x+y,z)$ *belongs to* $\mathcal{D}$.

Proof. Set $f(x,y,z) \stackrel{\text{def}}{=} z\dot{-}((z\dot{-}x)\dot{-}y)$. If $x+y \geq z$, then $(z\dot{-}x)\dot{-}y = 0$, which yields $z\dot{-}((z\dot{-}x)\dot{-}y) = z - 0 = z$ On the other hand, if $z > x+y \geq x$, then $z\dot{-}x = z - x > y > 0$. Hence $(z-x)\dot{-}y = ((z-x)-y) > 0$. But now $z > (z-(x+y)) > 0$, and so
$z\dot{-}((z\dot{-}x)\dot{-}y) = z - (z-(x+y)) = z - z + (x+y) = x+y$. □

This function is the key to proving several properties of $\mathcal{D}$ and $\mathcal{D}_\star$.

Proposition 2. *We have that* **(i)** $\min(x,y) \in \mathcal{D}$; **(ii)** $\mathcal{D}_\star$ *is boolean;* **(iii)** $\chi_=, \chi_< \in \mathcal{D}$; **(iv)** $\Gamma_{\max}, \Gamma_{\mathsf{C}} \in \mathcal{D}$; **(v)** *If* A *or* $\mathbb{N}^k \setminus A$ *is finite, then* $A \in \mathcal{D}_\star$.

Proof. Clearly $\min(x,y) = \min(x+0,y)$, thus **(i)**. Next, given χ_{R_1} and χ_{R_2} we have that

$$\chi_{R_1 \cap R_2} = \min(\chi_{R_1} + \chi_{R_2}, 1) \text{ and } \chi_{\mathbb{N}^k \setminus R} = 1\dot{-}\chi_{R_1}$$

hence **(ii)**. Since $x\dot{-}y = 0 \Leftrightarrow x \leq y$ and $y\dot{-}x = 0 \Leftrightarrow y \leq x$ yields **(iii)** in conjunction with **(ii)**. Next, $\max(x,y) = z \Leftrightarrow (y \leq x \wedge z = x) \vee (x < y \wedge z = y)$. Also, $\mathsf{C}(x,y,z) = w \Leftrightarrow (z = 0 \wedge w = x) \vee (z < 0 \wedge w = y)$, hence **(iv)**. **(v)** is proved by induction on $n = \min(|A|, |\mathbb{N}^k \setminus A|)$. We skip the details. □

Armed with PROPOSITION 2, we may prove the following lemma

Lemma 4. *Let* $\mathbf{r} \in \{<,=\}$, *and let* $1 \leq j < k \in \mathbb{N}$ *be arbitrary. Then, the following relations belong to* $\mathcal{D}_\star$:

$$\sum_{i=1}^{j} x_i \;\mathbf{r} \sum_{i=j+1}^{k} x_i$$

Proof. Observe first that the function $f(x,\vec{y}) = x\dot{-}\left(\sum_{i=1}^{k} y_i\right) \in \mathcal{D}$, by the length of $\vec{y}$ consecutive applications of composition. Note also that when $R(\vec{x}) \in \mathcal{D}_\star^k$, and $\vec{f} \in \mathcal{D}^m$, then the relation $S(\vec{y})$, defined by $S(\vec{y}) \stackrel{\text{def}}{\Leftrightarrow} R(f_1(\vec{y}), \ldots, f_k(\vec{y}))$, belongs to $\mathcal{D}_\star^m$.

We prove the lemma by induction on k. PROPOSITION 2 constitutes **induction start**. Note that

$\sum_{i=1}^{j} x_i = \sum_{i=j+1}^{k} x_i \Leftrightarrow \neg(\sum_{i=1}^{j} x_i < \sum_{i=j+1}^{k} x_i) \wedge \neg(\sum_{i=j+1}^{k} x_i < \sum_{i=1}^{j} x_i)$.

Hence, it is sufficient to perform the induction step when $\mathbf{r}$ is '<'. Moreover, as $\mathcal{D}_\star$ is boolean, we may invoke the induction hypothesis (i.h.) for $\mathbf{r} \in \{\geq, \leq, <, >, =\}$.

We proceed to the case $k+1$, with $2 \leq k$, and we shall first prove the special cases of $j = k$. Clearly

$$\sum_{i=1}^{k} x_i < x_{k+1} \Leftrightarrow 0 < x_{k+1}\dot{-}\left(\sum_{i=1}^{k} x_i\right) .$$

This shows that $\sum_{i=1}^{k} x_i < x_{k+1} \in \mathcal{D}_\star$ by the initial remarks.

Next consider the general case $1 \leq j < k$.

$$\sum_{i=1}^{j} x_i < \Big(\sum_{i=j+1}^{k} x_i \Big) + x_{k+1} \Leftrightarrow$$

$$\Big(\sum_{i=1}^{j} x_i < x_{k+1}\Big) \vee \Big(\underbrace{\Big(x_{k+1} \leq \sum_{i=1}^{j} x_i\Big)}_{\psi} \wedge \underbrace{\Big(\Big(\Big(\sum_{i=1}^{j} x_i\Big) \dot{-} x_{k+1}\Big) < \sum_{i=j+1}^{k} x_i\Big)}_{\phi}\Big) .$$

The conjunct marked ϕ above, needs special attention. Consider the function

$$g_j(x_1, \ldots, x_j, x_{k+1}) = (\sum_{i=1}^{j} x_i) \dot{-} x_{k+1} .$$

Now g_j is *not* in $\mathcal{D}$. However, we note that

$$g_j(\vec{x}, y) = (x_1 \dot{-} y) + (x_2 \dot{-} (y \dot{-} x_1)) + \cdots + (x_j \dot{-} (y \dot{-} x_1 \dot{-} x_2 \dot{-} \cdots \dot{-} x_{j-1})) \cdots)) \quad (\dagger)$$

Furthermore, when ψ is true, we have that $g_j(\vec{x}, y) = \big(\sum_i x_i\big) - y$. Importantly for us, the expression **(†)** is a sum of j summands, each summand being a $\mathcal{D}$-function of the variables involved. Hence

$$\psi \wedge \phi \Leftrightarrow \psi \wedge \underbrace{\Big(\sum_{i=1}^{j} \Big(x_i \dot{-} \Big(x_{k+1} \dot{-} \sum_{\ell=1}^{i-1} x_\ell \Big)\Big) < \sum_{i=j+1}^{k} x_i \Big)}_{\phi'} ,$$

which concludes the proof when $\mathbf{r}$ is '$<$', since the ϕ' is in $\mathcal{D}_\star$ by the i.h. □

3.1 Presburger Arithmetic and $\mathcal{D}_\star$

Let $\mathfrak{PrA}$ be the 1st-order language $\{0, \mathsf{S}, +, <, =\}$ with the intended structure $\mathfrak{N} \stackrel{\text{def}}{=} (\mathbb{N}, 0, \mathsf{S}, +, <)$—the natural numbers with the usual order, successor and addition. Terms, (atomic) formulae, and the *numerals* $\underline{m}$ are defined in the standard way. As usual we use abbreviations extensively, e.g. $t \leq s$ for $t = s \vee t < s$.

Many readers will no doubt have recognized $\mathfrak{PrA}$ as the language of *Presburger Arithmetic*, see e.g. Enderton [3, p. 188]. It is known that the theory $\mathfrak{N}$ is decidable (we return to this point in SECTION 4), and that it does *not* admit quantifier elimination.

We overload the symbol $\mathfrak{PrA}$ to also denote the set of $\mathfrak{PrA}$-formulae. For $\phi(\vec{x}) \in \mathfrak{PrA}$, define $R_\phi \subseteq \mathbb{N}^k$ by $R_\phi \stackrel{\text{def}}{=} \{\vec{m} \in \mathbb{N}^k | \mathfrak{N} \models \phi(\underline{\vec{m}})\}$, and set $\mathfrak{PrA}_\star \stackrel{\text{def}}{=} \{R_\phi | \phi \in \mathfrak{PrA}\}$. $\mathfrak{PrA}^{\text{qf}}$ denotes the set of *quantifier free* $\mathfrak{PrA}$-formulae; $\mathfrak{PrA}^{\Delta_0}$ the Δ_0-formulae. $\mathfrak{PrA}^{\text{qf}}_\star$ and $\mathfrak{PrA}^{\Delta_0}_\star$ are defined as expected.

It is well-known that for terms $t \in \mathfrak{PrA}$, we have $\mathfrak{N} \models t = \underline{m}$ for some m and that this m is unique. Also, any $\mathfrak{PrA}$-term t is clearly equivalent to some term $t = (\sum_{i \leq k} a_i x_i) + \underline{m}$, where the right side is shorthand for the term

$$\underbrace{(x_1 + \cdots + x_1)}_{a_1-\text{times}} + \cdots + \underbrace{(x_k + \cdots + x_k)}_{a_k-\text{times}} + \underline{m} \ .$$

Thus, any atomic formula $\phi(\vec{x}) \in \mathfrak{PrA}$ is equivalent to a formula of the form[6]

$$\sum a_i x_i + \underline{a} \ \mathbf{r} \sum b_i y_i + \underline{b} \ . \tag{‡}$$

The main result of this section is

Theorem 2. $\mathfrak{PrA}_\star^{qf} = \mathcal{D}_\star$.

Last section's LEMMA 4 provides us with most of the proof of $\mathfrak{PrA}_\star^{\mathrm{qf}} \subseteq \mathcal{D}_\star$.

Proof (of $\mathfrak{PrA}_\star^{qf} \subseteq \mathcal{D}_\star$). Let $R \in \mathfrak{PrA}_\star^{\mathrm{qf}}$. Then, by definition we have $R = R_\phi$ for some $\phi(\vec{x}) \in \mathfrak{PrA}^{\mathrm{qf}}$. Since $\mathcal{D}_\star$ is boolean, it is sufficient to prove the lemma for ϕ atomic.

Let $\phi(\vec{x})$ be atomic. Hence it is of the form specified in **(‡)** above. If we let R be the relation

$$R(\vec{x}, \vec{y}) \stackrel{\text{def}}{=} \sum a_i x_i + z \ \mathbf{r} \sum b_i y_i + w \ ,$$

we have by LEMMA 4 that $\chi_R \in \mathcal{D}$. But then

$$\chi_R(\vec{n}, \vec{m}, \mathsf{a}, \mathsf{b}) = 0 \Leftrightarrow \sum a_i n_i + a \ \mathbf{r} \sum b_i m_i + b \Leftrightarrow \mathfrak{N} \models \phi(\underline{\vec{n}}, \underline{\vec{m}}) \ ,$$

and so $\chi_R(\vec{x}, \vec{y}, \mathsf{a}, \mathsf{b}) = \chi_{R_\phi}$. Hence $\chi_{R_\phi} \in \mathcal{D}$. □

To facilitate the proof of the opposite inclusion, we first consider the language $\mathfrak{PrA}^{\dot{-}} \stackrel{\text{def}}{=} \mathfrak{PrA} \cup \{\dot{-}, -\}$, viz. $\mathfrak{PrA}$ augmented with new function symbols '$\dot{-}$' and '$-$', and with intended model $\mathfrak{Z} \stackrel{\text{def}}{=} (\mathbb{Z}, 0, \mathsf{S}, +, -, \dot{-}, <)$. Note that $\dot{-}$ is well-defined on all of $\mathbb{Z}^2$ by its original definition $\mathsf{max}(x - y, 0)$. We also just remark on the fact that for $\phi \in \mathfrak{PrA}$ variable free, we have $\mathfrak{Z} \models \phi \Leftrightarrow \mathfrak{N} \models \phi$.

Secondly, to every function $f \in \mathcal{D}^k$, there is a $\mathfrak{PrA}^{\dot{-}}$-term $t_f(\vec{x})$ such that, for all $\vec{n}, m \in \mathbb{N}$ we have $\mathfrak{Z} \models t_f(\underline{\vec{n}}) = \underline{m} \Leftrightarrow f(\vec{n}) = m$.

Lemma 5. *Let $\phi(\vec{x}) \in \mathfrak{PrA}^{\dot{-}}$ be atomic. Then, there is a $\phi'(\vec{x}) \in \mathfrak{PrA}^{qf}$ such that, for all $\vec{n} \in \mathbb{N}$ we have $\mathfrak{N} \models \phi'(\underline{\vec{n}}) \Leftrightarrow \mathfrak{Z} \models \phi(\underline{\vec{n}})$.*

Proof. For a $\mathfrak{PrA}^{\dot{-}}$-term t, we define $\mathrm{rk}_{\dot{-}}(\phi)$ to be the number of occurrences of the symbol $\dot{-}$ in t. Note that when $\phi \in \mathfrak{PrA}^{\dot{-}}$ is atomic it is of the form $t_1 \ \mathbf{r} \ t_2$, and that $\mathrm{rk}_{\dot{-}}(t_1) + \mathrm{rk}_{\dot{-}}(t_2) = 0$ implies that either $\phi \in \mathfrak{PrA}^{\mathrm{qf}}$, or it is equivalent to $\phi' \in \mathfrak{PrA}^{\mathrm{qf}}$ by basic arithmetical considerations.

The proof is by induction on $\mathrm{rk}_{\dot{-}}(t_1) + \mathrm{rk}_{\dot{-}}(t_2)$; the comments above effects the **induction start**.

Induction step: Let $\mathrm{rk}_{\dot{-}}(t_1) + \mathrm{rk}_{\dot{-}}(t_2) = \ell + 1$. At least one of the t_i must contain a sub-term s of the form $s_1 \dot{-} s_2$, and with with $\mathrm{rk}_{\dot{-}}(s) = 1$, i.e. s may be chosen to satisfy $\mathrm{rk}_{\dot{-}}(s_1) = \mathrm{rk}_{\dot{-}}(s_2) = 0$. Next, consider the terms defined

[6] We continue to use $\mathbf{r}$ as meta variable, ranging over $\{<, =\}$ if not otherwise specified.

by $t_i^- \stackrel{\text{def}}{=} t_i[s := s_1 - s_2]$ and $t_i^0 \stackrel{\text{def}}{=} t_i[s := \underline{0}]$, where $t_i[s := s']$ denotes the result of substituting s' for all occurrences of the sub-term s. We note that $\text{rk}_{\dot-}(t_i^-) = \text{rk}_{\dot-}(t_i^0) < \text{rk}_{\dot-}(t_i)$ for at least one i since we have removed *at least* one occurrence of $\dot-$ from one of the t_i in each construction. Thus we have $\text{rk}_{\dot-}(t_1^-) + \text{rk}_{\dot-}(t_2) = \text{rk}_{\dot-}(t_1^0) + \text{rk}_{\dot-}(t_2) \leq \ell$, and moreover

$$t_1(\vec{x}) \mathbf{\ r\ } t_2(\vec{x}) \Leftrightarrow \bigvee \begin{cases} s_1(\vec{x}) < s_2(\vec{x}) \wedge t_1^0(\vec{x}) \mathbf{\ r\ } t_2^0(\vec{x}) \\ s_2(\vec{x}) \leq s_1(\vec{x}) \wedge t_1^-(\vec{x}) \mathbf{\ r\ } t_2^-(\vec{x}) \end{cases} .$$

By the i.h., the relation $t_1(\vec{x}) \mathbf{\ r\ } t_2(\vec{x})$ is in $\mathfrak{PrA}_\star^{\text{qf}}$, since each disjunct above is a conjunction of atomic $\mathfrak{PrA}^{\dot-}$-terms of rank strictly less than $\ell + 1$. □

We are now ready to finish the proof of THEOREM 2.

Proof (of $\mathcal{D}_\star \subseteq \mathfrak{PrA}_\star^{qf}$). Let $R \in \mathcal{D}_\star^k$. Then $R = f^{-1}(0)$ for some $f \in \mathcal{D}^k$. Next, fix a $\mathfrak{PrA}^{\dot-}$-term t_f such that $\mathfrak{Z} \models t(\vec{x}) = y$ iff $f(\vec{x}) = y$. Apply LEMMA 5 to obtain $\phi \in \mathfrak{PrA}^{\text{qf}}$ satisfying $\mathfrak{N} \models \phi(\vec{x}, y) \Leftrightarrow \mathfrak{Z} \models t(\vec{x}) = y$. But then also $\vec{x} \in f^{-1}(0) \Leftrightarrow \mathfrak{N} \models \phi(\vec{x}, \underline{0})$, and thus $f^{-1}(0) = R_\phi \in \mathfrak{PrA}_\star^{\text{qf}}$. □

Considering that no increasing functions are available in $\mathcal{D}$, and perhaps even more striking, *only* composition is allowed, the class $\mathcal{D}$ really delivers more than one—at a first glance—would expect[7].

It is also of interest that when only composition is available, none of the standard linear initial functions add anything to $\mathcal{D}_\star$. More precisely

Theorem 3. $[\mathcal{I} \cup \mathcal{N} \cup \{\mathsf{max}, \mathsf{min}, \mathsf{C}, \mathsf{S}, \mathsf{P}, +, \dot-\}; \text{COMP}]_\star = \mathcal{D}_\star$. □

Space does not permit a proof, but by THEOREM 2 it is sufficient to construct $\mathfrak{PrA}_\star^{\text{qf}}$ formulae representing the graphs of the functions in this class—surely the reader can supply the details.

4 The Classes $\mathcal{BD}$ and G^1

In this section we turn our attention to the class $\mathcal{BD}$, the class obtained by adding the schema of bounded minimalisation to the closure operations of the class $\mathcal{D}$. Thus, define $\mathcal{BD} \stackrel{\text{def}}{=} [\mathcal{I} \cup \mathcal{N} \cup \{\dot-\}; \text{COMP}, \text{BMIN}]$. Obviously $\mathcal{D} \subseteq \mathcal{BD}$. It is straightforward to show that LEMMA 1 is true also of $f \in \mathcal{BD}$, in other words $(\vec{x}) \leq \mathsf{max}(x_i, c_f)$ for some fixed i and c_f, and $c_f = 0$ when $f(\mathbb{N}^k)$ is infinite. Also, $\mathcal{BD}_\star$ is boolean—since functions for intersection and complement are already in $\mathcal{D}$—and closed under quantifiers bounded by a variable—since all classes $\mathcal{F}_\star$ where $\mathcal{F}$ is closed under BMIN share this feature.

Define $\mathfrak{PrA}^{\Delta_V}$ by $\mathfrak{PrA}^{\text{qf}} \subseteq \mathfrak{PrA}^{\Delta_V}$, and $\phi \in \mathfrak{PrA}^{\Delta_V} \Rightarrow \exists_{z \leq y} \phi \in \mathfrak{PrA}^{\Delta_V}$. The '$V$' in Δ_V is meant to reflect the requirement that the quantified variable be bounded by a *variable*, and not a general *term*, and is also known as *finite* or *linear* quantification.

---

[7] In contrast, e.g. $[\mathcal{I} \cup \mathcal{N} \cup \{\mathsf{C}\}; \text{COMP}]_\star$ essentially consists of $\emptyset$, $\{0\}$, $\mathbb{N} \setminus \{0\}$ and $\mathbb{N}$, (and their products), and most other familiar functions, like S, P or $+$, even fail to produce a boolean set of relations.

Lemma 6. *If $f \in \mathcal{BD}$, then $\Gamma_f \in \mathfrak{PrA}_\star^{\Delta\vee}$.*

Proof. By induction on f. That $\mathcal{D}_\star = \mathfrak{PrA}_\star^{\mathrm{qf}}$ effects the **induction start**.

Case $h \circ \vec{g}$: Let i_j (for $1 \leq j \leq m$) be the strict top index of g_j if $g_j(\mathbb{N})$ is infinite, and $\max g_j(\mathbb{N})$ otherwise. Fix $\mathfrak{PrA}^{\Delta\vee}$-formulae $\phi_h(\vec{z}, y)$ and $\phi_j(\vec{x}, z_j)$ representing the graphs of h and the g_j's respectively. Then

$$(h \circ \vec{g})(\vec{x}) = y \;\Leftrightarrow\; \exists_{z_1 \leq t_1} \cdots \exists_{z_m \leq t_m} \Big(\Big(\bigwedge_{1 \leq i \leq m} \phi(\vec{x}, z_i) \Big) \wedge \phi_h(\vec{z}, y) \Big) \;,$$

where $t_j = x_{i_j}$ for unbounded g_j, and i_j else. Clearly, quantification bounded by a constant is merely a finite disjunction, and as such even $\mathfrak{PrA}_\star^{\mathrm{qf}}$ is closed under $\exists_{x \leq c}$ type quantifiers.

Case $\mu_{z \leq y}[g_1 = g_2]$: Let ϕ_i represent $g_i(\vec{x}) = y$, and define

$$\phi(\vec{x}, y, w) \;\Leftrightarrow\; \bigvee \begin{cases} w = y \;\wedge\; \neg \exists_{z \leq y} (\phi_1(\vec{x}, z) \wedge \phi_2(\vec{x}, z)) \\ \exists_{z \leq y} \Big(\bigwedge \begin{cases} w = z \wedge \phi_1(\vec{x}, z) \wedge \phi_2(\vec{x}, z) \\ \forall_{u \leq z} (\neg \phi_1(\vec{x}, u) \vee \neg \phi_2(\vec{x}, u) \vee u \neq z) \end{cases} \Big) \end{cases}$$

Then $\phi \in \mathfrak{PrA}^{\Delta\vee}$, and $\mathfrak{N} \models \phi(\vec{x}, y, w) \;\Leftrightarrow\; f(\vec{x}, y) = w$ as required. □

Corollary 1. $\mathcal{D}_\star \subsetneq \mathcal{BD}_\star = \mathfrak{PrA}_\star^{\Delta\vee}$. □

The class G^1 is defined by $G^1 \stackrel{\text{def}}{=} [\mathcal{I} \cup \{0, \mathsf{S}, +\}; \text{COMP}, \text{BMIN}]$, and Harrow proved that $G_\star^1 = \mathfrak{PrA}_\star$. Our next theorem is

Theorem 4. $\mathcal{BD}_\star = G_\star^1 = \mathfrak{PrA}_\star$.

We obtain THEOREM 4 via the original proof that the theory of $\mathfrak{PrA}$ is decidable. In 1930 Mojżes Presburger demonstrated this now well-known fact in[8] *Über die Vollständigket eines gewissen Systems der Arithmetik ganzer Zahlen, in welchem die Addition als einzige Operation hervortritt* [15] by proving that the theory of the intended structure $\mathfrak{N}^{\equiv} = (\mathbb{N}, 0, \mathsf{S}, +, <, \equiv_2, \equiv_3, \ldots)$ for the language $\mathfrak{PrA}^{\equiv} \stackrel{\text{def}}{=} \mathfrak{PrA} \cup \{\equiv_2, \equiv_3, \ldots\}$ does admit quantifier elimination. In particular, for any $\mathfrak{PrA}$-formula ϕ, there is a $\phi' \in \mathfrak{PrA}^{\equiv,\mathrm{qf}}$ such that

$$\mathfrak{N}^{\equiv} \models \phi' \;\Leftrightarrow\; \mathfrak{N}^{\equiv} \models \phi \;\Leftrightarrow\; \mathfrak{N} \models \phi \;.$$

The relation 'congruence modulo λ', is definable by

$$x \equiv_\lambda y \;\Leftrightarrow\; \exists_{u \leq x} \big(x = y + \underbrace{u + \cdots + u}_{\lambda\text{-terms}} \big) \vee \exists_{u \leq y} \big(y = x + \underbrace{u + \cdots + u}_{\lambda\text{-terms}} \big) \;.$$

Since the right side is clearly $\mathfrak{PrA}_\star^{\Delta\vee}$, by COROLLARY 1 these predicates belong to $\mathcal{BD}_\star$ for all $\lambda \in \mathbb{N}$. But we can do even better.

---

[8] The author has consulted the excellent translation by D. Jaquette, *On the Completeness of a Certain System of Arithmetic of Whole Numbers in Which Addition Occurs as the Only Operation* [16].

Lemma 7. *Let $p, q \in \mathbb{N}[\vec{x}]$ be linear polynomials, and let $\lambda \in \mathbb{N}$. Then, the relation $p(\vec{x}) \equiv_\lambda q(\vec{x})$ is in $\mathcal{BD}_\star$.*

Proof. First note that the unary function $\mathsf{rem}_\lambda(x) \stackrel{\text{def}}{=} \mathsf{rem}(x, \lambda) \in \mathcal{BD}$, for each fixed $\lambda \in \mathbb{N}$, since $\mathsf{rem}(x, \lambda) = \mu_{z \leq \lambda}[\chi_{\equiv_\lambda}(x, z) = 0]$.

Write $p(\vec{x}) = \sum_{i=1}^{k_p} a_i^p x_i + m_p$. Set $A_p = \sum_{i=1}^{k_p} a_i$, and note that A_p is independent of $\vec{x}$. Also, since $\mathsf{rem}_\lambda(x) < \lambda$, we have

$$s^p(\vec{x}) = \sum_{i=1}^{k_p} a_i \, \mathsf{rem}_\lambda(x_i) < A_p \lambda \ .$$

Similarly for $q(\vec{x})$. Then $p(\vec{x}) \equiv_\lambda q(\vec{x}) \ \Leftrightarrow \ s^p(\vec{x}) \equiv_\lambda s^q(\vec{x})$.

It remains to show that the relation $s^p(\vec{x}) \equiv_\lambda s^q(\vec{x})$ is in $\mathcal{BD}_\star$. Since bounded addition is in $\mathcal{D}$, we also have $\hat{s}^p(\vec{x}, z) \stackrel{\text{def}}{=} \mathsf{min}(s^p(\vec{x}), z)$ in $\mathcal{BD}$. But then, for $A = \mathsf{max}(\lambda A_p, \lambda A_q)$, we have $p(\vec{x}) \equiv_\lambda q(\vec{x}) \ \Leftrightarrow \ \hat{s}^p(\vec{x}, A) \equiv_\lambda \hat{s}^q(\vec{x}, A)$. □

Because LEMMA 7 yields a decision procedure for all atomic $\mathfrak{PrA}^{\equiv}$-formulae within $\mathcal{BD}_\star$, and since $\mathcal{BD}_\star$ is boolean, we have $\mathfrak{PrA}_\star^{\equiv,\mathrm{qf}} \subseteq \mathcal{BD}_\star$. Whence, via Presburger's original results

$$\mathcal{BD}_\star \supseteq \mathfrak{PrA}_\star^{\equiv,\mathrm{qf}} = \mathfrak{PrA}_\star \supseteq \mathfrak{PrA}_\star^{\Delta_V} = \mathcal{BD}_\star \ ,$$

which constitutes a proof of THEOREM 4. Note that this also proves that

Corollary 2. $\mathfrak{PrA}_\star^{\Delta_V} = \mathfrak{PrA}_\star$. □

5 The Classes $\mathcal{BDD}$ and G^2

Let $\mathfrak{PA} \stackrel{\text{def}}{=} \mathfrak{PrA} \cup \{\times\}$, be the language of *Peano Arithmetic*, and adopt the notations of the previous sections. The definition of the class G^2, minted in our notation, is $G^2 \stackrel{\text{def}}{=} [\mathcal{I} \cup \mathcal{N} \cup \{0, \mathsf{S}, +, \times\}; \text{COMP}, \text{BMIN}]$. Harrow proved that $\mathfrak{PA}_\star^{\Delta_0} = G_\star^2$, and that $\mathfrak{PA}_\star^{\Delta_0} = \mathfrak{PA}_\star^{\Delta_V}$. Off course $\mathfrak{PA}_\star^{\Delta_0} \subsetneq \mathfrak{PA}_\star$.

Let the class $\mathcal{BDD} \stackrel{\text{def}}{=} [\mathcal{I} \cup \mathcal{N} \cup \{\dot{-}, \lfloor \dot{:} \rfloor, \mathsf{rem}\}; \text{COMP}, \text{BMIN}]$; informally $\mathcal{BDD}$ is obtained by adding integer division and remainder to $\mathcal{BD}$. A natural question to ask is whether the inclusion of these new functions, which together make for an inverse to multiplication, makes $\mathcal{BDD}$ to G^2 what $\mathcal{BD}$ is to G^1.

The answer is yes.

Theorem 5. $\mathcal{BDD}_\star = \mathfrak{PA}_\star^{\Delta_0} = G_\star^2$. □

We only sketch the proof. Observe that

$$xy = z \ \Leftrightarrow \ (y > 0 \wedge x = \frac{z}{y}) \vee \phi \ \Leftrightarrow \ (y > 0 \wedge \mathsf{rem}(z, y) = 0 \wedge \left\lfloor \frac{z}{y} \right\rfloor = x) \vee \phi \ ,$$

where $\phi(x, y, z)$ is e.g. $y = 0 \wedge z = 0$, and so the graph of multiplication is computable in $\mathcal{BDD}_\star$.

The class $\mathcal{CA}$ of *constructive arithmetic* predicates as introduced by Smullyan (see [17]), is defined as the closure of the graphs of addition and multiplication under explicit transformations, boolean operations and quantification bounded by a variable. Hence $\mathcal{CA} \subseteq \mathcal{BDD}_\star$. It is straightforward to prove that $\mathcal{BDD}_\star \subseteq \mathfrak{PA}_\star^{\Delta_0}$, by a lemma analogous to LEMMA 6, especially since here we need not worry about the Δ_V-restriction and simply define the graphs of $\mathcal{BDD}$-functions within the framework of $\mathfrak{PA}_\star^{\Delta_0}$ (easy).

This is sufficient, since results by Bennett, Wrathall and Lipton in [1,18,14] imply the nontrivial identities $\mathcal{CA} \overset{\text{Ben}}{=} \mathcal{RUD} \overset{\text{Wra}}{=} \mathsf{LH} \overset{\text{Lip}}{=} \mathfrak{PA}_\star^{\Delta_0}$, which completes a proof of THEOREM 5. (Here $\mathcal{RUD}$ are Smullyan's *rudimentary* relations, and LH the *linear hierarchy*.)

Acknowledgments and final remarks. The author would like to thank Lars Kristiansen for valuable discussions, and the referees for their thorough reading and commenting on the article. In particular, SECTION 5 is new to the final version of this paper, and materialised after one of the referees pointed in the direction of Lipton [14]. The correct reference to the identity $\mathcal{CA} = \mathfrak{PA}_\star^{\Delta_0}$ proved invaluable. This theorem is widely—and falsely—attributed to Harrow, and the surrounding confusion initially stalled the proof of THEOREM 5.

Note however that it is not clear if $\mathcal{DD}_\star$ (where $\mathcal{DD}$ is '$\mathcal{BDD}$ without BMIN') is equal to $\mathfrak{PA}_\star^{\text{qf}}$, and so the question of whether the analogue to THEOREM 2 is true is an open problem. $\mathcal{DD}_\star^{\text{qf}} \subseteq \mathfrak{PA}_\star^{\text{qf}}$ is obvious, but the converse inclusion might very well be proper; expressing general polynomial equations in $\mathcal{BDD}_\star$ appears to make use of quantifiers in an essential way.

References

1. Bennett, J.H.: On Spectra, Ph.D. thesis, Princeton University (1962)
2. Clote, P.: Computation Models and Function Algebra. In: Handbook of Computability Theory, Elsevier, Amsterdam (1996)
3. Enderton, H.B.: A mathematical introduction to logic. Academic Press, Inc., San Diego (1972)
4. Grzegorczyk, A.: Some classes of recursive functions. In: Rozprawy Matematyczne, vol. IV, Warszawa (1953)
5. Harrow, K.: Sub-elementary classes of functions and relations, Ph.D.Thesis, New York University (1973)
6. Harrow, K.: Small Grzegorczyk classes and limited minimum. Zeitscr. f. math. Logik und Grundlagen d. Math. 21, 417–426 (1975)
7. Jones, N.D.: Logspace and PTIME characterized by programming languages. Theoretical Computer Science 228, 151–174 (1999)
8. Jones, N.D.: The expressive power of higher-order types or, life without CONS. J. Functional Programming 11, 55–94 (2001)
9. Kristiansen, L.: Neat function algebraic characterizations of logspace and linspace. Computational Complexity 14(1), 72–88 (2005)
10. Kristiansen, L.: Complexity-Theoretic Hierarchies Induced by Fragments of Gödel's T: Theory of Computing Systems (2007), `http://www.springerlink.com/content/e53154627x063685/`

11. Kristiansen, L., Barra, M.: The small Grzegorczyk classes and the typed λ-calculus. In: Cooper, S.B., Löwe, B., Torenvliet, L. (eds.) CiE 2005. LNCS, vol. 3526, pp. 252–262. Springer, Heidelberg (2005)
12. Kristiansen, L., Voda, P.J.: The surprising power of restricted programs and Gödel's functionals. In: Baaz, M., Makowsky, J.A. (eds.) CSL 2003. LNCS, vol. 2803, pp. 345–358. Springer, Heidelberg (2003)
13. Kristiansen, L., Voda, P.J.: Complexity classes and fragments of C. Information Processing Letters 88, 213–218 (2003)
14. Lipton, R.J.: Model theoretic aspects of computational complexity. In: Proc. 19th Annual Symp. on Foundations of Computer Science, Silver Spring MD, pp. 193–200. IEEE Computer Society, Los Alamitos (1978)
15. Presburger, M.: Über die Vollständigket eines gewissen Systems der Arithmetik ganzer Zahlen, in welchem die Addition als einzige Operation hervortritt. In: Sprawozdanie z I Kongresu Matematyków Słowańskich, pp. 92–101 (1930)
16. Presburger, M., Jaquette, D.: On the Completeness of a Certain System of Arithmetic of Whole Numbers in Which Addition Occurs as the Only Operation. History and Philosophy of Logic 12, 225–233 (1991)
17. Smullyan, R.M.: Theory of formal systems (revised edition). Princeton University Press, Princeton, New Jersey (1961)
18. Wrathall, C.: Rudimentary predicates and relative computation. SIAM J. Comput. 7(2), 194–209 (1978)

Finding Minimum 3-Way Cuts in Hypergraphs

Mingyu Xiao

Department of Computer Science and Engineering
The Chinese University of Hong Kong
myxiao@cse.cuhk.edu.hk

Abstract. The minimum 3-way cut problem in an edge-weighted hypergraph is to find a partition of the vertices into 3 sets minimizing the total weight of hyperedges with at least two endpoints in two different sets. In this paper we present some structural properties for minimum 3-way cuts and design an $O(dmn^3)$ algorithm for the minimum 3-way cut problem in hypergraphs, where n and m are the numbers of vertices and edges respectively, and d is the sum of the degrees of all the vertices. Our algorithm is the first deterministic algorithm finding minimum 3-way cuts in hypergraphs.

1 Introduction

Given a hypergraph $G = (V, E)$ with nonnegative hyperedge weights, and an integer k, a *k-way cut* for G is a subset of hyperedges whose deletion separates the graph into k nonempty components, and the *minimum k-way cut problem* is to find a k-way cut minimizing the total weight of it. In the literature k-way cuts are also referred as *k-cuts* or *multi-component cuts*. The minimum k-way cut problem is an extension of the classical *minimum cut* problem, and it has great applications in the area of VLSI system design, parallel computing systems, clustering, network reliability and finding cutting planes for the travelling salesman problems.

The minimum k-way cut problem in graphs was well studied during the past twenty years. First, Goldschmidt and Hochbaum [5] proved that this problem is NP-hard when k is part of the input and presented a polynomial algorithm for fixed k with running time $O(n^{k^2}T(n,m))$, where $T(n,m)$ is the time required to compute a minimum cut or maximum flow in an edge-weighted graph. Recently Kamidoi et al. [8] improved the running time bound to $O(n^{4k/(1-1.71/\sqrt{k})-34}$ $T(n,m))$. Karger and Stein [10] proposed a Monte Carlo algorithm with $O(n^{2(k-1)}$ $\log^3 n)$ time. It seems that all the above algorithms do not work in hypergraphs and no solid results in hypergraphs for $k > 2$ are known. When k is a small number, the problem has also drawn much attention.

The minimum 2-way cut problem is commonly known as the *minimum cut* problem. Another version of the minimum 2-way cut problem is the *minimum (s,t) cut* problem, which asks us to find a minimum cut that splits two given vertices s and t. These two problems are classical and fundamental problems in the subject of graph connectivity. For graphs, the minimum cut problem can

M. Agrawal et al. (Eds.): TAMC 2008, LNCS 4978, pp. 270–281, 2008.

be solved in $O(mn + n^2 \log n)$ time by Nagamochi and Ibaraki's algorithm [17] or Stoer and Wagner's algorithm [21], and the minimum (s,t) cut problem can be solved in $O(mn \log n^2/m)$ time by Goldberg and Tarjan's algorithm [4] and $O(\min\{n^{2/3}, m^{1/2}\}m \log(n^2/m) \log U)$ time by Goldberg and Rao's algorithm [3], where U is the maximum capacity of the edge. For hypergraphs, there are also some good results on them. Klimmek and Wagner [13] and Mak and Wong [16] extended Stoer and Wagner's algorithm [21] to hypergraphs and gave an $O(dn + n^2 \log n)$ algorithm for the minimum cut problem in hypergraphs, where d is the sum of the degrees of all the vertices. Lawler [14] showed that a minimum (s,t) cut in a hypergraph can be computed by using one maximum flow computation in an auxiliary digraph with $n+2m$ vertices and $2d+m$ edges. Then a minimum (s,t) cut in a hypergraph can be found in $\widetilde{O}(dm)$ time. In the remainder of the paper, we use $T(n,m)$ and $T(n,m,d)$ to denote the running times of computing a minimum (s,t) cut in a graph and a hypergraph respectively.

For the minimum 3-way cut problem in graphs, Kapoor [9] and Kamidoi et al. [7] showed that it can be solved by using $O(n^3)$ maximum flow computations. Burlet and Goldschmidt [1] and Nagamochi and Ibaraki [18] improved the result to $O(n^2)$. Furthermore, Nagamochi et al. [18], [19] proved that the minimum k-way cut problem can be solved in $\widetilde{O}(mn^k)$ time for $k = 4, 5, 6$. Unfortunately, we do not find many results in hypergraphs for $k \geq 3$. In fact, Nagamochi et al. [18] considered the minimum $\{3,4\}$-way cuts in hypergraphs, but their measure is different, which makes their problem easier. Currently the frequently used algorithms in VLSI system design to partition a hypergraph are some heuristic algorithms without any theoretic guarantee, such as hMETIS [11] and other algorithms based on multilevel partitioning framework [12].

Effective algorithms for minimum 3-way cuts in hypergraphs have potential applications in VLSI system design. A deeper understanding of the structure of 3-way cuts will help us investigating the structure of the general k-way cut problem. Motivated by these, in this paper we study the minimum 3-way cut problem in hypergraphs.

Roughly speaking, there are three techniques frequently used to design algorithm for finding minimum 3-way cuts, as well as minimum k-way cuts, in graphs. The first one is based on searching minimum (S,T) cuts. The first algorithm for the general k-way cut problem presented by Goldschmidt and Hochbaum [5] is an example. They proved that there are 4 vertices a, b, c, d such that the minimum $(\{a,b\},\{c,d\})$ cut is contained in a minimum 3-way cut. If we try all the $O(n^4)$ possibilities by taking each one as a subset of a 3-way cut and finding a such 3-way cut with weight minimized, then we can get a minimum 3-way cut by selecting a lightest one among all the $O(n^4)$ 3-way cuts. Later Kapoor [9] improved this method to $O(n^3)$ maximum flow computations. The second technique is enumerating all small 2-way cuts (We will simply call a 2-way cut a cut). We sort all cuts in the graph in the order of nondecreasing weights. If we try on each cut by taking it as a subset of a 3-way cut, finally we will meet a cut contained in a minimum 3-way cut and then get a minimum 3-way cut. Nagamochi et al. [18] proved that at least one of the first $O(n)$ small cuts is contained in a minimum 3-way cut and the first $O(n)$ cuts can be enunciated by using $O(n^2)$ maximum flow computations. So a minimum 3-way cut can be found in $O(n^2T(n,m))$ time. Levine [15] proved that a minimum 3-way cut can be found by considering the first $O(n)$ small cuts of all the 4/3-minimum cuts in 3 graphs. Since Karger and Stein's algorithms [10] can find all the 4/3-minimum cuts and a minimum cut in $O(n^2 \log n)$ and $O(m \log^3 n)$ time with high probability respectively, Levine got a Monte Carlo algorithm finding minimum 3-way cuts with time bound $O(mn \log^3 n)$. The last technique is based on 'divide-and-conquer', such as Kamidoi et al.'s algorithm [7]. We first find a proper cut of the graph that is noncrossing with a minimum 3-way cut, then we can find minimum 3-way cuts in two smaller graphs. The hard part of this method is that the proper cut is not always easy to get and sometimes we will turn to the second technique to find one.

All of the above algorithms can not be extended to the problem in hypergraphs directly. We look at one of the fastest algorithms for finding minimum 3-way cuts in graphs presented by Nagamochi et al. [18]. Their algorithm is based on the observation that at least one of the first r cuts $\{C_1, \cdots, C_r\}$ in the order of nondecreasing weights is contained in a minimum 3-way cut, where r is the first integer such that C_r is crossing with C_q $(1 \leq q < r)$. We give an example in Figure 1 to show that this theorem does not hold in hypergraphs. $G = (V, E)$ is a hypergraph with 6 vertices and 6 hyperedges, $V = \{a, b, c, d, e, f\}$, $E = \{abc, cdef, cd, de, ef, cf\}$, and $w(abc) = 4.5, w(cdef) = 2, w(cd) = w(de) = w(ef) = e(cf) = 1$. $[\{a, b, c, d\}, \{e, f\}]$ and $[\{a, b, c, f\}, \{d, e\}]$ are the first two small cuts and they are crossing, but none of them is contained in the minimum 3-way cut $[\{a\}, \{b\}, \{c, d, e, f\}]$. In this paper, based on the idea of Kamidoi et al.'s divided-and-conquer algorithm [7], we define a class of cuts called *k-way free cuts* (See Definition 3) and present a framework of designing algorithms for the minimum k-way cut problem in hypergraphs by finding k-way free cuts. Then we prove that minimum 3-way cuts in hypergraphs can be found by using $O(n^3)$ hypergraph minimum (s, t) cut computations.

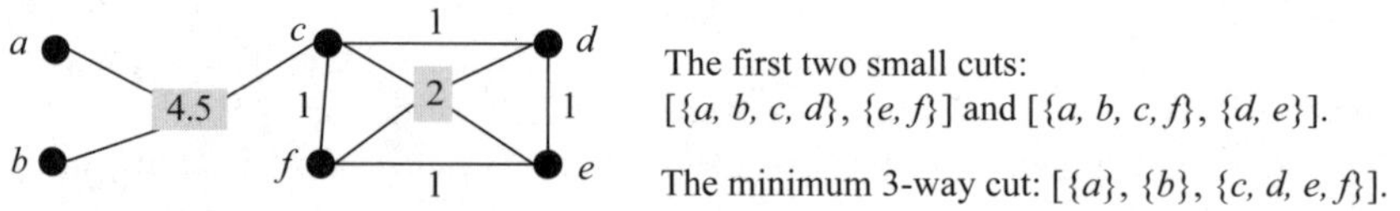

Fig. 1. The first two small cuts not being contained in a minimum 3-way cut

2 Preliminaries

Let $G = (V, E)$ be an edge-weighted hypergraph with $|V| = n$ vertices and $|E| = m$ hyperedges. For any hyperedge $e \in E$, $V(e)$ denotes the set of endpoints of e. An *induced sub hypergraph* $H = G[V']$ with $V' \subseteq V$ is a sub hypergraph with hyperedges whose endpoints in V'. For any hyperedge subset $C \subseteq E$, $w(C)$ denotes the total weight of the hyperedges in C. Let $X_1, X_2, \cdots, X_k \subset V$ be k $(2 \leq$

$k \leq n$) disjoint nonempty subsets of vertices, then $[X_1, X_2, \cdots, X_k]$ denotes the set of hyperedges crossing at least two different vertex sets of $\{X_1, X_2, \cdots, X_k\}$ (Hyperedges that the endpoints of each of them are in at least two different sets of $\{X_1, X_2, \cdots, X_k\}$). We also define another important hyperedge set: $e(X_1, X_2, \cdots, X_k) = \{e | e \in [X_1, X_2, \cdots, X_k] \& V(e) \subseteq \bigcup_{i=1}^{k} X_i\}$. There is an illustration for these two notations in Figure 2. Let $V = \{a, b, c, d, e\}$ and $E = \{ab, abc, abe, ae, bde, cd\}$. Then $[\{a, b\}, \{c\}, \{d\}] = \{abc, bde, cd\}$ and $e(\{a, b\}, \{c\}, \{d\}) = \{abc, cd\}$. When $\bigcup_{i=1}^{k} X_i = V$, i.e., $\{X_1, X_2, \cdots, X_k\}$ is k-partition of V, we say $[X_1, X_2, \cdots, X_k]$ is a *k-way cut* of the hypergraph and X_i is a *component* of the k-way cut. Over all the k-way cuts those with the minimum weight are called *minimum k-way cuts.* A 2-way cut $[X, \overline{X}]$ is also simply called a *cut* of the hypergraph, where $\overline{X} = V - X$. If we require a special vertex s in X and a special vertex t in $\overline{X}$, then such kind of cuts are called (s, t) cuts. Generally for any two disjoint sets of vertices S and T, a group of hyperedges is called an *(S, T) cut*, if deleting it splits S away from T. *Merging* some vertices means identifying these vertices as a new vertex while keeping all the hyperedges incident on them (we also need to update the hyperedges: combining 'parallel' hyperedges with weight being the sum of all the weights and deleting self-loops). For a vertex subset $X \subset V$, G_X stands for the graph generated by merging $\overline{X} = V - X$ into a single vertex. This notation is frequently used in this paper. Sometimes a singleton set $\{s\}$ is simply written as s, $w([X_1, X_2, \cdots, X_k])$ as $w(X_1, X_2, \cdots, X_k)$, and $w(e(X_1, X_2, \cdots, X_k))$ as $w_e(X_1, X_2, \cdots, X_k)$.

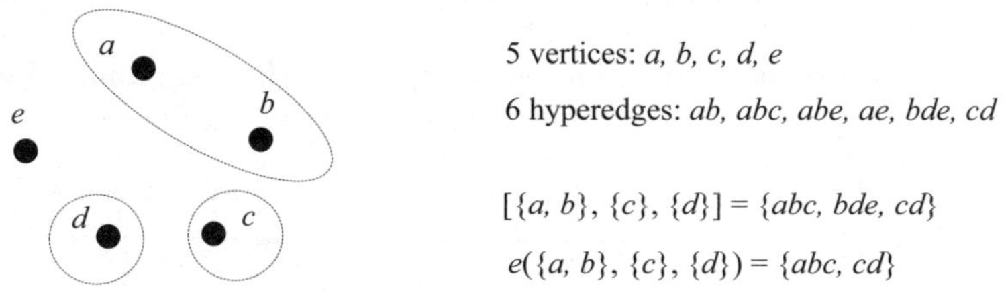

Fig. 2. An illustration for notations [] and $e()$

Definition 1. *Two cuts $[X, \overline{X}]$ and $[Y, \overline{Y}]$ are noncrossing if $X \subseteq Y$ or $\overline{X} \subseteq Y$. Otherwise, we say that they are crossing.*

Note that when $X = Y$ or $\overline{X} = Y$, the two cuts are the same. Some references exclude this case in their definition of noncrossing. The properties of noncrossing between two cuts are studied in many references [6], [2]. Here we extend the definition of noncrossing to describe a relation between a cut and a k-way cut.

Definition 2. *A cut $[X, \overline{X}]$ and a k-way cut $[Y_1, Y_2, \cdots, Y_k]$ are noncrossing if there exists $i \in \{1, 2, \cdots, k\}$ such that $X \subseteq Y_i$ or $\overline{X} \subseteq Y_i$. Otherwise, we say that they are crossing.*

Lemma 1. *Given a hypergraph G and two vertices u and v in G, let G' be the hypergraph after merging u and v into a new vertex, C'_k be a minimum k-way cut of G' and C_k be the corresponding hyperedge set of C'_k in G. If there is a minimum k-way cut C of G such that u and v are in a same component of C, then C_k is also a minimum k-way cut of G.*

Proof. Let C' be the corresponding hyperedge set of C in G'. u and v are in a same component of C, so C' is a k-way cut of G'. C'_k is a minimum k-way cut of G', and then $w(C_k) = w(C'_k) \leq w(C') = w(C)$. Thus, C_k is a minimum k-way cut of G.

Lemma 2. *Given a hypergraph G and a cut $C = [X, \overline{X}]$ of G, let C_X and $C_{\overline{X}}$ be a minimum k-way cut of G_X and $G_{\overline{X}}$ respectively, where G_X is the graph obtained by merging $\overline{X}$ into a single vertex and $G_{\overline{X}}$ is the graph obtained by merging X into a single vertex. And let C'_X and $C'_{\overline{X}}$ be the corresponding edge sets of C_X and $C_{\overline{X}}$ in G respectively. If there is a minimum k-way cut noncrossing with cut C, then either C'_X or $C'_{\overline{X}}$ is a minimum k-way cut of G.*

Proof. Assume that minimum k-way cut $C_k = [Y_1, Y_2, \cdots, Y_k]$ is noncrossing with cut C. Then there exists $i \in \{1, 2, \cdots, k\}$ such that $X \subseteq Y_i$ or $\overline{X} \subseteq Y_i$. According to Fact 1, when $X \subseteq Y_i$, we can merge X into a single vertex; when $\overline{X} \subseteq Y_i$, we can merge $\overline{X}$ into a single vertex. Thus, the lemma holds.

Lemma 2 provides a prospective divide-and-conquer method to find a minimum k-way cut in hypergraphs. If we get a proper cut which is noncrossing with a minimum k-way cut and each component of the cut contains at least two vertices, then we can find a minimum k-way cut by searching on two smaller graphs. This idea is one of the main ideas we used to find minimum 3-way cuts in this paper. For convenience, we give the following definition.

Definition 3. *A cut is k-way free, if there exists a minimum k-way cut noncrossing with it.*

Note that k-way free is just used to describe cuts, but not for k-way cuts for $k \geq 3$. To find a minimum k-way cut, we need to find proper k-way free cuts. In Section 3.2, we first present some structural properties of minimum 3-way cuts, which provide us some ways to get 3-way free cuts. The detailed algorithm and running time analysis are presented in Section 4. In Section 3.1, we first give some properties for hyperedges, which will be used in our main proofs.

3 Properties

3.1 Hyperedge's Properties

Lemma 3. *Let $\{X_1, X_2, \cdots, X_k\} \bigcup \{Y_1\}$ be a group of disjoint nonempty subsets of vertices in a hypergraph, then*

$$[X_1 + Y_1, X_2, \cdots, X_k] \supseteq [X_1, X_2, \cdots, X_k] + e(Y_1, X_2 + \cdots + X_k).$$

If $\{X_1, X_2, \cdots, X_k\} \bigcup \{Y_1\}$ *is a partition of the vertex set, then*

$$[X_1 + Y_1, X_2, \cdots, X_k] = [X_1, X_2, \cdots, X_k] + e(Y_1, X_2 + \cdots + X_k).$$

Lemma 4. *Let* $\{X_1, X_2, \cdots, X_k\} \bigcup \{Y_1, Y_2\}$ *be a group of disjoint nonempty subsets of vertices in a hypergraph, then*

$$[X_1+Y_1, X_2+Y_2, X_3, \cdots, X_k] \supseteq [X_1, \cdots, X_k]+e(Y_1+Y_2, X_3+\cdots+X_k)+e(Y_1, X_2)+e(X_1, Y_2).$$

Fact 3 and Fact 4 can be simply proved by enumerating all kinds of hyperedges and we just ignore the detailed proof steps here. These two facts are frequently used in the proofs in Section 3.2, and to make the proofs fluent we do not point out which fact being used each time. In fact, we only use the cases $k = 2$ and 3, which are somewhat intuitive.

3.2 Structural Properties

Lemma 5. *In a hypergraph, any minimum cut is 3-way free.*

Proof. We need to prove that for any minimum cut $C = [X, \overline{X}]$ of G, there is a minimum 3-way cut C_3 such that C and C_3 are noncrossing. Let $C_3' = [Y_1, Y_2, Y_3]$ be an arbitrary minimum 3-way cut, and $Z_{ij} = X_i \bigcap Y_j$, $(i = 1, 2, j = 1, 2, 3)$, where $X_1 = X$ and $X_2 = \overline{X}$ (See Figure 3). If C and C_3' are crossing, then $\exists j \in \{1, 2, 3\}$ such that Z_{1j} and Z_{2j} are nonempty sets. Without loss of generality, we assume that they are Z_{11} and Z_{21}. At least one of Z_{13} and Z_{23} is not an empty set. Without loss of generality, we further assume that $Z_{23} \neq \emptyset$.

Case 1: $Z_{22} \neq \emptyset$.
We will prove that $C_3 = [Y_1 + Z_{12} + Z_{13}, Z_{22}, Z_{23}]$ is a satisfied minimum 3-way cut (See Figure 3). It is easy to see that C_3 is a 3-way cut noncrossing with C. We only need to prove that C_3 is minimum.

$[Z_{11}, \overline{Z_{11}}] = [Z_{11}, X_2] + e(Z_{11}, Z_{12} + Z_{13})$ and $[X_1, X_2] = [Z_{11}, X_2] + e(Z_{12} + Z_{13}, X_2)$. $[X_1, X_2]$ is a minimum cut. $w(Z_{11}, \overline{Z_{11}}) \geq w(X_1, X_2)$. So

$$w_e(Z_{11}, Z_{12} + Z_{13}) \geq w_e(Z_{12} + Z_{13}, X_2). \quad (1)$$

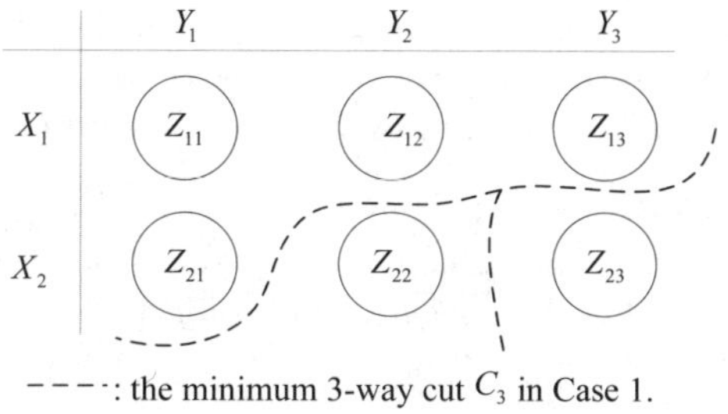

Fig. 3. Case 1 of the proof

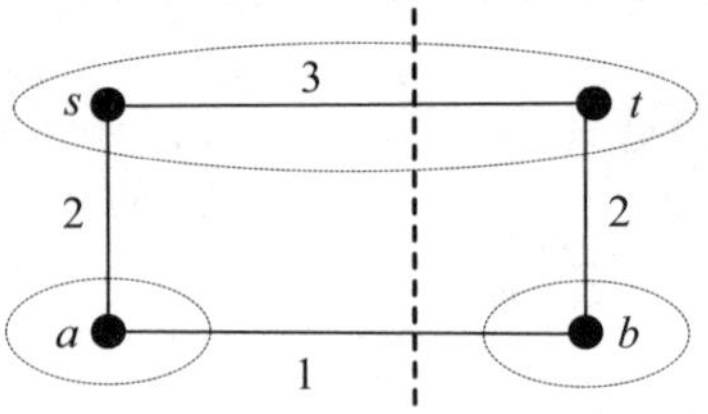

----: the minimum (s,t) cut. ◯: the minimum 3-way cut.

Fig. 4. A minimum (s,t) cut not being 3-way free

On the other hand, $C_3 = [Y_1 + Z_{12} + Z_{13}, Z_{22}, Z_{23}] = [Y_1, Z_{22}, Z_{23}] + e(Z_{12} + Z_{13}, Z_{22} + Z_{23})$ and $C'_3 = [Y_1, Y_2, Y_3] \supseteq [Y_1, Z_{22}, Z_{23}] + e(Y_1, Z_{12} + Z_{13})$. It is obvious that $e(Z_{12}+Z_{13}, Z_{22}+Z_{23}) \subseteq e(Z_{12}+Z_{13}, X_2)$ and $e(Z_{11}, Z_{12}+Z_{13}) \subseteq e(Y_1, Z_{12}+Z_{13})$. According to (1), we get that $w(C_3) \leq w(C'_3)$. C_3 is a minimum 3-way cut.

Case 2: $Z_{22} = \emptyset$.
For this case, $Z_{12} \neq \emptyset$. Let $C_3 = [Z_{11}, Z_{12}, Z_{21} + Y_3]$. Like what we do in **Case 1**, we will prove that C_3 is a satisfied minimum 3-way cut.

$[\overline{Z_{23}}, Z_{23}] = [X_1, Z_{23}] + e(Z_{21}, Z_{23})$ and $[X_1, X_2] = [X_1, Z_{23}] + e(X_1, Z_{21})$. $(\overline{Z_{23}}, Z_{23}) \geq w(X_1, X_2)$. So

$$w_e(Z_{21}, Z_{23}) \geq w_e(X_1, Z_{21}). \tag{2}$$

$C_3 = [Z_{11}, Z_{12}, Z_{21} + Y_3] = [Z_{11}, Z_{12}, Y_3] + e(Z_{11} + Z_{12}, Z_{21})$, where $e(Z_{11} + Z_{12}, Z_{21}) = e(Z_{11}, Z_{21}) + e(Z_{12}, Z_{21}) + e(Z_{11}, Z_{12}, Z_{21}) \subseteq e(Z_{12}, Z_{21}) + e(X_1, Z_{21})$. And $C'_3 = [Y_1, Y_2, Y_3] \supseteq [Z_{11}, Z_{12}, Y_3] + e(Z_{12}, Z_{21}) + e(Z_{21}, Y_3) \supseteq [Z_{11}, Z_{12}, Y_3] + e(Z_{12}, Z_{21}) + e(Z_{21}, Z_{23})$. According to (2), we know that $w(C_3) \leq w(C'_3)$.

We have discussed all the cases and then finished the proof.

Lemma 5 shows a noncrossing property between minimum cuts and minimum 3-way cuts. The noncrossing property between minimum (s,t) cuts and minimum 3-way cuts is a little weaker. Given two vertices s and t, the minimum (s,t) cut may not be 3-way free even in graphs. We give an example in Figure 4. $G = (V, E)$ is a graph with 4 vertices and 4 edges, $V = \{a, b, s, t\}$, $E = \{ab, bt, st, as\}$, and $w(ab) = 1, w(bt) = 2, w(st) = 3, w(as) = 2$. It is easy to see that $[a, b, \{s, t\}]$ is the unique minimum 3-way cut and $[\{a, s\}, \{b, t\}]$ is the unique minimum (s,t) cut in the graph. They are crossing. In the next two lemmas we show some further properties between minimum (s,t) cuts and minimum 3-way cuts.

Lemma 6. *Given a hypergraph G and two vertices s and t in it, if there is a minimum 3-way cut whose removal disconnects s from t (s and t are in two different components of the 3-way cut), then any minimum (s,t) cut is 3-way free.*

Proof. The proof is based on the proof of Lemma 5. Let $C'_3 = [Y_1, Y_2, Y_3]$ be a minimum 3-way cut, and s and t be two vertices in two different components of C_3'. Suppose $C = [X_1, X_2]$ is a minimum (s,t) cut, where $X_2 = \overline{X_1}$. Let $Z_{ij} = X_i \bigcap Y_j$, $(i = 1,2, j = 1,2,3)$. Without loss of generality, we also assume that $Z_{11}, Z_{21} \neq \emptyset$,$s \in X_1$, $t \in X_2$, and $t \in Y_3$. Then $t \in Z_{23}$ and s is in either Z_{11} or Z_{12}.

Case 1: $s \in Z_{11}$.
The proof just follows the proof of Lemma 5, because $[Z_{11}, \overline{Z_{11}}]$ and $[Z_{23}, \overline{Z_{23}}]$ are still (s,t) cuts and the two cases in the proof of Lemma 5 still work.

Case 2: $s \in Z_{12}$.
When $Z_{22} \neq \emptyset$, we exchange Y_1 and Y_2 and then it becomes **Case 1**. When $Z_{22} = \emptyset$, we only need to do **Case 2** in the proof of Lemma 5. For this case, $[Z_{23}, \overline{Z_{23}}]$ is still an (s,t) cut, so the proof is the same.

Lemma 7. *Given a hypergraph G and two disjoint vertex subsets $S, T \subset V$, if there is a minimum 3-way cut whose removal disconnects S from T, then any minimum (S,T) cut is 3-way free.*

Proof. Let $C_3' = [Y_1, Y_2, Y_3]$ be a minimum 3-way cut whose removal disconnects S from T, $C = [X_1, X_2]$ be a minimum (S,T) cut, and $Z_{ij} = X_i \bigcap Y_j$, $(i = 1,2, j = 1,2,3)$. Since removal of C_3' will disconnect S from T, either S or T is contained in a component of C_3'. Without loss of generality, we assume that $S \subseteq Y_1$. Furthermore, we assume that $S \subseteq X_1$ and $T \subseteq X_2$. Then $S \subseteq Z_{11}$ and $T \subseteq Z_{22} \bigcup Z_{23}$.

When T is also contained in a component of C_3', by Fact 1 we can safely merge S into a single vertex and merge T into a single vertex. The minimum 3-way cut in the remaining graph is still a minimum 3-way cut in the original graph. And by Lemma 6 we know that there is a minimum 3-way cut noncrossing with C. Otherwise, we let $T_1 = T \cap Y_2$ and $T_2 = T \cap Y_3$. T_1, $T_2 \neq \emptyset$. For this case, we can prove that $C_3 = [Y_1 + Z_{12} + Z_{13}, Z_{22}, Z_{23}]$ is a minimum 3-way cut noncrossing with C like what we do in **Case 1** of the proof of Lemma 5. The reason is that $T_1 \subseteq Z_{22} \neq \emptyset$ and $T_2 \subseteq Z_{23} \neq \emptyset$, and $[Z_{11}, \overline{Z_{11}}]$ is still an (S,T) cut.

If we want to use Lemma 2 to design algorithms, we first need to find a 3-free cut $[X, \overline{X}]$ such that neither X nor $\overline{X}$ only contains one vertex. The 3-way free cuts discussed above, including the minimum cuts and the minimum (s,t) cuts, may not have this property. For convenience, we give the following definition.

Definition 4. *A cut $[X, \overline{X}]$ is called a 2-size cut, if $|X| \geq 2$ and $|\overline{X}| \geq 2$. Over all the 2-size cuts those with the minimum weight are called minimum 2-size cuts.*

Lemma 8. *Let G be a hypergraph with more than 6 vertices. If G has a minimum 3-way cut such that each component of it has at least 2 vertices, then any minimum 2-size cut in G is 3-way free.*

Proof. Let the minimum 3-way cut be $C_3 = [Y_1, Y_2, Y_3]$, the minimum 2-size cut be $C = [X_1, X_2]$, and $Z_{ij} = X_i \bigcap Y_j$, $(i = 1,2, j = 1,2,3)$. We will prove the lemma by using Lemma 7.

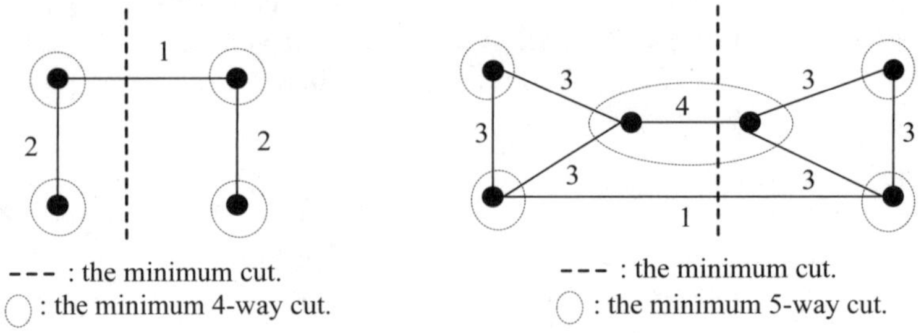

Fig. 5. Minimum cuts not being $\{4, 5\}$-way free

If there is a component of C_3 being contained in a component of C, say $Y_1 \subseteq X_1$, we can prove the lemma as the follows. Y_1 has at least two vertices a and b. X_2 has least two vertices c and d, and $c, d \in Y_2 \cup Y_3$. Let $S = \{a, b\}$ and $T = \{c, d\}$. Removal of C_3 will disconnect S from T and C is a minimum (S, T) cut. By Lemma 7 we know the lemma holds. Otherwise, none of Z_{ij} $(i = 1, 2, j = 1, 2, 3)$ is an emptyset. G has more than 6 vertices. At least one of Z_{ij} $(i = 1, 2, j = 1, 2, 3)$ contains more than one vertex. Without loss of generality, we assume that two different vertices $a, b \in Z_{11}$. $Z_{22}, Z_{23} \neq \emptyset$. Assume that $c \in Z_{22}$ and $d \in Z_{23}$. Let $S = \{a, b\}$ and $T = \{c, d\}$. We can also prove the lemma by using Lemma 7.

Remark. In Lemma 8, if graph G has right 6 vertices, the result may not hold. Here is an example. $G = (V, E)$ is a hypergraph with 6 vertices and 5 hyperedges, $V = \{a, b, c, d, e, f\}$, $E = \{ad, be, cf, abc, def\}$, and $w(ad) = w(be) = w(cf) = 2.5, w(abc) = w(def) = 4$. $[\{a, d\}, \{b, e\}, \{c, f\}]$ is the unique minimum 3-way cut and each component of it has two vertices. $[\{a, b, c\}, \{d, e, f\}]$ is the unique minimum 2-size cut. They are crossing.

Lemma 9. *A minimum cut may not be k-way free, $k \geq 4$.*

Two examples that minimum cuts are not being $\{4, 5\}$-way free are shown in Figure 5. In fact, Lemma 6, Lemma 7 and Lemma 8 also do not hold for minimum 4-way cuts.

4 The Algorithm

Using Lemma 2 and lemmas mentioned in Section 3.2, we can design various algorithms for the minimum 3-way cut problem in hypergraphs. But we should pay attention that when we meet a 3-way free cut $[X, \overline{X}]$ such that X or $\overline{X}$ contains only one vertex, then our algorithms may not work any more, because G_X or $G_{\overline{X}}$ will not be a smaller graph. At that time, we need to use Lemma 8.

Next, we use Lemma 5 and Lemma 8 to design a recursive algorithm. We iteratively find a minimum cut $C = [X, \overline{X}]$ as a 3-way free cut, and then find minimum 3-way cuts in smaller hypergraphs G_X and $G_{\overline{X}}$. When the 3-way free cut we found is not 2-size, we turn to Lemma 8. There is a 3-way free cut being 2-size or a minimum 3-way cut such that one component of it has only one vertex (Assume the graph has more than 6 vertices). For the former case, the recursive step can continue. And for the later case, we can find a solution directly by trying on each vertex. In fact, the algorithm only includes Step 1 and Step 4 of the following algorithm still works. The reason for it lies in Lemma 8. Step 2 and Step 3 can accelerate the algorithm sometimes but do not help in the running time analysis.

Algorithm Hyper3waycut(G):
0. Initially let the solution set S be the whole edge set E.
1. If the graph has less than or equal to 6 vertices, find the solution directly and return it; otherwise, do
2. Find a minimum cut $[X, \overline{X}]$ of the graph.
3. If both of X and $\overline{X}$ have more than one vertex, update $S \longleftarrow$ min $\{$**Hyper3waycut**(G_X), **Hyper3waycut**$(G_{\overline{X}})\}$, where G_X is the graph obtained by merging $\overline{X}$ into a single vertex and $G_{\overline{X}}$ is the graph obtained by merging X into a single vertex.
4. Otherwise, do
4.1 For each vertex v in the current graph, find a minimum cut C_v in induced subgraph $G[V - \{v\}]$ and update $S \longleftarrow \min\{S, C_v \cup [v, V - \{v\}]\}$.
4.2 Find a minimum 2-size cut $[X, \overline{X}]$ in the current graph and update $S \longleftarrow \min\{S,$ **Hyper3waycut**(G_X), **Hyper3waycut**$(G_{\overline{X}})\}$.
5. Return S.

In each iteration, we will find a 3-way free cut being 2-size in either Step 3 or Step 4.2 and divide the problem into two smaller problems. Now we consider how much time we will use in each iteration. First we compute a minimum cut in the hypergraph in Step 1. If one component of the cut has only one vertex, we will compute n minimum cuts in Step 4.1 and compute a minimum 2-size cut in Step 4.2. Via brute-force search, a minimum 2-size cut can be found by using $O(n^4)$ minimum (s,t) cut computations. In fact, we can improve it to $O(n^2)$ minimum (s,t) cut computations by enumerating small cuts. Vazirani and Yannakakis [22] presented an algorithm for finding all the cuts in a graph in the order of non-decreasing weights by building a 0-1 tree. And the delay between two successive outputs is at most $n-1$ minimum (s,t) cut computations. This algorithm still works in hypergraphs. A hypergraph has at most n cuts not being 2-size. So at most we need to compute the first $n+1$ outputs to get a minimum 2-size cut. We can find a minimum 2-size cut by using $O(n^2)$ minimum (s,t) cut computations in hypergraphs.

Suppose the current graph has n vertices and X has x vertices ($2 \leq x \leq n-2$). Then G_X has $x+1$ vertices and $G_{\overline{X}}$ has $n-x+1$ vertices. We get recurrence relation:

$$C(n) \leq C(x+1) + C(n-x+1) + O(n^2), \qquad (3)$$

where $C(n)$ is the number of minimum (s,t) cut computations used when **Hyper3waycut** computes on a hypergraph with n vertices. It is easy to verify that $C(n) = O(n^3)$ satisfies (3).

Theorem 1. *Algorithm* ***Hyper3waycut*** *finds a minimum* 3*-way cut in a hypergraph in* $O(n^3 T(n,m,d)) = \widetilde{O}(dmn^3)$ *time.*

5 Conclusion

In this paper, we have presented an $O(dmn^3)$ algorithm for finding minimum 3-way cuts in hypergraphs. As far as we know, it is the first deterministic algorithm for solving the minimum 3-way cut problem in hypergraphs.

k-way partitioning with minimum cost in hypergraphs is one of central problems in VLSI system design [20]. Effective algorithms for small k may have direct applications. But unfortunately most properties and algorithms for k-way cuts in graphs do not hold in hypergraphs, and currently there are not many known algorithms for hypergraphs except some heuristic algorithms without any theoretic guarantee. It would be interesting to find other fast algorithms for the k-way cut problem in hypergraphs, even when k is a small number.

Furthermore, to prove the correctness of our algorithm, we present some properties of hypergraphs and minimum k-way cuts, which may be helpful for us to understand the structure of the minimum k-way cuts and differences between the k-way cut problems in graphs and hypergraphs. We note that some good properties for 3-way cuts do not hold for 4-way cut cases, and the algorithm presented in this paper can not be simply extended to the minimum 4-way cut problem in hypergraphs. It seems that other approaches or properties are needed for the minimum k-way cut problem for $k \geq 4$.

References

1. Burlet, M., Goldschmidt, O.: A new and improved algorithm for the 3-cut problem. Operations Research Letters 21(5), 225–227 (1997)
2. Easley, R.F., Hartvigsen, D.: Crossing properties of multiterminal cuts. Networks 34(3), 215–220 (1999)
3. Goldberg, A.V., Rao, S.: Beyond the flow decomposition barrier. J. ACM. 45(5), 783–797 (1998): A preliminary version appeared in FOCS 1997
4. Goldberg, A.V., Tarjan, R.E.: A new approach to the maximum-flow problem. J. ACM 35(4), 921–940 (1988)
5. Goldschmidt, O., Hochbaum, D.: A polynomial algorithm for the k-cut problem for fixed k. Mathematics of Operations Research 19(1), 24–37 (1994): A preliminary version appeared in FOCS 1988
6. Gomory, R.E., Hu, T.C.: Multi-terminal network flows. J. SIAM. 9(4), 551–670 (1961)
7. Kamidoi, Y., Wakabayashi, S., Yoshida, N.: A divide-and-conquer approach to the minimum k-way cut problem. Algorithmica 32(2), 262–276 (2002)
8. Kamidoi, Y., Yoshida, N., Nagamochi, H.: A deterministic algorithm for finding all minimum k-way cuts. SIAM Journal on Computing 36(5), 1329–1341 (2006)

9. Kapoor, S.: On minimum 3-cuts and approximating k-cuts using cut trees. In: Proceedings of the 5th International IPCO Conference on Integer Programming and Combinatorial Optimization, Springer, London (1996)
10. Karger, D.R., Stein, C.: A new approach to the minimum cut problem. Journal of the ACM 43(4), 601–640 (1996): Preliminary portions appeared in SODA 1993 and STOC 1993
11. Karypis, G., Kumar, V.: hmetis: A hypergraph partitioning package version 1.5, user manual (1998), `http://glaros.dtc.umn.edu/gkhome/fetch/sw/hmetis/manual.pdf`
12. Karypis, G., Kumar, V.: Multilevel k-way hypergraph partitioning. VLSI Design 11(3), 285–300 (2000)
13. Klimmek, R., Wagner, F.: A simple hypergraph min cut algorithm, Internal Report B 96-02 Bericht FU Berlin Fachbereich Mathematik und Informatik (1995)
14. Lawler, E.L.: Cutsets and partitions of hypergraphs. Networks 3(3), 275–285 (1973)
15. Levine, M.S.: Fast randomized algorithms for computing minimum {3,4,5,6}-way cuts. In: Proceedings of the 11th annual ACM-SIAM symposium on Discrete algorithms (SODA 2000), Philadelphia, PA, USA. Society for Industrial and Applied Mathematics (2000)
16. Mak, W.-K., Wong, D.F.: A fast hypergraph min-cut algorithm for circuit partitioning. Integration, the VLSI Journal 30(1), 1–11 (2000)
17. Nagamochi, H., Ibaraki, T.: Computing edge connectivity in multigraphs and capacitated graphs. SIAM Journal on Discrete Mathematics 5(1), 54–66 (1992)
18. Nagamochi, H., Ibaraki, T.: A fast algorithm for computing minimum 3-way and 4-way cuts. Mathematical Programming 88(3), 507–520 (2000)
19. Nagamochi, H., Katayama, S., Ibaraki, T.: A Faster Algorithm for Computing Minimum 5-Way and 6-Way Cuts in Graphs. In: Asano, T., Imai, H., Lee, D.T., Nakano, S.-i., Tokuyama, T. (eds.) COCOON 1999. LNCS, vol. 1627, Springer, Heidelberg (1999)
20. Preas, B.T., Lorenzetti, M.: Physical Design Automation of VLSI Systems, Benjamin-Cummings, California (1988)
21. Stoer, M., Wagner, F.: A simple min-cut algorithm. Journal of the ACM 44(4), 585–591 (1997); A preliminary version appeared in ESA 1994
22. Vazirani, V.V., Yannakakis, M.: Suboptimal cuts: Their enumeration, weight and number. In: Kuich, W. (ed.) Automata, Languages and Programming. Proc. of the 19th International Colloquium, Springer, Berlin (1992)

Inapproximability of Maximum Weighted Edge Biclique and Its Applications

Jinsong Tan

Department of Computer and Information Science
School of Engineering and Applied Science
University of Pennsylvania, Philadelphia, PA 19104, USA
jinsong@seas.upenn.edu

Abstract. Given a bipartite graph $G = (V_1, V_2, E)$ where edges take on *both* positive and negative weights from set $\mathcal{S}$, the *maximum weighted edge biclique* problem, or $\mathcal{S}$-MWEB for short, asks to find a bipartite subgraph whose sum of edge weights is maximized. This problem has various applications in bioinformatics, machine learning and databases and its (in)approximability remains open. In this paper, we show that for a wide range of choices of $\mathcal{S}$, specifically when $\left|\frac{\min \mathcal{S}}{\max \mathcal{S}}\right| \in \Omega(\eta^{\delta-1/2}) \cap O(\eta^{1/2-\delta})$ (where $\eta = \max\{|V_1|, |V_2|\}$, and $\delta \in (0, 1/2]$), no polynomial time algorithm can approximate $\mathcal{S}$-MWEB within a factor of n^{ϵ} for some $\epsilon > 0$ unless $\mathsf{RP} = \mathsf{NP}$. This hardness result gives justification of the heuristic approaches adopted for various applied problems in the aforementioned areas, and indicates that good approximation algorithms are unlikely to exist. Specifically, we give two applications by showing that: 1) finding statistically significant biclusters in the SAMBA model, proposed in [18] for the analysis of microarray data, is n^{ϵ}-inapproximable; and 2) no polynomial time algorithm exists for the Minimum Description Length with Holes problem [4] unless $\mathsf{RP} = \mathsf{NP}$.

1 Introduction

Let $G = (V_1, V_2, E)$ be an undirected bipartite graph. A *biclique subgraph* in G is a complete bipartite subgraph of G and *maximum edge biclique* (MEB) is the problem of finding a biclique subgraph with the most number of edges. MEB is a well-known problem and received much attention in recent years because of its wide range of applications in areas including machine learning [14], management science [16] and bioinformatics, where it is found particularly relevant in the formulation of numerous biclustering problems for biological data analysis [5,2,18,19,17], and we refer readers to the survey by Madeira and Oliveira [13] for a fairly extensive discussion on this. Maximum edge biclique is shown to be NP-hard by Peeters [15] via a reduction from 3SAT. Its approximability status, on

M. Agrawal et al. (Eds.): TAMC 2008, LNCS 4978, pp. 282–293, 2008.

the other hand, remains an open question despite considerable efforts [7,8,12][1]. In particular, Feige and Kogan [8] conjectured that maximum edge biclique is hard to approximate within a factor of n^{ϵ} for some $\epsilon > 0$. In this paper, we consider a weighted formulation of this problem defined as follows

Definition 1. $\mathcal{S}$-Maximum Weighted Edge Biclique ($\mathcal{S}$-MWEB)
Instance: *A complete bipartite graph $G = (V_1, V_2, E)$ (throughout the paper, let $\eta = max\{|V_1|, |V_2|\}$ and $n = |V_1| + |V_2|$), a weight function $w_G : E \rightarrow \mathcal{S}$, where $\mathcal{S}$ is a set consisting of both positive and negative integers.*
Question: *Find a biclique subgraph of G where the sum of weights on edges is maximized.*

A few comments are in order. First note it is not a lose of generality but a technical convenience to require the graph be complete, one can always think of an incomplete bipartite graph as complete where non-edges are assigned weight 0. Also note we require that both positive and negative weights be in $\mathcal{S}$ at the same time because otherwise $\mathcal{S}$-MWEB becomes a trivial problem.

Our study of $\mathcal{S}$-MWEB is motivated by the problem of finding statistically significant biclusters in microarray data analysis in the SAMBA model [18] and the Minimum Description Length with Holes (MDLH) problem [3,4,10]; detailed discussion of the two problems can be found in Sect. 4. Our main technical contribution of this paper is to show that if $\mathcal{S}$ satisfies the condition $|\frac{\min \mathcal{S}}{\max \mathcal{S}}| \in \Omega(\eta^{\delta - 1/2}) \cap O(\eta^{1/2-\delta})$, where $\delta > 0$ is any arbitrarily small constant, then no polynomial time algorithm can approximate $\mathcal{S}$-MWEB within a factor of n^{ϵ} for some $\epsilon > 0$ unless $\mathsf{RP} = \mathsf{NP}$. This result enables us to answer open questions regarding the hardness of the SAMBA model and the MDLH problem. Since maximum edge biclique can be characterized as a special case of $\mathcal{S}$-MWEB with $\mathcal{S} = \{-\eta, 1\}$, the n^{ϵ}-inapproximability result also provides interesting insights into the conjectured n^{ϵ}-inapproximability [8] of maximum edge biclique.

The rest of the paper is organized in three sections. In Sect. 2, we present the main technical result by proving the aforementioned inapproximability of $\mathcal{S}$-MWEB. We give applications of this by answering hardness questions regarding two applied problems in Sect. 3. We conclude this work by raising a few open problems in the last section.

2 Approximating $\mathcal{S}$-Maximum Edge Biclique Is Hard

We start this section by giving two lemmas about CLIQUE, which will be used in establishing inapproximability for the biclique problems we consider later. Lemma 1 is a recent result by Zuckerman [20], obtained by a derandomization of results of Håstad [11]; Lemma 2 follows immediately from Lemma 1.

[1] Note it might be easy to confuse the MEB problem with the *Bipartite Clique* problem discussed by Khot in [12]. *Bipartite Clique*, which also known as *Balanced Complete Bipartite Subgraph* [8], aims to maximize the number of vertices of a *balanced* subgraph whereas MEB aims to maximize the total weights on edges in a (not necessarily balanced) subgraph.

Lemma 1. ([20]) *It is NP-hard to approximate CLIQUE within a factor of $n^{1-\epsilon}$, for any $\epsilon > 0$.*

Lemma 2. *For any constant $\epsilon > 0$, no polynomial time algorithm can approximate CLIQUE within a factor of $n^{1-\epsilon}$ with probability at least $\frac{1}{poly(n)}$ unless* $\mathsf{RP} = \mathsf{NP}$.

2.1 A Technical Lemma

We first describe the construction of a structure called $\{\gamma, \{\alpha, \beta\}\}$-Product, which will be used in the proof of our main technical lemma.

Definition 2. ($\{\gamma, \{\alpha, \beta\}\}$**-Product)**
Input: *An instance of $\mathcal{S}$-MWEB on complete bipartite graph $G = V_1 \times V_2$, where $\gamma \in \mathcal{S}$ and $\alpha < \gamma < \beta$; an integer N.*
Output: *Complete bipartite graph $G^N = V_1^N \times V_2^N$ constructed as follows: V_1^N and V_2^N are N duplicates of V_1 and V_2, respectively. For each edge $(i, j) \in G^N$, let $(\phi(i), \phi(j))$ be the corresponding edge in G. If $w_G(\phi(i), \phi(j)) = \gamma$, assign weight α or β to (i, j) independently and identically at random with expectation being γ, denote the weight by random variable X. If $w_G(\phi(i), \phi(j)) \neq \gamma$, then keep the weight unchanged. Call the weight function constructed this way $w(\cdot)$.*

For any subgraph H of G^N, denote by $w_\gamma(H)$ (resp., $w_{-\gamma}(H)$) the total weight of H contributed by former-γ-edges (resp., other edges). Clearly, $w(H) = w_\gamma(H) + w_{-\gamma}(H)$.

With a graph product constructed in this randomized fashion, we have the following lemma.

Lemma 3. *Given an $\mathcal{S}$-MWEB instance $G = (V_1, V_2, E)$ where $\gamma \in \mathcal{S}$, and a number $\delta \in (0, \frac{1}{2}]$; let $\eta = \max(|V_1|, |V_2|)$, $N = \eta^{\frac{\delta(3-2\delta)+3}{\delta(1+2\delta)}}$, $G^N = (V_1^N, V_2^N, E)$ be the $\{\gamma, \{\alpha, \beta\}\}$-product of G and $\mathcal{S}' = (\mathcal{S} \cup \{\alpha, \beta\}) - \{\gamma\}$. If*

1. $|\beta - \alpha| = O((N\eta)^{\frac{1}{2}-\delta})$; and

2. there is a polynomial time algorithm that approximates the $\mathcal{S}'$-MWEB instance within a factor of λ, where λ is some arbitrary function in the size of the $\mathcal{S}'$-MWEB instance

then there exists a polynomial time algorithm that approximates the $\mathcal{S}$-MWEB instance within a factor of λ, with probability at least $\frac{1}{poly(n)}$.

Proof. For notational convenience, we denote $\eta^{\frac{1}{2}-\delta}$ by $f(\eta)$ throughout the proof. Define random variable $Y = X - \gamma$, clearly $E[Y] = 0$. Suppose there is a polynomial time algorithm $\mathbb{A}$ that approximates $\mathcal{S}'$-MWEB within a factor of λ, we can then run $\mathbb{A}$ on G^N, the output biclique G_B^* corresponds to N^2 bicliques in G (not necessarily all distinct). Let G_A^* be the most weighted among these N^2 subgraphs of G, in the rest of the proof we show that with high probability, G_A^* is a λ-approximation of $\mathcal{S}$-MWEB on G.

Denote by $\mathbb{E}_1$ the event that G_B^* does not imply a λ-approximation on G. Let $\mathcal{H}$ be the set of subgraphs of G^N that do not imply a λ-approximation on G, clearly, $|\mathcal{H}| \leq 4^{N\eta}$. Let H' be an arbitrary element in $\mathcal{H}$, we have the following inequalities

$$\begin{aligned} Pr\{\mathbb{E}_1\} &\leq Pr\left\{\text{at least one element in } \mathcal{H} \text{ is a } \lambda\text{-approximation of } G^N\right\} \\ &\leq 4^{N\eta} \cdot Pr\left\{H' \text{ is a } \lambda\text{-approximation of } G^N\right\} \\ &= 4^{N\eta} \cdot Pr\{\mathbb{E}_2\} \end{aligned}$$

where $\mathbb{E}_2$ is the event that H' is a λ-approximation of G^N.

Let the weight of an optimal solution $U_1 \times U_2$ of G be K, denote by $U_1^N \times U_2^N$ the corresponding N^2-duplication in G^N. Let x_1 and x_2 be the number of former-γ-edges in H' and $U_1^N \times U_2^N$, respectively. Suppose $\mathbb{E}_2$ happens, then we must have

$$w_{-\gamma}(H') + x_1\gamma \leq N^2(\tfrac{K}{\lambda} - 1)$$
$$w_{-\gamma}(H') + w_\gamma(H') \geq \tfrac{1}{\lambda}(w_{-\gamma}(U_1^N \times U_2^N) + w_\gamma(U_1^N \times U_2^N))$$

where the first inequality follows from the fact that we only consider integer weights. Since $w_{-\gamma}(U_1^N \times U_2^N) = N^2K - x_2\gamma$, it implies

$$(w_\gamma(H') - x_1\gamma) - \frac{1}{\lambda}(w_\gamma(U_1^N \times U_2^N) - x_2\gamma) \geq N^2$$

so we have the following statement on probability

$$Pr\{\mathbb{E}_2\} \leq Pr\left\{(w_\gamma(H') - x_1\gamma) - \tfrac{1}{\lambda}(w_\gamma(U_1^N \times U_2^N) - x_2\gamma) \geq N^2\right\}$$

Let z_1 (resp., z_2 and z_3) be the number of edges in $E(H') - E(U_1^N \times U_2^N)$ (resp., $E(U_1^N \times U_2^N) - E(H')$ and $E(U_1^N \times U_2^N) \cap E(H')$) transformed from former-γ-edges in G. We have

$$\begin{aligned} & Pr\left\{(w_\gamma(H') - x_1\gamma) - \tfrac{1}{\lambda}(w_\gamma(U_1^N \times U_2^N) - x_2\gamma) \geq N^2\right\} \\ &= Pr\left\{\textstyle\sum_{i=1}^{z_1} Y_i - \frac{1}{\lambda}\sum_{j=1}^{z_2} Y_j + \frac{\lambda-1}{\lambda}\sum_{k=1}^{z_3} Y_k \geq N^2\right\} \\ &= Pr\left\{\textstyle\sum_{i=1}^{z_1} Y_i + \frac{1}{\lambda}\sum_{j=1}^{z_2} (-Y_j) + \frac{\lambda-1}{\lambda}\sum_{k=1}^{z_3} Y_k \geq N^2\right\} \\ &\leq Pr\left\{\textstyle\sum_{i=1}^{z_1} Y_i \geq \frac{N^2}{3}\right\} + Pr\left\{\frac{1}{\lambda}\sum_{j=1}^{z_2} (-Y_j) \geq \frac{N^2}{3}\right\} + Pr\left\{\frac{\lambda-1}{\lambda}\sum_{k=1}^{z_3} Y_k \geq \frac{N^2}{3}\right\} \\ &\leq Pr\left\{\textstyle\sum_{i=1}^{z_1} Y_i \geq \frac{N^2}{3}\right\} + Pr\left\{\sum_{j=1}^{z_2} (-Y_j) \geq \frac{N^2}{3}\right\} + Pr\left\{\sum_{k=1}^{z_3} Y_k \geq \frac{N^2}{3}\right\} \\ &\leq \textstyle\sum_{i\in\{1,2,3\}} \left(\exp\left(-2z_i\left(\frac{N^2}{3z_i(c_1 f(N\eta))}\right)^2\right)\right) && \text{(Hoeffding bound)} \\ &\leq 3 \cdot \exp\left(-c_2 \cdot \tfrac{N^{1+2\delta}}{\eta^{3-2\delta}}\right) && (z_i \leq \eta^2 N^2) \end{aligned}$$

where c_1, c_2 are constants ($c_2 > 0$). Now if we set $N = \eta^{\frac{3-2\delta}{1+2\delta}+\theta}$ for some θ, we have

$$Pr\{\mathbb{E}_1\} \leq 4^{N\eta} \cdot Pr\{\mathbb{E}_2\} \leq 3 \cdot \exp\left(\ln 4 \cdot \eta^{\frac{4}{(1+2\delta)}+\theta} - c_2 \cdot \eta^{(1+2\delta)\theta}\right)$$

For this probability to be bounded by $\frac{1}{2}$ as η is large enough, we need to have $\frac{4}{1+2\delta} + \theta < (1+2\delta)\theta$. Solving this inequality gives $\theta > \frac{2}{\delta(1+2\delta)}$. Therefore, for any $\delta \in (0, \frac{1}{2}]$, by setting $N = \eta^{\frac{\delta(3-2\delta)+3}{\delta(1+2\delta)}}$, we have $Pr\{\mathbb{E}_1\}$, i.e. the probability that the solution returned by $\mathbb{A}$ does not imply a λ-approximation of G, is bounded from above by $\frac{1}{2}$ once input size is large enough. This gives a polynomial time algorithm that approximates $\mathcal{S}$-MWEB within a factor of λ with probability at least $\frac{1}{2}$. □

This lemma immediately leads to the following corollary.

Corollary 1. *Following the construction in Lemma 3, if $\mathcal{S}'$-MWEB can be approximated within a factor of $n^{\epsilon'}$, for some $\epsilon' > 0$, then there exists a polynomial time algorithm that approximates $\mathcal{S}$-MWEB within a factor of n^{ϵ}, where $\epsilon = (1 + \frac{\delta(3-2\delta)+3}{\delta(1+2\delta)})\epsilon'$, with probability at least $\frac{1}{poly(n)}$.*[2]

Proof. Let $|G|$ and $|G^N|$ be the number of nodes in the $\mathcal{S}$-MWEB and $\mathcal{S}'$-MWEB problem, respectively. Since $\lambda = |G^N|^{\epsilon'} \leq |G|^{(1+\frac{\delta(3-2\delta)+3}{\delta(1+2\delta)})\epsilon'}$, our claim follows from Lemma 3. □

2.2 $\{-1, 0, 1\}$-MWEB

In this section, we prove inapproximability of $\{-1, 0, 1\}$-MWEB by giving a reduction from CLIQUE; in subsequence sections, we prove inapproximability results for more general $\mathcal{S}$-MWEB by constructing randomized reduction from $\{-1, 0, 1\}$-MWEB.

Lemma 4. *The decision version of the $\{-1, 0, 1\}$-MWEB problem is* NP*-complete.*

Proof. We prove this by describing a reduction from CLIQUE. Given a CLIQUE instance $G = (V, E)$, construct $G' = (V', E')$ such that $V' = V_1 \cup V_2$ where V_1, V_2 are duplicates of V in that there exist bijections $\phi_1 : V_1 \to V$ and $\phi_2 : V_2 \to V$. And

$$\begin{aligned}
E' &= E_1 \cup E_2 \cup E_3 \\
E_1 &= \{(u, v) \mid u \in V_1, v \in V_2 \text{ and } (\phi_1(u), \phi_2(v)) \in E\} \\
E_2 &= \{(u, v) \mid u \in V_1, v \in V_2, \phi_1(u) \neq \phi_2(v) \text{ and } (\phi_1(u), \phi_2(v)) \notin E\} \\
E_3 &= \{(u, v) \mid u \in V_1, v \in V_2, \text{ and } \phi_1(u) = \phi_2(v)\}
\end{aligned}$$

Clearly, G' is a biclique. Now assign weight 0 to edges in E_1, -1 to edges in E_2 and 1 to edges in E_3. We then claim that there is a clique of size k in G if and only if there is a biclique of total edge weight k in G'.

First consider the case where there is a clique of size k in G, let U be the set of vertices of the clique, then taking the subgraph induced by $\phi_1^{-1}(U) \times \phi_2^{-1}(U)$ in G' gives us a biclique of total weight k.

Now suppose that there is a biclique $U_1 \times U_2$ of total weight k in G'. Without loss of generality, assume U_1 and U_2 correspond to the same subset of vertices in

[2] Note we are slightly abusing notation here by always representing the size of a given problem under discussion by n. Here n refers to the size of $\mathcal{S}'$-MWEB (resp. $\mathcal{S}$-MWEB) when we are talking about approximation factor $n^{\epsilon'}$ (resp. n^{ϵ}). We adopt the same convention in the sequel.

V because if $(\phi_1(U_1) - \phi_2(U_2)) \cup (\phi_2(U_2) - \phi_1(U_1))$ is not empty, then removing $(U_1 - U_2) \cup (U_2 - U_1)$ will never decrease the total weight of the solution. Given $\phi_1(U_1) = \phi_2(U_2)$, we argue that there is no edge of weight -1 in biclique $U_1 \times U_2$; suppose otherwise there exists a weight -1 edge (i_1, j_2) ($i_1 \in U_1$, and $j_2 \in U_2$), then the corresponding edge (j_1, i_2) ($j_1 \in U_1$, and $i_2 \in U_2$) must be of weight -1 too and removing i_1, i_2 from the solution biclique will increase total weight by at least 1 because among all edges incident to i_1 and i_2, (i_1, i_2) is of weight 1, (i_1, j_2) and (i_2, j_1) are of weight -1 and the rest are of weights either 0 or -1.

Therefore, we have shown that if there is a solution $U_1 \times U_2$ of weight k in G', U_1 and U_2 correspond to the same set of vertices $U \in V$ and U is a clique of size k. It is clear that the reduction can be performed in polynomial time and the problem is NP, and thus NP-complete. □

Given Lemma 1, the following corollary follows immediately from the above reduction.

Theorem 1. *For any constant $\epsilon > 0$, no polynomial time algorithm can approximate problem $\{-1, 0, 1\}$-MWEB within a factor of $n^{1-\epsilon}$ unless* $\mathsf{P} = \mathsf{NP}$.

Proof. It is obvious that the reduction given in the proof of Lemma 4 preserves inapproximability exactly, and given that CLIQUE is hard to approximate within a factor of $n^{1-\epsilon}$ unless $\mathsf{P} = \mathsf{NP}$, the theorem follows. □

Theorem 2. *For any constant $\epsilon > 0$, no polynomial time algorithm can approximate $\{-1, 0, 1\}$-MWEB within a factor of $n^{1-\epsilon}$ with probability at least $\frac{1}{poly(n)}$ unless* $\mathsf{RP} = \mathsf{NP}$.

Proof. If there exists such a randomized algorithm for $\{-1, 0, 1\}$-MWEB, combining it with the reduction given in Lemma 4, we obtain an RP algorithm for CLIQUE. This is impossible unless $\mathsf{RP} = \mathsf{NP}$. □

2.3 $\{-1, 1\}$-MWEB

Lemma 5. *If there exists a polynomial time algorithm that approximates $\{-1, 1\}$-MWEB within a factor of n^{ϵ}, then there exists a polynomial time algorithm that approximates $\{-1, 0, 1\}$-MWEB within a factor of $n^{5\epsilon}$ with probability at least $\frac{1}{poly(n)}$.*

Proof. We prove this by constructing a $\{\gamma, \{\alpha, \beta\}\}$-Product from $\{-1, 0, 1\}$-MWEB to $\{-1, 1\}$-MWEB by setting $\gamma = 0$, $\alpha = -1$ and $\beta = 1$. Since $\delta = \frac{1}{2}$, according to Corollary 1, it is sufficient to set $N = \eta^4$ so that the probability of obtaining a $n^{5\epsilon}$-approximation for $\{-1, 0, 1\}$-MWEB is at least $\frac{1}{poly(n)}$. □

Theorem 3. *For any constant $\epsilon > 0$, no polynomial time algorithm can approximate $\{-1, 1\}$-MWEB within a factor of $n^{\frac{1}{5}-\epsilon}$ with probability at least $\frac{1}{poly(n)}$ unless* $\mathsf{RP} = \mathsf{NP}$.

Proof. This follows directly from Theorem 2 and Lemma 5. □

2.4 $\{-\eta^{\frac{1}{2}-\delta}, 1\}$-MWEB and $\{-\eta^{\delta-\frac{1}{2}}, 1\}$-MWEB

In this section, we consider the generalized cases of the $\mathcal{S}$-MWEB problem.

Theorem 4. *For any $\delta \in (0, \frac{1}{2}]$, there exists some constant ϵ such that no polynomial time algorithm can approximate $\{-\eta^{\frac{1}{2}-\delta}, 1\}$-MWEB within a factor of n^ϵ with probability at least $\frac{1}{poly(n)}$ unless* RP = NP. *The same statement holds for $\{-\eta^{\delta-\frac{1}{2}}, 1\}$-MWEB.*

Proof. We prove this by first construct a $\{\gamma, \{\alpha, \beta\}\}$-Product from $\{-1, 1\}$-MWEB to $\{-\eta^{\frac{1}{2}-\delta}, 1\}$-MWEB by setting $\gamma = -1$, $\alpha = -(N\eta)^{\frac{1}{2}-\delta}$ and $\beta = 1$. By Corollary 1, we know that for any $\delta \in (0, \frac{1}{2}]$, if there exists a polynomial time algorithm that approximates $\{-\eta^{\frac{1}{2}-\delta}, 1\}$-MWEB within a factor of n^ϵ, then there exists a polynomial time algorithm that approximates $\{-1, 1\}$-MWEB within a factor of $n^{(1+\frac{\delta(3-2\delta)+3}{\delta(1+2\delta)})\epsilon}$ with probability at least $\frac{1}{poly(n)}$. So invoking the hardness result in Theorem 3 gives the desired hardness result for $\{-\eta^{\frac{1}{2}-\delta}, 1\}$-MWEB.

The same conclusion applies to $\{-1, \eta^{\frac{1}{2}-\delta}\}$-MWEB by setting $\gamma = 1$, $\alpha = -1$ and $\beta = (N\eta)^{\frac{1}{2}-\delta}$. Since η is a constant for any given graph, we can simply divide each weight in $\{-1, \eta^{\frac{1}{2}-\delta}\}$ by $\eta^{\frac{1}{2}-\delta}$. □

Theorem 4 leads to the following general statement.

Theorem 5. *For any small constant $\delta \in (0, \frac{1}{2}]$, if $\left|\frac{\min \mathcal{S}}{\max \mathcal{S}}\right| \in \Omega(\eta^{\delta-1/2}) \cap O(\eta^{1/2-\delta})$, then there exists some constant ϵ such that no polynomial time algorithm can approximate $\mathcal{S}$-MWEB within a factor of n^ϵ with probability at least $\frac{1}{poly(n)}$ unless* RP = NP.

3 Two Applications

In this section, we describe two applications of the results establish in Sect. 3 by proving hardness and inapproximability of problems found in practice.

3.1 SAMBA Model Is Hard

Microarray technology has been the latest technological breakthrough in biological and biomedical research; in many applications, a key step in analyzing gene expression data obtained through microarray is the identification of a bicluster satisfying certain properties and with largest area (see the survey [13] for a fairly extensive discussion on this).

In particular, Tanay *et. al.* [18] considered the Statistical-Algorithmic Method for Bicluster Analysis (SAMBA) model. In their formulation, a complete bipartite graph is given where one side corresponds to genes and the other size corresponds to conditions. An edges (u, v) is assigned a real weight which could be either positive or negative, depending on the expression level of gene u in condition v, in a way such that heavy subgraphs corresponds to statistically significant biclusters. Two weight-assigning schemes are considered in their paper. In the first, or simple statistical model, a tight upper-bound on the probability of an observed biclusters in computed; in the second, or refined statistical model, the weights are assigned in a way such that a maximum weight biclique subgraph corresponds to a maximum likelihood bicluster.

The Simple SAMBA Statistical Model: Let $H = (V_1', V_2', E')$ be a subgraph of $G = (V_1, V_2, E)$, $\overline{E'} = \{V_1' \times V_2'\} - E'$ and $p = \frac{|E|}{|V_1||V_2|}$. The simple statistical model assumes that edges occur independently and identically at random with probability p. Denote by $BT(k, p, n)$ the probability of observing k or more successes in n binomial trials, the probability of observing a graph at least as dense as H is thus $p(H) = BT(|E'|, p, |V_1'||V_2'|)$. This model assumes $p < \frac{1}{2}$ and $|V_1'||V_2'| \ll |V_1||V_2|$, therefore $p(H)$ is upper bounded by

$$p^*(H) = 2^{|V_1'||V_2'|} p^{|E'|} (1-p)^{|V_1'||V_2'| - |E'|}$$

The goal of this model is thus to find a subgraph H with the smallest $p^*(H)$. This is equivalent to maximizing

$$-\log p^*(H) = |E'|(-1 - \log p) + (|V_1'||V_2'| - |E'|)(-1 - \log (1-p))$$

which is essentially solving a $\mathcal{S}$-MWEB problem that assigns either positive weight $(-1 - \log p)$ or negative weight $(-1 - \log (1-p))$ to an edge (u, v), depending on whether gene u express or not in condition v, respectively. The summation of edge weights over H is defined as the *statistical significance* of H.

Since $\frac{1}{\eta^2} \le p < \frac{1}{2}$, asymptotically we have $\frac{-1-\log(1-p)}{-1-\log p} \in \Omega(\frac{1}{\log \eta}) \cap O(1)$. Invoking Theorem 5 gives the following.

Theorem 6. *For the Simple SAMBA Statistical model, there exists some $\epsilon > 0$ such that no polynomial time algorithm, possibly randomized, can find a bicluster whose statistical significance is within a factor of n^ϵ of optimal unless* $\mathsf{RP} = \mathsf{NP}$.

The Refined SAMBA Statistical Model: In the refined model, each edge (u, v) is assumed to take an independent Bernoulli trial with parameter $p_{u,v}$, therefore $p(H) = (\prod_{(u,v)\in E'} p_{u,v})(\prod_{(u,v)\in \overline{E'}} (1 - p_{u,v}))$ is the probability of observing a subgraph H. Since $p(H)$ generally decreases as the size of H increases, Tanay *et al.* aims to find a bicluster with the largest (normalized) likelihood ratio $L(H) = \dfrac{(\prod_{(u,v)\in E'} p_c)(\prod_{(u,v)\in \overline{E'}} (1 - p_c))}{p(H)}$, where $p_c > \max_{(u,v)\in E} p_{u,v}$ is a constant probability and chosen with biologically sound assumptions. Note this is equivalent to maximizing the log-likelihood ratio

$$\log L(H) = \sum_{(u,v)\in E'} \log \frac{p_c}{p_{u,v}} + \sum_{(u,v)\in \overline{E'}} \log \frac{1-p_c}{1-p_{u,v}}$$

With this formulation, each edge is assigned weight either $\log \frac{p_c}{p_{u,v}} > 0$ or $\log \frac{1-p_c}{1-p_{u,v}} < 0$ and finding the most statistically significant bicluster is equivalent to solving $\mathcal{S}$-MWEB with $\mathcal{S} = \{\log \frac{1-p_c}{1-p_{u,v}}, \log \frac{p_c}{p_{u,v}}\}$. Since p_c is a constant and $\frac{1}{\eta^2} \le p_{u,v} < p_c$, we have $\frac{\log(1-p_c)-\log(1-p_{u,v})}{\log p_c - \log p_{u,v}} \in \Omega(\frac{1}{\log \eta}) \cap O(1)$. Invoking Theorem 5 gives the following.

Theorem 7. *For the Refined SAMBA Statistical model, there exists some $\epsilon > 0$ such that no polynomial time algorithm, possibly randomized, can find a bicluster whose log-likelihood is within a factor of n^ϵ of optimal unless* RP = NP.

3.2 Minimum Description Length with Holes (MDLH) Is Hard

Bu *et. al* [4] considered the Minimum Description Length with Holes problem (defined in the following); the 2-dimensional case is claimed NP-hard in this paper and the proof is referred to [3]. However, the proof given in [3] suffers from an error in its reduction[3], thus whether MDLH is NP-complete remains unsettled. In this section, by employing the results established in the previous sections, we show that no polynomial time algorithm exists for MDLH, under the slightly weaker (than P $\neq$ NP) but widely believed assumption RP $\neq$ NP.

We first briefly describe the Minimum Description Length summarization with Holes problem; for a detailed discussion of the subject, we refer the readers to [3,4].

Suppose one is given a k-dimensional binary matrix M, where each entry is of value either 1, which is of interest, or of value 0, which is not of interest. Besides, there are also k hierarchies (trees) associated with each dimension, namely $T_1, T_2, ..., T_k$, each of height $l_1, l_2, ..., l_k$ respectively. Define *level* $l = \max_i(l_i)$. For each T_i, there is a bijection between its leafs and the 'hyperplanes' in the ith dimension (e.g. in a 2-dimensional matrix, these hyperplanes corresponds to rows and columns). A *region* is a tuple $(x_1, x_2, ..., x_k)$, where x_i is a leaf node or an internal node in hierarchy T_i. Region $(x_1, x_2, ..., x_k)$ is said to *cover* cell $(c_1, c_2, ..., c_k)$ if c_i is a descendant of x_i, for all $1 \le i \le k$. A *k-dimensional l-level MDLH summary* is defined as two sets S and H, where 1) S is a set of regions covering all the 1-entries in M; and 2) H is the set of 0-entries covered (undesirably) by S and to be excluded from the summary. The *length* of a summary is defined as $|S| + |H|$, and the MDLH problem asks the question if there exists a MDLH summary of length at most K, for a given $K > 0$.

In an effort to establish hardness of MDLH, we first define the following problem, which serves as an intermediate problem bridging $\{-1, 1\}$-MWEB and MDLH.

---

[3] In Lemma 3.2.1 of [3], the reduction from CLIQUE to CEW is incorrect.

Definition 3. (Problem $\mathcal{P}$)
Instance: *A complete bipartite graph $G = (V_1, V_2, E)$ where each edge takes on a value in $\{-1, 1\}$, and a positive integer k.*
Question: *Does there exist an induced subgraph (a biclique $U_1 \times U_2$) whose total weight of edges is ω, such that $|U_1| + |U_2| + \omega \geq k$.*

Lemma 6. *No polynomial time algorithm exists for Problem $\mathcal{P}$ unless* RP = NP.

Proof. We prove this by constructing a reduction from $\{-1, 1\}$-MWEB to Problem $\mathcal{P}$ as follows: for the given input biclique $G = (V_1, V_2, E)$, make N duplicates of V_1 and N duplicates of V_2, where $N = (|V_1| + |V_2|)^2$. Connect each copy of V_1 to each copy of V_2 in a way that is identical to the input biclique, we then claim that there is a size k solution to $\{-1, 1\}$-MWEB if and only if there is a size N^2k solution to Problem $\mathcal{P}$.

If there is a size k solution to $\{-1, 1\}$-MWEB, then it is straightforward that there is a solution to Problem $\mathcal{P}$ of size at least N^2k. For the reverse direction, we show that if no solution to $\{-1, 1\}$-MWEB is of size at least k, then the maximum solution to Problem $\mathcal{P}$ is strictly less than N^2k. Note a solution $U_1^N \times U_2^N$ to Problem $\mathcal{P}$ consists of at most N^2 (not necessarily all distinct) solutions to $\{-1, 1\}$-MWEB, and each of them can contribute at most $(k - 1)$ in weight to $U_1^N \times U_2^N$, so the total weight gained from edges is at most $N^2(k - 1)$. And note the total weight gained from vertices is at most $N(|V_1| + |V_2|) = N\sqrt{N}$, therefore the weight is upper bounded by $N\sqrt{N} + N^2(k - 1) < N^2k$ and this completes the proof.

As a conclusion, we have a polynomial time reduction from $\{-1, 1\}$-MWEB to Problem $\mathcal{P}$. Since no polynomial time algorithm exists for $\{-1, 1\}$-MWEB unless RP = NP, the same holds for Problem $\mathcal{P}$. □

Theorem 8. *No polynomial time algorithm exists for MDLH summarization, even in the 2-dimension 2-level case, unless* RP = NP.

Proof. We prove this by showing that Problem $\mathcal{P}$ is a complementary problem of 2-dimensional 2-level MDLH.

Let the input 2D matrix M be of size $n_1 \times n_2$, with a tree of height 2 associated with each dimension. Without loss of generality, we only consider the 'sparse' case where the number of 1-entries is less than the number of 0-entries by at least 2 so that the optimal solution will never contain the whole matrix as one of its regions. Let S be the set of regions in a solution. Let R and C be the set of rows and columns not included in S. Let Z be the set of all zero entries in M. Let z be the total number of zero entries in the $R \times C$ 'leftover' matrix and let w be the total number of 1-entries in it. MDLH tries to minimize the following:

$$(n_1 - |R|) + (n_2 - |C|) + (|Z| - z) + w = (n_1 + n_2 + |Z|) - (|R| + |C| + z - w)$$

Since $(n_1 + n_2 + |Z|)$ is a fixed quantity for any given input matrix, the 2-dimensional 2-level MDLH problem is equivalent to maximizing $(|R|+|C|+z-w)$, which is precisely the definition of Problem $\mathcal{P}$.

Therefore, 2-dimensional 2-level MDLH is a complementary problem to Problem $\mathcal{P}$ and by Lemma 6 we conclude that no polynomial time algorithm exists for 2-dimensional 2-level MDLH unless $\mathsf{RP} = \mathsf{NP}$. □

4 Concluding Remarks

Maximum weighted edge biclique and its variants have received much attention in recently years because of it wide range of applications in various fields including machine learning, database, and particularly bioinformatics and computational biology, where many computational problems for the analysis of microarray data are closely related. To tackle these applied problems, various kinds of heuristics are proposed and experimented and it is not known whether these algorithms give provable approximations. In this work, we answer this question by showing that it is highly unlikely (under the assumption $\mathsf{RP} \neq \mathsf{NP}$) that good polynomial time approximation algorithm exists for maximum weighted edge biclique for a wide range of choices of weight; and we further give specific applications of this result to two applied problems. We conclude our work by listing a few open questions.

1. We have shown that $\{\Theta(-\eta^{\delta}), 1\}$-MWEB is n^{ϵ}-inapproximable for $\delta \in (-\frac{1}{2}, \frac{1}{2})$; also it is easy to see that (i) the problem is in P when $\delta \leq -1$, where the entire input graph is the optimal solution; (ii) for any $\delta \geq 1$, the problem is equivalent to MEB, which is conjectured to be n^{ϵ}-inapproximable [8]. Therefore it is natural to ask what is the approximability of the $\{-n^{\delta}, 1\}$-MWEB problem when $\delta \in (-1, -\frac{1}{2}]$ and $\delta \in [\frac{1}{2}, 1]$. In particular, can this be answered by a better analysis of Lemma 3?

2. We are especially interested in $\{-1, 1\}$-MWEB, which is closely related to the formulations of many natural problems [1,3,4,18]. We have shown that no polynomial time algorithm exists for this problem unless $\mathsf{RP} = \mathsf{NP}$, and we believe this problem is NP-complete, however a proof has eluded us so far.

References

1. Bansal, N., Blum, A., Chawla, S.: Correlation clustering. Machine Learning 56, 89–113 (2004)
2. Ben-Dor, A., Chor, B., Karp, R., Yakhini, Z.: Discovering local structure in gene expression data: The Order-Preserving Submatrix Problem. In: Proceedings of RECOMB 2002, pp. 49–57 (2002)
3. Bu, S.: The summarization of hierarchical data with exceptions. Master Thesis, Department of Computer Science, University of British Columbia (2004), `http://www.cs.ubc.ca/grads/resources/thesis/Nov04/Shaofeng_Bu.pdf`
4. Bu, S., Lakshmanan, L.V.S., Ng, R.T.: MDL Summarization with Holes. In: Proceedings of VLDB 2005, pp. 433–444 (2005)
5. Cheng, Y., Church, G.: Biclustering of expression data. In: Proceedings of ISMB 2000, pp. 93–103. AAAI Press, Menlo Park (2000)
6. Dawande, M., Keskinocak, P., Swaminathan, J.M., Tayur, S.: On Bipartite and multipartite clique problems. Journal of Algorithms 41(2), 388–403 (2001)

7. Feige, U.: Relations between average case complexity and approximation complexity. In: Proceedings of STOC 2002, pp. 534–543 (2002)
8. Feige, U., Kogan, S.: Hardness of approximation of the Balanced Complete Bipartite Subgraph problem. Technical Report MCS 2004-2004, The Weizmann Institute of Science (2004)
9. Garey, M.R., Johnson, D.S.: Computers and Intractability: A Guide to the Theory of NP-completeness. Freeman, San Francisco (1979)
10. Fontana, P., Guha, S., Tan, J.: Recursive MDL Summarization and Approximation Algorithms (preprint, 2007)
11. Håstad, J.: Clique is hard to approximate within $n^{1-\epsilon}$. Acta Mathematica 182, 105–142 (1999)
12. Khot, S.: Ruling out PTAS for Graph Min-Bisection, Densest Subgraph and Bipartite Clique. In: Proceedings of FOCS 2004, pp. 136–145 (2004)
13. Madeira, S.C., Oliveira, A.L.: Biclustering algorithms for biological data analysis: a survey. IEEE/ACM Transactions on Computational Biology and Bioinformatics 1, 24–45 (2004)
14. Mishra, N., Ron, D., Swaminathan, R.: On finding large conjunctive clusters. In: Proceedings of COLT 2003, pp. 448–462 (2003)
15. Peeters, R.: The maximum edge biclique problem is NP-complete. Discrete Applied Mathematics 131, 651–654 (2003)
16. Swaminathan, J.M., Tayur, S.: Managing Broader Product Lines Through Delayed Differentiation Using Vanilla Boxes. Management Science 44, 161–172 (1998)
17. Tan, J., Chua, K., Zhang, L., Zhu, S.: Complexity study on clustering problems in microarray data analysis. Algorithmica 48(2), 203–219 (2007)
18. Tanay, A., Sharan, R., Shamir, R.: Discovering statistically significant biclusters in gene expression data. Bioinformatics 18(1), 136–144 (2002)
19. Zhang, L., Zhu, S.: A New Clustering Method for Microarray Data Analysis. In: Proceedings of CSB 2002, pp. 268–275 (2002)
20. Zuckerman, D.: Linear Degree Extractors and the Inapproximability of Max Clique and Chromatic Number. In: Proceedings of STOC 2006, pp. 681–690 (2006)

Symbolic Algorithm Analysis of Rectangular Hybrid Systems*

Haibin Zhang and Zhenhua Duan

Institute of Computing Theory and Technology,
Xidian University
{hbzhang,ZhhDuan}@mail.xidian.edu.cn

Abstract. This paper investigates symbolic algorithm analysis of rectangular hybrid systems. To deal with the symbolic reachability problem, a restricted constraint system called *hybrid zone* is formalized. Hybrid zones are also applied to a symbolic model-checking algorithm for verifying some important classes of timed computation tree logic formulas. To present hybrid zones, a data structure called *difference constraint matrix* is defined. Using this structure, all reachability operations and model checking algorithms for rectangular hybrid systems are implemented. These enable us to deal with the symbolic algorithm analysis of rectangular hybrid systems in an efficient way.

Keywords: hybrid systems, model checking, temporal logic, reachability analysis.

1 Introduction

Hybrid systems are state-transition systems consisting of a non-trivial mixture of continuous activities and discrete events [6,8]. Most of verification tools [2,3,4] for hybrid systems deal with the reachability analysis and model checking [16,17]. In this paper, we will be concerned with a symbolic approach which is based on the manipulation of conjunctions of inequalities.

Many tools for timed automata use data structures and algorithms to manipulate timing constraints over clock variables called *clock zone* [11], which is a conjunction of inequalities that compare either a clock value or the difference between two clock values to an integer. In this paper, we are motivated to deal with the reachability of rectangular hybrid systems using some constraint system similar to clock zones. However, the immediate problem we encounter is that the constraint system is getting more complicated than clock zones. For linear hybrid systems, researchers have used *convex polyhedra* [10] as the basis for the symbolic manipulations. Using convex polyhedra, there is a basic operation "quantifier elimination" to compute reachability states, which has an exponential complexity [15]. As a subset of linear hybrid systems, we could also

* This research is supported by the NSFC under Grant No. 60373103 and 60433010, the SRFDP under Grant 20030701015.

M. Agrawal et al. (Eds.): TAMC 2008, LNCS 4978, pp. 294–305, 2008.

use convex polyhedra as a basic unit for the symbolic reachability analysis of rectangular hybrid systems. However, we use a more effective constraint system called *hybrid zone* in this paper[1], which is a conjunction of inequalities that compare either a variable value or a linear expression of two variables to a rational number. We prove that hybrid zones are closed over the reachability operations of rectangular hybrid systems.

The model checking problem for rectangular hybrid systems decides whether a rectangular hybrid system presented by a rectangular automaton meets a property given by a temporal logic formula. Unfortunately, this problem is proved to be undecidable. In [1], a semidecision model-checking algorithm for linear hybrid systems is given to check *timed computation tree logic* (TCTL) formulas. In the model-checking procedure, there are two important operations on state spaces of linear hybrid systems. They are *backward time closure* and *precondition*. We proved that our hybrid zones are also closed under those two operations for rectangular hybrid systems, which enables us to use the same algorithm to check properties of rectangular hybrid systems presented by some important classes of TCTL formulas.

To represent hybrid zones, we define a matrix data structure *difference constraint matrix* (DCM), which is indexed by the variables in the rectangular hybrid systems together, and each entry in the matrix represents an inequality in the hybrid zone. Using DCM, we implement all reachability operations and model checking algorithms for rectangular hybrid systems. Similar to DBM [5], after the DCM has been converted to canonical form, those reachability operations and model checking algorithms for rectangular hybrid systems can be implemented straightforwardly. Hence, the main computation is the operation for obtaining the canonical form of hybrid zones. Finding the canonical form can be automated by an algorithm with polynomial time complexity. However, if we use convex polyhedra, the quantifier elimination will lead to an exponential complexity.

This paper is organized as follows. The next section introduces rectangular automata. In Section 3, hybrid zones are defined. Section 4 investigates the model checking problem for rectangular hybrid systems. In Section 5, a data structure DCM is formalized. Conclusions are drawn in Section 6.

2 Rectangular Automata

Let $Y = \{y_1, \cdots, y_n\}$ be a set of variables. A rectangle $B \subseteq R^n$ over Y is defined by a conjunction of linear (in)equalities of the form $y_i \sim c$, where $\sim$ is $<$, $\leq$, $=$, $\geq$, or $>$, and c is a rational number [7,9]. For a rectangle $B \subseteq R^n$, we denote B_i by its projection onto the ith coordinate. The set of all n-dimensional rectangles is denoted by $\mathcal{B}_n$.

Definition 1. *A rectangular automaton [9] is a tuple* $(Q, X, init, E, jump, update, reset, act, inv)$*, where:*

[1] The definition of the hybrid zone is firstly formalized in our another paper [13], which is perfected in this paper.

- *Q is a finite set of locations.*
- *X is a finite set of real-numbered variables.*
- *$Init: Q \to \mathcal{B}_n$ is a labeling function that assigns an initial condition to each location $q \in Q$.*
- *$E \subseteq Q \times Q$ is a finite set of edges called transitions.*
- *$jump: E \to \mathcal{B}_n$ is a labeling function that assigns a jump condition to each edge $e \in E$, which relates values of variables before edge e.*
- *$update: E \to 2^{\{1,...,n\}}$ is a function that maps an element in $2^{\{1,...,n\}}$ to each edge $e \in E$. $update(e)$ indicates the variables $\{x_i | i \in update(e)\}$ whose values must be changed over edge e.*
- *$reset: E \to \mathcal{B}_n$ is a labeling function that assigns a reset condition to each edge $e \in E$, which relates values of the variables after edge e.*
- *$inv: Q \to \mathcal{B}_n$ is a labeling function that assigns an invariant condition to each location $q \in Q$.*
- *$act: Q \to \mathcal{B}_n$ is a labeling function that assigns a flow condition to each location $q \in Q$ being the form of $\dot{x} = a$ ($x \in X$, $a \in act(q)$).*

In this paper, we investigate the initialized rectangular automata with bounded nondeterminism [7]. In addition, we require that for $e = (q, q') \in E$, $jump(e)$ is bounded; and for each interval $act(q)_i$, either $act(q)_i \subset (-\infty, 0]$ or $act(q)_i \subset [0, \infty)$ while both of its end-points are integers.

A *valuation* is a function from $(x_1, ..., x_n)$ to R^n, where $x_i \in X$. We use V to denote the set of valuations. A state is a tuple (q, v), where $q \in Q$ and $v \in V$. We use S to denote the set of states. For $v \in V$ and $d \in R^+$, $v + l \cdot d$ denotes the valuation which maps each variable $x_i \in X$ to the value $v(x_i) + l_i \cdot d$. For $\theta \subseteq \{1, ..., n\}$, $B \in \mathcal{B}_n$, $[\theta \mapsto B]v$ denotes the valuation which maps each variable $x_i \in X$ ($i \in \theta$) to a value in B_i and agrees with v over $\{x_k | k \in \{1, ..., n\} \backslash \theta\}$. For $e = (q, q') \in E$, states (q, v) and (q', v'), if $v \in jump(e)$ and $v' \in reset(e)$, then state (q', v') is called a *transition successor* of state (q, v).

The semantics of rectangular automata can be given by an infinite transition system. A run of a rectangular automaton $\mathcal{H}$ is a finite or infinite sequence

$$\rho : s_0 \mapsto^{t_0}_{f_0} s_1 \mapsto^{t_1}_{f_1} s_2 \mapsto^{t_2}_{f_2} \cdots$$

of states $s_i = (q_i, v_i) \in S$, nonnegative reals $t_i \in R^+$, and $f_i \in act(q_i)$, such that for all $i \geq 0$:

1. For all $0 \leq t \leq t_i$, $v_i + f_i \cdot t \in inv(q_i)$.
2. The state s_{i+1} is a transition successor of the state $(q_i, v_i + f_i \cdot t_i)$.

The state s_{i+1} is called a *successor* of state s_i. The run ρ diverges if ρ is infinite and the infinite sum $\sum\limits_{i \geq 0} t_i$ diverges.

3 Reachability Analysis

Given two states s and s' of a hybrid system $\mathcal{H}$, the *reachability problem* asks whether there exists a run of $\mathcal{H}$ which starts at s and ends at s'. The reachability problem is central to the verification of hybrid systems. In the sequel, we will give a constraint system to deal with the symbolic reachability analysis of rectangular hybrid systems.

We can use the algorithm in Fig. 1 to deal with the symbolic reachability analysis of rectangular hybrid systems, which checks whether a rectangular automaton may reach a state satisfying a given state formula ϕ. This algorithm can be expressed in terms of the following three reachability operations: given two sets of valuations D and D', two locations $q, q' \in Q$ and the edge $e = (q, q') \in E$, *intersection* $D \wedge D'$ is the set of valuations in both D and D', *time progress in location* q $D^{\uparrow_q}$ is the set of valuations $\{u + l \cdot d | u \in D \wedge l \in act(q) \wedge d \in R^+\}$ that are reachable from some valuation in D by time progressing, *variable reset* $\mathbf{Reset}_e(D)$ is the set of valuations $\{[update(e) \mapsto reset(e)]u | u \in D\}$ that are reachable from some valuation in D over the transition e (In Fig. 1, $(q, D) \hookrightarrow (q_s, D_s)$ iff $q_s = q \wedge D_s = (D \wedge inv(q))^{\uparrow_q}$ or $q_s \neq q \wedge D_s = \mathbf{Reset}_{e'}(D \wedge jump(e')) \wedge e' = (q, q_s)$).

```
Passed:={}
Wait:={(q0, D0)}
while Wait≠ {}
        get (q, D) from Wait
        if (q, D) ⊨ φ then return 'YES'
        else if D ⊄ D' for all (q, D') ∈Passed
            add (q, D) to Passed
            Next:={(qs, Ds)| (q, D) ↪ (qs, Ds) ∧ Ds ≠ ∅}
            for all (qs', Ds') ∈Next do
                    put (qs', Ds') to Wait
return 'NO'
```

Fig. 1. An algorithm for symbolic reachability analysis

Hybrid zones

In the sequel, for each variable $x_i \in X$, l_i denotes the left end-point of $act(q)_i$, and r_i the right end-point, without clarification. We define a function $\widetilde{g}(a, b)$ as $\widetilde{g}(a, b) = gcd(a, b)$, if $a \cdot b \neq 0$, otherwise, $\widetilde{g}(a, b) = 1$, where $gcd(a, b)$ is the greatest common divisor of a and b.

Definition 2. *For a location q of a rectangular automaton, let l_i denote the left end-point of $act(q)_i$, and r_i the right end-point. A q-zone is the conjunction of inequalities:*

$$\bigwedge_{0<i\leq n} (x_i \prec c_{i0} \wedge -x_i \prec c_{0i}) \wedge \bigwedge_{0<i\neq j\leq n} a_{ij}x_i - b_{ij}x_j \prec c_{ij},$$

such that for $0 < i \neq j \leq n$

$$\begin{cases} a_{ij} = l_j/\widetilde{g}(l_j, r_i) & b_{ij} = r_i/\widetilde{g}(l_j, r_i), & if\ l_i \geq 0\ and\ l_j \geq 0 \\ a_{ij} = l_j/\widetilde{g}(l_j, l_i) & b_{ij} = l_i/\widetilde{g}(l_j, l_i), & if\ l_i \geq 0\ and\ r_j < 0 \\ a_{ij} = r_j/\widetilde{g}(r_j, r_i) & b_{ij} = r_i/\widetilde{g}(r_j, r_i), & if\ r_i < 0\ and\ l_j \geq 0 \\ a_{ij} = r_j/\widetilde{g}(r_j, l_i) & b_{ij} = l_i/\widetilde{g}(r_j, l_i), & if\ r_i < 0\ and\ r_j < 0 \end{cases}$$

A *hybrid zone* is a conjunction of inequalities that accords with some q–zone, where $q \in Q$ is a location of a rectangular automaton. By using a variable x_0 which is always 0, we can obtain a general form of a hybrid zone:

$$x_0 = 0 \wedge \bigwedge_{0 \leq i \neq j \leq n} a_{ij}x_i - b_{ij}x_j \prec c_{ij} \qquad (3-1)$$

where $a_{0k} = b_{0k} = a_{k0} = b_{k0} = 1$, for $0 \leq k \leq n$.

Given a hybrid zone D represented by $\bigwedge ax - by \prec c$, we call $(\prec, c)$ the *bound.* For two bounds $(\prec, c)$ and $(\prec', c')$, $(\prec, c)$ is tighter than $(\prec', c')$, denoted by $(\prec,\ c) \sqsubset (\prec', c')$, iff $c < c'$ or $c = c'$, $\prec$ is $<$ and $\prec'$ is $\leq$. Further, for $\prec$ and $\prec'$, we define $min(\prec, \prec')$ is $\leq$ if both $\prec$ and $\prec'$ are $\leq$, and $min(\prec, \prec')$ is $<$ if one of $\prec$ and $\prec'$ is $<$. For a constraint $ax - by \prec c$ in D, if $(\prec', c')$ is the tightest bound of linear expression $ax - by$ that can be deduced from the conjunction of other constraints in D, then we call $ax - by \prec'' c''$ the *canonical constraint*; and D with each constraint being canonical the *canonical* hybrid zone, where $(\prec'', c'') = (\prec, c)$ if $(\prec, c) \sqsubset (\prec', c')$, otherwise, $(\prec'', c'') = (\prec', c')$.

The following lemmas and theorems ensure that hybrid zones are closed over those three reachability operations of rectangular hybrid systems, which enables hybrid zones to be used as the basis for the state reachability analysis algorithm for rectangular hybrid systems in Fig. 1.

To prove that the projection of a hybrid zone onto a lower dimensional subspace is also a hybrid zone, we need to prove that any newly produced constraint from arbitrary two constraints contained in the hybrid zone must be redundant. For this, we give a sufficient condition that a hybrid zone has *q–property.* If a q-zone D has q-property, then any constraint newly produced from two constraints in D can be deduced from the conjunction of other constraints in D.

Given a location q of a rectangular automaton and a q-zone D, the canonical form of D is represented by (3-1). We say D has *q-property*, iff for $0 < i \neq j \neq k \leq n$, one of the following conditions is satisfied:

1. if $l_i \geq 0$, $l_j \geq 0$, $l_k \geq 0$, then
   $\widetilde{g}(l_j, r_i)l_k c_{ij} + (r_k r_i - l_k r_i)c_{0j} \leq \widetilde{g}(l_k, r_i)l_j c_{ik} + \widetilde{g}(l_j, r_k)r_i c_{kj}$
2. if $l_i \geq 0$, $r_j \geq 0$, $r_k < 0$, then
   $-\widetilde{g}(l_j, r_i)r_k c_{ij} + (r_k r_i - l_k r_i)c_{0j} \leq \widetilde{g}(r_i, r_k)l_j c_{ki} + \widetilde{g}(l_k, l_j)r_i c_{jk}$
3. if $l_i \geq 0$, $l_j < 0$, $l_k \geq 0$, then
   $\widetilde{g}(l_j, l_i)r_k c_{ij} + (r_k l_i - l_k l_i)c_{j0} \leq \widetilde{g}(l_j, l_k)l_i c_{kj} - \widetilde{g}(l_i, r_k)l_j c_{ki}$
4. if $l_i \geq 0$, $l_j < 0$, $l_k < 0$, then
   $-\widetilde{g}(l_j, l_i)l_k c_{ij} + (r_k l_i - l_k l_i)c_{j0} \leq \widetilde{g}(r_k, l_j)l_i c_{jk} - \widetilde{g}(l_k, l_i)l_j c_{ik}$
5. if $l_i < 0$, $l_j \geq 0$, $l_k \geq 0$, then
   $\widetilde{g}(r_j, r_i)r_k c_{ij} + (l_k r_i - r_k r_i)c_{0j} \leq \widetilde{g}(r_k, r_i)r_j c_{ik} - \widetilde{g}(l_k, r_j)r_i c_{jk}$

6. if $l_i < 0$, $r_j \geq 0$, $r_k < 0$, then
   $-\widetilde{g}(r_j, r_i)l_k c_{ij} + (l_k r_i - r_k r_i)c_{0j} \leq \widetilde{g}(r_i, l_k)r_j c_{ki} - \widetilde{g}(r_j, r_k)r_i c_{kj}$
7. if $l_i < 0$, $l_j < 0$, $l_k \geq 0$, then
   $\widetilde{g}(r_j, l_i)l_k c_{ij} + (l_k l_i - r_k l_i)c_{j0} \leq -\widetilde{g}(l_i, l_k)r_j c_{ki} - \widetilde{g}(r_k, r_j)l_i c_{jk}$
8. if $l_i < 0$, $l_j < 0$, $l_k < 0$, then
   $-\widetilde{g}(r_j, l_i)r_k c_{ij} + (l_k l_i - r_k l_i)c_{j0} \leq -\widetilde{g}(r_k, l_i)r_j c_{ik} - \widetilde{g}(r_j, l_k)l_i c_{kj}$

For a canonical q-zone D, and a variable x_r, D has q-property is a sufficient condition for $\exists x_r[D]$ being a q-zone with variables $X \setminus \{x_r\}$. Thus, the following lemma can be readily obtained.

Lemma 1. *Given a location q of a rectangular automaton, if D is a q-zone with q-property, x_r is a variable in X, then $\exists x_r[D]$ is a q-zone with q-property over variables $X \setminus \{x_r\}$.*

Proof. Suppose the canonical form of D is given by (3-1). Without loss of generation, let $r = n > 0$. Since D has q-property, then any inequality obtained from two inequalities in D by eliminating variable x_n are redundant. So, $\exists x_n[D]$ is the q-zone

$$x_0 = 0 \wedge \bigwedge_{0 \leq i \neq j < n} a_{ij}x_i - b_{ij}x_j \prec c_{ij}.$$

Since D has q-property, so $\exists x_n[D]$ has q-property over $X \setminus \{x_n\}$. □

Given a canonical hybrid zone D represented by (3-1), it is easy to prove that D is empty iff there exists a $k.(0 < k \leq n)$ such that $(min(\prec_{0k}, \prec_{k0}), c_{0k} + c_{k0}) \sqsubset (\leq, 0)$. For any location q of a rectangular automaton, a bounded rectangle can be easily expressed as a q-zone with q-property. This ensures that the assignment of values to variables in the initial condition and every variable constraint used in the invariant condition of a rectangular automaton location can be expressed as hybrid zones. Moreover the jump condition of a transition is a hybrid zone. The following three theorems ensure that hybrid zones keep closed over the three reachability operations.

Theorem 1. *Given a location q of a rectangular automaton, if D is a q-zone with q-property, then $D^{\uparrow_q}$ is a q-zone with q-property.*

Theorem 2. *Given a location q of a rectangular automaton, if D' is a bounded rectangle and D is a q-zone with q-property, then $D \wedge D'$ is a q-zone with q-property.*

The proofs of Theorem 1 and 2 can refer to our another paper [13]. To ensure $D \wedge D'$ having q-property is necessary and crucial, as after being intersected with D', the hybrid zone D will traverse an edge e with q as the left end location, $D \wedge D'$ having q-property is the sufficient condition that all the valuation that can be reached from $D \wedge D'$ by traversing the edge e can be represented by the hybrid zone. For the q-zone D, an edge $e' = (q, q')$, and a single variable set $update(e') = \{x_i\}$, by Lemma 1, $\mathbf{Reset}_{e'}[D] = \exists x_i[D] \wedge x_i \in reset(e')_i$ is a q'–zone. It is important that $\mathbf{Reset}_{e'}[D]$ has q'-property, as it will traverse another edge by time elapsing and intersected with a jump condition. This is ensured by the following lemma. The result can easily be extended to sets with more than one variable by induction, which is the conclusion of Theorem 3. The proofs of Lemma 2 and Theorem 3 can refer to our paper [13].

Lemma 2. *Given two locations q and q' of a rectangular automaton. Suppose D, given by*

$$x_0 = 0 \wedge \bigwedge_{0 \le i \ne j < n} a_{ij}x_i - b_{ij}x_j \prec c_{ij}$$

is a canonical q-zone with q-property, and q' is a location such that $act(q')_i = act(q)_i$, for $0 < i < n$. Then

$$x_0 = 0 \wedge x_0 - x_n \prec c_{0n} \wedge x_n - x_0 \prec c_{n0} \wedge \bigwedge_{0 \le i \ne j < n} a_{ij}x_i - b_{ij}x_j \prec c_{ij}$$

is a q'-zone with q'-property, where c_{0n} and c_{n0} are two constant rational numbers such that $c_{0n} + c_{n0} \ge 0$.

Theorem 3. *Given an edge $e = (q, q')$ of a rectangular automaton. If D is a q-zone with q-property, then $\boldsymbol{Reset}_e[D]$ is a q'-zone with q'-property.*

4 Model Checking

In this section, we address the model checking problem of rectangular hybrid systems, which is to check whether a given rectangular automaton satisfies a requirement expressed in the timed computing tree logic [1,14].

4.1 Timed Computation Tree Logic

Here, we introduce the TCTL presented in [1]. Let C be a set of clocks satisfying $C \cap X = \emptyset$, a *state predicate* is a linear formula over the set $C \cup X$. The syntax of TCTL is given by the grammar

$$\phi \ ::= \ \psi \mid \neg\phi \mid \phi_1 \vee \phi_2 \mid z.\phi_1 \mid \phi_1 \exists\mathcal{U}\phi_2 \mid \phi_1 \forall\mathcal{U}\phi_2$$

where ψ is a state predicate, '$\exists\mathcal{U}$' and '$\forall\mathcal{U}$' are temporal operators. The formula ϕ is *closed* if all occurrences of a clock $z \in C$ are within the scope of a reset quantifier z. In the following, some abbreviations are given: $\forall\Diamond\phi \overset{def}{=} true\forall\mathcal{U}\phi$, $\exists\Diamond\phi \overset{def}{=} true\exists\mathcal{U}\phi$, $\forall\Box\phi \overset{def}{=} \neg\exists\Diamond\neg\phi$, $\exists\Box\phi \overset{def}{=} \neg\forall\Diamond\neg\phi$. We also put timing constraints as subscripts on the temporal operators. For example, the formula $z.\exists\Diamond(\phi \wedge z \le 3)$ is abbreviated to $\exists\Diamond_{z\le3}\phi$.

Given a run $\rho = s_0 \mapsto^{t_0}_{f_0} s_1 \mapsto^{t_1}_{f_1} s_2 \mapsto^{t_2}_{f_2} \ldots$ of a rectangular automaton $\mathcal{H}$, we can construct a function $g : R^+ \to S$ such that for any $t \in R^+$, $g(t) = (q_i, v_i + f_i \cdot t')$, where i and $t' \in R^+$ satisfy $t = \sum_{j=0}^{i} t_j + t'$ and $s_i = (q_i, v_i)$. We use $\rho(t)$ to denote the state $g(t)$.

A *clock valuation* ν is a function from C to R^+. For any clock $z \in C$, we use $\nu[z := 0]$ to denote the clock valuation ν' such that $\nu'(z) = 0$ and $\nu'(z') = \nu(z')$, for all $z'(z' \neq z) \in C$. An *extended state* is a tuple (s, ν). The extended state (s, ν) satisfies the TCTL-formula ϕ, denoted $(s, \nu) \models \phi$, if

1. $(s, \nu) \models \psi$ iff $(s, \nu)(\psi)$.
2. $(s, \nu) \models \neg\phi$ iff $(s, \nu) \not\models \phi$.
3. $(s, \nu) \models \phi_1 \vee \phi_2$ iff $(s, \nu) \models \phi_1$ or $(s, \nu) \models \phi_2$.
4. $(s, \nu) \models z.\phi$ iff $(s, \nu[z := 0]) \models \phi$.
5. $(s, \nu) \models \phi_1 \exists \mathcal{U} \phi_2$ iff there exists a run ρ of rectangular automaton $\mathcal{H}$ with $\rho(0) = s$ and a time $t \in R^+$ such that $(\rho(t), \nu + t) \models \phi_2$ and for any $0 \leq t' \leq t$, $(\rho(t'), \nu + t') \models \phi_1 \vee \phi_2$.
6. $(s, \nu) \models \phi_1 \forall \mathcal{U} \phi_2$ iff for all divergent runs ρ of rectangular automaton $\mathcal{H}$ with $\rho(0) = s$ and a time $t \in R^+$ such that $(\rho(t), \nu + t) \models \phi_2$ and for any $0 \leq t' \leq t$, $(\rho(t'), \nu + t') \models \phi_1 \vee \phi_2$.

For a closed TCTL formula ϕ, a state $s \in S$ satisfies ϕ, denoted by $s \models \phi$, if $(s, \nu) \models \phi$ for all clock valuation ν. The rectangular automaton $\mathcal{H}$ satisfies ϕ, denoted by $\mathcal{H} \models \phi$, if all states of $\mathcal{H}$ satisfies ϕ. We use $|\phi|$ to denote the states of $\mathcal{H}$ that satisfies ϕ.

4.2 Model Checking

Given a closed TCTL-formula ϕ, a model-checking algorithm computes the characteristic set $|\phi|$. We use the algorithm in [1] to deal with the model-checking problem for rectangular hybrid systems with our hybrid zones.

The procedure is based on fixpoint characterizations of the TCTL-modalities in terms of a binary next operator $\triangleright$. Given two sets of states $\mathcal{R}$ and $\mathcal{R}'$, $\mathcal{R} \triangleright \mathcal{R}'$ is the set of the state s that have a successor $s' \in \mathcal{R}'$ such that all states between s and s' are contained in $\mathcal{R} \cup \mathcal{R}'$: $(q, v) \in \mathcal{R} \triangleright \mathcal{R}'$ iff

$$\begin{aligned} &\exists (q', v') \in \mathcal{R}', \exists t \in R^+, \exists e \in E.((q', v') \in (q', \mathbf{Reset}_e(\{v\}^{\uparrow_q})) \wedge \\ &\forall 0 \leq t' \leq t.((q, v + act(q) \cdot t) \in \mathcal{R} \cup \mathcal{R}')), \end{aligned}$$

that is, the $\triangleright$ operator is a *single-step until* operator. To define the $\triangleright$ operator syntactically, we introduce two notations.

Given an edge $e = (q, q') \in E$ and a set of valuations D, we define the *backward time elapsing* in location q $D^{\downarrow_q}$ and the *backward zone* $\mathbf{Bac}_e(D)$ as

$$\begin{aligned} &v' \in D^{\downarrow_q} \; iff \; \exists v \in D, \; \exists f \in act(q), \; \exists t \in R^+.(v = v' + f \cdot t \\ &\qquad \wedge \forall 0 \leq t' \leq t.((v' + f \cdot t') \in inv(q))), \\ &v \in \mathbf{Bac}_e(D) \; iff \; \exists v' \in D.(v' \in \mathbf{Reset}_e(\{v\})). \end{aligned}$$

Then, for two sets of states $\mathcal{R}$ and $\mathcal{R}'$, let $D = \{v | \exists q \in Q.((q, v) \in \mathcal{R}')\}$, we can define $\mathcal{R} \triangleright \mathcal{R}'$ as

$$(q, v') \in \mathcal{R} \triangleright \mathcal{R}' \; iff \; \exists \; e \in E, \; \exists \; v \in \mathbf{Bac}_e(D), \; \exists \; f \in act(q).(v' \in \{v\}^{\downarrow_q}).$$

Lemma 3. *Given a location q of a rectangular automaton, if D is a q-zone with q-property, then $D^{\downarrow_q}$ is a q-zone with q-property.*

Proof. Suppose q' is a location such that for all $k \in R$, $k \in act(q)$ if and only if $-k \in act(q')$. By the definition of q–zone and q–property, D is a q-zone iff D is a q'-zone, D has q-property iff D has q'-property. By the definition of *backward time elapsing*, $D^{\downarrow_q} = D^{\uparrow_{q'}}$. By Theorem 1, $D^{\uparrow_{q'}}$ is a q'-zone with q'-property, then $D^{\downarrow_q}$ is a q-zone with q-property.

Lemma 4. *Given an edge $e = (q, p)$ of a rectangular automaton, if D is a p-zone with p-property, then $\mathbf{Bac}_e[D]$ is a q-zone with q-property.*

Proof. Suppose *update*$(e) = \{i_1, ..., i_m\}$, by the definition of *backward zone*, $\mathbf{Bac}_e[D]$ is

$$\exists x_{i_1} \cdots \exists x_{i_m} [D \wedge \bigwedge_{j \in update(e)} x_j \in reset(e)_j].$$

By the definition of initialled rectangular automaton, for $\forall k \in \{1, ..., n\} \setminus update(e)$, $act(q)_k = act(p)_k$. By Lemma 1 and Theorem 2, $\mathbf{Bac}_e[D]$ is a q-zone with q-property.

Theorem 4. *Given two locations q and q' of a rectangular automaton, if D is a q-zone, D' is a q'-zone, $\mathcal{R} = (q, D)$ and $\mathcal{R}' = (q', D')$ are two state regions, then $D^* = \{v | (q, v) \in \mathcal{R} \rhd \mathcal{R}'\}$ is a q-zone.*

Proof. It is the immediate consequence of Lemma 3 and 4.

```
∃U(φ1, φ2)        {                          /* φ1, φ2 are two formulas */
    R1 = |φ1|;  R2 = |φ2|; R' = |φ2|;   R = |φ1| ▷ |φ2|;
    while(R ⊈ R')        {
        R' = R' ∪ R;
        R = R1 ▷ R;    }
    return  R';   }
∀□(φ)             {                          /* φ is a formula */
    R' = |φ|;    R = |φ| ∩ ¬(true ▷ ¬|φ|)
    while(R ⊈ R')                {
        R' = R' ∩ R;
        R = R ∩ ¬(true ▷ ¬R);    }
    return  R';   }
∀◇≤c(φ)           {                          /* φ is a formula */
    R = |φ|;    R* = |z > c|;   R' = |z > c| ∩ ((¬R) ▷ ¬|z > c|);     /*z ∈ C*/
    while(R' ⊈ R*)               {
        R* = R* ∩ R';
        R' = R' ∩ ((¬R) ▷ ¬R'); }
    return  R*; }
```

Fig. 2. The model checking procedures

The meaning of both TCTL- modalities $\forall\mathcal{U}$ and $\exists\mathcal{U}$ can be computed iteratively as fixpoints using the $\triangleright$ operator. By [1], the iterative fixpoint computation always terminates for rectangular hybrid systems. Theorem 4 ensures that all regions that are computed by the process are hybrid zones. Refer to the method presented in [1], Fig. 2 gives the model checking procedures for some important classes of TCTL-formulas,

5 Difference Constraint Matrix

To represent hybrid zones, we formalize a data structure *difference constraint matrix*. This matrix is indexed by the variables in X together with a special variable x_0 whose value is always 0. This variable plays exactly the same role as the variable x_0 in the previous section. Each entry $\mathcal{D}_{ij}$ $(i \neq j)$ in the matrix $\mathcal{D}$ has the form $(a_{ij}, b_{ij}, d_{ij}, \prec_{ij})$ and represents the inequality $a_{ij}x_i - b_{ij}x_j \prec_{ij} d_{ij}$, or $(a_{ij}, b_{ij}, \infty, <)$, if no bound is known for $a_{ij}x_i - b_{ij}x_j$. Each entry $\mathcal{D}_{ii}$ has the form $(1, 1, d_{ii}, \prec_{ii})$.

Given a hybrid zone D in its general form represented by (3-1). We can obtain the matrix $\mathcal{D}$ shown below:

- $\mathcal{D}_{ij} = (a_{ij}, b_{ij}, c_{ij}, \prec_{ij})$, for $0 \leq i \neq j \leq n$.
- $\mathcal{D}_{ii} = (1, 1, 0, \leq)$, for $0 \leq i \leq n$.

A DCM $\mathcal{D}$ with each $\mathcal{D}_{ij} = (a_{ij}, b_{ij}, d_{ij}, \prec_{ij})$ is *canonical* iff the hybrid zone represented by $\mathcal{D}$ is canonical, and $(min(\prec_{i0}, \prec_{0i}), (d_{i0} + d_{0i})) \not\subset (\prec_{ii}, d_{ii})$, for all $0 \leq i \leq n$. We describe five operations on canonical DCM. These operations correspond to the five operations defined on hybrid zones.

Intersection. Given a location q of a rectangular automaton, and two q-zones respectively represented by two DCMs $\mathcal{D}^1$ and $\mathcal{D}^2$, we define $\mathcal{D} = \mathcal{D}^1 \wedge \mathcal{D}^2$. Let $\mathcal{D}^1_{ij} = (a, b, c_1, \prec_1)$ and $\mathcal{D}^2_{ij} = (a, b, c_2, \prec_2)$. Then $\mathcal{D}_{ij} = (a, b, min(c_1, c_2), \prec)$, where $\prec$ is defined as follows:

- If $c_1 < c_2$, then $\prec = \prec_1$.
- If $c_2 < c_1$, then $\prec = \prec_2$.
- If $c_1 = c_2$ and $\prec_1 = \prec_2$, then $\prec = \prec_1$.
- If $c_1 = c_2$ and $\prec_1 \neq \prec_2$, then $\prec = <$.

Variable Reset. Given an edge $e = (q, q')$ of a rectangular automaton, and a q-zone represented by a DCM $\mathcal{D}$. Suppose $\mathcal{D}_{ij} = (a_{ij}, b_{ij}, c_{ij}, \prec_{ij})$ and $reset(e)_k = -x_k \prec_k \mu_k \wedge x_k \prec'_k \nu_k$, for $k \in update(e)$. We define $\mathcal{D}' = \mathbf{Reset}_e(\mathcal{D})$ as follows:

- $\mathcal{D}'_{ij} = (a_{ij}, b_{ij}, \infty, <)$, for $i \in update(e)$ or $j \in update(e)$ such that $i \neq j$ and $i \cdot j \neq 0$, where a_{ij} and b_{ij} satisfy the definition of q'−zone.
- $\mathcal{D}'_{ij} = \mathcal{D}_{ij}$, for $(i \neq j) \notin update(e)$.
- $\mathcal{D}'_{0i} = (1, 1, \mu_i, \prec_i)$, $\mathcal{D}'_{i0} = (1, 1, \nu_i, \prec'_i)$, for $i \in update(e)$ and $0 < i \leq n$.
- $\mathcal{D}'_{ii} = \mathcal{D}_{ii}$, for $0 \leq i \leq n$.

Elapsing of Time in Location *q*. Given a location q of a rectangular automaton, and a q-zone represented by a DCM $\mathcal{D}$, we define $\mathcal{D}' = \mathcal{D}^{\uparrow_q}$ as follows:

- $\mathcal{D}'_{i0} = (1, 1, \infty, <)$, $\mathcal{D}'_{0i} = \mathcal{D}_{0i}$, if $l_i \geq 0$; $\mathcal{D}'_{i0} = \mathcal{D}_{i0}$, $\mathcal{D}'_{0i} = (1, 1, \infty, <)$, if $r_i < 0$, for $i \neq 0$.
- $\mathcal{D}'_{ij} = \mathcal{D}_{ij}$, for $i = j = 0$ or $i \cdot j \neq 0$.

Backward Time Elapsing in Location *q*. Suppose $\mathcal{D}_{ij} = (a_{ij}, b_{ij}, c_{ij}, \prec_{ij})$, define $\mathcal{D}' = \mathcal{D}^{\downarrow_q}$ as follows:

- $\mathcal{D}'_{i0} = (1, 1, \infty, <)$, $\mathcal{D}'_{0i} = \mathcal{D}_{0i}$, if $l_i < 0$; $\mathcal{D}'_{i0} = \mathcal{D}_{i0}$, $\mathcal{D}'_{0i} = (1, 1, \infty, <)$, if $r_i \geq 0$, for any $i \neq 0$.
- $\mathcal{D}'_{ij} = \mathcal{D}_{ij}$, if $i = j = 0$ or $i \neq 0 \wedge j \neq 0$.

Backward Zone. Suppose $e = (q, q')$ is an edge of a rectangular automaton, and the canonical form of $\mathcal{D}' = \mathcal{D} \wedge \bigwedge_{i \in update(e)} reset(e)_i$ has each $\mathcal{D}'_{ij}$ as $(a_{ij}, b_{ij}, c_{ij}, \prec_{ij})$. We define $\mathcal{D}^* = \mathbf{Bac}_e(D)$ as follows:

- $\mathcal{D}^*_{ij} = (a_{ij}, b_{ij}, \infty, <)$, for $i \in update(e)$ or $j \in update(e)$.
- $\mathcal{D}^*_{ij} = \mathcal{D}'_{ij}$, for $i, j \notin update(e)$.

In each case the resulting DCM may fail in its canonical form. Thus, as a final step we need to reduce the DCM to the canonical form. Given a hybrid zone D represented by (3-1), to find the canonical form of a constraint $a_{ij}x_i - b_{ij}x_j \prec c_{ij}$ is equal to find the maximum of function $f(x_i, x_j) = a_{ij}x_i - b_{ij}x_j$ over the constraint system D. This is the issue of linear programming, and it can be automated by the algorithm in [12]. As clarified in [12], the running-time of this algorithm is $O(n^{3.5}L^2)$, where n is the dimension of the problem and L is the number of bits in the input. Hence, finding canonical form of hybrid zone D can be automated within a polynomial-time $O(n^{5.5}L^2)$.

6 Conclusion

In this paper, a hybrid zone with constraints less than n^2 and a data structure DCM are defined. After the DCM has been converted to canonical form, the manipulating operations of rectangular hybrid systems on hybrid zones can be implemented straightforwardly. Hence, the main computation is n^2 operations for obtaining canonical form at most. Finding the canonical form can be automated by an algorithm with polynomial time-complexity. In addition, we implement the reachability operations and model checking algorithms for rectangular hybrid systems using DCMs.

References

1. Alur, R., Courcoubetis, C., Halbwachs, N., Henzinger, T.A., Ho, P.-H., Nicollin, X., Olivero, A., Sifakis, J., Yovine, S.: The algorithmic analysis of hybrid systems. Theoretical Computer Science 138, 3–34 (1995)
2. Alur, R., Courcoubetis, C., Henzinger, T.A., Ho, P.-H.: Hybrid Automata: an Algorithmic Approach to the Specification and Verification of Hybrid Systems. In: Grossman, R.L., Ravn, A.P., Rischel, H., Nerode, A. (eds.) HS 1991 and HS 1992. LNCS, vol. 736, Springer, Heidelberg (1993)

3. Alur, R., Henzinger, T.A., Ho, P.-H.: Automatic Symbolic Verification of Embedded Systems. In: Proceedings of 1993 IEEE Real-Time System Symposium (1993)
4. Annichini, A., Asarin, E., Bouajjani, A.: Symbolic Techniques for Parametric Reasoning about Counter and Clock Systems. In: Emerson, E.A., Sistla, A.P. (eds.) CAV 2000. LNCS, vol. 1855, Springer, Heidelberg (2000)
5. Dill, D.L.: Timing Assumptions and Verification of Finite-State Concurrent Systems. LNCS, vol. 407, pp. 197–212 (1989)
6. Duan, Z.: Modeling of hybrid systems. Ph.D thesis, Department of Computer Science, University of Shefield, UK (February 1997)
7. Henzinger, T.A., Kopke, P.W., Puri, A., Varaiya, P.: What′s decidable about hybrid automata? J.Comput.Syst.Sci. 57, 94–124 (1998)
8. Henzinger, T.A., Majumdar, R.: Symbolic model checking for rectangular hybrid systems(Abstract). In: Schwartzbach, M.I., Graf, S. (eds.) ETAPS 2000 and TACAS 2000. LNCS, vol. 1785, pp. 146–156. Springer, Heidelberg (2000)
9. Kopke, P.W.: The thoery of Rectangular Hybrid Automata. Ph.D thesis, Cornell University (1996)
10. Wang, F.: Symbolic Parametric Safety Analysis of Linear Hybrid Systems with BDD-Like Data-Structures. In: Alur, R., Peled, D.A. (eds.) CAV 2004. LNCS, vol. 3114, pp. 295–307. Springer, Heidelberg (2004)
11. Behrmannl, G., Larsen, K.G., Pearson, J., Weise, C., Yi, W.: Efficient Timed Reachability Analysis Using Clock Diffierence Diagrams. In: Halbwachs, N., Peled, D.A. (eds.) CAV 1999. LNCS, vol. 1633, pp. 341–353. Springer, Heidelberg (1999)
12. Karmarkar, N.: A new polynomial-time algorithm for linear programming. Combinatorica 4(4), 373–395 (1984)
13. Zhang, H., Duan, Z.: Symbolic Reachability Analysis of Rectangular Hybrid Systems. In: ICCP 2006, Cluj-Napoca, Romania, pp. 240–248 (2006)
14. Alur, R., Courcoubetis, C., Dill, D.L.: Model-checking in dense real-time. Inform. Computat. 104, 2–34 (1993)
15. Anai, H., Weispfnening, V.: Reach Set Computations Using Real Quantifier Elimination. In: Di Benedetto, M.D., Sangiovanni-Vincentelli, A.L. (eds.) HSCC 2001. LNCS, vol. 2034, pp. 63–76. Springer, Heidelberg (2001)
16. Zhang, H., Duan, Z.: Symbolic Reachability Analysis of Multirate Hybrid Systems. Journal of Xi'an Jiao Tong University 41(4), 412–416 (2007)
17. Zhang, H., Duan, Z.: Decidability of The Dense Timed Interval Temporal Logic. Journal of Xidian University 34(3), 463–467 (2007)

On the OBDD Complexity of the Most Significant Bit of Integer Multiplication
(Extended Abstract)

Beate Bollig

LS2 Informatik, TU Dortmund,
44221 Dortmund, Germany
beate.bollig@uni-dortmund.de

Abstract. Integer multiplication as one of the basic arithmetic functions has been in the focus of several complexity theoretical investigations. Ordered binary decision diagrams (OBDDs) are the most common dynamic data structure for boolean functions. Among the many areas of application are verification, model checking, computer-aided design, relational algebra, and symbolic graph algorithms. In this paper it is shown that the OBDD complexity of the most significant bit of integer multiplication is exponential answering an open question posed by Wegener (2000).

Keywords: Computational complexity, integer multiplication, lower bounds, ordered binary decision diagrams.

1 Introduction and Result

Integer multiplication is certainly one of the most important functions in computer science and a lot of effort has been spent in designing good algorithms and small circuits and in determining its complexity. For one of the latest results see, e.g., [8]. Ordered binary decision diagrams (OBDDs) are the most common dynamic data structure for boolean functions. Although many even exponential lower bounds on the OBDD size of boolean functions are known and the lower bound methods how to obtain such bounds are simple, it is often a more difficult task to prove large lower bounds for some predefined and interesting functions. Despite the well-known lower bounds on the OBDD size of the so-called middle bit of multiplication ([7], [17]), until now the OBDD complexity of the most significant bit of multiplication has been unknown and Wegener [16] has asked whether its OBDD complexity is exponential. In the following we answer his question affirmatively.

1.1 Branching Programs or Binary Decision Diagrams

Besides boolean circuits and formulae, branching programs (BPs), sometimes also called binary decision diagrams (BDDs), are one of the standard representations for boolean functions. (For a history of results on branching programs see, e.g., the monograph of Wegener [16]).

M. Agrawal et al. (Eds.): TAMC 2008, LNCS 4978, pp. 306–317, 2008.

Definition 1. *A* branching program (BP) *on the variable set* $X_n = \{x_1, \ldots, x_n\}$ *is a directed acyclic graph with one source and two sinks labeled by the constants 0 and 1. Each non-sink node (or decision node) is labeled by a boolean variable and has two outgoing edges, one labeled by 0 and the other by 1.*

An input $b \in \{0,1\}^n$ *activates all edges consistent with* b*, i.e., the edges labeled by* b_i *which leave nodes labeled by* x_i*. A* computation path *for an input* b *in a* BP G *is a path of edges activated by the input* b *which leads from the source to a sink. A computation path for an input* b *which leads to the 1-sink is called* accepting path *for* b*.*

Let B_n *denote the set of all boolean functions* $f : \{0,1\}^n \rightarrow \{0,1\}$*. The* BP G *represents a function* $f \in B_n$ *for which* $f(b) = 1$ *iff there exists an accepting path for the input* b*.*

The size *of a branching program* G *is the number of its nodes and is denoted by* $|G|$*. The* branching program size *of a boolean function* f *is the size of the smallest* BP *representing* f*. The* length *of a branching program is the maximum length of a path.*

It is well known that the logarithm of the branching program size is essentially the same as the space complexity of the nonuniform variant of Turing machines. Hence, it is a fundamental open problem to prove superpolynomial lower bounds on the size of branching programs for explicitly defined boolean functions. In order to develop and strengthen lower bound techniques one considers restricted computation models. There are several possibilities to restrict BPs, among them restrictions on the multiplicity of variable tests or the order in which variables may be tested.

Definition 2. *i) A branching program is called* read-k-times (BPk) *if each variable is tested on each path at most* k *times.*

ii) A branching program is called s-oblivious *for a sequence of variables* $s = (s_1, \ldots, s_l)$*,* $s_i \in X_n$*, or short* oblivious*, if the set of decision nodes can be partitioned into disjoint sets* V_i*,* $1 \leq i \leq l$*, such that all nodes from* V_i *are labeled by* s_i *and the edges which leave* V_i*-nodes reach a sink or a* V_j*-node where* $j > i$*. The* length *of an* s*-oblivious branching program is the length of the sequence* s*.*

Besides the complexity theoretical viewpoint people have used branching programs in applications. Representations of boolean functions that allow efficient algorithms for many operations, in particular synthesis (combine two functions by a binary operation) and equality test (do two representations represent the same function?) are necessary. Bryant [6] introduced ordered binary decision diagrams (OBDDs) which have become the most popular data structure for boolean functions. Among the many areas of application are verification, model checking, computer-aided design, relational algebra, and symbolic graph algorithms.

Definition 3. *An* OBDD *is a branching program with a* variable order *given by a permutation* π *on the variable set. On each path from the source to the sinks, the variables at the nodes have to appear in the order prescribed by* π *(where* *some variables may be left out). A π-OBDD is an* OBDD *ordered according to π. The π-*OBDD *size of f denoted by π-*OBDD(f) *is the size of the smallest π-*OBDD *representing f. The* OBDD size *of f, sometimes also called* OBDD complexity *of f, (denoted by* OBDD(f)) *is the minimum of all π-*OBDD(f).

1.2 Integer Multiplication and Binary Decision Diagrams

Lower bounds for integer multiplication are motivated by the general interest in the complexity of important arithmetic functions.

Definition 4. *The boolean function* $\mathrm{MUL}_{i,n} \in B_{2n}$ *maps two n-bit integers $x = x_{n-1} \ldots x_0$ and $y = y_{n-1} \ldots y_0$ to the ith bit of their product, i.e.,* $\mathrm{MUL}_{i,n}(x, y) = z_i$, *where $x \cdot y = z_{2n-1} \ldots z_0$.*

The bit z_{2n-1} is the most significant bit of integer multiplication in the following sense. Let $(z_{2n-1}, \ldots, z_0)$ be the binary representation of the integer z, i.e., $z = \sum_{i=0}^{2n-1} z_i \cdot 2^i$. Since the bit z_{2n-1} has the highest value, for the approximation of the value of the product of two n-bit numbers x and y it is the most interesting one. On the other hand for space bounded models of computation the most significant bit of integer multiplication is the easiest one to compute in the sense that if it cannot be computed with size $s(n)$, then any other bit z_i, $2n-1 < i \leq n-1$, cannot be computed with size $s(n/2)$. and any other bit z_i, $n-1 < i \leq 0$, cannot be computed in size $s(i/2)$.

The middle bit of integer multiplication (the bit z_{n-1}) is the hardest bit to compute for space bounded models of computation in the sense that if it can be computed with size $s(n)$, then any other bit can be computed with size at most $s(2n)$. More precisely, any branching program for $\mathrm{MUL}_{2n-1,2n}$ can be converted into a branching program representing $\mathrm{MUL}_{i,n}$, $0 \leq i \leq 2n-1$, by relabeling the nodes and by replacing some inputs with the constant 0. Therefore, the first exponential lower bounds have been proved for $\mathrm{MUL}_{n-1,n}$. For OBDDs Bryant [7] has presented an exponential lower bound of $2^{n/8}$ and Gergov has extended the result for so-called nondeterministic linear-length oblivious branching programs [9]. Later Ponzio has shown that the complexity of this function is $2^{\Omega(\sqrt{n})}$ for read-once branching programs [12]. Progress in the analysis of $\mathrm{MUL}_{n-1,n}$ has been achieved by a new approach using universal hashing. Woelfel [17] has improved Bryant's lower bound to $\Omega(2^{n/2})$ and Bollig and Woelfel [3] have presented a lower bound of $\Omega(2^{n/4})$ for read-once branching programs. Exponential lower bounds have also been proved for more general read-once branching program models that allow limited nondeterminism and for models where some but not all variables may be tested multiple times (see, e.g., [2], [5], [18], [4]). Finally, Sauerhoff and Woelfel [13] have presented exponential lower bounds on the size of read-k-times branching programs representing the middle bit of multiplication.

Despite the well-known lower bounds for the middle bit of multiplication, until now the OBDD complexity of the most significant bit of multiplication has been unknown. Since the most significant bit is a monotone function it seems to be easier to compute than the middle bit. The known upper bounds on the OBDD size confirms this intuition. Amano and Maruoka [1] have presented an upper bound of $O(2^n)$ on the OBDD size of the most significant bit of multiplication, whereas the best known upper bound for the middle bit is $O(2^{6/5n})$. Furthermore, in the lower bound proofs on the OBDD size for $\mathrm{MUL}_{n-1,n}$ it has been shown that for an arbitrary variable order π there exists an assignment b to one of the input vectors such that the π-OBDD size for the resulting subfunction is exponential. In contrast it is not difficult to see that the OBDD size for any subfunction of $\mathrm{MUL}_{2n-1,n}$ where one of the input vectors is a constant is $O(n^2)$.

Computing the set of nodes that are reachable from some source $s \in V$ in a digraph $G = (V, E)$ is an important problem in computer-aided design, hardware verification, and model checking. Proving exponential lower bounds on the space complexity of a common class of OBDD-based algorithms for the reachability problem, Sawitzki [14] has presented the first exponential lower bound on the size of π-OBDDs representing the most significant bit for the variable order π where the variables are tested according to increasing significance, i.e. $\pi = (x_0, y_0, x_1, y_1, \ldots, x_{n-1}, y_{n-1})$. For the lower bounds on the space complexity of the OBDD-based algorithms he has used the assumption that the output OBDDs use the same variable order as the input OBDDs. But in contrast, practical algorithms usually run variable reordering heuristics on intermediate OBDD results in order to minimize their size. Therefore, it is interesting whether the OBDD size of the most significant bit of multiplication is exponential with respect to an arbitrary variable order.

In this paper we present the following result.

Theorem 1. $\mathrm{OBDD}(\mathrm{MUL}_{2n-1,n}) = \Omega(2^{n/720})$.

As a by-product we obtain almost the same lower bound on the π-OBDD size for the variable order $\pi = (x_0, y_0, x_1, y_1, \ldots, x_{n-1}, y_{n-1})$ as Sawitzki using a much simpler proof.

2 Preliminaries

2.1 Notation

In the rest of the paper we use the following notation.

Let $[x]_r^l$, $n-1 \geq l \geq r \geq 0$, denote the bits $x_l \ldots x_r$ of a binary number $x = (x_{n-1}, \ldots, x_0)$. For the ease of description we use the notation $[x]_r^l = z$ if $(x_l, \ldots, x_r)$ is the binary representation of the integer $z \in \{0, \ldots, 2^{l-r+1} - 1\}$. Sometimes, we identify $[x]_r^l$ with z if the meaning is clear from the context.

Let $\ell \in \{0, \ldots, 2^m - 1\}$, then $\overline{\ell}$ denotes the number $(2^m - 1) - \ell$.

2.2 Communication Complexity

In order to obtain lower bounds on the size of OBDDs one-way communication complexity has become a standard technique (see Hromkovič [10] and Kushilevitz and Nisan [11] for the theory of communication complexity and the results mentioned below).

The main subject is the analysis of the following (restricted) communication game. Consider a boolean function $f \in B_n$ which is defined on the variables in $X_n = \{x_1, \ldots, x_n\}$, and let $\Pi = (X_A, X_B)$ be a partition of X_n. Assume that Alice has only access to the input variables in X_A and Bob has only access to the input variables in X_B. In a one-way communication protocol, upon a given input x, Alice is allowed to send a single message (depending on the input variables in X_A) to Bob who must then be able to compute the answer $f(x)$. The *one-way communication complexity* of the function f denoted by $C(f)$ is the worst case number of bits of communication which need to be transmitted by such a protocol that computes f. It is easy to see that an OBDD G with respect to a variable order where the variables in X_A are tested before the variables in X_B can be transformed into a communication protocol and $C(f) \leq \lceil \log |G| \rceil$. Therefore, linear lower bounds on the communication complexity of a function f lead to exponential lower bounds on the OBDD complexity.

One central notion of communication complexity are fooling sets which play an important role for the lower bound proof used later on.

Definition 5. *Let $f : \{0,1\}^{|X_a|} \times \{0,1\}^{|X_B|} \rightarrow \{0,1\}$. A set $S \subseteq \{0,1\}^{|X_a|} \times \{0,1\}^{|X_B|}$ is called* fooling set *for f if $f(a,b) = c$ for all $(a,b) \in S$ and some $c \in \{0,1\}$ and if for different pairs $(a_1, b_1), (a_2, b_2) \in S$ at least one of $f(a_1, b_2)$ and $f(a_2, b_1)$ is unequal to c.*

Theorem 2. *If $f : \{0,1\}^{|X_a|} \times \{0,1\}^{|X_B|} \rightarrow \{0,1\}$ has a fooling set of size t, the communication complexity of f is bounded below by $\lceil \log t \rceil$.*

Because of our considerations above, the size t of a fooling set for f is a lower bound on the size of OBDDs representing f with respect to a variable order where the variables X_A are tested before the variables X_B. Because of the symmetric definition of fooling sets, t is also a lower bound on the size of OBDDs representing f with respect to a variable order where the variables X_B are tested before the variables X_A. The crucial thing to prove lower bounds on the OBDD complexity of a function is to obtain for all partitions of the variables lower bounds on the size of fooling sets for subfunctions of the given function.

Now we take a look at known results about the communication complexity of some functions. Let EQ: $\{0,1\}^n \times \{0,1\}^n \rightarrow \{0,1\}$ be defined by $\text{EQ}(a,b) = 1$ iff the vectors $a = (a_1, \ldots, a_n)$ and $b = (b_1, \ldots, b_n)$ are equal. It is well-known that $C(\text{EQ}) = n$. The same holds for the function IP: $\{0,1\}^n \times \{0,1\}^n \rightarrow \{0,1\}$ for which $\text{IP}(a,b) = 1$ iff $a_i \oplus b_i = 1$ for all $i \in \{1, \ldots, n\}$. Similar results can be obtained for the functions $\overline{\text{GT}} : \{0,1\}^n \times \{0,1\}^n \rightarrow \{0,1\}$, $\overline{\text{GT}^*} : \{0,1\}^n \times \{0,1\}^n \rightarrow \{0,1\}$, and $\overline{\text{GT}^{**}} : \{0,1\}^n \times \{0,1\}^n \rightarrow \{0,1\}$, where $\overline{\text{GT}}(a,b) = 1$ iff $[a]_1^n \leq [b]_1^n$, $\overline{\text{GT}^*}(a,b) = 1$ iff $\overline{\alpha} \leq [b]_1^n$, where α is the integer with binary representation a, and $\overline{\text{GT}^{**}}(a,b) = 1$ iff $[a]_1^n \leq \overline{\beta}$, where β is the integer with binary representation b. Furthermore, obviously the same results can be obtained if Alice gets exactly one of the variables a_i and b_i, $1 \leq i \leq n$. The reason is the following one. The addition function $\text{ADD}_{i,n} \in B_{2n}$ maps two n-bit integers $x = x_{n-1} \ldots x_0$ and $y = y_{n-1} \ldots y_0$ to the ith bit of their sum, i.e., $\text{ADD}_{i,n}(x,y) = s_i$, where $x + y = s_n \ldots s_0$. It is well-known that $\text{ADD}_{n,n}$

has a fooling set of size 2^n if for each i, $0 \le i \le n-1$, Alice gets exactly one of the variables x_i and y_i (variables of the same significance are symmetric variables for binary addition). $\overline{\mathrm{GT}}((a_{n-1}, \ldots, a_0), (b_{n-1}, \ldots, b_0))$ is equal to $\overline{\mathrm{ADD}_{n,n}}((a_{n-1}, \ldots, a_0), (\overline{b_{n-1}}, \ldots, \overline{b_0}))$, where $\overline{\mathrm{ADD}_{n,n}}(x, y)$ is the negated function $\mathrm{ADD}_{n,n}$.

The idea of Bryant's lower bound proof on the OBDD size of $\mathrm{MUL}_{n-1,n}$ [7] is the following. For each variable order, there is a subfunction of $\mathrm{MUL}_{n-1,n}$ which essentially equals the computation of the output bit at position m of the addition of two m-bit numbers x and y where $m \ge n/8$. The variable order is bad in the sense that among Alice's m variables is exactly one of the variables x_i and y_i.

3 An Exponential Lower Bound on the OBDD Complexity of the Most Significant Bit of Integer Multiplication

In this section, we determine the lower bound on the size of OBDDs for the representation of the most significant bit of multiplication mentioned above.

Besides Bryant's lower bound proof on the size of OBDDs representing the middle bit of multiplication we use the idea of the following reduction from multiplication to squaring presented by Wegener [15] where squaring computes the square of an n-bit input. For two n-bit numbers u and w the number $z := u \cdot 2^{2(n+1)} + w$ is defined. Then

$$z^2 = u^2 \cdot 2^{4(n+1)} + uw2^{2(n+1)+1} + w^2.$$

Since w^2 and uw are numbers of length $2n$, the binary representation of the product uw can be found in the binary representation of z^2.

In the following for the sake of simplicity we do not apply floor or ceiling functions to numbers even when they need to be integers whenever this is clear from the context and has no bearing on the essence of the proof.

We start our proof of the lower bound on the OBDD complexity of $\mathrm{MUL}_{2n-1,n}$ by the following observation (due to the lack of space we have to omit the proof).

Lemma 1. *A pair (x_i, y_{i+1}), $n/2 + 1 \le i \le (3/4)n - 2$, is called (x, y)-pair. Let S be the set of the first $|S|$ variables according to a variable order π. A pair (x_i, y_{i+1}) is called separated with respect to S iff $x_i \in S$ and $y_{i+1} \notin S$ or vice versa. If there exist a set S according to π such that there are at least m separated (x, y)-pairs with respect to S, the π-OBDD size of the most significant bit of integer multiplication is at least $2^{m/2}$.*

Now let π be an arbitrary variable order. The key observation for our lower bound proof is the following one.

Fact 1. *For a number $2^{n-1} + \ell 2^{(1/2)n}$, $\ell \le 2^{(1/6)n-1}$, the corresponding smallest number such that the product of the two numbers is at least 2^{2n-1} is $2^n - \ell 2^{(1/2)n+1} + 4\ell^2$. (Figure 1 shows the corresponding x- and y-inputs.)*

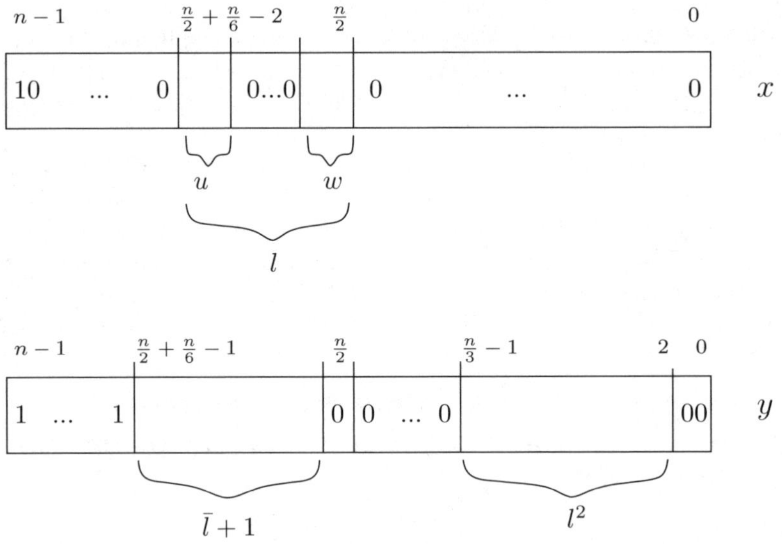

Fig. 1. The partition of the inputs x and y

In the following we take a closer look at the variables $x_{n/2}, \ldots, x_{n/2+n/6-2}$. For the ease of description we assume that $(n/6-1) \bmod 3 = 2$. We rename $[x]_{n/2}^{n/2+n/18-2}$ by $[w]_0^{m-1}$ and $[x]_{n/2+n/9}^{n/2+n/6-2}$ by $[u]_0^{m-1}$, where $m := (n/6-3)/3$. Figure 1 illustrates the partition of the input x.

Let S be the set of the first $|S|$ variables according to π where there are at least $(1/2)m$ variables from $\{w_0, \ldots, w_{m-1}\}$ for the first time. Let $I_S \subseteq \{0, \ldots, m-1\}$ be the set of indices i for which $w_i \in S$. Using simple counting arguments we can prove that there exists a distance parameter d such that there exists a set of pairs $P = \{(w_i, w_{i+d}) | i \in I_S \text{ and } (i+d) \notin I_S \text{ or } i \notin I_S \text{ and } (i+d) \in I_S, \text{ where } 0 \leq i < m/2 \leq i+d \leq m-1\}$ and $|P| \geq m/8$ (see [7] for a similar proof). Let I'' be the set of indices i, $0 \leq i < m/2$, where w_i belongs to a pair in P.

Case 1: There are at least $(2/5)m/8$ separated $(x_{n/2+i}, y_{n/2+i+1})$-pairs with respect to S, where $i \in I''$ or $i - d \in I''$. Using Lemma 1 we can conclude that the π-OBDD size of the most significant bit of integer multiplication is at least $2^{(1/5)m/8}$.

Case 2: There are less than $(2/5)m/8$ separated $(x_{n/2+i}, y_{n/2+i+1})$-pairs with respect to S, where $i \in I''$ or $i - d \in I''$. Let $I' \subseteq I''$ be the set of indices such that $(x_{n/2+i}, y_{n/2+i+1})$ and $(x_{n/2+i+d}, y_{n/2+i+d+1})$, $i \in I''$, are not separated with respect to S. Obviously, $|I'| \geq (3/5)m/8$.

Now we replace some of the variables in the following way.

- $y_{n-1}, \ldots, y_{n/2+n/6}$ are replaced by 1,
- $y_{n/2}, \ldots, y_{n/3}$, y_1, and y_0 are replaced by 0,
- x_{n-1} is replaced by 1,
- $x_{n-2}, \ldots, x_{n/2+n/6-1}$ are replaced by 0,

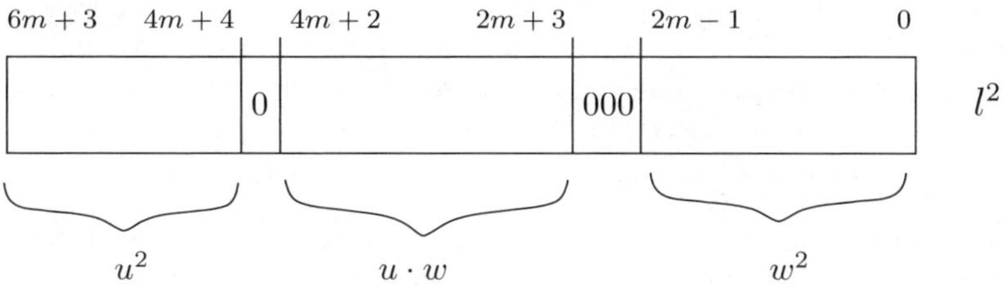

Fig. 2. The bit composition of the number l^2

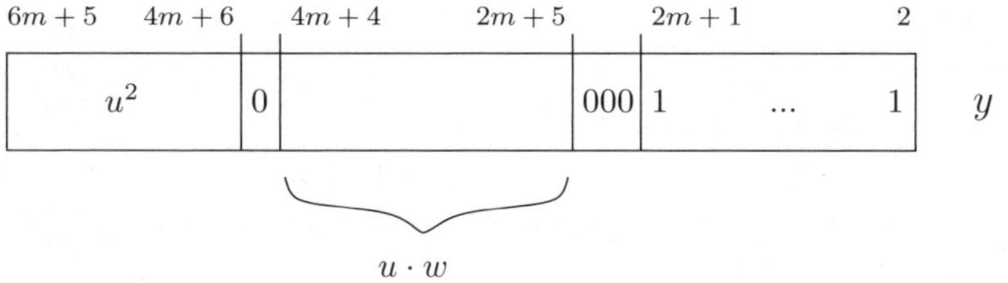

Fig. 3. The effect of the replacements of some of the y-variables

- $x_{n/2+n/9-1}, \ldots, x_{n/2+n/18-1}$ are replaced by 0,
- $x_0, \ldots, x_{n/2-1}$ are replaced by 0.

Figure 1 illustrates these replacements. Furthermore, u_0 and u_d are set to 1, all other u-variables are set to 0. The effect of these replacements is that $[u]_0^{m-1} = 2^d + 1 =: u$. The variables y_{4m+6}, y_{4m+d+7}, and $y_{4m+2d+6}$ are set to 1, the other variables y_j with $4m + 6 \leq j \leq 6m + 5$ are set to 0. The effect of these replacemenst is that $[y]_{4m+6}^{6m+5} = u^2$ (Figure 3 shows these replacements). The variables $y_{4m+5}, y_{2m+4}, y_{2m+3}$ and y_{2m+2} are set to 0, and the variables $y_{2m+1}, \ldots, y_2$ are set to 1. The effect of the last replacements is that $2^{2m} > [y]_2^{2m+1} > w^2$, where w is defined as the integer with binary representation $[w]_0^{m-1}$. Figure 3 illustrates these replacements. Now we take a closer look at the product $u \cdot w$, where u is equal to $2^d + 1$.

A pair $(w_{i+d}, y_{2m+5+2d+i})$, $i \in I'$, is called (w, y)-pair. A (w, y)-pair is called separated with respect to S iff $w_{i+d} \in S$ and $y_{2m+5+2d+i} \notin S$ or vice versa.

Case 2.1

In the following we prove the existence of a fooling set with at least $2^{(1/5)m/8}$ elements. For this reason we choose a subfunction of $\mathrm{MUL}_{2n-1,n}$ such that the computation of this subfunction resembles the computation of the function $\overline{\mathrm{GT}}_{(1/5)m/8}$.

There are at least $(2/5)m/8$ separated (w, y)-pairs with respect to S. W.l.o.g. we assume that there are at least $(1/5)m/8$ separated w-variables in S. Our aim is to prove that there exists a fooling set of size at least $(1/5)m/8$. The separated w-variables in S and their corresponding y-variables are called free. Furthermore, a variable $y_{n/2+i+d+1}$ for which the variable w_{i+d} is free is also called free. Remember that the variable $y_{n/2+i+d+1}$ is also in S because of the definition of I'. Let $\min I'$ be the minimal and $\max I'$ be the maximal element of I'. In the rest of the proof we choose for each variable $y_{n/2+i+d+1}$, where w_{i+d} is a free variable and $i \neq \min I'$, an assignment such that $y_{n/2+i+d+1} = w_{i+d} \oplus 1$ without further mentioning it.

- The variables $w_{\min I'+d}$, $y_{2m+5+2d+\min I'}$, and $y_{n/2+\min I'+d+1}$ are set to 1.
- The variables w_j and $y_{n/2+j+1}$, $0 \leq j < \min I' + d$, are set to 0.
- All other variables w_i which are not free are set to 0, their corresponding variables $y_{n/2+i+1}$ are set to 1.
- The other y-variables which are not free are replaced in the following way. The variable $y_{2m+5+\max I'+d+1}$ is set to 1, the remaining y-variables without the free variables to 0.

What is the effect of these replacements?
Remember that $[w]_0^{m-1} = [x]_{n/2}^{n/2+m-1}$. We only consider assignments to the variables for which the following holds. If $[x]_{n/2}^{n/2+3m+1} = \ell$ then $[y]_{n/2+1}^{n/2+3m+2} = \overline{\ell} + 1$. Now iff $[y]_2^{6m+3}$ represents a number r, where $r \geq \ell^2$, the product $x \cdot y$ is greater than 2^{2n-1}. We take a closer look at the variables $y_2, \ldots, y_{6m+3}$. Figure 2 shows the composition of the number ℓ^2. One effect of our replacements is that $[y]_{4m+6}^{6m+5} = u^2$ and $[y]_2^{2m+1} > w^2$. Therefore, iff $[y]_{2m+5}^{4m+4}$ represents a number r', where $r' \geq u \cdot w$, $[y]_2^{6m+5}$ represents a number r, where $r \geq \ell^2$. Since we have replaced the variable $y_{2m+5+\max I'+d+1}$ by 1 and because of our other replacements, $[y]_2^{6m+5}$ represents a number r, where $r \geq \ell^2$, iff for each separated (w, y)-pair, the assignment to the variable $y_{2m+5+2d+i}$ is at least as large as the assignment to the variable w_{i+d}. Therefore, the considered subfunction resembles the function $\overline{\mathrm{GT}_{1/5 \cdot m/8}}$.

In the rest of the proof we show that all possible assignments to the free variables w_{i+d} together with all possible assignments to the variables $y_{2m+5+2d+i}$, where $y_{2m+5+2d+i} = w_i$, are a fooling set of size at least $1/5 \cdot m/8$.

Together with the replacements to constants an assignment to the free w-variables can be seen as a number $2^{n-1} + \ell_1 2^{n/2}$, the corresponding assignment to the free y-variables together with the replacements to constants as number $2^n - \ell_1 2^{n/2+1} + c$, where $c > 4\ell_1^2$. Therefore, the product of the two numbers is larger than 2^{2n-1}. If we choose another assignment to the free variables w_{i+d} such that x can be seen as number $2^{n-1} + \ell_2 2^{n/2}$, $\ell_2 > \ell_1$, we get the following result

$$(2^{n-1} + \ell_2 2^{n/2}) \cdot (2^n - \ell_2 2^{n/2+1} + c) < 2^{2n-1},$$

because $c < 4\ell_2^2$. Therefore, we are done.

Case 2.2: In the following we prove the existence of a fooling set with at least $2^{(1/5)m/8}$ elements. For this reason we choose a subfunction of $\mathrm{MUL}_{2n-1,n}$ such that the computation of this subfunction resembles the computation of the function $\overline{\mathrm{GT}^{**}_{(1/5)m/8}}$.

There are less than $(2/5)m/8$ separated (w, y)-pairs with respect to S. Let $I \subseteq I'$ be the set of indices such that $(w_{i+d}, y_{2m+5+2d+i})$, $i \in I'$, are not separated with respect to S. Obviously, $|I| \geq (1/5)m/8$. Let $\min I$ and $\max I$ be the minimal resp. maximal element of I. For this reason we replace some of the variables in the following way.

- The variables $w_{\min I}$ and $y_{n/2+\min I+1}$ are set to 1, the variable $w_{\min I+d}$ is set to 0, the variable $y_{n/2+\min I+d+1}$ is set to 1,
- the variables w_i, $i < \min I$, are set to 0, the corresponding variables $y_{n/2+i+1}$ are set to 0,
- the variables w_i, $\min I < i < \max I$ and $i \notin I$ are set to 1, the corresponding variables $y_{n/2+i+1}$ are set to 0,
- all other variables w_j, $j \notin I$ and $j - d \notin I$, are set to 0, the corresponding variables $y_{n/2+j+1}$ are set to 1.

Furthermore, the variables $y_{2m+5+2d+i}$, $i \in I' \setminus I$, are replaced by 0. The variables $y_{2m+5+2d+i}$, $y_{n/2+i+1}$, and $y_{n/2+i+d+1}$, $i \in I$, are called free. The y-variables which are not free are replaced in the following way.

- The variables y_j, $2m + 5 + \min I \leq j \leq 2m + 5 + \max I$, are set to 1,
- the variables y_j, $2m + 5 + \min I + d \leq j \leq 2m + 5 + \max I + d$, are set to 1, and
- all other variables y_j with $2m + 5 \leq j \leq 6m + 5$ besides the free y-variables are set to 0.

The free y-variables are not separated from their corresponding w-variables, since we know that $w_i \in S$ and $y_{n/2+i+1} \in S$ or $w_i \notin S$ and $y_{n/2+i+1} \notin S$, where $i \in I$, because of the definition of I. The same holds for $w_{i+d}, y_{n/2+i+d+1}$, and $y_{2m+5+2d+i}$, where $i \in I$. In the rest of the proof we only consider assignments with the property that

- $y_{n/2+i+1} = w_i \oplus 1$,
- $y_{n/2+i+d+1} = w_{i+d} \oplus 1$, and
- $y_{2m+5+2d+i} = w_{i+d}$,

where $i \in I$, without further mentioning it.

In the following we prove that all possible assignments to the variables w_i, $i \in I$, together with the assignments to the variables w_{i+d}, $i \in I$, such that $w_{i+d} = w_i \oplus 1$ are a fooling set of size at least $(1/5)m/8$. Together with the replacements to constants our assignments to the variables w_i, $i \in I$ or $i - d \in I$, can be seen as a number $2^{n-1} + \ell 2^{n/2}$. The corresponding assignments to the y-variables can be interpreted as number $2^n - \ell 2^{n/2+1} + c$, where $c > 4\ell^2$. Therefore, the product of the two numbers is larger than 2^{2n-1}. To see this we decompose ℓ into $u \cdot 2^{2m+2} + w$, where $u = [u]_0^{m-1}$ and $w = [w]_0^{m-1}$. The number c can be decomposed into

$$[y]_{4m+6}^{6m+5} \cdot 2^{4m+6} + [y]_{2m+5}^{4m+4} \cdot 2^{2m+5} + [y]_2^{2m+1} \cdot 2^2.$$

As mentioned before, $[y]_{4m+6}^{6m+5} = u^2$ and $w^2 < [y]_2^{2m+1} < 2^{2m}$ (see Figure 3). The number $[w]_0^{m-1}$ can be decomposed into $[w]_{\min I+d}^{\max I+d} \cdot 2^{\min I+d} + [w]_{\min I}^{\max I} \cdot$

$2^{\min I}$. Let $w' := [w]_{\min I}^{\max I}$ and $w'' := [w]_{\min I+d}^{\max I+d}$. Now the number $[y]_{2m+5}^{4m+4}$ can be decomposed into $w'' \cdot 2^{2m+5+2d+\min I} + (2^{\max I - \min I+1} - 1) \cdot 2^{2m+5+d+\min I} + (2^{\max I - \min I+1} - 1) \cdot 2^{2m+5+\min I}$. Iff $w' + w'' \leq 2^{\max I - \min I+1} - 1$, the number $[y]_{2m+5}^{4m+4}$ is greater than $u \cdot w$ and altogether $[y]_2^{6m+5} > \ell^2$. Therefore, we can conclude $c > 4\ell^2$.

If $w' + w'' > 2^{\max I - \min I+1} - 1$, the number $[y]_{2m+5}^{4m+4}$ is less than $u \cdot w$. Therefore, x can be seen as number $2^{n-1} + \ell' 2^{n/2}$ and y as $2^n - \ell' 2^{n/2+1} + c$, where $c < 4\ell'^2$. Since

$$(2^{n-1} + \ell' 2^{n/2}) \cdot (2^n - \ell' 2^{n/2+1} + c) < 2^{2n-1}$$

we are done.

Altogether we have shown that for an arbitrary variable order π the π-OBDD size for the most significant bit of multiplication is at least $2^{(1/5)m/8}$. Considering the fact that $m := (n/6-3)/3 = n/18-1$ we obtain a lower bound of $2^{n/720-1} = \Omega(2^{n/720})$ on the OBDD complexity of $\mathrm{MUL}_{2n-1,n}$.

Using techniques from analytical number theory Sawitzki [14] has presented a lower bound of $2^{n/6}$ on the size of π-OBDDs representing the most significant bit of integer multiplication for the variable order π where the variables are tested according to increasing significance, i.e. $\pi = (x_0, y_0, x_1, y_1, \ldots, x_{n-1}, y_{n-1})$. Almost the same lower bound can be proved in an easier way and without analytical number theory using the fact that for a number $2^{n-1} + \ell 2^{(1/2)n}$, $\ell \leq 2^{(1/6)n-1}$, the corresponding smallest number such that the product of the two numbers is at least 2^{2n-1} is $2^n - \ell 2^{(1/2)n+1} + 4\ell^2$.

Furthermore, we only want to mention here that similar to Gergov's [9] generalization of Bryant's lower bound on the size of OBDDs for the middle bit of multiplication to arbitrary oblivious programs of linear length the result for the most significant bit of multiplication can be extended.

Acknowledgement

The author would like to thank Martin Sauerhoff for careful reading the first version of the paper and for valuable suggestions which helped to improve the paper.

References

1. Amano, K., Maruoka, A.: Better upper bounds on the QOBDD size of integer multiplication. Discrete Applied Mathematics 155, 1224–1232 (2007)
2. Bollig, B.: Restricted nondeterministic read-once branching programs and an exponential lower bound for integer multiplication. RAIRO Theoretical Informatics and Applications 35, 149–162 (2001)
3. Bollig, B., Woelfel, P.: A read-once branching program lower bound of $\Omega(2^{n/4})$ for integer multiplication using universal hashing. In: Proc. of 33rd STOC, pp. 419–424 (2001)

4. Bollig, B., Waack, S., Woelfel, P.: Parity graph-driven read-once branching programs and an exponential lower bound for integer multiplication. Theoretical Computer Science 362, 86–99 (2006)
5. Bollig, B., Woelfel, P.: A lower bound technique for nondeterministic graph-driven read-once branching programs and its applications. Theory of Computing Systems 38, 671–685 (2005)
6. Bryant, R.E.: Graph-based algorithms for Boolean manipulation. IEEE Trans. on Computers 35, 677–691 (1986)
7. Bryant, R.E.: On the complexity of VLSI implementations and graph representations of Boolean functions with application to integer multiplication. IEEE Trans. on Computers 40, 205–213 (1991)
8. Führer, M.: Faster integer multiplication. In: Proc. of 39th STOC, pp. 57–66 (2007)
9. Gergov, J.: Time-space trade-offs for integer multiplication on various types of input oblivious sequential machines. Information Processing Letters 51, 265–269 (1994)
10. Hromkovič, J.: Communication Complexity and Parallel Computing. Springer, Heidelberg (1997)
11. Kushilevitz, E., Nisan, N.: Communication Complexity. Cambridge University Press, Cambridge (1997)
12. Ponzio, S.: A lower bound for integer multiplication with read-once branching programs. SIAM Journal on Computing 28, 798–815 (1998)
13. Sauerhoff, M., Woelfel, P.: Time-space trade-off lower bounds for integer multiplication and graphs of arithmetic functions. In: Proc. of 33rd STOC, pp. 186–195 (2003)
14. Sawitzki, D.: Exponential lower bounds on the space complexity of OBDD-based graph algorithms. In: Proc. of LATIN. LNCS, vol. 3831, pp. 471–482 (2005)
15. Wegener, I.: Optimal lower bounds on the depth of polynomial-size threshold circuits for some arithmetic functions. Information Processing Letters 46(2), 85–87 (1993)
16. Wegener, I.: Branching Programs and Binary Decision Diagrams - Theory and Applications. SIAM Monographs on Discrete Mathematics and Applications (2000)
17. Woelfel, P.: New bounds on the OBDD-size of integer multiplication via universal hashing. Journal of Computer and System Science 71(4), 520–534 (2005)
18. Woelfel, P.: On the complexity of integer multiplication in branching programs with multiple tests and in read-once branching programs with limited nondeterminism. In: Proc. of 17th Computational Complexity, pp. 80–89 (2002)

Logical Closure Properties of Propositional Proof Systems
(Extended Abstract)

Olaf Beyersdorff*

Institut für Theoretische Informatik, Leibniz Universität Hannover, Germany
beyersdorff@thi.uni-hannover.de

Abstract. In this paper we define and investigate basic logical closure properties of propositional proof systems such as closure of arbitrary proof systems under modus ponens or substitutions. As our main result we obtain a purely logical characterization of the degrees of schematic extensions of *EF* in terms of a simple combination of these properties. This result underlines the empirical evidence that *EF* and its extensions admit a robust definition which rests on only a few central concepts from propositional logic.

1 Introduction

In their seminal paper [11] Cook and Reckhow gave a very general complexity-theoretic definition of the concept of a propositional proof system, focusing on efficient verification of propositional proofs. Due to the expressivity of Turing machines (or any other model of efficient computation) this definition includes a variety of rather unnatural propositional proof systems. In contrast, proof-theoretic research concentrates on propositional proof systems which, beyond efficient verification, satisfy a number of additional natural properties. Proof systems with nice structural properties are also exclusively used in practice (e.g. for automated theorem proving). Supported by this empirical evidence, we therefore formulate the thesis, that the Cook-Reckhow framework is possibly too broad for the study of natural proof systems of practical relevance. Motivated by these observations, we investigate the interplay of central logical closure properties of propositional proof systems, such as the ability to use modus ponens or substitutions in arbitrary proof systems.

Proof systems are compared with respect to their strength by simulations, and all equivalent systems form one degree of proof systems. Since in proof complexity we are mostly interested in the degree of a propositional proof system and not so much in specific representatives of this degree, we only study properties which are preserved inside a simulation degree. In particular, we think that it would be desirable to characterize the degrees of important proof systems

* Part of this work was done while at Humboldt University Berlin. Supported by DFG grant KO 1053/5-1.

M. Agrawal et al. (Eds.): TAMC 2008, LNCS 4978, pp. 318–329, 2008.

(e.g. resolution, cutting planes, or Frege) by meaningful and natural properties. Such results would provide strong confirmation for the empirical evidence, that these systems have indeed a natural and robust definition. One would expect that according to the general classification of propositional proof systems into logical systems (such as resolution, Frege, QBF), algebraic systems (polynomial calculus, Nullstellensatz) and geometric systems (cutting planes), these underlying principles should also be of logical, algebraic, and geometrical character, respectively.

As a first step of this more general program we exhibit a purely logical characterization of the degrees of schematic extensions of the extended Frege system *EF*. These schematic extensions enhance the extended Frege system by additional sets of polynomial-time decidable axiom schemes. Such systems are of particular importance: Firstly, because every propositional proof system is simulated by such an extension of *EF*, and secondly, because these systems admit a fruitful correspondence to theories of bounded arithmetic [8,14,16].

For our characterization we formalize closure properties such as modus ponens and substitutions in such a way that they are applicable for arbitrary propositional proof systems. We analyse the mutual dependence of these properties, providing in particular strong evidence for their independence. Our characterization of extensions of *EF* involves the properties modus ponens, substitutions, and reflection. This result tells us that the essence of extended Frege systems (and its generalizations) lies in the ability to use modus ponens and substitutions, and to prove the consistency of the system with short proofs (this property is known as reflection). Thus schematic extensions of *EF* are exactly those systems (up to p-equivalence) which can prove their consistency and are closed under modus ponens and substitutions. This result also allows the characterization of the existence of optimal propositional proof systems (which simulate every other system).

The paper is organized as follows. We start in Sect. 2 by recalling some background information on propositional proof systems and particularly Frege systems and their extensions. In Sect. 3 we define and investigate natural properties of proof systems which we use throughout this paper. A particularly important property for strong systems is the reflection property, which gives a propositional description of the correctness of a proof system. Different versions of such consistency statements are discussed in Sect. 4, leading in particular to a robust definition of the reflection property.

Section 5 contains the main result of this paper, consisting of a purely logical characterization of the degrees of schematic extensions of *EF*. This directly leads to a similar characterization of the existence of p-optimal proof systems. These results can also be explained in the more general context of the correspondence between strong propositional proof systems and arithmetic theories, of which we will sketch an axiomatic approach. Finally, in Sect. 6 we conclude with some open problems.

Most definitions and results of this paper can be given in two versions, one for simulations and the other, using slightly stronger assumptions, for the case of p-simulations between proof systems (cf. Sect. 2). For brevity we will restrict this exposition to the efficient case of p-simulations.

Due to space limitations we only sketch proofs or omit them in this extended abstract.

2 Propositional Proof Systems

Propositional proof systems were defined in a very general way by Cook and Reckhow [11] as polynomial-time functions P which have as their range the set TAUT of all propositional tautologies. A string π with $P(\pi) = \varphi$ is called a P-proof of the tautology φ. By $P \vdash_{\leq m} \varphi$ we indicate that there is a P-proof of φ of size $\leq m$. If Φ is a set of propositional formulas we write $P \vdash_* \Phi$ if there is a polynomial p such that $P \vdash_{\leq p(|\varphi|)} \varphi$ for all $\varphi \in \Phi$. If $\Phi = \{\varphi_n \mid n \geq 0\}$ is a sequence of formulas we also write $P \vdash_* \varphi_n$ instead of $P \vdash_* \Phi$.

Proof systems are compared according to their strength by simulations, introduced in [11] and [18]. A proof system S *simulates* a system P (denoted by $P \leq S$) if there exists a polynomial p such that for all tautologies φ and P-proofs π of φ there is an S-proof π' of φ with $|\pi'| \leq p(|\pi|)$. If such a proof π' can even be computed from π in polynomial time we say that S *p-simulates* P and denote this by $P \leq_p S$. Systems P and S, that mutually (p-)simulate each other, are called *(p-)equivalent*, denoted by $P \equiv_{(p)} S$. A proof system is *(p-)optimal* if it (p-)simulates all proof systems.

A prominent example of a class of proof systems is provided by *Frege systems* which are usual textbook proof systems based on axioms and rules. In the context of propositional proof complexity these systems were first studied by Cook and Reckhow [11], and it was proven there that all Frege systems, i.e., systems using different axiomatizations and rules, are p-equivalent. A different characterization of Frege systems is provided by *Gentzen's sequent calculus* [12], that is historically one of the first and best analysed proof systems. The sequent calculus is widely used, both for propositional and first-order logic, and it is straightforward to verify that Frege systems and the propositional sequent calculus LK p-simulate each other [11].

Augmenting Frege systems by the possibility to abbreviate complex formulas by propositional variables, we arrive at the *extended Frege proof system* EF. The extension rule might further reduce the proof size, but it is not known whether EF is really stronger than ordinary Frege systems. Both Frege and the extended Frege system are very strong systems for which no non-trivial lower bounds to the proof size are currently known (cf. [6]).

It is often desirable to further strengthen the proof system EF by additional axioms. This can be done by allowing a polynomial-time computable set Φ as new axioms, i.e., formulas from Φ as well as their substitution instances may be freely used in EF-proofs. These schematic extensions of EF are denoted by $EF + \Phi$. In this way, we obtain proof systems of arbitrary strength (cf. Theorem 15). More detailed information on Frege systems and its extensions can be found in [9,16].

3 Closure Properties of Proof Systems

In this section we define and investigate natural properties of propositional proof systems that are satisfied by many important proof systems. One of the most common rules is modus ponens, which serves as the central rule in Frege systems. Carrying out modus ponens in a general proof system might be formalized as:

Definition 1. *A proof system P is* closed under modus ponens *if there exists a polynomial-time computable algorithm that takes as input P-proofs $\pi_1, \ldots, \pi_k$ of propositional formulas $\varphi_1, \ldots, \varphi_k$ together with a P-proof π_{k+1} of the implication $(\varphi_1 \to (\varphi_2 \to \ldots (\varphi_k \to \varphi_{k+1}) \ldots))$ and outputs a P-proof of φ_{k+1}.*

Defining closure under modus ponens by requiring $k = 1$ in the above definition seems to lead to a too restrictive notion. Namely, in some applications we need to use modus ponens polynomially many times (cf. Theorem 11). In this case, the above definition with $k = 1$ would only guarantee an exponential upper bound on the size of the resulting proof, whereas Definition 1 results only in a polynomial increase.

If π is a Frege proof of a formula φ, then we can prove substitution instances $\sigma(\varphi)$ of φ by applying the substitution σ to every formula in the proof π. This leads us to the general concept of closure of a proof system under substitutions.

Definition 2. *P is* closed under substitutions *if there exists a polynomial-time procedure that takes as input a P-proof of a formula φ as well as a substitution instance $\sigma(\varphi)$ of φ and computes a P-proof of $\sigma(\varphi)$.*

It also makes sense to consider other properties like closure under conjunctions or disjunctions. A particularly simple property is the following: a proof system *evaluates formulas without variables* if formulas using only constants but no propositional variables have polynomial-size proofs. As this is true even for truth-table evaluations, all proof systems simulating the truth-table system evaluate formulas without variables.

We can classify properties of proof systems like those above along the following lines. Some properties are *monotone* in the sense that they are preserved from weaker to stronger systems, i.e., if $P \leq Q$ and P has the property, then also Q satisfies the property. Evaluation of formulas without variables is such a monotone property. Other properties might not be monotone but still *robust* in the sense that the property is preserved when we switch to a p-equivalent system. Since we are interested in the degree of a proof system and not in the particular representative of that degree, it is desirable to investigate only robust or even monotone properties. It is straightforward to verify that closure under modus ponens and closure under substitutions are robust properties.

We remark that Frege systems and their extensions have very good closure properties.

Proposition 3. *The Frege system F, the extended Frege system EF, and all extensions $EF + \Phi$ by polynomial-time computable sets of axioms $\Phi \subseteq \mathrm{TAUT}$ are closed under modus ponens and under substitutions.*

It is interesting to ask whether these properties of propositional proof systems are independent from each other. With respect to this question we observe the following.

Proposition 4. *Assume that the extended Frege proof system is not optimal. Then there exist proof systems which are closed under substitutions but not under modus ponens.*

Proof. (Idea) We use the assumption of the non-optimality of *EF* to obtain polynomial-time constructable sequences φ_n and ψ_n of tautologies, such that $EF \vdash_* \varphi_n$, but $EF \not\vdash_* \psi_n$. We then encode the implications $\varphi_n \to \psi_n$ into an extension of *EF*, thus obtaining a system Q that is closed under substitutions, but not under modus ponens, because $Q \vdash_* \varphi_n$, $Q \vdash_* \varphi_n \to \psi_n$, and $Q \not\vdash_* \psi_n$. □

Candidates for proof systems that are closed under modus ponens but not under substitutions come from extensions of Frege systems by polynomial-time computable sets $\Phi \subseteq$ TAUT as new axioms. Clearly these systems are closed under modus ponens. In [3], however, we exhibit a suitable hypothesis, involving disjoint NP-pairs, which guarantees that these proof systems are not even closed under substitutions by constants for suitable choices of Φ.

4 Consistency Statements

Starting with this section, we will use the correspondence of propositional proof systems to theories of bounded arithmetic. Bounded arithmetic is the general denomination of a whole collection of weak fragments of Peano arithmetic, that are defined by adding a controlled amount of induction to a set of basic axioms (cf. [14]). One of the most prominent examples of these arithmetic theories is Buss' theory S^1_2, defined in [8]. In addition to the usual ingredients, the language L of S^1_2 uses a number of technical symbols to allow a smooth formalization of syntactic concepts.

A central ingredient of the correspondence of arithmetic theories to propositional proof systems is the translation of first-order arithmetic formulas into propositional formulas [10,19]. An L-formula in prenex normal form with only bounded existential quantifiers is called a Σ^b_1-formula. These formulas describe NP-predicates in the sense that the class of all Σ^b_1-definable subsets of $\mathbb{N}$ coincides with the class of all NP-subsets of $\mathbb{N}$ (cf. [25,8]). Likewise, Π^b_1-formulas only have bounded universal quantifiers and describe coNP-predicates. A Π^b_1-formula $\varphi(x)$ is translated into a sequence $\|\varphi(x)\|^n$ of propositional formulas containing one formula per input length for the number x, such that $\varphi(x)$ is true, i.e., $\mathbb{N} \models (\forall x)\varphi(x)$, if and only if $\|\varphi(x)\|^n$ is a tautology where $n = |x|$ (cf. [16]). We use $\|\varphi(x)\|$ to denote the set $\{\|\varphi(x)\|^n \mid n \geq 1\}$.

The consistency of a proof system P is described by the *consistency statement* $Con(P) = (\forall \pi)\neg Prf_P(\pi, \bot)$, where $Prf_P(\pi, \varphi)$ is a suitable arithmetic formula describing that π is a P-proof of φ. The formula Prf_P can be chosen such that Prf_P is provably equivalent in S^1_2 both to a Σ^b_1- and a Π^b_1-formula (such formulas are called Δ^b_1-formulas with respect to S^1_2, cf. [16]).

A somewhat stronger formulation of consistency is given by the *reflection principle* of a propositional proof system P, which is defined by the arithmetic formula

$$RFN(P) = (\forall\pi)(\forall\varphi)Prf_P(\pi,\varphi) \rightarrow Taut(\varphi) \ ,$$

where $Taut$ is a Π_1^b-formula formalizing propositional tautologies. Therefore $Con(P)$ and $RFN(P)$ are $\forall\Pi_1^b$-formulas, i.e., these formulas are in prenex normal form with unbounded $\forall$-quantifiers followed by bounded $\forall$-quantifiers and can therefore be translated via $\|.\|$ into sequences of propositional formulas.

The two consistency notions $Con(P)$ and $RFN(P)$ are compared by the following well-known observation, contained e.g. in [16]:

Proposition 5. *Let P be a proof system that is closed under substitutions and modus ponens and evaluates formulas without variables, and assume that these properties are provable in S_2^1. Then $S_2^1 \vdash RFN(P) \leftrightarrow Con(P)$.*

Very often propositional descriptions of the reflection principle are needed. These can be simply obtained by translating $RFN(P)$ to a sequence of propositional formulas using the translation $\|.\|$.

Definition 6. *A propositional proof system P has the* reflection property *if there exists a polynomial-time algorithm that on input 1^n outputs a P-proof of $\|RFN(P)\|^n$.*

There is a subtle problem with Definition 6 which is somewhat hidden in the definition. Namely, the formula Prf_P describes the computation of some Turing machine computing the function P. However, the provability of the formulas $\|RFN(P)\|^n$ with polynomial-size P-proofs might depend on the actual choice of the Turing machine computing P. Let us illustrate this with the following example.

Proposition 7. *If EF is not p-optimal, then there exists a proof system $Q \equiv_p EF$ such that S_2^1 does not prove the reflection principle of Q, i.e., S_2^1 does not prove the formula $(\forall\pi)(\forall\varphi)Prf_Q(\pi,\varphi) \rightarrow Taut(\varphi)$ for some suitable choice of the Turing machine that computes Q and is used for the formula Prf_Q.*

Proof. (Sketch) If EF is not p-optimal, then there exists a proof system R such that $R \not\leq_p EF$. We define the system P as $EF + \|RFN(R)\|$ and the system Q as

$$Q(\pi) = \begin{cases} \varphi & \text{if } \pi = 0\pi' \text{ and } \pi' \text{ is an } EF\text{-proof of } \varphi \\ P(\pi') & \text{if } \pi = 1\pi' \text{ and } P(\pi') \in \{\top,\bot\} \\ \top & \text{otherwise.} \end{cases}$$

It is easily checked that EF and Q are $\leq_p$-equivalent. We have to show that S_2^1 does not prove the formula $RFN(Q)$ where for the predicate Prf_Q we use the canonical Turing machine M according to the above definition of Q, i.e., on input $0\pi'$ the machine M checks whether π' is a correct EF-proof and on input $1\pi'$ the machine M evaluates $P(\pi')$. Assume on the contrary that $S_2^1 \vdash RFN(Q)$. Because of line 2 of the definition of Q this means that S_2^1 can prove that there is no P-proof of $\bot$, i.e., S^1_2 proves the consistency statement of P. The system P is closed under substitutions by constants and modus ponens. Therefore $Con(P)$ and $RFN(P)$ are equivalent in S^1_2 by Proposition 5. Hence S^1_2 not only proves $Con(P)$, but also $RFN(P)$, which gives a p-simulation of P by EF (cf. Definition 13 and Theorem 14 below). This, however, contradicts the choice of P, and hence $S^1_2 \not\vdash RFN(Q)$. □

Note that $S^1_2 \vdash RFN(EF)$ (cf. [18]), contrasting $S^1_2 \not\vdash RFN(Q)$ in the above proposition. This observation tells us that we should understand the meaning of Definition 6 in the following, more precise way:

Definition 8. *A propositional proof system P has the* robust reflection property *if there exists a deterministic polynomial-time Turing machine M computing the function P such that for some Δ^b_1-formalization Prf_P of the computation of M with respect to S^1_2 we have a polynomial-time algorithm that constructs on input 1^n a P-proof of the formula $\|(\forall\pi)(\forall\varphi)Prf_P(\pi,\varphi) \rightarrow Taut(\varphi)\|^n$.*

For this definition of reflection we can show the robustness of the reflection principle under p-simulations:

Proposition 9. *Let P and Q be p-equivalent proof systems. Then P has the robust reflection property if and only if Q has the robust reflection property.*

Proof. (Idea) If Q proves its reflection principle with respect to the Turing machine M, then P can prove its reflection for the Turing machine $M \circ N$, where the machine N computes a p-simulation of Q by P. □

It is known that strong propositional proof systems like EF and its extensions have the reflection property [18]. In contrast, weak systems like resolution do not have reflection. Pudlák [21] proved that the cutting planes system CP requires nearly exponential-size refutations of some *canonical* formulation of the formulas $RFN(CP)$. Atserias and Bonet [1] obtained the same result for the resolution system Res. This, however, does not exclude the possibility that we have short proofs for the reflection principle of resolution or CP with respect to some other formalization of Prf_{CP}, Prf_{Res}, or $Taut$. It therefore remains as an open question whether these systems have the robust reflection property.

Alternatively, we can view robust reflection as a condition on the *canonical disjoint* NP*-pair* $(Ref(P), Sat^*)$ of a proof system P, introduced by Razborov [22]. Its first component $Ref(P) = \{(\varphi, 1^m) \mid P \vdash_{\leq m} \varphi\}$ contains information about proof lengths in P, and the second component $Sat^* = \{(\varphi, 1^m) \mid \neg\varphi \in \mathrm{SAT}\}$ is a padded version of SAT. The link of the canonical pair with the reflection property was already noted by Pudlák [21]. We can extend this idea to obtain a characterization of robust reflection for weak proof systems.

Proposition 10. *Let P be the resolution or cutting planes system. Then P has the robust reflection property if and only if the canonical pair of P is p-separable, i.e., there exists a polynomial-time decidable set S such that $Ref(P) \subseteq S$ and $S \cap Sat^* = \emptyset$.*

Proof. (Idea) Robust reflection for P means that we can efficiently generate P-proofs for the disjointness of $(\mathit{Ref}(P), \mathit{Sat}^*)$ with respect to some propositional representations of its components. Using feasible interpolation for P [17,7,20], we get a polynomial-time computable separator for $(\mathit{Ref}(P), \mathit{Sat}^*)$.

Conversely, if the canonical P-pair is p-separable, then it can be given a simple propositional description, for which we can devise short P-proofs of the disjointness of the pair. This is possible, as all p-separable disjoint NP-pairs are equivalent (via suitable reductions for pairs [13]). We can then choose a simple p-separable pair, prove its disjointness in P, and translate these proofs into proofs for the disjointness of $(\mathit{Ref}(P), \mathit{Sat}^*)$ (cf. [2] for the details of this approach). □

As it is conjectured that none of the canonical pairs of natural proof systems is p-separable [21], Proposition 10 indicates the absence of robust reflection for weak systems that satisfy the interpolation property.

5 Characterizing the Degree of Extended Frege Systems

Using the results from the previous section, we will now exhibit a characterization of the degrees of schematic extensions of EF.

Theorem 11. *For all proof systems $P \geq_p EF$ the following conditions are equivalent:*

1. *P is p-equivalent to a proof system of the form $EF + \|\varphi\|$ with a true Π_1^b-formula φ.*
2. *P is p-equivalent to a proof system of the form $EF + \|\Phi\|$ with a polynomial-time decidable set of true Π_1^b-formulas Φ.*
3. *P has the robust reflection property and is closed under modus ponens and substitutions.*

Proof. Item 1 trivially implies item 2. For the implication $2 \Rightarrow 3$ let $P \equiv_p EF + \|\Phi\|$. Then the closure properties of $EF + \|\Phi\|$ are transferred to P. Systems of the form $EF + \|\Phi\|$ are known to have the reflection property (cf. [19] and Theorem 14 below). By Proposition 9 robust reflection for $EF + \|\Phi\|$ is transferred to P.

The main part of the proof is the implication $3 \Rightarrow 1$. Its proof involves a series of results that are also of independent interest. The first step is an efficient version of the deduction theorem for EF:

1. Efficient Deduction theorem for EF. *There exists a polynomial-time procedure that takes as input an EF-proof of a formula ψ from a finite set of tautologies Φ as extra assumptions, and produces an EF-proof of the implication $(\bigwedge_{\varphi \in \Phi} \varphi) \rightarrow \psi$.*

A similar deduction theorem was shown for Frege systems by Bonet and Buss [4,5]. For stronger systems like EF we just remark that there are different ways to formalize deduction. These deduction properties seem to be quite powerful, as they allow the characterization of the existence of optimal and even polynomially bounded proof systems [3].

In the second step we compare schematic extensions of EF and strong proof systems with sufficient closure properties.

2. Simulation of extensions of EF by sufficiently closed systems. *Let P be a proof system such that $EF \leq_p P$ and P is closed under substitutions and modus ponens. Let Φ be some polynomial-time decidable set of tautologies such that P-proofs of all formulas from Φ can be constructed in polynomial time. Then $EF + \Phi \leq_p P$.*

The idea of the proof of this simulation is the following: if $EF + \Phi \vdash_{\leq m} \varphi$, then there are substitution instances $\psi_1, \ldots, \psi_k$ of formulas from Φ such that we have an EF-proof of φ from $\psi_1, \ldots, \psi_k$. Using the deduction theorem for EF, we get a polynomial-size EF-proof of $(\bigwedge_{i=1}^{k} \psi_i) \rightarrow \varphi$. By the hypotheses $P \geq_p EF$ and $P \vdash_* \Phi$, together with the closure properties of P, we can transform this proof into a polynomial-size P-proof of φ.

Item 2 is most useful in the following form:

3. *If the proof system $P \geq_p EF$ has the robust reflection property and P is closed under substitutions and modus ponens, then we get the p-simulation $EF + \|RFN(P)\| \leq_p P$.*

The converse simulation extends a result of Krajíček and Pudlák [18], namely that every proof system P is p-simulated by the system $EF + \|RFN(P)\|$.

4. Simulation of arbitrary systems by extensions of EF. *Let P be an arbitrary proof system, and let Φ be some polynomial-time decidable set of tautologies. If $EF + \Phi$-proofs of $\|RFN(P)\|^n$ can be generated in polynomial time, then $P \leq_p EF + \Phi$.*

After these preparations we can now prove the implication $3 \Rightarrow 1$. Let P be a proof system which has the robust reflection property and is closed under modus ponens and substitutions. We choose the formula φ as $RFN(P)$. Then $EF + \|\varphi\| \leq_p P$ by the above item 3. The converse simulation $P \leq_p EF + \|\varphi\|$ follows from item 4. □

The equivalence of items 1 and 2 in the above corollary expresses some kind of compactness for extensions of EF: systems of the form $EF + \|\Phi\|$ are always equivalent to a system $EF + \|\varphi\|$ with a single arithmetic formula φ. The equivalence to item 3 shows that these systems have a robust logical definition, independent of the particular axiomatization chosen for EF.

We will now apply Theorem 11 to characterize the existence of optimal and p-optimal proof systems. These problems were posed by Krajíček and Pudlák [18] and have been intensively studied during the last years [15,23,24]. We call a set A *printable* if there exists a polynomial-time algorithm which on input 1^n outputs all words from A of length n.

Corollary 12. *The following conditions are equivalent:*

1. *P is a p-optimal propositional proof system.*
2. *$P \geq_p EF$ and P is closed under modus ponens and substitutions. Further, for every printable set of tautologies, P-proofs can be constructed in polynomial time.*

Using non-constructive versions of the conditions in item 2 we get a similar characterization of the existence of optimal proof systems.

Probably the strongest available information on EF and its extensions stems from the connection of these systems to theories of bounded arithmetic. Actually, also Theorem 11 can be derived as a consequence from this more general context. In the remaining space we will sketch an axiomatic approach to this correspondence, as suggested by Krajíček and Pudlák [19]. The correspondence works for pairs (T, P) of arithmetic theories T and propositional proof systems P. It can be formalized as follows:

Definition 13. *A propositional proof system P is called* regular *if there exists an L-theory T such that the following two properties are fulfilled for (T, P).*

1. *Let $\varphi(x)$ be a Π_1^b-formula such that $T \vdash (\forall x)\varphi(x)$. Then there exists a polynomial-time computable function which on input 1^n outputs a P-proof of $\|\varphi(x)\|^n$.*
2. *T proves the correctness of P, i.e., $T \vdash RFN(P)$. Furthermore, P is the strongest proof system for which T proves the correctness, i.e., $T \vdash RFN(Q)$ for a proof system Q implies $Q \leq_p P$.*

Probably the most important instance of the general correspondence is the relation between S_2^1 and EF. Property 1 of the correspondence, stating the simulation of S_2^1 by EF, is essentially contained in [10], but for the theory PV instead of S_2^1. Examining the proof of this result, it turns out that the theorem is still valid if both the theory S_2^1 and the proof system EF are enhanced by further axioms. Property 2 of the correspondence between S_2^1 and EF was established by Krajíček and Pudlák [19]. Again, this result can be generalized to extensions of S_2^1 and EF by additional axioms. Combining these results, we can state:

Theorem 14 (Cook [10], Buss [8], Krajíček, Pudlák [19]). *Let Φ be a polynomial-time decidable set of true Π_1^b-formulas. Then the proof system $EF + \|\Phi\|$ is regular and corresponds to the theory $S_2^1 + \Phi$.*

Using these results, we can exhibit sufficient conditions for the regularity of a propositional proof system. From the definition of a regular system it is clear that regular proof systems have the reflection property. Furthermore, a combination of the properties of proof systems introduced in Sects. 3 and 4 guarantees the regularity of the system, namely:

Theorem 15. *If P is a proof system such that $EF \leq_p P$ and P has the robust reflection property and is closed under substitutions and modus ponens, then P is regular and corresponds to the theory $S_2^1 + RFN(P)$.*

Proof. The hypotheses on P imply $EF + \|RFN(P)\| \equiv_p P$ by Theorem 11. We will now check the axioms of the correspondence for $S^1_2 + RFN(P)$ and P. Suppose φ is a $\forall\Pi^b_1$-formula such that $S^1_2 + RFN(P) \vdash \varphi$. By Theorem 14 we can construct $EF + \|RFN(P)\|$-proofs of $\|\varphi\|^n$ in polynomial time. As we already know that $EF + \|RFN(P)\|$ is p-simulated by P, we obtain polynomial-size P-proofs of $\|\varphi\|^n$. This proves part 1 of the correspondence.

It remains to verify the second part. Clearly $S^1_2 + RFN(P) \vdash RFN(P)$. Finally, assume $S^1_2 + RFN(P) \vdash RFN(Q)$ for some proof system Q. By Theorem 14 this implies that we can efficiently construct proofs of $\|RFN(Q)\|$ in the system $EF + \|RFN(P)\|$. Applying items 3 and 4 from the proof of Theorem 11 we infer $Q \leq_p EF + \|RFN(P)\| \leq_p P$. □

6 Conclusion

The results of this paper suggest that logical closure properties can be used to give robust definitions of strong proof systems such as EF and its extensions. Continuing this line of research, it is interesting to ask, whether we can also characterize the degrees of weak systems like resolution or cutting planes in terms of similar closure properties. In particular, these weak systems are known to satisfy the feasible interpolation property [17]. Can interpolation in combination with other properties be used to characterize the degrees of weak systems? Pudlák [21] provides strong evidence that interpolation and reflection are mutually exclusive properties. Which other combinations of such properties are possible? Further investigation of these questions will hopefully contribute to a better understanding of propositional proof systems.

Acknowledgements. I am indebted to Jan Krajíček for many helpful and stimulating discussions on the topic of this paper. I also thank Sebastian Müller for detailed suggestions on how to improve the presentation of the paper.

References

1. Atserias, A., Bonet, M.L.: On the automatizability of resolution and related propositional proof systems. Information and Computation 189(2), 182–201 (2004)
2. Beyersdorff, O.: Classes of representable disjoint NP-pairs. Theoretical Computer Science 377, 93–109 (2007)
3. Beyersdorff, O.: The deduction theorem for strong propositional proof systems. In: Arvind, V., Prasad, S. (eds.) FSTTCS 2007. LNCS, vol. 4855, pp. 241–252. Springer, Heidelberg (2007)
4. Bonet, M.L.: Number of symbols in Frege proofs with and without the deduction rule. In: Clote, P., Krajíček, J. (eds.) Arithmetic, Proof Theory and Computational Complexity, pp. 61–95. Oxford University Press, Oxford (1993)
5. Bonet, M.L., Buss, S.R.: The deduction rule and linear and near-linear proof simulations. The Journal of Symbolic Logic 58(2), 688–709 (1993)
6. Bonet, M.L., Buss, S.R., Pitassi, T.: Are there hard examples for Frege systems? In: Clote, P., Remmel, J. (eds.) Feasible Mathematics II, pp. 30–56. Birkhäuser, Basel (1995)

7. Bonet, M.L., Pitassi, T., Raz, R.: Lower bounds for cutting planes proofs with small coefficients. The Journal of Symbolic Logic 62(3), 708–728 (1997)
8. Buss, S.R.: Bounded Arithmetic. Bibliopolis, Napoli (1986)
9. Buss, S.R.: An introduction to proof theory. In: Buss, S.R. (ed.) Handbook of Proof Theory, pp. 1–78. Elsevier, Amsterdam (1998)
10. Cook, S.A.: Feasibly constructive proofs and the propositional calculus. In: Proc. 7th Annual ACM Symposium on Theory of Computing, pp. 83–97 (1975)
11. Cook, S.A., Reckhow, R.A.: The relative efficiency of propositional proof systems. The Journal of Symbolic Logic 44, 36–50 (1979)
12. Gentzen, G.: Untersuchungen über das logische Schließen. Mathematische Zeitschrift 39, 68–131 (1935)
13. Grollmann, J., Selman, A.L.: Complexity measures for public-key cryptosystems. SIAM Journal on Computing 17(2), 309–335 (1988)
14. Hájek, P., Pudlák, P.: Metamathematics of First-Order Arithmetic. In: Perspectives in Mathematical Logic, Springer, Berlin (1993)
15. Köbler, J., Messner, J., Torán, J.: Optimal proof systems imply complete sets for promise classes. Information and Computation 184, 71–92 (2003)
16. Krajíček, J.: Bounded Arithmetic, Propositional Logic, and Complexity Theory. Encyclopedia of Mathematics and Its Applications, vol. 60. Cambridge University Press, Cambridge (1995)
17. Krajíček, J.: Interpolation theorems, lower bounds for proof systems and independence results for bounded arithmetic. The Journal of Symbolic Logic 62(2), 457–486 (1997)
18. Krajíček, J., Pudlák, P.: Propositional proof systems, the consistency of first order theories and the complexity of computations. The Journal of Symbolic Logic 54, 1079–1963 (1989)
19. Krajíček, J., Pudlák, P.: Quantified propositional calculi and fragments of bounded arithmetic. Zeitschrift für mathematische Logik und Grundlagen der Mathematik 36, 29–46 (1990)
20. Pudlák, P.: Lower bounds for resolution and cutting planes proofs and monotone computations. The Journal of Symbolic Logic 62, 981–998 (1997)
21. Pudlák, P.: On reducibility and symmetry of disjoint NP-pairs. Theoretical Computer Science 295, 323–339 (2003)
22. Razborov, A.A.: On provably disjoint NP-pairs. Technical Report TR94-006, Electronic Colloquium on Computational Complexity (1994)
23. Sadowski, Z.: On an optimal propositional proof system and the structure of easy subsets of TAUT. Theoretical Computer Science 288(1), 181–193 (2002)
24. Sadowski, Z.: Optimal proof systems, optimal acceptors and recursive presentability. Fundamenta Informaticae 79(1–2), 169–185 (2007)
25. Wrathall, C.: Rudimentary predicates and relative computation. SIAM Journal on Computing 7(2), 149–209 (1978)

Graphs of Linear Clique-Width at Most 3*

Pinar Heggernes, Daniel Meister, and Charis Papadopoulos

Department of Informatics, University of Bergen, 5020 Bergen, Norway
pinar.heggernes@ii.uib.no, daniel.meister@ii.uib.no,
charis.papadopoulos@ii.uib.no

Abstract. We give the first characterisation of graphs of linear clique-width at most 3, and we give a polynomial-time recognition algorithm for such graphs.

1 Introduction

Clique-width is an important graph parameter that is useful for measuring the computational complexity of NP-hard problems. In particular, all problems that can be expressed in a certain kind of monadic second order logic can be solved in linear time on graphs whose clique-width is bounded by a constant [5]. The clique-width of a graph is defined as the smallest number of labels that are needed for constructing the graph using the graph operations 'vertex creation', 'union', 'join' and 'relabel'. The related graph parameter *linear clique-width* is obtained by restricting the allowed clique-width operations to only 'vertex creation', 'join' and 'relabel'. Both parameters are NP-hard to compute, even on complements of bipartite graphs [7].

The relationship between clique-width and linear clique-width is similar to the relationship between treewidth and pathwidth, and the two pairs of parameters are related [7,10,12]. However, clique-width can be viewed as a more general concept than treewidth since there are graphs of bounded clique-width but unbounded treewidth, whereas graphs of bounded treewidth have bounded clique-width. While treewidth is widely studied and well understood the knowledge on clique-width is still limited. The study of the more restricted parameter linear clique-width is a step towards a better understanding of clique-width. For example, NP-hardness of clique-width is obtained by showing that linear clique-width is NP-hard to compute [7].

In this paper, we contribute to the study of linear clique-width. The main result that we report is the first characterisation of graphs that have linear clique-width at most 3 and the first polynomial-time algorithm to decide whether a graph has linear clique-width at most 3. We also show that such graphs are both cocomparability and weakly chordal graphs, and we present a decomposition scheme for them. Such a decomposition scheme is useful for designing algorithms especially tailored for this graph class. For bounded linear clique-width, until now only graphs of linear clique-width at most 2 could be recognised in polynomial

* This work is supported by the Research Council of Norway through grant 166429/V30. Some proofs are omitted in this version. They can be found in [11].

M. Agrawal et al. (Eds.): TAMC 2008, LNCS 4978, pp. 330–341, 2008.

time [8,3], and it was an open question whether the same is true for graphs of linear clique-width at most 3 [9,10]. For bounded clique-width, graphs of clique-width at most 2 [6] and at most 3 [2] can be recognized in polynomial time. Whereas graphs of clique-width at most 2 are characterised as the class of cographs [6], no characterisation is known for graphs of clique-width at most 3. Furthermore, from the proposed algorithm in [2] there is even no straightforward way of deciding whether a graph of clique-width at most 3 has linear clique-width at most 3. Before giving the mentioned results, we define a graph reduction operation that preserves linear clique-width.

2 Graph Preliminaries and Linear Clique-Width

We consider undirected finite graphs with no loops or multiple edges. For a graph $G = (V, E)$, we denote its vertex and edge set by $V(G) = V$ and $E(G) = E$, respectively. Usually, graphs are non-empty, and may be empty only in case it is explicitly mentioned. For a vertex set $S \subseteq V$, the *subgraph of G induced by S* is denoted by $G[S]$. We denote by $G - S$ the graph $G[V \setminus S]$ and by $G{-}v$ the graph $G[V \setminus \{v\}]$.

The *neighbourhood* of a vertex x in G is $N_G(x) = \{v : xv \in E\}$ and its *degree* is $|N_G(x)|$. The *closed neighbourhood* of x is $N_G[x] = N_G(x) \cup \{x\}$. Vertex x is *isolated* in G if $N_G(x) = \emptyset$ and *universal* if $N_G[x] = V(G)$. Two vertices x, y of G are called *false twins* if $N_G(x) = N_G(y)$.

An induced cycle on k vertices is denoted by C_k and an induced path on k vertices is denoted by P_k. The graph consisting of only two disjoint edges is denoted by $2K_2$, the complement of the graph consisting of two disjoint P_3's is denoted by co-$(2P_3)$.

Let G and H be two vertex-disjoint graphs. The *(disjoint) union* of G and H, denoted by $G \oplus H$, is the graph with vertex set $V(G) \cup V(H)$ and edge set $E(G) \cup E(H)$. The *join* of G and H, denoted by $G \otimes H$, is the graph obtained from the union of G and H and adding all edges between vertices of G and vertices of H.

Let $\mathcal{H}$ be a family of graphs. A graph is *$\mathcal{H}$-free* if none of its induced subgraphs is in $\mathcal{H}$. The class of *cographs* is defined recursively as follows: a single vertex is a cograph, the disjoint union of two cographs is also a cograph, and the complement of a cograph is a cograph. The class of cographs coincides with the class of P_4-free graphs [3].

The notion of clique-width was first introduced in [4]. The *clique-width* of a graph G, denoted by $\mathrm{cwd}(G)$, is defined as the minimum number of labels needed to construct G, using the following operations:

(i) Creation of a new vertex v with label i, denoted by $i(v)$;
(ii) Disjoint union, denoted by $\oplus$;
(iii) Changing all labels i to j, denoted by $\rho_{i \to j}$;
(iv) Adding edges between all vertices with label i and all vertices with label j, $i \neq j$, denoted by $\eta_{i,j} (= \eta_{j,i})$.

An expression built by using the above four operations is called a *clique-width expression.* If k labels are used in a clique-width expression then it is called a k-*expression.* We say that a k-expression t *defines* a graph G if G is equal to the graph obtained by using the operations in t in the order given by t.

The *linear clique-width* of a graph, denoted by $\mathrm{lcwd}(G)$, is introduced in [10] and defined by restricting the disjoint union operation (ii) of clique-width. In a linear clique-width expression all clique-width operations are allowed, but whenever the $\oplus$ operation is used, at least one of the two operands must be an expression defining a graph on a single vertex. The restricted version of operation (ii) becomes redundant if we allow operation (i) to automatically add the vertex to the graph as an isolated vertex when it is created. For simplicity, we adopt this notation in this paper. Hence whenever a vertex v is created with operation $i(v)$, it is added to the graph as an isolated vertex v with label i, which means that we never use operation $\oplus$ in our linear clique-width expressions. With this convention, the linearity of a linear clique-width expression becomes clearly visible.

For every k-expression that defines a graph G there is a parse tree that encodes the composition of G starting with single vertices followed by interleaved operations of relabeling, edge insertion, and disjoint union. If t is a linear clique-width expression then its parse tree is path-like. Thus one can view the relationship between clique-width and linear clique-width analogous to the relationship between treewidth and pathwidth. Note that the difference between clique-width and linear clique-width of a graph can be arbitrarily large. For example cographs and trees have bounded clique-width [6] but unbounded linear clique-width [10].

3 A Graph Reduction Operation That Preserves Linear Clique-Width

Some graph operations preserve clique-width. *Substituting* a vertex by a graph is the replacement of a vertex v of G with a graph H such that every vertex of H is adjacent to its neighbours in H and the neighbours of v in G. The *modular decomposition* is the reverse operation of obtaining a graph recursively by substitution. A *module* in a graph is a set of vertices that have the same neighbourhood outside of the module. A *prime* graph with respect to modular decomposition is a graph that cannot be obtained by non-trivial substitution. The clique-width of a graph G is equal to the maximum of the clique-width of all prime induced subgraphs of G [6]. Thus for clique-width it is enough to consider the prime graphs appearing in the modular decomposition. For linear clique-width this is not true, since cographs have unbounded linear clique-width whereas they are completely decomposable with respect to modular decomposition and they have no prime graph [1].

In other words modules do not affect the clique-width of a graph and the scheme provided by modular decomposition gives an efficient way of considering only the clique-width of the decomposed subgraphs. Motivated by the above property on clique-width, in this section we give an analogous result for linear clique-width. In particular we are able to show that for certain types of modules this nice property holds when restricted to linear clique-width of graphs.

Definition 1. *A set of vertices M in a graph G is a* maximal independent-set module *of G if M is an inclusion-maximal set of vertices that is both an independent set and a module of G.*

For a graph G, we define a binary relation for false twins u and v, denoted by $u \sim_{\text{ft}} v$. Since $\sim_{\text{ft}}$ is an equivalence relation, the corresponding equivalence classes partition the vertex set. It is not difficult to verify that the equivalence classes are exactly the maximal independent-set modules of G. We define the graph $G/\sim_{\text{ft}}$ as follows: there is a vertex in $G/\sim_{\text{ft}}$ for every maximal independent-set module of G, and two vertices of $G/\sim_{\text{ft}}$ are adjacent if and only if the corresponding maximal independent-set modules in G contain adjacent vertices. Clearly $G/\sim_{\text{ft}}$ is isomorphic to an induced subgraph of G. An independent-set module is called *trivial* if it contains a single vertex.

Lemma 1. *For a graph G, the independent-set modules of $G/\sim_{\text{ft}}$ are trivial.*

Lemma 2. *For a graph G,* $\text{lcwd}(G) = \text{lcwd}(G/\sim_{\text{ft}})$.

Proof. Since $G/\sim_{\text{ft}}$ is isomorphic to an induced subgraph of G, the inequality $\text{lcwd}(G/\sim_{\text{ft}}) \leq \text{lcwd}(G)$ is immediate. For showing $\text{lcwd}(G) \leq \text{lcwd}(G/\sim_{\text{ft}})$, let M_x for $x \in V(G/\sim_{\text{ft}})$ be the maximal independent-set module of G corresponding to x. Let a be a linear clique-width k-expression for $G/\sim_{\text{ft}}$ where $k = \text{lcwd}(G/\sim_{\text{ft}})$ and let ℓ_x be the label of x when adding x in the expression a. Define a linear clique-width k-expression a' for G as: for every vertex x of $G/\sim_{\text{ft}}$ place the vertices of M_x at the occurrence of the addition of x and give them the same label of x. That is, we replace the appearance of $\ell_x(x)$ in a with $\ell_x(v_1) \cdots \ell_x(v_{|M_x|})$, where $v_i \in M_x$. M_x is a module in G and every vertex of M_x will obtain the same neighbourhood as x in the graph defined by a'. And since every vertex of M_x receives the same label in a', M_x is an independent set in the graph defined by a'. Hence $\text{lcwd}(G) \leq \text{lcwd}(G/\sim_{\text{ft}})$.

4 Characterisation of Graphs of Linear Clique-Width at Most 3

We introduce a graph decomposition scheme such that graphs of linear clique-width at most 3 are exactly the graphs that are completely decomposable. Using this characterisation, we show that graphs of linear clique-width at most 3 are both cocomparability and weakly-chordal graphs, which particularly means that graphs of linear clique-width at most 3 do not contain long cycles or complements of long cycles as induced subgraphs.

As a first step, we give a characterisation of graphs of linear clique-width at most 2. The only known characterisation of this graph class is as the class of $\{2K_2, P_4, \text{co-}(2P_3)\}$-free graphs [8].

Definition 2. *A graph G is a* simple cograph *if G satisfies one of the following conditions:*

(1) G is an edgeless graph.
(2) G can be partitioned into two graphs A and B such that A is a simple cograph and B is an edgeless graph and $G = A \otimes B$ or $G = A \oplus B$.

Recall that cographs are graphs that are completely decomposable into edgeless graphs with respect to operations 'join' and 'union'. Hence simple cographs are defined as the graphs that are completely decomposable in the same way, with the restriction that one operand is always an edgeless graph in the 'join' and 'union' operations.

Theorem 1. *For a graph G, the following statements are equivalent:*

(1) G is a simple cograph.
(2) G is $\{2K_2, P_4, \text{co-}(2P_3)\}$-free.
(3) G can be reduced to a graph on a single vertex by repeatedly deleting an isolated vertex, a universal vertex or a false twin vertex.

Note that characterisation (3) of Theorem 1 implies a simple and linear-time recognition algorithm for graphs of linear clique-width at most 2.

Similar to graphs of linear clique-width at most 2, we want to give a decomposition scheme for graphs of linear clique-width at most 3. We define a set of operations and show that graphs of linear clique-width at most 3 are exactly the graphs that are completely decomposable into edgeless graphs by using these operations. Unlike for simple cographs, the operations are complex. We define the decomposition as a class of formal expressions. An *lc3-expression* is inductively defined as follows, where $d \in \{l, r\}$.

(d1) (A) is an lc3-expression, where A is a non-empty set of vertices.
(d2) Let T be an lc3-expression and let A be a (possibly empty) set of vertices not containing a vertex appearing in T, then $(d[T], A)$ is an lc3-expression.
(d3) Let T be an lc3-expression and let A_1, A_2, A_3, A_4 be disjoint (and possibly empty) sets of vertices not containing a vertex appearing in T, then $(A_1|A_2, d[T], A_3|A_4)$ is an lc3-expression.
(d4) Let T be an lc3-expression and let A_1, A_2, A_3, A_4 be disjoint (and possibly empty) sets of vertices not containing a vertex appearing in T, and let p be one of the following number sequences: 123, 132, 312, 321, 12, 32 (not allowed are 213, 231, 21, 23, 13, 31), then $T \odot (A_1|A_2, \bullet)$, $T \odot (\bullet, A_1|A_2)$ and $T \circ (A_1, A_2|A_3, A_4, d[p])$ are lc3-expressions where $\odot \in \{\otimes, \oplus\}$.

This completes the definition of lc3-expressions. These expressions define graphs. Important and necessary is that these expressions allow to fix a vertex partition for the defined graphs. The *graph defined by an lc3-expression and its associated vertex partition* is obtained according to the following inductive definition. Let T be an lc3-expression. Then, $G(T)$ is the following graph, where T' always means an lc3-expression and A, A_1, A_2, A_3, A_4 are sets of vertices and $d \in \{l, r\}$:

(i1) Let $T = (A)$, then $G(T)$ is the edgeless graph on vertex set A; the vertex partition associated with $G(T)$ is $(A, \emptyset)$.

(i2) Let $T = T' \odot (A_1|A_2, \bullet)$ for $\odot \in \{\otimes, \oplus\}$, and let $G(T')$ with vertex partition (B, C) be given, then $G(T)$ is obtained from $G(T')$ by adding the vertices in A_1 and A_2 and the edges $B \odot (A_1 \cup A_2)$ and $A_2 \otimes C$; the vertex partition associated with $G(T)$ is $((B \cup A_1 \cup A_2), C)$.

(i3) The case $T = T' \odot (\bullet, A_1|A_2)$ is similar to the previous one, where the edges are $C \odot (A_1 \cup A_2)$ and $A_2 \otimes B$ and the vertex partition is $(B, (C \cup A_1 \cup A_2))$.

(i4) Let $T = T' \circ (A_1, A_2|A_3, A_4, d[p])$, and let $G(T')$ with vertex partition (B, C) be given, then $G(T)$ is obtained from $G(T')$ by adding the vertices in $A_1 \cup A_2 \cup A_3 \cup A_4$; the set of edges is dependent on p: we have three start sets, A_2, B, C, and three additional sets, A_3, A_1, A_4, that correspond to A_2, B, C, respectively; the first join is executed between the two specified start sets, then the two additional sets are added to the sets involved in the first join; the second join is executed between a now enlarged set and a start set, then the third additional set is added to its corresponding start set and the last join operation (if there is one) is executed; so the result depends on the order of the operations; the numbers stand for: 1 means $A_2 \otimes B$, 2 means $B \otimes C$, 3 means $A_2 \otimes C$; note that the start sets become bigger after each join operation; the vertex partition associated with $G(T)$ is $((B \cup A_1 \cup A_2 \cup A_3), (C \cup A_4))$ or $((B \cup A_1), (C \cup A_4 \cup A_2 \cup A_3))$ for $d = l$ or $d = r$, respectively.

(i5) Let $T = (d[T'], A)$, and let $G(T')$ with vertex partition (B, C) be given, then $G(T)$ is the graph defined by $T' \otimes (\bullet, A|\emptyset)$ or $T' \otimes (A|\emptyset, \bullet)$ for $d = l$ or $d = r$, respectively; the vertex partition associated with $G(T)$ is $(B \cup C, A)$.

(i6) Let $T = (A_1|A_2, d[T'], A_3|A_4)$, and let $G(T')$ with vertex partition (B, C) be given, then $G(T)$ is the graph defined by $T' \circ (A_1, A_3|A_4, A_2, d[p'])$ where $p' =_{\text{def}} 132$ or $p' =_{\text{def}} 312$ for $d = l$ or $d = r$, respectively; the vertex partition associated with $G(T)$ is $((A_1 \cup A_2 \cup B \cup C), (A_3 \cup A_4))$.

Definition 3. *A graph G is an* lc3-graph *if there is an lc3-expression T such that $G = G(T)$.*

It is not difficult to show that induced graphs of lc3-graphs are also lc3-graphs. Note that this is not trivial, since a given lc3-expression might require considerable changes to obtain an lc3-expression for the induced subgraph.

Before showing that lc3-graphs are exactly the graphs of linear clique-width at most 3, we give two structural properties. For simplicity, we assume that empty graphs are simple cographs.

Lemma 3. *Let T be an lc3-expression, and let (B, C) be the vertex partition associated with $G(T)$. Then, the subgraph of $G(T)$ induced by C is a simple cograph.*

Proof. We give a construction, that works on the inductive definition of lc3-expressions. Let G_C denote the subgraph of $G(T)$ induced by vertex partition set C. If $T = (A)$ for some set A of vertices, then C is empty, and then G_C is

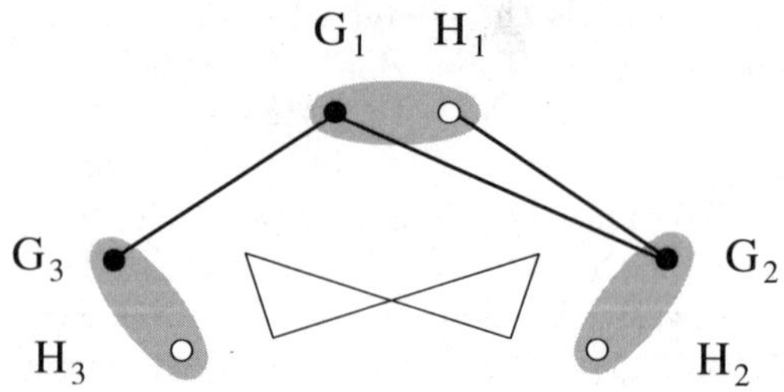

Fig. 1. The result of an lc3-composition of the graphs $G_1, G_2, G_3, H_1, H_2, H_3$. The thick lines from graphs G_1 and H_1 represent joins. The bow tie between $G_2 \oplus H_2$ and $G_3 \oplus H_3$ means either a join or no edge at all.

an empty graph. Empty graphs are simple cographs. If $T = (d[T'], A)$ or $T = (A_3|A_4, d[T], A_1|A_2)$ for $d \in \{l, r\}$, T' an lc3-expression and A_1, A_2, A_3, A_4 sets of vertices, then G_C is an edgeless graph on vertex set A or $A_1 \cup A_2$, respectively. If $T = T' \odot (A_1|A_2, \bullet)$ where $\odot \in \{\oplus, \otimes\}$, then C induces G_C also in $G(T')$, and we obtain the result by application of the induction hypothesis. Now, let (B', C') be the vertex partition associated with $G(T')$. If $T = T' \circ (A_1, A_2|A_3, A_4, l[p])$ and A_4 is non-empty, then G_C is equal to $G_{C'} \oplus G(T)[A_4]$, where $G_{C'}$ denotes the subgraph of G induced by C'. According to induction hypothesis, $G_{C'}$ is a simple cograph, and since $G(T)[A_4]$ is an edgeless graph, G_C is a simple cograph. Using the same definitions for the case $T = T \odot (\bullet, A_1|A_2)$ where $\odot \in \{\oplus, \otimes\}$, G_C is equal to $G_{C'} \odot G(T)[A_1 \cup A_2]$, which is a simple cograph. Finally, let $T = T' \circ (A_1, A_2|A_3, A_4, r[p])$. Depending on p, G_C is one of the following graphs:

- $(G_{C'} \otimes G(T)[A_2 \cup A_3]) \oplus G(T)[A_4]$
- $(G_{C'} \otimes G(T)[A_2]) \oplus G(T)[A_3 \cup A_4]$
- $(G_{C'} \oplus G(T)[A_4]) \otimes G(T)[A_2 \cup A_3]$.

All these graphs are simple cographs, and we conclude the proof.

The second property of lc3-graphs is a closure property for a special composition operation. Let G_1, G_2, G_3 be an edgeless graph, a simple cograph and an lc3-graph (in arbitrary assignment) and let H_1, H_2, H_3 be edgeless graphs. Graphs may be empty but at least two of them are non-empty. Then, the graph that is obtained from these six graphs and the additional edges as in Figure 1 is an lc3-graph. The bow tie in Figure 1 means either a join between $G_2 \oplus H_2$ and $G_3 \oplus H_3$ or no edge at all. We call this operation *complete* or *incomplete lc3-composition* depending on whether the bow tie represents a join operation ('complete') or a union operation ('incomplete').

Lemma 4. *Let G_1, G_2, G_3 be an edgeless graph, a simple cograph and an lc3-graph and let H_1, H_2, H_3 be edgeless graphs, where at least two of the six graphs are non-empty. Then, both the complete and the incomplete lc3-compositions of these graphs yield lc3-graphs.*

Proof. Let G_2 be an edgeless graph, G_3 be a simple cograph, G_1 be an lc3-graph. Let T be an lc3-expression for G_1. Let $T' =_{\text{def}} (l[T], \emptyset)$. Obtain T^* by adding operations of the form $\oplus(\bullet, \emptyset|A)$ and $\otimes(\bullet, \emptyset|A)$ to T' to obtain an lc3-expression for $G_1 \otimes G_3$. This can be done by reversing the construction procedure described in the proof of Lemma 3. Then, the expressions

$$T^* \oplus (V(H_1)|\emptyset, \bullet) \oplus (\bullet, V(H_3)|\emptyset) \otimes (\bullet, V(H_2)|V(G_2))$$
$$T^* \oplus (V(H_1)|\emptyset, \bullet) \oplus (\bullet, V(H_3)|\emptyset) \oplus (\bullet, V(H_2)|V(G_2))$$

define lc3-expressions for the complete and incomplete lc3-composition of the six graphs. In case that G_2 is an edgeless graph, G_1 is a simple cograph and G_3 is an lc3-graph, we obtain lc3-expressions for the lc3-compositions in a similar way. Let G_1 be an edgeless graph. Similar to the construction above, we obtain an lc3-expression T for $G_2 \oplus G_3$. Then,

$$T \circ (V(H_2), V(G_1)|V(H_1), V(H_3), l[p]) \quad \text{or}$$
$$T \circ (V(H_1), V(G_1)|V(H_1), V(H_2), l[p])$$

for an appropriate choice of p is an lc3-expression for the complete or incomplete lc3-composition. The case of G_3 being an edgeless graph is analogous.

Our main theorem is proven by considering linear clique-width 3-expressions. Linear clique-width expressions may contain redundant operations with no affect to the defined graph. As an auxiliary result, the following lemma shows a normalisation result for linear clique-width expressions. We use the following notation. Let $t = t_1 \cdots t_q$ be a linear clique-width k-expression. By $G[t_1 \cdots t_i]$, we denote the graph that is defined by subexpression $t_1 \cdots t_i$ of t. We say that vertices with the same label in $G[t_1 \cdots t_i]$ belong to the same label class of $G[t_1 \cdots t_i]$.

Lemma 5. *Let $k \geq 1$ and let G be a graph that has a linear clique-width k-expression. Then, G has a linear clique-width k-expression $a = a_1 \cdots a_r$ such that the following holds for all join and relabel operations a_i in a:*

(1) $G[a_1 \cdots a_i]$ does not contain an isolated vertex
(2) $G[a_1 \cdots a_{i-1}]$ contains vertices of the two label classes involved in a_i.

Proof. Let $b = b_1 \cdots b_s$ be a linear clique-width k-expression for G. Let b_i be a join or relabel operation and suppose that $G[b_1 \cdots b_i]$ contains an isolated vertex, say x. Let c be the label of x in $G[b_1 \cdots b_i]$. If b_i is a join operation then c is not one of the two join labels. We obtain $b' = b'_1 \cdots b'_s$ from b by deleting the vertex creation operation for x in b and adding the operation $c(x)$ right after operation b_i. Then, $G[b'_1 \cdots b'_i] = G[b_1 \cdots b_i]$. Iterated application of this operation shows existence of a linear clique-width k-expression having the first property. For the second property, let $d = d_1 \cdots d_t$ be a linear clique-width k-expression that has the first property. Let $d_i = \eta_{c,c'}$ be a join operation. If $G[d_1 \cdots d_i]$ does not contain a vertex with label c or c', d_i does not add an edge to $G[d_1 \cdots d_{i-1}]$, i.e.,

$G[d_1 \cdots d_{i-1}] = G[d_1 \cdots d_i]$. We obtain d' from d by deleting operation d_i. Let $d_i = \rho_{c \to c'}$ be a relabel operation, and suppose that one of the two involved label classes is empty in $G[d_1 \cdots d_{i-1}]$. If the label class corresponding to c is empty, we obtain d' from d by just deleting operation d_i. If the class corresponding to c is non-empty but the class corresponding to c' is empty, we obtain d' from d by first exchanging c and c' in all operations $d_{i+1}, \ldots, d_t$ and then deleting d_i. It is clear that $G[d_1 \cdots d_j]$ and $G[d'_1 \cdots d'_{j-1}]$ correspond to each other for every $j \in \{i+1, \ldots, t\}$ with the exception that the classes corresponding to c and c' are exchanged. Repeated application of the modification completes the proof.

Theorem 2. *A graph has linear clique-width at most 3 if and only if it is an lc3-graph.*

The main idea of the proof of the 'only-if' part is to partition a given normalised linear clique-width 3-expression into subexpressions between two relabel operations. The modifications of the described graph are expressed by appropriate combinations of lc3-expression operations. For the converse, it is mainly sufficient to give linear clique-width 3-expressions for each lc3-expression operation.

Using the new characterisation of graphs of linear clique-width at most 3, we can show that these graphs are cocomparability graphs. A graph is called *cocomparability graph* if directions can be assigned to the edges of its complement to obtain a directed graph that is transitive. For a graph $G = (V, E)$, a vertex ordering σ is called *cocomparability ordering* if for every triple u, v, w of vertices such that $u \prec_\sigma v \prec_\sigma w$, $uw \in E$ implies $uv \in E$ or $vw \in E$. A graph is cocomparability if and only if it has a cocomparability ordering [13].

Proposition 1. *Lc3-graphs are cocomparability graphs.*

Proof. We show the statement by induction over the definition of lc3-expressions. Let $G = (V, E)$ be an lc3-graph with lc3-expression T. We show that there is a cocomparability ordering for G that respects the vertex partition associated with $G(T)$. First, let $T = (A)$ for a set A of vertices. Then, $G(T)$ is an edgeless graph, and every vertex sequence for G is a cocomparability ordering for G. Furthermore, every vertex sequence for G respects the vertex partition $(A, \emptyset)$. Now, let T be more complex. For the rest of the proof, let T' be an lc3-expression with associated vertex partition (B, C), let σ' be a cocomparability ordering for $G(T')$ that respects partition (B, C), which means that the vertices in B appear consecutively and the vertices in C appear consecutively in σ', let A, A_1, A_2, A_3, A_4 be sets of vertices, $d \in \{l, r\}$ and p an appropriate number sequence. We consider only the more interesting cases here.

- $T = T' \odot (A_1|A_2, \bullet)$ for $\odot \in \{\oplus, \otimes\}$. We obtain σ from σ' by adding the vertices in A_1 to the left of the vertices in B and the vertices in A_2 between the vertices in B and C. Then, σ is a cocomparability ordering for G and the vertices in $B \cup A_1 \cup A_2$ appear consecutively.
- $T = (l[T'], A)$. The vertices in A are adjacent to only the vertices in C. We obtain σ from σ' by adding the vertices in A to the right of the vertices in C. Then, σ is a cocomparability ordering for G that respects the partition $(B \cup C, A)$.

- $T = T' \circ (A_1, A_2|A_3, A_4, l[12])$. The vertices in B and A_2 are adjacent and the vertices in $B \cup A_1$ and C are adjacent. We obtain σ by placing the vertices in the following order: A_3, A_2, B, A_1, C, A_4, and the vertices in B and C appear in order determined by σ'. Then, σ is a cocomparability ordering for G and respects the vertex partition $((B \cup A_1 \cup A_2 \cup A_3), (C \cup A_4))$.
- $T = T' \circ (A_1, A_2|A_3, A_4, r[12])$. We place the vertices in order A_4, A_3, A_2, C, B, A_1.

The remaining cases are $(A_1|A_2, d[T'], A_3|A_4)$ and $T = T' \circ (A_1, A_2|A_3, A_4, d[p])$ where $p \in \{123, 132, 312, 321\}$. We begin with the situation in Figure 1. Suppose cocomparability orderings for the graphs $G_1, G_2, G_3, H_1, H_2, H_3$ are given. We define three vertex orderings for the depicted graph. The vertices of every partition graph appear consecutively and in order defined by the given orderings. The order of the partition graphs is as follows:

$$H_2, G_3, G_2, H_3, G_1, H_1 \ ; \quad H_3, H_1, G_1, G_3, G_2, H_2 \ ; \quad H_2, G_1, G_2, H_1, G_3, H_3 \, .$$

It is easy to check that all three vertex orderings actually define cocomparability orderings for the graph (scheme) depicted in Figure 1. Furthermore, for every $i \in \{1, 2, 3\}$, there is a cocomparability ordering such that the vertices of $G_i \oplus H_i$ appear consecutively at the end of the ordering. Depending on p, the pairs $(B, A_1), (C, A_4), (A_2, A_3)$ are matched to $(G_1, H_1), (G_2, H_2), (G_3, H_3)$, and depending on d and the particular case, one of the vertex orderings is chosen to achieve the correct vertex partition. This completes the proof.

Proposition 1 shows that lc3-graphs do not contain induced cycles of length more than 4, since such cycles are no cocomparability graphs. This also holds the complements of such cycles. Graphs that neither contain induced cycles of length at least 5 nor their complements as induced subgraphs, are called *weakly-chordal.* The proof of the following proposition requires a careful but not difficult argumentation.

Proposition 2. *Lc3-graphs are weakly-chordal.*

5 Recognition of Graphs of Linear Clique-Width at Most 3

The results of the previous section provide an efficient recognition algorithm for lc3-graphs. This algorithm, called Lc3-graphRecognition, is given in Figure 2. The main loop decomposes the input graph using the reverse of lc3-composition as long as possible. When such a decomposition is not possible, a vertex partition has to be determined. Algorithm Simplify applies the opposite strategy, since it starts already with a vertex partition. We call a vertex *almost-universal* if it has exactly one non-neighbour. The following lemma summarises properties of Simplify.

Lemma 6. *Let $G = (V, E)$ be a graph, and let (B, C) be a vertex partition for G. Let G not have false twin vertices in B and in C. If* Simplify *applied to G and (B, C) ...*

```
Algorithm LC3-GRAPHRECOGNITION
Input    a graph G = (V, E)
Result   an answer ACCEPT or REJECT

set G := G/~ft;
while  no return  do
   if  G is edgeless  then  return ACCEPT
   else-if  G is the result of an lc3-composition of at least two non-empty graphs  then
      if  all partition graphs are simple cographs  then  return ACCEPT
      else  set G to the partition graph that is not a simple cograph  end if
   else-if  there is a vertex u such that
            SIMPLIFY on G−u with partition (N_G(u), V(G) \ N_G[u]) returns a graph G'  then
      set G := G'
   else-if  there is a pair u, v of non-adjacent vertices where the degree of v is |V(G)| − 2 such that
            SIMPLIFY on G − {u, v} with partition (N_G(u), V(G) \ (N_G(u) ∪ {u, v}))
            returns a graph G'  then
      set G := G'
   else  return REJECT  end if
end while.

Algorithm SIMPLIFY
Input    a graph G and a vertex partition (B, C) for G
Result   an answer ACCEPT or REJECT or a graph G', where ACCEPT returns a graph on a single vertex

while  no return  do
   if  G is edgeless  then  return ACCEPT
   else-if  B = ∅ or C = ∅  then  return G
   else-if  G has a vertex u such that
            N_G(u) = ∅  or  N_G(u) = B  or  N_G(u) = C  or
            N_G[u] = V(G)  or  N_G[u] = B  or  N_G[u] = C  then
      set G := G−u and (B, C) := (B \ {u}, C \ {u})
   else-if  there is a pair u, v of non-adjacent vertices where v is almost-universal such that
            u, v ∈ B  and  N_G(u) = B \ {u, v}  or
            u, v ∈ C  and  N_G(u) = C \ {u, v}  then
      set G := G − {u, v} and (B, C) := (B \ {u, v}, C \ {u, v})
   else-if  B or C is singleton set containing vertex u  and
             |V(G) \ N_G[u]| ≠ 2  or  V(G) \ N_G[u] = {x, z} and the sets
             N_G(x) \ {z} and N_G(z) \ {x} can be ordered by inclusion  then
      set G := G−u and (B, C) := (V(G) \ N_G[u], N_G(u))
   else-if  G is the result of an lc3-composition of at least two non-empty graphs
            that respects the given vertex partition  then
      if  all partition graphs are simple cographs  then  return ACCEPT
      else  return the partition graph that is not a simple cograph  end if
   else  return REJECT  end if
end while.
```

Fig. 2. The recognition algorithm for lc3-graphs

(1) *returns* ACCEPT *then G is an lc3-graph and there is an lc3-expression T for G that associates $G(T)$ with vertex partition (B, C) or (C, B).*
(2) *returns* REJECT *then G is not an lc3-graph or there is no lc3-expression T for G that associates $G(T)$ with vertex partition (B, C) or (C, B).*
(3) *returns a graph G' then G is an lc3-graph and there is an lc3-expression T for G that associates $G(T)$ with vertex partition (B, C) or (C, B) if and only if G' is an lc3-graph. Furthermore, G' is a module and subgraph of G that is induced by vertices only in B or in C.*

Theorem 3. *There is an $\mathcal{O}(n^2m)$-time algorithm for recognising lc3-graphs.*

Correctness. If the algorithm accepts, an lc3-expression can be constructed for the input graph. Important subroutines are the proofs of Lemmata 4 and 2. The converse is proven by induction. Let G be the input graph, and let T be an lc3-expression for G. For space reasons, we restrict G to have only trivial independent-set modules and not to be the result of an lc3-composition. The last operation in T is not of the forms (d1) and (d3) and the complex operation of (d4). Let the last operation be of the form (d2). There is a vertex u such that $G-u$ is an lc3-graph with lc3-expression T' that associates $G(T')$ with the vertex partition $(N_G(u), V(G)\setminus N_G[u])$ or its reverse. Due to Lemma 6, Simplify outputs a proper subgraph of G, that is accepted. Let the last operation of T be of the form $\odot(A_1|A_2, \bullet)$ or $\odot(\bullet, A_1|A_2)$. If $\odot = \oplus$ then A_1 is empty. The vertex in A_2, say u, defines vertex partition $(V(G) \setminus N_G[u], N_G(u))$ or $(N_G(u), V(G) \setminus N_G[u])$ for $G-u$, and this partition corresponds to the vertex partition associated with $G(T)-u$. So, G is accepted. Let $\odot = \otimes$. If A_2 is empty, the case is similar to the previous one. If A_1 is empty, the vertex in A_2 is universal. Thus, A_1 and A_2 are non-empty. The vertex in A_2 is adjacent to all vertices but the vertex in A_1, and the vertex in A_1 defines a vertex partition for G. The algorithm accepts.

Running time. Graph $G/\sim_{\text{ft}}$ can be computed in linear time. Every **while**-loop execution is done with a smaller graph, so that there are at most n **while**-loop executions. A single **while**-loop execution takes time $\mathcal{O}(nm)$: linear time for checking for result of an lc3-composition and at most $2n$ applications of Simplify, that requires linear time each call. This shows the total $\mathcal{O}(n^2m)$ running time.

References

1. Brandstädt, A., Le, V.B., Spinrad, J.P.: Graph Classes: A Survey. SIAM Monographs on Discrete Mathematics and Applications (1999)
2. Corneil, D.G., Habib, M., Lanlignel, J.-M., Reed, B.A., Rotics, U.: Polynomial time recognition of clique-width ≤ 3 graphs. In: Gonnet, G.H., Viola, A. (eds.) LATIN 2000. LNCS, vol. 1776, pp. 126–134. Springer, Heidelberg (2000)
3. Corneil, D.G., Perl, Y., Stewart, L.K.: A linear recognition algorithm for cographs. SIAM J. Comput. 14, 926–934 (1985)
4. Courcelle, B., Engelfriet, J., Rozenberg, G.: Handle-rewriting hypergraph grammars. J. Comput. System Sci. 46, 218–270 (1993)
5. Courcelle, B., Makowsky, J.A., Rotics, U.: Linear time solvable optimization problems on graphs of bounded clique-width. Theory Comput. Syst. 33, 125–150 (2000)
6. Courcelle, B., Olariu, S.: Upper bounds to the clique width of graphs. Discrete Applied Mathematics 101, 77–114 (2000)
7. Fellows, M.R., Rosamond, F.A., Rotics, U., Szeider, S.: Clique-width Minimization is NP-hard. In: Proceedings of the thirty-eighth annual ACM Symposium on Theory of Computing, STOC 2006, pp. 354–362 (2006)
8. Gurski, F.: Characterizations for co-graphs defined by restricted NLC-width or clique-width operations. Discrete Mathematics 306, 271–277 (2006)
9. Gurski, F.: Linear layouts measuring neighbourhoods in graphs. Discrete Mathematics 306, 1637–1650 (2006)
10. Gurski, F., Wanke, E.: On the relationship between NLC-width and linear NLC-width. Theoretical Computer Science 347, 76–89 (2005)
11. Heggernes, P., Meister, D., Papadopoulos, Ch.: Graphs of small bounded linear clique-width. Technical report, no.362, Institute for Informatics, University of Bergen, Norway (2007)
12. Hliněný, P., Oum, S., Seese, D., Gottlob, G.: Width parameters beyond tree-width and their applications. Computer Journal (to appear)
13. Kratsch, D., Stewart, L.: Domination on cocomparability graphs. SIAM Journal on Discrete Mathematics 6, 400–417 (1993)

A Well-Mixed Function with Circuit Complexity $5n \pm o(n)$: Tightness of the Lachish-Raz-Type Bounds

Kazuyuki Amano[1] and Jun Tarui[2]

[1] Dept of Comput Sci, Gunma Univ, Tenjin 1-5-1, Kiryu, Gunma 376-8515, Japan
amano@cs.gunma-u.ac.jp

[2] Dept of Info & Comm Eng, Univ of Electro-Comm, Chofu, Tokyo 182-8585, Japan
tarui@ice.uec.ac.jp

Abstract. A Boolean function on n variables is called k-mixed if for any two different restrictions fixing the same set of k variables must induce different functions on the remaining $n-k$ variables. In this paper, we give an explicit construction of an $n - o(n)$-mixed Boolean function whose circuit complexity over the basis U_2 is $5n + o(n)$. This shows that a lower bound method on the size of a U_2-circuit that uses the property of k-mixed, which gives the current best lower bound of $5n - o(n)$ on a U_2-circuit size (Iwama, Lachish, Morizumi and Raz [STOC '01, MFCS '02]), has reached the limit.

1 Introduction and Results

A Boolean function on n variables is called k-mixed if for any two different restrictions fixing the same set of k variables must induce different functions on the remaining $n - k$ variables. The notion of a k-mixed Boolean function, which was introduced by Jukna [7], plays an important role on deriving explicit lower bounds on several computational models.

The best known $5n - o(n)$ lower bound on the size of a Boolean circuit over the basis U_2 can be established for *any* $n - o(n)$-mixed Boolean function [5,6]. The basis U_2 is the set of all Boolean functions over two variables except for the XOR function and its complement. It is also known that the size of any read-once branching program computing a k-mixed Boolean function is at least 2^k (see e.g., [12,10]). This result was used to show the existence of a function in P having read-once branching program size not smaller than $2^{n-O(\sqrt{n})}$ [10] (see also [2] for an improved result).

In this paper, we focus on the complexity of Boolean circuits over the basis U_2. Deriving a good lower bound for an explicit Boolean function in such a general circuit model is one of the central problems in computer science. In 1991, Zwick [14] gave a lower bound of $4n - O(1)$ for a certain family of symmetric Boolean functions. After a decade, Lachish and Raz [9] introduced a new family of Boolean functions called *strongly-two-dependent* and improved the lower bound to $4.5n - o(n)$. Shortly after, Iwama and Morizumi [5] proved a lower

M. Agrawal et al. (Eds.): TAMC 2008, LNCS 4978, pp. 342–350, 2008.

bound of $5n - o(n)$ for the same family of Boolean functions, which is the current best (see also [6]). Since it is easily shown that the property of k-mixed is stronger than that of strongly-two-dependent, the above mentioned lower bound of $5n - o(n)$ can be established for every k-mixed function with $k = n - o(n)$.

A simple and explicit construction of an $n - 3\sqrt{n}$-mixed Boolean function was given by Savický and Zák [10]. Their function is of the form $h(x) = x_{\phi(x)}$, where $\phi : \{0,1\}^n \to \{1, 2, \ldots, n\}$ is a kind of weighted sum of inputs. If we consider the circuit complexity of their function, a straightforward implementation yields an upper bound of $O(n \log n)$. It seems to be difficult to improve this upper bound without changing the definition of the function, since we need to manipulate n numbers each has $O(\log n)$-digits in order to compute the index function $\phi(\cdot)$. Hence it would be natural to expect that a lower bound method that uses the property of k-mixed [6,9,5], which gives the current best lower bound on a U_2-circuit size of $5n - o(n)$, can further be extended for obtaining a higher lower bound.

The main result of this paper is to show that this is not the case since there is a well-mixed function with circuit complexity $5n + o(n)$. Precisely, we give a construction of an explicit Boolean function f_n on n variables such that

(i) f_n is $n - t(n)$-mixed for any $t(n) = \omega(\sqrt{n} \log^2 n)$, and
(ii) f_n can be computed by a circuit of size $5n + o(n)$ over the basis U_2.

The circuit complexity of our function over the basis U_2 is optimal up to a lower order term. The result also shows that a lower bound method that uses the property of k-mixed (Iwama, Lachish, Morizumi and Raz [6,9,5]) has reached the limit.

The organization of the paper is as follows. In Section 2, we give some preliminaries. In Section 3, we describe the definition and the analysis of our function. Finally, we give some concluding remarks in Section 4.

2 Notations and Definitions

For a natural number n, $[n]$ denotes the set $\{1, 2, \ldots, n\}$. For a binary sequence $x = x_{k-1} \cdots x_0$, $(x)_2$ denotes the integer represented by x, i.e., $(x)_2 = \sum_{i=0}^{k-1} 2^i x_i$.

Throughout the paper, we consider a Boolean function on the variable set $X_n = \{x_1, \ldots, x_n\}$. Let B_2 denote the set of all (sixteen) Boolean functions over two variables, and let U_2 denote $B_2 - \{\oplus, \equiv\}$, i.e., U_2 contains all Boolean functions over two variables except for the XOR function and its complement. For a basis $B \subseteq B_2$, *a Boolean circuit* over the basis B is a directed acyclic graph with nodes of in-degree 0 or 2. Nodes of in-degree 0 are called *input-nodes*, and each one of them is labeled by a variable in X_n or a constant 0 or 1. Nodes of in-degree 2 are called *gate-nodes*, and each one of them has two inputs and an output, and is labeled by a function in B. For a basis B and for a Boolean function f, the circuit complexity of f over the basis B, which is denoted by $\mathsf{Size}_B(f)$, is the minimum number of gate-nodes in a circuit over the basis B that computes f.

Let f be a Boolean function on X_n. A *partial assignment* is a function $\sigma : X_n \to \{0, 1, *\}$, where $\sigma(x_i) = 0$ or 1 means that the input variable x_i is fixed to the corresponding constant, and $\sigma(x_i) = *$ means x_i remains free. For a partial assignment σ, the support of σ, denoted by $\mathsf{Sup}(\sigma)$, is $\mathsf{Sup}(\sigma) = \{x_i \mid \sigma(x_i) \neq *\}$. Let $f|_\sigma$ denote the subfunction (over $X_n \backslash \mathsf{Sup}(\sigma)$) obtained by restricting all variables x_i in $\mathsf{Sup}(\sigma)$ to $\sigma(x_i)$.

All logarithms in the paper are base 2.

3 The Function and Proofs

In this section, we give an explicit construction of a well-mixed Boolean function whose circuit complexity over the basis U_2 is $5n \pm o(n)$.

Definition 1. *For $k \in \mathbf{N}$, a Boolean function f on X_n is called k-mixed, if for every $V \subseteq X_n$ such that $|V| = k$ and for any two distinct partial assignments α, β with support V yield different subfunctions, i.e., $f|_\alpha \neq f|_\beta$.*

We now describe the definition of our function. Our function is of the form $f_n(x) = x_z$ where z is computed from a weighted sum of *block parities* of input variables.

Definition 2. *For an integer n, let $p[n]$ be the smallest prime greater than or equal to n. Note that the Bertrand-Chebyshev theorem (see e.g., [4, p.373]), which states that there is a prime p such that $n \leq p < 2n$ for every $n > 1$, guarantees that $p[n] < 2n$. Let $w_n : \mathbf{N} \to [n]$ be defined as follows. For an integer s, let $\tilde{s}$ be the unique integer satisfying $\tilde{s} \equiv s \pmod{p[n]}$ and $1 \leq \tilde{s} \leq p[n]$. Then,*

$$w_n(s) = \begin{cases} \tilde{s}, & 1 \leq \tilde{s} \leq n, \\ 1, & \textit{otherwise}. \end{cases}$$

Let $b = \lceil \frac{n}{\lceil \log^2 n \rceil} \rceil$. For an index $i \in [n]$, the value $\lceil \frac{i}{\lceil \log^2 n \rceil} \rceil$ is called a weight of the index i and is denoted by $\mathsf{wgt}(i)$. We divide $[n]$ into b blocks $D_1, \ldots, D_b$ such that $j \in D_i$ iff $\mathsf{wgt}(j) = i$.

For a total assignment x to X_n and $i \in [b]$, let $\mathsf{PAR}(x, i)$ denote the parity of variables whose weight is i, i.e.,

$$\mathsf{PAR}(x, i) = \bigoplus_{j \in D_i} x_j.$$

For every n, the Boolean function $f_n : \{0,1\}^n \to \{0,1\}$ is defined as $f_n(x) = x_z$, where

$$z = w_n\left(\sum_{i=1}^{b} i \cdot \mathsf{PAR}(x, i)\right). \tag{1}$$

We should note that our function is a modification of the function introduced by Savický and Zák [10], which was shown to be $n - 3\sqrt{n}$-mixed. The difference between their function and ours is that they defined $z := w_n(\sum_{i=1}^n i \cdot x_i)$ instead of Eq. (1) in Definition 2.

Below we prove that the function f_n defined above is $n - t(n)$ mixed for every $t(n) = \omega(\sqrt{n}\log^2 n)$ (Theorem 3), and then prove that the circuit complexity of f_n over the basis U_2 is $5n + o(n)$ (Theorem 6).

Theorem 3. *For any $t = t(n) = \omega(\sqrt{n}\log^2 n)$, the function f_n is $(n-t)$-mixed.*

As for the result of Savický and Zák [10], we use the following theorem due to da Silva and Hamidoune [11] (see also e.g., [1]).

Theorem 4. *Let p be a prime number and let z and h be two integers. Let $h \le z \le p$ and let $A \subseteq Z_p$ such that $|A| = z$. Let B be the set of all sums of h distinct elements of A. Then $|B| \ge \min(p, hz - h^2 + 1)$.*

The outline of the proof of Theorem 3 is similar to the proof of Theorem 2.5 in [10]. However, we need more detailed analysis since the effect of a partial assignment to our function is more complex than that to the function introduced by Savický and Zák [10].

Proof. (of Theorem 3) Let $I \subseteq [n]$ with $|I| = n - t$ and let u and v be two different partial assignments both of whose support is I. To show $f_n|_u \ne f_n|_v$, it is sufficient to show that there are two total assignments x and y such that $f_n(x) \ne f_n(y)$, where x is an extension of u and y is an extension of v. Let u^0 (v^0, resp.) be a total assignment obtained from u (v, resp.) by additionally assigning every variables in $[n]\backslash I$ to the constant 0.

Let J be an arbitrary maximal subset of $[n]\backslash I$ such that every two elements of J have distinct weights, i.e., $\mathsf{wgt}(i) \ne \mathsf{wgt}(j)$ for every $i, j \in J$ with $i \ne j$. Note that $|J| = \omega(\sqrt{n})$ since there are at most $O(\log^2 n)$ indices having same weight.

Intuitively, Theorem 4 guarantees that, for every assignment to the variable set $[n]\backslash J$, it is possible to set the remaining inputs J in such a way that the weighted sum of parities in Eq. (1) belongs to any given value residue class modulo $Z_{p[n]}$.

Now we divide the set of indices J into two classes J_s and J_d defined as

$$J_s = \{j \in J \mid \mathsf{PAR}(u^0, \mathsf{wgt}(j)) = \mathsf{PAR}(v^0, \mathsf{wgt}(j))\},$$
$$J_d = \{j \in J \mid \mathsf{PAR}(u^0, \mathsf{wgt}(j)) \ne \mathsf{PAR}(v^0, \mathsf{wgt}(j))\}.$$

We divide the proof of the theorem into two cases depending on the sizes of J_s and J_d.

In the rest of the proof, the symbol $\equiv$ means that the congruence modulo $p[n]$.

Case A. $|J_s| \ge |J_d|$.
In this case we have $|J_s| = \omega(\sqrt{n})$. Let $\tilde{u}$ ($\tilde{v}$, resp.) be a partial assignment obtained from u (v, resp.) by additionally assigning every variables in $([n]\backslash I)\backslash J_s$ to 0.

Let $S(u) = \sum_{i=1}^{b} i \cdot \mathsf{PAR}(u^0, i)$ and $S(v) = \sum_{i=1}^{b} i \cdot \mathsf{PAR}(v^0, i)$. Let Δ be an element of Z_p such that $\Delta \equiv S(v) - S(u)$.

Case A-1. $\Delta \not\equiv 0$.
We extend $\tilde{u}$ and $\tilde{v}$ to u' and v' by setting some positions not fixed in $\tilde{u}$ and $\tilde{v}$. We first choose any $j \in J_s \backslash \{1\}$. Let $\ell = w_n(j + \Delta)$. Since $\ell \equiv j + \Delta \not\equiv j$ or $\ell = 1 \neq j$, we have $j \neq \ell$. If $\ell \in J_s$, in u' and v' we set the position j to 0 and set the position ℓ to 1. This ensures that $u'_j = v'_j \neq u'_\ell = v'_\ell$. If $\ell \notin J_s$, we set the position j of both u' and v' in such a way that $u'_j = v'_j \neq v'_\ell$.

We have $|J_s| - 2 = \omega(\sqrt{n})$ positions that are still not specified in both u' and v'. We denote A by the set of these positions. For an index $j \in [n]$, the *contribution* of j, denoted by c_j, is defined as

$$c_j = \begin{cases} \mathsf{wgt}(j), & \text{if } \mathsf{PAR}(u^0, \mathsf{wgt}(j)) = 0, \\ p[n] - \mathsf{wgt}(j), & \text{if } \mathsf{PAR}(u^0, \mathsf{wgt}(j)) = 1. \end{cases}$$

Intuitively, setting 1 to an unassigned variable x_j increases the total weight of x (i.e., $\sum_{i=1}^{b} i \cdot \mathsf{PAR}(x, i)$) by c_j. Since $\mathsf{wgt}(j) \leq b = \lceil \frac{n}{\lceil \log^2 n \rceil} \rceil < p[n]/2$ for every j, all indices in A have distinct contributions. Let $h = \lfloor |A|/2 \rfloor$. Since $|A|h - h^2 + 1 = \omega(n) \geq p[n]$, Theorem 4 guarantees that there is a set $H \subseteq A$ of size h such that

$$\sum_{i \in H} c_i \equiv j - \sum_{i=1}^{b} i \cdot \mathsf{PAR}(u', i)$$

Let x be an extension of u' such that $x_i = 1$ for every $i \in H$ and $x_i = 0$ for every $i \in A \backslash H$. Then, we have

$$\sum_{i=1}^{b} i \cdot \mathsf{PAR}(x, i) \equiv j.$$

This implies that $f_n(x) = u'_j$. Let y be an assignment extending v such that the bits with indices in J_s have the same value in x and y. This implies that

$$\omega_n \left(\sum_{i=1}^{b} i \cdot \mathsf{PAR}(y, i) \right) = \omega_n \left(\sum_{i=1}^{b} i \cdot \mathsf{PAR}(x, i) + \Delta \right) = \ell,$$

and hence $f_n(y) = v'_\ell \neq f_n(x)$.

Case A-2. $\Delta \equiv 0$.
Let $j \in I$ with $u_j \neq v_j$. Since $|J_s| = \omega(\sqrt{n})$, Theorem 4 guarantees that there is an extension x of u such that $\sum_{i=1}^{b} i \cdot \mathsf{PAR}(x, i) \equiv j$. Let y be an assignment extending v such that the bits with indices in J_s have the same value in x. Then, we have $f_n(x) = u_j \neq v_j = f_n(y)$.

Case B. $|J_d| > |J_s|$.
We divide the set J_d into two sets $J_{d,0}$ and $J_{d,1}$ defined as

$$J_{d,0} = \{j \in J_d \mid \mathsf{PAR}(u^0, \mathsf{wgt}(j)) = 0 \wedge \mathsf{PAR}(v^0, \mathsf{wgt}(j)) = 1\},$$
$$J_{d,1} = \{j \in J_d \mid \mathsf{PAR}(u^0, \mathsf{wgt}(j)) = 1 \wedge \mathsf{PAR}(v^0, \mathsf{wgt}(j)) = 0\}.$$

Without loss of generality, we can assume that $|J_{d,0}| \geq |J_{d,1}|$ (otherwise we swap u with v). Note that $|J_{d,0}| \geq |J_d|/2 = \omega(\sqrt{n})$. Let $\tilde{u}$ ($\tilde{v}$, resp.) be a partial assignment obtained from u (v, resp.) by additionally assigning every variables in $([n]\backslash I)\backslash J_{d,0}$ to 0.

Let $S(u) = \sum_{i=1}^{b} i \cdot \mathsf{PAR}(u^0, i)$ and $S(v) = \sum_{i=1}^{b} i \cdot \mathsf{PAR}(v^0, i)$. Let Δ be an element of Z_p such that $\Delta \equiv S(v) - S(u)$.

As for Case A-1, we will extend $\tilde{u}$ and $\tilde{v}$ to u' and v' by setting some positions in $J_{d,0}$. The difference to Case A-1 is that if we set the value 1 to an unassigned variable whose weight is k, then the total weight of $\tilde{u}$ increases by k and that of $\tilde{v}$ decreases by k.

Let k be an arbitrary element in Z_p that satisfies $\omega_n(S(u)+k) \in J_{d,0}\backslash\{1\}$, and let $j = \omega_n(S(u) + k)$ for a chosen k. Let $\ell = \omega_n(S(v) - k) = \omega_n(S(u) + \Delta - k)$. Without loss of generality, we can assume that $j \neq \ell$. This is because that since $j = \ell$ implies $S(u)+k \equiv S(u)+\Delta-k$, the number of choices of k that forces $j = \ell$ is at most one. Hence we can always avoid such a k since $|J_{d,0}|$ is sufficiently large. If $\ell \in J_{d,0}$, in u' and v' we set the position j to 0 and set the position ℓ to 1. If $\ell \notin J_{d,0}$, we set the position j of both u' and v' in such a way that $u'_j = v'_j \neq v_\ell$.

The rest of the proof is an analogous to Case A-1. Since we have at least $|J_{d,0}| - 2 = \omega(\sqrt{n})$ unassigned variables, Theorem 4 guarantees that we can always extend u' to x such that $\sum_{i=1}^{b} i \cdot \mathsf{PAR}(x, i) \equiv S(u) + k$. Let y be a total assignment obtained from v' by applying the same extension as for u'. Then we have $\sum_{i=1}^{b} i \cdot \mathsf{PAR}(y, i) \equiv S(v) - k$, and therefore $f_n(x) = x_j \neq y_\ell = f_n(y)$. This completes the proof of Theorem 3. □

Now we proceed to the analysis of the complexity of our function. We use a circuit called *decoder* whose definition is as follows.

Definition 5. *An n-to-2^n decoder* $\mathsf{Decode}_n : \{0,1\}^n \to \{0,1\}^{2^n}$ *is the function that takes an n-bit binary input x and outputs $d_0 d_1 \cdots d_{2^n-1}$ such that $d_i = 1$ iff $(x)_2 = i$.*

It is well known that $\mathsf{Size}_{U_2}(\mathsf{Decode}_n) = 2^n + O(n2^{n/2})$ (see e.g., [13, p. 75]).

Theorem 6. *The function f_n can be computed by a circuit of size at most $5n + o(n)$ over the basis U_2.*

Proof. We break down the computation of f_n into five steps as follows :

(i) Compute $\mathsf{PAR}(x, i)$ for each $i = 1, \ldots, b$. Recall that b denotes the number of blocks of the input variables.

(ii) Compute the binary representation of $i \cdot \mathsf{PAR}(x, i)$ for each $i = 1, \ldots, b$.
(iii) Compute the binary representation of $\sum_{i=1}^{b} i \cdot \mathsf{PAR}(x, i)$.
(iv) Compute the binary representation of $z = w_n \left(\sum_{i=1}^{b} i \cdot \mathsf{PAR}(x, i)\right)$.
(v) Output x_z.

In step (i), since a parity gate can be implemented by using three U_2-gates ($g_1 = \overline{x_1} \wedge x_2$, $g_2 = x_1 \wedge \overline{x_2}$, $g_3 = x_1 \oplus x_2 = g_1 \vee g_2$), all $\mathsf{PAR}(x, i)$'s can be computed by a circuit of size $3(n - b) < 3n$ over the basis U_2.

Let $(i_q, \ldots, i_1)$ be the binary representation of an $i \in [b]$, where $q = O(\log n)$. Then the binary representation of $i \cdot \mathsf{PAR}(x, i)$ is obviously

$$(i_q \wedge \mathsf{PAR}(x, i), \ldots, i_1 \wedge \mathsf{PAR}(x, i)). \tag{2}$$

Since i_j's are not depending on an input, i.e., i_j's can be considered as constants, we need no gates in step (ii). In fact, we can replace $i_j \wedge \mathsf{PAR}(x, i)$ in Eq. (2) by the constant 0 for each j with $i_j = 0$, and by $\mathsf{PAR}(x, i)$ for each j with $i_j = 1$.

In step (iii), we need $b - 1$ $O(\log n)$-bit adders, which can be realized using at most $b \cdot O(\log n) = o(n)$ gates over the basis U_2, since the addition of two k bit numbers can be implemented by a circuit of size $O(k)$ (see, e.g., [13]).

In step (iv), we only need several basic arithmetic operations on $O(\log n)$-digit numbers. The number of gates needed is obviously a polynomial in $O(\log n) = o(n)$.

In step (v), we use the *n-way multiplexer* (a.k.a. *storage access function*) M_n whose definition is as follows: Let $q = \lceil \log n \rceil$. The function M_n takes $q + n$ binary inputs and is defined as

$$M_n(z_1, \ldots, z_q, x_0, \ldots, x_{n-1}) = x_{(z_q \cdots z_1)_2}.$$

If $(z_q \cdots z_1)_2 \geq n$, then the output of M_n is unspecified. It is well known that M_n can be computed by a circuit of size $2n + o(n)$ over the basis U_2 when n is a power of two [8] (see also [13, p.77]). Below we describe the construction due to [8] and verify that $\mathsf{Size}_{U_2}(M_n) = 2n + o(n)$ for every n.

By using the identity [8], we have

$$M_n(z_1, \ldots, z_q, x_0, \ldots, x_{n-1}) = M_{2^{\lfloor q/2 \rfloor}}(z_1, \ldots, z_{\lfloor q/2 \rfloor}, x'_0, \ldots, x'_{2^{\lfloor q/2 \rfloor}-1}),$$

where, for each $i = 0, \ldots, 2^{\lfloor q/2 \rfloor} - 1$,

$$x'_i = \bigvee_{t=0}^{2^{\lceil q/2 \rceil}-1} (d_t \wedge x_{2^{\lceil q/2 \rceil} i + t}), \tag{3}$$

and d_t is the t-th output of $\mathsf{Decode}_{\lceil q/2 \rceil}$ applied to $(z_{\lfloor q/2 \rfloor + 1}, \ldots, z_q)$. Here we put $x_j = 0$ for every $j \geq n$ in Eq. (3). Thus, M_n can be realized using n AND gates and at most n OR gates, and additional $O(2^{q/2}) = o(n)$ gates those are used to compute the function $M_{2^{\lfloor q/2 \rfloor}}$ and $\mathsf{Decode}_{\lceil q/2 \rceil}$.

Overall, the total size of our circuit for the function f_n is $(3+2)n + o(n) = 5n + o(n)$. This completes the proof of Theorem 6. □

In summary, we have:

Theorem 7. *There is a sequence of Boolean functions* $\{f_n\}_n$*, which is given by Definition 2, such that (i)* f_n *is a Boolean function on* n *variables, (ii) for every* $t = t(n) = \omega(\sqrt{n}\log^2 n)$*,* f_n *is* $n-t$*-mixed, and (iii) the circuit complexity of* f_n *over the basis* U_2 *satisfies*

$$5n - o(n) \leq \mathsf{Size}_{U_2}(f_n) \leq 5n + o(n).$$

4 Concluding Remarks

In the paper, we gave an explicit construction of an $n-o(n)$-mixed Boolean function with circuit complexity $5n \pm o(n)$. Our results shows that a lower bound method on the size of a U_2-circuit that uses the property of k-mixed has reached the limit. This gives a strong motivation to find another property of Boolean functions that can be used for deriving a higher lower bound, which is an interesting and challenging open problem.

References

1. Alon, N., Nathanson, M.B., Ruzsa, I.: The Polynomial Method and Restricted Sums of Congruence Classes. J. Number Theory 56, 404–417 (1996)
2. Andreev, A., Baskakov, J., Clementi, A., Rolim, J.: Small Pseudo-Random Sets yields Hard Functions: New Tight Explicit Lower Bounds for Branching Programs. In: Wiedermann, J., Van Emde Boas, P., Nielsen, M. (eds.) ICALP 1999. LNCS, vol. 1644, pp. 179–189. Springer, Heidelberg (1999); Also in ECCC TR97-053 (1997)
3. Forster, C.C., Stockton, F.D.: Counting Responders in an Associative Memory. IEEE Trans. Comput. C-20(12), 1580–1583 (1971)
4. Hardy, G.H., Wright, E.M.: An Introduction to the Theory of Numbers, 5th edn. Oxford University Press, Oxford (1978)
5. Iwama, K., Morizumi, H.: An Explicit Lower Bound of $5n - o(n)$ for Boolean Circuits. In: Proc. 27th MFCS, pp. 353–364 (2002)
6. Iwama, K., Lachish, O., Morizumi, H., Raz, R.: An Explicit Lower Bound of 5n − o(n) for Boolean Circuits (manuscript, 2005), (preliminary versions are [9] and [5]), `http://www.wisdom.weizmann.ac.il/~ranraz/publications/`
7. Jukna, S.P.: Entropy of Contact Circuits and Lower Bounds on Their Complexity. Theoret. Comput. Sci. 57, 113–129 (1988)
8. Klein, P., Paterson, M.S.: Asymptotically Optimal Circuit for a Storage Access Function. IEEE Trans. Comput. 29(8), 737–738 (1980)
9. Lachish, O., Raz, R.: Explicit Lower Bound of $4.5n - o(n)$ for Boolean Circuits. In: Proc. 33rd STOC, pp. 399–408 (2001)
10. Savický, P., Zák, S.: A Large Lower Bound for 1-Branching Programs. ECCC TR96-036 Rev.01 (1996)
11. Dias da Silva, J.A., Hamidoune, Y.O.: Cyclic Spaces for Grassmann Derivatives and Additive Theory. Bull. London. Math. Soc. 26, 140–146 (1994)

12. Simon, J., Szegedy, M.: A New Lower Bound Theorem for Read Only Once Branching Programs and its Applications. In: Advances in Computational Complexity Theory, DIMACS Series, vol. 13, pp. 183–193 (1993)
13. Wegener, I.: The Complexity of Boolean Functions, Willey-Teubner (1987)
14. Zwick, U.: A $4n$ Lower Bound on the Combinatorial Complexity of Certain Symmetric Boolean Functions over the Basis of Unate Dyadic Boolean Functions. SIAM J. Comput. 20, 499–505 (1991)

A Logic for Distributed Higher Order π-Calculus*

Zining Cao

Department of Computer Science and Technology
Nanjing University of Aero. & Astro., Nanjing 210016, P.R. China
caozn@nuaa.edu.cn

Abstract. In this paper, we present a spatial logic for distributed higher order π-calculus. In order to prove that the induced logical equivalence coincides with distributed context bisimulation, we present some new bisimulations, and prove the equivalence between these new bisimulations and distributed context bisimulation. Furthermore, we present a variant of this spatial logic and prove that it gives a logical characterisation of distributed bisimulations.

1 Introduction

Higher order π-calculus was proposed and studied intensively in Sangiorgi's dissertation [13]. In higher order π-calculus, processes and abstractions over processes of arbitrarily high order, can be communicated. Some interesting equivalences for higher order π-calculus, such as barbed equivalence, context bisimulation and normal bisimulation, were presented in [13]. Barbed equivalence can be regarded as a uniform definition of bisimulation for a variety of concurrent calculi. Context bisimulation is a very intuitive definition of bisimulation for higher order π-calculus, but it is heavy to handle, due to the appearance of universal quantifications in its definition. In the definition of normal bisimulation, all universal quantifications disappeared, therefore normal bisimulation is a very economic characterisation of bisimulation for higher order π-calculus. The coincidence between the three weak equivalences was proved [13,12,9].

In [3] we design a distributed higher order π-calculus which allows the observer to see the distributed nature of processes. For this distributed higher order π-calculus, the processes under observation are considered to be distributed in nature and locations are associated with parallel components which represent sites. So, the observer cannot only test the process by communicating with it but also can observe or distinguish that part of the distributed process which reacted to the test. Furthermore, three general notions of bisimulation related to this observation of distributed systems are introduced. The equivalence between the distributed bisimulations is proved, which generalizes the equivalence between

* This work was supported by the National Natural Science Foundation of China under Grant 60473036.

M. Agrawal et al. (Eds.): TAMC 2008, LNCS 4978, pp. 351–363, 2008.

barbed equivalence, normal bisimulation and context bisimulation for distributed process calculus.

Providing a logical characterisation for various equivalence relations has been one of the major research topics in the development of process theories. A logical characterisation not only allows us to reason about process behaviours, but also helps to understand the properties of the operators.

For CCS, a famous modal logic is Hennessy-Milner logic [10], for which logical equivalence is known to characterise bisimilarity. For higher order π-calculus, the logical characterisation is in general difficult to give. The main problem lies in the fact that universal quantifications appear in the definition of bisimulations for higher order π-calculus. For instance, to equal two processes with higher order input prefixing, one has to equal the residuals under all possible process substitutions. Some Hennessy-Milner style logics for bisimulation in *CHOCS* were presented in [1,2].

Spatial logic was presented in [7]. Spatial logic extends classical logic with connectives to reason about the structure of the processes. The additional connectives belong to two families. Intensional operators allow one to inspect the structure of the process. A formula $A_1|A_2$ is satisfied whenever we can split the process into two parts satisfying the corresponding subformula A_i, $i = 1, 2$. In presence of restriction in the underlying model, a process P satisfies formula $n®A$ if we can write P as $(\nu n)P'$ with P' satisfying A. Finally, formula 0 is only satisfied by the inaction process. Connectives $|$ and ® come with adjunct operators, called guarantee ($\triangleright$) and hiding ($\oslash$) respectively, that allow one to extend the process being observed. In this sense, these can be called contextual operators. P satisfies $A_1 \triangleright A_2$ whenever the spatial composition (using $|$) of P with any process satisfying A_1 satisfies A_2, and P satisfies $A \oslash n$ if $(\nu n)P$ satisfies A. Some spatial logics have an operator for fresh name quantification [6].

Existing spatial logics for concurrency are intensional [11], in the sense that they induce an equivalence that coincides with structural congruence, which is much finer than bisimilarity. In [8], Hirschkoff studied an extensional spatial logic. This logic only has spatial composition adjunct ($\triangleright$), revelation adjunct ($\oslash$), a simple temporal modality, and an operator for fresh name quantification. For π-calculus, this extensional spatial logic was proved to induce the same separative power as strong early bisimilarity.

There are lots of works of spatial logics for π-calculus and *Mobile Ambients*. But as far as we know, there is no spatial logic for higher order π-calculus up to now.

In this paper, we present a spatial logic for distributed higher order π-calculus, called WL, which comprises two barb detecting predicates $\Downarrow$ and $\Downarrow_\mu$, a simple temporal modality $\langle\rangle$ and a spatial composition adjunct $\triangleright$. We show that this equivalence induced by this spatial logic coincides with distributed context bisimulation. To establish this result, our strategy is to first give a simpler bisimulation which is equivalent to distributed context bisimulation, and then present a logical characterisation for this simpler bisimulation. Consequently, this logic is also a characterisation for context bisimulation. Similar to [8], we exploit the characterisation of distributed context bisimulation in the terms of distributed barbed equivalence. However, in the definition of distributed barbed equivalence, the testing context is an arbitrary process. It is difficult to give the logical characterisation of barbed equivalence directly. Therefore we first give a variant of normal bisimulation, $\approx_{nr}^{d}$, where in the case of higher order output action, we only need to test the remainder process with some kind of finite processes. Then we give a similar variant of contextual barbed bisimulation, $\approx_{nb}^{d}$, where a process only need to be tested with some special finite processes. The bisimulations $\approx_{nr}^{d}$and $\approx_{nb}^{d}$ are proved to coincide with weak context bisimulation. Furthermore, we prove that WL is a characterisation of $\approx_{nb}^{d}$. Thus WL is a logic characterisation of weak context bisimulation. Finally, we present a variant of WL, named WSL, which does not have $\Downarrow_{\mu}$, but contains $\oslash$. We prove that WSL induces an equivalence that coincides with weak context bisimulation. Thus WSL is also a logic characterisation of context bisimulation.

This paper is organized as follows: In Section 2, we introduce a distributed higher order π-calculus. In Section 3, we present some distributed bisimulations and give the equivalence between these bisimulations. In Section 4, we give a spatial logic for weak distributed context bisimulation, and prove that the logical induced equivalence coincides with weak distributed context bisimulation. The paper is concluded in Section 5.

2 Syntax and Labelled Transition System of Distributed Higher Order π-Calculus

In this section, a distributed higher order π-calculus is given, which was first presented in [3]. We now introduce the concept of distributed process. Given a location set Loc, w.l.o.g., let Loc be the set of natural numbers, the syntax for distributed processes allows the parallel composition of located process $\{P\}_i$, which may share a defined name, using the construct $(\nu a)-$. The class of the distributed processes is denoted as DPr, ranged over by $K, L, M, N, \ldots$.

The formal definition of distributed process is given as follows:

$M ::= \{P\}_i \mid M_1|M_2 \mid (\nu a)M$, where $i \in Loc$ and P is a process of higher order π-calculus.

Intuitively, $\{P\}_i$ represents process P residing on location i. The actions of such a process will be observed to occur "within location i". $M_1|M_2$ represents the parallel of two distributed systems M_1 and M_2. $(\nu a)M$ is the restriction operator, which makes name a local to system M.

For example, $\{(\nu l)(\overline{b}\langle l.0|!\tau.0\rangle.0|\overline{l}.0)\}_0|\{b(U).U\}_1$, $(\nu l)(\{\overline{l}.0\}_0|\{l.0|!\tau.0\}_1)$, $\{\overline{b}\langle !\tau.0\rangle.0\}_0|\{b(U).U\}_1$ and $\{0\}_0|\{!\tau.0\}_1$ are distributed processes. Here intuitively $\{(\nu l)(\overline{b}\langle l.0|!\tau.0\rangle.0|\overline{l}.0)\}_0|\{b(U).U\}_1$ represents $(\nu l)(\overline{b}\langle l.0|!\tau.0\rangle.0|\overline{l}.0)$ and $b(U).U$ running on locations 0 and 1 respectively, $(\nu l)(\{\overline{l}.0\}_0|\{l.0|!\tau.0\}_1)$ represents the parallel of $\{\overline{l}.0\}_0$ and $\{l.0|!\tau.0\}_1$ with a private name l.

In each distributed process of the form $(\nu a)M$ the occurrence of a is a bound within the scope of M. An occurrence of a in M is said to be free iff it does not lie within the scope of a bound occurrence of a. The set of names occurring free in M is denoted $fn(M)$. An occurrence of a name in M is said to be bound if it is not free, we write the set of bound names as $bn(M)$. $n(M)$ denotes the set of names of M, i.e., $n(M) = fn(M) \cup bn(M)$. We use $n(M, N)$ to denote $n(M) \cup n(N)$. Higher order input prefix $\{a(U).P\}_i$ binds all free occurrences of U in P. The set of variables occurring free in M is denoted as $fv(M)$. We write the set of bound variables in M as $bv(M)$. A distributed process is closed if it has no free variable; it is open if it may have free variables. DPr^c is the set of all closed distributed processes.

Processes M and N are α-convertible, $M \equiv_\alpha N$, if N can be obtained from M by a finite number of changes of bound names and bound variables.

Structural congruence is a congruence relation including the following rules: $\{P\}_i|\{Q\}_i \equiv \{P|Q\}_i$; $M|N \equiv N|M$; $(L|M)|N \equiv L|(M|N)$; $M|0 \equiv M$; $(\nu a)0 \equiv 0$; $(\nu a)(\nu b)M \equiv (\nu b)(\nu a)M$; $(\nu a)(M|N) \equiv M|(\nu a)N$ if $a \notin fn(M)$; $M \equiv N$ if $M \equiv_\alpha N$.

The distributed actions are given by

$I\alpha ::= \{\tau\}_{i,j} \mid \{l\}_i \mid \{\overline{l}\}_i \mid \{a\langle E\rangle\}_i \mid \{\overline{a}\langle E\rangle\}_i \mid \{(\nu\widetilde{b})\overline{a}\langle E\rangle\}_i$

We write $bn(I\alpha)$ for the set of names bound in $I\alpha$, which is $\{\widetilde{b}\}$ if $I\alpha$ is $\{(\nu\widetilde{b})\overline{a}\langle E\rangle\}_i$ and $\emptyset$ otherwise. $n(I\alpha)$ denotes the set of names that occur in $I\alpha$.

We give the operational semantics of distributed processes in Table 1. We have omitted the symmetric of the parallelism and communication rules. The main character of Table 1 is that the label $I\alpha$ on the transition arrow is of the

Table 1.

$$ALP: \frac{M \xrightarrow{I\alpha} M'}{N \xrightarrow{I\alpha} N'} M \equiv N, M' \equiv N' \qquad TAU: \frac{P \xrightarrow{\tau} P'}{\{P\}_i \xrightarrow{\{\tau\}_{i,i}} \{P'\}_i}$$

$$OUT1: \frac{P \xrightarrow{\overline{l}} P'}{\{P\}_i \xrightarrow{\{\overline{l}\}_i} \{P'\}_i} \qquad IN1: \frac{P \xrightarrow{l} P'}{\{P\}_i \xrightarrow{\{l\}_i} \{P'\}_i}$$

$$OUT2: \frac{P \xrightarrow{(\nu\widetilde{b})\overline{a}\langle E\rangle} P'}{\{P\}_i \xrightarrow{\{(\nu\widetilde{b})\overline{a}\langle E\rangle\}_i} \{P'\}_i} \qquad IN2: \frac{P \xrightarrow{a\langle E\rangle} P'}{\{P\}_i \xrightarrow{\{a\langle E\rangle\}_i} \{P'\}_i}$$

$$PAR: \frac{M \xrightarrow{I\alpha} M'}{M|N \xrightarrow{I\alpha} M'|N} bn(I\alpha) \cap fn(N) = \emptyset$$

$$COM1: \frac{M \xrightarrow{\{\overline{l}\}_i} M' \quad N \xrightarrow{\{l\}_j} N'}{M|N \xrightarrow{\{\tau\}_{i,j}} M'|N'}$$

$$COM2: \frac{M \xrightarrow{\{(\nu\widetilde{b})\overline{a}\langle E\rangle\}_i} M' \quad N \xrightarrow{\{a\langle E\rangle\}_j} N'}{M|N \xrightarrow{\{\tau\}_{i,j}} (\nu\widetilde{b})(M'|N')} \widetilde{b} \cap fn(N) = \emptyset$$

$$RES: \frac{M \xrightarrow{I\alpha} M'}{(\nu a)M \xrightarrow{I\alpha} (\nu a)M'} a \notin n(I\alpha)$$

$$OPEN: \frac{M \xrightarrow{\{(\nu\widetilde{c})\overline{a}\langle E\rangle\}_i} M'}{(\nu b)M \xrightarrow{\{(\nu b,\widetilde{c})\overline{a}\langle E\rangle\}_i} M'} a \neq b,\ b \in fn(E) - \widetilde{c}$$

form $\{\alpha\}_i$ or $\{\tau\}_{i,j}$, where α is an input or output action, i and j are locations. From the point of distributed view, $\{\alpha\}_i$ can be regarded as an input or output action performed on location i, and $\{\tau\}_{i,j}$ can be regarded as a communication, where the output happens on location i and the input happens on location j. In the following, we view $\{\alpha\}_i$ as a distributed input or output action, and view $\{\tau\}_{i,j}$ as a distributed communication.

Remark: In the above table, the labelled transition system of $P \stackrel{\alpha}{\longrightarrow} P'$ was given in operational semantics of higher order π-calculus [13]. Transition $M \stackrel{\{\alpha\}_i}{\longrightarrow} M'$ means that distributed process M performs an action on location i, then continues as M'; and transition $M \stackrel{\{\tau\}_{i,j}}{\longrightarrow} M'$ means that after a communication between locations i and j, distributed process M continues as M'.

3 Bisimulations for Distributed Higher Order π-Calculus

3.1 Distributed Bisimulations

In our distributed calculus, processes contain locations, and actions in the scope of locations will be observed on those locations. For such distributed processes, some new distributed bisimulations, distributed context bisimulation and distributed normal bisimulation, will be proposed. The new semantics we give to distributed processes additionally takes the distribution in space into account.

For distributed calculus, two distributed processes are equated if not only the actions, but also the locations, where the actions happen, can be matched against each other. So we get the following definition of distributed context bisimulation. In the following, we abbreviate $M\{E/U\}$ as $M\langle E\rangle$.

We first give the weak distributed context bisimulation, weak distributed normal bisimulation and the weak distributed reduction bisimulation. Then we give their congruence property and the equivalence between these three bisimulations.

Before giving the weak distributed bisimulations, let us compare communication $\{\tau\}_{i,i}$ with $\{\tau\}_{i,j}$ firstly, where $i \neq j$. For weak distributed bisimulations, it seems natural to view $\{\tau\}_{i,i}$ as an invisible communication and $\{\tau\}_{i,j}$ as a visible communication, since for an observer who can distinguish between sites, $\{\tau\}_{i,i}$ is an internal communication on location i, and $\{\tau\}_{i,j}$ represents an external communication between two different locations i and j. Therefore in the following, we regard $\{\tau\}_{i,i}$ as a private event on location i, and $\{\tau\}_{i,j}$ as a visible event between locations i and j.

For example, let us consider a system consisting of a satellite and an earth station. It is clear that in this system, the satellite is physically far away from the earth station. If the program controlling the satellite has to be changed, either because of a program error or because the job of the satellite is to be changed, then a new program will be sent to the satellite. The satellite is ready to receive a new program. After reception it acts according to this program until it is "interrupted" either by a new job or because a program error has occurred or because the program has finished.

In this example, the satellite and the earth can be specified in our distributed calculus syntax as follows:

$Sat \stackrel{def}{=} \{!a(U).U|satsys\}_S$, where $_S$ represents location *satellite*.

$Earth \stackrel{def}{=} \{\overline{a}\langle newprg_1\rangle.\overline{a}\langle newprg_2\rangle...\}_E$, where $_E$ represents location *earth*.

Now the system is specified as:

$Sys \stackrel{def}{=} (\nu a)(\{!a(U).U|satsys\}_S|\{\overline{a}\langle newprg_1\rangle.\overline{a}\langle newprg_2\rangle...\}_E)$.

The earth can send a new program to the satellite, then in the satellite, this program interacts with the old satellite system:

$Sys \xrightarrow{\{\tau\}_{S,E}} (\nu a)(\{!a(U).U|newprg_1|satsys\}_S|\{\overline{a}\langle newprg_2\rangle...\}_E) \stackrel{\{\tau\}_{S,S}}{\Longrightarrow} (\nu a)(\{!a(U).U|newsatsys\}_S|\{\overline{a}\langle newprg_2\rangle...\}_E)$.

In this transition sequence, $\{\tau\}_{S,E}$ is a communication between locations *satellite* and *earth*, $\{\tau\}_{S,S}$ is an internal action on location *satellite*. Intuitively, we can consider $\{\tau\}_{S,E}$ as an external communication and it is visible; meanwhile, $\{\tau\}_{S,S}$ can be viewed as an internal action of satellite and it is private. Therefore, from the point of view of distributed calculus, we can neglect internal communication such as $\{\tau\}_{S,S}$ but $\{\tau\}_{S,E}$ is treated as a visible action.

Firstly we give the definition of weak distributed context bisimulation. The difference from strong distributed context bisimulation is that in the case of weak bisimulation we neglect $\{\tau\}_{i,i}$ since from the point of distributed view $\{\tau\}_{i,i}$ happens internally on location i.

In the following, we use $M \stackrel{\varepsilon}{\Longrightarrow} M'$ to abbreviate $M \xrightarrow{\{\tau\}_{i_1,i_1}} ... \xrightarrow{\{\tau\}_{i_n,i_n}} M'$, and use $M \stackrel{I\alpha}{\Longrightarrow} M'$ to abbreviate $M \stackrel{\varepsilon}{\Longrightarrow}\xrightarrow{I\alpha}\stackrel{\varepsilon}{\Longrightarrow} M'$.

Definition 1. Weak distributed context bisimulation

Let M, $N \in DPr^c$, we write $M \approx^d_{cxt} N$, if there is a symmetric relation R, such that $M\ R\ N$ implies:

(1) whenever $M \stackrel{\varepsilon}{\Longrightarrow} M'$, there exists N' such that $N \stackrel{\varepsilon}{\Longrightarrow} N'$ and $M'\ R\ N'$;

(2) whenever $M \stackrel{\{\tau\}_{i,j}}{\Longrightarrow} M'$, there exists N' such that $N \stackrel{\{\tau\}_{i,j}}{\Longrightarrow} N'$ with $M'\ R\ N'$, where $i \neq j$;

(3) whenever $M \stackrel{\{l\}_i}{\Longrightarrow} M'$, there exists N' such that $N \stackrel{\{l\}_i}{\Longrightarrow} N'$ with $M'\ R\ N'$;

(4) whenever $M \stackrel{\{\overline{l}\}_i}{\Longrightarrow} M'$, there exists N' such that $N \stackrel{\{\overline{l}\}_i}{\Longrightarrow} N'$ with $M'\ R\ N'$;

(5) whenever $M \stackrel{\{a\langle E\rangle\}_i}{\Longrightarrow} M'$, there exists N' such that $N \stackrel{\{a\langle E\rangle\}_i}{\Longrightarrow} N'$ and $M'\ R\ N'$;

(6) whenever $M \stackrel{\{(\nu\widetilde{b})\overline{a}\langle E\rangle\}_i}{\Longrightarrow} M'$, there exist N', F, $\widetilde{c}$, such that $N \stackrel{\{(\nu\widetilde{c})\overline{a}\langle F\rangle\}_i}{\Longrightarrow} N'$, and for any distributed process $C(U)$ with $fn(C(U)) \cap \{\widetilde{b},\widetilde{c}\} = \emptyset$, $(\nu\widetilde{b})(M'|C\langle E\rangle)\ R\ (\nu\widetilde{c})(N'|C\langle F\rangle)$.

Definition 2. Weak distributed normal bisimulation

Let M, $N \in DPr^c$, we write $M \approx^d_{nor} N$, if there is a symmetric relation R, such that $M\ R\ N$ implies:

(1) whenever $M \stackrel{\varepsilon}{\Longrightarrow} M'$, there exists N' such that $N \stackrel{\varepsilon}{\Longrightarrow} N'$ and $M'\ R\ N'$;

(2) whenever $M \overset{\{\tau\}_{i,j}}{\Longrightarrow} M'$, there exists N' such that $N \overset{\{\tau\}_{i,j}}{\Longrightarrow} N'$ with $M'\ R\ N'$, where $i \neq j$;

(3) whenever $M \overset{\{l\}_i}{\Longrightarrow} M'$, there exists N' such that $N \overset{\{l\}_i}{\Longrightarrow} N'$ with $M'\ R\ N'$;

(4) whenever $M \overset{\{\overline{l}\}_i}{\Longrightarrow} M'$, there exists N' such that $N \overset{\{\overline{l}\}_i}{\Longrightarrow} N'$ with $M'\ R\ N'$;

(5) whenever $M \overset{\{a\langle\overline{m}.0\rangle\}_i}{\Longrightarrow} M'$, there exists N' such that $N \overset{\{a\langle\overline{m}.0\rangle\}_i}{\Longrightarrow} N'$ and $M'\ R\ N'$, where m is a fresh name;

(6) whenever $M \overset{\{(\nu\widetilde{b})\overline{a}\langle E\rangle\}_i}{\Longrightarrow} M'$, there exist N', F, $\widetilde{c}$, such that $N \overset{\{(\nu\widetilde{c})\overline{a}\langle F\rangle\}_i}{\Longrightarrow} N'$, and for a fresh name m and fresh location j, $(\nu\widetilde{b})(M'|\{!m.E\}_j)\ R\ (\nu\widetilde{c})(N'|\{!m.F\}_j)$.

In the following, we present some new bisimulations, and prove that they are equivalent to distributed context bisimulation. These results are used to give a logical charatersation of $\approx^d_{cxt}$.

Now we first give a variant of distributed normal bisimulation, where parallel of finitary copies is used as a limit form of replication. We prove it coincides with distributed normal bisimulation. In the following, we write $\Pi_k M$ to denote the parallel composition of k copies of M, i.e., $\Pi_k M \overset{def}{=} M|\Pi_{k-1}M$ and $\Pi_0 M \overset{def}{=} 0$. For example, $\Pi_3 M$ represents $M|M|M$.

Definition 3. A symmetric relation $R \subseteq DPr^c \times DPr^c$ is a weak distributed limit normal bisimulation if $M\ R\ N$ implies:

(1) whenever $M \overset{\varepsilon}{\Longrightarrow} M'$, there exists N' such that $N \overset{\varepsilon}{\Longrightarrow} N'$ and $M'\ R\ N'$;

(2) whenever $M \overset{\{\tau\}_{i,j}}{\Longrightarrow} M'$, there exists N' such that $N \overset{\{\tau\}_{i,j}}{\Longrightarrow} N'$ with $M'\ R\ N'$, where $i \neq j$;

(3) whenever $M \overset{\{l\}_i}{\Longrightarrow} M'$, there exists N' such that $N \overset{\{l\}_i}{\Longrightarrow} N'$ with $M'\ R\ N'$;

(4) whenever $M \overset{\{\overline{l}\}_i}{\Longrightarrow} M'$, there exists N' such that $N \overset{\{\overline{l}\}_i}{\Longrightarrow} N'$ with $M'\ R\ N'$;

(5) whenever $M \overset{\{a\langle\overline{m}.0\rangle\}_i}{\Longrightarrow} M'$, there exists N' such that $N \overset{\{a\langle\overline{m}.0\rangle\}_i}{\Longrightarrow} N'$ and $M'\ R\ N'$, where m is a fresh name;

(6) whenever $M \overset{\{(\nu\widetilde{b})\overline{a}\langle E\rangle\}_i}{\Longrightarrow} M'$, there exist N', F, $\widetilde{c}$ such that $N \overset{\{(\nu\widetilde{c})\overline{a}\langle F\rangle\}_i}{\Longrightarrow} N'$ and $(\nu\widetilde{b})(M'|\{\Pi_k m.E\}_j)\ R\ (\nu\widetilde{c})(N'|\{\Pi_k m.F\}_j)$ for all $k \in \{0, 1, 2, ...\}$, where m is a fresh name.

We write $M \approx^d_{nr} N$ if M and N are weakly distributed limit normal bisimilar.

In [8], the logical characterisation of strong early bisimulation was exploited in terms of strong barbed equivalence. The extensional spatial logic [8] was proved to capture strong barbed equivalence firstly, then since strong early bisimulation coincides with strong barbed equivalence, this spatial logic is also a logical characterisation of strong early bisimulation. But the proof techniques in [8] does not work for weak bisimulation. In [8], the problem of the spatial logical equivalence of weak early bisimulation was presented as an open question.

To give a logical charatersation of $\approx^d_{cxt}$, we will give a distributed variant of barbed equivalence, where testing contexts are some special processes.

Definition 4. For each name or co-name μ, the observability predicate $\downarrow_\mu$ is defined by

(1) $M \downarrow_l$ if there exists M' such that $M \stackrel{\{l\}_i}{\longrightarrow} M'$;

(2) $M \downarrow_{\bar{l}}$ if there exists M' such that $M \stackrel{\{\bar{l}\}_i}{\longrightarrow} M'$;

(3) $M \downarrow_a$ if there exist E, M' such that $M \stackrel{\{a\langle E\rangle\}_i}{\longrightarrow} M'$;

(4) $M \downarrow_{\bar{a}}$ if there exist $\widetilde{b}, E, M'$ such that $M \stackrel{\{(\nu\widetilde{b})\bar{a}\langle E\rangle\}_i}{\longrightarrow} M'$.

Definition 5. A symmetric relation $R \subseteq DPr^c \times DPr^c$ is a weak distributed normal barbed bisimulation if $M\ R\ N$ implies:

(1) $M|C\ R\ N|C$ for any C in the form of $\{0\}_i$, $\{\bar{l}.0\}_i$, $\{l.0\}_i$, $\{\bar{a}\langle\overline{m}.0\rangle.0\}_i$, $\{a(U).\Pi_k m.U\}_i$, where k is any natural number;

(2) whenever $M \stackrel{\varepsilon}{\Longrightarrow} M'$ there exists N' such that $N \stackrel{\varepsilon}{\Longrightarrow} N'$ and $M'\ R\ N'$;

(3) whenever $M \stackrel{\{\tau\}_{i,j}}{\Longrightarrow} M'$, there exists N' such that $N \stackrel{\{\tau\}_{i,j}}{\Longrightarrow} N'$ with $M'\ R\ N'$, where $i \neq j$;

(4) if $M \Downarrow_\mu$, then also $N \Downarrow_\mu$, where $M \Downarrow_\mu$ means $\exists M'$, $M \stackrel{\varepsilon}{\Longrightarrow} M' \downarrow_\mu$.

We write $M \approx_{nb}^d N$ if M and N are weakly distributed normal barbed bisimilar.

It is worth to note that all testing contexts in the above definition are finite processes. Hence it is possible to give the characteristic formulas for these testing contexts.

3.2 Equivalence of Distributed Bisimulations

Now we study the equivalence between $\approx_{cxt}^d$, $\approx_{nor}^d$ and $\approx_{nr}^d$.

Lemma 1. For any $M, N \in DPr^c$, $M \approx_{cxt}^d N \Rightarrow M \approx_{nr}^d N$.

Proof : It is trivial by the definition of $\approx_{cxt}^d$ and $\approx_{nr}^d$.

Lemma 2. For any $M, N \in DPr^c$, $M \approx_{nr}^d N \Rightarrow M \approx_{nor}^d N$.

Proposition 1. For any $M, N \in DPr^c$, $M \approx_{cxt}^d N \Leftrightarrow M \approx_{nr}^d N \Leftrightarrow M \approx_{nor}^d N$.

Proof : By the equivalence between $\approx_{cxt}^d$ and $\approx_{nor}^d$ [3], $M \approx_{nor}^d N \Leftrightarrow M \approx_{cxt}^d N$. Furthermore, by Lemmas 1 and 2, we have that the proposition holds.

The following proposition states that $\approx_{nr}^d$ coincides with $\approx_{nb}^d$.

Proposition 2. For any $M, N \in DPr^c$, $M \approx_{nr}^d N \Leftrightarrow M \approx_{nb}^d N$.

Proposition 3. For any $M, N \in DPr^c$, $M \approx_{cxt}^d N \Leftrightarrow M \approx_{nr}^d N \Leftrightarrow M \approx_{nor}^d N \Leftrightarrow M \approx_{nb}^d N$.

Proof : By Propositions 1 and 2.

In fact, a symmetric relation R satisfying conditions (1), (2) and (3) of Definition 5 is also equivalent to $\approx_{cxt}^d$.

4 A Logic for Distributed Higher Order π-Calculus

In this section, we present a logic to reason about distributed higher order π-calculus. Roughly speaking, this logic extends propositional logic with four connectives: (1) formula $\Downarrow$ is an atom formula, which means that the process can communicate through some channel; (2) formula $\Downarrow_\mu$ is an atom formula, which means that the process can communicate through channel μ; (3) formula $\langle\rangle A$ means that the process can perform several $\{\tau\}_{i,j}$ actions, where $i = j$, and transform to a process that satisfies A; (4) formula $\langle i,j\rangle A$ means that process can perform $\{\tau\}_{i,j}$ and transform to a process that satisfies A; (5) if M satisfies $A_1 \triangleright A_2$, then the parallel composition of processes M and N satisfies formula A_2 if N satisfies formula A_1.

To prove that the equivalence induced by this logic coincides with distributed context bisimulation, we present a variant of distributed normal bisimulation, $\approx^d_{nr}$, where replication is replaced by a series of parallel composition. Then we prove the equivalence between $\approx^d_{nr}$ and $\approx^d_{nor}$. Furthermore, we present a simplified version of distributed contextual barbed bisimulation, $\approx^d_{nb}$. In the definition of distributed contextual barbed bisimulation, two equivalent processes should be tested by any context. So it is not convenient to verify the equivalence between two processes. In the definition of $\approx^d_{nb}$, to equate two processes we only need to test them by some special contexts. We prove that $\approx^d_{nb}$ coincides with $\approx^d_{nr}$. Moreover, since $\approx^d_{cxt}$, $\approx^d_{nor}$ and $\approx^d_{nr}$ are equivalent, to give a logical characterisation of $\approx^d_{cxt}$, it is enough to prove that this logic captures $\approx^d_{nb}$. Since the special contexts appearing in the definition of $\approx^d_{nb}$ are finite, we can give characterisation formulas for all these processes. Thus that the equivalence induced by this logic coincides with $\approx^d_{nb}$ can be proved.

4.1 Syntax and Semantics of Logic *WL*

Now we introduce a spatial logic called WL.

Definition 6. Syntax of WL
$A ::= \Downarrow \mid \Downarrow_x \mid \Downarrow_{\overline{x}} \mid \neg A \mid \wedge_{i\in I} A_i \mid \langle\rangle A \mid \langle i,j\rangle A \mid A_1 \triangleright A_2$

Definition 7. Semantics of WL
$M \models \Downarrow$ iff $\exists M'.\ M \stackrel{\varepsilon}{\Longrightarrow} M'$, and $M' \downarrow_x$ or $M' \downarrow_{\overline{x}}$ for some x;
$M \models \Downarrow_x$ iff $\exists M'.\ M \stackrel{\varepsilon}{\Longrightarrow} M' \downarrow_x$;
$M \models \Downarrow_{\overline{x}}$ iff $\exists M'.\ M \stackrel{\varepsilon}{\Longrightarrow} M' \downarrow_{\overline{x}}$;
$M \models \neg A$ iff $M \not\models A$;
$M \models \wedge_{i\in I} A_i$ iff $M \models A_i$ for any $i \in I$;
$M \models \langle\rangle A$ iff $\exists M'.\ M \stackrel{\varepsilon}{\Longrightarrow} M'$ and $M' \models A$;
$M \models \langle i,j\rangle A$ iff $\exists M'.\ M \stackrel{\{\tau\}_{i,j}}{\Longrightarrow} M'$ and $M' \models A$;
$M \models A_1 \triangleright A_2$ iff $\forall N.\ N \models A_1$ implies $M|N \models A_2$.

Definition 8. M and N are logically equivalent with respect to WL, written $M =_{WL} N$, iff for any formula A, $M \models A$ iff $N \models A$.

4.2 WL is a Logical Characterisation of $\approx_{cxt}^{d}$

In this section we prove the equivalence between $\approx_{nb}^{d}$ and $=_{WL}$. Since $\approx_{nb}^{d}$ coincides with $\approx_{cxt}^{d}$, we have the equivalence between $\approx_{cxt}^{d}$ and $=_{WL}$.

Proposition 4. For any $M, N \in DPr^c$, $M \approx_{nb}^{d} N$ implies $M =_{WL} N$.

Proof : We prove by structural induction on A that whenever $M \approx_{nb}^{d} N$ and $M \models A$, we have $N \models A$. The case corresponding to the adjunct operator $\triangleright$ follows from congruence properties of $\approx_{nb}^{d}$ with respect to parallel composition.

To prove the converse proposition, we need the following definition, which gives the characteristic formulas for the testing contexts in the definition of $\approx_{nb}^{d}$.

Definition 9

(1) $WF_{\{0\}_i} \stackrel{def}{=} \not\Downarrow$, where $\not\Downarrow \stackrel{def}{=} \neg \Downarrow$;

(2) $WF_{\{ch(x)\}_i} \stackrel{def}{=} \Downarrow_x \wedge [j,i]\bot \wedge ([j,i]\bot \succ \langle j,i\rangle WF_{\{0\}_i})$, where $A \succ B \stackrel{def}{=} \neg(A \triangleright \neg B)$, $[i,j]A \stackrel{def}{=} \neg\langle i,j\rangle\neg A$.

(3) $WF_{\{ch(\overline{x})\}_i} \stackrel{def}{=} \Downarrow_{\overline{x}} \wedge [i,j]\bot \wedge ([i,j]\bot \succ \langle i,j\rangle WF_{\{0\}_i})$.

(4) $WF_{\{l.0\}_i} \stackrel{def}{=} WF_{\{ch(l)\}_i}$;

(5) $WF_{\{\overline{l}.0\}_i} \stackrel{def}{=} WF_{\{ch(\overline{l})\}_i}$;

(6) $WF_{\{\overline{a}\langle 0\rangle.0\}_i} \stackrel{def}{=} WF_{\{ch(\overline{a})\}_i} \wedge (WF_{\{ch(a)\}_j} \triangleright [i,j](\not\Downarrow_{\overline{a}} \rightarrow WF_{\{0\}_i}))$;

(7) $WF_{\{\overline{a}\langle\overline{m}.0\rangle.0\}_i} \stackrel{def}{=} WF_{\{ch(\overline{a})\}_i} \wedge (WF_{\{ch(a)\}_j} \succ [i,j](\not\Downarrow_{\overline{a}} \rightarrow WF_{\{\overline{m}.0\}_i}))$;

(8) $WF_{\{\Pi_0 m.0\}_i} \stackrel{def}{=} WF_{\{0\}_i}$;

(9) $WF_{\{\Pi_k m.0\}_i} \stackrel{def}{=} WF_{\{\overline{m}.0\}_j} \triangleright \langle j,i\rangle WF_{\{\Pi_{k-1} m.0\}_i}$;

(10) $WF_{\{\Pi_0 m.\overline{n}.0\}_i} \stackrel{def}{=} WF_{\{0\}_i}$;

(11) $WF_{\{\Pi_k m.\overline{n}.0\}_i} \stackrel{def}{=} \not\Downarrow_{\overline{n}} \wedge (WF_{\{\overline{m}.0\}_j} \triangleright \langle j,i\rangle)^k WF_{\{\Pi_k \overline{n}.0\}_i} \wedge WF_{\{\overline{m}.0\}_j} \triangleright \langle j,i\rangle (WF_{\{n.0\}_h} \triangleright \langle i,h\rangle WF_{\{\Pi_{k-1} m.\overline{n}.0\}_i})$, where $k \neq 0$ $(B \triangleright \langle i,j\rangle)^i C \stackrel{def}{=} B \triangleright \langle i,j\rangle (B \triangleright \langle i,j\rangle)^{i-1} C$, $(B \triangleright \langle i,j\rangle)^1 C \stackrel{def}{=} B \triangleright \langle i,j\rangle C$;

(12) $WF_{\{a(U).\Pi_k m.U\}_i} \stackrel{def}{=} \not\Downarrow_m \wedge WF_{\{\overline{a}\langle 0\rangle.0\}_j} \triangleright \langle j,i\rangle WF_{\{\Pi_k m.0\}_i} \wedge WF_{\{\overline{a}\langle\overline{n}.0\rangle.0\}_j} \triangleright \langle j,i\rangle WF_{\{\Pi_k m.\overline{n}.0\}_i}$.

Intuitively, formula WF_M captures the class of processes that are bisimilar to process M. The following lemma states this formally:

Lemma 3. The above formulas have the following interpretation:

(1) $M \models WF_{\{0\}_i}$ iff $M \approx_{cxt}^{d} \{0\}_i$;

(2) $M \models WF_{\{ch(a)\}_i}$ where a is a higher order name iff $M \approx_{cxt}^{d} \{a(U).C(U)\}_i$, and there exist $\widetilde{b}, E, N$, such that $(\nu\widetilde{b})(\{C\langle E\rangle\}_i|N) \approx_{cxt}^{d} \{0\}_i$;

(3) $M \models WF_{\{ch(\overline{a})\}_i}$ where a is a higher order name iff $M \approx_{cxt}^{d} (\nu\widetilde{b})(\{\overline{a}\langle E\rangle.Q\}_i)$, and there exists $N(U)$, such that $(\nu\widetilde{b})(N\langle E\rangle|\{Q\}_i) \approx_{cxt}^{d} \{0\}_i$;

(4) $M \models WF_{\{l.0\}_i}$ iff $M \approx_{cxt}^{d} \{l.0\}_i$;

(5) $M \models WF_{\{\overline{l}.0\}_i}$ iff $M \approx_{cxt}^{d} \{\overline{l}.0\}_i$;

(6) $M \models WF_{\{\overline{a}\langle 0\rangle.0\}_i}$ iff $M \approx^d_{cxt} \{\overline{a}\langle 0\rangle.0\}_i$;
(7) $M \models WF_{\{\overline{a}\langle\overline{m}.0\rangle.0\}_i}$ iff $M \approx^d_{cxt} \{\overline{a}\langle\overline{m}.0\rangle.0\}_i$;
(8) $M \models WF_{\{\Pi_k m.0\}_i}$ iff $M \approx^d_{cxt} \{\Pi_k m.0\}_i$, where $\Pi_k M$ denotes the parallel composition of k copies of M;
(9) $M \models WF_{\{\Pi_k m.\overline{n}.0\}_i}$ iff $M \approx^d_{cxt} \{\Pi_k m.\overline{n}.0\}_i$, where $m \neq n$;
(10) $M \models WF_{\{a(U).\Pi_k m.U\}_i}$ iff $M \approx^d_{cxt} \{a(U).\Pi_k m.U\}_i$.

Lemma 4. $D \models WF_C \Leftrightarrow C \approx^d_{cxt} D$, where C is in the form of $\{0\}_i$, $\{\overline{l}.0\}_i$, $\{l.0\}_i$, $\{\overline{a}\langle\overline{m}.0\rangle.0\}_i$, or $\{a(U).\Pi_k m.U\}_i$, where k is an arbitrary natural number.

Proof : It is clear by Lemma 3.

Proposition 5. For any $M, N \in DPr^c$, $M =_{WL} N$ implies $M \approx^d_{nb} N$.

The following proposition states that WL captures $\approx^d_{cxt}$.

Proposition 6. For any $M, N \in DPr^c$, $M \approx^d_{cxt} N \Leftrightarrow M =_{WL} N$.

Proof : By Proposition 3, 4 and 5.

4.3 A Variant of *WL*

We have proved that $\approx^d_{cxt}$ coincides with $=_{WL}$. In the following we consider a variant of WL, called WSL, in which we remove $\Downarrow$ and $\Downarrow_\mu$, and add $\Downarrow_{in}$, $\Downarrow_{out}$ and $\oslash$. We show that $\Downarrow$ and $\Downarrow_\mu$ can be defined by using $\Downarrow_{in}$, $\Downarrow_{out}$ and $\oslash$, then it is clear that WSL is a spatial logical characterisation of $\approx^d_{cxt}$.

Definition 10. Syntax and semantics of WSL

Formulas of WSL are defined by the following grammar:

$A ::= \Downarrow_{in} \mid \Downarrow_{out} \mid \neg A \mid \wedge_{i\in I} A_i \mid \langle\rangle A \mid \langle i, j\rangle A \mid A_1 \triangleright A_2 \mid A \oslash n$

Semantics of WSL is similar to WL except that the definition of $\Downarrow_\mu$ is eliminated and the definitions of $\Downarrow_{in}$, $\Downarrow_{out}$ and $\oslash$ are added as follows:

$M \models \Downarrow_{in}$ iff $\exists M'.\ M \stackrel{\varepsilon}{\Longrightarrow} M' \downarrow_\mu$ where μ is in the form of l or a;

$M \models \Downarrow_{out}$ iff $\exists M'.\ M \stackrel{\varepsilon}{\Longrightarrow} M' \downarrow_\mu$ where μ is in the form of $\overline{l}$ or $\overline{a}$;

$M \models A \oslash n$ iff $(\nu n)M \models A$.

We abbreviate $A \oslash n_1 \oslash n_2 \oslash ... \oslash n_k$ as $A \oslash \widetilde{n}$, where $\widetilde{n} = \{n_1, n_2, ..., n_k\}$.

Definition 11. M and N are logically equivalent with respect to WSL, written $M =_{WSL} N$, iff for any formula A, $M \models A$ iff $N \models A$.

Lemma 5. $M \models \Downarrow \Leftrightarrow M \models \Downarrow_{in} \vee \Downarrow_{out}$

Proof : It is trivial.

Lemma 6. (1) $M \models \Downarrow_\mu \Leftrightarrow M \models \Downarrow_{in} \wedge (WF_0 \oslash \widetilde{m} \oslash \mu) \wedge \neg(WF_0 \oslash \widetilde{m})$, where μ is in the form of l or a, $\widetilde{m} = fn(M) - \{\mu\}$;

(2) $M \models \Downarrow_\mu \Leftrightarrow M \models \Downarrow_{out} \wedge (WF_0 \oslash \widetilde{m} \oslash \overline{\mu}) \wedge \neg(WF_0 \oslash \widetilde{m})$, where μ is in the form of $\overline{l}$ or $\overline{a}$, $\overline{\mu} = l$ if $\mu = \overline{l}$, $\overline{\mu} = a$ if $\mu = \overline{a}$, and $\widetilde{m} = fn(M) - \{\overline{\mu}\}$.

Now we can give the equivalence between $\approx^d_{cxt}$ and $=_{WSL}$.

Proposition 7. For any $M, N \in DPr^c$, $M \approx^d_{cxt} N \Leftrightarrow M =_{WSL} N$.

Proof : Similar to Proposition 4, we prove by structural induction on A that whenever $M \approx_{cxt}^{d} N$ and $M \models A$, we have $N \models A$. Thus $M \approx_{cxt}^{d} N \Rightarrow M =_{WSL} N$. By Lemmas 5, 6 and Proposition 6, we have $M =_{WSL} N \Rightarrow M \approx_{cxt}^{d} N$. Hence the proposition holds.

5 Conclusions

In this paper, we have defined a logic WL, which comprises some spatial operators. We have shown that the induced logical equivalence coincides with weak distributed context bisimulation for distributed higher order π-calculus. As far as we know, this is the first spatial logical characterisation of distributed context bisimulation.

In [8], Hirschkoff studied a contextual spatial logic for the π-calculus, which lacks the spatial operators to observe emptiness, parallel composition and restriction, and only has composition adjunct and hiding. The induced logical equivalence was proved to coincide with strong early bisimulation. The proof involves the definition of non-trivial formulas, including characteristic formulas for restriction-free processes up to bisimulation and characteristic formulas for barbed bisimulation. In [8], Hirschkoff pointed out that his techniques do not apply directly if we consider weak early bisimulation, and studying spatial logical equivalence of weak early bisimulation was viewed as a challenging question. In this paper, to prove that $=_{WL}$ coincides with $\approx_{cxt}^{d}$, we present two simplified notions of observable equivalence on distributed higher order processes named $\approx_{nr}^{d}$ and $\approx_{nb}^{d}$. Then $\approx_{nr}^{d}$ and $\approx_{nb}^{d}$ was proved to coincide with $\approx_{cxt}^{d}$. Moreover, since the testing contexts in the definition of $\approx_{nb}^{d}$ are finite processes, we can give the characteristic formulas for such testing contexts. Thus the equivalence between $=_{WL}$ and $\approx_{cxt}^{d}$ is proved.

References

1. Amadio, R.M., Dam, M.: Reasoning about Higher-order Processes. In: Mosses, P.D., Schwartzbach, M.I., Nielsen, M. (eds.) CAAP 1995, FASE 1995, and TAPSOFT 1995. LNCS, vol. 915, pp. 202–216. Springer, Heidelberg (1995)
2. Baldamus, M., Dingel, J.: Modal Characterization of Weak Bisimulation for Higher-order Processes. In: Bidoit, M., Dauchet, M. (eds.) CAAP 1997, FASE 1997, and TAPSOFT 1997. LNCS, vol. 1214, pp. 285–296. Springer, Heidelberg (1997)
3. Cao, Z.: Bisimulations for a Distributed Higher Order π -Calculus. In: Jones, C.B., Liu, Z., Woodcock, J. (eds.) ICTAC 2007. LNCS, vol. 4711, pp. 94–108. Springer, Heidelberg (2007)
4. Cao, Z.: More on bisimulations for higher-order π -calculus. In: Aceto, L., Ingólfsdóttir, A. (eds.) FOSSACS 2006 and ETAPS 2006. LNCS, vol. 3921, pp. 63–78. Springer, Heidelberg (2006)
5. Castellani, I.: Process Algebras with Localities, ch. 15. In: Bergstra, J., Ponse, A., Smolka, S. (eds.) Handbook of Process Algebra, pp. 945–1045. North-Holland, Amsterdam (2001)

6. Caires, L., Cardelli, L.: A Spatial Logic for Concurrency (Part II). Theoretical Computer Science 322(3), 517–565 (2004)
7. Caires, L., Cardelli, L.: A Spatial Logic for Concurrency (Part I). Information and Computation 186(2), 194–235 (2003)
8. Hirschkoff, D.: An Extensional Spatial Logic for Mobile Processes. In: Gardner, P., Yoshida, N. (eds.) CONCUR 2004. LNCS, vol. 3170, pp. 325–339. Springer, Heidelberg (2004)
9. Jeffrey, A., Rathke, J.: Contextual equivalence for higher-order π-calculus revisited. In: Proceedings of Mathematical Foundations of Programming Semantics, Elsevier, Amsterdam (2003)
10. Milner, R., Parrow, J., Walker, D.: Modal logics for mobile processes. Theoretical Computer Science 114(1), 149–171 (1993)
11. Sangiorgi, D.: Extensionality and Intensionality of the Ambient Logic. In: Proc. of the 28th POPL, pp. 4–17. ACM Press, New York (2001)
12. Sangiorgi, D.: Bisimulation in higher-order calculi. Information and Computation 131(2) (1996)
13. Sangiorgi, D.: Expressing mobility in process algebras: first-order and higher-order paradigms. Ph.D thesis, University of Einburgh (1992)

Minimum Maximal Matching Is NP-Hard in Regular Bipartite Graphs

M. Demange[1] and T. Ekim[2,*]

[1] ESSEC Business School, Avenue Bernard HIRSH, BP 105, 95021 Cergy Pontoise cedex France
demange@essec.fr

[2] Boğaziçi University, Department of Industrial Engineering, 34342, Bebek-Istanbul, Turkey
tinaz.ekim@boun.edu.tr

Abstract. Yannakakis and Gavril showed in [10] that the problem of finding a maximal matching of minimum size (MMM for short), also called Minimum Edge Dominating Set, is NP-hard in bipartite graphs of maximum degree 3 or planar graphs of maximum degree 3. Horton and Kilakos extended this result to planar bipartite graphs and planar cubic graphs [6]. Here, we extend the result of Yannakakis and Gavril in [10] by showing that MMM is NP-hard in the class of k-regular bipartite graphs for all $k \geq 3$ fixed.

Keywords: stable marriage, unmatched pairs, Minimum Maximal Matching, regular bipartite graphs.

1 Introduction

Given a graph, a *matching* is a set of edges which are pairwise non-adjacent. A matching M is said to be *maximal* if no other edge can be added to M while keeping the property of being a matching. We say that an edge in a matching M *dominates* all edges adjacent to it. Also, vertices contained in edges of M are said to be *saturated* by M. In a given graph, a matching saturating all vertices is called *perfect*. Note that in a maximal matching M of a graph $G = (V, E)$, every edge in E is necessarily dominated. The problem of finding a maximal matching of minimum size is called Minimum Maximal Matching (MMM) or Minimum (Independent) Edge Dominating Set (see [5] for the equivalence of these problems as well as all other graph theoretical definitions not given here).

MMM, NP-hard in general, is extensively studied due to its wide range of applications. See [10] for the description of an application related to a telephone switching network built to route phone calls from incoming lines to outgoing trunks. The worst case behavior of such a network (minimum number of calls routed when the system is saturated) can be evaluated by a minimum maximal matching.

* The research of Tınaz Ekim is funded by the B.U. Research Fund, Grant 08A301, whose support is greatly acknowledged.

M. Agrawal et al. (Eds.): TAMC 2008, LNCS 4978, pp. 364–374, 2008.

A second application appears in the so-called *stable marriage* problem which involves a set of *institutions* and *applicants*. Suppose that institutions, respectively applicants, have preference lists on each other set. The objective is to find a matching between institutions and applicants which is *stable*; that is, a matching where there is no pair of institution-applicant which are not matched but which would be both better off if matched to each other. The problem to solve is to find a stable matching in a bipartite graph where one part of the bipartition represents institutions and the other part applicants, and where the preferences of each institution and each applicant are given. In [4], Gale and Shapley give an algorithm to determine a stable matching between medical school graduate students and hospitals. Their algorithm works in a greedy fashion; in a successive way, institutions tentatively admit applicants and admitted applicants tentatively accept the admissions.

It is easy to see that a stable matching is necessarily a maximal matching since otherwise there would be at least one more pair of institution-applicant where both prefer to be matched to each other compared to the current situation. However, the well-known Gale-Shapley algorithm finding a stable matching in a bipartite graph may leave unmatched some institutions and applicants; this is clearly not a desired situation. One can easily see that the number of unmatched institutions and applicants is bounded above by the number of the remaining vertices after the removal of a minimum maximal matching (since a stable matching is necessarily maximal).

In some problems, the number of applicants can be very large; it is the case for instance during the university admission examination in Turkey where each year over two million students apply. Here, the national examination office places students to universities by giving priority to the preferences of students with best scores (see [2] for further information). Departing from the idea that the examination score can not be the only criterion in the student placement procedure, after the first placement results, it may be convenient to establish shortlists (of the same lenght) based on the initial preferences of students and universities (which are for now represented by examination scores). These shortlists can be obtained by constructing a *stable many-to-many matching* where each student and each university has the same number of choices; this is represented by a regular bipartite graph (see [1] for the description of such an algorithm). Once we have the shortlists, students and universities can have a second round to express their new preferences which are this time based on interviews and a better knowledge of each other; criteria other then exam scores for institutions or university reputations for students are now taken into account. A stable matching obtained from this regular bipartite graph with new preferences will be the ultimate placement result. So this motivates the problem of studying the tractability of minimum maximal matching restricted to regular bipartite graphs.

In [10], Yannakakis and Gavril show that MMM is NP-hard in several classes of graphs including bipartite (or planar) graphs with maximum degree 3. In [6], Horton and Kilakos obtained some extensions of these results including the NP-hardness of MMM in planar bipartite graphs and planar cubic graphs.

In the present paper, we give another strengthening of the result of Yannakakis and Gavril in [10] by showing that MMM is NP-hard in k-regular bipartite graphs for any fixed $k \geq 3$; this shows the NP-hardness of finding the best upper bound on the number of unmatched pairs given by a stable matching based on shortlists. Note that in the case $k = 2$, MMM is polynomially solvable since 2-regular graphs consist of disconnected cycles and therefore any maximal matching is in fact a maximum matching. Our result implies in particular that MMM is NP-hard already in cubic bipartite graphs which are strictly contained in the class of bipartite graphs with maximum degree 3. Besides, one may observe that the generalization of this result in cubic bipartite graphs to k-regular bipartite graphs for $k > 3$ is far from being straightforward. Note also that the class of k-regular bipartite graphs for $k \geq 3$ has no inclusion relationship with the class of planar bipartite graphs while having a non-empty intersection (the Heawood graph, also called (3,6)-cage graph, is a 3-regular bipartite graph which is non-planar [5]). These results contribute to narrowing down the gap between P and NP-complete with respect to MMM.

Other polynomial time solvable cases for MMM can be found in [8] for trees, in [7] for block graphs and in [9] for bipartite permutation graphs and cotrianglated graphs.

Throughout the paper, we consider only connected bipartite graphs; all results remain valid for not necessarily connected bipartite graphs simply by repeating the described procedures for each connected component.

The paper is organized in the following way. In Section 2, we show that if MMM is NP-hard in k-regular bipartite graphs then it is also NP-hard in $(k+1)$-regular bipartite graphs. In Section 3, the NP-hardness of MMM in 3-regular bipartite graphs is established; to this purpose, we first reduce a restricted 3-SAT problem to MMM in bipartite graphs with vertices of degree 2 or 3 (inspired from [10]), and then the later problem to MMM in 3-regular bipartite graphs by using the gadget R_k for $k = 2$ already described in Section 2. Finally, we put together the results of Sections 2 and 3 to conclude that MMM is NP-hard in k-regular bipartite graphs for all fixed $k \geq 3$.

2 MMM in k-Regular Bipartite Graphs $\propto$ MMM in $(k+1)$-Regular Bipartite Graphs

Let $B_k = (V_k, E_k)$ be a k-regular bipartite graph. We shall explain a polynomial time reduction from MMM in k-regular bipartite graphs to MMM in $(k+1)$-regular bipartite graphs. It has two steps; in the first step, called degree increasing step, we increase by 1 the degree of all vertices in B_k by branching to each of them a *Degree Increasing* gadget DI_k which itself contains vertices of degree k and $k+1$. This transformation is followed by the addition of a *Regularity* gadget R_k to each DI_k. The obtained graph, denoted by $\widetilde{B}_k$, contains vertices of degree k and $k+1$. The second step, called regularity step, consists of making $\widetilde{B}_k$ $(k+1)$-regular by first taking $k+1$ copies of $\widetilde{B}_k$ and then branching Regularity gadgets R_k's to the vertices of degree k in such a way that, in the final graph B_{k+1}, all vertices have degree $k+1$. To conclude, we show that if one can solve MMM in B_{k+1}, then one can also solve MMM in B_k.

2.1 Degree Increasing Step

Let us first discuss about some properties of the gadget DI_k depicted in Figure 1. DI_k is bipartite as can be observed in the bipartition shown in Figure 1 by black (referred to as sign $-$) and white (referred to as sign $+$) vertices. DI_k has one vertex in Level 1, $k-1$ vertices in Level 2, k vertices in Level 3 and two vertices in Level 4; let i_j denote the j'th vertex (from left) in Level i in a DI_k. The edges of DI_k are the following: there are all possible edges between Levels 1 and 2 and between Levels 3 and 4; every vertex in Level 2 has exactly $k-1$ edges towards the vertices of Level 3 which are distributed as equally as possible (hence each vertex of Level 3 receives exactly $k-2$ edges from Level 2 except one vertex which receives $k-1$). We suppose without loss of generality that this is the vertex 3_1 in Level 3 which receives $k-1$ edges from Level 2. The dashed edge e_0 is the connection edge to the original graph B_k whereas the other $k+1$ dashed edges $e_1, \ldots, e_{k+1}$ will be used to connect DI_k to a Regularity gadget R_k.

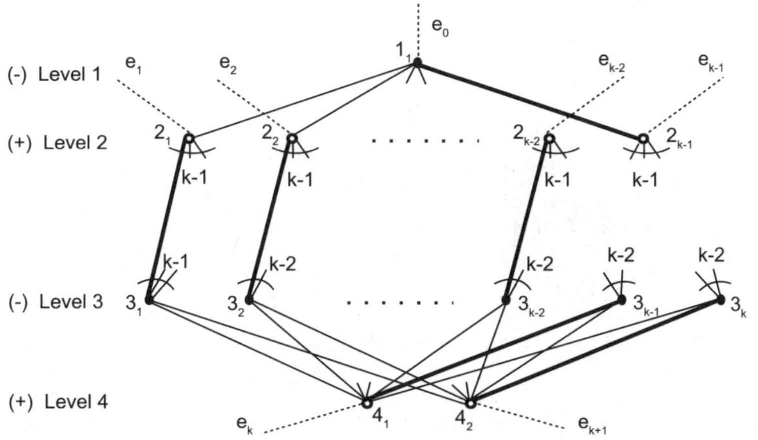

Fig. 1. DI_k: Degree Increasing gadget of order k

Suppose we have branched a gadget DI_k to each vertex of the k-regular graph B_k using the edge e_0. Now, we branch a Regularity Gadget R_k (see Figure 2) to each DI_k using the $k+1$ (dashed) connection edges $e_1, \ldots, e_{k+1}$. The graph obtained in this manner is denoted by $\widetilde{B}_k = (\widetilde{V}_k, \widetilde{E}_k)$ and depicted in Figure 3.

The gadget R_k has $k+1$ vertices in each one of Level 1, Level 2 and Level 3, and k vertices in Level 4; the edges of R_k are such that a k-regular bipartite graph is induced by the vertices of Level 1 and 2, a 1-regular bipartite graph (a perfect matching) is induced by the vertices of Level 2 and 3, and all possible edges exist between Levels 3 and 4.

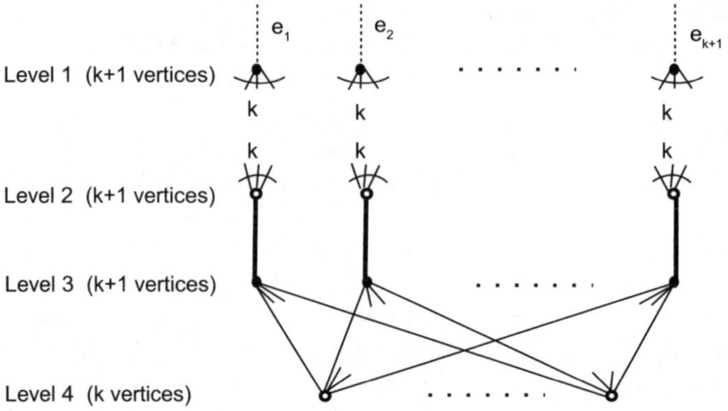

Fig. 2. R_k: Regularity gadget of order k

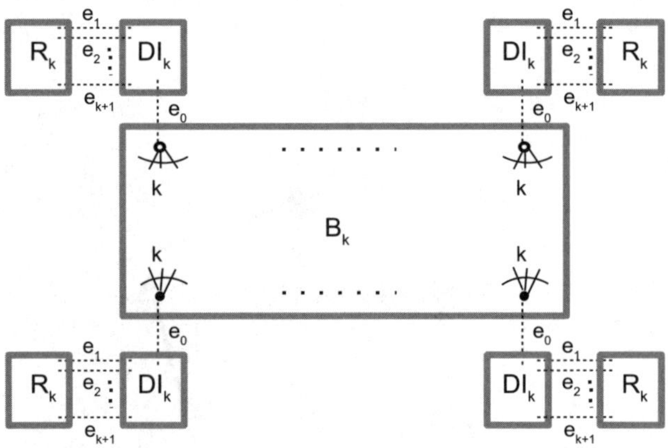

Fig. 3. $\widetilde{B}_k = (\widetilde{V}_k, \widetilde{E}_k)$: The graph obtained at the end of the first step

It is important to note that the connection edges $e_0, e_1, \ldots, e_{k+1}$ are not included in DI_k, nor in R_k.

Let $n_k = |V_k|$ be the number of vertices in B_k. The following property is a direct consequence of the construction of $\widetilde{B}_k$:

Property 1. $\widetilde{B}_k$ is bipartite; moreover all of its vertices are of degree $k+1$, except k vertices in each one of the n_k gadgets DI_k which are of the same sign and of degree k.

Proof. A bipartition of $\widetilde{B}_k$ follows from the fact that both DI_k and R_k are bipartite; moreover all end-vertices of the connection edges $e_1 \ldots, e_{k+1}$ in DI_k, respectively in R_k, have the same sign in the bipartition of DI_k, respectively of R_k.

The property on the degrees of vertices in $\widetilde{B}_k$ is a direct consequence of its definition; the connection edges e_0's ensure that the degree of all vertices of

B_k is $k+1$; all vertices in R_k's have degree $k+1$ since the connection edges $e_1, \ldots, e_{k+1}$ are present; all vertices in DI_k's are of degree $k+1$ except the vertices $1_1, 3_2, 3_3, \ldots, 3_k$ in each copy of DI_k which are of degree k. □

Other properties follow from the above definitions.

Property 2. There is a unique minimum maximal matching (of size $k+1$) in R_k; it leaves non-dominated all connection edges.

Proof. First, notice that at least $k+1$ edges are needed to dominate all edges in R_k since there is an induced perfect matching of size $k+1$ between Levels 2 and 3 (shown by bold edges in Figure 2). In addition, this matching dominates all edges in R_k, hence it is a minimum maximal matching in R_k. Besides, suppose that there is another maximal matching M in R_k which does not contain $p > 0$ edges of the matching between Levels 2 and 3. This requires that $\min(p, k)$ other edges between Levels 3 and 4, and at least $\lceil p/2 \rceil$ other edges between Levels 1 and 2 have to be in M (since there are $p \times k$ edges to be dominated between Levels 1 and 2, and one edge can dominate at most $2k-1$ edges (plus itself)). It follows that this matching has at least $k+1-p+\min(p,k)+\lceil p/2 \rceil > k+1$ edges and consequently there is no other minimum maximal matching in R_k. □

Let H_k be the graph which consists of a $DI_k = (V_D, E_D)$ connected to a $R_k = (V_R, E_R)$ with the connection edges $e_1, \ldots, e_{k+1}$ and where all edges incident to e_0 (that is all edges between Levels 1 and 2 in DI_k) are removed. In other words, H_k is a subgraph of $\tilde{B}_k$ induced by the vertex set $(V_D \setminus \{1_1\}) \cup V_R$. The following property shows that, even if an edge e_0 connecting a couple of gadgets DI_k and R_k (which are already connected) to B_k is included in a maximal matching of $\tilde{B}_k$, we still need at least $2(k+1)$ additional edges to dominate all edges of these gadgets DI_k and R_k.

Property 3. There are at least $2(k+1)$ edges in any maximal matching M_H of H_k.

Proof. We set $M_H = M_H^R \cup M_H^D \cup M'_H$ where $M_H^R = M_H \cap E_R$ and $M_H^D = M_H \cap E_D$ and hence $M'_H \subseteq \{e_1, \ldots, e_{k+1}\}$. Firstly, we note that $|M_H^D| \geq k$. In fact, the k vertices of DI_k in Level 3 are of degree at least k and an edge incident to one of them can dominate at most one edge which is incident to another vertex of Level 3. As a consequence, if $|M_H^D| = k-1$ then these $k-1$ edges can dominate at most all edges incident to $k-1$ vertices of Level 3 and $k-1$ edges incident to the remaining vertex of Level 3; hence at least one edge remains non-dominated.

Let us now consider the edges in M_H^R; suppose $p \geq 0$ of them are adjacent to one of the connection edges $e_1, \ldots, e_{k+1}$. If $p = 0$, then $|M_H^R| = k+1$ since there is an induced matching of size $k+1$ (between Levels 2 and 3) in R_k. In this case, none of the connection edges $e_1, \ldots, e_{k+1}$ is dominated by an edge in M_H^R and moreover, they form an induced matching of size $k+1$; thus $|M_H^D \cup M'_H| \geq k+1$ and therefore $|M_H| \geq 2(k+1)$. If $p \geq 1$, note that M_H^R contains at least k additional edges to dominate all edges between Levels 3 and 4. Indeed, either k edges saturating all vertices of Level 4 are in M_H^R; or at least one vertex of

Level 4 is not saturated by an edge in M_H^R and consequently at least $k+1$ edges are in M_H^R. In both cases we have $|M_H^R| \geq k+p$. Now, if $p \geq 2$ then $|M_H| \geq |M_H^R| + |M_H^D| \geq 2(k+1)$. Finally, if $p=1$, we consider two cases mentioned above. If M_H^R contains k edges dominating all vertices of Level 4, then there is at least one edge which is not yet dominated between Levels 1 and 2; thus, at least one more edge (in M_H^R or M'_H) is necessary in addition to $k+1$ edges in M_H^R, this gives $|M_H| \geq 2(k+1)$. If there is at least one vertex of Level 4 which is not saturated, then $|M_H^R| \geq k+1+1$ and therefore $|M_H| \geq 2(k+1)$, which concludes the proof. □

The following result summarizes the above discussion on the degree increasing step.

Lemma 1. *There is a maximal matching M_0 of size at most m_0 in B_k if and only if there is a maximal matching M_1 of size at most $m_1 = m_0 + 2n_k(k+1)$ in $\widetilde{B}_k$ where $n_k = |V_k|$ and all vertices in copies of DI_k are saturated.*

Proof. ($\Longrightarrow$) The matching M_1 contains all (at most) m_0 edges in M_0 which dominate all edges in B_k. In addition, the matchings of size $(k+1)$ in each DI_k (there are n_k of them) as depicted in Figure 1 by bold edges, and the matchings of size $(k+1)$ in each R_k (there are n_k of them) as depicted in Figure 2 by bold edges dominate together all edges in $\widetilde{B}_k$. Note that edges $e_0, e_1, \ldots, e_{k+1}$ are dominated since the matchings in DI_k's saturate their end-vertices in DI_k's. This matching of size $m_1 = m_0 + 2n_k(k+1)$ is clearly maximal.

($\Longleftarrow$) Suppose we have a maximal matching M_1 of size at most $m_1 = m_0 + 2n_k(k+1)$ in $\widetilde{B}_k$. Let V_{e_0} be the set of vertices in B_k which are incident to an edge of type $e_0 \in M_1$. By Property 3, we know that $|M_1| \geq 2n_k(k+1) + |V_{e_0}| + |M_1 \cap E_k|$; hence $m_0 \geq |V_{e_0}| + |M_1 \cap E_k|$. Besides, the edges of B_k not dominated by $M_1 \cap E_k$ are incident to a vertex in V_{e_0} and the partial subgraph of B_k formed by these edges is a bipartite graph with V_{e_0} as a vertex cover (of its edges). Consequently, any maximal matching M'_1 of this subgraph has at most $|V_{e_0}|$ edges. Therefore $M'_1 \cup (M_1 \cap E_k)$ is a maximal matching of B_k of size at most m_0. □

2.2 Regularity Step

Property 1 states that all vertices in the graph $\widetilde{B}_k$ obtained at the end of the first step has degree $k+1$ except k vertices of the same sign in each one of the n_k copies of DI_k. Now, we describe a procedure which allows to make $\widetilde{B}_k$ a $(k+1)$-regular bipartite graph. To this purpose, we first take $k+1$ copies of $\widetilde{B}_k$, denoted by $(k+1)\widetilde{B}_k$. Then, for each vertex $x \in \widetilde{V}_k$ of degree k, we consider the $k+1$ copies $x_1, \ldots, x_{k+1}$ of x in $(k+1)\widetilde{B}_k$; clearly, they are all of degree k and there is a bipartition of $(k+1)\widetilde{B}_k$ where all of $x_1, \ldots, x_{k+1}$ have the same sign. We branch a regularity gadget R_k to vertices $x_1, \ldots, x_{k+1}$ for every vertex x of degree k in $\widetilde{B}_k$; this new graph is denoted by B_{k+1}. We have the following.

Theorem 1. *There is a maximal matching M_0 of size at most m_0 in B_k if and only if there is a maximal matching M_2 of size at most $m_2 = (k+1)m_1 + k(k+1)n_k$ in B_{k+1} where $m_1 = m_0 + 2n_k(k+1)$ and $n_k = |V_k|$.*

Proof. ($\Longrightarrow$) By Property 1, in $\widetilde{B}_k$, there are k vertices of degree k in each one of the n_k gadgets DI_k. This results in kn_k copies of the regularity gadget R_k in B_{k+1}. By Property 2, we know that we can dominate all edges in these kn_k copies of R_k by $k(k+1)n_k$ edges. Moreover, we know by Lemma 1 that one can find a minimum maximal matching M_1 of size $m_1 = m_0 + 2n_k(k+1)$ in $\widetilde{B}_k$ where all vertices in different copies of DI_k are saturated. Therefore, by taking $k+1$ copies of M_1 (one by each $\widetilde{B}_k$) in M_2, we obtain a matching of size $m_2 = (k+1)m_1 + k(k+1)n_k$ which also dominates all connection edges between DI_k's and newly added R_k's.

($\Longleftarrow$) There are kn_k independent copies of R_k's in B_{k+1} which are added during the regularity step; by Property 2, there are at least $k(k+1)n_k$ edges of M_2 which dominate all edges in these R_k's (but not necessarily their connection edges $e_1, \ldots, e_{k+1}$). Consequently, the remaining at most $(k+1)(m_0+2n_k(k+1))$ edges in M_2 dominate all edges in the $k+1$ independent copies of $\widetilde{B}_k$ and their connection edges to the newly added kn_k copies of R_k. It follows from Lemma 1 that there is a maximal matching M_0 of size at most m_0 in B_k. □

3 MMM in 3-Regular Bipartite Graphs Is NP-Hard

We will make use of the following problem in the sequel.

Restricted 3-SAT: A set of clauses $C_1, \ldots, C_p$ with variables $x_1, \ldots x_n$ where there are at most three literals (negative or positive variables) per clause, such that every variable occurs twice and its negation once.

Lemma 2. *The decision version of MMM in bipartite graphs with vertices of degree 2 or 3 is NP-complete.*

Proof. The reduction is a modified version of the reduction of Restricted 3-SAT to the decision version of MMM in bipartite graphs with vertices of maximum degree 3 [10]. We construct a graph $B = (V, E)$ as follows: for each variable x_i, there is a gadget G_1^i as depicted in Figure 4 where the edges (d_{j1}, d'_{j1}) and (d_{j2}, d'_{j2}) represent the two clauses C_{j1}, C_{j2} in which x_i appears in positive form, and (d_{j3}, d'_{j3}) the clause C_{j3} in which x_i appears in negative form. Edges $(a_i^{j1}, d_{j1}), (b_i^{j2}, d_{j2})$ and (c_i^{j3}, d'_{j3}) are the connection edges between subgraphs representing variables and clauses. In this way, all variables and clauses are represented; nevertheless, the vertices d_j (respectively d'_j) corresponding to clauses C_j which consist of 3-positive (respectively 3-negative) variables will have degree four. We avoid such a situation by representing this kind of clauses by the gadget G_2^j of Figure 4. For a clause $C_j = (x_r, x_k, x_h)$ (respectively $C_j = (\bar{x}_r, \bar{x}_k, \bar{x}_h)$), the connection edges l_1, l_2, l_3 are incident to a_r^j or b_r^j, a_k^j or b_k^j, and a_h^j or b_h^j (respectively c_r^j, c_k^j and c_h^j); a connection edge is incident to a vertex of type a_i^j

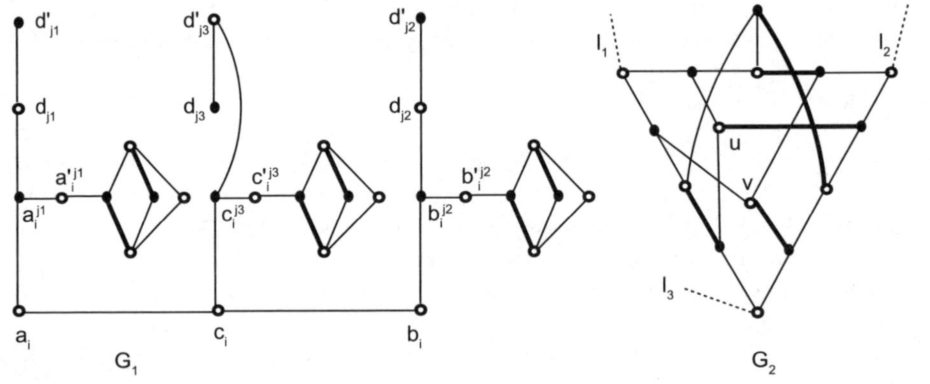

Fig. 4. Gadgets G_1 and G_2

if the first occurrence of variable x_i is in the clause C_j, and to a vertex of type b_i^j if the second occurrence of variable x_i is in the clause C_j. Now, we notice that the constructed graph B is bipartite with vertices of degree 2 or 3.

Suppose a satisfying assignment of Restricted 3-SAT is given; let p_3 be the number of clauses with 3-positive or 3-negative variables, then a maximal matching M of size $9n + 5p_3$ in B is obtained as follows:

$$\begin{aligned} M = & \{(a_i^{j1}, d_{j1}), (b_i^{j2}, d_{j2}), (c_i, c_i^{j3}) | x_i = 1\} \cup \{6 \text{ bold edges in } G_1^i | x_i = 1\} \\ \cup & \{(a_i, a_i^{j1}), (b_i, b_i^{j2}), (c_i^{j3}, d'_{j3}) | x_i = 0\} \cup \{6 \text{ bold edges in } G_1^i | x_i = 0\} \\ \cup & \quad \{5 \text{ bold edges in } G_2^j | C_j \text{ has 3-positive or 3-negative variables}\}. \end{aligned}$$

Conversely, let M be a maximal matching of size $9n + 5p_3$ in B, a truth assignment for Restricted 3-SAT can be obtained in polynomial time. First, notice that M must contain at least 5 edges per gadget G_2^j even though all connection edges l_1, l_2, l_3 are in M (since there are three disjoint paths of length 4 between vertices u and v). In particular, if at least one connection edge is in M, then 5 additional edges suffice to dominate all edges of G_2^j (In Figure 4, the bold edges of G_2^j show these 5 edges in case l_1 is in M). On the other hand, if none of the connection edges is in M, then at least 6 edges from G_2^j should be in M.

Secondly, one can observe that in G_1^i, we need at least 6 edges to dominate all edges containing non-labeled vertices (such a set of 6 edges in G_1 is shown in bold in Figure 4). Moreover, there is no such a set of 6 edges which also dominates at least one edge from $\{(a_i^{j1}, a'^{j1}_i), (c_i^{j3}, c'^{j3}_i), (b_i^{j2}, b'^{j2}_i)\}$. It follows that vertices a_i^{j1}, c_i^{j3} and b_i^{j2} must be saturated for all variables x_i in every maximal matching of B, so in particular in M. This implies that there are at least three more edges per variable in M. Since $|M| = 9n + 5p_3$ and at least $6n$ edges are necessarily used to dominate all edges containing non-labeled vertices as well as at least $5p_3$ edges to dominate all edges in G_2^j's, there are exactly three additional edges per variable which should saturate a_i^{j1}, c_i^{j3} and

b_i^{j2}; therefore neither (a_i, c_i) nor (c_i, b_i) is in M for any x_i. Let us now have a closer look at these three remaining edges per variable in M. If $(c_i, c_i^{j3}) \notin M$, then necessarily $(a_i, a_i^{j1}), (b_i, b_i^{j2}), (c_i^{j3}, d'_{j3}) \in M$. If $(c_i, c_i^{j3}) \in M$, then we can suppose without loss of generality that $(a_i^{j1}, d_{j1}), (b_i^{j2}, d_{j2}) \in M$ because if not, M can be transformed in polynomial time into such a maximal matching since $|M| = 9n + 5p_3$ implies that for every j we have $(d_j, d'_j) \notin M$.

We define a truth assignment by setting $x_i = 0$ in the first case and $x_i = 1$ in the second case. As $\forall j, (d_j, d'_j) \notin M$, there is at least one edge in M which is incident to d_j or d'_j and which determine(s) the variable(s) satisfying the clause j. One can notice that if for some i, we have $(a_i, a_i^{j1}), (c_i, c_i^{j3}), (b_i, b_i^{j2}) \in M$ then variable x_i can be set equally to 0 or 1; in fact, this simply means that there is a truth assignment of the Restricted 3-SAT instance where variable x_i does not satisfy any clause. □

Theorem 2. *The decision version of MMM in 3-regular bipartite graphs is NP-complete.*

Proof. The reduction is from MMM in bipartite graphs with degree 2 or 3. Let B be a bipartite graph with vertices of degree 2 or 3. We take 3 copies of B and for each vertex x of degree 2 in B, we branch a Regularity gadget (see Figure 2) of order $k = 2$ (R_2) to its three copies x_1, x_2 and x_3. Let us denote this new graph by B'; it is easy to see that B' is a 3-regular bipartite graph. Now, it follows from the discussion in the proof of Theorem 1 that there is a maximal matching M of size m in B if and only if there is a maximal matching M' of size $m' = 3m + 6|V^2|$ in B' where V^2 is the set of vertices of degree 2 in B. □

4 Final Remarks

The following is a corollary of Theorems 1 and 2:

Corollary 1. *MMM is NP-hard in k-regular bipartite graphs for all fixed $k \geq 3$.*

A simple computation shows that our reduction can be done in polynomial time if k is fixed. In fact, if the number of vertices in a k-regular bipartite graph is n_k, then our reduction (in Theorem 1) gives a $(k+1)$-regular bipartite graph with $n_k(10k^2 + 14k + 5)$ vertices. It follows that the graph obtained (by successive applications of Theorem 1 starting from parameter $k' = 3$ until $k' = k$) to show the NP-hardness of MMM in k-regular bipartite graphs will have a number of vertices which is a multiplier of $k!$.

Our result narrowed down the gap between P and NP for MMM in bipartite graphs. This theoretical progress opens the doors of several research directions. Several approximation algorithms with performance guarantee for MMM are known in the literature (MMM is 2-approximable even if the edges are weighted [3]); one could hope to derive algorithms with better performance guarantee in the case of regular bipartite graphs. The sharpness of such approximation ratios in special cases can be discussed knowing that this allows to bound the number of unmatched pairs in a stable matching.

References

1. Alkan, A.: Current Trends in Economics: Theory and Applications, pp. 30–39. Springer, Heidelberg (1999)
2. Balinski, M., Sönmez, T.: A Tale of Two Mechanisms: Student placement. Journal of Economic Theory 84, 73–94 (1999)
3. Fujito, T., Nagamochi, H.: A 2-approximation algorithm for the minimum weight edge dominating set problem. Discrete Applied Mathematics 118, 199–207 (2002)
4. Gale, D., Shapley, I.S.: College admissions and the stability of marriage. Amer. Math. Montly 69, 9–14 (1962)
5. Harary, F.: Graph Theory. Addison-Wesley, Reading (1994)
6. Horton, J.D., Kilakos, K.: Minimum edge dominating sets. SIAM J. Disc. Math. 6(3), 375–387 (1993)
7. Hwang, S.F., Chang, G.J.: The edge domination problem. Discuss. Math. Graph. Theory 15(1), 51–57 (1995)
8. Mitchell, S.L., Hedetniemi, S.T.: Edge domination in trees. In: Proc. of the 8th Southeastern Conference on Combinatorics, Graph Theory and Computing (Louisiana State Univ., Baton Rouge, La., 1977), pp. 489–509 (1977)
9. Srinivasan, A., Madhukar, K., Nagavamsi, P., Pandu Rangan, C., Chang, M.-S.: Edge domination on bipartite permutation graphs and cotriangulated graphs. Inform. Process. Lett. 56(3), 165–171 (1995)
10. Yannakakis, M., Gavril, F.: Edge dominating sets in graphs. SIAM J. Appl. Math. 38, 364–372 (1980)

A Topological Study of Tilings

Grégory Lafitte[1] and Michael Weiss[2,⋆]

[1] Laboratoire d'Informatique Fondamentale de Marseille (LIF), CNRS – Aix-Marseille Université, 39, rue Joliot-Curie, F-13453 Marseille Cedex 13, France

[2] Centre Universitaire d'Informatique, Université de Genève, Battelle bâtiment A, 7 route de Drize, 1227 Carouge, Switzerland

Abstract. To tile consists in assembling colored tiles on $\mathbb{Z}^2$ while respecting color matching. Tilings, the outcome of the tile process, can be seen as a computation model. In order to better understand the global structure of tilings, we introduce two topologies on tilings, one *à la* Cantor and another one *à la* Besicovitch. Our topologies are concerned with the whole set of tilings that can be generated by any tile set and are thus independent of a particular tile set. We study the properties of these two spaces and compare them. Finally, we introduce two infinite games on these spaces that are promising tools for the study of the structure of tilings.

1 Introduction

Wang was the first to introduce in [Wan61] the study of tilings with colored tiles. A tile is a unit size square with colored edges. Two tiles can be assembled if their common edge has the same color. A finite set of tiles is called a tile set. To tile consists in assembling tiles from a tile set on the grid $\mathbb{Z}^2$.

One of the first famous problems on tilings was the domino problem: can one decide whether given a tile set, there exists a tiling of the plane generated by this tile set? Berger proved the undecidability of the domino problem by constructing an aperiodic set of tiles, *i.e.*, a tile set that can generate only non-periodic tilings [Ber66]. Simplified proofs can be found in [Rob71] and later [AD96]. The main argument of this proof was to simulate the behavior of a given Turing machine with a tile set, in the sense that the Turing machine M stops on an instance ω if and only if the tile set $\tau_{\langle M,\omega\rangle}$ does not tile the plane. Hanf and later Myers [Mye74, Han74] have strengthened this and constructed a tile set that has only non-recursive tilings.

Later, tilings have been studied for different purposes: some researchers have used tilings as a tool for studying mathematical logical problems [AD96], others have studied the different kinds of tilings that one tile set can produce [CK97, DLS01, Rob71], or defined tools to quantify the regular structure of a tiling [Dur99]. One of the most striking facts concerning tilings is that tilings constitute a Turing equivalent computation model. This computation model is particularly

⋆ This author has been supported by the FNS grant 200020-105515.

M. Agrawal et al. (Eds.): TAMC 2008, LNCS 4978, pp. 375–387, 2008.

relevant as a model of computation on the plane. Notions of reductions that have led to notions of universality for tilings and completeness for tile sets have been introduced in [LW07]. It is difficult to quantify this completeness property and other ones as periodicity, quasiperiodicity: one would like to be able to measure how common such a property is, in order to determine when and how they occur, or to give a size to the different sets of tilings with a certain property, or to say if a tile set is more likely to produce tilings with a certain property. One of our ultimate goals is to be able to say that if a set of tilings (generated from a given tile set or a family of tile sets) is large enough, then it necessarily contains a tiling with such and such properties. Naturally, topological tools on tilings would be the first step in this direction; first step that we aim at developing in this paper.

We introduce two metrics on tilings which have the particularity to be independent of a particular tile set, *i.e.*, we can measure a distance between two tilings that are not necessarily generated by the same tile set. These two metrics are similar to two traditional metrics used in the study of cellular automata: the so-called Cantor and Besicovitch metrics. The former gives more importance to the local structure around $(0, 0)$ and the later measures the asymptotic difference of information contained in the two tilings. They give rise to two natural topologies on the set of tilings.

The topological study of subsets of reals is inherently linked to the study of infinite games. In some of these games, such as Banach-Mazur games, a strong connection exists between the existence of a winning strategy and the co-meager property of the set on which the game is played. This connection allows one to show that some sets are meager. Others games, such as Gale-Stewart games, yield a hierarchy of winning strategies for one of the two players - we say in this case that the game is determined - depending on the structure of the sets on which the games are played. Having such a game-topological study of tilings, instead of subsets of reals, can lead to a better understanding of the structure of the tilings generated by a tile set.

This paper is organized as follows: we first recall basic notions on tilings, define the two topologies that we use and prove basic properties of these topologies. Then we study in a deeper way the structure of our topological spaces. We conclude by introducing two types of games on tilings, which help us to prove in a simpler way results of the previous section, and open a new direction for the study of the structure of tilings.

2 Topologies on Tilings

2.1 Tilings

In this paper we use the following terminologies: a *tile set* S is an initial subset $\{1, \ldots, n\}$ of $\mathbb{N}$. To map consists in placing the numbers of S on the grid $\mathbb{Z}^2$. A mapping generated by S is called a *S-mapping*. It is associated to a mapping function $f_A \in S^{\mathbb{Z}^2}$ that gives the tile of S at position (x, y) in A. We call $\mathfrak{M}$ the set of all mappings, *i.e.*, $\mathfrak{M} \equiv \{\{1, \ldots, n\}^{\mathbb{Z}^2}\}_{n \geq 1}$.

An S-pattern without constraints, or just *S-pattern*, is an S-mapping defined on a finite subset of $\mathbb{Z}^2$.

A *Wang tile* is an oriented unit size square with colored edges from C, where C is a finite set of colors. A *Wang tile set*, or just *tile set*, is a finite set of different tiles. To tile consists in placing the tiles of a given tile set on the grid $\mathbb{Z}^2$ such that two adjacent tiles share the same color on their common edge. A tiling P generated by a tile set τ is called a τ-tiling. It is associated to a tiling function f_P where $f_P(x, y)$ gives the tile at position (x, y) in P.

An S-mapping $A \in \mathfrak{M}$ is a Wang tiling if there exist a Wang tile set τ, a τ-tiling P and a bijective function $h : S \rightarrow \tau$ such that $h \circ f_A(x, y) = f_P(x, y)$. By this, we mean that A *works* as P. We define $\mathfrak{T}$ as the subset of mappings of $\mathfrak{M}$ which are Wang tilings.

Different kinds of tile sets and tilings have been identified: a periodic tiling is a tiling such that there exist two integers a and b such that for any (x, y), the tiles at position (x, y) and $(x+a, y+b)$ are the same; a tile set is periodic if it generates a periodic tiling; a tiling is finite if it is a pattern; a tiling P is universal if for any tile set τ, P simulates at least one τ-tiling. For more precisions on simulation and universality we refer the reader to [LW07].

Different tools are used to quantify the regular structure of a tiling. One of these is the quasiperiodic function. For a tiling P, the quasiperiodic function of P, denoted g_P, is the function that given an integer n, gives the smallest integer s which has the following property: if m is a square pattern of P of size n (the length of the sides of the square), then m appears in any square of size s in $\mathbb{Z}^2$. Thus, the quasiperiodic function of a tiling quantifies the regularity of appearance of the patterns in the tiling. Some tilings do not have a quasiperiodic function defined for every n. We say that a tiling is quasiperiodic if it has a quasiperiodic function defined for every n. An important result in [Dur99] is that any tile set, that can tile the plane, can generate a quasiperiodic tiling of the plane.

2.2 A Besicovitch Topology

The first metric we introduce is a metric similar to the cellular automata metric *à la* Besicovitch. A $\{n \times n'\}$ τ-pattern m can be seen has a sequence of numbers placed in a rectangle of size $n \times n'$; we have the number k in position (x, y) if $f_m(x, y) = t_k$ where t_k is the kth tile of τ. Then intuitively any reordering of the tiles of τ gives the same pattern. Thus, we would like to say that the distance between two patterns m and m' is c if the proportion of different tiles between m and m' is at most c up to a reordering of the tiles. We formalize this notion in the following definitions, by defining a metric for any pattern without constraints:

Definition 1. *Let S and S' be two initial subsets of $\mathbb{N}$ such that $|S| \leq |S'|$. Let A be a $\{n \times n\}$ S-pattern, B be a $\{n \times n\}$ S'-pattern, and g be a one-to-one function from S to S'. We define the metric related to g by:* $\delta_g(A, B) = \frac{\#\{(x,y) \mid g \circ f_A(x,y) \neq f_B(x,y)\}}{n^2}$.

If A is an S-pattern and B is an S'-pattern such that $|S| \geq |S'|$*, then* $\delta_g(A, B)$ *is defined to be equal to* $\delta_g(B, A)$*.*

We define the absolute metric by: $\delta(A, B) = \min_{g \in S'^S}\{\delta_g(A, B)\}$*.*

From this definition of distance between patterns, we have that δ is symmetric and satisfies the triangle inequality.

Now that we have defined a metric between patterns, we can generalize it to tilings of the whole plane, since a tiling can be seen as the infinite union of patterns of ever increasing sizes:

Definition 2. *Let A be an S-tiling and let B be an S'-tiling. For any function* $g \in S'^S$*, we define the tiling metric* d_g *by:* $d_g(A, B) = \limsup_{i \to \infty} \delta_g(A_i, B_i)$*, where the* A_i *(*resp. B_i*) are the* $\{n \times n\}$ *S-patterns (*resp. *S'-patterns) centered around* $(0, 0)$ *in A (*resp. *B).*

We define the absolute tiling metric d by: $d(A, B) = \min_{g \in S'^S}\{d_g(A, B)\}$*.*

Since δ satisfies the triangle inequality, and since reflexivity and symmetry are obvious, then d is a pseudometric on $\mathfrak{M}$. The natural way to obtain a metric on $\mathfrak{M}$ is to introduce an equivalence relation $\equiv_d$ defined by: $A \equiv_d B \Leftrightarrow d(A, B) = 0$. One can see that $\equiv_d$ is an equivalence relation. We can now consider the quotient space $\mathfrak{M}/\equiv_d$ where a typical element is the equivalence class $[A] = \{\, B \,|\, d(A, B) = 0 \,\}$. By adding to this space the metric d we obtain a metric space that we call $\mathfrak{M}_B$. In this paper, $[A]$ denotes the equivalence class of the particular mapping A; an element of $\mathfrak{M}_B$ is designated by a capital letter in bold, *e.g.*, $\mathbf{A}$; and we say "let $A \in \mathfrak{M}_B$" in the sense that we consider a mapping of the equivalence class of A, *i.e.*, a mapping in $[A]$. Similarly, we define the space $\mathfrak{T}/_{\equiv_d}$ where a typical element is $[P] = \{\, Q \,|\, d(P, Q) = 0 \,\}$. By adding to this space the metric d we obtain the metric space $\mathfrak{T}_B$. Of course, we have $\mathfrak{T}_B \subset \mathfrak{M}_B$. An element of $\mathfrak{T}_B$ is an equivalence class of $\mathfrak{M}_B$ that can contain mappings that are not Wang tilings, but which work "almost" like Wang tilings since the local constraint is respected almost everywhere.

From this definition, we have the two following results: for any two mappings $A, B \in \mathfrak{M}$, the distance between A and B is in $[0, 1[$ and for any mapping $C \in \mathfrak{M}$ (*resp.* any tiling $P \in \mathfrak{T}$) and any $\epsilon \in [0, 1[$, there exists a tiling D (*resp.* Q) such that $d(C, D) \geq 1 - \epsilon$ (*resp.* $d(P, Q) \geq 1 - \epsilon$). Therefore, for any tiling A, we can build a tiling B such that the distance between A and B is almost one.

To obtain a topological space, since $\mathfrak{M}_B$ is a metric space, we use the natural topology induced by the metric where the open sets are the balls $B(A, r)$, where A is a mapping. We use the subset topology on $\mathfrak{T}_B$.

The Besicovitch metric is one of the traditional metrics used for tilings. The other one is the Cantor one which gives more importance to the finite patterns centered around the origin.

2.3 A Cantor Metric

We define and study another *traditional* metric adapted for tilings, a metric *à la* Cantor. The metric studied above, is a metric that allows one to understand the behavior of a tiling in the whole $\mathbb{Z}^2$ grid. The distance between two tilings is small if their behavior is close. Another way to measure distance between tilings is to consider the greatest common pattern centered around $(0,0)$ that they share. We first define the function p as the function $p : \mathbb{N} \to \mathbb{Z}^2$ such that $p(0) = (0,0)$, $p(1) = (0,1)$, $p(2) = (1,1)$, $p(3) = (1,0)\ldots$ and p keeps having the behavior of a spiral afterward. This function allows us to enumerate the tiles of a given tiling. We define the *prefix-patterns* of a tiling P:

Definition 3. *Let P be a tiling. The* prefix-pattern *of size n of P is the pattern m defined on the finite subset $D = \{p(0),\ldots,p(n)\} \subsetneq \mathbb{Z}^2$ with the pattern function $f_m(x,y) = f_P(x,y)$ if $(x,y) \in D$. We denote the prefix-pattern m by a finite sequence of tiles ordered by the function p: $m = \{f_m \circ p(0), f_m \circ p(1),\ldots, f_m \circ p(n-1)\}$. If m is a prefix-pattern of size n, then $m.t_1$, the concatenation of m and the tile t_1, is the prefix-pattern of size $n+1$ such that $m.t_1 = \{f_m \circ p(0), f_m \circ p(1),\ldots, f_m \circ p(n-1), t_1\}$.*

The set of prefix-patterns can be enumerated by a tree T. The rules of the construction of T are the following: at level 0 in T, we have the empty prefix-pattern; at level 1 we have an unique prefix-pattern $\{1\}$. Then a pattern m at the level i, composed of j different tiles, has $j+1$ sons: $m.1, m.2,\ldots, m.(j+1)$.

One can see that for any tile set S and any prefix-pattern $m = \{m_1,\ldots,m_n\}$ generated by S, there exists an unique bijective function $e_m : S \to \{1,\ldots,|S|\}$ such that the prefix-pattern $e_m(m) \stackrel{\text{def}}{=} \{e_m(m_1), e_m(m_2),\ldots, e_m(m_n)\}$ is an element of T. $e_m(m)$ is said to be the *canonical form* of m. In T, an infinite branch corresponds to a mapping of the plane. Thus, to any S-mapping A there exists a unique bijective function e_A such that the set $e_A(A) \stackrel{\text{def}}{=} \{e_A \circ f_A \circ p(0), e_A \circ f_A \circ p(1),\ldots\}$ corresponds to an infinite branch of T. $e_A(A)$ is said to be the *canonical form* of A. We say that m is a prefix-pattern of A if $e_m(m)$ is a prefix-pattern of $e_A(A)$. We can now define a metric *à la* Cantor:

Definition 4. *Let A be an S-mapping and B be an S'-mapping. We define the Cantor metric δ_C as: $\delta_C(A,B) = 2^{-i}$ where i is the size of the greatest common prefix-pattern of $e_A(A)$ and $e_B(B)$, i.e., the highest level in T where $e_A(A)$ and $e_B(B)$ are equal.*

If $e_A(A) = e_B(B)$ then $\delta_C(A,B) = 0$.

We can see that δ_C is a pseudometric on $\mathfrak{M}$. In fact, d_C is a hypermetric, *i.e.*, a metric such that for any three mappings A, B, C, $d_C(A,C) \leq \max\{d_C(A,B), d_C(B,C)\}$. This is a stronger version of the inequality of the triangle. One can note that in a hypermetric space, any point of an open ball is center of this ball.

To obtain a metric on $\mathfrak{M}$, we say that two tilings P and Q are equivalent, $P \equiv_C Q$, if their Cantor distance is null. Thus, two tilings are equivalent if they represent the same tiling up to a color permutation. We denote $\mathfrak{M}_C$ the space of equivalence classes $\mathfrak{M}/_{\equiv_C}$ equipped with the metric δ_C.

Similarly, we define $\mathfrak{T}/_{\equiv_C}$ that we denote $\mathfrak{T}_C$. The metric δ_C is a metric on $\mathfrak{T}_C$. From this, we can define a topology on $\mathfrak{T}_C$: we say that the set $U_m = \{\, P \,|\, m \text{ is a prefix-pattern of } P\,\}$ is a clopen set for any prefix-pattern m. This topology gives rise to a different understanding of the topology of tilings than the topological space $\mathfrak{M}_B$ since it gives more importance to the local structure centered around $\{0,0\}$. Since there is a finite set of pattern of a given size, then we can cover $\mathfrak{M}_B$ and $\mathfrak{T}_B$ with a finite set of open sets. Therefore, $\mathfrak{M}_B$ and $\mathfrak{T}_B$ are precompact, *i.e.*, for all r, there does not exist a finite set of open balls of radius r that covers these spaces.

Since we have a Hausdorff space, because $\mathfrak{M}_C$ is a metric space, and since we have a basis of clopen sets, then $\mathfrak{M}_C$ is a $0-$*dimensional* space.

We finish the definition of the Cantor space by stating some obvious facts about the Cantor metric: if A and B are two mappings, then $d_C(A,B) \in [0,1/2]$, and for any mapping A, there exists a mapping B such that $d_C(A,B) = 1/2$.

2.4 Basic Properties

The space $\mathfrak{M}_C$ is well-defined since two mappings at distance 0 are in fact the same tiling up to a reordering of the tiles, or, to say it differently, have the same canonical form. The space $\mathfrak{M}_B$ is slightly different, since two tilings at distance 0 can be different. We have to redefine properties for the mappings of $\mathfrak{M}_B$: if a tiling class contains a tiling with a certain property, then all the class has this property, since in fact, all other tilings of the class have "almost" the property. Thus, we can define the following tiling classes: if $\mathbf{P} \in \mathfrak{T}_B$ is a tiling class, we say that $\mathbf{P}$ is: *periodic* if $\mathbf{P}$ contains a periodic tiling, *quasiperiodic* if $\mathbf{P}$ contains a quasiperiodic tiling, *finite* if $\mathbf{P}$ contains a finite tiling (*i.e.*, a pattern), *universal* if $\mathbf{P}$ contains a universal tiling and a tiling with a quasiperiodic function f if $\mathbf{P}$ contains a tiling with a quasiperiodic function f and if $\mathbf{P}$ does not contain a tiling with a quasiperiodic function $g < f$.

We obtain now a space that works almost like the space of all Wang tilings. We can see how the different classes work. The following proposition states that the distance between two periodic tilings of $\mathfrak{T}_B$ is a rational number and gives a characterization of the classes of periodic tilings.

Proposition 1. *If P and Q are two different periodic tilings, then there exist $n, m \in \mathbb{N}^*$ such that $d(P,Q) = n/m$. Therefore, if $\mathbf{P} \in \mathfrak{M}_B$ is periodic, then it contains one and only one periodic tiling up to a reordering of the tiles.*

The following proposition shows that two quasiperiodic tilings which belong to the same equivalence class have the same quasiperiodic function:

Proposition 2. *If P and Q are two quasiperiodic tilings such that $P \equiv_d Q$, then $g_P = g_Q$, where g_P and g_Q are the quasiperiodic functions of P and Q.*

We recall a basic notion of tilings: *extraction.* Consider an infinite set of τ-patterns $\{m_1, m_2, \ldots\}$ of ever increasing sizes. We can see them as an infinite tree with the root representing the empty pattern and where a pattern m is a direct son of a pattern n if n is a subpattern of m, and if there does not exist a pattern $m' \neq m$ such that n is a subpattern of m' and m' is a subpattern of m. Thus, we obtain an infinite tree with finite degree. Therefore, by Koenig's lemma, we have at least one infinite branch. In our tree, this branch represents a tiling Q of the plane. This tiling Q is said to be an extraction of the set $\{m_1, m_2, \ldots\}$. Now, we say that Q is extracted from P if there exists an infinite set $\{m_1, m_2, \ldots\}$ of patterns of P such that Q is an extraction of $\{m_1, m_2, \ldots\}$.

Notions of universality and completeness for tilings have been introduced in [LW07]. We relate them to our topological space:

Proposition 3. *Let* $\mathbf{P} \in \mathfrak{M}_B$ *be a universal* (resp. *quasiperiodic, periodic*) *tiling. Then for any tiling* $A \in \mathbf{P}$, *we can extract from* A *a universal* (resp. *quasiperiodic, periodic*) *tiling* A'.

The previous result shows again that belonging to an equivalence class with a certain property is almost like having this property.

Now we study the different distances that we can obtain between Wang tilings and mappings in our two Besicovitch spaces. We have the following properties:

Theorem 1. *i) There exist a mapping* $A \in \mathfrak{M}_B \setminus \mathfrak{T}_B$ *and* $\epsilon > 0$ *such that the ball* $B(A, \epsilon)$ *does not contain any Wang tilings;*

ii) For any n, *there exists an infinite subset* H *of* $\mathfrak{T}_B$ *such that for any two tilings* P *and* Q *of* H, $d(P, Q) \geq 1 - 1/n^2$.

As a corollary, we have that the spaces $\mathfrak{M}_B$ and $\mathfrak{T}_B$ are not precompact. We can remark that that there exist prefix-patterns that can not be represented by the local constraint of a Wang tile set. From this, we obtain the following proposition:

Proposition 4. *There exists an open set in* $\mathfrak{M}_C$ *that does not contain any Wang tiling.*

3 Properties of Our Topologies

3.1 Properties of the Metric Spaces

We first study some basic notions of our spaces to have a better understanding of how they work. First of all, since we have metrizable spaces, we have that our spaces are completely Hausdorff and that there are no isolated points neither in $\mathfrak{M}_B$ nor in $\mathfrak{T}_B$. This seems natural for $\mathfrak{M}_B$, and is more interesting in the case of $\mathfrak{T}_B$.

Proposition 5. $\mathfrak{M}_B$, $\mathfrak{T}_B$, $\mathfrak{M}_C$ *and* $\mathfrak{T}_C$ *are all perfectly normal Hausdorff and perfect.*

The set of tilings is uncountable. But there exist tile sets that generate an uncountable set of tilings but only a countable set of equivalence classes in $\mathfrak{M}_B$. From this, and with the fact that any equivalence class contains an uncountable set of mappings, there arises the question of the cardinality of $\mathfrak{M}_B$. The following proposition shows that even the equivalence classes of $\mathfrak{T}_B$ are uncountable.

Proposition 6. *$\mathfrak{T}_B$ has the cardinality of the continuum.*

The next theorem is important for the understanding of Wang tilings. The metric used is strongly related to the information contained in the tiling. Thus, two tilings are close if the information contained in them is similar. So, theorem 2 can be stated as follows: a countable set of Wang tilings can not approach all the information that Wang tilings can generate. Then, by generalizing this theorem, we obtain a nice corollary.

Theorem 2. *There does not exist a countable set of Wang tilings which is dense in $\mathfrak{T}_B$.*

Corollary 1. *i) For any countable set of Wang tilings H, there exist a tiling P and a natural number n such that $d(P, Q) \geq 1/n$ for any $Q \in H$;*
ii) There exist a Wang tiling P and an integer n such that for any tile set τ, there exists at least one τ-tiling Q such that $d(P, Q) \geq 1/n$.

The next proposition shows how different the two topological spaces $\mathfrak{T}_B$ and $\mathfrak{T}_C$ are. This comes from the fact that one takes a glimpse at the whole tiling since the other one just look at it with blinkers.

Proposition 7. *There exists a countable set of Wang tilings that is dense in τ_C.*

From this, we have that $\mathfrak{M}_B$ is separable, and since it is completely metrizable, we have that $\mathfrak{M}_B$ is a Polish space. The next theorem shows that for any two tilings $A, B \in \mathfrak{T}_B$, there exists a continuous path $c : [0, 1] \to \mathfrak{T}_B$ such that $c(0) = A$ and $c(1) = B$:

Theorem 3. *$\mathfrak{T}_B$ is a path-connected space.*

3.2 Topological Properties

We now study the topological structure of our spaces. Since they are metric spaces, then natural topologies are induced on them by the metrics. We define the following sets:

i) Mapping(S) = $\{ [A] \mid A$ is an S-mapping $\}$,
ii) Wang(S) = $\{ [P] \mid P$ is a Wang S-tiling $\}$.

And for any tile set τ, we define the set: $Wang(\tau) = \{ [P] \mid P$ is a τ-tiling $\}$.

The following theorem shows that the set of mappings or tilings generated by a tile set is a closed set. Then we give a characterization of the set of Wang tilings that can produce a tile set:

Theorem 4. *Let S and τ be two tile sets. Then $Wang(\tau)$ and $Mapping(S)$ are closed sets and $Wang(\tau)$ is either a closed discrete set, or a closed non-discrete nowhere-dense set.*

Corollary 2. *i) $\mathfrak{T}_B$ is meager in $\mathfrak{M}_B$;*
ii) $\mathfrak{M}_B \setminus \mathfrak{T}_B$ is dense in $\mathfrak{M}_B$.

We now show that our spaces are Baire spaces:

Theorem 5. *$\mathfrak{T}_B$, $\mathfrak{T}_C$, $\mathfrak{M}_B$ and $\mathfrak{M}_C$ are complete metric spaces, and thus, are Baire spaces.*

In the following section, we introduce games on our topological spaces as tools for the study of the structure of tilings.

4 Games on Tilings

Since the tilings computation model is equivalent to the Turing machines, tilings make possible a geometrical point of view of computability. The different topological tools studied in this paper have shown some interesting aspects of the behavior of computability in tiling spaces. As we have seen, the set of tilings generated by a tile set τ, gives rise to complex subsets of $\mathfrak{M}_B$ and $\mathfrak{M}_C$. These sets can even be uncountable. A natural next step for studying these sets is to consider infinite games on tilings.

Several kinds of infinite games exist and are used in many different fields. Games have been studied for computation models such as pushdown automata (see Serre's PhD thesis [Ser05] for detailed survey). Considering the tilings computation model, we now give definitions for two types of infinite games on tilings.

The first one, Banach-Mazur games [Oxt57], is a play on the topological structure of the space. Different definitions of Banach-Mazur games exist. We propose this one:

Definition 5. *Let X be a topological space and Y a family of subsets of X such that:*

i) any member of Y has nonempty interior;
ii) any nonempty open subset of X contains a member of Y.

Let C be a subset of X. The game proceeds as follows: Player I chooses a subset Y_1 of Y. Player II chooses a subset Y_2 of Y such that $Y_2 \subseteq Y_1$. Then Player I chooses a subset Y_3 of Y such that $Y_3 \subseteq Y_2$ and so on. At the end of the infinite game, we obtain a decreasing sequence of sets: $X \supseteq Y_1 \supseteq Y_2 \supseteq \dots$ such that Player I has chosen the sets with odd indexes and Player II has chosen the sets with even indexes. Player II wins the game if $\bigcup_{n\geq 1} Y_i \subseteq X$.

The study of the different subsets of X such that Player II has a winning strategy is the main application of Banach-Mazur games. Of course, if $C = X$ Player II has a winning strategy. The question is: how big C has to be to allow Player II to have a winning strategy? This gives rise to classical theorem concerning Banach-Mazur games on topological spaces which states: a subset C of X is meager if and only if Player II has a winning strategy for the game on $\{X, X \setminus C\}$. We propose a Banach-Mazur game on the space $\mathfrak{M}_C$:

Definition 6. *Let X be a subset of $\mathfrak{M}_C$ and $C \subseteq X$. The game $\{X, C\}$ is defined as follows: Player I chooses a prefix-pattern m_1 such that m_1 is a prefix-pattern* *of a mapping of X. Player II chooses a prefix-pattern m_2 such that $m_1 \subset m_2$ and such that m_2 is a prefix-pattern of a mapping of X and so on. At the end of the game, we obtain a sequence of prefix-patterns $m_1 \subset m_2 \subset m_3 \ldots$ from which we extract a unique mapping A. Player II wins the game if $A \in C$.*

This amounts to playing the classical Banach-Mazur game with $X = \mathfrak{M}_C$ and $Y \subseteq \{ U_m \,|\, U_m$ is an open set in $\mathfrak{M}_C \}$. In our Cantor space, choosing a prefix-pattern amounts to choosing an open set, since the open sets of $\mathfrak{M}_C$ can be defined from prefix-patterns. Using this tool, we show that the set of Wang tilings is meager in the set of mappings for our topology $\mathfrak{M}_C$:

Theorem 6. *$\mathfrak{T}_C$ is meager in $\mathfrak{M}_C$.*

Proof. We use the game $\{\mathfrak{M}_C, \mathfrak{M}_C \setminus \mathfrak{T}_C\}$. We now have to show that Player II has a winning strategy, *i.e.*, Player II can always chooses integers in such a way that the final mapping can not be a Wang tiling.

This is true since whatever plays Player I at the first round, then Player II can choose a prefix-pattern which does not respect any possible local constraint generated by Wang tilings.

This trivial proof shows the convenience of using games to prove that some subsets are meager. To obtain the same kind of results for $\mathfrak{M}_B$ we define a Banach-Mazur game more adapted to the topology of $\mathfrak{M}_B$:

Definition 7. *Let X be a subset of $\mathfrak{M}_B$ and $C \subseteq X$. A Banach-Mazur game on $\{X, C\}$ is defined as follows: Player I chooses a mapping A_1 of X and an integer n_1; Player II chooses a mapping A_2 of X such that $d(A_1, A_2) \leq 1/n_1$ and an integer $n_2 \geq n_1$; Player I chooses a mapping A_3 of X such that $d(A_3, A_2) \leq 1/n_2$ and an integer $n_3 \geq n_2$, and so on. Player II wins the game if $lim_{i\to\infty} A_i \in C$.*

This game is still equivalent to a classical Banach-Mazur game, and since $\mathfrak{M}_B$ is a topological space, we still have that Player II has a winning strategy if and only if C is co-meager. We now prove the same result for $\mathfrak{M}_B$:

Theorem 7. *$\mathfrak{T}_B$ is meager in $\mathfrak{M}_B$.*

Proof. We will show that Player II has a winning strategy in the game $\{\mathfrak{M}_B, \mathfrak{M}_B \setminus \mathfrak{T}_B\}$. To show this, we first prove the following result: for any tiling P and any open ball $B(P, 1/n)$, there exist a mapping A and an integer $m > n$ such that $B(A, 1/m) \subset B(P, 1/n)$ and $B(A, 1/m) \cap \mathfrak{T}_B = \emptyset$.

Let P be a tiling and n an integer. The idea is to insert error patterns in P. We can build a pattern of size six generated by two tiles such that it can not be represented by a Wang pattern since its construction would imply that the two tiles that compose it are equal. Thus, at least one tile of this pattern can not be represented by Wang tiles. We introduce it in P in such a way that the new mapping A that we obtain is at distance $1/2n$ of P. Because of the error patterns, A can not be a Wang tiling. In the error pattern we have at least one of the six tiles that can not be represented by a Wang tile. Therefore if Q

is a Wang tiling, then $d(Q, A) \geq 1/12n$. Thus, $B(A, 1/12n) \subset B(P, 1/n)$ and $B(A, 1/12n) \cap \mathfrak{T}_B = \emptyset$.

With this result, we can see that the strategy of Player II will be to choose the tiling A and the integer $12n_1$ to be sure to win. Thus, $\mathfrak{T}_B$ is meager.

We introduce another type of games for the study of the complexity of the set of tilings generated by a given tile set: games *à la* Gale-Stewart. Here is a general definition of these games:

Definition 8. *Let A be a nonempty set and $X \subseteq A^{\mathbb{N}}$. We associate with X the following game: Player I chooses an element a_1 of A, Player II chooses an element a_2 of A and so on. Player I wins if $\{a_n\}_{n \in \mathbb{N}} \in X$. We denote $G(A, X)$ this game.*

A traditional question about a game $G(A, X)$ is to know whether one of the two players has a winning strategy, or in the terminology of games, if the game is determined. In [Mar75], Martin has shown that any Borel set is determined. Thus, we have to go beyond the Borelian hierarchy to find subsets complicated enough not to be determined. We would like to use these games on tilings to obtain similar structural complexity results for the set of tilings generated by a tile set. In that direction we give the following definition:

Definition 9. *Let $H \subseteq \mathfrak{M}_B$ and $X \subseteq H$. The Gale-Stewart game $G(H, X)$ is defined as follows: Player I chooses a tile a_1 such that $\{a_1\}$ is a prefix-pattern of a tiling of H; Player II chooses a tile a_2 such that $\{a_1, a_2\}$ is a prefix-pattern of a tiling of H and so on. Player I wins if the tiling $\{a_1, a_2, \ldots\} \in X$.*

If one of the two players has a winning strategy we say that $G(H, X)$ is determined *or that X is* determined in *H. We say that τ is* determined *if Wang(τ) is determined in T_τ, where T_τ is the set of all τ-tilings and τ-patterns, and that τ is* completely undetermined *if for any subset $X \in$ Wang(τ), X is undetermined in Wang(τ).*

The question is to know which kinds of games on tilings are determined, and which ones are not. We give a glimpse in that direction by showing that there exist tile sets determined and other ones completely undetermined:

Theorem 8. *i) There exists a determined tile set;*
ii) There exists a completely undetermined tile set.

Proof. i) To show this, we just have to find a tile set simple enough to generate a determined game. The tile set EASY5composed of a unicolor blue tile and four tile with three sides blue and one red satisfies the theorem. Player I has a winning strategy in the game $G(T_{\text{EASY5}}, \text{Wang}(\text{EASY5}))$: in this game, the goal of player I is to obtain a tiling of the plane, and the goal of player II, while respecting the local constraint of EASY5, is to obtain a situation where Player I can not move anymore. Player I can force the play of Player II by playing always one of the tile with a colored edge to force Player II to play the symmetric of this tile. Thus, the game is determined.

ii) Since Hanf and Myers [Mye74, Han74], we know that there exist tile sets that generate only non-recursive tilings. Let τ be one of them; we consider the game $G(\text{Wang}(\tau), X)$ where X is a subset of $\text{Wang}(\tau)$. If this game is determined, then there exists a winning strategy for one of the two players. Without loss of generality, suppose Player I has a winning strategy. Therefore, whatever Player II plays, Player I has a recursive process that allows him to choose a tile to go in a winning position: this strategy can generate a τ-tiling in a recursive way. This is a contradiction.

5 Concluding Remarks

The topologies and games introduced in this paper have made possible some descriptions of the structure of Wang tilings. This is a first step in the direction of measuring the largeness or meagerness of sets of tilings. One of the many remaining questions is to be able to measure how common universality is.

To reach these goals, the topological study of tilings through games appears as a promising approach.

References

[AD96] Allauzen, C., Durand, B.: The Classical Decision Problem. In: Appendix A: "Tiling problems", pp. 407–420. Springer, Heidelberg (1996)

[Ber66] Berger, R.: The undecidability of the domino problem. Memoirs of the American Mathematical Society 66, 1–72 (1966)

[CD04] Cervelle, J., Durand, B.: Tilings: recursivity and regularity. Theoretical Computer Science 310(1-3), 469–477 (2004)

[CK97] Culik II, K., Kari, J.: On aperiodic sets of Wang tiles. In: Foundations of Computer Science: Potential - Theory - Cognition, pp. 153–162 (1997)

[DLS01] Durand, B., Levin, L.A., Shen, A.: Complex tilings. In: Proceedings of the Symposium on Theory of Computing, pp. 732–739 (2001)

[Dur99] Durand, B.: Tilings and quasiperiodicity. Theoretical Computer Science 221(1-2), 61–75 (1999)

[Dur02] Durand, B.: De la logique aux pavages. Theoretical Computer Science 281(1-2), 311–324 (2002)

[GS53] Gale, F., Stewart, F.M.: Infinite games with perfect information. Ann. Math. Studies 28, 245–266 (1953)

[Han74] Hanf, W.P.: Non-recursive tilings of the plane. I. Journal of Symbolic Logic 39(2), 283–285 (1974)

[LW07] Lafitte, G., Weiss, M.: Universal Tilings. In: Thomas, W., Weil, P. (eds.) STACS 2007. LNCS, vol. 4393, pp. 367–380. Springer, Heidelberg (2007)

[Mar75] Martin, D.A.: 1975. Annals of Math. 102, 363–371 (1975)

[Mye74] Myers, D.: Non-recursive tilings of the plane. II. Journal of Symbolic Logic 39(2), 286–294 (1974)

[Oxt57] Oxtoby, J.C.: Contribution to the theory of games, Vol. III. Ann. of Math. Studies 39, 159–163 (1957)

[Rob71] Robinson, R.: Undecidability and nonperiodicity for tilings of the plane. Inventiones Mathematicae 12, 177–209 (1971)

[Ser05] Serre, O.: Contribution á l'étude des jeux sur des graphes de processus á pile, PhD thesis, Université Paris VII (2005)

[Wan61] Wang, H.: Proving theorems by pattern recognition II. Bell System Technical Journal 40, 1–41 (1961)

[Wan62] Wang, H.: Dominoes and the ∀∃∀-case of the decision problem. In: Proceedings of the Symposium on Mathematical Theory of Automata, pp. 23–55. (1962)

A Denotational Semantics for Total Correctness of Sequential Exact Real Programs

Thomas Anberrée

Division of Computer Science, University of Nottingham in Níngbō, P.R. China
Thomas.Anberree@nottingham.edu.cn
http://www.cs.bham.ac.uk/~txa

Abstract. We provide a domain-based denotational semantics for a sequential language for exact real number computation, equipped with a non-deterministic test operator. The semantics is only an approximate one, because the denotation of a program for a real number may not be precise enough to tell which real number the program computes. However, for many first-order common functions $f : \mathbb{R}^n \to \mathbb{R}$, there exists a program for f whose denotation is precise enough to show that the program indeed computes the function f. In practice such programs possessing a faithful denotation are not difficult to find.

1 Introduction

We provide a denotational model for a functional programming language for exact real number computation. A well known difficulty in real number computation is that the tests $x = y$ and $x \leq y$ are undecidable and hence cannot be used to control the execution flow of programs. One solution, proposed by Boehm and Cartwright, is to use a non-deterministic test [2]. For any two rational numbers $p < q$ and any real number x, at least one of the relations $p < x$ or $x < q$ can be determined to hold; thus, operators $\mathrm{rtest}_{p,q}$ are used, whose evaluation never diverges when x is a real number:

1. $\mathrm{rtest}_{p,q}(x)$ evaluates to true or to false,
2. $\mathrm{rtest}_{p,q}(x)$ may evaluate to true iff $x < q$ and
3. $\mathrm{rtest}_{p,q}(x)$ may evaluate to false iff $p < x$.

Since a program can in general produce different results in different runs, Escardó and Marcial-Romero took the view in previous work that programs of real-number type denote sets of real numbers, and the question arose as to which power domains would be suitable for modelling the behaviour of rtest [7,8]. It was shown that, among the known power domains, only the Hoare power domains are suitable. However, a semantics based on Hoare power domains only accounts for partial correctness of programs (If a program computes a real number then it is the one given by its denotation).

We argue that, although Smyth power domains cannot faithfully model the $\mathrm{rtest}_{p,q}$ operators, they can nevertheless provide an approximation which is sufficiently precise to define many common first-order functions, and has the further

M. Agrawal et al. (Eds.): TAMC 2008, LNCS 4978, pp. 388–399, 2008.

advantage over other choices that total correctness of programs is expressed in the model (If the denotation of a program is a real number, then the program computes this real number). Writing $[\![M]\!]$ and $[M]$ for the denotational and operational meaning of a program M, respectively, we show that, in general, one has $[\![M]\!] \sqsubseteq [M]$, but not necessarily $[\![M]\!] = [M]$. However, if $[\![M]\!]$ represents a real number, then $[\![M]\!]$ is maximal, and hence $[\![M]\!] = [M]$. To illustrate the usefulness of the approximate semantics in practice, it can be shown that many programs of interest have a total denotation, which allows to establish their total correctness without resorting to operational methods.

Some background in domain theory is assumed [1,6]. In this paper, a domain is a continuous, directed-complete poset and all domains are seen as topological spaces, with the Scott topology. Continuity of functions between domains is to be understood relatively to the Scott topologies.

2 The Language

The language we consider is an extension of the functional language PCF [9,10] with a type for real numbers and related basic constructors. Types are given by

$$\sigma = \texttt{real} \mid \texttt{nat} \mid \texttt{bool} \mid \sigma \to \sigma. \tag{1}$$

Types `real`, `nat` and `bool` are called *ground types.* The typing judgements and reductions rules are recapitulated in Table 1 and Table 2 respectively. The expressions $|M| = \bot$, $|L| < 1$ and $0 < |L|$, to the right of certain reduction rules, are decidable, side conditions indicating when these rules are applicable (*cf.* Definition 2 below).

2.1 Approximating Real Numbers with Intervals

The constructors $\mathtt{bound}_a$, $p+$ and $p\times$ are used to approximate real numbers with intervals and operate on these approximations. The index a ranges over

$$\mathcal{R}_{\mathbb{Q}} = \{[s,t] \mid s \leq t \text{ and } s,t \in \mathbb{Q}\}$$

and p over the rational numbers. The set $\mathcal{R}_{\mathbb{Q}} \cup \{(-\infty,+\infty)\}$ forms a basis of the interval domain [5,3].

Definition 1. *The interval domain $\mathcal{R}$ is the set of all non-empty, closed and bounded real intervals ordered by reverse inclusion, along with a least element $\bot_{\mathcal{R}} = (-\infty,+\infty)$. We write $x \sqsubseteq y$ for $x \supseteq y$.*

The interval domain is a continuous Scott domain in which the supremum of a directed set is its intersection. The real numbers are embedded in $\mathcal{R}$ via the continuous map $r \mapsto \{r\} = [r,r]$.

Table 1. Typing judgements

$$\frac{}{\Gamma, x\colon \sigma \vdash x\colon \sigma} \qquad \frac{\Gamma \vdash M\colon \sigma \to \tau \quad \Gamma \vdash N\colon \sigma}{\Gamma \vdash M(N)\colon \tau}$$

$$\frac{\Gamma, x\colon \sigma \vdash M\colon \tau}{\Gamma \vdash (\lambda x\colon \sigma.M)\colon \sigma \to \tau} \qquad \frac{\Gamma \vdash M\colon \sigma \to \sigma}{\Gamma \vdash \mathtt{Y}_\sigma(M)\colon \sigma}$$

$$\frac{}{\Gamma \vdash \mathtt{true}\colon \mathtt{bool}} \qquad \frac{}{\Gamma \vdash \mathtt{false}\colon \mathtt{bool}}$$

$$\frac{\Gamma \vdash L\colon \mathtt{nat}}{\Gamma \vdash (0 =)\,(L) : \mathtt{bool}}$$

$$\frac{\Gamma \vdash L\colon \mathtt{bool}, \quad \Gamma \vdash M\colon \gamma, \quad \Gamma \vdash N\colon \gamma}{\Gamma \vdash \mathtt{if}_\gamma(L, M, N)\colon \gamma}\ \gamma \in \{\mathtt{bool}, \mathtt{nat}, \mathtt{real}\}$$

$$\frac{}{\Gamma \vdash \underline{n}\colon \mathtt{nat}} \qquad \text{for all } n \in \mathbb{N}$$

$$\frac{\Gamma \vdash M\colon \mathtt{nat}}{\Gamma \vdash \mathtt{succ}(M)\colon \mathtt{nat}} \qquad \frac{\Gamma \vdash M\colon \mathtt{nat}}{\Gamma \vdash \mathtt{pred}(M)\colon \mathtt{nat}}$$

$$\frac{\Gamma \vdash M\colon \mathtt{real}}{\Gamma \vdash \mathtt{bound}_a(M)\colon \mathtt{real}} \qquad a \in \mathcal{R}_\mathbb{Q}, \text{ where } \mathcal{R}_\mathbb{Q} = \{[p, q] \mid p \le q, \quad p, q \in \mathbb{Q}\}$$

$$\frac{\Gamma \vdash M\colon \mathtt{real}}{\Gamma \vdash p + (M)\colon \mathtt{real}} \qquad p \in \mathbb{Q}$$

$$\frac{\Gamma \vdash M\colon \mathtt{real}}{\Gamma \vdash p \times (M)\colon \mathtt{real}} \qquad p \in \mathbb{Q}$$

$$\frac{\Gamma \vdash L\colon \mathtt{real}}{\Gamma \vdash \mathtt{rtest}(L)\colon \mathtt{bool}}$$

Γ ranges over *contexts*, that is finite sequences $x_1\colon \sigma_1, \ldots, x_k\colon \sigma_k$ where x_i are pairwise distinct variables and σ_i are types.

The operation $\mathrm{bound}_a\colon \mathcal{R} \to \mathcal{R}$ that appears as index in the reduction rule (15) of Table 2 is continuous. It is the direct image function obtained from the function $\mathrm{bound}_a\colon \mathbb{R} \to \mathbb{R}$, $r \mapsto \max(\underline{a}, \min(r, \overline{a}))$ where $\underline{a}$ and $\overline{a}$ are the upper and lower bounds of the interval a (*i.e.* $a = [\underline{a}, \overline{a}]$). Notice that we have

Table 2. Reduction relation

$$(1)\ \frac{}{\mathtt{Y}_\sigma(M) \to M(\mathtt{Y}_\sigma(M))} \qquad (2)\ \frac{}{(\lambda x\colon \sigma.M)(N) \to M[N/x]}$$

$$(3)\ \frac{M \to M'}{M(N) \to M'(N)}$$

$$(4)\ \frac{}{\mathtt{if}(\mathtt{true}, M, N) \to M} \qquad (5)\ \frac{}{\mathtt{if}(\mathtt{false}, M, N) \to N}$$

$$(6)\ \frac{B \to B'}{\mathtt{if}(B, M, N) \to \mathtt{if}(B', M, N)}$$

$$(7)\ \frac{}{(0 =)(\underline{0}) \to \mathtt{true}} \qquad (8)\ \frac{}{(0 =)(\underline{n+1}) \to \mathtt{false}} \qquad (9)\ \frac{L \to L'}{(= 0)(L) \to (= 0)(L')}$$

$$(10)\ \frac{}{\mathtt{succ}(\underline{n}) \to \underline{n+1}} \qquad (11)\ \frac{M \to M'}{\mathtt{succ}(M) \to \mathtt{succ}(M')}$$

$$(12)\ \frac{}{\mathtt{pred}(\underline{0}) \to \underline{0}} \qquad (13)\ \frac{}{\mathtt{pred}(\underline{n+1}) \to \underline{n}} \qquad (14)\ \frac{M \to M'}{\mathtt{pred}(M) \to \mathtt{pred}(M')}$$

$$(15)\ \frac{}{\mathtt{bound}_a\left(\mathtt{bound}_b(M)\right) \to \mathtt{bound}_{\mathrm{bound}_a(b)}(M)}$$

$$(16)\ \frac{M \to M'}{\mathtt{bound}_a(M) \to \mathtt{bound}_a(M')}\,|M| = \bot$$

$$(17)\ \frac{}{p+\left(\mathtt{bound}_a(M)\right) \to \mathtt{bound}_{p+a}\left(p+(M)\right)} \qquad (18)\ \frac{M \to M'}{p+(M) \to p+(M')}\,|M| = \bot$$

$$(19)\ \frac{}{p\times\left(\mathtt{bound}_a(M)\right) \to \mathtt{bound}_{p\times a}\left(p\times(M)\right)} \qquad (20)\ \frac{M \to M'}{p\times(M) \to p\times(M')}\,|M| = \bot$$

$$(21)\ \frac{}{\mathtt{rtest}(L) \to \mathtt{true}}\,|L| < 1 \qquad (22)\ \frac{}{\mathtt{rtest}(L) \to \mathtt{false}}\,0 < |L|$$

$$(23)\ \frac{L \to L'}{\mathtt{rtest}(L) \to \mathtt{rtest}(L')}$$

$\mathrm{bound}_a(\bot_{\mathcal{R}}) = \bot_{\mathcal{R}}$ and hence $a \sqsubseteq \mathrm{bound}_a(x)$, for all $x \in \mathcal{R}$. Furthermore, if a and x are consistent ($a \cap x \neq \emptyset$), then $\mathrm{bound}_a(x) = a \cap x = a \sqcup x$.

Similarly, $p+$ and $p\times\colon \mathcal{R} \to \mathcal{R}$ are the continuous functions obtained from addition and multiplication of real numbers.

As an example, here are a few reduction steps from a program zero for 0.

$$\text{zero} = \mathtt{Y}_{\mathtt{real}}\left(\left(\lambda z\colon \mathtt{real}.\frac{1}{2}\times\left(\mathtt{bound}_{[-1,1]}(z)\right)\right)\right)$$

$$\rightarrow \left(\lambda z\colon \mathtt{real}.\frac{1}{2}\times\left(\mathtt{bound}_{[-1,1]}(z)\right)\right)(\text{zero}) \rightarrow \frac{1}{2}\times\left(\mathtt{bound}_{[-1,1]}(\text{zero})\right)$$

$$\rightarrow \mathtt{bound}_{\left[-\frac{1}{2},\frac{1}{2}\right]}\left(\frac{1}{2}\times(\text{zero})\right)$$

$$\rightarrow^4 \mathtt{bound}_{\left[-\frac{1}{2},\frac{1}{2}\right]}\left(\mathtt{bound}_{\left[-\frac{1}{4},\frac{1}{4}\right]}\left(\frac{1}{2}\times\left(\frac{1}{2}\times(\text{zero})\right)\right)\right)$$

$$\rightarrow \mathtt{bound}_{\left[-\frac{1}{4},\frac{1}{4}\right]}\left(\frac{1}{2}\times\left(\frac{1}{2}\times(\text{zero})\right)\right)$$

$$\rightarrow^5 \mathtt{bound}_{\left[-\frac{1}{4},\frac{1}{4}\right]}\left(\mathtt{bound}_{\left[-\frac{1}{8},\frac{1}{8}\right]}\left(\frac{1}{2}\times\left(\frac{1}{2}\times\left(\frac{1}{2}\times(\text{zero})\right)\right)\right)\right)$$

$$\rightarrow \mathtt{bound}_{\left[-\frac{1}{8},\frac{1}{8}\right]}\left(\frac{1}{2}\times\left(\frac{1}{2}\times\left(\frac{1}{2}\times(\text{zero})\right)\right)\right) \rightarrow \textit{etc.}$$

It is not hard to see that there is only one, infinite reduction path from program zero, of the form

$$\text{zero} \rightarrow \ldots \rightarrow \mathtt{bound}_{\left[-\frac{1}{2},\frac{1}{2}\right]}\frac{1}{2}\,\text{zero} \rightarrow \ldots \rightarrow \mathtt{bound}_{\left[-\frac{1}{2^n},\frac{1}{2^n}\right]}\frac{1}{2}\times\ldots\times\frac{1}{2}\,\text{zero} \rightarrow \ldots$$

The reduction relation reduces a program of type `real` to another one that possibly displays a better approximation of its result, in the form of intervals a, at the head of programs such as $\mathtt{bound}_a(M)$, but the evaluation process never terminates. We say that a is the *immediate output* of program $\mathtt{bound}_a\,(M)$.

Definition 2 (Immediate outputs of programs). *The* immediate output *of a program of type* `real` *is an element of* $\mathcal{R}$ *and is defined by:*

$\lvert\mathtt{bound}_a(N)\rvert = a$	*for any term* $N\colon \mathtt{real}\rightarrow\mathtt{real}$ *and any* $a\in\mathcal{R}_{\mathbb{Q}}$,
$\lvert M\rvert = \bot_{\mathcal{R}}$	*for any program not of the form* $\mathtt{bound}_a(N)$.

We also define the immediate output of programs of type `bool` *and* `nat`, *as an element of* $\mathbb{B}_\bot$ *or* $\mathbb{N}_\bot$*:* $\lvert\mathtt{true}\rvert = \textit{true}$, $\lvert\mathtt{false}\rvert = \textit{false}$, $\lvert\underline{n}\rvert = n$ *and* $\lvert M\rvert = \bot$ *in other cases.*

The immediate output of a program of ground type is the information that can be read at once from the head of a program without further evaluation.

Definition 3. *The supremum* $\bigsqcup\lvert M_n\rvert$ *of immediate outputs of programs along a reduction path* (M_n) *is called the* output of the path.

At type `real`, the output of a path is an element of the interval domain $\mathcal{R}$. For example the output of the reduction path from program zero given above, is $[0,0]$, that is $\{0\}$, which represents the real number 0 in $\mathcal{R}$. The output of any reduction path exists because a program can only reduce to a program with equal or greater immediate output.

It is easy to see that there exists only one reduction path from any given term with no occurrence of `rtest`. In this deterministic setting, the *operational meaning* $[M]$ of a program M of ground type could simply be defined as the output of the unique reduction path from it, that is as an element of $\mathcal{R}$, $\mathbb{N}_\perp$ or $\mathbb{B}_\perp$. However, the expressivity of the language without `rtest` is very weak [4].

2.2 Redundant Test: The rtest Constructor

We have already motivated the introduction of the `rtest` operator in Section 1. Reduction rules for the `rtest` constructor are the only ones with overlapping premises, which renders the operational semantics non-deterministic, in the sense that one term may reduce to several ones (Table 2). We only introduce one constructor $\texttt{rtest}_{0,1}$, simply written `rtest`, because the other ones are definable in the language, *e.g.* $\mathrm{rtest}_{p,q}(x) = \mathrm{rtest}_{0,1}\left(\frac{x-p}{q-p}\right)$.

The purpose of rule (23) is to eventually enable the applicability of one or both rules (21) and (22). Only when none of (21) and (22) become applicable, should rule (23) be allowed to be applied infinitely often.

2.3 Fair Reduction Paths

To avoid applying rule (23) infinitely many times in a row when doing otherwise is possible, one could give priority to the application of rules (21) and (22) over the application of rule (23), by stipulating that rule (23) should not be applied whenever one of the two other rules can be applied. But this would make `rtest` able to distinguish between programs computing the same real number. For example, suppose that half is a program for computing $\frac{1}{2}$, *i.e.* all reduction paths from half have output $[\frac{1}{2}, \frac{1}{2}]$; for example, take half $= \frac{1}{2} +$ zero. The term $\texttt{rtest}(\texttt{bound}_{[0,\frac{1}{2}]}(\mathrm{half}))$ would then only reduce to `true` whereas the term $\texttt{rtest}(\texttt{bound}_{[\frac{1}{2},1]}(\mathrm{half}))$ would only reduce to `false`, although both terms $\texttt{bound}_{[0,\frac{1}{2}]}(\mathrm{half})$ and $\texttt{bound}_{[\frac{1}{2},1]}(\mathrm{half})$ compute the number $\frac{1}{2}$. We wish to remain as general as possible and not put artificial constraints on the operational behaviour of the language. It should be up to a particular implementation to make specific decisions to avoid undesirable paths. In order to rule out undesirable paths without imposing too much constraint, we stipulate that only *fair paths* should be allowed. Roughly, a fair path is one along which rules (21) and (22) are eventually applied when possible.

Definition 4. *A* computation *is a maximal, fair reduction path.*

Computations are the only paths that any implementation of the language should consider. What is important, is that there exists a strategy that allows to produce maximal fair paths. Even if we only consider computations from a fixed program, these may have different outputs. For example, the program $\texttt{rtest}(\text{half})$ reduces to both `true` and `false`.

Definition 5. *The* outputs of a program *is the set of outputs of all* computations *from this program:*

$$\text{Outputs}\,(M) = \Big\{ \bigsqcup_n |M_n| \;\Big|\; (M_n) \textit{ is a computation from } M \Big\}.$$

Since there is at least one fair path from a given term M, the set Outputs(M) is not empty. A program $M\colon \texttt{real}$ *computes* a real number r if $\text{Outputs}(M) = [r, r]$. For example, consider the program $\text{abs}\colon \texttt{real} \to \texttt{real}$ recursively defined by

$$\begin{aligned} \text{abs}\,(x) = \texttt{if}\,\texttt{rtest}_{-1,0}(x)\,\texttt{then}\; -x \\ \texttt{else if}\,\texttt{rtest}_{0,1}(x)\,\texttt{then}\;\tfrac{1}{2}\,\texttt{bound}_{[-1,1]}(\text{abs}\,(2x)) \\ \texttt{else}\; x. \end{aligned} \tag{2}$$

It can be shown that, for any program $M\colon \texttt{real}$ that computes some real number r, the program abs (M) computes the absolute value of r.

2.4 Operational Meaning of Programs of Ground Types

To account for the multi-valuedness of programs, both the operational meaning $[M]$ and the denotation $[\![M]\!]$ of a program M of ground type are defined in a power domain $\mathcal{P}D$. As mentioned in the introduction, our choice is to use Smyth power domains because it allows us to prove the correctness of programs.

Definition 6 (Smyth power domains). *Given a domain D, its* Smyth power domain $\mathcal{P}^{\mathcal{S}}\,D$ *is the set of non empty, Scott compact and upper subsets of D, ordered by reverse inclusion (It is itself a domain).*

The Hasse diagram of the Smyth power domain $\mathcal{P}^{\mathcal{S}}\,\mathbb{B}_\perp$ is represented in Figure 1.

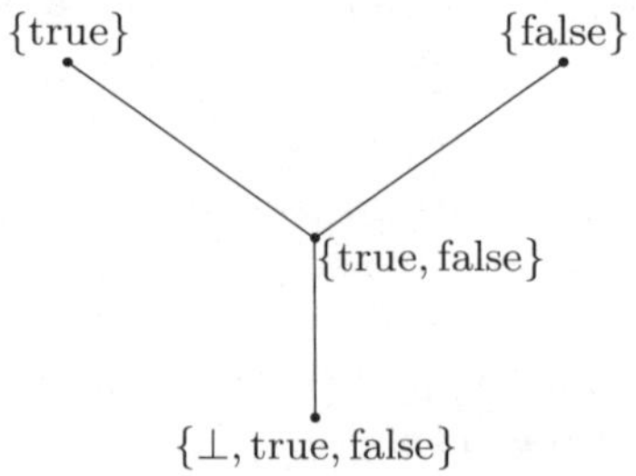

Fig. 1. Smyth power domain $\mathcal{P}^{\mathcal{S}}\,\mathbb{B}_\perp$ over the lifted booleans $\mathbb{B}_\perp$

Definition 7 (Interpretation of ground types.). *The interpretation $[\![\gamma]\!]$ of a ground type γ is given by:* $[\![\texttt{nat}]\!] = \mathcal{P}^{\mathcal{S}}\,\mathbb{N}_\perp$, $[\![\texttt{bool}]\!] = \mathcal{P}^{\mathcal{S}}\,\mathbb{B}_\perp$, $[\![\texttt{real}]\!] = \mathcal{P}^{\mathcal{S}}\,\mathcal{R}$.

Definition 8. *The* operational meaning $[M]$ *of a closed term* M *of ground type* γ *is the greatest element in the domain* $[\![\gamma]\!]$ *that contains the set of outputs of* M*:*

$$[M] = \bigcap \{K \in [\![\gamma]\!] \mid K \supseteq \mathrm{Outputs}(M)\}\,. \tag{3}$$

Notice that, for $\gamma = \mathtt{real}$, Definition 8 relies on the fact that, for any closed term M of ground type γ, the set $\{K \in [\![\gamma]\!] \mid K \supseteq \mathrm{Outputs}\,(M)\}$ is directed. This follows from the fact that, in continuous Scott domains (bounded complete domains), the intersection of finitely many Scott compact upper sets is Scott compact [1].

It is not always possible to recover the set of outputs of a program from its operational meaning. However, for a program M of ground type whose operational meaning is a finite set of maximal elements of $\mathbb{N}_\perp$, $\mathbb{B}_\perp$ or $\mathcal{R}$, we have that $\mathrm{Outputs}\,(M) = [M]$. In particular, if a program M computes a real number r, then $[M] = \{[r, r]\}$ in $\mathcal{P}^{\mathcal{S}}\,\mathcal{R}$. The main motivation of our work is to define a denotational semantics such that, in such cases, $[\![M]\!]$ coincides with $[M]$ and hence with $\mathrm{Outputs}\,(M)$, at least for enough programs M for the denotational semantics to be useful.

3 An Approximate Semantics for Total Correctness

The interpretations of ground types were given in Definition 7. The interpretation of a higher type $\sigma \to \tau$ is given by $[\![\sigma \to \tau]\!] = D_{\sigma \to \tau} = \big[\,[\![\sigma]\!] \to [\![\tau]\!]\,\big]$, the domain of Scott continuous functions from $[\![\sigma]\!]$ to $[\![\tau]\!]$. The denotation of a typing judgement $x_1 \colon \sigma_1, \ldots, x_k \colon \sigma_k \vdash M \colon \sigma$ is a continuous function from $[\![\sigma_1]\!] \times \ldots \times [\![\sigma_k]\!]$ to $[\![\sigma]\!]$ in the category of bounded-complete continuous domains. Table 3 on the following page gives the recursive definition of the denotations of typing judgements. The fact that the denotations of terms are continuous comes from classical results of domain theory (see *e.g.* [10]) and the fact that the denotations of basic constructors other than `rtest` are given as functions known to be continuous. We only have space to give details on the interpretation of `rtest`.

3.1 Denotation of `rtest`

Theorem 1. *The function* $\mathrm{ttest} \colon \mathcal{R} \to \mathcal{P}^{\mathcal{S}}\,\mathbb{B}_\perp$ *defined as follows is continuous.*

$$\mathrm{ttest}(l) = \begin{cases} \{\mathrm{true}\} & \textit{if}\ \bar{l} < 0 \\ \{\mathrm{false}\} & \textit{if}\ 1 < \underline{l} \\ \{\mathrm{true}, \mathrm{false}\} & \textit{if}\ \bar{l} \not< 0, 1 \not< \underline{l}\ \textit{and}\ l \not\sqsubseteq [0,1] \\ \{\perp_{\mathbb{B}_\perp}\} & \textit{if}\ l \sqsubseteq [0,1] \end{cases} \tag{4}$$

The denotation of `rtest` is the continuous function $\mathrm{ttest}^* \colon \mathcal{P}^{\mathcal{S}}\,\mathcal{R} \to \mathcal{P}^{\mathcal{S}}\,\mathbb{B}_\perp$ that maps K to $\bigcup \{\uparrow \mathrm{ttest}(k) \mid k \in K\}$.

Table 3. Denotation of typing judgements

$$[\![x_1 : \sigma_1, \ldots, x_k : \sigma_k \vdash x_i : \sigma_i]\!]\,(\vec{d}\,) = d_i$$

$$[\![\Gamma \vdash (\lambda x : \sigma.M) : \sigma \to \tau]\!]\,(\vec{d}\,)(e) = [\![\Gamma, x : \sigma \vdash M : \tau]\!]\,(\vec{d}, e)$$

$$[\![\Gamma \vdash \mathtt{Y}_\sigma(M) : \sigma]\!]\,(\vec{d}\,) = \mu_{D_\sigma}\left([\![\Gamma \vdash M : \sigma \to \sigma]\!]\,(\vec{d}\,)\right)$$

$$[\![\Gamma \vdash M(N) : \tau]\!]\,(\vec{d}\,) = [\![\Gamma \vdash M : \sigma \to \tau]\!]\,(\vec{d}\,)\Big([\![\Gamma \vdash N : \sigma]\!]\,(\vec{d}\,)\Big)$$

$$[\![\Gamma \vdash (0 =)(M) : \mathtt{bool}]\!]\,(\vec{d}\,) = (\mathcal{P}^{\mathcal{S}}(0 =))\,([\![\Gamma \vdash M : \mathtt{nat}]\!])$$

$$[\![\Gamma \vdash \mathtt{succ}(M) : \mathtt{nat}]\!]\,(\vec{d}\,) = (\mathcal{P}^{\mathcal{S}}(\mathrm{succ}))\,([\![\Gamma \vdash M : \mathtt{nat}]\!])$$

$$[\![\Gamma \vdash \mathtt{pred}(M) : \mathtt{nat}]\!]\,(\vec{d}\,) = (\mathcal{P}^{\mathcal{S}}(\mathrm{pred}))\,([\![\Gamma \vdash M : \mathtt{nat}]\!])$$

$$[\![\Gamma \vdash \mathtt{true} : \mathtt{bool}]\!]\,(\vec{d}\,) = \{\mathrm{true}\}$$

$$[\![\Gamma \vdash \mathtt{false} : \mathtt{bool}]\!]\,(\vec{d}\,) = \{\mathrm{false}\}$$

$$[\![\Gamma \vdash \underline{n} : \mathtt{nat}]\!]\,(\vec{d}\,) = \{n\}$$

$$[\![\Gamma \vdash \mathtt{rtest}(L) : \mathtt{bool}]\!]\,(\vec{d}\,) = \mathrm{ttest}^*\left([\![\Gamma \vdash L : \mathtt{real}]\!]\,(\vec{d}\,)\right)$$

$$[\![\Gamma \vdash \mathtt{if}_\gamma(B, M, N) : \gamma]\!]\,(\vec{d}\,)$$
$$= \begin{cases} [\![\Gamma \vdash M : \gamma]\!]\,(\vec{d}\,) & \text{if } [\![\Gamma \vdash B : \mathtt{bool}]\!]\,(\vec{d}\,) = \{\mathrm{true}\} \\ [\![\Gamma \vdash N : \gamma]\!]\,(\vec{d}\,) & \text{if } [\![\Gamma \vdash B : \mathtt{bool}]\!]\,(\vec{d}\,) = \{\mathrm{false}\} \\ \left([\![\Gamma \vdash M : \gamma]\!]\,(\vec{d}\,)\right) \cup \left([\![\Gamma \vdash N : \gamma]\!]\,(\vec{d}\,)\right) & \text{if } [\![\Gamma \vdash B : \mathtt{bool}]\!]\,(\vec{d}\,) = \{\mathrm{true}, \mathrm{false}\} \\ \bot_{D_\gamma} & \text{otherwise} \end{cases}$$

$$[\![\Gamma \vdash \mathtt{bound}_a(M) : \mathtt{real}]\!]\,(\vec{d}\,) = (\mathcal{P}^{\mathcal{S}}(\mathrm{bound}_a))\,([\![\Gamma \vdash M : \mathtt{real}]\!])$$

$$[\![\Gamma \vdash p + (M) : \mathtt{real}]\!]\,(\vec{d}\,) = (\mathcal{P}^{\mathcal{S}}(p+))\,([\![\Gamma \vdash M : \mathtt{real}]\!])$$

$$[\![\Gamma \vdash p \times (M) : \mathtt{real}]\!]\,(\vec{d}\,) = (\mathcal{P}^{\mathcal{S}}(p\times))\,([\![\Gamma \vdash M : \mathtt{real}]\!])$$

The left side of Figure 2 represents the values taken by the function ttest on the interval domain. The black wedges indicate to which area the boundary lines belong. In particular

$$\mathrm{ttest}\,([0,0]) = \mathrm{ttest}\,([1,1]) = \{\mathrm{true}, \mathrm{false}\}\,. \tag{5}$$

Similarly, the right side represents the values taken by `rtest` when seen as a function rtest from $\mathcal{R}$ to $\mathcal{P}(\mathbb{B}_\bot)$, the power set of $\mathbb{B}_\bot$, in the following way:

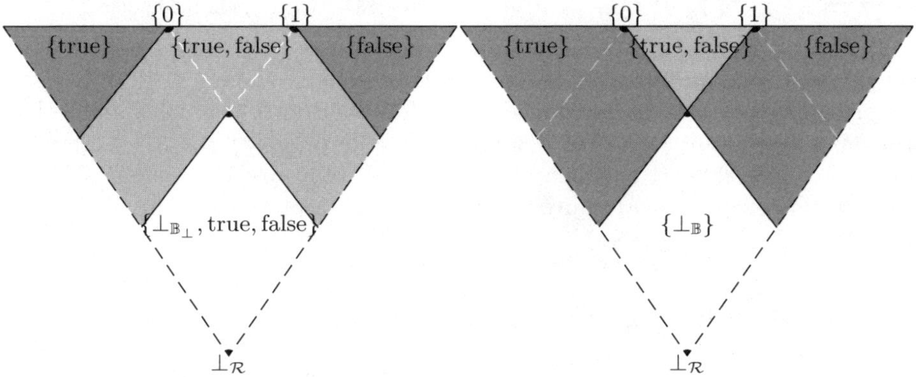

Fig. 2. Values of ttest(l) for $l \in \mathcal{R}$

given a program M: `real` such that $[\![M]\!] = \uparrow\{x\} = \{y \in \mathcal{R} \mid x \sqsubseteq y\}$ for some interval $x \in \mathcal{R}$, the figure indicates the set of outputs of the program `rtest` (M). On the areas that are of different colours from one figure to the other, we have that ttest(x) strictly contains rtest(x). In particular rtest $([0,0]) = \{\text{true}\}$ and rtest$([1,1]) = \{\text{false}\}$. It follows that for some programs M: `real`, even some computing real numbers, one has $[\![\texttt{rtest}\,(M)]\!] \sqsubset [\texttt{rtest}\,(M)]$.

Our main technical result states that adequacy between the operational and denotational semantics holds for programs of ground type with maximal denotation.

Theorem 2. *For any closed term M of ground type, $[\![M]\!] \sqsubseteq [M]$.*

In particular, if M has a maximal denotation, then equality holds. Our proof makes use of a logical relation and is adapted from the proof of adequacy for PCF as found in *e.g.* [10]. In itself, Theorem 2 does not say much about our semantics, because the semantics that assigns bottom to every term also satisfies the conclusion of the theorem. However, our model is "close enough" to the operational semantics, as we shall see in the next two sections.

4 Example

For many common first-order computable functions, it is not difficult to find programs whose denotation is total (sends maximal elements to maximal elements). We only have space to provide one example (but see conclusion). The denotation of the closed term we gave earlier (page 394) for the absolute value function is the map abs: $\mathcal{P}^{\mathcal{S}}\mathcal{R} \to \mathcal{P}^{\mathcal{S}}\mathcal{R}$ considered in the next theorem.

Theorem 3. *The following recursively defined function* abs: $\mathcal{P}^{\mathcal{S}}\mathcal{R} \to \mathcal{P}^{\mathcal{S}}\mathcal{R}$ *represents the absolute value map on real numbers:*

$$\begin{array}{rlll} \text{abs}\,(x) = \text{cases } x \le 0 & \to & -x & \\ -1 \le x & \to & \text{cases } x \le 1 & \to \ \text{bound}_{[-1,1]}\left(\frac{1}{2}\,\text{abs}(2x)\right) \\ & & \qquad 0 \le x & \to \ x \end{array}$$

where $\text{cases } \begin{array}{lll} x \leq q & \rightarrow & M_0 \\ p \leq x & \rightarrow & M_1 \end{array}$ *is a notation for* if $\text{ttest}_{p,q}(x)$ then M_0 else M_1.

Let us say that a program M *defines* a real number r if $[\![M]\!] = \{[r,r]\}$. Using the fact that $\{[r,r]\}$ is maximal in $\mathcal{P}^{\mathcal{S}}\,\mathcal{R}$, it follows from Theorems 2 and 3 that, whenever a program M : `real` defines $r \in \mathbb{R}$, the program abs(M) defines the absolute value $|r|$ of r, and hence that the program abs(M) *computes* $|r|$: all computations from abs(M) output $[|r|,|r|]$, which represents $|r|$ in the interval domain.

5 Conclusion

The undecidability of (in)equality tests on real numbers from a constructive point of view creates problems in the design of control-flow mechanisms for programming languages for exact real-number computation. A good operational solution, based on constructive analysis, has been proposed by Boehm and Cartwright [2] and developed by Marcial-Romero and Escardó [8] to some extent. A main contribution of [8] is that the Hoare power domain of the interval domain can be used to reason about partial correctness in a natural way, combining real analysis and domain theory. A main problem left open in the same work is the development of a denotational semantics that would allow for total correctness proofs. Marcial-Romero and Escardó ruled out the Smyth and Plotkin power domains for such a semantics, by proving that there is no continuous interpretation of Boehm and Cartwright's `rtest` construction in those power domains. In this work, we have shown that the best continuous approximation of `rtest` in the Smyth power domain allows one to develop total correctness proofs based on real analysis and denotational semantics, in a natural way, avoiding termination proofs based on operational semantics. Technically, our main contribution is the formulation and proof of a suitable computational adequacy theorem for the approximate semantics, which we expressed as $[\![M]\!] \sqsubseteq [M]$, for any program M of ground type.

Acknowledgement. This work is part of my PhD Thesis, under the supervision of Martín Escardó, to whom I express all my gratitude. Detailed proofs and many other examples of programs will appear in my PhD dissertation, a draft of which is already available on my web page. The dissertation also contains a universality result (all computable first-order total functions are definable in the language); this will be the subject of another paper.

References

1. Abramsky, S., Jung, A.: Domain theory. In: Abramsky, S., Gabbay, D.M., Maibaum, T.S.E. (eds.) Handbook of Logic in Computer Science, vol. 3, pp. 1–168. Clarendon Press (1994)
2. Boehm, H.J., Cartwright, R.: Exact real arithmetic: Formulating real numbers as functions. In: Research topics in functional programming, pp. 43–64. Addison-Wesley Longman Publishing Co., Inc., Boston, MA, USA (1990)

3. Escardó, M.H.: PCF extended with real numbers: A domain-theoretic approach to higher-order exact real number computation. PhD thesis, Department of Computing, Imperial College, University of London (1996)
4. Farjudian, A.: Sequentiality in Real Number Computation. PhD thesis, School of Computer Science, University of Birmingham (2004)
5. Di Gianantonio, P.: A Functional Approach to Computability on Real Numbers. PhD thesis, University of Pisa, Udine (1993)
6. Gierz, G., Hofmann, K.H., Keimel, K., Lawson, J.D., Mislove, M., Scott, D.S.: Continuous Lattices and Domains. Encyclopedia of Mathematics and its Applications, vol. 93. Cambridge University Press, Cambridge (2003)
7. Marcial-Romero, J.R.: Semantics of a Sequential Language for Exact Real-Number Computation. PhD thesis, School of Computer Science, University of Birmingham U.K. (2004)
8. Marcial-Romero, J.R., Escardó, M.H.: Semantics of a sequential language for exact real-number computation. Theoretical Computer Science 379(1-2), 120–141 (2007)
9. Plotkin, G.D.: LCF considered as a programming language. Theor. Comput. Sci. 5(3), 225–255 (1977)
10. Streicher, T.: Domain-Theoretic Foundations of Functional Programming. Imperial College Press, London (2006)

Weak Bisimulations for the Giry Monad (Extended Abstract)

Ernst-Erich Doberkat*

Chair for Software Technology
Technical University of Dortmund
doberkat@acm.org

Abstract. The existence of bisimulations for objects in the Kleisli category associated with the Giry monad of subprobabilities over Polish spaces is studied. We first investigate these morphisms and show that the problem can be reduced to the existence of bisimulations for objects in the base category of stochastic relations using simulation equivalent congruences. This leads to a criterion for two objects related through the monad to be bisimilar.

1 Introduction

The Giry monad [8] provides a categorical approach to the measure-theoretic foundations of Probability Theory. Permitting to represent a variety of models in computer Science, it has recently attracted some attention with respect to the foundations of concurrency theory (like bisimulations) [11,1,3,5], Kripke models for modal logics with applications to model checking [15,4], and even to modelling software architectures [2]. The monograph [6] provides an overview over the recent development. E. Moggi proposes in his paper *Notions of Computations and Monads* [10] a monadic model of computation. Functor $\mathbf{T}$ acts as a type constructor, so that if A is a type, then $\mathbf{T}A$ is the object of computations of type A. Assuming that $\mathbf{T}$ is the functorial part of a monad, the Kleisli category for this monad is identified as the category of programs, and morphisms in this category are the programs.

The present paper is about these morphisms in the Giry monad; they are usually referred to as *stochastic relations*. It studies the problem under which conditions two stochastic relations are bisimilar in the Kleisli category for the Giry monad. Thus, given two stochastic relations $\mathcal{K}$ and $\mathcal{L}$, we ask under which conditions we can find a stochastic relation $\mathcal{M}$ and two Kleisli morphisms $\Phi : \mathcal{K} \curvearrowright \mathcal{M}$ and $\Psi : \mathcal{L} \curvearrowright \mathcal{M}$, hence we want to know under which conditions a span of morphisms in the Kleisli category for the Giry monad exists. The strong case, i.e., the case that F and G both are replaced by Borel measurable maps (and Kleisli composition by the composition of maps) has been studied extensively in different settings [6,7], and here general criteria are known. The case of weak morphisms, however, did not find sufficient attention yet.

This problem is interesting for a variety of reasons. First, it is known that there is an intimate connection between bisimilarity and Kripke models for modal logics, not

* Research funded in part by Deutsche Forschungsgemeinschaft, grant DO 263/8-3, *Algebraische Eigenschaften stochastischer Relationen*.

M. Agrawal et al. (Eds.): TAMC 2008, LNCS 4978, pp. 400–409, 2008.

only in the case when the logics is interpreted coalgebraically through a set based functor [13] but also when the subprobability based functor is used [7]. It can be shown that the weak probabilistic case is by no means as clearly cut as the strong case, hence the interest is somewhat pronounced here. Then, bisimilarity is a basic notion in the theory of concurrency, consequently, a better knowledge about an interesting special case is welcome. From a technical point of view it is important to see that weak morphisms are closely related to strong ones, which gives some insight into the inner working of the Kleisli construction for the Giry monad.

The problem is tackled here in terms of congruences with an approach similar to the one proposed in [4]. Congruences are introduced in Section 3, after having recalled in Section 2 some familiar constructions with the Giry monad. In Section 4 we reduce the problem to finding a weak morphism to the strong case through a construction that resembles a tensor product and analyse congruences that occur in this case. We know that the existence of simulation equivalent congruences implies the existence of a bisimulation for the strong case, hence we study simulation equivalence in the context of the transformed problem. After some transformations involving the corresponding factor spaces, we show that two stochastic relations are weakly bisimilar if we can find simulation equivalent congruences on the base spaces that are compatible with the Kleisli composition. The conclusing Section 5 wraps it all up and proposes further work.

2 Some Constructions Associated with the Giry Monad

Given a measurable space $(X, \mathcal{A})$, $\mathbf{S}(X, \mathcal{A})$ is the space of all subprobability measures on $(X, \mathcal{A})$. It is again a measurable space with the $* - \sigma$-algebra $\mathcal{A}^{\bullet}$. This is the smallest σ-algebra on $\mathbf{S}(X, \mathcal{A})$ which makes all the evaluations $\mu \mapsto \mu(Q)$ with $Q \in \mathcal{A}$ measurable. Defining for another measurable space $(Y, \mathcal{B})$ and an $\mathcal{A}$-$\mathcal{B}$-measurable map $f : X \to Y$ the map $\mathbf{S}(f) \to \mathbf{S}(Y)$ through $\mathbf{S}(f)(\mu)(B) := \mu(f^{-1}[B])$ constitutes a $\mathcal{A}^{\bullet}$-$\mathcal{B}^{\bullet}$-measurable map. The subprobability functor $\mathbf{S}$ is hence an endofunctor on the category of measurable spaces. This functor is the functorial part of a monad, the *Giry monad* [8]. The *Kleisli construction* [9] for this monad is particularly interesting: If $(X, \mathcal{A})$ and $(Y, \mathcal{B})$ are measurable spaces, then a stochastic relation $K : (X, \mathcal{A}) \rightsquigarrow (Y, \mathcal{B})$ is a map $K : X \to \mathbf{S}(Y, \mathcal{B})$ which is $\mathcal{A}$-$\mathcal{B}^{\bullet}$-measurable, stochastic relations are just the Kleisli morphisms for the Giry monad. The reader may wish to compare this with Kleisli morphisms for the powerset monad: they are exactly the relations, and in fact there are many interesting and not so obvious similarities, see, e.g. [6]. In Probability Theory, stochastic relations are known as subprobability kernels, the algebraic context is usually not being considered there.

The *Kleisli composition* of the stochastic relations $K : (X, \mathcal{A}) \rightsquigarrow (Y, \mathcal{B})$ and $L : (Y, \mathcal{B}) \rightsquigarrow (Z, \mathcal{C})$ (for some measurable space $(Z, \mathcal{C})$) is the stochastic relation $L{*}K : (X, \mathcal{A}) \rightsquigarrow (Z, \mathcal{C})$ defined through $(x \in X, G \in \mathcal{C})$

$$(L{*}K)(x)(G) := \int_Y L(y)(G)\ K(x)(dy).$$

In Probability Theory, this operation is usually referred to as the *convolution of the kernels K and L*.

Define the coproduct $\mathcal{K}+\mathcal{K}'$ of the stochastic relations $K = ((X, \mathcal{A}), (Y, \mathcal{B}), K)$ and $K' = ((X', \mathcal{A}'), (Y', \mathcal{B}'), K')$ through $\big((X+X', \mathcal{A}+\mathcal{A}'), (Y+Y', \mathcal{B}+\mathcal{B}'), K+K'\big)$, where

$$(K + K')(z) := \begin{cases} \big(\mathbf{S}\,(\mathrm{inj}_Y) \circ K\big)(x), & \text{if } z = \mathrm{inj}_X(x), \\ \big(\mathbf{S}\,(\mathrm{inj}_{Y'}) \circ K'\big)(x'), & \text{if } z = \mathrm{inj}_{X'}(x'). \end{cases}$$

Here e.g., inj_Y is the injection $Y \to Y + Y'$ from Y into the sum $Y + Y'$. Thus we have e.g. for $x \in X$ and the measurable set $T \subseteq Y + Y'$ the identity $\big(K + K'\big)(\mathrm{inj}_X(x))(T) = K(x)(T \cap Y)$. It is not difficult to see that this defines the coproduct in the category of stochastic relations, anticipating the definition of strong morphisms from Definition 3.

If X is a *Polish space*, i.e., a Hausdorff space with a countable dense subset for which a complete metric exists, then X becomes a measurable space with the *Borel sets* $\mathcal{B}(X)$ as a σ-algebra. $\mathcal{B}(X)$ is the smallest σ-algebra that contains the open (or, equivalently, the closed) sets of X. We suppress $\mathcal{B}(X)$ usually from the notation of a Polish space X. Then $\mathbf{S}\,(X)$ is a Polish space as well [12], taking the topology of weak convergence as a topology, and it is well known that the $* - \sigma$-algebra are just the Borel sets for this topology.

Let again $(X, \mathcal{A})$ and $(Y, \mathcal{B})$ be measurable spaces, and denote by $\mathcal{A} \otimes \mathcal{B}$ the product σ-algebra on $X \times Y$. Define for a stochastic relation $K : (X, \mathcal{A}) \rightsquigarrow (Y, \mathcal{B})$ and for a subprobability $\mu \in \mathbf{S}\,(Y, \mathcal{B})$ on $(X \times Y, \mathcal{A} \otimes \mathcal{B})$ the measure

$$\big(\mu \otimes K\big)(H) := \int_X K(x)(H_x)\, \mu(dx)$$

(here $H_x := \{y \in Y \mid \langle x, y \rangle \in H\}$ is the vertical cut of H at $x \in X$). It is folklore in Probability Theory that this defines a measure on the product space. The following Lemma will be helpful in expressing the Kleisli product in a sometimes more convenient manner.

Lemma 1. *$x \mapsto K(x) \otimes L$ defines a stochastic relation $K(\cdot) \otimes L : (X, \mathcal{A}) \rightsquigarrow (Y \times Z, \mathcal{B} \otimes \mathcal{C})$ such that $\big(L{*}K\big)(x)(G) = \mathbf{S}\,(\pi_{Y \times Z, Z})\,(K(x) \otimes L)(G)$ for all $G \in \mathcal{C}$.* □

Now let X and Y be Polish spaces, and assume that a measure $\mu \in \mathbf{S}\,(X \times Y)$ is given. Then it is well known that μ can be represented through a measure on X and a stochastic relation $K : X \rightsquigarrow Y$ (see [12, Theorem V.8.1] or [6, Section 1.5.3]):

Theorem 1. *Given $\mu \in \mathbf{S}\,(X \times Y)$, then $\mu = \mathbf{S}\,(\pi_{X \times Y, X})\,(\mu) \otimes K$ for some stochastic relations $K : X \rightsquigarrow Y$. K is unique $\mathbf{S}\,(\pi_{X \times Y, X})\,(\mu)$-almost everywhere. K is called a* disintegration *of μ with respect to X, Y.* □

Thus, whenever $\mu \in \mathbf{S}\,(X \times Y)$ and $H \in \mathcal{B}(X \times Y)$, we can write

$$\mu(H) = \int_X K(x)(H_x)\, \mathbf{S}\,(\pi_{X \times Y, X})\,(\mu)(dx).$$

The Change of Variables Formula will be a helpful technical tool. It uses the basic fact stated above that a measurable map between measurable spaces induces a measurable map between the corresponding subprobabilities. We formulate it for convenience for Polish spaces, but it holds for general measurable spaces as well.

Lemma 2. *Let X and Y be Polish spaces, and assume that $f : X \to Y$ is Borel measurable. Then*

$$\int_B g(y)\, \mathbf{S}(f)(\mu)(dy) = \int_{f^{-1}[B]} (g \circ f)(x)\, \mu(dx)$$

whenever $B \subseteq Y$ is a Borel set. □

3 Smooth Relations

If $(X, \mathcal{A})$ is a measurable space, then an equivalence relation ϱ on X is called *smooth* (or *countably generated*) iff there exists a sequence $(Q_n)_{n\in\mathbb{N}}$ of sets in $\mathcal{A}$ such that

$$x\,\varrho\,x' \text{ iff } \forall n \in \mathbb{N} : [x \in Q_n \Leftrightarrow x' \in Q_n].$$

The sequence $(Q_n)_{n\in\mathbb{N}}$ is said to *determine* ϱ. For example, when one constructs an equivalence relation from the sets on which the formulas of a modal logic are valid [4], these sets then form a determining sequence. In this case two states are equivalent iff they cannot be separated by the logic.

Concerning the measurable structure of the factor space, it is well known that if X is Polish, then the factor space X/ϱ is analytic (i.e., homeomorphis to a continuous image of a Polish space), whenever ϱ is smooth, provided the factor carries the largest σ-algebra that makes the factor map $\mathfrak{e}_\varrho : x \mapsto [x]_\varrho$ $\mathcal{B}(X)$-measurable [14, Exercise 5.1.14]. As a rule, the factor spaces are analytic spaces, even if the base spaces are Polish, see [7, Example 2.7] for a discussion.

Call a set $B \subseteq X$ *ϱ-invariant* iff $B = \bigcup\{[x]_\varrho \mid x \in B\}$, so that $x \in B$ and $x\,\varrho\,x'$ implies $x' \in B$. The *ϱ-invariant $\mathcal{A}$-measurable sets* $\mathcal{INV}(\mathcal{A}, \varrho) := \{C \in \mathcal{A} \mid C$ is ϱ-invariant$\}$ form a σ-algebra for a measurable space $(X, \mathcal{A})$. This is a collection of results for smooth equivalence relations that will be used silently throughout.

Lemma 3. *Let X be a Polish space, ϱ be a smooth equivalence relation with the Borel sets $(Q_n)_{n\in\mathbb{N}}$ as determining sequence. Then*

a. $\mathcal{INV}(\mathcal{B}(X), \varrho) = \sigma(\{Q_n \mid n \in \mathbb{N}\})$,
b. $C \in \mathcal{B}(X/\varrho)$ *iff* $\mathfrak{e}_\varrho^{-1}[C] \in \mathcal{INV}(\mathcal{B}(X), \varrho)$, *and* $\mathfrak{e}_\varrho^{-1}[\mathfrak{e}_\varrho[A]] = A$ *for each* $A \in \mathcal{INV}(\mathcal{B}(X), \varrho)$,
c. $x\,\varrho\,x'$ *iff* $[x \in Q \Leftrightarrow x' \in Q]$ *holds for all* $Q \in \mathcal{INV}(\mathcal{B}(X), \varrho)$.
d. The equivalence classes $[x]_\varrho$ *are exactly the atoms of the σ-algebra* $\mathcal{INV}(\mathcal{B}(X), \varrho)$.

As a consequence of the characterization for the atoms of $\mathcal{INV}(\mathcal{B}(X), \varrho)$ we note that the $\mathcal{INV}(\mathcal{B}(X), \varrho)$-measurable real functions are constant on the equivalence classes.

Corollary 1. *Let X and ϱ be as in Lemma 3. If $f : X \to \mathbb{R}$ is $\mathcal{INV}(\mathcal{B}(X), \varrho)$-measurable, then $\varrho \subseteq \mathsf{ker}(f)$.* □

A pair (α, β) of smooth equivalence relations α on X and β on Y (with both X and Y Polish) is called a *congruence* for the stochastic relation $K : X \rightsquigarrow Y$ iff $K(x)(B) = K(x')(B)$ whenever $x\,\alpha\,x'$ and $B \subseteq Y$ is a β-invariant Borel set. Thus if α cannot distinguish x from x', and if β cannot discern the elements of B, then $K(x)$ and $K(x')$ both assign the same probability to B. Given a congruence (α, β) for $K : X \rightsquigarrow Y$, one constructs the factor relation $K_{(\alpha,\beta)} : X/\alpha \rightsquigarrow Y/\beta$ upon setting

$$K_{(\alpha,\beta)}([x]_\alpha)(G) := K(x)(\mathfrak{e}_\beta^{-1}[G]) = (\mathbf{S}(\mathfrak{e}_\beta) \circ K)(x)(G)$$

for $x \in X, G \in \mathcal{B}(Y/\beta)$. This is a characterization of congruences in terms of the relation involved [5, Lemma 2.8]:

Proposition 1. *The following statements are equivalent for the Polish spaces X and Y*

a. (α, β) *is a congruence for* $K : X \rightsquigarrow Y$.
b. $K : \mathcal{INV}(\mathcal{B}(X), \alpha) \to \mathcal{INV}(\mathcal{B}(Y), \beta)$ *is a stochastic relation.* □

For the investigation of bisimilarity, the notion of simulation equivalent congruences will be important. A necessary condition for the bisimilarity of stochastic relations is the existence of equivalent congruences on them. This condition is technically a bit involved because it requires the notion of the mutual generation of smooth equivalence relations; this concept is called *spawning*, it is discussed in detail in [4], so is the closely related concept of simulation equivalent congruences.

Lemma 4. *Let α and β be smooth equivalence relations on the Polish spaces X resp. Y, and assume that $\mathfrak{p} : X/\alpha \to Y/\beta$ is a map between the equivalence classes. Put $\sqrt{\mathfrak{p}}(A) := \bigcup\{\mathfrak{p}([x]_\alpha) \mid x \in A\}$ for $A \subseteq X$. Then $\sqrt{\mathfrak{p}} : \mathcal{INV}(\mathcal{B}(X), \alpha) \to \mathcal{INV}(\mathcal{B}(Y), \beta)$. If $\mathfrak{p}$ is injective, then $\sqrt{\mathfrak{p}}$ is a Boolean σ-morphism.* □

It is useful to notice that $\sqrt{\mathfrak{p}}(A) = \mathfrak{e}_\beta^{-1}[\mathfrak{p}[\mathfrak{e}_\alpha[A]]]$.

Definition 1. *Let α and β be smooth equivalence relations on the Polish spaces X resp. Y, and assume that $\mathfrak{p} : X/\alpha \to Y/\beta$ is an injective map between the equivalence classes. We say that α* spawns *β* via *$(\mathfrak{p}, \mathcal{A}_0)$ iff $\mathcal{A}_0$ is a countable generator $\mathcal{INV}(\mathcal{B}(X), \alpha)$ such that $\{\sqrt{\mathfrak{p}}(A) | A \in \mathcal{A}_0\}$ is a generator of $\mathcal{INV}(\mathcal{B}(Y), \beta)$.*

Thus if α spawns β, then the measurable structure induced by α on X is all we need for constructing the measurable structure induced by β on Y: the map $\mathfrak{p}$ can be made to carry over the generator $\mathcal{A}_0$ from $\mathcal{INV}(\mathcal{B}(X), \alpha)$ to $\mathcal{INV}(\mathcal{B}(Y), \beta)$ and to transport the atoms from one σ-algebra to the other. This is of particular interest since the atoms constitute the equivalence classes. It can be shown that the specific generator in question is not important because the map between the classes is injective, so that any generator will do.

Definition 2. *Let $K : X \rightsquigarrow Y$ and $K' : X' \rightsquigarrow Y'$ be stochastic relations over analytic spaces on which congruences (α, β) resp. (α', β') are defined.*

a. Call (α, β) proportional to *(α', β') (written as $(\alpha, \beta) \propto (\alpha', \beta')$) iff α spawns α' via $(\mathfrak{p}, \mathcal{A}_0)$, β spawns β' via $(\mathfrak{q}, \mathcal{B}_0)$ such that both $\mathcal{A}_0$ and $\mathcal{B}_0$ are closed under finite intersections and*

$$\forall x \in X \forall x' \in \mathfrak{p}([x]_\alpha) \forall B \in \mathcal{B}_0 : K(x)(B) = K'(x')(\sqrt{\mathfrak{q}}(B)).$$

b. *Call these congruences* simulation equivalent *iff both* $(\alpha, \beta) \propto (\alpha', \beta')$ *and* $(\alpha', \beta') \propto (\alpha, \beta)$

We require the generators to be closed under finite intersections since this is — by the famous π-λ-Theorem of Measure Theory [6, Theorem 1.1] — a helpful condition to ensure uniqueness of a measure. Simulation equivalent congruences behave in identical fashion on the class structure induced by the respective congruences.

If α and α' are smooth equivalence relations on X resp. X', define for $\bar{x}, \bar{x}' \in X+X'$ the equivalence relation $\alpha + \alpha'$ on the coproduct $X + X'$ through $[\mathrm{inj}_X(x)]_{\alpha+\alpha'} := [x]_\alpha$ whenever $x \in X$, and $[\mathrm{inj}_{X'}(x')]_{\alpha+\alpha'} := [x']_{\alpha'}$, whenever $x' \in X'$. Thus $(X + X')/(\alpha + \alpha')$ is isomorphic to $X/\alpha + X'/\alpha'$, hence both spaces will be identified. Simulation equivalence is preserved through coproducts. The proof is fairly straighforward.

Lemma 5. *Let $\mathcal{K}$ and $\mathcal{L}$ be stochastic relations with simulation equivalent congruences (χ, α) and (υ, β) and assume that $\mathcal{K}'$ and $\mathcal{L}'$ are stochastic relations with simulation equivalent congruences (χ', α') and (υ', β'). Then $(\chi+\chi', \alpha+\alpha')$ and $(\upsilon+\upsilon', \beta+\beta')$ are simulation equivalent congruences for $\mathcal{K} + \mathcal{K}'$ and $\mathcal{L} + \mathcal{L}'$.* □

4 Weak Bisimilarity

We will define weak and strong morphisms now in order to be able to define bisimilarity. This property is introduced through a span of morphisms, thus we take here a coalgebraic point of view. In what follows, all spaces are Polish.

Definition 3. *Let $\mathcal{K} = (X, Y, K)$ and $\mathcal{L} = (A, B, L)$ be stochastic relations, then a pair $(f, g) : \mathcal{K} \to \mathcal{L}$ is called a* strong morphism *iff $f : X \to A$ and $g : Y \to B$ are surjective Borel maps, such that $L \circ f = \mathbf{S}(g) \circ K$.*

Thus strong morphisms are based on measurable Borel maps. In contrast, weak morphisms are based on Kleisli morphisms.

Definition 4. *Let $\mathcal{K} = (X, Y, K)$ and $\mathcal{L} = (A, B, L)$ be stochastic relations, then a pair $(F, G) : \mathcal{K} \curvearrowright \mathcal{L}$ is called a* weak morphism *iff $F : X \rightsquigarrow A$ and $G : Y \rightsquigarrow B$ are Kleisli morphisms with $L{*}F = G{*}K$.*

The condition $L{*}F = G{*}K$ entails for each $x \in X$ and each Borel set $P \in \mathcal{B}(B)$

$$(L{*}F)(x)(P) = \int_A L(a)(P)\ F(x)(da) = \int_Y G(y)(P)\ K(x)(dy) = (G{*}K)(x)(P)$$

If $(f, g) : \mathcal{K} \to \mathcal{L}$ is a strong morphisms, then $L(f(x))(P) = K(x)(g^{-1}[P])$ holds. Clearly, strong morphisms are special cases of weak ones. This is so because $\delta_f : x \mapsto \delta_{f(x)}$ is a stochastic relation $X \rightsquigarrow Y$, whenever $f : X \to Y$ is a Borel map, and because $\int_Y h(y)\ \delta_f(x)(dy) = h(f(x))$, whenever $f : Y \to \mathbb{R}$ is a measurable map. Some properties of weak morphisms are discussed in [5].

The coproduct permits to reduce co-spans of weak morphisms to weak morphisms.

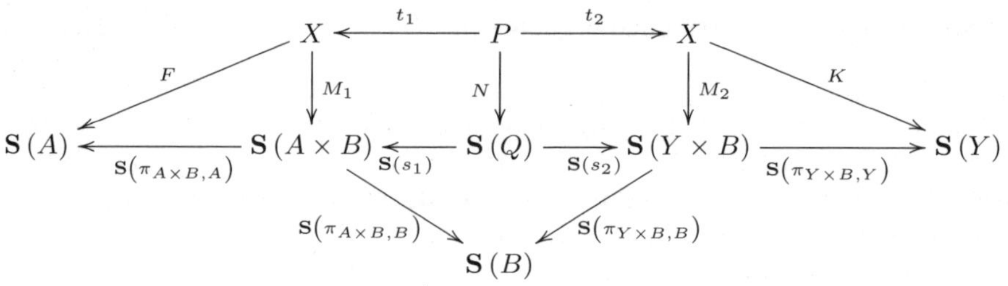

Fig. 1. Derived diagram for the proof of Proposition 3

Lemma 6. *Let $(F, G) : \mathcal{K} \curvearrowright \mathcal{N}, (F', G') : \mathcal{K}' \curvearrowright \mathcal{N}$ be a span of weak morphisms. Then $(F + F', G + G') : \mathcal{K} + \mathcal{K}' \curvearrowright \mathcal{N} + \mathcal{N}$ constitutes a weak morphism.* □

Proposition 2. *A weak morphism $(F, G) : (X, Y, K) \curvearrowright (A, B, L)$ is characterized through the stochastic relations $M_1 : X \rightsquigarrow A \times B$ and $M_2 : X \rightsquigarrow Y \times B$ with these properties:*

a. $\mathbf{S}(\pi_{A\times B,B}) \circ M_1 = \mathbf{S}(\pi_{Y\times B,B}) \circ M_2$.
b. The disintegration of $M_1(x)$ with respect to A, B is independent of $x \in X$ and coincides with L.
c. $K = \mathbf{S}(\pi_{Y\times B,Y}) \circ M_2$.
d. The disintegration of $M_2(x)$ with respect to Y, B is independent of $x \in X$.

Then G is the disintegration of $M_2(x)$ with respect to Y, B, and $F = \mathbf{S}(\pi_{A\times B,A}) \circ M_1$. □

Using this construction, we derive a first criterion for the weak bisimilarity of two stochastic relations.

Proposition 3. *Construct $M_1 : X \rightsquigarrow A \times B$ and $M_2 : X \rightsquigarrow Y \times B$ from the weak morphism $(F, G) : (X, Y, K) \curvearrowright (A, B, L)$. If M_1 and M_2 are strongly bisimilar, then K and L are weakly bisimilar.*

Proof. The construction yields the diagram of commuting maps displayed in Fig. 1 for a suitable stochastic relation $N : P \rightsquigarrow Q$. Chasing L (which is hidden through disintegration in M_1) and K, the assertion is easily established. □

This reduction result encourages to look for conditions that ensure the derived relations M_1 and M_2 being strongly bisimilar. For this, fix smooth equivalence relations α on A, β on B, χ on X and υ on Y and a weak morphism $(F, G) : (X, Y, K) \curvearrowright (A, B, L)$ such that (χ, υ) is a congruence for K, (α, β) is a congruence for L, (χ, α) is a congruence for F and finally, (υ, β) is a congruence for G. We assume that (χ, α) and (χ, υ) are simulation equivalent such that $\mathfrak{p} : A/\alpha \to Y/\upsilon$ is a bijection with inverse $\mathfrak{q}$. The maps $\mathfrak{p}$ and $\mathfrak{q}$ are used to transport the structures to and fro in the sense of Definition 1 with $\mathcal{A}_0$ and $\mathcal{Y}_0$ as the respective $\cap$-stable generators for $\mathcal{INV}(\mathcal{B}(A), \alpha)$ and $\mathcal{INV}(\mathcal{B}(Y), \upsilon)$. We assume that the identity on X/χ maps the factor structure onto itself so that $K(x)(W_0) = F(x')(\sqrt{\mathfrak{q}}(W_0))$ holds for every $W_0 \in \mathcal{Y}_0$ whenever $x \,\chi\, x'$. Recall from both $\mathcal{A}_0$ and $\mathcal{Y}_0$ are closed under finite intersections, see Definition 2.

Define the product $\alpha \times \beta$ of the equivalence relations α and β through $\langle a, b\rangle\langle a', b'\rangle$ $(\alpha \times \beta)$ iff $a\,\alpha\,a'$ and $b\,\beta\,b'$, then $\alpha\times\beta$ and, similarly, $\upsilon\times\beta$ are smooth equivalence relations on $A\times B$ resp. $Y\times B$ such that e.g. $\mathcal{INV}\,(\mathcal{B}(A\times B), \alpha\times\beta) = \mathcal{INV}\,(\mathcal{B}(A), \alpha)\otimes \mathcal{INV}\,(\mathcal{B}(B), \beta)$, see [6, Lemma 5.13]. Let $\mathcal{B}_0$ be a $\cap$-stable countable generator of $\mathcal{INV}\,(\mathcal{B}(B), \beta)$, then $\{A_0 \times B_0 \mid \mathcal{A}_0 \in A_0, B_0 \in \mathcal{B}_0\}$ is a countable and $\cap$-stable generator of $\mathcal{INV}\,(\mathcal{B}(A\times B), \alpha\times\beta)$. A similar characterizsation holds for the σ-algebra $\mathcal{INV}\,(\mathcal{B}(Y\times B), \upsilon\times\beta)$.

We claim that $(\chi, \alpha\times\beta)$ is simulation equivalent to $(\chi, \upsilon\times\beta)$. It suffices to demonstrate that these equations hold: $M_2(x)(\sqrt{\mathfrak{p}}(A_0) \times B_0) = M_1(x')(A_0 \times B_0)$, and $M_1(x)(\sqrt{\mathfrak{q}}(Y_0) \times B_0) = M_2(x')(Y_0 \times B_0)$, whenever $A_0 \in \mathcal{A}_0, B_0 \in \mathcal{B}_0, Y_0 \in \mathcal{Y}_0$ and $x, x' \in X$ with $x\,\chi\,x'$ are given. It is enough to establish the first equality, since the maps $\mathfrak{p}$ and $\mathfrak{q}$ are symmetric, the second one will follow in the same way.

The proof is broken into several steps. The first step shows that

$$M_2(x)(\sqrt{\mathfrak{p}}(A_0) \times B_0) = \int_{\mathfrak{e}_\upsilon[\sqrt{\mathfrak{p}}(A_0)]} G_{(\upsilon,\beta)}(t)(\mathfrak{e}_\beta\,[B_0])\; K_{(\chi,\upsilon)}([x]_\chi)(dt).$$

Because (υ, β) is a congruence for G, we know from Proposition1 that $y \mapsto G(y)(B_0)$ is $\mathcal{INV}\,(\mathcal{B}(Y), \upsilon)$-measurable, since $B_0 \in \mathcal{INV}\,(\mathcal{B}(B), \beta)$. From Corollary 1 we infer that this map is constant on the atoms of $\mathcal{INV}\,(\mathcal{B}(Y), \upsilon)$, which are just the equivalence classes, thus we may conclude that $G(y)(B_0) = \big(G_{(\upsilon,\beta)} \circ \mathfrak{e}_\upsilon\big)(y)(\mathfrak{e}_\upsilon\,[B_0])$. Apply the Change of Variables Formula, and observe that for a Borel set $T \in \mathcal{B}(Y/\upsilon)$ the equality $\mathfrak{e}_\upsilon^{-1}\,[T] = \sqrt{\mathfrak{p}}(A_0)$ is equivalent to $T = \mathfrak{e}_\upsilon\,[\sqrt{\mathfrak{p}}(A_0)]$, because $\sqrt{\mathfrak{p}}(A_0) \in \mathcal{INV}\,(\mathcal{B}(Y), \upsilon)$. Recalling the construction of the factor relation, it is evident that the measures $\mathbf{S}\,(\mathfrak{e}_\upsilon)\,(K(x))$ and $K_{(\chi,\upsilon)}([x]_\chi)$ coincide on $\mathcal{B}(Y/\upsilon)$, yielding the equation.

Because (α, β) and (υ, β) are simulation equivalent, we may conclude that $G_{(\upsilon,\beta)}(t)(\mathfrak{e}_\beta\,[B_0]) = L_{(\alpha,\beta)}(\mathfrak{q}(t))(\mathfrak{e}_\beta\,[B_0])$. Using change of variables, and observing $\mathfrak{q}^{-1}\,[\mathfrak{q}\,[\mathfrak{e}_\upsilon\,[\sqrt{\mathfrak{p}}(A_0)]]] = \mathfrak{e}_\upsilon\,[\sqrt{\mathfrak{p}}(A_0)]$, because $\mathfrak{q}$ is injective, we continue and show in a second step

$$M_2(x)(\sqrt{\mathfrak{p}}(A_0) \times B_0) = \int_{\mathfrak{q}[\mathfrak{e}_\upsilon[\sqrt{\mathfrak{p}}(A_0)]]} L_{(\alpha,\beta)}(s)(\mathfrak{e}_\beta\,[B_0])\; \mathbf{S}\,(\mathfrak{q})\,\big(K_{(\chi,\upsilon)}([x]_\chi)\big)(ds)$$

Now let $G \in \mathcal{B}(A/\alpha)$, then $\mathbf{S}\,(\mathfrak{q})\,\big(K_{(\chi,\upsilon)}([x]_\chi)\big)(G) = \mathbf{S}\,(\mathfrak{e}_\alpha)\,\big(F(x')\big)(G)$ This is derived from the definition of the factor relation, from the assumption that $x\,\chi\,x'$, and from the assumption that $\mathfrak{q}$ is injective, since $\sqrt{\mathfrak{q}}\circ\mathfrak{e}_\upsilon^{-1}\circ\mathfrak{q}^{-1} = \big(\mathfrak{e}_\alpha^{-1}\circ\mathfrak{q}\circ\mathfrak{e}_\upsilon\big)\circ\mathfrak{e}_\upsilon^{-1}\circ\mathfrak{q}^{-1} = \mathfrak{e}_\alpha^{-1}$. From $\mathfrak{e}_\alpha^{-1} \circ \mathfrak{q} \circ \mathfrak{e}_\upsilon \circ \mathfrak{e}_\upsilon^{-1} \circ \mathfrak{p} \circ \mathfrak{e}_\alpha = id$ we obtain

$$\begin{aligned}
M_2(x)(\sqrt{\mathfrak{p}}(A_0) \times B_0) &= \int_{\mathfrak{q}[\mathfrak{e}_\upsilon[\sqrt{\mathfrak{p}}(A_0)]]} L_{(\alpha,\beta)}(s)(\mathfrak{e}_\beta\,[B_0])\; \mathbf{S}\,(\mathfrak{e}_\alpha)\,\big(F(x')\big)(ds)\\
&= \int_{\mathfrak{e}_\alpha^{-1}[\mathfrak{q}[\mathfrak{e}_\upsilon[\sqrt{\mathfrak{p}}(A_0)]]]} L_{(\alpha,\beta)}(\mathfrak{e}_\alpha(a))(\mathfrak{e}_\beta\,[B_0])\; F(x')(da)\\
&= \int_{A_0} L(a)(B_0)\; F(x')(da)\\
&= M_1(x')(A_0 \times B_0),
\end{aligned}$$

as desired. As an aside, it is noted that the proof does not use the assumption that (F, G) is a weak morphism.

The considerations above establish the main result now.

Theorem 2. *Let (χ, υ) and (α, β) be congruences for the stochastic relations $\mathcal{K} = (X, Y, K)$ resp. $\mathcal{L} = (A, B, L)$ and assume that there exists a weak morphism $(F, G) : \mathcal{K} \curvearrowright \mathcal{L}$ such that (χ, α) and (υ, β) are simulation equivalent congruences for (X, A, F) resp. (Y, B, G). Then $\mathcal{K}$ and $\mathcal{L}$ are weakly bisimilar.*

Proof. Define the relations $M_1 : X \rightsquigarrow A \times B$ through $M_1(x) := F(x) \otimes L$ resp. $M_2 : X \rightsquigarrow Y \times B$ through $M_2(x) := K(x) \otimes G$. The discussion above shows that the congruences $(\chi, \alpha \times \beta)$ and $(\chi, \upsilon \times \beta)$ are simulation equivalent. Thus we obtain from the general criterion in [4, Theorem 4.8] that M_1 and M_2 are strongly bisimilar. From Proposition 3we infer the claim. □

In the category of stochastic relations with strong morphisms, a bisimulation is constructed through a cospan. This construction is helpful, e.g., when showing that Kripke models for a modal logic are bisimilar, provided they are behavioral equivalent [6, 6.2.3]. Thus the question arises whether a similar construction is possible as well when strong morphisms are replaced by weak ones. We did reduce a cospan of weak morphisms to a weak morphism on a coproduct in Lemma 6, and this construction will be helpful now for answering the question above.

Proposition 4. *Let $(F, G) : \mathcal{K} \curvearrowright \mathcal{N}$ and $(F', G') : \mathcal{K}' \curvearrowright \mathcal{N}$ be a cospan of weak morphisms, and assume that (χ, υ) and (χ', υ') is a congruence on $\mathcal{K}$ resp. $\mathcal{K}'$, and (α, β) is a congruence on $\mathcal{N}$ such that (χ, α) is simulation equivalent to (υ, β), (χ', α) is simulation equivalent to (υ', β). Then $\mathcal{K}$ and $\mathcal{K}'$ are weakly bisimilar.*

Proof. We infer from Lemma 5 that $(\chi + \chi', \alpha + \alpha)$ is simulation equivalent to $(\upsilon + \upsilon', \beta + \beta')$, so we find through Theorem 2 a stochastic relation $\mathcal{M} = (C, D, M)$ and weak morphisms $(I, J) : \mathcal{M} \curvearrowright \mathcal{K} + \mathcal{K}'$ and $(I', J') : \mathcal{M} \curvearrowright \mathcal{N} + \mathcal{N}$ (using the notation from Theorem 2 with $\mathcal{N} = (A, B, C)$). Discarding the latter morphisms, we define the stochastic relation $I_1 : C \rightsquigarrow X$ through $I_1(c)(G) := I(c)(G \cap X)$ with $G \in \mathcal{B}(X)$; the relations $I_2 : C \rightsquigarrow X'$, $J_1 : D \rightsquigarrow Y$ and $J_2 : D \rightsquigarrow Y'$ are defined similarly. An easy integration argument shows that $K' * I_2 = J_2 * M$, holds. □

Thus we have a general criterion for two stochastic relations being weakly bisimilar. It is technically more involved than the corresponding criterion for strong morphisms (which simply states that two relations are strongly bisimilar if there exists simulation equivalent congruences for them). Both statements have in common that they do not need an external instance (like a modal logic) for a decision of whether or not they are bisimilar, and both use the existence of a cospan, albeit in very different ways.

5 Conclusion

We derive a criterion for weak bisimilarity in the Giry monad. Technically, this was done through a reduction argument together with an appliction of a general criterion for strong bisimilarity. This shows a close relationship between Kleisli morphisms and morphisms in the base category which has not yet been sufficiently exploited in this special case or in the general case, as it seems. Looking again into tensored categories [1] for this purpose may be useful.

References

1. Abramsky, S., Blute, R., Panangaden, P.: Nuclear and trace ideal in tensored *-categories. J. Pure Appl. Alg. 143(1–3), 3–47 (1999)
2. Doberkat, E.-E.: Pipelines: Modelling a software architecture through relations. Acta Informatica 40, 37–79 (2003)
3. Doberkat, E.-E.: Eilenberg-Moore algebras for stochastic relations. Information and Computation 204, 1756–1781 (2006)
4. Doberkat, E.-E.: Stochastic relations: congruences, bisimulations and the Hennessy-Milner theorem. SIAM J. Computing 35(3), 590–626 (2006)
5. Doberkat, E.-E.: Kleisli morphisms and randomized congruences for the Giry monad. J. Pure Appl. Alg. 211, 638–664 (2007)
6. Doberkat, E.-E.: Stochastic Relations. Foundations for Markov Transition Systems. Chapman & Hall/CRC Press, Boca Raton, New York (2007)
7. Doberkat, E.-E., Schubert, C.: Coalgebraic logic for stochastic right coalgebras. Technical report, Chair for Software-Technology, University of Dortmund (September 2007)
8. Giry, M.: A categorical approach to probability theory. In: Categorical Aspects of Topology and Analysis, J. Pure Appl. Alg., 915, 68–85. (1981)
9. Mac Lane, S.: Categories for the Working Mathematician. In: Graduate Texts in Mathematics, Springer, Berlin (1997)
10. Moggi, E.: Notions of computation and monads. Information and Computation 93, 55–92 (1991)
11. Panangaden, P.: Probabilistic relations. In: Baier, C., Huth, M., Kwiatkowska, M., Ryan, M. (eds.) Proc. PROBMIV, pp. 59–74 (1998)
12. Parthasarathy, K.R.: Probability Measures on Metric Spaces. Academic Press, New York (1967)
13. Schröder, L., Pattinson, D.: Modular Algorithms for Heterogeneous Modal Logics. In: Arge, L., Cachin, C., Jurdziński, T., Tarlecki, A. (eds.) ICALP 2007. LNCS, vol. 4596, pp. 459–471. Springer, Heidelberg (2007)
14. Srivastava, S.M.: A Course on Borel Sets. In: Graduate Texts in Mathematics, Springer, Berlin (1998)
15. Zhou, C.: Complete Deductive Systems for Probabilistic Logic with Application to Harsany Type spaces. PhD thesis, Department of Mathematics, University of Indiana (2007)

Approximating Border Length for DNA Microarray Synthesis

Cindy Y. Li[1], Prudence W.H. Wong[1], Qin Xin[2], and Fencol C.C. Yung[3]

[1] Department of Computer Science, University of Liverpool, UK
{cindyli,pwong}@liverpool.ac.uk
[2] Simula Research Lab, Norway
xin@simula.no
[3] Department of Computer Science, University of Hong Kong, Hong Kong
ccyung@graduate.hku.hk

Abstract. We study the border minimization problem (BMP), which arises in microarray synthesis to place and embed probes in the array. The synthesis is based on a light-directed chemical process in which unintended illumination may contaminate the quality of the experiments. Border length is a measure of the amount of unintended illumination and the objective of BMP is to find a placement and embedding of probes such that the border length is minimized. The problem is believed to be NP-hard. In this paper we show that BMP admits an $O(\sqrt{n}\log^2 n)$-approximation, where n is the number of probes to be synthesized. In the case where the placement is given in advance, we show that the problem is $O(\log^2 n)$-approximable. We also study a related problem called agreement maximization problem (AMP). In contrast to BMP, we show that AMP admits a constant approximation even when placement is not given in advance.

1 Introduction

DNA microarrays [9] have become a very important research tool which have proved to benefit areas including gene discovery, disease diagnosis, and multi-virus discovery. They are used for performing a large number of hybridization experiments simultaneously. Besides their prevalent use to measure the amount of gene expression [21] in a cell, microarray is an efficient tool for making a qualitative statement about the presence or absence of biological target sequences in a sample. A DNA microarray ("chip") is a plastic or glass slide which consists of thousands of (about 60,000) short DNA sequences known as *probes*. DNA microarray design raises a number of challenging combinatorial problems, such as probe selection [10,14,18,22], deposition sequence design [17,19] and probe placement and synthesis [12,3,4,5,15,16]. In this paper, we focus on the probe placement and synthesis problem.

Probes are synthesized on the microarray through the process called *very large-scale immobilized polymer synthesis* (VLSIPS) [8]. In each step, light is selectively allowed through a *mask* to expose *spots* in the microarray in order to activate the nucleotides in the spots. The patterns of the masks used and the sequence of the deposition nucleotides in the illumination define the ultimate sequence of nucleotides of the array spot. A mask consists of masked (blocking light) and unmasked (allowing light) regions and induces

M. Agrawal et al. (Eds.): TAMC 2008, LNCS 4978, pp. 410–422, 2008.

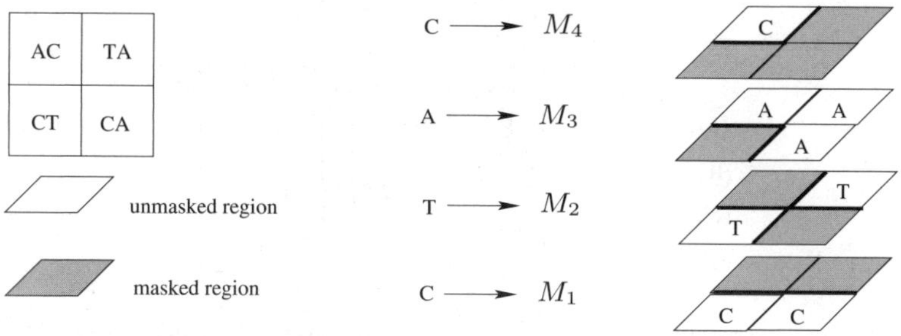

Fig. 1. Synthesis of a 2×2 microarray. The deposition sequence D = CTAC corresponds to the sequence of four masks M_1, M_2, M_3, and M_4. The masked regions are shaded. The borders between the masked and unmasked regions are represented by bold lines.

deposition of a particular nucleotide (A, C, G or T) at its exposed array *spots*. The *deposition sequence* D corresponding to the sequence of masks is a supersequence of all probes in the array (see example in Figure 1).

DNA microarray synthesis consists of two components, namely *probe placement* and *probe embedding*. Given a set of probes to be synthesized, probe placement is to place each probe to a unique spot in the microarray and probe embedding is the sequence of masked and unmasked steps used in the synthesis. For example, in Figure 2, the deposition sequence is $(\text{ACGT})^3$ and the sequence (a) $\text{A}(-)^4\text{C}(-)^5\text{T}$ is a possible embedding of the probe ACT, where " $-$ " represents a space.

We distinguish two types of synthesis, namely, *synchronous* and *asynchronous* synthesis. In synchronous synthesis, each deposition nucleotide can only be deposited to the i-th position of the probes for a particular i. In asynchronous synthesis, there is no such restriction, allowing arbitrary embeddings. For example, Figure 1 shows an asynchronous synthesis in which M_2 deposits a nucleotide to the second position of the sequence CT and the first position of TA. Asynchronous synthesis is more flexible, yet more difficult to optimize. In this paper we focus on asynchronous synthesis.

Due to diffraction, internal reflection and scattering, spots on the *border* between masked and unmasked regions are often subject to unintended illumination [8]. This uncertainty produces unpredicted probes that can compromise experimental results. As microarray chip is expensive to synthesize, it is usual that as many probes as possible are placed in a chip (i.e., as many entries are used), while unintended illumination has to be minimized. The magnitude of unintended illumination can be measured by the *border length* of the masks used, which is the number of borders shared between masked and unmasked regions, e.g., in Figure 1, the border length of M_1, M_3, M_4 is 2 and M_2 is 4.

To reduce the amount of unintended illumination, one can exploit freedom in placing probes in the microarray during probe placement and choosing different probe embeddings. The *Border Minimization Problem (BMP)* [12] is to find a placement of the probes on the microarray together with their embeddings in such a way that the sum of border lengths over all masks is minimized. It has been stated in [3,4] that the problem is believed to be NP-hard because of the exponential number of possible placements

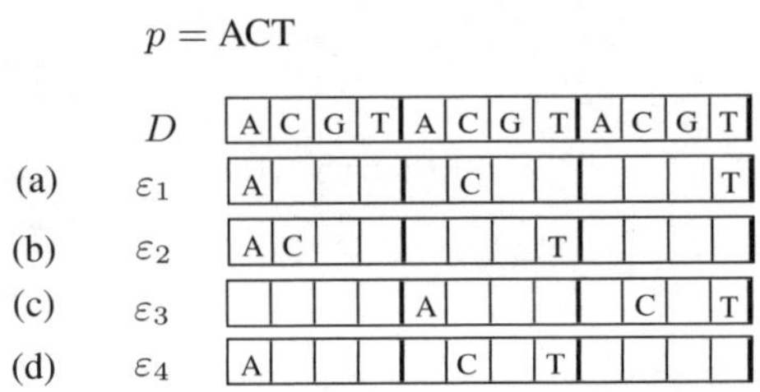

Fig. 2. Different embeddings of probe p = ACT into deposition sequence $D = (\mathrm{ACGT})^3$

(although we are not aware of an NP-hardness proof). For this reason, we focus on approximation algorithms for BMP in this paper.

Previous work. The BMP problem has attracted a lot of attention [12,3,4,5,15,16] and most work is experimental in nature. As far as we know, no polynomial time approximation algorithm is known for BMP with non-trivial performance guarantee.

BMP was first formally defined by Hannenhalli et al. [12]. They focused on synchronous synthesis and the only concern becomes probe placement. Their algorithm computes an approximated travelling salesman path (TSP) in the complete graph with nodes representing probes and edge costs representing the Hamming distance between the probes. The TSP path is then placed on the microarray in a certain way called *threading*. Experiments shows that threading is effective in reducing border length. Since then, other algorithms [15,16,4] have been proposed to improve the experimental results.

Asynchronous probe embedding was introduced by Kahng et al. [15]. They studied a special case that the deposition sequence D is given and the embeddings of all but one probes are known. A polynomial time dynamic programming algorithm was proposed to compute the optimal embedding of this single probe whose neighbors are already embedded. This algorithm is used as the basis for several heuristics [3,4,5,15,16] that are shown experimentally to reduce unintended illumination in terms of border length.

On the other hand, there are few theoretical results. In [15], lower bounds on the total border length for synchronous and asynchronous BMP problem were given, which are based on Hamming distance, and Longest Common Subsequence (LCS), respectively. The asynchronous dynamic programming mentioned above computes the optimal embedding of a single probe in time $O(\ell|D|)$, where ℓ is the length of a probe and D is the deposition sequence. The algorithm can be extended to an exponential time algorithm to find the optimal embedding of all n probes in $O(2^n \ell^n |D|)$ time.

Our contribution. In this paper, we study approximation of BMP in asynchronous synthesis. This is the first result with proved performance guarantee. The main result is an $O(\sqrt{n}\log^2 n)$-approximation, where n is the number of probes in the microarray. This is based on an approximation algorithm for the variant when the placement of probes is given in advance (called P-BMP problem). We show that P-BMP is $O(\log^2 n)$-approximable. We further show that if the array is one-dimensional, P-BMP can be solved optimally in polynomial time and there is a constant approximation for BMP. On the other hand, we show that BMP can be defined as the maximum agreement problem (AMP) with a different objective called "agreement". Minimizing the border length is equivalent to maximizing the agreement. Yet we are able to devise $O(1)$-approximation algorithms for AMP regardless of whether the placement is given in advance or not.

Organization of the paper. In Section 2, we give some definitions and notations. In Sections 3 and 4, we present and analyze approximation algorithms for BMP and AMP, respectively. Finally we give a conclusion and discuss future work in Section 5.

2 Preliminaries

We are given a set of n length-ℓ probes $\mathcal{P} = \{p_1, p_2, \ldots, p_n\}$, a $\sqrt{n} \times \sqrt{n}$ array (for simplicity, we assume that $\sqrt{n}$ is an integer). For any sequence p_i, we denote the t-th character of a sequence p_i by $p_i[t]$. The probes in $\mathcal{P}$ are to be placed on the $\sqrt{n} \times \sqrt{n}$ array. We represent this array by a grid graph $G = (V, E)$. Two grid vertices (x_1, y_1) and (x_2, y_2) are said to be *neighbor* if $|x_1 - x_2| + |y_1 - y_2| = 1$. For each vertex $v \in V$, we denote the set of neighbors of v by $\mathcal{N}(v)$.

Placement and embedding. A *placement* of the probes is a bijective function $\phi : \mathcal{P} \rightarrow V$ that maps each probe to a unique vertex in the grid G. An *embedding* of a set of probes $\mathcal{P}$ into a deposition sequence D is denoted by $\varepsilon = \{\varepsilon_1, \varepsilon_2, \ldots, \varepsilon_n\}$. For $1 \leq i \leq n$, ε_i is a length-$|D|$ sequence such that (1) $\varepsilon_i[t]$ is either $D[t]$ or a space "$-$"; and (2) removing all spaces from ε_i gives p_i. The hamming distance between ε_i and ε_j measures the border length between p_i and p_j if they are neighbors in a certain placement. We define this quantity as the *conflict* between the embeddings of p_i and p_j, denoted by $\text{conf}_\varepsilon(p_i, p_j)$. Note that $\text{conf}_\varepsilon(p_i, p_j) \leq 2\ell$. We define the *share* between the embeddings of p_i and p_j as $2\ell - \text{conf}_\varepsilon(p_i, p_j)$, and denote it by $\text{share}_\varepsilon(p_i, p_j)$.

Border length and agreement. The *border length* of a placement ϕ and an embedding ε is defined as the sum of conflicts between the embeddings of probes that are neighbors in the placement ϕ in G:

$$\text{BL}(\phi, \varepsilon) = \frac{1}{2} \sum_{\substack{p_i, p_j : \\ \phi(p_j) \in \mathcal{N}(\phi(p_i))}} \text{conf}_\varepsilon(p_i, p_j) \,. \tag{1}$$

The objective of the BMP problem is to find a placement ϕ and an embedding ε, so that $\text{BL}(\phi, \varepsilon)$ is minimized. We denote the optimal placement and the corresponding optimal embedding by ϕ^* and ε^*, respectively. We further define the counter part of border length, the *agreement*, which is the sum of shares between the embeddings of probes that are neighbors in the placement ϕ in G:

$$\text{A}(\phi, \varepsilon) = \frac{1}{2} \sum_{\substack{p_i, p_j : \\ \phi(p_j) \in \mathcal{N}(\phi(p_i))}} \text{share}_\varepsilon(p_i, p_j) \tag{2}$$

The *Maximum Agreement Problem* (AMP) is to find a placement ϕ and an embedding ε, so that $\text{A}(\phi, \varepsilon)$ is maximized. Since $\text{A}(\phi, \varepsilon) = 4\ell(n - \sqrt{n}) - \text{BL}(\phi, \varepsilon)$, minimizing the border length $\text{BL}(\phi, \varepsilon)$ is equivalent to maximizing the agreement $\text{A}(\phi, \varepsilon)$.

Common subsequence and common supersequence. The border length is closely related to the common subsequence and common supersequence between neighboring sequences in the placement. Consider any two length-ℓ sequences p, q. We denote the longest common subsequence and shortest common supersequence of two sequences p and q by $LCS(p, q)$ and $SCS(p, q)$, respectively, and the corresponding length as $|LCS(p, q)|$ and $|SCS(p, q)|$, respectively. $SCS(p, q)$ can be obtained by finding $LCS(p, q)$ and inserting into p the characters in q that are not in $LCS(p, q)$ while preserving the order in q. Therefore, $|SCS(p, q)| = 2\ell - |LCS(p, q)|$. For any embedding ε, the maximum number of common deposition nucleotides between p and q is $|LCS(p, q)|$, in other words, $\text{conf}_\varepsilon(p, q) \geq 2(\ell - |LCS(p, q)|)$ and $\text{share}_\varepsilon(p, q) \leq 2|LCS(p, q)|$. We define the *LCS distance* to be $2(\ell - |LCS(p, q)|)$, denoted by $\text{dist}(p, q)$. In other words, $\text{dist}(p, q)$ is a lower bound of $\text{conf}_\varepsilon(p, q)$ for any embedding ε.

Multiple sequence alignment (MSA) and Weighted MSA (WMSA). As we will see in later sections, a variant of BMP problem, named P-BMP (BMP problem in which the placement is given), can be polynomial time reducible to WMSA. As a consequence, we can apply the approximation results on WMSA to P-BMP, which we can further use as a building block for the approximation for BMP. We first review the MSA and WMSA problems. MSA and WMSA have been studied extensively [2,7,11,20]. Let Σ be the set of characters and $S = \{S_1, S_2, \ldots, S_k\}$ be a set of k sequences, with maximum length m, over Σ. An alignment of S is a matrix $S' = (S'_1, S'_2, \ldots, S'_k)$ such that $|S'_i| = m'$ and S'_i is formed by inserting spaces into S_i. For a given distance function $\delta(a, b)$ where $a, b \in \Sigma \cup \{-\}$, the *pair-wise score* of S'_i and S'_j is defined as $\sum_{1 \leq y \leq m'} \delta(S'_i[y], S'_j[y])$. Given a weight function $w(i, j)$ for the pair of sequences S_i and S_j, the *weighted sum-of-pair* (SP) score $\text{SP}(S', w) = \frac{1}{2} \sum_{1 \leq i,j \leq k} w(i, j) \sum_{1 \leq y \leq m'} \delta(S'_i[y], S'_j[y])$. The WMSA problem is to find an alignment S' such that $\text{SP}(S', w)$ is minimized. WMSA has been proved to be NP-complete. An $O(\log^2 n)$-approximation algorithm [23] has been given via a reduction to the minimum routing cost tree problem (MRCT) [1].

Minimum routing cost tree problem (MRCT). In this problem, a graph with weighted edges is given. For a spanning tree of the graph, the *routing cost* between two vertices is the sum of weights of the edges on the unique path between the two vertices in the spanning tree. The *routing cost* of the spanning tree is defined as the sum of routing cost between every pair of two vertices. The MRCT problem is to find a spanning tree whose routing cost is minimum. The results in [1] state that there is a polynomial time reduction from WMSA to MRCT. Each sequence in the input of WMSA corresponds to a vertex in the input graph of MRCT. The edge weight between two vertices is set to be the weighted edit distance between the two corresponding sequences. The reduction result states that (1) there is a routing spanning tree T whose routing cost is at most $O(\log^2 n)$ times $\sum_{i,j} w(i, j)d(i, j)$, where $d(i, j)$ is the edit distance between the two sequences i and j; and (2) there is an alignment S' whose $\text{SP}(S', w)$ is at most the routing cost of T. Note that $\sum_{i,j} w(i, j)d(i, j)$ is a lower bound on the weighted SP score. Therefore, the following lemma follows.

Lemma 1. [23] *There is an $O(\log^2 n)$-approximation algorithm for the WMSA problem, where n is the number of sequences to be aligned.*

3 The BMP Problem

In this section, we study the BMP problem. We are to find a placement and an embedding for the given probe set. An $O(\sqrt{n}\log^2 n)$-approximation algorithm is given for BMP (Section 3.2), which is based on an approximability result for a variant of BMP, named P-BMP (Section 3.1). At the end of this section, we also discuss the case when the array is one-dimensional and we show that BMP admits better results in this case.

3.1 P-BMP: Finding Embedding When Placement Is Given

In this section, we study the P-BMP problem, a variant of BMP with a placement given in advance. The concern becomes to find an embedding. We show that P-BMP is $O(\log^2 n)$-approximable by giving a reduction to the weighted multiple sequence alignment problem (WMSA), for which there is an $O(\log^2 n)$-approximation algorithm [23].

Lemma 2. *There is a polynomial time reduction from P-BMP to WMSA.*

Proof. Let I be an instance of the P-BMP problem with a given placement ϕ. We construct an instance I' for WMSA such that there is a solution for I with border length X if and only if there is a solution for I' with a weighted SP score of X.

Construction of I'. We first show the construction of I'. The input sequence for WMSA is the same as the input probe set $\mathcal{P}$. The weight $w(i, j)$ is defined as follows:

$$w(i,j) = \begin{cases} 1 & \text{if } \phi(p_j) \in \mathcal{N}(\phi(p_i)), \\ 0 & \text{otherwise.} \end{cases}$$

The distance function $\delta(a, b)$, for $a, b \in \Sigma \cup \{-\}$, is defined as follows:

$$\delta(a,b) = \begin{cases} 0 & \text{if } a = b, \\ 1 & \text{if } a \neq b \text{ and } (a = \text{“} - \text{”} \text{ or } b = \text{“} - \text{”}), \\ \infty & \text{otherwise.} \end{cases}$$

Note that the edit distance of p_i and p_j in WMSA is the same as dist(p_i, p_j) in BMP.

Solution for I implies solution for I'. Suppose we have an embedding ε for I. Note that $\varepsilon = \{\varepsilon_1 \cdots \varepsilon_n\}$ is an alignment for $\mathcal{P}$ and the pairwise score of ε_i and ε_j equals $\text{conf}_\varepsilon(p_i, p_j)$. So, $\text{SP}(\mathcal{P}', w) = \frac{1}{2}\sum_{1\le i,j\le n} w(i,j) \sum_{1\le y\le |D|} \delta(\varepsilon_i[y], \varepsilon_j[y]) = \frac{1}{2}\sum_{1\le i,j\le n} w(i,j)\text{conf}_\varepsilon(p_i,p_j) = \frac{1}{2}\sum_{p_i,p_j:\phi(p_j)\in\mathcal{N}(\phi(p_i))} \text{conf}_\varepsilon(p_i,p_j) = \text{BL}(\phi, \varepsilon)$. The second last equality is due to the definition of $w(i, j)$, which is based on ϕ.

Solution for I' implies solution for I. On the other hand, suppose we have a solution for I', i.e., an alignment $\mathcal{P}' = (p'_1 \cdots p'_n)$ for $\mathcal{P}$ and $|p'_i| = m'$, for some m'. In the alignment $\mathcal{P}'$, each column contains the same character or "$-$" because of the definition of the distance function $\delta(a, b)$. We denote the resulting matrix as $\varepsilon = (\varepsilon_1 \cdots \varepsilon_n)$. It can be seen that ε is an embedding for $\mathcal{P}$ and the hamming distance between ε_i and ε_j equals the pair-wise score of p'_i and p'_j. Then $\text{BL}(\phi, \varepsilon) = \frac{1}{2}\sum_{p_i,p_j:\phi(p_j)\in\mathcal{N}(\phi(p_i))} \text{conf}_\varepsilon(p_i,p_j) = \frac{1}{2}\sum_{p_i,p_j:\phi(p_j)\in\mathcal{N}(\phi(p_i))} \sum_{1\le y\le|D|} \delta(p'_i[y], p'_j[y]) = \frac{1}{2}\sum_{1\le i,j\le n} w(i,j)\sum_{1\le y\le|D|} \delta(p'_i[y], p'_j[y]) = \text{SP}(\mathcal{P}', w)$. Note that the second last equality holds for the same reason as above. Therefore, the lemma follows. □

Corollary 1. *The P-BMP problem is $O(\log^2 n)$-approximable.*

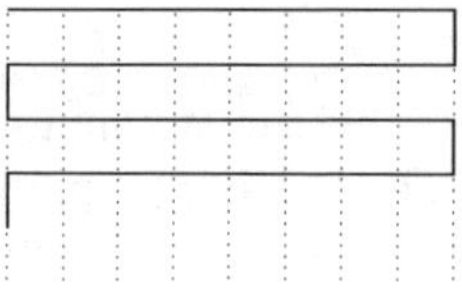

Fig. 3. Row-by-row threading of a TSP (solid edges) on a grid. Solid and dotted edges connect neighbors in the placement that are and are not, respectively, neighbors on the TSP.

3.2 BMP: Finding Placement and Embedding

In this section, we study the BMP problem in which we are to find both the placement as well as the embedding. We give an $O(\sqrt{n}\log^2 n)$-approximation, which makes use of the approximability result for P-BMP (Section 3.1). To make use of the result for P-BMP, we need a certain placement, the choice of which is guided by some travelling salesman path (TSP) on a particular graph (to be defined). Note that finding the minimum TSP is NP-hard, yet there is a polynomial time $O(1)$-approximation [6].

The algorithm PLACE&EMBED. The approximation algorithm PLACE&EMBED is shown in Algorithm 1. The graph G_c constructed in the algorithm is a weighted complete graph with vertices representing $\mathcal{P}$ and edge weight representing dist() between the two vertices. A travelling salesman path (TSP) is obtained from G_c, which we *"thread"* on the grid G in a row-by-row fashion to form a placement [12]: the TSP is placed from left to right on the first row, right to left on the second, and then alternate in the same way in the remaining rows (see Figure 3 for an example). We then employ the approximation algorithm in Section 3.1. We denote the placement and embedding computed by PLACE&EMBED as $\tilde{\phi}$ and $\tilde{\varepsilon}$, respectively.

Algorithm 1. PLACE&EMBED: Approximation algorithm for BMP.

Input: Probe set $\mathcal{P} = \{p_1, p_2, \ldots, p_n\}$ to be placed on a $\sqrt{n} \times \sqrt{n}$ array.
Output: A placement $\tilde{\phi}$ and an embedding $\tilde{\varepsilon}$ for $\mathcal{P}$.
1: Construct the weighted complete graph G_c.
2: Find an approximate TSP $\tilde{Q}$ for G_c using algorithm in [6].
3: Thread $\tilde{Q}$ in a row-by-row fashion to obtain a placement $\tilde{\phi}$.
4: Run the approximation algorithm for P-BMP in Section 3.1 (i.e., by reducing the P-BMP instance to an WMSA instance) to obtain an embedding $\tilde{\varepsilon}$.

Theorem 1. *Algorithm* PLACE&EMBED *is an $O(\sqrt{n}\log^2 n)$-approximation for BMP.*

To analyze the performance of PLACE&EMBED, we need some notations. Recall that we define for any sequences p, q, $\text{dist}(p, q) = 2(\ell - |LCS(p, q)|)$. We overload the notation dist() for any subgraph of G_c. For any subgraph H of G_c, we define the LCS distance of H, denoted by $\text{dist}(H)$, to be the sum of LCS distances of neighboring probes in H, i.e., $\text{dist}(H) = \frac{1}{2}\sum_{p,\, q\,:\, q \in \mathcal{N}(p) \text{ in } H} \text{dist}(p, q)$.

As mentioned before in Section 2, $\text{dist}(p, q)$ is the minimum conflict between probes p and q. Yet the embeddings needed to achieve $\text{dist}(p, q)$ may not be compatible with each other in a particular placement. For example, consider the placement ϕ in Figure 1, $\text{dist}(\phi) = 8$ since $\text{dist}(p, q) = 2$ for every neighboring pair p, q. Yet the minimum border length is 10 with CTAC as the deposition sequence, and embeddings $(- -$ AC, $-$TA$-$, CT $- -$, C $-$ A$-)$. We summarize this as follows.

Observation 1. *Given a placement ϕ, $dist(\phi) \leq BL(\phi, \varepsilon)$, for any embedding ε.*

Observation 1 implies that for the optimal placement ϕ^* and embedding ε^*, $\text{dist}(\phi^*) \leq \text{BL}(\phi^*, \varepsilon^*)$. To approximate BMP, it suffices to bound the border length by $\text{dist}(\phi^*)$. On the other hand, we make an observation about a graph H_1 and its subgraph H_2. The observation is true since any neighbors in H_2 are also neighbors in H_1.

Observation 2. *Consider any graph H_1 and a subgraph H_2 of it. $dist(H_2) \leq dist(H_1)$.*

Corollary 2. *Suppose Q^* is the optimal TSP for G_c. Then, we have $dist(Q^*) \leq dist(\phi^*)$.*

Proof. ϕ^* can be viewed as threading a TSP Q in a row-by-row fashion. By Observation 2, $\text{dist}(Q) \leq \text{dist}(\phi^*)$. As Q^* is the optimal TSP, $\text{dist}(Q^*) \leq \text{dist}(Q) \leq \text{dist}(\phi^*)$. □

It is known that TSP can be approximated by $3/2$ (Lemma 3). So, $\text{dist}(\tilde{Q}) \leq 3\,\text{dist}(Q^*)/2$.

Lemma 3. [6] *The travelling salesman problem admits a $3/2$-approximation if the weight satisfies the triangle inequality.*

Lemma 4. *(i) $dist(\tilde{\phi}) \leq 2\sqrt{n}\, dist(\tilde{Q})$; and (ii) $BL(\tilde{\phi}, \tilde{\varepsilon}) \leq O(\log^2 n)\, dist(\tilde{\phi})$.*

Proof (Sketch). (i) Suppose $\tilde{Q} = \{u_1, u_2, \ldots, u_n\}$. Note that the LCS distance dist() satisfies the triangular inequality, i.e., $\text{dist}(u_i, u_j) \leq \sum_{i \leq k < j} \text{dist}(u_k, u_{k+1})$. Neighboring probes on $\tilde{Q}$ are also neighbors in $\tilde{\phi}$ but not vice versa. For any two probes u_i and u_j which are neighbors in $\tilde{\phi}$, we have $1 \leq |j - i| < 2\sqrt{n}$. When we sum up $\text{dist}(\tilde{\phi})$, $\text{dist}(u_k, u_{k+1})$, for any k, may be counted more than once, but no more than $2\sqrt{n}$ times. Therefore, $\text{dist}(\tilde{\phi}) \leq 2\sqrt{n}\,\text{dist}(\tilde{Q})$.

(ii) In Step 4 of PLACE&EMBED, we reduce the P-BMP instance with $\tilde{\phi}$ as the placement to an WMSA instance. Lemma 2 asserts that the border length of the embedding obtained is the same as the weighted SP score of the alignment. Furthermore, we have seen in Section 2 that approximation for WMSA can be found by the approximation for MRCT and the resulting routing tree has a routing cost, and thus, the weighted SP score, at most $O(\log^2 n)$ times the total weighted edit distance in WMSA. In the proof of Lemma 2, we note that the weighted edit distance of two sequences is the same as dist() of the two sequences. So, $\text{BL}(\tilde{\phi}, \tilde{\varepsilon}) \leq O(\log^2 n)\,\text{dist}(\tilde{\phi})$. □

Proof (Theorem 1). By Lemmas 4, 3, and Corollary 2, we have $\text{BL}(\tilde{\phi}, \tilde{\varepsilon}) \leq O(\sqrt{n}\log^2 n)\,\text{dist}(\tilde{Q}) \leq O(\sqrt{n}\log^2 n)\,\text{dist}(Q^*) \leq O(\sqrt{n}\log^2 n)\,\text{dist}(\phi^*)$. Furthermore, Observation 1 holds for all placements, and hence for ϕ^*, in other words, $\text{dist}(\phi^*) \leq \text{BL}(\phi^*, \varepsilon^*)$. Therefore, $\text{BL}(\tilde{\phi}, \tilde{\varepsilon}) \leq O(\sqrt{n}\log^2 n)\,\text{BL}(\phi^*, \varepsilon^*)$. □

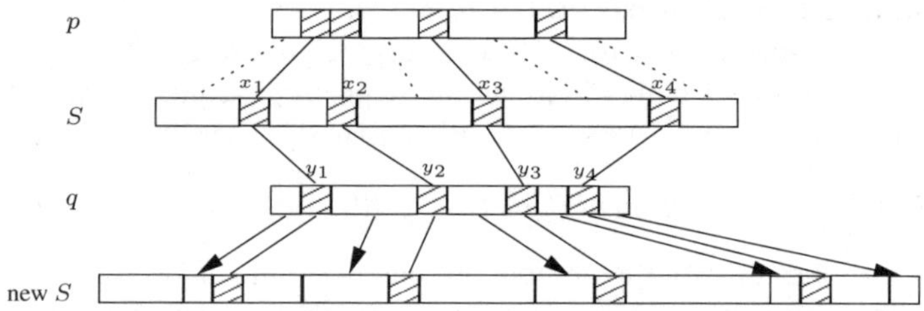

Fig. 4. An illustration of EXTEND. Shaded squares refer to characters in $LCS(p, q)$. Characters in q but not in $LCS(p, q)$ are inserted into S so that the order preserves as in q (see the arrows).

3.3 One Dimensional Array

In this section, we study the special case on an 1D array. Intuitively, the problem is easier than the 2D case. We show that P-BMP on an 1D array can be solved optimally in polynomial time while BMP on an 1D array admits an $O(1)$-approximation.

P-BMP on 1D array. The algorithm EMBED1D shown in Algorithm 2 makes use of a procedure called EXTEND. EXTEND takes two sequences p and q, and a supersequence S of p as input and returns a supersequence of S and q. Let $c = |LCS(p, q)|$, $x_1, x_2, \ldots, x_c$ be the indices of S corresponding to p that belongs to $LCS(p, q)$, and $y_1, y_2, \ldots, y_c$ be the indices of q that belongs to $LCS(p, q)$. EXTEND then extends S by inserting characters in q but not in $LCS(p, q)$: characters between $q[y_{k-1}]$ and $q[y_k]$ are inserted right before $S[x_k]$ and characters beyond $q[y_c]$ are appended to the end of S. EXTEND keeps track of the indices of the new S that correspond to q (see Figure 4).

Algorithm 2. EMBED1D: Optimal embedding for P-BMP on 1D array.

Input: Probe set $\mathcal{P} = \{p_1, p_2, \ldots, p_n\}$, placed on a 1D array in that order.
Output: An embedding ε with minimum border length.

1: Set $D = p_1$.
2: For $i > 1$, call the procedure EXTEND with p_{i-1}, p_i and D as the input to obtain a new D.
3: For each p_i, set ε_i such that $\varepsilon[y] = D[y]$ if $D[y]$ corresponds to a character in p_i kept track by EXTEND, and $\varepsilon[y] =$ " $-$ " otherwise.

Theorem 2. EMBED1D *finds an optimal embedding for the P-BMP problem on 1D array in polynomial time.*

Proof. We first observe that D constructed in each iteration by EXTEND is a common supersequence of $p_1, \ldots, p_i$. This is clear from the way EXTEND finds $LCS(p_{i-1}, p_i)$ and inserts characters into D. It also implies that the number of nucleotides shared by p_{i-1} and p_i is maintained as $|LCS(p_{i-1}, p_i)|$, which is the maximum possible. Note that this property does not change by later steps. Hence, the border length of the final embedding is the minimum. As for time complexity, the bottleneck is finding the longest common subsequences of two sequences, which is known to take polynomial time [13]. This is done for $n - 1$ times only. Therefore, EMBED1D also takes polynomial time. □

BMP on 1D array. Similar to the case on 2D array, we find a placement by finding an approximate TSP on the weighted complete graph G_c and then find an embedding by EMBED1D. This algorithm gives a $3/2$-approximation for BMP on 1D array.

Theorem 3. *There is a polynomial time algorithm for BMP on 1D array with approximation ratio* $3/2$.

4 The Maximum Agreement Problem (AMP)

In this section, we study the counter part of BMP, which we called maximum agreement problem (AMP) (recall definition in Section 2). In contrast to BMP, AMP admits constant approximations, whether the placement is given in advance or not.

4.1 Approximation for P-AMP

We first study the P-AMP problem, a variant of AMP with a placement already given.

Algorithm AEMBED**.** The algorithm AEMBED (EMBED for Agreement) makes use of procedure EXTEND in Section 3.3. The order of probes to be considered is determined by a certain tree T with the bottom rightmost probe in G being the root. To construct T, for each probe p, we assign a parent to the probe, denoted by $parent(p)$. We denote by $r(p)$ and $b(p)$ the right and bottom neighbors of probe p, respectively. The probes in the rightmost column and bottommost column has $r(p)$ = NULL and $b(p)$ = NULL, respectively. We set $parent(p)$ to $r(p)$ or $b(p)$ depending on whether $|LCS(p, r(p))|$ or $|LCS(p, b(p))|$ is larger. Details of AEMBED is shown in Algorithm 3. The embedding found is denoted by $\hat{\varepsilon}$. Figure 5 shows an example.

Analysis. To analyze the performance of AEMBED, we first observe that in the final embedding $\hat{\varepsilon}$, the number of nucleotides shared by a probe and its parent equals to the length of their LCS (by a similar argument as the proof of Theorem 2). We then bound the performance of AEMBED as follows.

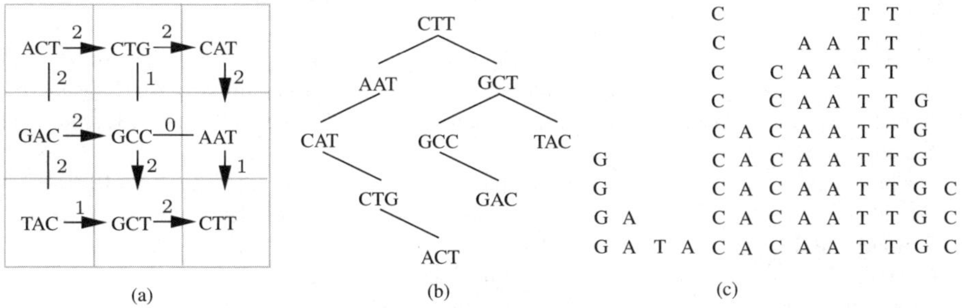

Fig. 5. (a) A set of probes placed on a 3×3 grid G. The values represent the length of LCS between the two neighboring probes. An arrow from p to q means $parent(p) = q$. (b) The tree constructed by AEMBED with root CTT. (c) How the deposition sequence D changes iteratively. The sequences are drawn in a way the characters align with the final D.

Algorithm 3. AEMBED: Approximate algorithm for P-AMP.

Input: Probe set $\mathcal{P} = \{p_1, p_2, \ldots, p_n\}$ placed on a $\sqrt{n} \times \sqrt{n}$ array according to a placement ϕ.
Output: An embedding $\hat{\varepsilon}$ for $\mathcal{P}$.

1: Construct a tree T by assigning parent to each probe p: if $|LCS(p, r(p))| \geq |LCS(p, b(p))|$ set $parent(p) = r(p)$ else set $parent(p) = b(p)$.
2: Set D to be the bottom rightmost probe in the grid G.
3: Traverse T in a pre-order fashion: for each probe p traversed, call the procedure EXTEND with $parent(p)$, p and D as input.
4: For each p_i, set $\hat{\varepsilon}_i$ such that $\hat{\varepsilon}[y] = D[y]$ if $D[y]$ corresponds to a character in p_i kept track by EXTEND, and $\hat{\varepsilon}[y] =$ " $-$ " otherwise.

Theorem 4. AEMBED *is a polynomial-time* 2*-approximation algorithm for P-AMP.*

Proof. For the given placement ϕ and the optimal embedding ε^*, the optimal agreement is: $\mathrm{A}(\phi, \varepsilon^*) = \sum_{p \in \mathcal{P}} (\mathrm{share}_{\varepsilon^*}(p, r(p)) + \mathrm{share}_{\varepsilon^*}(p, b(p)))$. We assume $\mathrm{share}_{\varepsilon^*}(p, q) = 0$ if $q = \mathrm{NULL}$. As mentioned in Section 2, for any embedding, the share between the embeddings of probes p, q is at most $2|LCS(p, q)|$. Thus, $2|LCS(p, r(p))| \geq \mathrm{share}_{\varepsilon^*}(p, r(p))$ and $2|LCS(p, b(p))| \geq \mathrm{share}_{\varepsilon^*}(p, b(p))$. Note that $\mathrm{share}_{\hat{\varepsilon}}(p, parent(p)) = 2\max\{\, |LCS(p, r(p))|, |LCS(p, b(p))|\} \geq \frac{1}{2}(\mathrm{share}_{\varepsilon^*}(p, r(p)) + \mathrm{share}_{\varepsilon^*}(p, b(p)))$. Therefore, $\mathrm{A}(\phi, \hat{\varepsilon}) = \sum_{p \in \mathcal{P}} \mathrm{share}_{\hat{\varepsilon}}(p, parent(p)) \geq \frac{1}{2}\mathrm{A}(\phi, \varepsilon^*)$. Finally, AEMBED runs in polynomial time as the bottleneck is finding LCS between two sequences. □

4.2 Approximation for AMP

In this section, we study the general AMP problem to find both the placement and the embedding to maximize the agreement. We prove that the algorithm APLACE&EMBED as shown in Algorithm 4 has an asymptotic approximation ratio of 4.

Algorithm 4. APLACE&EMBED: Approximation algorithm for AMP.

Input: Probe set $\mathcal{P} = \{p_1, p_2, \ldots, p_n\}$ to be placed on a $\sqrt{n} \times \sqrt{n}$ array.
Output: A placement $\check{\phi}$ and an embedding $\check{\varepsilon}$ for $\mathcal{P}$.

1: Partition $\mathcal{P}$ into four disjoint groups $\mathcal{A}$, $\mathcal{C}$, $\mathcal{G}$ and $\mathcal{T}$: a probe belongs to $\mathcal{A}$ if the number of A in the probe is the maximum over the number of other characters (similarly for $\mathcal{C}$, $\mathcal{G}$ and $\mathcal{T}$).
2: Thread the probes in group $\mathcal{A}$ on the array in a row-by-row fashion, followed by threading of probes in $\mathcal{C}$, $\mathcal{G}$, and $\mathcal{T}$ to form the placement $\check{\phi}$.
3: For probes in $\mathcal{A}$, align them such that the maximum number of A are aligned while different characters are not aligned. This forms a partial embedding $\check{\varepsilon}_a$ with deposition sequence D_a. Similarly, find $\check{\varepsilon}_c$, $\check{\varepsilon}_g$, $\check{\varepsilon}_t$ and D_c, D_g, D_t.
4: Combine D_a, D_c, D_g, and D_t to form D (append one after the other).
5: Extend the embeddings $\check{\varepsilon}_a$, $\check{\varepsilon}_c$, $\check{\varepsilon}_g$, $\check{\varepsilon}_t$ according to D by inserting " $-$ " in the columns corresponding to other groups. The union of the extended embeddings is the resulting embedding $\check{\varepsilon}$.

Theorem 5. *The asymptotic approximation ratio of* APLACE&EMBED *is* 4.

Proof. Consider the optimal placement ϕ^* and embedding ε^*. For every pair of neighboring probes p, q, $\text{share}_\varepsilon(p, q) \leq 2\ell$. There are a total of $2(n - \sqrt{n})$ pairs of neighbors on the grid in total. So, the optimal agreement $A(\phi^*, \varepsilon^*) \leq 4\ell(n - \sqrt{n})$. On the other hand, consider $\check{\phi}$ and $\check{\varepsilon}$ returned by APLACE&EMBED. According to the way we partition the probes into group, for any two probes p, q in a group, the number of nucleotides that can be shared is at least $\ell/4$. Hence, $\text{share}_{\check{\varepsilon}}(p, q) \geq 2(\ell/4) = \ell/2$. As we seen above, there are altogether $2(n - \sqrt{n})$ pairs of neighbors in the grid. We may not share any nucleotide when the pair belongs to different groups. According to the way we thread the groups, there are at most $3\sqrt{n} + 3$ such pairs ($\sqrt{n}$ pairs of vertical neighbors between consecutive groups and 3 pairs of neighbors that are the last one in a group and the first one in the next group). As a result, we have at least $2n - 5\sqrt{n} - 3$ pairs each with $\text{share}_{\check{\varepsilon}}()$ at least $\ell/2$. Therefore, $A(\check{\phi}, \check{\varepsilon}) \geq \ell(n - 2.5\sqrt{n} - 1.5)$. Then $A(\check{\phi}, \check{\varepsilon})/A(\phi^*, \varepsilon^*)$ tends to 4 as $A(\phi^*, \varepsilon^*)$ tends to infinity. So, the asymptotic approximation ratio of APLACE&EMBED is 4. □

5 Concluding Remarks

To summarize, we study the border minimization problem which is believed to be NP-hard with no known NP-hardness proof. An open question is to derive an NP-hardness proof. Another interesting open question is to improve the approximation ratio and/or derive inapproximability result. As mentioned before, there is an exponential time algorithm to compute the optimal BMP solution. Improving the exponential time algorithm could be useful in practice and is of theoretical interest.

References

1. Bartal, Y.: Probabilistic approximation of metric spaces and its algorithmic applications. In: Proc. 37th FOCS, pp. 184–193 (1996)
2. Bonizzoni, P., Vedova, G.D.: The complexity of multiple sequence alignment with SP-score that is a metric. Theoretical Computer Science 259(1–2), 63–79 (2001)
3. Carvalho Jr., S.A., Rahmann, S.: Improving the layout of oligonucleotide. microarrays: Pivot partitioning. In: Proc. 6th WABI, pp. 321–332 (2006)
4. Carvalho Jr., S.A., Rahmann, S.: Microarray layout as quadratic assignment problem. In: Proc. GCB, pp. 11–20 (2006)
5. Carvalho Jr., S.A., Rahmann, S.: Improving the design of genechip arrays by combining placement and embedding. In: Proc. 6th CSB, pp. 54–63 (2007)
6. Christofides, N.: Worst-case analysis of a new heuristic for the travelling salesman problem. Technical Report 388, Graduate School of Industrial Administration, Carnegie-Mellon University, Pittsburgh, PA (ND33) (1976)
7. Feng, D.F., Doolittle, R.F.: Approximation algorithms for multiple sequence alignment. Theoretical Computer Science 182(1), 233–244 (1987)
8. Fodor, S., Read, J.L., Pirrung, M.C., Stryer, L., Lu, A.T., Solas, D.: Light-directed, spatially addressable parallel chemical synthesis. Science 251(4995), 767–773 (1991)
9. Gerhold, D., Rushmore, T., Caskey, C.T.: DNA chips: promising toys have become powerful tools. Trends in Biochemical Sciences 24(5), 168–173 (1999)

10. Gąsieniec, L., Li, C.Y., Sant, P., Wong, P.W.H.: Randomized probe selection algorithm for microarray design. Journal of Theoretical Biology 248(3), 512–521 (2007)
11. Gusfield, D.: Efficient methods for multiple sequence alignment with guaranteed error bounds. Bulletin of Mathematical Biology 55(1), 141–154 (1993)
12. Hannenhalli, S., Hubell, E., Lipshutz, R., Pevzner, P.A.: Combinatorial algorithms for design of DNA arrays. Advances in Biochemical Engineering/Biotechnology 77, 1–19 (2002)
13. Hirschberg, D.S.: A linear space algorithm for computing maximal common subsequences. Communications of the ACM 18(6), 341–343 (1975)
14. Kaderali, L., Schliep, A.: Selecting signature oligonucleotides to identify organisms using DNA arrays. Bioinformatics 18, 1340–1349 (2002)
15. Kahng, A.B., Mandoiu, I.I., Pevzner, P.A., Reda, S., Zelikovsky, A.: Scalable heuristics for design of DNA probe arrays. Journal of Computational Biology 11(2/3), 429–447 (2004)
16. Kahng, A.B., Mandoiu, I.I., Reda, S., Xu, X., Zelikovsky, A.: Computer-aided optimization of DNA array design and manufacturing. IEEE Transactions on Computer-Aided Design of Integrated Circuits and Systems 25(2), 305–320 (2006)
17. Kasif, S., Weng, Z., Detri, A., Beigel, R., DeLisi, C.: A computational framework for optimal masking in the synthesis of oligonucleotide microarrays. Nucleic Acids Research 30(20), e106 (2002)
18. Li, F., Stormo, G.: Selection of optimal DNA oligos for gene expression arrays. Bioinformatics 17(11), 1067–1076 (2001)
19. Rahmann, S.: The shortest common supersequence problem in a microarray production setting. Bioinformatics 19(suppl. 2), 156–161 (2003)
20. Reinert, K., Lenhof, H.P., Mutzel, P., Mehlhorn, K., Kececioglu, J.D.: A branch-and-cut algorithm for multiple sequence alignment. In: Proc. 1st RECOMB, pp. 241–250 (1997)
21. Slonim, D.K., Tamayo, P., Mesirov, J.P., Golub, T.R., Lander, E.S.: Class prediction and discovery using gene expression data. In: Proc. 4th RECOMB, pp. 263–272 (2000)
22. Sung, W.K., Lee, W.H.: Fast and accurate probe selection algorithm for large genomes. In: Proc. 2nd CSB, pp. 65–74 (2003)
23. Wu, B.Y., Lancia, G., Bafna, V., Chao, K.M., Ravi, R., Tang, C.Y.: A polynomial-time approximation scheme for minimum routing cost spanning trees. SIAM Journal on Computing 29(3), 761–778 (1999)

On a Question of Frank Stephan

Klaus Ambos-Spies, Serikzhan Badaev, and Sergey Goncharov

[1] University of Heidelberg, Im Neuenheimer Feld 294, D-69120 Heidelberg, Germany
ambos@math.uni-heidelberg.de
http://www.math.uni-heidelberg.de/logic/personen.html
[2] Kazakh National University, 71 Al-Farabi ave., Almaty 050038, Kazakhstan
badaev@kazsu.kz
[3] Sobolev's Math. Institute, 4 Koptyug ave., Novosibirsk 630090, Russia
s.s.goncharov@math.nsc.ru
http://mmfd.nsu.ru/mmf/persons/gonchar/

Abstract. For many TxtEX-learnable computable families of recursively enumerable sets, all their computable numberings are equivalent with respect to the reduction via the functions recursive in the halting problem. We show that this holds for every TxtEX-learnable computable family of recursively enumerable sets, but, in general, the converse is not true.

Keywords: computable family of sets, TxtEX learning, equivalent numberings.

1 Introduction

It is well-known (see [1]) that computable classes of finite sets, finite classes of recursively enumerable sets, and some classes of the graphs of recursive functions are TxtEX-learnable. On the other hand, the computable numberings of any of these classes are pairwise equivalent with respect to the reduction by $\mathbf{0}'$-recursive functions. It might be that these observations led Frank Stephan to propose the following conjecture to one of the authors of this paper:

For every computable family $\mathcal{A}$ of r.e. sets, the following are equivalent.

(i) $\mathcal{A}$ is TxtEX-learnable.
(ii) All computable numberings of $\mathcal{A}$ are $\mathbf{0}'$-equivalent.

Our aim is to show that one of the directions of this statement is true (namely the (easy) direction $(i) \Rightarrow (ii)$) while the converse fails.

We follow the monograph [4] of Yu.L. Ershov in Russian and the survey papers [2], [3] for the terminology and notations accepted in the theory of numberings. A mapping $\alpha : \mathbb{N} \longrightarrow \mathcal{L}$ of the set $\mathbb{N}$ of natural numbers onto a family $\mathcal{L}$ of recursively enumerable (r.e. for short) sets is called a *computable numbering* of $\mathcal{L}$ if the set $\{\langle x, n\rangle : x \in \alpha(n)\}$ is r.e., and a family $\mathcal{L}$ of subsets of $\mathbb{N}$ is called *computable* if it has a computable numbering. In other words, a computable family $\mathcal{L}$ is a *uniformly r.e. class* of sets, and every computable numbering α of $\mathcal{L}$ defines a uniform r.e. sequence $\alpha(0), \alpha(1), \ldots$ of the members of $\mathcal{L}$. A

M. Agrawal et al. (Eds.): TAMC 2008, LNCS 4978, pp. 423–432, 2008.

numbering α is called *reducible* (*$\mathbf{0}'$-reducible*) to a numbering β if $\alpha = \beta \circ f$ for some recursive (recursive relative to the halting problem) function f. Numberings α, β are called *equivalent* (*$\mathbf{0}'$-equivalent*) if they are reducible ($\mathbf{0}'$-reducible) to each other.

A language in this paper is just an r.e. set. The class of all r.e. subsets of $\mathbb{N}$ is denoted by $\mathcal{E}$, and $W_0, W_1, \ldots$ stands for the standard computable numbering of the family $\mathcal{E}$. A *text for a set* L is a function $t : \mathbb{N} \longrightarrow \mathbb{N}$ such that $\mathrm{range}(t) = L$.

Definition 1. *A* learner *is a partial function* $M : \mathbb{N}^* \longrightarrow \mathbb{N}$.

Notice that a learner is not required to be recursive. We call a learner M *computable* if the function M is partial recursive.

Definition 2. *A learner M* learns *or* identifies *a language L if for every text t for L,*

1. $\lim_n M(t \upharpoonright n)$ exists (we say in this case that M converges on the text t);
2. $L = W_{\lim_n M(t\upharpoonright n)}$.

If $\mathcal{L}$ is a family of languages, we say that M learns *or* identifies *$\mathcal{L}$ if M identifies every language $L \in \mathcal{L}$. $\mathcal{L}$ is* TxtEX-learnable *if it is identified by some computable learner.*

TxtEX denotes the class of all TxtEX-learnable families, [5].

Remark 1. If a computable family $\mathcal{L}$ of languages is TxtEX-learnable then it is identifiable by some primitive recursive learner.

2 Computable Numberings of TxtEX-Learnable Families

In this section we will show that any two computable numberings of a computable TxtEX-learnable family of languages are $\mathbf{0}'$-equivalent. Our proof will use the following representation of computable numberings by Lachlan (see [6] for the details).

Let $\mathcal{L}$ be a computable family of r.e. sets. We say that an r.e. set A *represents* $\mathcal{L}$ if $\mathcal{L} = \{W_x : x \in A\}$. Now if A represents $\mathcal{L}$ then any recursive function f enumerating A induces a computable numbering α_f of $\mathcal{L}$ where $\alpha_f(x) = W_{f(x)}$. Moreover, if f and g are recursive functions enumerating A then the corresponding numberings α_f and α_g of $\mathcal{L}$ are equivalent. So, up to equivalence, any r.e. set A representing $\mathcal{L}$ induces a unique computable numbering of $\mathcal{L}$ in the just described way. Conversely, for any computable numbering α of $\mathcal{L}$, there is an r.e. set A representing $\mathcal{L}$ such that the numbering induced by A is equivalent to α. The latter follows from the following well-known fact of the theory of numberings: $\alpha : \mathbb{N} \longrightarrow \mathcal{L} \subseteq \mathcal{E}$ is a computable numbering iff α is reducible to the standard numbering $W = \langle W_e : e \geq 0 \rangle$, so A can be chosen as the range of a function which reduces α to W.

In the following we will refer to the above observations on representations as Lachlan's representation theorem. The following lemma gives a criterion for the numberings induced by two representations of a computable family to be $\mathbf{0}'$-equivalent.

Lemma 1. *Let A and B be r.e. sets such that*

$$\{W_a : a \in A\} = \{W_b : b \in B\} \tag{1}$$

and let α and β be the numberings induced by the sets A and B respectively. Then α is $\mathbf{0}'$-reducible to β if and only if there exists a recursive function $g : \mathbb{N}\times\mathbb{N} \to \mathbb{N}$ such that, for any $a \in A$,

$$\lim_s g(a,s)\downarrow \ \& \ \lim_s g(a,s) \in B \ \& \ W_a = W_{\lim_s g(a,s)}. \tag{2}$$

Proof. The proof is a straightforward modification of the proof of Lemma 2.2 from [6].

Theorem 1. *Computable numberings of a computable* TxtEX*-learnable family of languages are pairwise $\mathbf{0}'$-equivalent.*

Proof. Let $\mathcal{C}$ be a computable TxtEX-learnable family, let M be a computable learner of $\mathcal{C}$, and let A and B be any r.e. sets such that

$$\mathcal{C} = \{W_a : a \in A\} = \{W_b : b \in B\}.$$

By Lachlan's representation theorem and by Lemma 1, it suffices to define a recursive function g satisfying (2).

In order to do so, we will show that there is a recursive function

$$\mathit{state} : \mathbb{N}^3 \to \{0,1\}$$

such that, for any numbers $a \in A$ and $b \in B$ the following hold.

$$W_a = W_b \Rightarrow \lim_s \mathit{state}(a,b,s)\downarrow \ \& \ \lim_s \mathit{state}(a,b,s) = 1 \tag{3}$$

$$W_a \neq W_b \Rightarrow \lim_s \mathit{state}(a,b,s)\downarrow \ \& \ \lim_s \mathit{state}(a,b,s) = 0 \tag{4}$$

Then the function g defined by letting $g(a,s)$ be the least $b \in B_s$ such that $\mathit{state}(a,b,s) = 1$ (if there is such a b and by letting $g(a,s) = 0$ otherwise) will have the desired properties. Namely, given $a \in A$, by (1), we may fix b minimal such that $b \in B$ and $W_a = W_b$. Then, given a stage s_0 such that $b \in B_{s_0}$, by (3) and (4), we may fix $s_1 \geq s_0$ such that, for $s \geq s_1$, $\mathit{state}(a,b,s) = 1$ and $\mathit{state}(a,b',s) = 0$ for all $b' < b$. So, for any stage $s \geq s_1$, $g(a,s) = b$.

The function $\mathit{state}(a,b,s)$ is defined by induction on s. Simultaneously with $\mathit{state}(a,b,s)$ we define a finite string $\sigma(a,b,s)$ over $\mathbb{N}$ and we let $\mathit{content}(\sigma(a,b,s))$ be the set of numbers occuring in $\sigma(a,b,s)$.

For $s = 0$ we let $\sigma(a,b,0) = \lambda$ and $\mathit{state}(a,b,0) = 1$. For the definition of $\sigma(a,b,s+1)$ and $\mathit{state}(a,b,s+1)$ we distinguish the following cases.

Case 1: $\mathit{content}(\sigma(a,b,s)) \not\subseteq W_{a,s} \cap W_{b,s}$.

Then let $\sigma(a,b,s+1) = \sigma(a,b,s)$ and $\mathit{state}(a,b,s+1) = 0$.

Case 2: $content(\sigma(a, b, s)) \subseteq W_{a,s} \cap W_{b,s}$ and

$$M(\sigma(a, b, s)) = M(\sigma(a, b, s)W_{a,s}) = M(\sigma(a, b, s)W_{b,s}).$$

(Here $W_{a,s}$ is viewed as the string of the elements of $W_{a,s}$ in the order of enumeration (and, similarly, for $W_{b,s}$).)

Then let $\sigma(a, b, s+1) = \sigma(a, b, s)$ and $state(a, b, s+1) = 1$.

Case 3: Otherwise, i.e., $content(\sigma(a, b, s)) \subseteq W_{a,s} \cap W_{b,s}$ and $M(\sigma(a, b, s)) \neq M(\sigma(a, b, s)W_{a,s})$ or $M(\sigma(a, b, s)) \neq M(\sigma(a, b, s)W_{b,s})$.

Then let $\sigma(a, b, s+1) = \sigma(a, b, s)W_{a,s}W_{b,s}$ if $M(\sigma(a, b, s)) \neq M(\sigma(a, b, s)W_{a,s})$ and $\sigma(a, b, s+1) = \sigma(a, b, s)W_{b,s}W_{a,s}$ otherwise. In either case let $state(a, b, s+1) = 0$.

This completes the definition of σ and *state*.

To show that the function *state* satisfies (3) and (4), fix $a \in A$ and $b \in B$. Note that, by definition,

$$\sigma(a, b, s) \sqsubseteq \sigma(a, b, s+1), \tag{5}$$

$$content(\sigma(a, b, s)) \subseteq W_{a,s} \cup W_{b,s}, \tag{6}$$

and

$$\sigma(a, b, s) \sqsubset \sigma(a, b, s+1) \Rightarrow content(\sigma(a, b, s)) \subseteq W_{a,s} \cap W_{b,s}. \tag{7}$$

Next we will show that there are only finitely many stages s such that Case 3 applies in the definition of $\sigma(a, b, s+1)$ and $state(a, b, s+1)$. For a contradiction assume that this happens infinitely often. Since, for $s < s'$ such that Case 3 applies to $s+1$ and $s'+1$, $content(\sigma(a, b, s+1)) = W_{a,s} \cup W_{b,s}$ and $content(\sigma(a, b, s+1)) \subseteq content(\sigma(a, b, s')) \subseteq W_{a,s'} \cap W_{b,s'}$, it follows that $W_a \cup W_b \subseteq W_a \cap W_b$ whence $W_a = W_b$. So the infinite sequence $\lim_s \sigma(a, b, s)$ is an enumeration (i.e., a text) of both W_a and W_b. Moreover, if Case 3 holds at stage $s+1$ then the extension $\sigma(a, b, s+1)$ of $\sigma(a, b, s)$ is chosen in such a way that $M(\sigma(a, b, s)) \neq M(\sigma)$ for some σ with $\sigma(a, b, s) \sqsubset \sigma \sqsubseteq \sigma(a, b, s+1)$. So on the sequence $\lim_s \sigma(a, b, s)$ the learner M changes its mind infinitely often, hence does not learn W_a contrary to assumption.

Now, since Case 3 applies only finitely often and since $\sigma(a, b, s)$ is only extended at a stage $s+1$ at which Case 3 applies, we may fix s_0 such that $\sigma(a, b, s) = \sigma(a, b, s_0)$ for all $s > s_0$ and Case 3 does not apply after stage s_0. Distinguish the following two cases.

First assume that $content(\sigma(a, b, s_0)) \not\subseteq W_a \cap W_b$. Then, by (6), $W_a \neq W_b$ and Case 1 applies to all stages $s > s_0$ whence $\lim_s state(a, b, s) = 0$. So (4) holds.

Finally assume that $content(\sigma(a, b, s_0)) \subseteq W_a \cap W_b$. Then, for $s_1 > s_0$ with $content(\sigma(a, b, s_0)) \subseteq W_{a,s_1} \cap W_{b,s_1}$, Case 2 applies to all stages $s > s_1$ whence $\lim_s state(a, b, s) = 1$. It remains to show that $W_a = W_b$ in this case. Now, by definition, $M(\sigma(a, b, s_0)) = M(\sigma(a, b, s_0)W_{a,s})$ and $M(\sigma(a, b, s_0)) = M(\sigma(a, b, s_0)W_{b,s})$, so given the enumeration of W_a obtained by the initial segments $\sigma(a, b, s_0)W_{a,s}$ $(s \geq s_1)$, the learner M does not make any changes in his prediction after stage s_1. Since M learns W_a this implies that $M(\sigma(a,b,s_0))$ is an index of W_a. By a similar argument, $M(\sigma(a,b,s_0))$ is an index of W_b too, whence $W_a = W_b$.

This completes the proof.

3 A Counterexample

Now we will give a counterexample to the converse part of the conjecture of Frank Stephan. Indeed, we will prove the following stronger statement.

Theorem 2. *There exists a computable family* $\mathcal{L} \notin \mathrm{TxtEX}$ *whose computable numberings are pairwise equivalent.*

Proof. We will only sketch the proof by discussing the strategies for meeting the requirements, the conflicts among these strategies and how these conflicts are resolved.

We will construct a computable numbering α such that for the family $\mathcal{L} = \{\alpha(x) : x \in \mathbb{N}\}$ of r.e. sets the following hold.

(a) All computable numberings of $\mathcal{L}$ are equivalent to α.
(b) $\mathcal{L}$ is not TxtEX-learnable.

Indeed, we will build a *positive* numbering α, i.e., a numbering α such that the relation $\alpha(x) = \alpha(y)$ is r.e. in x and y. Any positive numbering is minimal under reduction, [4], and, therefore, in order to ensure (a) it suffices to reduce all computable numberings of $\mathcal{L}$ to that numbering α. Moreover, in order to ensure (b), by Remark 1 it suffices to guarantee that no primitive recursive learner identifies $\mathcal{L}$.

Requirements. Let M_0, M_1, ... be a computable sequence of all primitive recursive learners, and let γ_0, γ_1, ... be a uniformly computable sequence of all possible computable numberings of computable families of r.e. sets. We fix any uniform approximation $\gamma_k^s(x)$ for this sequence. Then the numbering α has to meet the following requirements for all $k, e \in \mathbb{N}$:

$$\mathcal{P} : \ \alpha \text{ is a computable positive numbering.}$$

$$\mathcal{R}_k : \ \text{If } \gamma_k \text{ is a numbering of } \mathcal{L} \text{ then } \gamma_k \text{ is reducible to } \alpha.$$

$$\mathcal{M}_e : \ \text{There exists an } \alpha\text{-index } m \text{ such that } M_e \text{ fails to learn the set } \alpha(m).$$

We will refer to $\mathcal{R}_k$ as the kth *reduction requirement* and to $\mathcal{M}_e$ as the eth *nonlearning requirement.* The priority ordering among the requirements $\mathcal{R}_k$ and $\mathcal{M}_e$ is defined as usual by giving requirements with smaller index higher priority and by giving $\mathcal{R}_n$ higher priority than $\mathcal{M}_n$.

We identify each α-index n with a triple of numbers, $n = \langle e, i, j\rangle$, where the individual components of n have the following meaning:

- e means that the language $\alpha(n)$ might be used for diagonalizing against the learner M_e, i.e., for meeting the nonlearning requirement $\mathcal{M}_e$,
- i denotes the attempt number for trying to diagonalize against M_e (due to our strategy for meeting the higher priority reduction requirements, a single attempt might not suffice),
- j means that $\alpha(n)$ is the jth candidate in the ith attempt for diagonalizing against M_e.

We denote the components e, i, j of a triple $n = \langle e, i, j\rangle$ by $\pi_1(n), \pi_2(n)$, and $\pi_3(n)$ respectively. Moreover, we refer to the sets $\alpha(n)$ with $\pi_1(n) = e$ and $\pi_2(n) = i$ as the sets in *section* (e, i). (So sets in section (e, i) are reserved for the ith attempt for meeting $\mathcal{M}_e$.)

Strategies. For meeting the requirements $\mathcal{P}$ and $\mathcal{R}_k$ we have to take some precautions in the enumerations $\alpha^s(n)$ of the sets $\alpha(n)$.

Strategy for meeting $\mathcal{P}$. In order to make α positive we ensure, for all stages s and numbers m and n,

$$\alpha^s(m) = \alpha^s(n) \Rightarrow \forall\, t \geq s\ (\alpha^t(m) = \alpha^t(n)) \tag{8}$$

and

$$\alpha(m) = \alpha(n) \Rightarrow \exists t\ (\alpha^t(m) = \alpha^t(n)). \tag{9}$$

So, in particular, $\alpha(m) = \alpha(n)$ if and only if $\alpha^s(m) = \alpha^s(n)$ for some stage s. Obviously, this implies that $\{(m, n) : \alpha(m) = \alpha(n)\}$ is r.e.

Strategy for meeting $\mathcal{R}_k$. Let $b(n) = 2n$ and let $a(n) = 2n + 1$. Initially, we let

$$\alpha^0(n) = \{b(n)\} \tag{10}$$

and call $b(n)$ the *base element* of $\alpha(n)$. So, for $n \neq m$, $\alpha^0(n) \neq \alpha^0(m)$ and $\alpha^0(n)$ can be positively distinguished from $\alpha^0(m)$ by its base element. Numbers $a(n)$ may be enumerated into some sets $\alpha(m)$ later, where $a(s)$ will not enter any set $\alpha(m)$ before stage s.

Sets in different sections will be distinguishable by their base elements, i.e., the base element of a set $\alpha(n)$ will never be put into any set $\alpha(m)$ in a different section. For sets $\alpha(m)$ and $\alpha(n)$ in the same section (e, i), however, our strategy for meeting the nonlearning requirements may force us to enumerate the base element of $\alpha(n)$ into $\alpha(m)$. If this happens, the role of the base element of $\alpha(n)$ will be played by some new number $a(s)$ put into $\alpha(n)$ before $b(n)$ enters $\alpha(m)$ - unless we make $\alpha(n)$ and $\alpha(m)$ agree.

To be more precise, the enumeration of the sets in a given section (e, i) will obey the following rules. At any stage s there will be a distinguished set, $\alpha(\langle e, i, j_{e,i}(s)\rangle)$, called the *active set*, where $j_{e,i}(0) = 1$ and $j_{e,i}(s+1) \geq j_{e,i}(s)$ is defined at stage $s+1$. Then, at any stage $s \geq 1$,

- the sets $\alpha(\langle e, i, j\rangle)$ for $j < j_{e,i}(s)$ have been previously merged with the *primary set* of the section, $\alpha(\langle e, i, 0\rangle)$, i.e., $\alpha^s(\langle e, i, j\rangle) = \alpha^s(\langle e, i, 0\rangle)$ for $j < j_{e,i}(s)$;

- $\alpha^s(\langle e,i,j_{e,i}(s)\rangle)$ and $\alpha^s(\langle e,i,0\rangle)$ can be positively distinguished from each other, namely the base element $b(\langle e,i,j_{e,i}(s)\rangle)$ of $\alpha(\langle e,i,j_{e,i}(s)\rangle)$ is not in $\alpha^s(\langle e,i,0\rangle)$ and *some* element of $\alpha^{s-1}(\langle e,i,0\rangle)$, denoted by $\hat{b}(\langle e,i,0\rangle,s-1)$ below, is not in $\alpha^s(\langle e,i,j_{e,i}(s)\rangle)$; and
- the sets $\alpha(\langle e,i,j\rangle)$ for $j > j_{e,i}(s)$ are still in their initial states and can be positively distinguished from all other sets by their base elements, i.e., $\alpha^s(\langle e,i,j\rangle) = \{b(\langle e,i,j\rangle)\}$ and $b(\langle e,i,j\rangle) \notin \alpha^s(m)$ for all $m \neq \langle e,i,j\rangle$.

This will be achieved by initially letting $j_{e,i}(0) = 1$ and $\hat{b}(\langle e,i,0\rangle,0) = b(\langle e,i,0\rangle)$, and by allowing only the following procedures to be applied to the sets of section (e,i) at a stage $s+1>0$.

- COPY$(e,i,s+1)$.

$$\alpha^{s+1}(\langle e,i,j_{e,i}(s)\rangle) = \alpha^s(\langle e,i,j_{e,i}(s)\rangle) \cup [\alpha^s(\langle e,i,0\rangle) \setminus \{\hat{b}(\langle e,i,0\rangle,s)\}].$$

- MERGE$(e,i,s+1)$. For all $j \leq j_{e,i}(s)$ let

$$\alpha^{s+1}(\langle e,i,j\rangle) = \alpha^s(\langle e,i,0\rangle) \cup \alpha^s(\langle e,i,j_{e,i}(s)\rangle) \cup \{a(s+1)\}$$

  and set $\hat{b}(\langle e,i,0\rangle,s+1) = a(s+1)$ and $j_{e,i}(s+1) = j_{e,i}(s)+1$.

where these procedures are applied alternatingly starting with COPY and where at most one of the procedures is applied at any stage $s+1>0$. (If not stated otherwise, a parameter will maintain its value at stage $s+1$.)

Note that the above will ensure that (8) and (9) are satisfied whence α is positive.

Now, in order to show how the above will help us to meet the reduction requirements, fix k. Assuming that γ_k is a numbering of $\mathcal{L}$ we have to give a reduction function g_k from γ_k to α. Given x, we will define $g_k(x)$ as follows.

1. Wait for a stage $s_1 > x$ and a number n such that $b(n) \in \gamma_k^{s_1}(x)$, say $n = \langle e,i,j\rangle$. Distinguish the following two cases.
2. If $j \geq j_{e,i}(s_1)$ then let $g_k(x) = n$.
3. If $j < j_{e,i}(s_1)$ then wait for the least stage $s_2 \geq s_1$ such that $\gamma_k^{s_2}(x) = \alpha^{s_2}(\langle e,i,0\rangle)$ or $j_{e,i}(s_1) < j_{e,i}(s_2)$ or $\gamma_k^{s_2}(x) = \alpha^{s_2}(\langle e,i,j_{e,i}(s_1)\rangle)$. In the former two cases let $g_k(x) = \langle e,i,0\rangle$, in the latter case let $g_k(x) = \langle e,i,j_{e,i}(s_1)\rangle$.

Note that if we wait for a stage s_1 and n as above forever then $\gamma_k(x) \notin \mathcal{L}$, hence γ_k is not a numbering of $\mathcal{L}$. Similarly, if there is no stage $s_2 \geq s_1$ such that $j_{e,i}(s_1) < j_{e,i}(s_2)$ then only finitely many operations are applied to section (e,i). So there will be $s' \geq s_1$ at which all sets in the section have reached their final (finite) state, and $\gamma_k(x)$ will disagree from all of these sets. So, again, γ_k is not a numbering of $\mathcal{L}$. So we may conclude, that g_k is total. (In the actual construction we will define $g_k(x)$ according to the above procedure only if, for s_1 and n as above, $\pi_1(n) \geq k$. If $\pi_1(n) < k$ then the value of $g_k(x)$ will be specified depending on the outcomes of the strategies for meeting the nonlearning requirements $\mathcal{M}_0,\dots,\mathcal{M}_{k-1}$.)

It remains to argue that $\gamma_k(x) = \alpha(g_k(x))$. Note that if we let $g_k(x) = n$, $n = \langle e, i, j\rangle$, at stage s then

$$b(n) \in \gamma_k^s(x) \;\&\; \forall\, m \neq n\; (b(n) \notin \alpha^s(m))$$

(namely in case of 2. and in the third case of 3.) or

$$j = 0 \;\&\; [\; (b(n) \in \gamma_k^s(x) \cap \alpha^s(\langle e,i,0\rangle) \;\&\; b(n) \notin \alpha^s(\langle e,i,j_{e,i}(s)\rangle)) \text{ or }$$
$$\gamma_k^s(x) = \alpha^s(\langle e,i,0\rangle)\;].$$

Now, in the former case, if $\alpha(n)$ is not merged with $\alpha(\langle e,i,0\rangle)$, then $b(n)$ will positively distinguish $\alpha(n)$ from all sets $\alpha(m)$ with $m \neq n$ in the limit. So it suffices to show that if the first case applies and $\alpha(n)$ is merged with $\alpha(\langle e,i,0\rangle)$ or if the second case applies then $\gamma_k(x) = \alpha(\langle e,i,0\rangle)$. Here a problem may arise by a COPY operation applied at a stage $t > s$. Then the current approximation $\gamma_k^t(x)$ of $\gamma_k(x)$ may be a subset of both, $\alpha^t(\langle e,i,0\rangle)$ and $\alpha^t(\langle e,i,j_{e,i}(t-1)\rangle)$, and in the limit the latter two sets may disagree. (So it could turn out that $\gamma_k(x) = \alpha(\langle e,i,j_{e,i}(t-1)\rangle) \neq \alpha(\langle e,i,0\rangle)$.) We can prevent this from happening by requiring that any COPY operation has to wait for *k-confirmation.*

Here we say that stage $s+1$ is *k-confirmed* (w.r.t. section (e,i)) if, for any x and j such that $g_k^s(x) = \langle e,i,j\rangle$ for some $j < j_{e,i}(s)$, $\alpha^s(\langle e,i,0\rangle) \subseteq \gamma_k^s(x)$.

Note that if $s+1$ is k-confirmed then (for x and j as above), in particular, $\hat{b}(\langle e,i,0\rangle, s) \in \gamma_k^s(x)$, so if COPY is applied at stage s, still $\gamma_k^s(x) \not\subseteq \alpha^{s+1}(\langle e,i,j_{e,i}(s)\rangle)$. Hence if we limit the COPY operation to k-confirmed stages then the reduction g_k will be correct.

But what effect does this limitation may have on the $\mathcal{M}_e$ strategies? For a single k such that γ_k is a numbering of $\mathcal{L}$ there will be infinitely many k-confirmed stages (provided that we sufficiently slow down the enumeration of g_k), so that waiting for confirmation will not interfere with the intended action on section (e,i). For k such that γ_k is not a numbering of $\mathcal{L}$, however, we may wait for a k-confirmation forever.

So any attack on the nonlearning requirement $\mathcal{M}_e$ will be provided with a guess at which of the higher priority reduction requirements have a correct hypothesis. An attack based on guess $(i_0,\dots,i_e) \in \{0,1\}^{e+1}$ will wait for k-confirmation for those $k \leq e$ such that $i_k = 1$. If the guess is correct, waiting for confirmation will not interfere with the nonlearning startegy. On the other hand, if in the course of the construction it seems that an attack erroneously ignores some k, then the attack is abandoned and the sets in the section (e,i) used by the attack are all identified (by letting $\alpha(\langle e,i,j\rangle) = \{b(\langle e,i,j'\rangle) : j' \geq 0\} \cup \{a(n) : n \geq 0\}$). So the reduction g_k will be trivially correct on section (e,i).

Strategy for meeting $\mathcal{M}_e$. Given a section (e,i), in the ith attempt for meeting $\mathcal{M}_e$ we build a (finite or infinite) sequence of strings over $\mathbb{N}$, namely

$$\sigma_{e,i}^0 \sqsubset \sigma_{e,i}^1 \sqsubset \cdots \sqsubset \sigma_{e,i}^{j^*}$$

$(j^* \geq 0)$ or

$$\sigma_{e,i}^0 \sqsubset \sigma_{e,i}^1 \sqsubset \sigma_{e,i}^2 \sqsubset \dots$$

such that -if the ith attempt is the successful one- either

$$\begin{array}{c} content(\sigma^{j^*}_{e,i}) \subseteq \alpha(\langle e,i,0\rangle) \cap \alpha(\langle e,i,j^*+1\rangle) \ \& \\ M_e \text{ fails to learn } \alpha(\langle e,i,0\rangle) \text{ or } \alpha(\langle e,i,j^*+1\rangle) \end{array} \tag{11}$$

or

$$\begin{array}{c} \bigcup_{j\in\mathbb{N}} content(\sigma^{j}_{e,i}) = \alpha(\langle e,i,0\rangle) \ \& \\ M_e \text{ fails to learn } \alpha(\langle e,i,0\rangle) \text{ from the text } \sigma_{e,i} = \lim_{j\to\infty} \sigma^{j}_{e,i}. \end{array} \tag{12}$$

This is achieved by induction on $j \geq 1$, where string $\sigma^{j}_{e,i}$ is defined in the j-*cycle* given below ($\sigma^{0}_{e,i}$ is the empty string). Cycle j will affect only $\alpha(\langle e,i,0\rangle)$ and $\alpha(\langle e,i,j\rangle)$. If cycle j is started at stage s_j then $\alpha^{s_j}(\langle e,i,0\rangle) = content(\sigma^{j-1}_{e,i}) \cup \{\hat{b}(\langle e,i,0\rangle, s_j)\}$, j is active at stage s_j, i.e., $j = j_{e,i}(s_j)$, and $\alpha^{s_j-1}(\langle e,i,j\rangle)$ is still in its initial state.

Cycle j ($j \geq 1$).

1. Wait for the least stage $s'_j > s_j$ which is k-confirmed for $k \leq e$ such that $i_k = 1$ where $(i_0, \dots, i_e)$ is the guess underlying the attack.
   Perform COPY(e, i, s'_j).

2. Wait for the least stage $s''_j > s'_j$ such that

$$M_e(\sigma^{j-1}_{e,i}\hat{b}(\langle e,i,0\rangle, s_j)^{s''_j - s'_j}) \neq M_e(\sigma^{j-1}_{e,i}) \tag{13}$$

   or

$$M_e(\sigma^{j-1}_{e,i} b(\langle e,i,j\rangle)^{s''_j - s'_j}) \neq M_e(\sigma^{j-1}_{e,i}). \tag{14}$$

   Then MERGE(e, i, s''_j), let

$$\sigma^{j}_{e,i} = \begin{cases} \sigma^{j-1}_{e,i}\ \hat{b}(\langle e,i,0\rangle, s_j)^{s''_j - s'_j}\ b(\langle e,i,j\rangle) & \text{if (13) holds} \\ \sigma^{j-1}_{e,i}\ b(\langle e,i,j\rangle)^{s''_j - s'_j}\ \hat{b}(\langle e,i,0\rangle, s_j) & \text{otherwise,} \end{cases}$$

   and start cycle $j+1$.

The success of this strategy (if based on the correct guess, hence not stuck in step 1 of any cycle) is shown as follows. If the strategy gets stuck in step 2 of cycle j then, for $j^* = j-1$, (11) holds. Namely, the learner M_e will make the same prediction for the text $\sigma^{j-1}_{e,i}\hat{b}(\langle e,i,0\rangle, s_j)^{\omega}$ of $\alpha(\langle e,i,0\rangle) = \alpha^{s_j}(\langle e,i,0\rangle)$ and for the text $\sigma^{j-1}_{e,i} b(\langle e,i,j\rangle)^{\omega}$ of $\alpha(\langle e,i,j\rangle) = \alpha^{s_j}(\langle e,i,j_{e,i}(s_j)\rangle)$ though these sets differ. If all cycles are completed then (12) holds since M_e makes infinitely many changes on the text $\sigma_{e,i} = \lim_{j\to\infty} \sigma^{j}_{e,i}$ for $\alpha(\langle e,i,0\rangle)$.

This completes our discussion of the strategies and the basic ideas underlying the proof.

Acknowledgments. The second author's research was partially supported by the State Grants of Kazakhstan "Best Teacher of Higher Education" for 2006 and 2007. The third author was partially supported by the Grant RFBR-08-01-00336 and the Grant for Leading Scientific Schools SS-4413 for 2006.

References

1. Jain, S., Osherson, D., Royer, J.S., Sharma, A. (eds.): Systems That Learn. An Introduction to Learning Theory, 2nd edn. The MIT Press, Cambridge Massachusetts (1999)
2. Ershov, Y. L.: Theory of Numberings. In: Handbook of Computability Theory, pp. 473–503. North-Holland, Amsterdam (1999)
3. Badaev, S.A., Goncharov, S.S.: The Theory of Numberings: Open Problems. In: Cholak, P.A., Lempp, S., Lerman, M., Shore, R.A. (eds.) Computability Theory and its Applications. Current Trends and Open Problems, pp. 23–38, Amer. Math. Soc., Providence (2000)
4. Ershov, Y.L.: Theory of Numberings. Nauka, Moscow (1977)
5. Gold, E.M.: Language Identification in the Limit. Information and Control 10, 447–474 (1967)
6. Badaev, S.A., Goncharov, S.S., Podzorov, S., Podzorov, S.Y., Sorbi, A.: Algebraic Properties of Rogers Semilattices of Arithmetical Numberings. In: Cooper, S.B., Goncharov, S.S. (eds.) Computability and Models, pp. 45–77. Kluwer / Plenum Publishers, New York (2003)

A Practical Parameterized Algorithm for the Individual Haplotyping Problem MLF*

Minzhu Xie[1,2], Jianxin Wang[1,**], and Jianer Chen[1]

[1] School of Information Science and Engineering,
Central South University
[2] College of Physics and Information Science,
Hunan Normal University
xieminzhu@hotmail.com,jxwang@mail.csu.edu.cn
http://netlab.csu.edu.cn

Abstract. The individual haplotyping problem Minimum Letter Flip (MLF) is a computational problem that, given a set of aligned DNA sequence fragment data of an individual, induces the corresponding haplotypes by flipping minimum SNPs. There has been no practical exact algorithm to solve the problem. In DNA sequencing experiments, due to technical limits, the maximum length of a fragment sequenced directly is about 1kb. In consequence, with a genome-average SNP density of 1.84 SNPs per 1 kb of DNA sequence, the maximum number k_1 of SNP sites that a fragment covers is usually small. Moreover, in order to save time and money, the maximum number k_2 of fragments that cover a SNP site is usually no more than 19. Based on the properties of fragment data, the current paper introduces a new parameterized algorithm of running time $O(nk_2 2^{k_2} + mlogm + mk_1)$, where m is the number of fragments, n is the number of SNP sites. The algorithm solves the MLF problem efficiently even if m and n are large, and is more practical in real biological applications.

Keywords: SNP (single-nucleotide polymorphism); haplotype; MLF (Minimum Letter Flip); NP-hardness; parameterized algorithm.

1 Introduction

A *single nucleotide polymorphism* (SNP) is a single base mutation of a DNA sequence that occurs in at least 1% of the population. SNPs are the predominant form of human genetic variation, and more than 3 million SNPs are distributed throughout the human genome [1, 2]. Detection of SNPs is used in identifying

* This research was supported in part by the National Natural Science Foundation of China under Grant Nos. 60433020 and 60773111, the Program for New Century Excellent Talents in University No. NCET-05-0683, the Program for Changjiang Scholars and Innovative Research Team in University No. IRT0661, and the Scientific Research Fund of Hunan Provincial Education Department under Grant No.06C526.
** The corresponding author.

M. Agrawal et al. (Eds.): TAMC 2008, LNCS 4978, pp. 433–444, 2008.

biomedically important genes for the diagnosis and therapy of human hereditary diseases, in identification of individual and descendant, and in the analysis of genetic relations of populations.

In humans and other diploid organisms, chromosomes are paired up. A *haplotype* describes the SNP sequence of a chromosome. In Fig. 1, the haplotypes of the individual are "ATACG" and "GCATG".

Because two chromosomes paired up are not completely identical, the haplotype contains more information than the genotype. *Haplotyping*, i.e. identification of chromosome haplotypes, plays an important role in SNP applications [3]. Stephens *et al.* [4] identified 3899 SNPs that were present within 313 genes from 82 unrelated individuals of diverse ancestry. Their analysis of the pattern of haplotype variation strongly supports the recent expansion of human populations. Based on linkage studies of SNPs and the association analysis between haplotypes and type 2 diabetes, Horikawa *et al.* [5] localized the gene *NIDDM1* to the distal long arm of chromosome 2 and found 3 SNPs in *CAPN10* associated with type 2 diabetes.

A..T..A..C..G.

G..C..A..T..G.

Fig. 1. SNPs

Haplotyping has been time-consuming and expensive using biological techniques. Therefore, effective computational techniques have been in demand for solving the haplotyping problem. A number of combinatorial versions of the haplotyping problem have been proposed. In the current paper, we will be concentrated on an important version *Minimum Letter Flips* (MLF) [6, 7] that comes from the *individual haplotyping problem* [8]: Given a set of aligned SNP sequence fragment data from the two copies of a chromosome, find a minimum number of SNPs to correct so that there exist two haplotypes compatible with the corrected fragments. The MLF problem is also called the *Minimun Error Correction* (MEC) problem [9, 10]. The problem is NP-hard [9] and there has been no practical exact algorithm to solve the problem [6, 10, 11].

By carefully studying related properties of fragment data, we have found the following fact. In all sequencing centers, due to technical limits, the sequencing instruments such as ABI 3730 and MageBACE can only sequence DNA fragments whose maximum length is about 1000 nucleotide bases. In consequence, with a genome-average SNP density of 1.84 SNPs per 1 kb of DNA sequence [12], the maximum number k_1 of SNP sites that a fragment covers is small. Moreover, in order to save time and money, the maximum number k_2 of fragments that cover a SNP site is usually no more than 19 [1, 13, 14].

Based on the observation above, the current paper proposes a new algorithm of time $O(nk_2 2^{k_2} + mlogm + mk_1)$, where m is the number of fragments and n is the number of SNP sites. The algorithm solves the problem efficiently even if m and n are large, and is more practical in real biological applications.

2 The Individual Haplotyping MLF Problem

For a pair of chromosomes, a SNP site where both haplotypes have the same nucleotide is called a *homozygous* site, and a SNP site where both haplotypes

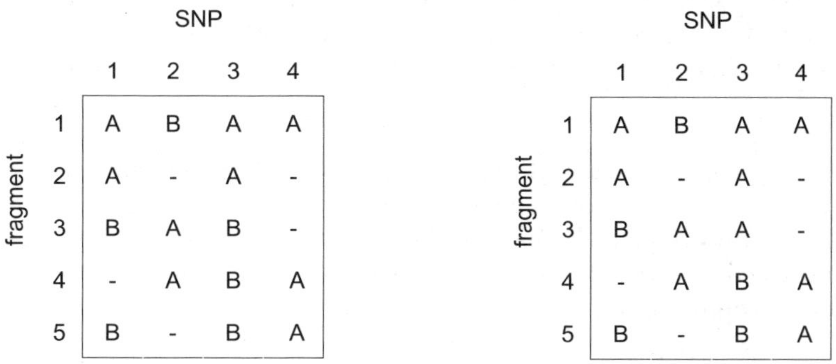

(a) Error-free (b) $M_{3,3}$ is an error read

Fig. 2. SNP Matrices

have different nucleotides is called a *heterozygous* site. To reduce the complexity, a haplotype can be represented as a string over a two-letter alphabet {A, B} rather than the four-letter alphabet {A, C, G, T} , where 'A' denotes the major allele and 'B' denotes the minor. In Fig. 1, the haplotypes can be represented by "ABABB" and "BAAAB".

Only considering SNPs, the aligned DNA sequence fragment data of an individual can be represented as an $m \times n$ SNP matrix M over the alphabet {A,B,-}, in which n columns represent a sequence of SNPs according to the order of sites in a chromosome and m rows represent m fragments. Fig. 2 shows two 5×4 SNP matrices. In the matrix M, the ith row's value at the jth column is denoted by $M_{i,j}$, which equals the ith fragment's value at the jth SNP site. If the value of the ith fragment at the jth SNP site misses (i.e. there is a hole in the fragment) or the ith fragment doesn't cover the jth SNP site, then $M_{i,j}$ takes the value "$-$" (the value "$-$" will be called the *empty value*).

The following are some definitions related to the SNP matrix M.

We say that the ith row *covers* the jth column if there are two indices k and r such that $k \leq j \leq r$, and both $M_{i,k}$ and $M_{i,r}$ are not empty. For example, in Fig. 2(a), row 2 covers columns 1, 2 and 3.

The set of (ordered) rows covering the jth column is denoted by *rowset*(j). The first and the last column that the ith row covers are denoted by $left(i)$ and $right(i)$ respectively.

If $M_{i,j} \neq$ " $-$ ", $M_{k,j} \neq$ " $-$ " and $M_{i,j} \neq M_{k,j}$, then the ith and kth rows of M are said to *conflict* at column j. If the ith and kth rows of M do not conflict at any column then they are *compatible.*

A SNP matrix M is *feasible* if its rows can be partitioned into two classes such that the rows in each class are all compatible.

Obviously, a set of DNA sequence fragment data without error corresponds to a feasible SNP matrix, with each of the compatible classes in the matrix corresponding to one of a pair of haplotypes. As in Fig. 2(a), it is easy to see fragments 1 and 2 are from a chromosome and fragments 3, 4 and 5 are from another chromosome. So it is easy to infer that a haplotype is "ABAA" and another is "BABA".

However, because of contaminants and read errors during the sequencing process, the SNP matrix corresponding to the fragment data is in general not feasible. This fact has resulted in the following combinatorial optimization problem [6, 9, 10]:

> *Minimum Letter Flips* (MLF): Given a SNP matrix M, flip (or correct) a minimum number of elements ("A" into "B" and vice versa) so that the resulting matrix is feasible, i.e. the corrected SNP fragments can be divided into two disjoint sets of pairwise compatible fragments, with each set determining a haplotype.

With a SNP matrix M, we will denote MLF(M) a solution to the MLF problem (i.e. the minimum number of elements have to be flipped to make M feasible). As for the SNP matrix M in Fig. 2(b), MLF(M)=1 because with no element flipped, M is not feasible, and once $M_{3,3}$ is flipped to "B", M becomes feasible.

The MLF problem is NP-hard [9]. Wang *et al.* [10] proposed a branch-and-bound algorithm whose time complexity is $O(2^m)$, which is impractical when the number of fragments m is large. There has been no more efficient exact algorithm [6, 10, 11]. Therefore it is of practical importance to introduce more efficient algorithms.

3 Parameterized MLF Algorithm

Currently the main method to identify SNPs is DNA direct sequencing [15]. The prevailing method to sequence a DNA fragment is the Sanger's technology of DNA sequencing with chain-termination inhibitors [16], which cannot sequence any DNA fragment longer than 1200 bases due to the technique limits. A recent research [17] revealed that two copies of the human genome differ from one another by approximately 0.5% of nucleotide sites, and the number of SNP sites in a fragment read is very small and no more than 10 according to the available data in spite of the varying distribution density along a chromosome [3, 18].

In order to save money and time, in DNA sequencing experiments, the fragment coverage is also small. In Celera's whole-genome shotgun assembly of the human genome, the fragment average coverage is 5.11 [1], and in the human genome project of the International Human Genome Sequencing Consortium, the fragment average coverage is 4.5 [14]. Huson *et al.* [13] have analyzed the fragment data of the human genome project of Celera's, and given a fragment coverage plot. Although the fragment covering rate is not the same at all sites along the whole genome, the plot shows that most sites are covered by 5 fragments, and that the maximum of fragments covering a site is no more than 19. Therefore, for an SNP site, compared with the total number of fragments, the number of fragments covering the SNP site is very small.

Based on the observations above, we introduce the following parameterized condition.

Definition 1. *The* (k_1, k_2) *parameterized condition*: the number of SNP sites covered by a single fragment is bounded by k_1, and each SNP site is covered by no more than k_2 fragments.

Accordingly, in a SNP matrix satisfying the (k_1, k_2) parameterized condition, each row covers at most k_1 columns and each column is covered by at most k_2 rows. For an $m \times n$ SNP matrix M, the parameters k_1 and k_2 can be obtained by scanning all rows of M. In the worst case, $k_1 = n$ and $k_2 = m$. But as to the fragment data of Celera's human genome project, k_2 is no more than 19 [13].

For a solution to the MLF problem for a SNP matrix M, after flipping the corresponding elements, all the rows of M can be partitioned into two classes H_0 and H_1, such that all pairs of rows in the same class are compatible.

Definition 2. Let R be a subset of (ordered) rows of a SNP matrix M. A *partition function* P on R maps each row in R to one of the values $\{0, 1\}$. Suppose that R contains $h > 0$ rows, then a partition function P on R can be coded by an h-digit binary number in $\{0, 1\}$, where the i-th digit is the P value of the ith row in R. For completeness, if $R = \varnothing$, we also define a unique partition function P, which is coded by -1.

For briefness, a partition function defined on *rowset*(j) is called a partition function at column j. For a SNP matrix M satisfying the (k_1, k_2) parameterized condition, there are at most 2^{k_2} different partition functions at column j.

Let R be a set of rows of the matrix M, and P be a partition function on R. For a subset R' of R, the partition function P' on R' obtained by restricting P on the subset R' is called the *projection* of P on R', and P is called an *extension* of P' on R.

Definition 3. Fix a j. Let P be a partition function on a row set R. Define $V_E[P, j]$ to be any subset S of elements of M that satisfies the following conditions:

(1) For each element $M_{r,k}$ in S: $1 \le k \le j$.

(2) After flipping the elements of S, there is a partition (H_0, H_1) of all rows of M such that any two rows in the same class do not conflict at any column from 1 to j, and for any row $i \in R$, row i is in the class H_q if and only if $P(i) = q$, for $q \in \{0, 1\}$.

Definition 4. Fix a j. Let P be a partition function at column j. Define $S_E[P, j]$ to be any $V_E[P, j]$ that minimizes the number of elements in $V_E[P, j]$. And $E[P, j]$ is defined as the number of elements in $S_E[P, j]$.

From Definitions 3 and 4, it is easy to verify that the following equation holds true:

$$\mathrm{MLF}(M) = \min_{P:\, P \text{ is a partition function at column } n} (E[P, n]) \tag{1}$$

Let P be a partition function at column j, for $k \in \{0, 1\}$, $N_A(P, j, k)$ denotes the number of the rows whose P value is k, and whose value at column j is

```
CompFlips(j, P, F)    // F denotes F(P, j)
      // A_s[k] denotes N_A(P, j, k), B_s[k] denotes N_B(P, j, k), k = 0, 1
1.    A_s[0] = B_s[0] = A_s[1] = B_s[1] = 0; k = 0; P_tmp = P; F = ∅;
2.    for i=the first row .. the last row of rowset(j) do
2.1.      k =the least significant bit of P_tmp ;    //k = P(i)
2.2.      P_tmp right shift 1 bit;
2.3.      if M_{i,j} ='A' then A_s[k] = A_s[k] + 1;
2.4.      if M_{i,j} ='B' then B_s[k] = B_s[k] + 1;
3.    for k = 0..1 do
3.1.      if A_s[k] > B_s[k] then Minor[k] ='B';
3.2.      else Minor[k] ='A';
4.    P_tmp = P;
5.    for i = the first row .. the last row of rowset(j) do
5.1.      k = the least significant bit of P_tmp ; P_tmp right shift 1 bit;
5.2.      if M_{i,j} = Minor[k] then F = F ∪ M_{i,j};
```

Fig. 3. Function **CompFlips**

'A', i.e. $N_A(P, j, k) = |\ \{i \mid P(i) = k \wedge M_{i,j} = \text{'}A\text{'}\}\ |$; similarly, $|\ \{i \mid P(i) = k \wedge M_{i,j} = \text{'}B\text{'}\}\ |$ is denoted by $N_B(P, j, k)$. If $N_A(P, j, k) > N_B(P, j, k)$, then $Minor(P, j, k) = \text{'}B\text{'}$, otherwise $Minor(P, j, k) = \text{'}A\text{'}$.

Under the condition that the rows covering column j are partitioned by P and the rows partitioned into the same class don't conflict at column j, the values at column j of some rows have to be flipped to avoid confliction. The number of the elements flipped at column j is minimal if and only if those $M_{i,j}$, whose value equal to $Minor(P, j, P(i))$, are flipped. We denote the set of the flipped elements above by $F(P, j)$, i.e. $F(P, j) = \{M_{i,j} \mid M_{i,j} = Minor(P, j, P(i))\}$.

The function **CompFlips** to compute $F(P, j)$ is given in Fig. 3, whose time complexity is $O(k_2)$.

So $S_E[P, 1]$ and $E[P, 1]$ can be obtained as follows.

$$S_E[P, 1] = F(P, 1) \tag{2}$$

$$E[P, 1] = |F(P, 1)|, \quad \text{i.e. the number of elements in } F(P, 1) \tag{3}$$

In order to present our algorithm, we need to extend the above concepts from one column to two columns as follows. Let the set of all rows that cover both columns j_1 and j_2 be $R_c(j_1, j_2)$.

Definition 5. Fix a j. Let P' be a partition function defined on $R_c(j, j+1)$. Define $S_B[P', j]$ to be any $V_E[P', j]$ that minimizes the number of elements in $V_E[P', j]$. And $B[P', j]$ is defined as the number of elements in $S_B[P', j]$.

Given j and a partition function P' on $R_c(j, j+1)$. If $E[P, j]$ and $S_E[P, j]$ are known for every extension P of P' on *rowset*(j), then $B[P', j]$ and $S_B[P', j]$ can be calculated by the following equations:

$$B[P', j] = \min_{P:\, P \text{ is an extension of } P' \text{ on } rowset(j)} (E[P, j]) \tag{4}$$

$$S_B[P', j] = S_E[P, j] \mid P \text{ minimizes } E[P, j] \tag{5}$$

Inversely, for each partition function P on *rowset*(j), since $R_c(j-1, j)$ is a subset of *rowset*(j), the project P' of P on $R_c(j-1, j)$ is unique. Once $B[P', j-1]$ and $S_B[P', j-1]$ are known, $E[P, j]$ and $S_E[P, j]$ can be calculated according to the following equations, whose correctness can be proven similarly as Equation (2) and (3).

$$S_E[P, j] = S_B[P', j-1] \cup F(P, j) \tag{6}$$

$$E[P, j] = B[P', j-1] + \mid F(P, j) \mid \tag{7}$$

Based on the equations above, the solution to the MLF problem for a SNP matrix M can be obtained as follows: firstly, $E[P, 1]$ and $S_E[P, 1]$ can be obtained according to Equations (2) and (3) for all partition functions P at column 1; secondly, $B[P', 1]$ and $S_B[P', 1]$ can be obtained by using Equations (4) and (5) for all partition functions P' on $R_c(1, 2)$; thirdly, $E[P, 2]$ and $S_E[P, 2]$ can be obtained by using Equations (6) and (7) for all partition functions P on *rowset*(2); and so on, at last $E[P, n]$ and $S_E[P, n]$ can be obtained for all partition functions P at column n. Once $E[P, n]$ and $S_E[P, n]$ for all possible P are known, a solution to the MLF problem for M can be obtained by using Equation (1). Please see Fig. 4 for the details of our P-MLF algorithm.

Theorem 1. *For an $m \times n$ SNP matrix M, if M satisfies the (k_1, k_2) parameterized condition, then the P-MLF algorithm solves the MLF problem correctly in time $O(nk_2 2^{k_2} + mlogm + mk_1)$ and space $O(mk_1 2^{k_2} + nk_2)$.*

Proof. The P-MLF algorithm is based on Eqs. (1)-(7). Eqs. (2)-(7) have been proven in the above discussion and the correctness of Eq. (1) is obvious. Given an $m \times n$ SNP matrix M satisfying the (k_1, k_2) parameterized condition, consider the following storage structure: each row keeps the first and the last column that the row covers, i.e. its *left* and *right* value, and its values at the columns from its left column to its right column. In such a storage structure, M takes space $O(mk_1)$. It is easy to see that *rowset* takes space $O(nk_2)$, H takes space $O(n)$, E and B take space $O(2^{k_2})$, and S_E and S_B take space $O(mk_1 2^{k_2})$. In summary, the space complexity of the algorithm is $O(mk_1 2^{k_2} + nk_2)$.

Now we discuss the time complexity of the P-MLF algorithm. In Step 1, sorting takes time $O(mlogm)$. All *rowsets* can be obtained by scanning the rows only once, which takes time $O(mk_1)$. Since M satisfies the (k_1, k_2) parameterized condition, there are no more than k_2 rows covering a column. Therefore, the

Algorithm P-MLF
input: an $m \times n$ SNP matrix M
output: a solution to the MLF problem for M
1. **initiation**: sort the rows in M in ascending order such that for any two rows i_1 and i_2, if $i_1 < i_2$, then $left(i_1) \leq left(i_2)$; for each column j, calculate an ordered set *rowset*(j) and the number $H[j]$ of the rows that cover column j;
2. **for** $P = 0..2^{H[1]} - 1$ **do** // P is coded by a binary number
2.1. **CompFlips**$(1, P, S_E[P])$; //calculate $S_E[P, 1]$ using Eq. (2)
2.2. $E[P] =| S_E[P] |$; //calculate $E[P, 1]$ using Eq. (3),
3. j=1;
4. **while** $j < n$ **do** //recursion based on Eqs. (4)-(7)
// MAX denotes the maximum integer
4.1. calculate N_C, the number of rows that cover both columns j and $j+1$, and a vector **Bits** such that **Bits**$[i]$=1 denotes the ith row of *rowset*(j) covers column $j+1$;
4.2. **for** $P' = 0..2^{N_C} - 1$ **do** $B[P']$=MAX;
4.3. **for** $P = 0..2^{H[j]} - 1$ **do**
4.3.1. calculate the project P' of P on $R_c(j, j+1)$ using **Bits**. //Eqs. (4) and (5)
4.3.2. **if** $B[P'] > E[P]$ **then** $B[P'] = E[P]$; $S_B[P'] = S_E[P]$;
4.4. $j++$; //next column
4.5. **for** $P' = 0..2^{N_C} - 1$ **do**
4.5.1. **for** each extensions P of P' on *rowset*(j) **do**
4.5.1.2. **CompFlips**(j, P, F);
4.5.1.3. $S_E[P] = S_B[P'] \cup F$; $E[P] = B[P'] + | F |$; //Eqs. (6) and (7) //Eq. (1)
5. output the minimum $E[P]$ and the corresponding $S_E[P]$ ($P = 0..2^{H[n]}-1$).

Fig. 4. P-MLF Algorithm

function **CompFlips** takes time $O(k_2)$, and for each column j, $H[j] \leq k_2$. In consequence, Step 2 takes time $O(k_2 2^{k_2})$. In Step 4.1, scanning *rowset*(j) and *rowset*$(j+1)$ simultaneously can obtain N_C and **Bits**, and takes time $O(k_2)$. Step 4.2 takes time $O(2^{k_2})$, and Step 4.3 takes time $O(k_2 2^{k_2})$. In Step 4.5, for each P', there are $2^{H[j]-N_C}$ extensions of P' on *rowset*(j). Given P', an extension of P' can be obtained by a bit-or operation in time $O(1)$. In all, Step 4.5 takes time $O(k_2 2^{N_c} 2^{k_2-N_c})$. Then Step 4 is iterated $n-1$ times and takes time $O(nk_2 2^{k_2})$. Step 5 takes time $O(2^{k_2})$. In summary, the time complexity of the algorithm is $O(nk_2 2^{k_2} + mlogm + mk_1)$. This completes the proof. □

4 Experimental Results

To the best of our knowledge, real DNA sequence fragment data in the public domain are not available, References [15, 19, 20] used computer-generated

Table 1. Comparison of performances of two algorithms

Parameters			$Tavg(s)$ [a]		$Tmax(s)$ [a]		R [a]	
n	m	e	B-MLF	P-MLF	B-MLF	P-MLF	B-MLF	P-MLF
		0.01	0.04	0.002	0.13	0.011	0.985	0.984
16	32	0.03	0.40	0.002	2.34	0.011	0.978	0.980
		0.05	5.62	0.002	56.8	0.013	0.971	0.971
		0.01	186.9	0.002	738.5	0.018	0.962	0.963
22	44	0.03	200.1	0.003	1503	0.019	0.958	0.958
		0.05	532.3	0.003	3214	0.029	0.953	0.952
		0.01	-	0.013	>96 hours	0.030	-	0.949
50	100	0.03	-	0.018	>96 hours	0.037	-	0.945
		0.05	-	0.025	>96 hours	0.038	-	0.943

[a] All experiments are repeated 100 times except for B-MLF with $n = 50$ and $m = 100$. $Tavg$ (the average time), $Tmax$ (the maximum running time) and R (the reconstruction rate) are over the repeated experiments with the same parameters.

simulated data under some realistic assumptions to compare the algorithms on the haplotypes assembly problem. Accordingly, we generated artificial fragment data in the same way and using the same parameters as in the above references.

In order to make the generated data have the same statistical features as the real data, a widely used shotgun assembly simulator Celsim [21] is adopted. At first a haplotype h_1 of length n is generated at random, then another haplotype h_2 of the same length is generated by flipping every char of h_1 with the probability of d, then Celsim is invoked to generated m fragments whose lengths are between $lMin$ and $lMax$. At last the output fragments are processed to plant reading errors with probability e and empty values with probability p. In the DNA sequencing experiments, fragment coverage rate c is about 10 [13]. In our experiments, the parameters are as follows: fragment coverage rate $c = 10$, the difference rate between two haplotypes d =10%, the minimal length of fragment $lMin = 3$, the maximal length of fragment $lMax = 7$ and empty values probability $p = 2\%$.

The length of haplotype n, the number of fragment m ($m = 2 \times n \times c/(lMax + lMin)$) and the reading error probability e are varied to compare the performance of Wang et al.'s algorithms B-MLF [10] and our algorithms P-MLF. Please refer to [20] and [21] for the details about how to generate artificial data.

P-MLF is implemented in C++ and B-MLF comes from Wang [10]. We ran our experiments on a Linux server (4 Intel Xeon 3.6GHz CPU and 4GByte RAM). In the experiments, we compare the running time and the reconstruction rate of haplotypes [10] between both the algorithms. The reconstruction rate of haplotypes is defined as the ratio of the number of the SNP sites that are correctly inferred by an algorithm to the total number of the SNP sites of the haplotypes.

From Table 1, we can see when n and m increase, the running time of B-MLF increases sharply. When $n = 50$ and $m = 100$, B-MLF cannot work out a solution in 4 days, but the running time of P-MLF is still small and less than 1 second. We also can see when the reading error probability e is small both the algorithms have good performance in reconstructing the haplotypes.

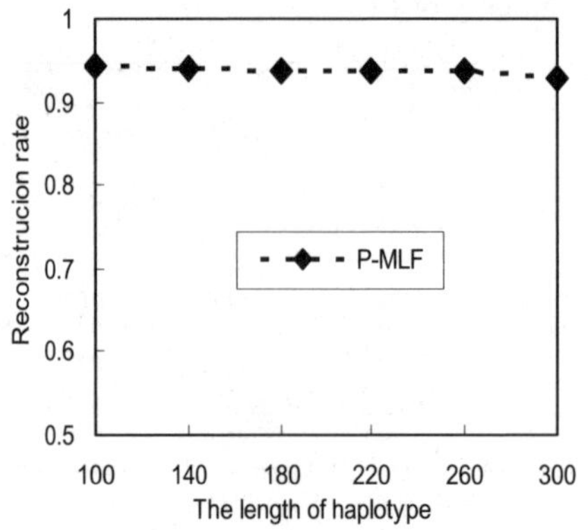

(a) Reconstruction rate

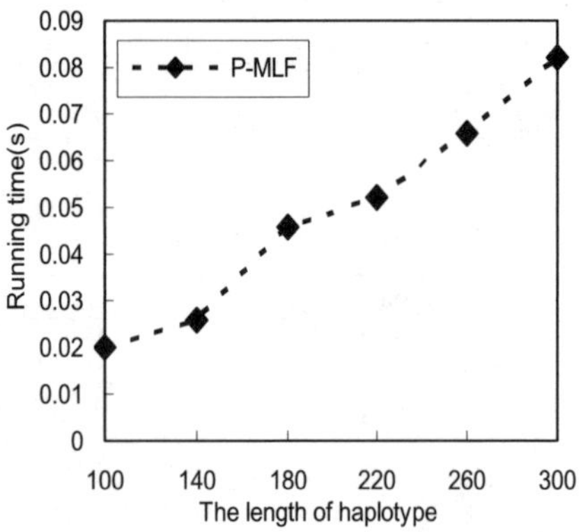

(b) Running time

Fig. 5. The performance of P-MLF

From Fig. 5, we can also see that P-MLF still has good performance when the length of the haplotypes increases to 300.

5 Conclusion

Haplotyping plays a more and more important role in some regions of genetics such as locating of genes, designing of drugs and forensic applications. MLF is an important computational model to infer the haplotypes of an individual from one's aligned SNPs fragment data by flipping the minimum number of SNPs. MLF has been proven to be NP-hard. To solve the problem, Wang et al. [10] proposed a branch-and-bound algorithm. Because it is of $O(2^m)$ time complexity, Wang et al.'s alogrithm is impractical when the number of fragments m is large. Based on the fact that the maximum number of fragments covering a SNP site is small (usually no more than 19 [13]), the current paper introduced a new parameterized algorithm P-MLF to solve the MLF problem. With the fragments of maximum length k_1 and the maximum number k_2 of fragments covering a SNP site, the P-MLF algorithm can solve the MLF problem in time $O(nk_2 2^{k_2} + mlogm + mk_1)$ and in space $O(mk_1 2^{k_2} + nk_2)$. Compared with other exact algorithms, the P-MLF algorithm has good scalability and is more efficient in practice. As for other MLF type computational models such as MEC/GI (MEC with Genotype Information) [10], the P-MLF algorithm can also be applicable with a little change to ensure that the haplotypes reconstructed by the algorithm make up the input genotype.

Acknowledgement. We thank Dr. Rui-Sheng Wang and Prof. Gene Myers for their kindly providing us with the source codes of B-MLF and Celsim respectively.

References

[1] Venter, J.C., Adams, M.D., Myers, E.W., et al.: The sequence of the human genome. Science 291(5507), 1304–1351 (2001)
[2] The International HapMap Consortium: A haplotype map of the human genome. Nature 437(7063) 1299–1320 (2005)
[3] Gabriel, S.B., Schaffner, S.F., Nguyen, H., et al.: The structure of haplotype blocks in the human genome. Science 296(5576), 2225–2229 (2002)
[4] Stephens, J.C., Schneider, J.A., Tanguay, D.A., et al.: Haplotype variation and linkage disequilibrium in 313 human genes. Science 293(5529), 489–493 (2001)
[5] Horikawa, Y., Oda, N., Cox, N.J., et al.: Genetic variation in the gene encoding calpain-10 is associated with type 2 diabetes mellitus. Nature Genetics 26(2), 163–175 (2000)
[6] Greenberg, H.J., Hart, W.E., Lancia, G.: Opportunities for combinatorial optimization in computational biology. INFORMS J. Comput. 16(3), 211–231 (2004)
[7] Zhao, Y.Y., Wu, L.Y., Zhang, J.H., Wang, R.S., Zhang, X.S.: Haplotype assembly from aligned weighted snp fragments. Computational Biology and Chemistry 29(4), 281–287 (2005)
[8] Lancia, G., Bafna, V., Istrail, S., Lippert, R., Schwartz, R.: Snps problems, complexity and algorithms. In: Meyer auf der Heide, F. (ed.) ESA 2001. LNCS, vol. 2161, pp. 182–193. Springer, Heidelberg (2001)
[9] Lippert, R., Schwartz, R., Lancia, G., Istrail, S.: Algorithmic strategies for the single nucleotide polymorphism haplotype assembly problem. Brief. Bioinform 3(1), 1–9 (2002)
[10] Wang, R.S., Wu, L.Y., Li, Z.P., Zhang, X.S.: Haplotype reconstruction from snp fragments by minimum error correction. Bioinformatics 21(10), 2456–2462 (2005)
[11] Bonizzoni, P., Vedova, G.D., Dondi, R., Li, J.: The haplotyping problem: an overview of computational models and solutions. J. Comp. Sci. Technol. 18(6), 675–688 (2003)
[12] Chen, C., Wang, J., Cohen, B.: The strength of selection on ultraconserved elements in the human genome. The American Journal of Human Genetics 80(4), 692–704 (2007)
[13] Huson, D.H., Halpern, A.L., Lai, Z., Myers, E.W., Reinert, K., Sutton, G.G.: Comparing assemblies using fragments and mate-pairs. In: Gascuel, O., Moret, B.M.E. (eds.) WABI 2001. LNCS, vol. 2149, pp. 294–306. Springer, Heidelberg (2001)
[14] International Human Genome Sequencing Consortium: Initial sequencing and analysis of the human genome. Nature 409(6822), 860–921 (2001)
[15] Wernicke, S.: On the algorithmic tractability of single nucleotide polymorphism (SNP) analysis and related problems. Ph. d. thesis, Univ. Tübingen (2003)
[16] Sanger, F., Nicklen, S., Coulson, A.R.: Dna sequencing with chain-terminating inhibitors. PNAS 74(12), 5463–5467 (1977)
[17] Levy, S., Sutton, G., Ng, P.C., et al.: The diploid genome sequence of an individual human. PLoS Biology 5(10), October 2007, e254–e254 (2007)
[18] Hinds, D.A., Stuve, L.L., Nilsen, G.B., Halperin, E., Eskin, E., Ballinger, D.B., Frazer, K.A., Cox, D.R.: Whole-genome patterns of common dna variation in three human populations. Science 307(5712), 1072–1079 (2005)
[19] Hüffner, F.: Algorithm engineering for optimal graph bipartization. In: Nikoletseas, S.E. (ed.) WEA 2005. LNCS, vol. 3503, pp. 240–252. Springer, Heidelberg (2005)

[20] Panconesi, A., Sozio, M.: Fast hare: a fast heuristic for single individual snp haplotype reconstruction. In: Jonassen, I., Kim, J. (eds.) WABI 2004. LNCS (LNBI), vol. 3240, pp. 266–277. Springer, Heidelberg (2004)

[21] Myers, G.: A dataset generator for whole genome shotgun sequencing. In: Lengauer, T., Schneider, R., Bork, P., Brutlag, D.L., Glasgow, J.I., Mewes, H.W., Zimmer, R. (eds.) Proc. ISMB, California, pp. 202–210. AAAI Press, Menlo Park (1999)

Improved Algorithms for Bicluster Editing

Jiong Guo[1,⋆], Falk Hüffner[1,⋆], Christian Komusiewicz[1,⋆⋆], and Yong Zhang[2]

[1] Institut für Informatik, Friedrich-Schiller-Universität Jena
Ernst-Abbe-Platz 2, D-07743 Jena, Germany
{guo,hueffner,ckomus}@minet.uni-jena.de
[2] Department of Mathematical Sciences, Eastern Mennonite University
Harrisonburg, VA 22802, USA
yong.zhang@emu.edu

Abstract. The NP-hard Bicluster Editing is to add or remove at most k edges to make a bipartite graph $G = (V, E)$ a vertex-disjoint union of complete bipartite subgraphs. It has applications in the analysis of gene expression data. We show that by polynomial-time preprocessing, one can shrink a problem instance to one with $4k$ vertices, thus proving that the problem has a linear kernel, improving a quadratic kernel result. We further give a search tree algorithm that improves the running time bound from the trivial $O(4^k + |E|)$ to $O(3.24^k + |E|)$. Finally, we give a randomized 4-approximation, improving a known approximation with factor 11.

1 Introduction

Data clustering is a classical task, where the goal is to partition a data set into *clusters* such that elements within a cluster are similar, while between clusters there is less similarity. This similarity is often modeled as a graph: Each vertex represents a data point, and two vertices are connected by an edge iff the entities that they represent have some (context-specific) similarity. If the data were perfectly clustered, this would result in a *cluster graph*, that is, a graph where every connected component is a clique. However, for real-world data, there is typically noise in the data. A simple clustering model is then the Cluster Editing problem [4, 19]: find a minimum set of edges to add or delete to make the graph a cluster graph.

Cluster Editing is NP-hard [15]; a number of approaches have been recently suggested to deal with this. After a series of improvements, the best known polynomial-time approximation is by a factor of 2.5 [2, 21]. Another technique is that of *fixed-parameter (FPT) algorithms* [7, 9, 17]. The idea is to accept the superpolynomial running time that seems to be inherent to NP-hard problems, but to restrict the combinatorial explosion to a *parameter* that is expected to be small. For Cluster Editing, the number of editing operations k is a suitable

⋆ Supported by the Deutsche Forschungsgemeinschaft, Emmy Noether research group PIAF (fixed-parameter algorithms), NI 369/4.
⋆⋆ Supported by a PhD fellowship of the Carl-Zeiss-Stiftung.

M. Agrawal et al. (Eds.): TAMC 2008, LNCS 4978, pp. 445–456, 2008.

parameter, since for data with not too much noise it should be low. Several fixed-parameter algorithms for CLUSTER EDITING have been suggested (see also Hüffner et al. [14] for a survey on FPT techniques in graph-modeled clustering). The search tree algorithm of Gramm et al. [10] with a running time bound of $O(2.27^k + n^3)$ has been experimentally evaluated [6]. A recent manuscript [5] claims a running time of $O(1.83^k + n^3)$ by using a different branching strategy and reports further experimental results.

An important tool of FPT algorithmics is *kernelization* [7, 9, 17]. A kernelization is a polynomial-time preprocessing that reduces an instance to a size that depends only on the parameter k, and not on the input size $|G|$ anymore. Clearly, such a preprocessing is useful for basically any approach to solving the problem, be it exact, approximative, or heuristic. For CLUSTER EDITING, after a series of improvements [8, 10, 18], a kernel of only $4k$ vertices is known [11].

In some settings, the standard clustering model is not satisfactory. An important example is clustering of gene expression data, where under a number of conditions the level of expression of a number of genes is measured. This yields a bipartite similarity graph. Here, clustering only genes or only conditions often does not yield sufficient insight; we would like to find subsets of genes and subsets of conditions that together behave in a consistent way. This is called *biclustering* [16, 20]. A simple formulation of biclustering analogous to CLUSTER EDITING is BICLUSTER EDITING. Here, as a consistency condition for a cluster, we demand that it forms a *biclique*, that is, a complete bipartite subgraph. With bipartite graphs, we mean two-colorable graphs. Further, we do not allow any clusters to overlap.

BICLUSTER EDITING
Instance: A bipartite graph $G = (V, E)$ and an integer $k \geq 0$.
Question: Can we delete and add at most k edges in G such that it becomes a *bicluster graph*, that is, a graph where every connected component is a biclique?

Further applications of biclustering arise in collaborative filtering, information retrieval, and data mining. Despite its importance, there are fewer results for BICLUSTER EDITING than for CLUSTER EDITING. Amit [3] proved the NP-hardness and gave a factor-11 approximation based on the relaxation of a linear program. Using a simple branching strategy, the problem can be solved in $O(4^k + m)$ time [18], where m is the number of edges in the graph. Protti et al. [18] showed how to construct a problem kernel with $4k^2 + 6k$ vertices.

Contributions. Following the work recently done for CLUSTER EDITING, our aim is to improve FPT and approximation algorithms also for its sister problem BICLUSTER EDITING. We first improve the size of the problem kernel from $4k^2 + 6k$ to $4k$ vertices (Sect. 2). The methods used are similar to those of Guo [11]. If the input graph is not already a bipartite graph, we can still get a $6k$-vertex kernel by similar means. Next, we show that the trivial $O(4^k + m)$ time branching algorithm can be improved to $O(3.24^k + m)$ time by a more refined branching strategy (Sect. 3). Finally, we give a randomized approximation algorithm with an expected approximation factor of 4, using similar techniques as those introduced for CLUSTER EDITING [1].

Preliminaries. We consider only undirected graphs $G = (V, E)$ with $n := |V|$ and $m := |E|$. Since singleton vertices do not play an interesting role in our problems, we assume that $n \in O(m)$. Let P_4 denote an induced path comprising 4 vertices. Furthermore, let $ijkl$ denote a P_4 in which i and l have degree 1 and j and k have degree 2. The neighborhood of a vertex v is denoted by $N(v)$, and the closed neighborhood $N(v) \cup \{v\}$ is denoted by $N[v]$. We furthermore extend this notation to vertex sets, that is, for a vertex set S, $N(S) := (\bigcup_{v \in S} N(v)) \setminus S$. For a vertex v, $N_2(v) := N(N(v)) \setminus \{v\}$ denotes the set of vertices that have distance exactly 2 from v.

Due to lack of space, several proofs are deferred to a full version of this paper.

2 Linear Problem Kernel

In this section, we present a kernelization algorithm for BICLUSTER EDITING that produces a kernel consisting of at most $4k$ vertices, improving the kernel consisting of $O(k^2)$ vertices given by Protti et al. [18]. This kernelization follows the idea of the kernelization algorithm for CLUSTER EDITING in [11] that also produces a kernel consisting of at most $4k$ vertices. However, since here we are dealing with bipartite graphs and bicliques, the concrete handling of the data reduction rules and the argumentation of the kernel size are different from the one for CLUSTER EDITING. The first step is to introduce a useful structure.

Definition 1. *A set S of vertices is called a* critical independent set *if all vertices in S have the same open neighborhood and S is maximal under this property.*

Observe that every critical independent set is an independent set. The connection between critical independent sets and BICLUSTER EDITING is given by the following lemma.

Lemma 1. *For any critical independent set I, there is an optimal solution of* BICLUSTER EDITING *in which any two vertices v_1 and v_2 from I end up in the same biclique.*

We apply the following two data reduction rules; the second one works on critical independent sets.

Rule 1. Remove all connected components that are bicliques from the graph.

Rule 2. Consider a critical independent set R. Let $S := N(R)$ and $T := N(S) \setminus R$. If $|R| > |T|$, then remove arbitrary vertices from R until $|R| = |T|$.

Rule 1 is clearly correct and can be carried out in $O(m)$ time. A situation in which Rule 2 can be applied is illustrated in Fig. 1. Next, we prove the correctness of Rule 2.

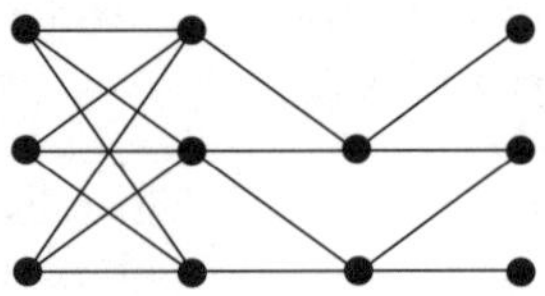

Fig. 1. Example for the application of Rule 2

Lemma 2. *Rule 2 is correct and works in $O(n^2)$ time.*

Proof. To prove the correctness of Rule 2, we first claim that, as long as $|R| \geq |T|$, there is always an optimal solution constructing a bicluster graph that contains a biclique B with $R \cup S \subseteq B \subseteq R \cup S \cup T$ and, thus, deleting or inserting no edge incident to R. Since the input graph G and the graph resulting by one application of Rule 2 to G differ only in the size of R, the correctness of Rule 2 follows.

To show the claim, let $U := N(T) \setminus S$. First, observe that R will not be "split" (following from Lemma 1), that is, there is always an optimal solution leaving a biclique B with $R \subseteq B$. Second, we prove that no vertex outside of $R \cup S \cup T$ can be in B, that is, $B \subseteq R \cup S \cup T$. To see this, if a vertex $u \notin R \cup S \cup T$ is in B, then obviously u is from U. However, to add u to B needs at least $|R|$ edge insertions. Thus, as long as $|R| \geq |T|$, adding u to B is never better than putting u in a biclique different from B, which requires at most $|T|$ edge deletions. Finally, we show that no vertex from S can be outside of B, that is, $R \cup S \subseteq B$. This is easy to see, since every vertex $u \in S$ has only neighbors in R and T. By $|R| \geq |T|$ and $B \subseteq R \cup S \cup T$, including u in B requires at most $|T|$ edge modifications, namely, deleting all edges between u and $N(u) \cap T$ and adding edges between u and $T \setminus N(u)$. In comparison, excluding u from B needs at least $|R|$ edge deletions, since $R \subseteq N(u)$.

Concerning the running time, one can compute all critical independent sets in $O(m)$ time [12]. Then, we determine the sets S and T for all independent sets R, which can be done in $O(n^2)$ time. To check the applicability of Rule 2, one iterates over all critical independent sets and uses the already computed information about S and T to decide if the precondition of Rule 2 is fulfilled by R. Note that after one application of Rule 2, one has only to consider the critical independent sets whose vertices are in S and T and to change the sizes of their S's and T's. Therefore, each application of Rule 2 can be carried out in $O(n)$ time. Rule 2 can be applied at most n times, which gives the total running time $O(n^2)$. □

With these two rules we can now prove a kernel consisting of at most $4k$ vertices.

Theorem 1. Bicluster Editing *on bipartite graphs admits a $4k$-vertex problem kernel.*

Proof. Let G denote a bipartite graph on which Rules 1 and 2 have been exhaustively applied. Furthermore, let F be a bicluster editing set with $|F| \leq k$

and let G' be the resulting bicluster graph after applying the edge modifications in F to G. We partition the vertices in G' into two sets, X containing the endpoints of the edges in F, and Y the rest. Clearly, $|X| \leq 2k$. It remains to upper-bound $|Y|$. Suppose G' consists of l biclusters, $B_1, \ldots, B_l$. It is easy to see that for every $i \in \{1, \ldots, l\}$, the unaffected vertices from one partition in B_i must have the same neighborhood in G. Hence, for every $i \in \{1, \ldots, l\}$, the vertices in $B_i \cap Y$ form at most two critical independent sets in G. Let R be one critical independent set in B_i, $S := N_G(R)$, and $T := N_G(S) \setminus R$. Due to Rule 2, $|R| \leq |T|$. Due to Rule 1, all vertices of T are in X. Some of them are in B_i after gaining some edges between them and the vertices in S and the others are in other bicliques after losing all edges between them and S. Therefore, summing up over all critical independent sets in all bicliques, we can conclude that $|Y| \leq |X|$, giving the claimed number of vertices in the kernel. □

Protti et al. [18] have considered BICLUSTER EDITING on general graphs as well and presented a problem kernel with $O(k^2)$ vertices. With a slight modification of Rule 2, we can improve this result to a $6k$-vertex problem kernel. The main difference to the kernelization for bipartite graphs lies in the edges between the vertices of S: If the vertices in S have much more edges between them than the size of R, it could be better to keep them and to remove some edges between R and S. To take this into account, we make a partition of the vertices in S as described below.

Modified Rule 2. Consider a critical independent set R, let $S := N(R)$ and $T := N(S) \setminus R$. Further partition S into two sets, S_1 the set of vertices without neighbors in S and $S_2 := S \setminus S_1$. If $|R| > |S_2| + |T|$, then reduce R until $|R| = |S_2| + |T|$.

The correctness proof of the modified Rule 2 is almost the same as the one for Lemma 2, namely, showing that in case $|R| \geq |S_2| + |T|$ there is always an optimal solution creating a biclique B, such that $R \cup S \subseteq B \subseteq R \cup S \cup T$. The only difference concerns the vertices in S_2. Since they have neighbors in S_2, including them in B requires not only deleting and adding edges between them and T but also deleting edges between them and their neighbors in S_2. However, if $|R| \geq |S_2| + |T|$, then it is never better to exclude them from B than to include them in B.

Theorem 2. BICLUSTER EDITING *on general graphs admits a* $6k$*-vertex problem kernel.*

3 Fixed-Parameter Algorithm

In this section, we present a search tree algorithm for BICLUSTER EDITING in bipartite graphs that is based on the forbidden subgraph characterization of BICLUSTER EDITING and has a running time of $O(3.24^k + m)$, improving upon the trivial search tree algorithm with a running time of $O(4^k + m)$ [18].

Let $ijkl$ be a P_4 in G. The trivial search tree algorithm for BICLUSTER EDITING branches on $ijkl$ in 4 cases: one case corresponds to adding edge $\{i, l\}$; the other three cases correspond to removing one of the three edges of $ijkl$. The improvement of the running time of our algorithm is achieved by applying a refined branching strategy on larger subgraphs that contain a P_4. In this branching strategy, we distinguish two main cases. For the first case, we show that an improved branching can be achieved. For the second case, we show that it can be solved in polynomial time. In the following, we describe this branching strategy.

Clearly, branching is only performed as long as G is not a bicluster graph. Therefore, we assume that G contains a P_4. Furthermore, we deal with each connected component separately. Therefore, without loss of generality assume that G is connected. We distinguish two main cases.

Case 1: There is a connected subgraph of size 5 *of* G *that contains a* P_4 *and has 4 edges.* Let G' be such a subgraph, and let $ijkl$ denote a P_4 contained in G'. Since G' is connected and contains 5 vertices and 4 edges, it must contain a vertex u that is connected to exactly one vertex in $ijkl$. Hence, the resulting graph is either a P_5—in case u is adjacent to i or l—or a so-called *fork*—in case u is adjacent to j or k. We describe the branching strategy for P_5's in detail. Branching on forks works analogously. Both branchings are depicted in Fig. 2.

Let $ijklu$ be the P_5 that we branch on. In the first branch, we delete the edge $\{k, l\}$, the parameter is decreased by 1. In the second branch, we delete the edge $\{j, k\}$ and the parameter is decreased by 1. Since we have already considered deleting $\{k, l\}$ or $\{j, k\}$, we can mark these two edges as *permanent*, that is, we may not delete these edges in the remaining branches. To destroy $jklu$, we must either delete $\{l, u\}$ or add $\{j, u\}$. But after performing either of these two edge modifications the graph still contains $ijkl$, and $\{j, k\}$ and $\{k, l\}$ are marked as permanent. Hence, for each of these two branches, we create two subbranches, one in which $\{i, l\}$ is added, and one in which $\{i, j\}$ is deleted. In total, we have 6 branches, two branches in which the parameter is decreased by 1, and 4 branches in which the parameter is decreased by 2. To estimate the size of the search tree, we use the concept of *branching vectors* [17]. The branching vector of this branching is $(1, 1, 2, 2, 2, 2)$. The branching on a fork works analogously.

Case 2: Otherwise. Since Case 1 did not apply, every connected subgraph of size 5 that contains a P_4 has at least 5 edges. We show that in this case, no branching is needed because G can be turned into a biclique by adding one edge.

Lemma 3. *Let* $G = (V_1, V_2, E)$ *be a fork-free and* P_5*-free connected bipartite graph, and let* $ijkl$ *be a* P_4 *in* G*. Then, adding edge* $\{i, l\}$ *transforms* G *into a biclique.*

Proof. W.l.o.g. assume that $\{i, k\} \subseteq V_1$ and $\{j, l\} \subseteq V_2$. We prove the lemma by showing that with the exception of $\{i, l\}$ all edges are present in G.

First, we show that k is adjacent to all vertices in V_1, and that j is adjacent to all vertices in V_2. Clearly, if i has a neighbor u, then u is a neighbor of k. Otherwise, $G[\{i, j, k, l, u\}]$ is a P_5, and G is thus not P_5-free. Also, every vertex $u \in$

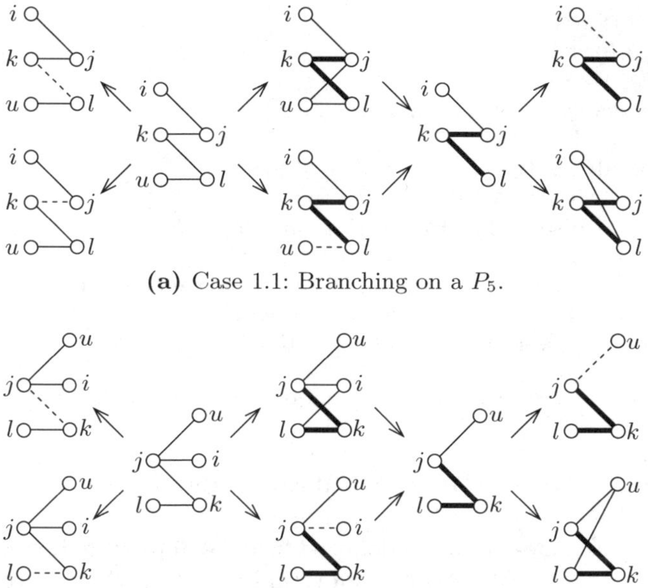

(a) Case 1.1: Branching on a P_5.

(b) Case 1.2: Branching on a fork.

Fig. 2. Branching on subgraphs of size 5 that contain a P_4 and exactly 4 edges. *Dashed lines* are deleted edges; *bold lines* are permanent edges.

$N(k) \setminus \{l\}$ is a neighbor of i. Otherwise, $G[\{i, j, k, l, u\}]$ is a fork, and G thus not fork-free. Therefore $N(i) = N(k) \setminus \{l\}$. Analogously, we can show that $N(l) = N(j) \setminus \{i\}$. Furthermore, every vertex $v \in N_2(i)$ is adjacent to j. Otherwise, suppose that there is a vertex $v \in N_2(i)$ that is not adjacent to j and let u be the common neighbor of i and v. Then, the subgraph $G[\{i, k, l, u, v\}]$ is a fork, because u is also adjacent to k (since $u \in N(i) \subset N(k)$), and u is not adjacent to l (since $u \notin N(j) \supset N(l)$). Hence, G is not fork-free in this case. Analogously, we can show that k is adjacent to all vertices in $N_2(l)$. With this it becomes obvious that k is adjacent to all vertices in V_1, and j is adjacent to all vertices in V_2.

Now we show that every pair of vertices $u \in V_1 \setminus \{i\}$ and $v \in V_2 \setminus \{l\}$ must be pairwise adjacent. Suppose, there are two vertices $u \in V_1 \setminus \{i\}$ and $v \in V_2 \setminus \{l\}$ that are not pairwise adjacent. Since i is adjacent to all vertices in $V_1 \setminus \{l\}$, it is adjacent to v. Analogously, one can show that j and l are adjacent to u. Therefore, $G[\{i, j, l, v, u\}]$ is a P_5 if v and u are not adjacent, contradicting the fact that G is P_5-free.

Since all vertices in $V_1 \setminus \{i\}$ are adjacent to all vertices $V_2 \setminus \{l\}$, it is clear that adding edge $\{i, l\}$ transforms G into a biclique. □

In the following theorem, we bound the running time of the described search tree algorithm, when it is combined with kernelization.

Theorem 3. Bicluster Editing *can be solved in* $O(3.24^k + m)$ *time.*

```
ApproxBicluster(G = (V1, V2, E))
  G' := (∅, ∅, ∅)
  while V1 ∪ V2 ≠ ∅:
        randomly select a pivot vertex i ∈ V1 ∪ V2
        C := {i} ∪ N(i)
        for all j ∈ {v ≠ i | N(v) ∩ N(i) ≠ ∅} :
              if N(j) = N(i) : add j to C
              else : add j to C with probability 1/2
        transform G[C] into an isolated biclique
        G' := G' ∪ G[C]      ▷ add G[C] as new component to G'
        G := G[V \ C]        ▷ remove C from G
  output set of edge modifications from G to G'
```

Fig. 3. A randomized factor-4 approximation algorithm for Bicluster Editing

4 Randomized 4-Approximation Algorithm

We present a polynomial time randomized factor-4 approximation algorithm for Bicluster Editing that is based on a technique introduced by Ailon et al. [1]. This improves the previously best factor-11 approximation algorithm by Amit [3]. The basic strategy of the algorithm is to randomly pick a pivot vertex v, and then to randomly destroy all P_4's that contain v. In doing so, we create an isolated biclique that contains v, since a connected component in which one vertex does not appear in a P_4 is a biclique. This procedure is applied until the graph is a bicluster graph. In the following, we describe how the P_4's containing the pivot vertex v are destroyed.

Given a pivot vertex i, we create a vertex set C that initially contains $N[i]$. In the end this set C contains the vertices that are in the same biclique as i in the final bicluster graph. First, we add all vertices that are in the same critical independent set as i.

Then we randomly decide for each vertex w that is adjacent to at least one vertex of $N(i)$ whether w should be added to C. Since w is adjacent to neighbors of $N(i)$ but is not in the same critical independent set as i, there must be a P_4 that contains i and w. By randomly deciding whether i and w end up in the same biclique, we randomly decide which edge modification is made to destroy the P_4. After this is done for all such vertices, we output C and cut C from G.

This is done until G has no vertex. The pseudo-code of the algorithm is shown in Fig. 3. In order to apply the method of Ailon et al. [1], the algorithm must guarantee that after an edit operation is made on an edge, this edge is never again modified during the course of the algorithm, and that for a given P_4 each edit operation has the same probability. In our case this probability is $\frac{1}{4}$, which leads to an approximation factor of 4.

To prove this upper bound on the approximation factor, we first need the notion of a *fractional packing*.

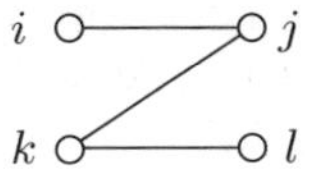

pivot	$Pr[E_{ij}]$	$Pr[E_{il}]$	$Pr[E_{kj}]$	$Pr[E_{kl}]$
i	0	0	$\frac{1}{2}$	$\frac{1}{2}$
j	0	$\frac{1}{2}$	0	$\frac{1}{2}$
k	$\frac{1}{2}$	$\frac{1}{2}$	0	0
l	$\frac{1}{2}$	0	$\frac{1}{2}$	0
$i \vee j \vee k \vee l$	$\frac{1}{4}$	$\frac{1}{4}$	$\frac{1}{4}$	$\frac{1}{4}$

Fig. 4. The probabilities of edge modifications in $ijkl$ in case one of $\{i, j, k, l\}$ is chosen as pivot. The event that there is an edge modification between two vertices i and j is denoted by E_{ij}.

Definition 2. *Let $G = (V_1, V_2, E)$ be a bipartite graph, P the set of P_4's of G, and $w : P \to \mathbb{R}^+$ a weight function. The function w is called a* fractional packing *of P if $\forall i \in V_1, j \in V_2 : \sum_{\{p \in P | \{i,j\} \in p\}} w(p) \leq 1$.*

In the following lemma, we show that given a fractional packing w, the sum of the weights $w(p)$ of all $p \in P$ is a lower bound on the cost of optimal solutions.

Lemma 4. *Let $G = (V_1, V_2, E)$ be a bipartite graph, C_{Opt} the cost of an optimal solution of* Bicluster Editing *of G, and P the set of P_4's. If a weight function $w : P \to \mathbb{R}^+$ is a fractional packing of P, then $\sum_{p \in P} w(p) \leq C_{\text{Opt}}$.*

Using Lemma 4, we can show an upper bound on the approximation factor of ApproxBicluster, as follows. First, we show that the sum of the probabilities of making edge modifications in all P_4's caused by choosing one of their vertices as pivot equals the expected cost of the solution that is output by ApproxBicluster. Then, we show that dividing these probabilities by 4 also yields a fractional packing and thus that the expected cost of the output solution is at most 4 times the cost of an optimal solution.

Theorem 4. ApproxBicluster *is a randomized factor-4 approximation algorithm for* Bicluster Editing, *running in $O(n^2)$ time.*

Proof. Obviously, the output of ApproxBicluster is a solution of Bicluster Editing. We thus prove the theorem by bounding the approximation factor of the expected cost of the output of ApproxBicluster. Let C be the cost of a solution that is output by ApproxBicluster, and let C_{Opt} be the cost of an optimal solution. We prove the theorem by showing that the expected cost $E[C] \leq 4 \cdot C_{\text{Opt}}$.

Clearly, the algorithm inserts or removes edges only between vertices that appear in a P_4. An edit operation is performed in a P_4 $ijkl$ only if i, j, k, and l are still in the graph and one of them is chosen as pivot. Let A_{ijkl} denote the event that one vertex in $\{i, j, k, l\}$ is chosen as pivot when all of them are still in the graph. Furthermore, let π_{ijkl} denote the probability that event A_{ijkl} occurs during the execution of ApproxBicluster. In case that event A_{ijkl} occurs, we perform exactly one edge edit operation between the vertices of $\{i, j, k, l\}$. Therefore, the expected cost $E[C]$ of ApproxBicluster is $\sum_{p \in P} \pi_p$.

We complete the proof by first showing that the weight function w that is obtained by assigning the weight $\frac{\pi_{ijkl}}{4}$ to the P_4 $ijkl$ is a fractional packing of the P_4's of G, and then showing that this leads to the claimed expected approximation factor.

Let p be a P_4, and let $i \in V_1$, $j \in V_2$ be two vertices in p. Let E_{ij} denote the event that there is an edit operation between i and j.

As Fig. 4 shows, the probability $Pr[E_{ij} \mid A_p] = \frac{1}{4}$. Therefore,

$$Pr[E_{ij} \wedge A_p] = Pr[E_{ij} \mid A_p] \cdot Pr[A_p] = \frac{1}{4}\pi_p.$$

Furthermore, note that after event E_{ij} occurred, at least one of i and j is removed from the graph, and thus no further editing between i and j takes place. Therefore, for distinct P_4's p and p', the events $E_{ij} \wedge A_p$ and $E_{ij} \wedge A_{p'}$ are disjoint and thus

$$\sum_{\{p \in P \mid \{i,j\} \in p\}} Pr[E_{ij} \wedge A_p] = \sum_{\{p \in P \mid \{i,j\} \in p\}} \frac{1}{4}\pi_p \leq 1.$$

With this, it becomes obvious that assigning the weight $\frac{1}{4}\pi_p$ to every $p \in P$ results in a fractional packing of the P_4's of G. As shown by Lemma 4, this means that $\sum_{p \in P} \frac{1}{4}\pi_p \leq C_{\text{Opt}}$. Therefore,

$$E[C] = \sum_{p \in P} \pi_p \leq 4 \cdot C_{\text{Opt}},$$

which proves the upper bound on the approximation factor.

Now we prove the running time of the algorithm. In a preprocessing step, we compute the critical independent sets of the graph, which can be performed in $O(n+m)$ time [12]. This is done so that the test in line 6 of the algorithm can be performed in constant time. Determining which vertices end up in C takes $O(m)$ time overall, since the test for membership in the same critical independent set can now be performed in constant time and each edge is visited at most once: either the edge is cut or it is part of the isolated biclique that is removed from G and added to G'. Finally, the number of added edges is in $O(n^2)$, which results in the claimed running time bound. □

Note that the running time of Theorem 4 can be improved to $O(m)$ when the output is merely a list of the bicliques and the vertices they contain. Otherwise, a linear running time cannot be achieved, because the output size cannot be bounded by $O(m)$.

5 Outlook

We have improved kernelization, parameterized algorithm, and approximation algorithm for Bicluster Editing. It is probably possible to further improve the bound of the FPT algorithm, albeit only at the cost of a more complicated case distinction. Further improvement of the approximation factor and derandomization of the result of Theorem 4 should be possible using similar techniques as for Cluster Editing [1, 21, 22].

Together, the improvements make non-heuristic algorithm implementations much more feasible. In particular useful seems the kernelization, which for example is guaranteed to reduce an instance with 1000 vertices and $k = 50$ to only 200 vertices, without losing optimality (note that here the quadratic kernelization [18] does not give any useful bound). It is conceivable that in many cases the kernelized instance can be solved by the branching algorithm from Sect. 3 within reasonable time. Further, it would be interesting to see whether the observed approximation quality of the approximation algorithm from Sect. 4 improves by the preprocessing.

Variants of Bicluster Editing are also of interest, for example considering weights, allowing the deletion of vertices instead of adding and deleting edges, or the generation of a prespecified number of bicliques (the corresponding variations for Cluster Editing have received some attention, see e. g. [11, 13]).

References

[1] Ailon, N., Charikar, M., Newman, A.: Aggregating inconsistent information: ranking and clustering. In: Proc. 37th STOC, pp. 684–693. ACM Press, New York (2005)

[2] Ailon, N., Charikar, M., Newman, A.: Proofs of conjectures in Aggregating inconsistent information: Ranking and clustering. Technical Report TR-719-05, Department of Computer Science, Princeton University (2005)

[3] Amit, N.: The bicluster graph editing problem. Master's thesis, Tel Aviv University, School of Mathematical Sciences (2004)

[4] Bansal, N., Blum, A., Chawla, S.: Correlation clustering. Machine Learning 56(1–3), 89–113 (2004)

[5] Böcker, S., Briesemeister, S., Bui, Q.B.A., Truß, A.: PEACE: Parameterized and exact algorithms for cluster editing. Manuscript, Lehrstuhl für Bioinformatik, Friedrich-Schiller-Universität Jena (September 2007)

[6] Dehne, F.K.H.A., Langston, M.A., Luo, X., Pitre, S., Shaw, P., Zhang, Y.: The cluster editing problem: Implementations and experiments. In: Bodlaender, H.L., Langston, M.A. (eds.) IWPEC 2006. LNCS, vol. 4169, Springer, Heidelberg (2006)

[7] Downey, R.G., Fellows, M.R.: Parameterized Complexity. Springer, Heidelberg (1999)

[8] Fellows, M.R., Langston, M.A., Rosamond, F.A., Shaw, P.: Efficient parameterized preprocessing for cluster editing. In: Csuhaj-Varjú, E., Ésik, Z. (eds.) FCT 2007. LNCS, vol. 4639, pp. 312–321. Springer, Heidelberg (2007)

[9] Flum, J., Grohe, M.: Parameterized Complexity Theory. Springer, Heidelberg (2006)

[10] Gramm, J., Guo, J., Hüffner, F., Niedermeier, R.: Graph-modeled data clustering: Exact algorithms for clique generation. Theory of Computing Systems 38(4), 373–392 (2005)

[11] Guo, J.: A more effective linear kernelization for Cluster Editing. In: Chen, B., Paterson, M., Zhang, G. (eds.) ESCAPE 2007. LNCS, vol. 4614, Springer, Heidelberg (2007)

[12] Hsu, W., Ma, T.: Substitution decomposition on chordal graphs and applications. In: Hsu, W.-L., Lee, R.C.T. (eds.) ISA 1991. LNCS, vol. 557, pp. 52–60. Springer, Heidelberg (1991)

[13] Hüffner, F., Komusiewicz, C., Moser, H., Niedermeier, R.: Fixed-parameter algorithms for cluster vertex deletion. In: Lin, G. (ed.) COCOON. LNCS, vol. 4598, pp. 711–722. Springer, Heidelberg (2007)

[14] Hüffner, F., Niedermeier, R., Wernicke, S.: Fixed-parameter algorithms for graph-modeled data clustering. In: Clustering Challenges in Biological Networks, World Scientific, Singapore (to appear, 2008)

[15] Křivánek, M., Morávek, J.: NP-hard problems in hierarchical-tree clustering. Acta Informatica 23(3), 311–323 (1986)

[16] Madeira, S.C., Oliveira, A.L.: Biclustering algorithms for biological data analysis: a survey. IEEE/ACM Transactions on Computational Biology and Bioinformatics 1(1), 24–45 (2004)

[17] Niedermeier, R.: Invitation to Fixed-Parameter Algorithms. Oxford Lecture Series in Mathematics and Its Applications, vol. 31. Oxford University Press, Oxford (2006)

[18] Protti, F., da Silva, M.D., Szwarcfiter, J.L.: Applying modular decomposition to parameterized bicluster editing. In: Bodlaender, H.L., Langston, M.A. (eds.) IWPEC 2006. LNCS, vol. 4169, pp. 1–12. Springer, Heidelberg (2006); To appear under the title "Applying modular decomposition to parameterized cluster editing problems" in Theory of Computing Systems.

[19] Shamir, R., Sharan, R., Tsur, D.: Cluster graph modification problems. Discrete Applied Mathematics 144(1–2), 173–182 (2004)

[20] Tanay, A., Sharan, R., Shamir, R.: Biclustering algorithms: A survey. In: Aluru, S. (ed.) Handbook of Computational Molecular Biology, pp. 26–1 – 26–17. Chapman Hall/CRC Press (2006)

[21] van Zuylen, A.Z., Williamson, D.P.: Deterministic algorithms for rank aggregation and other ranking and clustering problems. In: Kaklamanis, C., Skutella, M. (eds.) WAOA 2007. LNCS, vol. 4927, pp. 260–273. Springer, Heidelberg (2007)

[22] van Zuylen, A.Z., Hegde, R., Jain, K., Williamson, D.P.: Deterministic pivoting algorithms for constrained ranking and clustering problems. In: Proc. 18th SODA, pp. 405–414. SIAM, Philadelphia (2007)

Generation Complexity Versus Distinction Complexity

Rupert Hölzl and Wolfgang Merkle

Institut für Informatik,
Ruprecht-Karls-Universität,
Heidelberg, Germany
{hoelzl,merkle}@math.uni-heidelberg.de

Abstract. Among the several notions of resource-bounded Kolmogorov complexity that suggest themselves, the following one due to Levin [Le] has probably received most attention in the literature. With some appropriate universal machine U understood, let the Kolmogorov complexity of a word w be the minimum of $|d| + \log t$ over all pairs of a word d and a natural number t such that U takes time t to check that d determines w. One then differentiates between generation complexity and distinction complexity [A, Sip], where the former asks for a program d such that w can actually be computed from d, whereas the latter asks for a program d that distinguishes w from other words in the sense that given d and any word u, one can effectively check whether u is equal to w.

Allender et al. [A] consider a notion of solvability for nondeterministic computations that for a given resource-bounded model of computation amounts to require that for any nondeterministic machine N there is a deterministic machine that exhibits the same acceptance behavior as N on all inputs for which the number of accepting paths of N is not too large. They demonstrate that nondeterminism is solvable for computations restricted to polynomially exponential time if and only if for any word the generation complexity is at most polynomial in the distinction complexity. We extend their work and a related result by Fortnow and Kummer [FK] as follows. First, nondeterminism is solvable for linearly exponential time bounds if and only if generation complexity is at most linear in distinction complexity. Second, nondeterminism is solvable for polynomial time bounds if and only if the conditional generation complexity of a word w given a word y is at most linear in the conditional distinction complexity of w given y; hence, in particular, the latter condition implies that **P** is equal to **UP**. Finally, in the setting of space bounds it holds unconditionally that generation complexity is at most linear in distinction complexity.

In general, the Kolmogorov complexity of a word w is the length $|d|$ of a shortest program d such that d determines w effectively. In a setting of unbounded computations, this approach leads canonically to the usual notion of plain Kolmogorov complexity and its prefix-free variant. In a setting of resource-bounded computations though, there are several notions of Kolmogorov complexity that are in some sense natural – and none of them is considered canonical.

M. Agrawal et al. (Eds.): TAMC 2008, LNCS 4978, pp. 457–466, 2008.

A straight-forward approach is to cap the execution time and/or used space by simply not allowing descriptions that take too long or too much space for producing the word we want to describe. This notion has the disadvantage that for a fixed resource-bound there is no canonical notion of universal machine. Another approach, which has received considerable attention in the literature, was introduced by Levin [Le], where, in contrast to the notion just mentioned, arbitrarily long computations are allowed, but a large running time increases in some way the complexity value. More precisely, with some appropriate universal machine U understood, in Levin's model the Kolmogorov complexity of a word d is the minimum of $|d|+\log t$ over all pairs of a word d and a natural number t such that U takes time t to check that w is the word determined by d. As for other notions of resource-bounded Kolmogorov complexity, here one can differentiate between generation complexity and distinction complexity [A, Sip], where the former asks for a program d such that w can actually be computed from d, whereas the latter asks for a program d that distinguishes w from other words in the sense that given d and any word u, one can effectively check whether u is equal to w.

The question of how generation and distinction complexity relate to each other in the setting of Levin's notion of resource-bounded Kolmogorov complexity has been investigated by Allender et al. [A]. They consider a notion of solvability for nondeterministic computations that — for a given resource-bounded model of computation — amounts to require that for any nondeterministic machine N there is a deterministic machine that exhibits the same acceptance behavior as N on all inputs for which the number of accepting paths of N is not too large, e.g., is at most logarithmic in the number of all possible paths. Their main result then asserts that nondeterminism is solvable for computations restricted to polynomially exponential time if and only if for any word the generation complexity is at most polynomial in the distinction complexity.

We extend the work of Allender et al. [A] and a related result by Fortnow and Kummer [FK] as follows. First, nondeterminism is solvable for linearly exponential time bounds if and only if generation complexity is at most linear in distinction complexity. Second, nondeterminism is solvable for polynomial time bounds if and only if the conditional generation complexity of a word w given a word y is at most linear in the conditional distinction complexity of w given y; as a consequence, the latter condition implies in particular that **P** is equal to **UP**. Combining the result on polynomial time bounds with a result by Fortnow and Kummer [FK] about Kolmogorov complexity defined in terms of fixed polynomial time bounds, one obtains that in the model just mentioned conditional generation and distinction complexity are close if and only if they are close in Levin's model. Finally, in the setting of space bounds, more precisely, for complexity measures Ks and KDs that logarithmically count the used space instead of the running time used on a program, it holds unconditionally that generation complexity is at most linear in distinction complexity.

The notion of generation complexity considered below differs from Levin's original notion insofar as one has to generate only single bits of the word to be generated but not the word as a whole. This variant has already been used by Allender et al. [A]; their results mentioned above, as well as the results demonstrated below extend to Levin's original model by almost identical proofs.

For a complexity class **C**, we will refer by **C**-machine to any machine M that uses a model of computation and obeys a time- or space-bound such that M witnesses $\mathrm{L}(M) \in \mathbf{C}$ with respect to the standard definition of **C**. For example, an **NE**-machine is a nondeterministic machine that runs in linearly exponential time.

The individual bits of a word x will be denoted by x_1 to $x_{|x|}$. We fix an appropriate universal machine U that receives as input encoded tuples of words and e.g. (x, y, z) will be encoded by $\tilde{x}01\tilde{y}01\tilde{z}$ where the word $\tilde{u}$ is obtained by doubling every symbol in u, i.e., $\tilde{u} = u_1u_1u_2u_2 \ldots u_{|u|}u_{|u|}$.

Logarithms to base 2 are denoted by log, and often a term of the form $\log t$ will indeed denote the least natural number s such that $t \leq 2^s$.

1 Known Results

Definition 1 (Levin [Le], Allender et al. [A]). *Time-bounded generation complexity* Kt(.) *and distinction complexity* KDt(.) *are defined by*

$$\mathrm{Kt}(x) = \min\left\{ |d| + \log t \;\middle|\; \begin{array}{c} \forall b \in \{0,1,*\} \colon \forall i \leq |x|+1 \colon U(d,i,b) \\ \text{runs for } t \text{ steps and accepts iff } (x*)_i = b \end{array} \right\},$$

$$\mathrm{KDt}(x) = \min\left\{ |d| + \log t \;\middle|\; \begin{array}{c} \forall y \in \Sigma^{|x|} \colon U(d,y) \text{ runs for} \\ t \text{ steps and accepts iff } x = y \end{array} \right\}.$$

Observe that in the definition of Kt-complexity the symbol $*$ has to be generated as an end marker for the word x.

Remark 2. The notion of Kt-complexity introduced in Definition 1 was proposed by Allender et al. in [A] as a variation of Levin's original definition, where the latter requires to generate whole words instead of individual bits. Levin's original definition has the advantage of assuring that for all x, it holds that $\mathrm{KDt}(x) \leq \mathrm{Kt}(x) + \log |x|$.

In connection with Theorem 14 we also use the following conditional complexity notions.

Definition 3. *The conditional time-bounded distinction complexity* Kt(.|.) *and conditional generation complexity* KDt(.|.) *are defined by*

$$\mathrm{Kt}(x|y) = \min\left\{ |d| + \log t \;\middle|\; \begin{array}{c} \forall b \in \{0,1,*\} \colon \forall i \leq |x|+1 \colon U(d,y,i,b) \\ \text{runs for } t \text{ steps and accepts iff } (x*)_i = b \end{array} \right\},$$

$$\mathrm{KDt}(x|y) = \min\left\{ |d| + \log t \;\middle|\; \begin{array}{c} \forall z \in \Sigma^{|x|} \colon U(d,y,z) \text{ runs for} \\ t \text{ steps and accepts iff } z = x \end{array} \right\}.$$

We will shortly review Theorem 17 from Allender et al. [A] before we will state our extensions.

Definition 4 (Allender et al. [A]). *We say that* **FewEXP** search instances are **EXP**-solvable *if, for every* **NEXP**-*machine* N *and every* k, *there is an* **EXP**-*machine* M *with the property that if* N *has fewer than* $2^{|x|^k}$ *accepting paths on input* x, *then* M *produces on input* x *some accepting path as output if there is one.*

We say that **FewEXP** decision instances are **EXP**-solvable *if, for every* **NEXP**-*machine* N *and every* k, *there is an* **EXP**-*machine* M *with the property that if* N *has fewer than* $2^{|x|^k}$ *accepting paths on input* x, *then* M *accepts* x *if and only if* N *accepts* x.

We say that **FewEXP** decision instances are **EXP**/poly-solvable *if, for every* **NEXP**-*machine* N *and every* k, *there is an* **EXP**-*machine* M *having access to advice of polynomial length, such that if* N *has fewer than* $2^{|x|^k}$ *accepting paths on input* x, *then* M *accepts* x *if and only if* N *accepts* x.

The notion of solvability can be equivalently characterized in terms of promise problems [CHV, FK]. This will be discussed further in connection with Theorem 17 by Fortnow and Kummer.

Remark 5. Note that by definition **EXP**-solvability of **FewEXP** decision instances implies **FewEXP** = **EXP**, but it is unknown whether the reverse implication holds as well. This is because the definition of **EXP**-solvability does *not* require the considered machines to have a limited number of accepting paths on *all* inputs.

Theorem 6 (Allender et al. [A]). *The following statements are equivalent:*

1 *For all* x, $\mathrm{Kt}(x) \in (\mathrm{KDt}(x))^{\mathrm{O}(1)}$.
2 **FewEXP** *search instances are* **EXP**-*solvable.*
3 **FewEXP** *decision instances are* **EXP**-*solvable.*
3′ **FewEXP** *decision instances are* **EXP**/poly-*solvable.*
4 *For all* $A \in \mathbf{P}$ *and for all* $y \in A^{=l}$ *it holds that*

$$\mathrm{Kt}(y) \in (\log |A^{=l}| + \log l)^{\mathrm{O}(1)}.$$

In words this means, that if generating is "not much more difficult" than distinguishing, then witnesses for certain nondeterministic computations with few witnesses can be found deterministically, and vice versa.

2 Tools

In what follows we will use a corollary of the following result by Buhrman et al., which have also been used by Allender et al. We omit proofs and more detailed discussion due to space considerations.

Lemma 7 (Buhrman et al. [BFL]). *Let n be large enough and let*

$$A := \{x_1, x_2, \ldots, x_{|A|}\} \subseteq \{l, l+1, \ldots, l+n-1\}.$$

Then for all $x_i \in A$ and at least half of the prime numbers $p \leq 4 \cdot |A| \cdot \log^2 n$ it holds for all $j \neq i$ that $x_i \not\equiv x_j (\operatorname{mod} p)$.

Corollary 8. *Let $A \subseteq \Sigma^*$, $y \in \Sigma^*$ and $l \in \mathbb{N}$. Let*

$$A_{y,l} := A \cap \{x \mid y \sqsubseteq x \wedge |x| = l\}.$$

Then it holds that

$$\forall y \in A_{y,l} \colon \mathrm{KDt}^{A_{y,l}}(x) \leq 2 \log |A_{y,l}| + \mathrm{O}(\log l)$$

In particular, if there is a machine that on input y, l and x decides in polynomial time whether x is in the set $A_{y,l}$, then

$$\forall x \in A_{y,l} \colon \mathrm{KDt}(x|y) \leq 2 \log |A_{y,l}| + \mathrm{O}(\log l).$$

3 New Results

We will now state our variants of Theorem 6 by Allender et al., which are demonstrated by similar proofs.

Definition 9. *We say that* **FewE** search instances are **E**-solvable *if, for every* **NE**-*machine N and every k, there is an* **E**-*machine M with the property that if N has fewer than $2^{k \cdot |x|}$ accepting paths on input x, then M produces on input x some accepting path as output if there is one.*

We say that **FewE** decision instances are **E**-solvable *if, for every* **NE**-*machine N and every k, there is an* **E**-*machine M with the property that if N has fewer than $2^{k \cdot |x|}$ accepting paths on input x, then M accepts x if and only if N accepts x.*

We say that **FewE** decision instances are **E**/lin-solvable *if, for every* **NE**-*machine N and every k, there is an* **E**-*machine M having access to advice of linear length, such that if N has fewer than $2^{k \cdot |x|}$ accepting paths on input x, then M accepts x if and only if N accepts x.*

We say that **UE** decision instances are **E**-solvable *if, for every* **NE**-*machine N and every k, there is an* **E**-*machine M with the property that if N has at most one accepting path on input x, then M accepts x if and only if N accepts x.*

Theorem 10. *The following statements are equivalent:*

1 *For all words x, $\mathrm{Kt}(x) \in \mathrm{O}(\mathrm{KDt}(x))$.*
2 **FewE** *search instances are* **E**-*solvable.*
3 **FewE** *decision instances are* **E**-*solvable.*
3′ **UE** *decision instances are* **E**-*solvable.*
3″ **FewE** *decision instances are* **E**/lin-*solvable.*

4 *For all* $A \in \mathbf{P}$ *it holds that for* $A_{y,l} := A \cap \{x \mid y \sqsubseteq x \wedge |x| = l\}$ *and for all* $x \in A_{y,l}$

$$\mathrm{Kt}(x) \in \mathrm{O}(\log |A_{y,l}| + \log l + |y|).$$

We omit the proof of Theorem 10 due to space constraints.

Remark 11. Theorem 10 remains valid by essentially the same proof when formulated for Levin's original notion of Kt instead of the variant of Allender et al.

Corollary 12. *If for all* x, $\mathrm{Kt}(x) \in \mathrm{O}(\mathrm{KDt}(x))$, *then* $\mathbf{UE} = \mathbf{E}$.

Proof. According to the theorem, the assumption implies that **UE** decision instances are **E**-solvable. Since a language in **UE** contains only such instances, the claim follows. □

The equivalence stated in Theorem 10 can be extended to the setting of polynomial time bounds when considering conditional complexities

Definition 13. *We say that* **FewP** search instances are **P**-solvable *if, for every* **NP** *machine* N *and every* k *there is a* **P** *machine* M *with the property that if* N *has fewer than* $|x|^k$ *accepting paths on input* x, *then* M *produces on input* x *some accepting path as output if there is one.*

We say that **FewP** decision instances are **P**-solvable *if, for every* **NP** *machine* N *and every* k *there is a* **P** *machine* M *with the property that if* N *has fewer than* $|x|^k$ *accepting paths on input* x, *then* M *accepts* x *if and only if* N *accepts* x.

We say that **UP** decision instances are **P**-solvable *if, for every* **NP**-*machine* N *and every* k, *there is a* **P**-*machine* M *with the property that if* N *has at most one accepting path on input* x, *then* M *accepts* x *if and only if* N *accepts* x.

Theorem 14. *The following statements are equivalent:*

1 *For all words* x *and* y, $\mathrm{Kt}(x|y) \in \mathrm{O}(\mathrm{KDt}(x|y))$.
2 **FewP** *search instances are* **P**-*solvable.*
3 **FewP** *decision instances are* **P**-*solvable.*
3′ **UP** *decision instances are* **P**-*solvable.*
4 *For all* $A \in \mathbf{P}$ *it holds that for* $A_{y,l} := A \cap \{x \mid y \sqsubseteq x \wedge |x| = l\}$ *and for all* $x \in A_{y,l}$

$$\mathrm{Kt}(x|y) \in \mathrm{O}(\log |A_{y,l}| + \log l).$$

Proof. (1 ⇒ 4)**:** We have access to y through the conditioning. If we also have access to l, we can decide membership of x in $A_{y,l}$ in polynomial time. To do this, we first check whether $y \sqsubseteq x$ and whether y has the correct length l. If yes, compute the value $A(x) = A_{y,l}(x)$ using the fact that $A \in \mathbf{P}$. Corollary 8 then yields

$$\mathrm{KDt}(x|y) \leq 2 \log |A_{y,l}| + \mathrm{O}(\log l).$$

Using assumption 1 the claim follows.

(4 ⇒ 2): Let N be any nondeterministic machine running in polynomial time n^k, where we can assume that N branches binarily. Let L denote $\mathrm{L}(N)$. Let

$$D := \{yx \mid x \in \{0,1\}^{|y|^k} \text{codes an accepting computation of } N \text{ on } y\}.$$

Obviously, $D \in \mathbf{P}$. Now fix any y such that M on input y has at most $|y|^k$ accepting paths. Then the set $D_y := D \cap \{yx \mid |x| = |y|^k\}$ contains at most $|y|^k$ words and by assumption 4 it follows that

$$\begin{aligned} \forall yx \in D_y \colon \mathrm{Kt}(yx|y) &\in \mathrm{O}(\log |D_y| + \log(|y| + |y|^k)) \\ &= \mathrm{O}(\log(|y|)) \end{aligned}$$

So, in order to find an accepting path of M on input y, if there is one, it suffices to search through all words x with $\mathrm{Kt}(x|y) \leq \mathrm{O}(\log |y|)$. This can be done in polynomial time, so that $L \in \mathbf{P}$ as was to be shown. Note that it causes no problems that we have to deal with conditional complexity here. This is because when we are searching for an accepting path x for a word y we obviously have access to y.

(2 ⇒ 3): This is trivial.

(3 ⇒ 3′): This is trivial.

(3′ ⇒ 1): Let us assume that we have a KDt-description d, finite conditioning information y and a described word x such that the universal machine U accepts the triple (d, y, x) in t_{KDt} steps. Assuming that this is the optimal description for x we would have $\mathrm{KDt}(x|y) = |d| + \log t_{\mathrm{KDt}}$. Since we can in t_{KDt} steps only access the first t_{KDt} bits of y (with the encoding of tupels introduced with the universal machine U) we can use $y[1..t_{\mathrm{KDt}}]$ instead of y for the rest of the proof and can therefore w.l.o.g. assume $|y| < t_{\mathrm{KDt}}$. By a similar argument we can assume $|d| < t_{\mathrm{KDt}}$.

Consider a variant $U_{\mathcal{UP}}$ of the universal machine U where $U_{\mathcal{UP}}$ is given inputs of the form $(\hat{d}, \hat{y}, 1^{\hat{t}}, \hat{\imath}, \hat{b}, \hat{n})$. On any such input, if $\hat{n} > \hat{t}$, then reject. Otherwise guess a word $\hat{x} \in \{0,1\}^{\hat{n}}$ and check whether $\hat{x}_i = \hat{b}$. If yes, $U_{\mathcal{UP}}$ behaves for $\hat{t}$ steps like U on input $(\hat{d}, \hat{y}, \hat{x})$, that is $U_{\mathcal{UP}}$ accepts iff $U(\hat{d}, \hat{y}, \hat{x})$ accepts in these $\hat{t}$ steps.

For our fixed triple (d, y, x) this computation takes $t_{\mathcal{UP}}$ steps, with $t_{\mathcal{UP}} \in \Theta(|x| + t_{\mathrm{KDt}})$. Note that the running time of the simulation is coded unarily into its input. Let's call the coded number t_{coded}. So we have $t_{\mathrm{KDt}} = t_{\mathrm{coded}}$. The execution of a distinguishing description for x on U takes at least $|x|$ steps (otherwise x could not even be read completely, and therefore it could not be correctly distinguished). So we have $t_{\mathcal{UP}} \in \Theta(t_{\mathrm{KDt}})$.

This computation is now a nondeterministic one that already correctly recognizes the given bit of x. Because no part of the input pentuple P is longer than t_{coded}, we have that the program has length $\Theta(t_{\mathrm{coded}}) = \Theta(t_{\mathrm{KDt}})$ and that therefore the running time $t_{\mathcal{UP}}$ of the nondeterministic computation is in $\Theta(|P|)$.

Since d is a *distinguishing* description for the word $x \in \{0,1\}^n$, for all i on input $(d, y, 1^{t_{\mathrm{coded}}}, i, x_i, |x|)$ there is a unique accepting path of $U_{\mathcal{UP}}$ and none on input $(d, y, 1^{t_{\text{coded}}}, i, \bar{x}_i, |x|)$. By assumption 3′ there is a deterministic machine M that for all such inputs has the same acceptance and rejection behaviour as N and works in some fixed polynomial time bound.

The input for M together with an encoding of M is a generating program for x. It only remains to prove that this program is small enough and computes fast enough, compared to the KDt-program. This then implies $\mathrm{Kt}(x|y) \leq \mathrm{O}(\mathrm{KDt}(x|y))$, as desired.

Let us first inspect the program length. The input for M consists of t_{coded}, $|x|$ (both encoded in binary), $\ulcorner M \urcorner$, d. Since t_{KDt} counted logarithmically for KDt we have $\log t_{\text{coded}} \leq \mathrm{KDt}(x|y)$. $\mathrm{KDt}(x|y)$ is always greater than $\log |x|$, for the same argument as above. One fixed M works for all appropriate inputs and its encoding therefore has constant length. Obviously, $d \leq \mathrm{KDt}(x|y)$. Furthermore, y is given as part of the input to M, but does not add to $\mathrm{Kt}(x|y)$.

Let us now inspect the running time of the code. The nondeterministic machine used a running time in $\Theta(|P|)$. After the conversion to a deterministic procedure we have by assumption a running time of $(|P|)^{\mathrm{O}(1)}$. In other words: A polynomial overhead might have been introduced relative to the nondeterministic running time $t_{\mathcal{UP}}$. Therefore we have $t_{\mathbf{P}} \in \mathrm{O}((t_{\mathcal{UP}})^c) = \mathrm{O}(t_{\mathrm{KDt}}^c)$, hence $\log(t_{\mathbf{P}}) \in \mathrm{O}(\log t_{\mathrm{KDt}})$.

All this, together with the fact that running time counts logarithmically, results in the required inequality $\mathrm{Kt}(x|y) \leq \mathrm{O}(\mathrm{KDt}(x|y))$. □

Remark 15. For the same reason as in Remark 11, this proof would also work if we considered Levin's original definition of Kt.

Corollary 16. *If for all x and y, $\mathrm{Kt}(x|y) \in \mathrm{O}(\mathrm{KDt}(x|y))$, then* $\mathbf{UP} = \mathbf{P}$.

Proof. According to the theorem, the assumption implies that **UP** decision instances are **P**-solvable. Since a language in **UP** contains only such instances, the claim follows. □

Fortnow and Kummer [FK, Theorem 24] proved an equivalence related to Theorem 14 in the setting of the "traditional" polynomially time-bounded Kolmogorov complexities C^t and CD^t [LiV, Chapter 7] where for example

$$\mathrm{C}^t(x) = \min\{|d| \mid U(d) \text{ runs on input } x \text{ for } t(|x|) \text{ steps and outputs } x\}.$$

Theorem 17 (Fortnow, Kummer). *The following two statements are equivalent:*

1. **UP** *decision instances are* **P***-solvable.*
2. *For any polynomial t there are a polynomial t' and a constant $c \in \mathbb{N}$ such that for all x and y it holds that $C^{t'}(y|x) \leq CD^t(y|x) + c$.*

Remark 18. In fact Fortnow and Kummer formulated their equivalence in terms of promise problems. Instead of the first statement in the theorem they used the assertion that the promise problem (1SAT, SAT) is in **P**, where for a promise problem (Q, R) to be in **P** means that there is a **P**-machine that accepts all $x \in Q \cap R$ and rejects all $x \in \Sigma^* - R$.

Their formulation of the first statement is indeed equivalent to the one used above, because (1SAT, SAT) is complete for $\mathcal{UP}$, as witnessed by a parsimonious version of Cook's Theorem due to Simon [Sim, Theorem 4.1].

The following corollary is immediate from Theorems 14 and 17.

Corollary 19. *The following two statements are equivalent.*

1. *For all x and y, $\mathrm{Kt}(x|y) \in \mathrm{O}(\mathrm{KDt}(x|y))$.*
2. *For any polynomial t there is a polynomial t' and a constant $c \in \mathbb{N}$ such that for all x and y it holds that $C^{t'}(y|x) \leq CD^t(y|x) + c$.*

In analogy to the time-bounded case one can define the following two notions of space-bounded Kolmogorov complexity.

Definition 20. *The space-bounded distinction complexity* Ks *and generation complexity* KDs *are defined by*

$$\mathrm{Ks}(x) = \min\left\{ |d| + \log s \;\middle|\; \begin{array}{c} \forall b \in \{0,1,*\}\colon \forall i \leq |x|+1\colon U(d,i,b) \\ \text{runs in space } s \text{ and accepts iff } (x*)_i = b \end{array} \right\},$$

$$\mathrm{KDs}(x) = \min\left\{ |d| + \log \max(s, |x|) \;\middle|\; \begin{array}{c} \forall y \in \Sigma^{|x|}\colon U(d,y) \text{ runs in} \\ \text{space } s \text{ and accepts iff } x = y \end{array} \right\}.$$

Here U is a machine with a two-way read-only input tape where only the space on the work tapes is counted.

Remark 21. For the definition of KDs it is relevant how the candidate y is provided to U and if the space for y is counted. Here we chose to *do* count the space for y which accounts for the term $\max |x|$ in the definition of KDs. This then implies the inequality $\log |x| \leq \mathrm{KDs}(x)$, which is analogous to the corresponding statement for KDt and will be used in the proof of Theorem 22.

Theorem 22. *For almost all x, it holds that $\mathrm{Ks}(x) \leq 5 \cdot \mathrm{KDs}(x)$.*

Proof. Let N be a nondeterministic machine which on input (d, s, i, b, n) guesses a word $y \in \{0,1\}^n$, simulates the computation of $U(d, y)$ while limiting the used space to s, and then accepts iff $y_i = b$ and $U(d, y)$ accepts. In particular, if d is a *distinguishing* description for a word $x \in \{0,1\}^n$, then for all sufficiently large s and for all $i \leq n$ there is an accepting path of N on input $(d, s, i, x_i, |x|)$ but none on $(d, s, i, \bar{x}_i, |x|)$.

By the Theorem of Savitch there is a deterministic machine M that has the same acceptance behavior as N and uses space at most s^2; observe in this connection that s is specified in the input of N and M, hence doesn't have to be computed by M.

Given a word x, fix a pair d and s such that d is a distinguishing program for x, it holds that $|d| + \log s \leq \mathrm{KDs}(x)$, and U uses space at most s on input (d, x).

The specification of d, s, $|x|$ and M therefore constitutes a Ks-program for x which runs in space s^2. By choice of d and s we have

$$|d| + \log s + \log |x| \leq 2\mathrm{KDs}(x).$$

Furthermore, the space s^2 used in the computation of M counts only logarithmically, where $2 \cdot \log s \leq 2 \cdot \mathrm{KDs}(x)$. Taking into account that M has to be specified and that some additional information is needed to separate the components of the Ks-program for x, we obtain $\mathrm{Ks}(x) \leq 5 \cdot \mathrm{KDs}(x)$ for all sufficiently large x. □

References

[A] Allender, E., Koucký, M., Ronneburger, D., Roy, S.: Derandomization and distinguishing complexity. In: Proc. 18th Annual IEEE Conference on Computational Complexity, pp. 209–220. IEEE Computer Society Press, Los Alamitos (2003)

[BFL] Buhrman, H., Fortnow, L., Laplante, S.: Resource-bounded Kolmogorov complexity revisited. SIAM Journal on Computing 31(3), 887–905 (2002)

[CHV] Cai, J., Hemachandra, L.A., Vyskoč, J.: Promises and fault-tolerant database access. In: Ambos-Spies, K., Homer, S., Schöning, U. (eds.) Complexity Theory, pp. 227–244. Cambridge University Press, Cambridge (1993)

[FK] Fortnow, L., Kummer, M.: On resource-bounded instance complexity. Theoretical Computer Science 161, 123–140 (1996)

[Le] Levin, L.A.: Randomness conservation inequalities: Information and independence in mathematical theories. Information and Control 61, 15–37 (1984)

[LiV] Li, M., Vitányi, P.: An Introduction to Kolmogorov Complexity and Its Applications. Springer, Heidelberg (1997)

[Sim] Simon, J.: On some central problems in computational complexity. Technical Report TR75-224, Cornell University (1975)

[Sip] Sipser, M.: A complexity theoretic approach to randomness. In: Proc. 15th ACM Symp. Theory Comput, pp. 330–335. ACM Press, New York (1983)

Balancing Traffic Load Using One-Turn Rectilinear Routing

Stephane Durocher[1], Evangelos Kranakis[2], Danny Krizanc[3], and Lata Narayanan[4]

[1] School of Computer Science, University of Waterloo, Waterloo, Ontario, Canada
sdurocher@cs.uwaterloo.ca
[2] School of Computer Science, Carleton University, Ottawa, Ontario, Canada
kranakis@scs.carleton.ca
[3] Department of Mathematics and Computer Science, Wesleyan University, Middletown, Connecticut, USA
dkrizanc@wesleyan.edu
[4] Department of Computer Science, Concordia University, Montréal, Québec, Canada
lata@cse.concordia.ca

Abstract. We consider the problem of load-balanced routing, where a dense network is modelled by a continuous square region and origin and destination nodes correspond to pairs of points in that region. The objective is to define a routing policy that assigns a continuous path to each origin-destination pair while minimizing the traffic, or *load*, passing through any single point. While the average load is minimized by straight-line routing, such a routing policy distributes the load non-uniformly, resulting in higher load near the center of the region. We consider one-turn rectilinear routing policies that divert traffic away from regions of heavier load, resulting in up to a 33% reduction in the maximum load while simultaneously increasing the path lengths by an average of less than 28%. Our policies are simple to implement, being both local and oblivious. We provide a lower bound that shows that no one-turn rectilinear routing policy can reduce the maximum load by more than 39% and we give a polynomial-time procedure for approximating the optimal randomized policy.

1 Introduction

The problem of routing in multi-hop wireless networks has received extensive attention in the last decade [1,2,12,14,18]. Many of the proposed routing protocols attempt to find shortest paths between pairs of nodes, or try to bound the *stretch factor* of the paths, while trying to ensure that the paths are loop-free. This approach takes into account a single packet traversing the network and tries to optimize performance for this packet. A more global and realistic view would consider the performance of the protocol under the assumption of many traffic flows in the network. In this situation, there can often be *congestion* created by several packets that need to be forwarded by the same intermediate nodes at the same time. This congestion is very likely to influence the latency experienced by

M. Agrawal et al. (Eds.): TAMC 2008, LNCS 4978, pp. 467–478, 2008.

a packet. A routing protocol should therefore attempt to avoid creating highly congested nodes. Not only does this improve packet latency, it would also improve the lifetime of a wireless network, where heavily loaded nodes may run out of battery power and disconnect the network.

In this paper, we investigate routing protocols for wireless networks with the aim of minimizing the congestion experienced at nodes. We consider a multi-hop ad hoc network consisting of identical location-aware nodes, uniformly and densely deployed within a given planar region. Furthermore, we assume that the traffic pattern is uniform point-to-point communication, i.e., each node has the same number of packets to send to every other node in the network. This is sometimes called the *all-to-all* communication pattern. A routing policy must define, for every ordered pair of nodes (u, v), a path in the network to get from u to v. The *load* at a given node v is the number of paths that pass through v. The average (maximum) load for a network with respect to a particular routing policy is the average (respectively maximum) load over all nodes in the network. The fundamental question we wish to answer is: *what routing policy minimizes the maximum load in the network?*

It seems intuitively evident that for nodes within a convex planar region, shortest path routing should cause maximum load near the *geometric center*. Indeed, this has been proved analytically for disks (see for example [16]) and squares and rectangles (see Section 3). This suggests that if load balancing is a fundamental concern then a good routing policy should redirect some of the traffic away from the geometric center and other areas of high load. However, load balancing cannot be the *only* concern: taking unnecessarily long paths just to bypass the center can drastically increase the stretch factor and the average load of nodes in the network, and can therefore be very inefficient in terms of energy consumption. Furthermore, it is critical that the forwarding strategy required to implement the routing policy be simple and have low memory requirements. Ideally, the routing policy should be *oblivious* (the route between u and v depends only on the identities or locations of u and v) and the forwarding strategy should be *local* (the forwarding node can make its decision based only on itself and its neighbors, and the packet header contains only the address of the destination).

In the setting of nodes uniformly distributed in a given planar convex region, very little research has been done on finding a simple routing policy that achieves both a reasonable stretch factor and a minimum value of maximum congestion. In [6], an algorithm achieving a good tradeoff between stretch factor and load balance is shown for the special case when all nodes are located in a narrow strip of width at most 0.86 times the transmission radius. The analysis is not specific to the all-to-all communication pattern. Popa *et al.* [16] address the all-to-all routing problem for the case when the region containing the nodes is a unit disk. They establish quantitatively the *crowded center* effect for shortest-path routing as a nearly-quadratic function that peaks at the center of the disk and present a theoretical approach that is guaranteed to find paths minimizing the maximum load. They also give a practical solution (curveball routing) whose performance compares favorably to the optimum. No theoretical bounds are given on the stretch factor of the routes for either strategy.

In this paper, we investigate the problem of load-balanced routing when the nodes are uniformly and densely packed in a square or rectangular region. As in [16], our approach is to look at the unit square (and the $k \times 1$ rectangle) as a *continuous space* rather than formed by discrete nodes. This makes it possible to analyze the average and maximum load induced by a routing policy, without regard to the topology of the actual network. At the same time, the results should predict the behavior of a network with very densely and uniformly deployed nodes. Shortest-path routing corresponds to *straight-line routing* in this setting. We derive the average and maximum load for straight-line routing in a unit square and confirm the crowded-center effect for squares and rectangles. In keeping with the goal of minimizing congestion while ensuring a reasonable stretch factor, we investigate the class of *rectilinear* routing policies that assign to each origin-destination pair of nodes one of the two possible rectilinear paths containing only one turn. It is not difficult to show that all such one-turn rectilinear strategies have a maximum stretch factor of $\sqrt{2}$. Furthermore, they are simple and realistic in the ad hoc network setting; the routing policy is oblivious and the forwarding algorithm is local. We propose and analyze several simple rectilinear strategies, the best of which reduces the maximum load by about 33% compared to the straight-line policy. We also characterize the optimal randomized rectilinear policy as the solution to an optimization problem and provide an efficient procedure for approximating it.

1.1 Overview of Results

Our main contributions are summarized below:

- We derive an exact expression for the load induced by a straight-line routing policy at an arbitrary point in the unit square. We show that the average and maximum load for the straight-line routing policy are 0.5214 and 1.1478 respectively.
- We show that the average load for *every* one-turn rectilinear routing policy is 2/3. The maximum and average stretch factor for such policies are shown to be $\sqrt{2}$ and 1.2737 respectively.
- We propose several one-turn rectilinear routing policies and derive their maximum load. The best of these, called the *diagonal rectilinear policy*, achieves a maximum load of 0.7771, which represents a 33% improvement over straight-line routing.
- We prove a lower bound of 0.7076 on the the maximum load for *any* one-turn rectilinear policy.
- We characterize the optimal randomized rectilinear policy as the solution to an optimization problem and provide an efficient procedure for approximating it. Numerical results suggest that the maximum load for the best possible rectilinear policy is close to 0.74.

Detailed proofs for some results are omitted due to space restrictions.

1.2 Related Work

In this section, we briefly describe other efforts to address the congestion problem. Several studies confirm the crowded center effect for shortest path routing [8,11,15,16]. In [15,16], the load at the center of a circular area is derived analytically, by modelling the area as a continuous region, as in this paper, rather than as formed by discrete nodes. The node distribution resulting from a random waypoint mobility model in an arbitrary convex domain is analyzed in [10]; this is related to the load probability density for straight-line routing.

The tradeoffs between congestion and stretch factor in wireless networks has been studied in [13] and [7]. For instance, for growth-bounded wireless networks, Gao and Zhang [7] show routing algorithms that simultaneously achieve a stretch factor of c and a load balancing ratio of $O((n/c)^{1-1/k})$ where k is the growth rate. (The load balancing ratio is defined to be the ratio between the maximum load on any node induced by the algorithm versus that created by the optimal algorithm.) They also derive an algorithm for unit disk graphs with bounded density and show that if the density is constant, shortest path routing has a load balancing ratio of $\Theta(\sqrt{n})$. The communication patterns considered are arbitrary, the lower bound does not derive from the all-to-all communication pattern, and the routing algorithms are not oblivious.

The all-to-all communication pattern has been studied extensively in the context of interconnection networks, and particularly in WDM optical networks. In this context, [4] defined the forwarding index of a communication network with respect to a specific routing algorithm to be the maximum number of paths going through any vertex in the graph. The forwarding index of the network itself is the minimum over all possible routing algorithms for the network. This notion was extended to the maximum load on an edge [9], which is more appropriate to wired networks. However, for wireless networks, the node forwarding index captures the load on a wireless node better. While the node forwarding index for specific networks, including the ring and torus networks has been derived exactly [4], it has not been studied for two-dimensional grid networks, which would perhaps be a good approximation for the dense wireless networks of interest to us. Our results in Section 6 provide an approximation for the forwarding index in grid graphs for the class of one-turn rectilinear routing schemes.

There does not appear to be much work on routing with a view to reducing the congestion for the all-to-all communication pattern in specific planar regions, the model of interest in this paper. As stated earlier, [6] looks at nodes contained in a narrow strip and [16] addresses the problem for the unit disk. Busch et al. [3] analyze routing on embedded graphs via a random intermediate point located near the perpendicular bisector of the origin and destination; we consider the generalization of this strategy to convex regions in Section 4.4. Popa *et al.* [16] give expressions for the maximum and average load induced by straight-line routing in unit disks, and propose a practical algorithm called curveball routing whose performance is close to the optimum for disks. They also provide experimental results on greedy routing versus curveball routing in square- and rectangular-shaped areas, and show that curveball routing achieves a reduction in load in such areas, but they do not provide any theoretical results.

2 Definitions

2.1 Routing Policies and Traffic Load

Given a convex region $A \subseteq \mathbb{R}^2$, a *routing policy* P assigns a route to every origin-destination pair $(u, v) \in A^2$, where the *route* from u to v, denoted $\mathrm{route}_P(u, v)$, is a plane curve segment contained in A, whose endpoints are u and v. For a given routing policy P on a region A, the traffic load at a point p is proportional to the number of routes that pass through p. Formally,

Definition 1. *Given a routing policy P on a region A, the* load *at point p is*

$$\lambda_P(p) = \iint_A f_P(p, u, v)\, du\, dv, \qquad \textit{where } f_P(p, u, v) = \begin{cases} 1 \textit{ if } p \in \mathrm{route}_P(u, v), \\ 0 \textit{ otherwise.} \end{cases}$$

The *average load* of routing policy P on region A is given by

$$\lambda_{\mathrm{avg}}(P) = \frac{1}{\mathrm{Area}(A)} \int_A \lambda_P(p)\, dp, \tag{1}$$

where $\mathrm{Area}(A) = \int_A dp$ denotes the area of region A. The average length of a route determined by policy P between two points in A is given by

$$\mathrm{length}_{\mathrm{avg}}(P) = \frac{1}{\mathrm{Area}(A)^2} \iint_A \mathrm{length}(\mathrm{route}_P(p, q))\, dq\, dp. \tag{2}$$

Since $\mathrm{length}(\mathrm{route}_P(u, v)) = \int_A f_P(p, u, v)\, dp$, Proposition 1 follows from (1) and (2):

Proposition 1. *Given routing policy P on a region A,*

$$\lambda_{\mathrm{avg}}(P) = \mathrm{Area}(A) \cdot \mathrm{length}_{\mathrm{avg}}(P). \tag{3}$$

In addition to average load, a routing policy P on a region A is also characterized by its *maximum load*, given by

$$\lambda_{\max}(P) = \max_{p \in A} \lambda_P(p). \tag{4}$$

2.2 Straight-Line Routing Policy

The *straight-line routing policy*, denoted S, assigns to every pair (u, v) the route consisting of the line segment between u and v. In straight-line routing,

$$\mathrm{length}(\mathrm{route}_S(p, q)) = ||p - q|| = \sqrt{(p_x - q_x)^2 + (p_y - q_y)^2}. \tag{5}$$

Since the line segment from u to v is the shortest route from u to v, it follows that straight-line routing minimizes (2). Consequently, for any convex region A and any routing policy $P \neq S$,

$$\lambda_{\text{avg}}(S) \leq \lambda_{\text{avg}}(P). \tag{6}$$

The *average stretch factor* and *maximum stretch factor* of routing policy P on region A are respectively given by

$$\text{str}_{\text{avg}}(P) = \frac{1}{\text{Area}(A)^2} \iint_A \frac{\text{length}(\text{route}_P(p,q))}{\text{length}(\text{route}_S(p,q))}\, dq\, dp \tag{7}$$

$$\text{str}_{\text{max}}(P) = \max_{\{p,q\} \subseteq A} \frac{\text{length}(\text{route}_P(p,q))}{\text{length}(\text{route}_S(p,q))}. \tag{8}$$

2.3 One-Turn Rectilinear Routing Policies

Recent related work on this problem has considered the case when region A is a disk [16]. In this paper, we consider the case when region A is bounded by a square or a rectangle. As we show in Section 3, the load in straight-line routing on a square or a rectangle is maximized at its center. The maximum load can be decreased by redirecting routes that pass near the center to regions of lower traffic. This motivates the examination of one-turn rectilinear routing policies which we now define.

A *monotonic rectilinear routing policy* assigns to every pair (u, v) a route consisting of a monotonic rectilinear path from u to v, *i.e.*, a path comprised of a series of axis-parallel line segments such that any axis-parallel line intersects the path at most once. A *one-turn rectilinear routing policy* assigns to every pair (u, v) a monotonic rectilinear path consisting of one horizontal line segment and one vertical line segment joining u to v via an intermediate point w. Point w may coincide with u or v.

For any monotonic rectilinear routing policy P,

$$\text{length}(\text{route}_P(p,q)) = |p_x - q_x| + |p_y - q_y|. \tag{9}$$

In general, there are two possible one-turn rectilinear routes from a given origin (u_x, u_y) to a given destination (v_x, v_y). We refer to these as *row-first* and *column-first*, where the row-first route passes through the intermediate point (v_x, u_y) and the column-first route passes through the intermediate point (u_x, u_y).

3 Straight-Line Routing on a Square

In this section we examine the load of straight-line routing on the unit square. These values serve as milestones against which the optimality of all other routing policies on the unit square are compared.

3.1 Average Load

By Proposition 1, the average load in the unit square under straight-line routing is equal to the expected distance between two points selected at random in the square. This value is a box integral with the following solution [17]:

$$\text{length}_{\text{avg}}(S) = \frac{2+\sqrt{2}+5\ln(1+\sqrt{2})}{15} \approx 0.5214. \tag{10}$$

By (6), the average load (and maximum load) of any routing policy on the unit square is bounded from below by (10).

3.2 Load at an Arbitrary Point

Since straight-line routing is symmetric in the x- and y-dimensions, we derive the load at an arbitrary point p located in an octant of the unit square. The load at an arbitrary point in the unit square is then easily found using the appropriate coordinate transformation.

Theorem 1. *Given a point $p = (p_x, p_y)$ such that $1/2 \le p_y \le p_x \le 1$, the load at p using straight-line routing is given by*

$$\begin{aligned}\lambda_S(p) = {} & (1-p_x)p_x \left[\Big|_{\theta=0}^{\alpha} g_1(\theta) \right] + (1-p_x)^2 p_y \left[\Big|_{\theta=\alpha}^{\pi/2-\beta} g_4(\theta) \right] \\ & + (1-p_x)p_y^2 \left[\Big|_{\theta=\alpha}^{\pi/2-\beta} g_3(\theta) \right] + (1-p_y)p_y \left[\Big|_{\theta=\pi/2-\beta}^{\pi/2} g_2(\theta) \right] \\ & + (1-p_y)p_y \left[\Big|_{\theta=\pi/2}^{\pi/2+\gamma} g_2(\theta) \right] - (1-p_y)^2(1-p_x) \left[\Big|_{\theta=\pi/2+\gamma}^{\pi-\delta} g_3(\theta) \right] \\ & + (1-p_y)(1-p_x)^2 \left[\Big|_{\theta=\pi/2+\gamma}^{\pi-\delta} g_4(\theta) \right] - p_x(1-p_x) \left[\Big|_{\theta=\pi-\delta}^{\pi} g_1(\theta) \right], \end{aligned} \tag{11}$$

where expressions for α, β, γ, δ, and g_1 through g_4 are omitted for lack of space.

Expression (11) has a closed-form polylogarithmic representation (free of any trigonometric terms). The complete expression is not reproduced here due to the large number of terms but can be easily reconstructed from (11).

3.3 Maximum Load

We now derive the maximum load for straight-line routing on the unit square and show that this value is realized at the center of the square.

Theorem 2. *The maximum load for straight-line routing on the unit square is*

$$\lambda_{\max}(S) = \frac{1}{\sqrt{2}} + \frac{3}{8}\ln(\sqrt{2}+1) - \frac{1}{8}\ln(\sqrt{2}-1) \approx 1.1478, \tag{12}$$

realized uniquely at the center of the square.

4 One-Turn Rectilinear Routing on a Square

In this section we consider various one-turn rectilinear routing policies on the unit square and compare these against straight-line routing. Our objective in designing these policies was to reduce the maximum load by redirecting routes for particular regions of origin-destination pairs away from high-traffic areas and towards low-traffic areas while maintaining a low stretch factor.

4.1 Average Load

Theorem 3. *The average load for any monotonic rectilinear routing policy on the unit square is* 2/3.

Proof. By Proposition 1 and (9), the average load is equal to the average ℓ_1 distance between two points in the unit square. This value is

$$\lambda_{\text{avg}}(P) = \int_0^1\int_0^1\int_0^1\int_0^1 |u_x - v_x| + |u_y - v_y| \, dv_y \, dv_x \, du_y \, du_x = \frac{2}{3}. \; \square \qquad (13)$$

4.2 Average Stretch Factor

It is straightforward to see that the maximum stretch factor for any monotonic rectilinear routing policy is $\sqrt{2}$. We now consider the average stretch factor.

Theorem 4. *The average stretch factor for any monotonic rectilinear routing policy P on the unit square is*

$$\text{str}_{\text{avg}}(P) = \frac{1}{6}\left(10\ln(2+\sqrt{2}) + 2\sqrt{2} - 4 - 5\ln(2)\right) \approx 1.2737. \qquad (14)$$

4.3 Diagonal Rectilinear Routing

We define a routing policy in terms of the partition of the unit square induced by its two diagonals. Let R_1 through R_4 denote the four regions of the partition such that R_1 is at the bottom of the square and the regions are numbered in clockwise order. If the origin lies in R_1 or R_3, the row-first route is selected. Otherwise, the column-first route is selected. We refer to this routing policy, denoted P_D, as *diagonal rectilinear routing.*

As we did in Section 3.2, we derive the load at an arbitrary point p located in an octant of the unit square since P_D is symmetric in the x- and y-dimensions. The load at an arbitrary point in the unit square is then easily found using the appropriate coordinate transformation.

Theorem 5. *Given a point $p = (p_x, p_y)$ such that $0 \leq p_y \leq p_x \leq 1/2$, the load at p using diagonal rectilinear routing is*

$$\lambda_{P_D}(p) = 2p_x^3 - 5p_x^2 + \frac{7}{2}p_x - 2p_xp_y + \frac{3}{2}p_y - 3p_y^2 + 2p_y^3. \qquad (15)$$

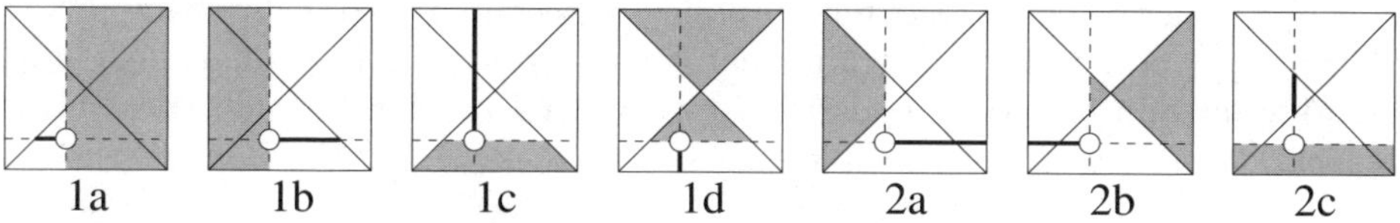

Fig. 1. Illustration in support of Theorem 5. The white dot denotes point p.

Proof. Let $u = (u_x, u_y)$ denote the origin and let $v = (v_x, v_y)$ denote the destination. The relative positions of u, v, and p can be divided into seven cases such that the load at p corresponds to the sum of the measure of the regions of possible origin-destination combinations in each case. In Cases 1a through 1d, u lies in region R_1 or R_3 and, consequently, the row-first route is selected. In Cases 2a through 2c, u lies in region R_2 or R_4 and, consequently, the column-first route is selected. See Fig. 1. The theorem follows by summing the contribution to load in each of these cases. Details are omitted for lack of space. □

It can be shown that (15) is maximized when $p_x = \frac{5}{6} - \frac{1}{3}\sqrt{3 - \frac{1}{2}\sqrt{11}} \approx 0.4472$ and $p_y = \frac{2}{3} - \frac{\sqrt{11}}{6} \approx 0.1139$, with load $\lambda_{\max}(P_D) = \frac{1}{27}\left[\sqrt{11} - \frac{31}{2}\right] \approx 0.7771$.

4.4 Additional Policies Considered

We describe additional one-turn rectilinear routing policies considered. In each case, the maximum load was shown to be strictly greater than that of diagonal rectilinear routing. Recall that all one-turn rectilinear routing policies have equal average load (Theorem 3).

Equal Distribution. A simple initial strategy to consider is to assign to each origin-destination pair (u, v) the row-first route. For any point $p \in [0, 1]^2$, $\lambda_{P_R}(p) = \lambda_P(p)$, where P_R denotes the row-first routing policy and P denotes any policy that assigns the pairs (u, v) and (v, u) different one-turn rectilinear routes for all u and v.

Outer Turn. Consider the routing policy that selects the one-turn rectilinear route whose intermediate point is furthest from the center of the square. If the two intermediate points are equidistant from the origin, then a route is assigned as in the equal distribution policy.

Grid-Based Regions. Divide the unit square into nine rectangular regions whose boundaries intersect the x- and y-axes at 0, k, $1 - k$, and 1, respectively, for some fixed $k \in [0, 1/2]$. There are three types of regions: corner, mid-boundary, and one central region. If the origin is located in a mid-boundary region and the destination is in a non-adjacent corner region then select the one-turn rectilinear route that avoids passing through the central region. Similarly, a route from a corner regions to a non-adjacent mid-boundary region must avoid passing through the central region. For all other origin-destination pairs, routes are assigned as in the equal distribution policy. A second grid-based routing policy is defined by adding the constraint that a route from a mid-boundary region to an adjacent mid-boundary region must also avoid passing through the central region.

Line Division. Let l denote the line passing through the origin u and destination v. If l does not pass through the center of the square, c, select the one-turn rectilinear route whose intermediate point is opposite l from c. If l passes through c, then a route is assigned as in the equal distribution policy.

Random Intermediate Point. A random intermediate point connected to the origin and destination by straight-line routes results in load exactly twice that of straight-line routing. Perhaps a better strategy is that described by Busch et al. [3] which can be generalized to convex regions; an intermediate point is selected at random on the perpendicular bisector of the origin and destination. Note that although this policy involves a single turn, it is not a rectilinear routing policy.

Summary. Table 1 summarizes bounds on average and maximum load for each of the above routing policies and straight-line routing. We derived the load at an arbitrary point in the unit square for five of these policies; the corresponding plots are illustrated in Fig. 2. The diagonal rectilinear routing policy, P_D, achieves the lowest maximum load, significantly lower than the maximum load of straight-line routing and not much greater than the lower bound.

Table 1. Comparing routing policies on the unit square

routing policy	λ_{avg}	λ_{max}
straight-line S	0.5214	1.1478
diagonal P_D	2/3	0.7771
equal dist. P_R	2/3	1
outer turn	2/3	≥ 0.8977

routing policy	λ_{avg}	λ_{max}
grid-based (1)	2/3	7/8
grid-based (2)	2/3	0.8541
line division	2/3	$\geq 7/8$
lower bound		0.7076

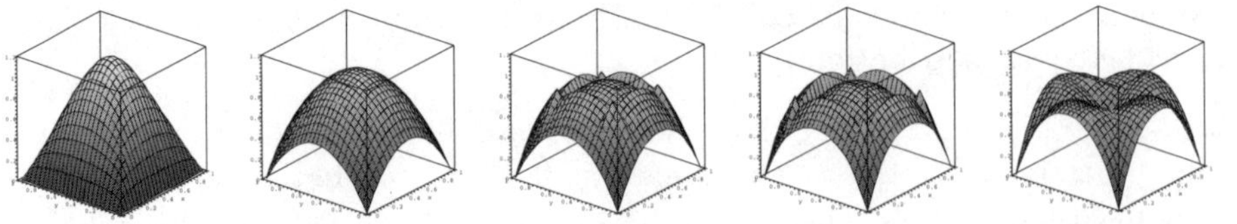

Fig. 2. These plots display $\lambda_P(p)$ for $p \in [0, 1]^2$ for five routing policies: (left to right) straight-line S, equal distribution P_R, grid-based (two policies), and diagonal P_D

5 Lower Bounds on Load for One-Turn Rectilinear Routing Policies

Naturally, no monotonic rectilinear routing policy can have a maximum load less than the average load of 2/3. In this section we establish a stronger lower bound on the maximum load of any one-turn rectilinear routing policy.

Theorem 6. *No one-turn rectilinear routing policy can guarantee a maximum load less than* 0.7076.

6 Optimal Randomized One-Turn Rectilinear Routing Policies

In this section we give a characterization of the optimal randomized one-turn rectilinear strategy as the solution of an optimization problem and provide an efficient procedure for approximating it. A deterministic one-turn rectilinear strategy is equivalent to a function $P : [0,1]^4 \to \{0,1\}$ where $P(u,v,s,t) = 1$ iff the route from (u,v) to (s,t) uses the column-first path. (If $u = s$ or $v = t$, we define $P(u,v,u,t) = P(u,v,s,v) = 1$.) We can generalize this to randomized rectilinear schemes by considering $Q : [0,1]^4 \to [0,1]$ where if $Q(u,v,s,t) = q$ then a packet travelling from point (u,v) to (s,t) takes the the column-first path with probability q and the row-first path with probability $1-q$. A formula for the expected $\lambda(x,y)$ at a point (x,y) can be found easily.

The optimal strategy is given by the solution to the following optimization problem: $\min_Q \max_{(x,y)} \lambda(x,y)$. While we can't directly solve this problem we can approximate it by considering finer and finer partitions of the square into n^2 $1/n$ by $1/n$ subsquares and giving a strategy for all packets routing between each pair of subsquares. Now our problem is equivalent to finding a randomized one-turn rectilinear routing strategy for an $n \times n$ grid that minimizes the number of packets using any particular node of the grid under an all-to-all communication pattern.

Let p_{ijkl}, $1 \le i,j,k,l \le n$, be the probability that a packet starting in subsquare (i,j) going to subsquare (k,l) uses the column-first path and let the maximum expected load at any point in subsquare (r,s) be $\lambda(r,s)$.

An upper bound on $\lambda(r,s)$ is easily derived and our problem now reduces to $\min_{p_{ijkl}} \max_{r,s} \lambda(r,s)$, which is equivalent to the following linear program with n^4+1 variables and $2n^4+n^2$ constraints (solvable in polynomial time):

Minimize z
Subject to
$0 \le p_{ijkl} \le 1, \quad 1 \le i,j,k,l \le n,$
$z - \lambda(r,s) \ge 0, \quad 1 \le r,s \le n.$

Table 2 shows an upper bound on the maximum load achieved by the strategy obtained by using an $n \times n$ grid to approximate the unit square for $2 \le n \le 12$. The results indicate that the optimal strategy achieves a maximum load

Table 2. Approximations to the optimal randomized strategy using $n \times n$ grids

n	2	3	4	5	6	7	8	9	10	11	12
max. load	1.0000	0.8889	0.8264	0.8009	0.7813	0.7759	0.7650	0.7610	0.7530	0.7499	0.7446

of approximately 0.74. The solutions were found using the CVXOPT convex optimization package [5]. We were unable to obtain results for larger n due to memory limitations.

References

1. Bose, P., Morin, P., Stojmenovic, I., Urrutia, J.: Routing with guaranteed delivery in ad hoc wireless networks. Wireless Networks 7, 609–616 (2001)
2. Broch, J., Johnson, D., Maltz, D.: The dynamic source routing protocol for mobile ad hoc networks (1998): Internet-draft, draft-ietf-manet-dsr-00.txt
3. Busch, C., Magdon-Ismail, M., Xi, J.: Oblivious routing on geometric networks. In: Proc. ACM SPAA, vol. 17 (2005)
4. Chung, F.R.K., Coffman Jr., E.G., Reiman, M.I., Simon, B.: The forwarding index of communication networks. IEEE Trans. Inf. Th. 33(2), 224–232 (1987)
5. Dahl, J.: Cvxopt: A python package for convex optimization. In: Proc. Eur. Conf. Op. Res. (2006)
6. Gao, J., Zhang, L.: Load balanced short path routing in wireless networks. In: Proc. IEEE INFOCOM, vol. 23, pp. 1099–1108 (2004)
7. Gao, J., Zhang, L.: Tradeoffs between stretch factor and load balancing ratio in routing on growth restricted graphs. In: Proc. ACM PODC, pp. 189–196 (2004)
8. Gupta, P., Kumar, P.R.: The capacity of wireless networks. IEEE Trans. Inf. Th. 46, 388–404 (2000)
9. Heydemann, M.C., Meyer, J.C., Sotteau, D.: On forwarding indices of networks. Disc. App. Math. 23, 103–123 (1989)
10. Hyytiä, E., Lassila, P., Virtamo, J.: Spatial node distribution of the random waypoint mobility model with applications. IEEE Trans. Mob. Comp. 6, 680–694 (2006)
11. Jinyang, L., Blake, C., Couto, D.D., Lee, H., Morris, R.: Capacity of ad hoc wireless networks. In: Proc. ACM MOBICOM (2001)
12. Kranakis, E., Singh, H., Urrutia, J.: Compass routing on geometric networks. In: Proc. CCCG, pp. 51–54 (1999)
13. Meyer auf der Heide, F., Schindelhauer, C., Volbert, K., Grunwald, M.: Energy, congestion, and dilation in radio networks. In: Proc. ACM SPAA, pp. 230–237 (2002)
14. Park, V., Corson, S.: A highly adaptive distributed routing algorithm for mobile wireless networks. In: Proc. IEEE INFOCOM, pp. 1405–1413 (1997)
15. Pham, P.P., Perreau, S.: Performance analysis of reactive shortest path and multipath routing mechanism with load balance. In: Proc. IEEE INFOCOM, pp. 251–259 (2003)
16. Popa, L., Rostami, A., Karp, R.M., Papadimitriou, C., Stoica, I.: Balancing traffic load in wireless networks with curveball routing. In: Proc. ACM MOBIHOC (2007)
17. Santaló, L.A.: Integral Geometry and Geometric Probability. Cambridge University Press, Cambridge (2004)
18. Stojmenovic, I.: Position based routing in ad hoc networks. IEEE Comm. Mag. 40, 128–134 (2002)

A Moderately Exponential Time Algorithm for Full Degree Spanning Tree

Serge Gaspers, Saket Saurabh, and Alexey A. Stepanov

Department of Informatics, University of Bergen, N-5020 Bergen, Norway
{serge,saket}@ii.uib.no, ljosha@ljosha.org

Abstract. We consider the well studied FULL DEGREE SPANNING TREE problem, a NP-complete variant of the SPANNING TREE problem, in the realm of moderately exponential time exact algorithms. In this problem, given a graph G, the objective is to find a spanning tree T of G which maximizes the number of vertices that have the same degree in T as in G. This problem is motivated by its application in fluid networks and is basically a graph-theoretic abstraction of the problem of placing flow meters in fluid networks. We give an exact algorithm for FULL DEGREE SPANNING TREE running in time $\mathcal{O}(1.9172^n)$. This adds FULL DEGREE SPANNING TREE to a very small list of "non-local problems", like FEEDBACK VERTEX SET and CONNECTED DOMINATING SET, for which non-trivial (non brute force enumeration) exact algorithms are known.

1 Introduction

The problem of finding a spanning tree of a connected graph arises at various places in practice and theory, like the analysis of communication or distribution networks, or modeling problems, and can be solved efficiently in polynomial time. On the other hand, if we want to find a spanning tree with some additional properties like maximizing the number of leaves or minimizing the maximum degree of the tree, the problem becomes NP-complete. This paper deals with one of the NP hard variants of SPANNING TREE, namely FULL DEGREE SPANNING TREE from the view point of moderately exponential time algorithms.

> FULL DEGREE SPANNING TREE (FDST): Given an undirected connected graph $G = (V, E)$, find a spanning tree T of G which maximizes the number of vertices of *full degree*, that is the vertices having the same degree in T as in G.

The FDST problem is motivated by its applications in water distribution and electrical networks [15,16,17,18]. Pothof and Schut [18] studied this problem in the context of water distribution networks where the goal is to determine or control the flows in the network by installing and using a small number of flow meters. It turns out that to measure flows in all pipes, it is sufficient to find a full degree spanning tree T of the network and install flow meters (or pressure

M. Agrawal et al. (Eds.): TAMC 2008, LNCS 4978, pp. 479–489, 2008.

gauges) at each vertex of T that does not have full degree. We refer to [1,4,11] for a more detailed description of various applications of FDST.

The FDST problem has attracted a lot of attention recently and has been studied extensively from different algorithmic paradigms, developed for coping with NP-completeness. Pothof and Schut [18] studied this problem first and gave a simple heuristic based algorithm. Bhatia et al. [1] studied it from the view point of approximation algorithms and gave an algorithm of factor $\mathcal{O}(\sqrt{n})$. On the negative side, they show that FDST is hard to approximate within a factor of $\mathcal{O}(n^{\frac{1}{2}-\epsilon})$, for any $\epsilon > 0$, unless $coR = NP$, a well known complexity-theoretic hypothesis. Guo et al. [10] studied the problem in the realm of parameterized complexity and observed that the problem is W[1]-complete. The problem which is dual to FDST is also studied in the literature, that is the problem of finding a spanning tree that minimizes the number of vertices not having full degree. For this dual version of the problem, Khuller et al [11] gave an approximation algorithm of factor $2 + \epsilon$ for any fixed $\epsilon > 0$, and Guo et al. [10] gave a fixed parameter tractable algorithm running in time $4^k n^{\mathcal{O}(1)}$. FDST has also been studied on special graph classes like planar graphs, bounded degree graphs and graphs of bounded treewidth [4]. The goal of this paper is to study Full Degree Spanning Tree in the context of moderately exponential time algorithms, another coping strategy to deal with NP-completeness. We give a $\mathcal{O}(1.9172^n)$ time algorithm breaking the trivial $2^n n^{\mathcal{O}(1)}$ barrier.

Exact exponential time algorithms have an old history [5,14] but the last few years have seen a renewed interest in the field. This has led to the advancement of the state of the art on exact algorithms and many new techniques based on Inclusion-Exclusion, Measure & Conquer and various other combinatorial tools have been developed to design and analyze exact algorithms [2,3,7,8,12]. Branch & Reduce has always been one of the most important tools in the area but its applicability was mostly limited to 'local problems' (where the decision on one element of the input has direct consequences for its neighboring elements) like Maximum Independent Set, SAT and various other problems, until recently. In 2006, Fomin et al.[9] devised an algorithm for Connected Dominating Set (or Maximum Leaf Spanning Tree) and Razgon [19] for Feedback Vertex Set combining sophisticated branching and a clever use of measure. Our algorithm adheres to this machinery and adds an important real life problem to this small list. We also need to use an involved measure, which is a function of the number of vertices and the number of edges to be added to the spanning tree, to get the desired running time.

2 Preliminaries

Let G be a graph. We use $V(G)$ and $E(G)$ to denote the vertices and the edges of G respectively. We simply write V and E if the graph is clear from the context. For $V' \subseteq V$ we define an *induced subgraph* $G[V'] = (V', E')$, where $E' = \{uv \in E : u, v \in V'\}$.

Let $v \in V$, we denote by $N(v)$ the *neighborhood* of v, namely $N(v) = \{u \in V : uv \in E\}$. The *closed neighborhood* $N[v]$ of v is $N(v) \cup \{v\}$. In the same way we define $N[S]$ for $S \subseteq V$ as $N[S] = \cup_{v \in S} N[v]$ and $N(S) = N[S] \setminus S$. We define the *degree* of vertex v in G as the number of vertices adjacent to v in G. Namely, $d_G(v) = |\{u \in V(G) : uv \in E(G)\}|$.

Let G be a graph and T be a spanning tree of G. A vertex $v \in V(G)$ is a *full degree* vertex in T, if $d_G(v) = d_T(v)$. We define a *full degree spanning tree* to be a spanning tree with the maximum number of full degree vertices. One can similarly define *full degree spanning forest* by replacing tree with forest in the earlier definition.

A set $I \subseteq V$ is called an *independent set* for G if no vertex v in I has a neighbor in I.

3 Algorithm for Full Degree Spanning Tree

In this Section we give an exact algorithm for the FDST problem.

Given an input graph $G = (V, E)$, the basic idea is that if we know a subset S of V for which there exists a spanning tree T where all the vertices in S have full degree then, given this set S, we can construct a spanning tree T where all the vertices in S have full degree in polynomial time. Our first observation towards this is that all the edges incident to the vertices in S, that is

$$E_S = \{uv \in E \text{ such that } u \in S \text{ or } v \in S\} \tag{1}$$

induce a forest. For our polynomial time algorithm we start with the forest (V, E_S) and then complete this forest into a spanning tree by adding edges to connect the components of the forest. The last step can be done by using a slightly modified version of the SPANNING TREE algorithm of Kruskal [13] that we denote by `poly_fdst`(G, S).

The rest of the section is devoted to finding a largest subset of vertices S for which we can find a spanning tree where the vertices of S have full degree.

Our algorithm follows a branching strategy and as a partial solution keeps a set of vertices S for which there exists a spanning tree where the vertices in S have full degree. The standard branching step chooses a vertex v that could be included in S and then recursively tries to find a solution by including v in S and not including v in S. But when v is not included in S, it cannot be removed from further consideration as cycles involving v might be created later on in (V, E_S) by adding neighbors of v to S. Hence we resort to a coloring scheme for the vertices, which can also be thought of as a partition of the vertex set of the input graph. At any point of the execution of the algorithm, the vertices are partitioned as below:

1. *Selected* S: The set of vertices which are decided to be of full degree.
2. *Discarded* D: The set of vertices which are not required to be of full degree.
3. *Undecided* U: The set of vertices which are neither in S nor D, that is those vertices which are yet to be decided. So, $U = V \setminus (S \cup D)$.

Next we define a generalized form of the FDST problem based on the above partition of the vertex set. But before that we need the following definition.

Definition 1. *Given a vertex set $S \subseteq V$, we define the* partial spanning tree *of G induced by S as $T(S) = (N[S], E_S)$ where E_S is defined as in Equation (1).*

For our generalized problem, we denote by $G = (S, D, U, E)$ the graph (V, E) with vertex set $V = S \cup D \cup U$ partitioned as above.

> Generalized Full Degree Spanning Tree (GFDST): Given an instance $G = (S, D, U, E)$ such that $T(S)$ is connected and acyclic, the objective is to find a spanning forest which maximizes the number of vertices of U of full degree under the constraint that all the vertices in S have full degree.

If we start with a graph G, an instance of FDST, with the vertex partition $S = D = \emptyset$ and $U = V$ then the problem we will have at every intermediate step of the recursive algorithm is GFDST. Also, note that a full degree spanning forest of a connected graph can easily be extended to a full degree spanning tree and that a full degree spanning tree is a full degree spanning forest.

As suggested earlier our algorithm is based on branching and will have some reduction rules that can be applied in polynomial time, leading to a refined partitioning of the vertices. Before we come to the detailed description of the algorithm, we introduce a few more important definitions. For given sets S, D and U, we say that an edge is

(a) *unexplored* if one of its endpoints is in U and the other one in $U \cup D$,
(b) *forced* if at least one of its endpoints is in S, and
(c) *superfluous* if both its endpoints are in D.

The basic step of our algorithm chooses an undecided vertex $u \in U$ and considers two subcases that it solves recursively: either u is *selected*, that is u is moved from U to S, or u is *discarded*, that is moved from U to D. But the main idea is to choose a vertex in a way that the connectivity of $T(S)$ is maintained in both recursive calls. To do so we choose u from $U \cap N[N[S]]$. This brings us to the following definition.

Definition 2. *The vertices in $U \cap N[N[S]]$ are called* candidate *vertices.*

On the other hand, if S is not empty and the graph does not contain a candidate vertex, then D can be partitioned into two sets: (a) those vertices in D that have neighbors in S and (b) those that have neighbors in U. Superfluous edges (with both endpoints in D) are removed by reduction rule **R1** making G disconnected in this case, and then the algorithm is executed on each connected component.

Now we are ready to describe the algorithm in details. We start with a procedure for reduction rules in the next subsection and prove that these rules are correct.

3.1 Reduction Rules

Given an instance $G = (S, D, U, E)$ of GFDST, a reduced instance of G is computed by the following procedure.

`Reduce`$(G = (S, D, U, E))$

R1 If there is a superfluous edge e, then return `Reduce`$((S, D, U, E \setminus \{e\}))$
R2 If there is a vertex $u \in D \cup U$ such that $d(u) = 1$, then remove the unique edge e incident on it and return `Reduce`$((S, D, U, E \setminus \{e\}))$.
R3 If there is an undecided vertex $u \in U$ such that $T(S \cup \{u\})$ contains a cycle, then discard u, that is return `Reduce`$((S, D \cup \{u\}, U \setminus \{u\}, E))$.
R4 If there is a candidate vertex u that is incident to at most one vertex in $U \cup D$, then select u, and return `Reduce`$((S \cup \{u\}, D, U \setminus \{u\}, E))$.
R5 If $S = \emptyset$ and there exists a vertex $u \in U$ of degree 2, then select u and return `Reduce`$((S \cup \{u\}, D, U \setminus \{u\}, E))$.
R6 If there is a candidate vertex u of degree 2, then select u and return `Reduce`$((S \cup \{u\}, D, U \setminus \{u\}, E))$.
Else return G

Now we argue about the correctness of the reduction rules, more precisely that there exists a spanning forest of G such that a maximum number of vertices preserve their degree and the partitionning of the vertices into the sets S, D and U of the graph resulting from a call to `Reduce`$(G = (S, D, U, E))$ is respected. Note that the reduction rules are applied in the order of their appearance. The correctness of **R1** follows from the fact that discarded vertices are not required to have full degree.

For the correctness of reduction rule **R2**, consider a vertex $u \in D \cup U$ of degree 1 with unique neighbor w. Let $G' = (S, D, U, E \setminus \{uw\})$ be the graph resulting from the application of the reduction rule. Note that the edge uw is not part of any cycle and that a full degree spanning forest of G can be obtained from a full degree spanning forest of G' by adding the edge uw. As Algorithm `poly_fdst`(G, S) adds edges to make the obtained spanning forest into a spanning tree, the edge uw is added to the final solution.

For the correctness of reduction rule **R3**, it is enough to observe that if for a subset $S \subseteq V$, there exists a spanning tree T such that all the vertices of S have full degree then $T(S)$ is a forest.

We prove the correctness of **R4**, **R5** and **R6** by the following lemmata.

Lemma 1. *Let $G = (V, E)$ be a graph and T be a full degree spanning forest for G. If $v \in V$ is a vertex of degree $d_G(v) - 1$ in T, then there exists a full degree spanning forest T' such that v has degree $d_G(v)$ in T'.*

Proof. Let $u \in V$ be the neighbor of v such that uv is not an edge of T. Note that both u and v do not have full degree in T, are not adjacent and belong to the same tree in T. The last assertion follows from the fact that if u and v belong to two different trees of T then one can safely add uv to T and obtain a forest T' that has a larger number of full degree vertices, contradicting that T

is a full degree spanning forest. Now, adding the edge uv to T creates a unique cycle passing through u and v. We obtain the new forest T' by removing the other edge incident to u on the cycle, say uw, $w \neq v$. So, $T' = T \setminus \{uw\} + \{uv\}$. The number of full degree vertices in T' is at least as high as in T as v becomes a full degree vertex and at most one vertex, w, could become non full degree. □

We also need a generalized version of Lemma 1.

Lemma 2. *Let $G = (S, D, U, E)$ be a graph and T be a full degree spanning forest for G such that the vertices in S have full degree. Let $v \in U$ a candidate vertex such that its neighbors in $D \cup U$ are not incident to a forced edge. If v has degree $d_G(v) - 1$ in T, then there exists a full degree spanning forest T' such that v has degree $d_G(v)$ in T' and the vertices in S have full degree.*

Proof. The proof is similar to the one of Lemma 1. The only difference is that we need to show that the vertices of S remain of full degree and for that we need to show that all the edges of $T(S)$ remain in T'. To this observe that all the edges incident to the neighbors of v in $D \cup U$ in T do not belong to edges of $T(S)$, that is they are not forced edges. So if uv is the unique edge incident to v missing in T then we can add uv to T and remove the other non-forced edge on u from the unique cycle in $T + \{uv\}$ and get the desired T'. □

Now consider reduction rule **R4**. If u is a candidate vertex with unique neighbor w in $D \cup U$ then (a) $u \in N(S)$ and (*b*) all the edges incident to w are not forced, otherwise reduction rule **R2** or **R3** would have applied. Now the correctness of the reduction rule follows from Lemma 2. The correctness proof of reduction rule **R6** is similar. Here u belongs to $N[N[S]] \cap U$ but all the edges incident to its unique neighbor in $V \setminus N[S]$ are not forced and again Lemma 2 comes into play. To prove the correctness of reduction rule **R5**, we need to show that there exists a spanning forest where u has full degree. Suppose not and let T be any full degree spanning forest of G. Without loss of generality, suppose that u has degree 1 in T (if u is an isolated vertex in T, then add one edge incident to u to T; this does not create any cycle in T and does not decrease the number of vertices of full degree in T). Let v be the unique neighbor of u in T. But since $S = \emptyset$, there are no forced edges and we can apply Lemma 2 again and conclude.

This finishes the correctness proof of the reduction rules. Before we go into the details of the algorithm we would like to point out that all our reduction rules preserve the connectivity of $T(S)$.

3.2 Algorithm

In this section we describe our algorithm in details. Given an instance $G = (S, D, U, E)$ of GDPST, our algorithm recursively solves the problem by choosing a vertex $u \in U$ and including u in S or in D and then returning as solution the one which has maximum sized S. The algorithm has various cases based on the number of unexplored edges incident to u.

Algorithm `fdst`(G), described below, returns a super-set S^* of S corresponding to the full degree vertices in a full degree spanning forest respecting the initial choices for S and D. After this, `poly_fdst`(G, S^*) returns a full degree spanning tree of G as described in the beginning of the section. The description of the algorithm consists of the application of the reduction rules and a sequence of cases. A case consists of a condition (first sentence) and a procedure to be executed if the condition holds. The first case which applies is used in the algorithm. Thus, inside a given case, the conditions of all previous cases are assumed to be false.

`fdst`$(G = (S, D, U, E))$

Replace G by `Reduce`(G).

Case 1: U is a set of isolated vertices. Return $S \cup U$.

Case 2: $S = \emptyset$. Choose a vertex $u \in U$ of degree at least 3. Return the largest set among `fdst`$((S \cup \{u\}, D, U \setminus \{u\}, E))$ and `fdst`$((S, D \cup \{u\}, U \setminus \{u\}, E))$.

Case 3: G has at least 2 connected components, say $G_1, G_2, \cdots, G_k$. Return $\bigcup_{i=1}^{k}$ `fdst`$((S \cap V(G_i), D \cap V(G_i), U \cap V(G_i), E \cap E(G_i)))$.

Case 4: There is a candidate vertex u with at least 3 unexplored incident edges. Make two recursive calls: `fdst`$((S \cup \{u\}, D, U \setminus \{u\}, E))$ and `fdst`$((S, D \cup \{u\}, U \setminus \{u\}, E))$, and return the largest obtained set.

Case 5: There is a candidate vertex u with at least one neighbor v in U and exactly two unexplored incident edges. Make two recursive calls: `fdst`$((S \cup \{u\}, D, U \setminus \{u\}, E))$ and `fdst`$((S, D \cup \{u, v\}, U \setminus \{u, v\}, E))$, and return the largest obtained set.

From now on let v_1 and v_2 denote the discarded neighbors of a candidate vertex u (see Figure 1).

Case 6: Either v_1 and v_2 have a common neighbor $x \neq u$; or
v_1 (or v_2) has a neighbor $x \neq u$ that is a candidate vertex; or
v_1 (or v_2) has a neighbor x of degree 2.
Make two recursive calls: `fdst`$((S \cup \{u\}, D, U \setminus \{u\}, E))$ and `fdst`$((S, D \cup \{u\}, U \setminus \{u\}, E))$, and return the largest obtained set.

Case 7: Both v_1 and v_2 have degree 2. Let w_1 and w_2 ($w_1 \neq w_2$) be the other (different from u) neighbors of v_1 and v_2 in U respectively. Make recursive calls as usual, but also explore all the possibilities for w_1 and w_2 if $u \in S$. When u is in S, recurse on all possible ways one can add a subset of $A = \{w_1, w_2\}$ to S. That is make recursive calls `fdst`$((S, D \cup \{u\}, U \setminus \{u\}, E))$ and `fdst`$((S \cup \{u\} \cup X, D \cup (A - X), U \setminus (\{u\} \cup A), E))$ for each independent set $X \subseteq A$, and return the largest obtained set.

Case 8: At least one of $\{v_1, v_2\}$ has degree ≥ 3. Let $\{u, w_1, w_2, w_3\} \subseteq N(\{v_1, v_2\})$ and let $A = \{w_1, w_2, w_3\}$. Make recursive calls `fdst`$((S, D \cup \{u\}, U \setminus \{u\}, E))$ and `fdst`$((S \cup \{u\} \cup X, D \cup (A - X), U \setminus (\{u\} \cup A), E))$ for each independent set $X \subseteq A$, and return the largest obtained set.

---

4 Correctness and Time Complexity of the Algorithm

We prove the correctness and the time complexity of Algorithm `fdst` in the following theorem.

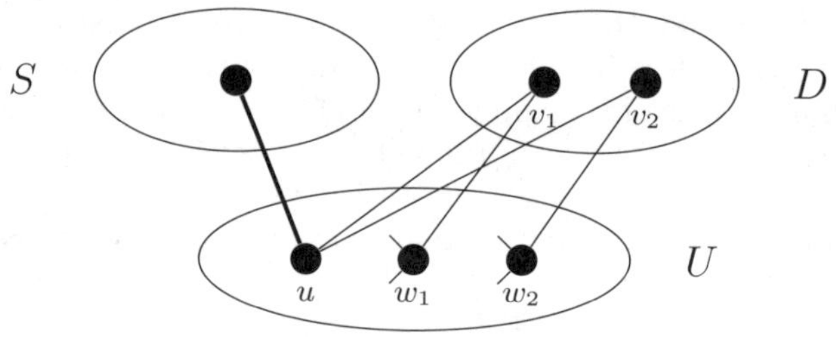

Fig. 1. Illustration of Case 7. Cases 6 and 8 are similar.

Theorem 1. *Given an input graph $G = (S, D, U, E)$ on n vertices such that $T(S)$ is connected and acyclic, Algorithm* **fdst** *returns a maximum size set S^*, $S \subseteq S^* \subseteq S \cup U$ such that there exists a spanning forest for G where all the vertices in S^* have full degree in time $\mathcal{O}(1.9172^n)$.*

Proof. The correctness of the reduction rules is described in Section 3.1. The correctness of **Case 1** follows, as any isolated vertex belonging to U has full degree in any spanning forest. Similarly, the correctness of **Case 3** follows from the fact that any spanning forest of G is a spanning forest of each of the connected components of G. The remaining cases, except **Case 5**, of Algorithm `fdst` are branching steps where the algorithm chooses a vertex $u \in U$ and tries both possibilities: $u \in S$ or $u \in D$. Sometimes the algorithm branches further by looking at the local neighborhood of u and trying all possible ways these vertices can be added to either S or D. Since all possibilities are tried to add vertices of U to D or S in **Cases 3**, **4** and **6** to **8**, these cases are correct and do not need any further justifications. The correctness of **Case 5** requires special attention. Here we use the fact that there exists a full degree spanning forest with all the vertices in S having full degree, such that either $u \in S$ or u and its neighbor $v \in U$ are in D. We prove the correctness of this assertion by contradiction. Suppose all the full degree spanning forests such that all the vertices in S are of full degree have u of non full degree and v of full degree. But notice that $u \in N(S)$ (see **R6**) and all the neighbors of u in $D \cup U$ do not have any incident forced edges. Now we can use Lemma 2 to get a spanning forest which contains u and is a full degree spanning forest with all the vertices in S having full degree.

Now we move on to the time complexity of the algorithm. The measure of subproblems is generally chosen as a function of structure, like vertices, edges or other graph parameters, which change during the recursive steps of the algorithm. In our algorithm, this change is reflected when vertices are moved to either S or D from U. The second observation is that any spanning tree on n vertices has at most $n - 1$ edges and hence when we select a vertex in S we increase the number of edges in $T(S)$ and decrease the number of edges we can add to $T(S)$. Finally we also gain when the degree of a vertex becomes two because reduction rules apply as soon as the degree 2 vertex becomes a candidate vertex. Our measure is precisely a function of these three parameters and is defined as follows:

$$\mu(G) = \eta|U_2| + |U_{\geq 3}| + \alpha m', \tag{2}$$

where U_2 is the subset of undecided vertices of degree 2, $U_{\geq 3}$ is the subset of undecided vertices of degree at least 3, $m' = n - 1 - |E(T(S))|$ is the number of edges that can be added to the spanning tree and $\alpha = 0.372$ and $\eta = 0.5$ are numerically obtained constants to optimize the running time. We write μ instead of $\mu(G)$ if G is clear from the context. We prove that the problem can be solved for an instance of size μ in time $\mathcal{O}(x^\mu)$ where $x < 1.60702$. As $\mu \leq 1.372n$, the final running time of the algorithm will be $\mathcal{O}(x^{1.372n}) = \mathcal{O}(1.9172^n)$. Denote by $P[\mu]$ the maximum number of times the algorithm is called recursively on a problem of size μ (i. e. the number of leaves in the search tree). Then the running time $T(\mu)$ of the algorithm is bounded by $P[\mu] \cdot n^{\mathcal{O}(1)}$ because in any node of the search tree, the algorithm executes only a polynomial number of steps. We use induction on μ to prove that $P[\mu] \leq x^\mu$. Then $T(\mu) = x^\mu \cdot n^{\mathcal{O}(1)}$, and since the polynomial is suppressed by rounding the exponential base, we have $T(\mu) = \mathcal{O}(1.60702^\mu)$. Clearly, $P[0] = 1$. Suppose that $P[k] \leq x^k$ for every $k < \mu$ and consider a problem of size μ. We remark that in all the branching steps all the candidate vertices in U have degree at least 3, otherwise reduction rules **R5** or **R6** would have applied.

Case 2: In this case, the number of vertices in $U_{\geq 3}$ decreases by one in both recursive calls and the number of edges in $T(S)$ increases by at least 3 in the first recursive call. Thus,

$$P[\mu] \leq P[\mu - 1 - 3\alpha] + P[\mu - 1].$$

Case 3: Here we branch on different connected components of G, and hence

$$P[\mu(G)] \leq \sum_{i=1}^{k} P[\mu(G_i)].$$

Case 4: This case has the same recurrence as **Case 2** as the number of vertices in $U_{\geq 3}$ decreases by one in both recursive calls and the number of edges in $T(S)$ increases by at least 3 in the first recursive call.

Case 5: When the algorithm adds u to S, the number of vertices in $U_{\geq 3}$ decreases by one and the number of edges in $T(S)$ increases by 2 while in the other case, $|U_{\geq 3}|$ decreases by two as both u and v are candidate vertices. So we get:

$$P[\mu] \leq P[\mu - 1 - 2\alpha] + P[\mu - 2].$$

Case 6: When the algorithm adds u to S, reduction rule **R3** or **R6** applies to x. We obtain the following recurrences, based on the degree of x:

$$P[\mu] \leq P[\mu - 1 - \eta - 2\alpha] + P[\mu - 1],$$

$$P[\mu] \leq P[\mu - 2 - 2\alpha] + P[\mu - 1].$$

Case 7: In this case we distinguish two subcases based on the degrees of w_1 and w_2. Our first subcase is when either w_1 or w_2 has degree 3 and the other subcase is when both w_1 and w_2 have degree at least 4. (Note that because of **Case 5**, v_1 and v_2 do not have a common neighbor and do not have a neighbor of degree 2). Suppose w_1 has degree 3. When the algorithm adds u to D, the edges uv_1 and uv_2 are removed (**R1**), the degree of v_1 is reduced to 1 and then reduction rule **R2** is applied and makes w_1 of degree 2. So, in this subcase, μ decreases by $2 - \eta$. The analysis of the remaining branches is standard and we get the following recurrence:

$$P[\mu] \leq P[\mu - 3 - 2\alpha] + 2P[\mu - 3 - 5\alpha] + P[\mu - 3 - 8\alpha] + P[\mu - 2 + \eta].$$

For the other subcase we get the following recurrence:

$$P[\mu] \leq P[\mu - 3 - 2\alpha] + 2P[\mu - 3 - 6\alpha] + P[\mu - 3 - 10\alpha] + P[\mu - 1].$$

Case 8: This case is similar to **Case 7** and we get the following recurrence:

$$P[\mu] \leq P[\mu-4-2\alpha]+3P[\mu-4-5\alpha]+3P[\mu-4-8\alpha]+P[\mu-4-11\alpha]+P[\mu-1].$$

In each of these recurrences, $P[\mu] \leq x^{\mu}$ which completes the proof of the theorem. □

The bottleneck of the analysis is the second recurrence in **Case 7**. Therefore, an improvement of this case would lead to a faster algorithm.

5 Conclusion

In this paper we have given an exact algorithm for the Full Degree Spanning Tree problem. The most important feature of our algorithm is the way we exploit connectivity arguments to reduce the size of the graph in the recursive steps of the algorithm. We think that this idea of combining connectivity while developing Branch & Reduce algorithms could be useful for various other non-local problems and in particular for other NP-complete variants of the Spanning Tree problem. Although the theoretical bound we obtained for our algorithm seems to be only slightly better than a brute-force enumeration algorithm, practice shows that Branch & Reduce algorithms perform usually better than the running time proved by a worst case analysis of the algorithm. Therefore we believe that this algorithm, combined with good heuristics, could be useful in practical applications.

One problem which we would like to mention here is Minimum Maximum Degree Spanning Tree, where, given an input graph G, the objective is to find a spanning tree T of G such that the maximum degree of T is minimized. This problem is a generalization of the famous Hamiltonian Path problem for which no algorithm faster than $2^n n^{\mathcal{O}(1)}$ is known. It remains open to find even a $2^n n^{\mathcal{O}(1)}$ time algorithm for the Minimum Maximum Degree Spanning Tree problem.

References

1. Bhatia, R., Khuller, S., Pless, R., Sussmann, Y.J.: The full degree spanning tree problem. Networks 36(4), 203–209 (2000)
2. Björklund, A., Husfeldt, T.: Inclusion–Exclusion Algorithms for Counting Set Partitions. In: The proceedings of FOCS 2006, pp. 575–582 (2006)
3. Björklund, A., Husfeldt, T., Kaski, P., Koivisto, M.: Fourier meets Möbius: Fast Subset Convolution. In: The proceedings of STOC 2007, pp. 67–74 (2007)
4. Broersma, H., Koppius, O.R., Tuinstra, H., Huck, A., Kloks, T., Kratsch, D., Müller, H.: Degree-preserving trees. Networks 35(1), 26–39 (2000)
5. Christofides, N.: An Algorithm for the Chromatic Number of a Graph. Computer Journal 14(1), 38–39 (1971)
6. Fomin, F.V., Gaspers, S., Pyatkin, A.V.: Finding a Minimum Feedback Vertex Set in Time $O(1.7548^n)$. In: Bodlaender, H.L., Langston, M.A. (eds.) IWPEC 2006. LNCS, vol. 4169, pp. 184–191. Springer, Heidelberg (2006)
7. Fomin, F.V., Grandoni, F., Kratsch, D.: Measure and Conquer: Domination - A Case Study. In: Caires, L., Italiano, G.F., Monteiro, L., Palamidessi, C., Yung, M. (eds.) ICALP 2005. LNCS, vol. 3580, pp. 191–203. Springer, Heidelberg (2005)
8. Fomin, F.V., Grandoni, F., Kratsch, D.: Measure and Conquer: A simple $O(2^{0.288n})$ Independent Set Algorithm. In: The proceedings of SODA 2006, pp. 18–25 (2006)
9. Fomin, F.V., Grandoni, F., Kratsch, D.: Solving Connected Dominating Set Faster Than 2^n. In: Arun-Kumar, S., Garg, N. (eds.) FSTTCS 2006. LNCS, vol. 4337, pp. 152–163. Springer, Heidelberg (2006)
10. Guo, J., Niedermeier, R., Wernicke, S.: Fixed-Parameter Tractability Results for Full-Degree Spanning Tree and Its Dual. In: Bodlaender, H.L., Langston, M.A. (eds.) IWPEC 2006. LNCS, vol. 4169, pp. 203–214. Springer, Heidelberg (2006)
11. Khuller, S., Bhatia, R., Pless, R.: On Local Search and Placement of Meters in Networks. SIAM Journal of Computing 32(2), 470–487 (2003)
12. Koivisto, M.: An $O(2^n)$ Algorithm for Graph Colouring and Other Partitioning Problems via Inclusion-Exclusion. In: The proceedings of FOCS 2006, pp. 583–590 (2006)
13. Kruskal, J.B.: On the Shortest Spanning Subtree and the Traveling Salesman Problem. The proceedings of the American Mathematical Society 7, 48–50 (1956)
14. Lawler, E.L.: A Note on the Complexity of the Chromatic Number Problem. Information Processing Letters 5(3), 66–67 (1976)
15. Lewinter, M.: Interpolation Theorem for the Number of Degree-Preserving Vertices of Spanning Trees. IEEE Transaction Circ. Syst. 34, 205 (1987)
16. Ormsbee, L.E.: Implicit Network Calibration, Journal of Water Resources. Planning and Management 115(2), 243–257 (1989)
17. Ormsbee, L.E., Wood, D.J.: Explicit Pipe Network Calibratio. Journal of Water Resources, Planning and Management 112(2), 166–182 (1986)
18. Pothof, I.W.M., Schut, J.: Graph-theoretic approach to identifiability in a water distribution network. Memorandum, vol. 1283, Universiteit Twent (1995)
19. Razgon, I.: Exact Computation of Maximum Induced Forest. In: Arge, L., Freivalds, R. (eds.) SWAT 2006. LNCS, vol. 4059, pp. 160–171. Springer, Heidelberg (2006)

Speeding up Dynamic Programming for Some NP-Hard Graph Recoloring Problems

Oriana Ponta[1,⋆], Falk Hüffner[2,⋆], and Rolf Niedermeier[2]

[1] Mathematisches Institut, Ruprecht-Karls-Universität Heidelberg, Im Neuenheimer Feld 288, D-69120 Heidelberg, Germany
oriana@urz.uni-heidelberg.de

[2] Institut für Informatik, Friedrich-Schiller-Universität Jena, Ernst-Abbe-Platz 2, D-07743 Jena, Germany
{hueffner,niedermr}@minet.uni-jena.de

Abstract. A vertex coloring of a tree is called *convex* if each color induces a connected component. The NP-hard CONVEX RECOLORING problem on vertex-colored trees asks for a minimum-weight change of colors to achieve a convex coloring. For the non-uniformly weighted model, where the cost of changing a vertex v to color c depends on both v and c, we improve the running time on trees from $O(\Delta^\kappa \cdot \kappa n)$ to $O(3^\kappa \cdot \kappa n)$, where Δ is the maximum vertex degree of the input tree T, κ is the number of colors, and n is the number of vertices in T. In the uniformly weighted case, where costs depend only on the vertex to be recolored, one can instead parameterize on the number of *bad* colors $\beta \leq \kappa$, which is the number of colors that do not already induce a connected component. Here, we improve the running time from $O(\Delta^\beta \cdot \beta n)$ to $O(3^\beta \cdot \beta n)$. For the case where the weights are integers bounded by M, using fast subset convolution, we further improve the running time with respect to the exponential part to $O(2^\kappa \cdot \kappa^4 n^2 M \log^2(nM))$ and $O(2^\beta \cdot \beta^4 n^2 M \log^2(nM))$, respectively. Finally, we use fast subset convolution to improve the exponential part of the running time of the related 1-CONNECTED COLORING COMPLETION problem.

1 Introduction

The issue of recoloring vertex-colored graphs by a minimum-cost set of color changes in order to achieve a desired property of the color classes such as being connected has recently received considerable attention; approximation as well as fixed-parameter algorithms have been developed for the corresponding NP-hard problems [2, 7, 8, 15, 16, 20]. Here, we focus on exact fixed-parameter algorithms [9, 11, 18] for two prominent types of these problems, significantly improving on the associated exponential running time factors. The two types of problems we investigate are as follows. First, we study vertex-colored trees and the task is to recolor some of its vertices such that each color class forms

⋆ Supported by the Deutsche Forschungsgemeinschaft, Emmy Noether research group PIAF (fixed-parameter algorithms), NI 369/4.

M. Agrawal et al. (Eds.): TAMC 2008, LNCS 4978, pp. 490–501, 2008.

a connected component. The problem was introduced by Moran and Snir [15], and it concerns the major part of this work. The second type of problem is not only concerned with trees, but with general graphs. However, here we cannot recolor freely, but a subset of vertices is uncolored and the task is to complete the coloring such that each color class forms a connected component [8].

Convex Recoloring. Most of this work deals with convex recoloring problems on trees. The most general version, that is, *non-uniformly weighted*, is defined as follows.

CONVEX RECOLORING
Instance: A tree $T = (V, E)$ with a vertex coloring $C : V \to \mathcal{C}$ and a weight function $w : V \times \mathcal{C} \to \mathbb{Q}_+$, where $w(v, C(v)) = 0$ for all $v \in V$.
Task: Find a convex coloring $C' : V \to \mathcal{C}$ with minimum weight $w(C') := \sum_{v \in V} w(v, C'(v))$.

We defined CONVEX RECOLORING only for trees. There are some positive results for the special case of paths [15, 16], but there do not seem to be positive results for general graphs (Moran et al. [17] considered the slightly more general class of galled networks).

Let κ be the number of colors $|\mathcal{C}|$ and n the number of vertices $|V|$. Let $\beta \leq \kappa$ be the number of *bad* colors, that is, colors that do not already induce a connected component.

CONVEX RECOLORING was introduced by Moran and Snir [15], who showed that the decision version is NP-complete, even for unweighted paths. They also gave an algorithm for non-uniformly weighted CONVEX RECOLORING running in $O(\Delta^\kappa \kappa n)$ time, where Δ is the maximum degree of the input graph. They further gave an algorithm running in $O((\kappa/\log\kappa)^\kappa \cdot \kappa n^4)$ time, thus showing that the problem is fixed-parameter tractable with respect to the parameter κ. For the uniformly weighted case, they showed that κ can be replaced by the potentially smaller parameter β in these running times.

For the unweighted case, Razgon [20] gave an $256^k \cdot n^{O(1)}$ time algorithm, where k is the number of vertices recolored. This can be related to other results by noting that $k \geq \beta/2$ (every color change can make at most two bad colors good). Bodlaender and Weyer [5] considered a different parameter, namely the *separation of colors* ℓ, which is the maximum number of colors separated by a vertex, where we say that a vertex v separates a color c if there is a path between two vertices of color c that passes through v. They presented a running time of $O(3^\ell \cdot \ell^3 n)$. Since they showed that $\ell \leq k+1$, this also improves Razgon's result to $O(3^k \cdot kn)$. Bar-Yehuda et al. [2] further improved the bound to $O(2^k \cdot kn + n^2)$ by doing a better analysis of a variant of the dynamic programming algorithm of Moran and Snir [15]. Bodlaender et al. [6] showed a problem kernel with $O(k^6)$ vertices, which was later improved to $O(k^2)$ vertices [7]. Finally, Moran and Snir [16] gave a factor-3 approximation for the uniformly weighted case running in $O(\kappa n^2)$ time. This was improved to a $(2 + \epsilon)$-approximation running in $O(n^2 + n(1/\epsilon)^2 4^{1/\epsilon})$ time [2].

Bachoore and Bodlaender [1] gave an $O(4^k n)$ time algorithm for the variant where only the leaves are precolored.

Connected Coloring Completion. The second type of problem we study has only recently been introduced by Chor et al. [8]; accordingly, so far less results are known for this problem. As CONVEX RECOLORING, it is motivated by applications in bioinformatics.

1-CONNECTED COLORING COMPLETION
Instance: A graph $G = (V, E)$ with k uncolored vertices $U \subseteq V$ and a vertex coloring $C : V \setminus U \to \mathcal{C}$.
Task: Find a convex coloring $C' : V \to \mathcal{C}$ that extends C, that is, for all $v \in V \setminus U : C'(v) = C(v)$.

Chor et al. [8] also considered the more general r-CONNECTED COLORING COMPLETION, where the goal is to find a coloring where each color induces at most r connected components. They showed that 1-CONNECTED COLORING COMPLETION is NP-hard, even for only two colors, but can be solved in $O(8^k \cdot k + 2^k \cdot kn)$ time on an n-vertex graph. They further showed that for the parameter treewidth, r-CONNECTED COLORING COMPLETION is fixed-parameter tractable for $r = 1$ but W[1]-hard for $r \geq 2$.

Our contributions. The main purpose of this paper can be seen in "engineering" dynamic programs for weighted CONVEX RECOLORING problems and for (unweighted) 1-CONNECTED COLORING COMPLETION with respect to their exponential running time factors. To this end, we make use of two main technical tricks investigated in greater depth in the following two sections. First, we observe how a method for tree problems originally going back to Maffioli [14] (which meanwhile has found several applications, see, e.g., [4, 5]) also helps to significantly speed up and somewhat simplify dynamic programming algorithms for weighted convex recoloring problems. Second, we show how a recent general breakthrough result of Björklund et al. [3] concerning a more efficient computation of subset convolutions can be tailored towards applying it to recoloring problems.[1] More specifically, for non-uniformly weighted CONVEX RECOLORING we improve a previous exponential factor of Δ^κ to 3^κ and further on to 2^κ, and for uniformly weighted CONVEX RECOLORING we improve a previous exponential factor of Δ^β to 3^β and further on to 2^β; herein, Δ denotes the maximum vertex degree in the tree. Note that the improvements from exponential base 3 to 2 come along with increased polynomial factors in the running time. Finally, we also adapt the subset convolution trick to 1-CONNECTED COLORING COMPLETION in order to improve the previous exponential factor of 8^k to 4^k.

[1] Lingas and Wahlen [13] recently presented an application in the context of subgraph homeomorphism problems.

2 Fine-Grained Dynamic Programming

The major part of this work is concerned with improvements for non-uniformly and uniformly weighted Convex Recoloring based on a more efficient dynamic programming strategy. The essence of the underlying trick can be traced back to work of Maffioli [14]. We start with the somewhat less technical case concerning a dynamic program for *non-uniformly weighted* Convex Recoloring with respect to the parameter "number of colors" and then extend our findings to *uniformly weighted* Convex Recoloring with respect to the parameter "number of bad colors".

2.1 Non-uniformly Weighted Convex Recoloring

In this section, we show how to improve the running time of the dynamic programming by Moran and Snir [15] from $O(\Delta^\kappa \cdot \kappa n)$ to $O(3^\kappa \cdot \kappa n)$, where κ is the number of colors and n is the number of vertices in the input graph. The dynamic programming works bottom-up from the leaves of the tree. The improvement comes from not considering all children of an inner vertex at once, but rather taking them into account one-by-one. This is a classical trick for dynamic programming on trees (see e. g., [14, 5, 4]). A more detailed presentation of our result is given in the thesis of Ponta [19].

We designate an arbitrary vertex r of T as the root. For each vertex $v \in V$, we denote by T_v the subtree induced by v and all descendants of v. For a vertex v with children $w_1, \ldots, w_p$ in an arbitrary but fixed order, we denote by $T_{v,i}$ the subtree induced by v, the first i children $w_1, \ldots, w_i$ of v, and all descendants of $w_1, \ldots, w_i$. Note that $T_{v,0}$ contains only the vertex v and that $T_{v,p}$ equals T_v.

The basic structure of Moran and Snir's original algorithm is preserved. The algorithm visits the vertices in postorder. We start by determining the trivial convex recolorings for the leaves of the tree and proceed with the computation of weights of convex recolorings of subtrees T_v for internal vertices v in a bottom-up fashion. A solution for T_v is constructed using the previously computed solutions for the subtrees induced by the children of v. The way a solution for the extended problem is computed differs from Moran and Snir's algorithm and is the key to the running time improvement.

For the description of the algorithm, we need two dynamic programming tables denoted by opt and opt_r. Let $C'[T_v]$ be the set of colors appearing in the subtree T_v.

Definition 1. *Let $v \in V$ and $\mathcal{D} \subseteq \mathcal{C}$ be a set of colors. A recoloring C' is a* $(T_v, \mathcal{D})$-coloring *if it is a convex recoloring of T_v such that $C'[T_v] = \mathcal{D}$. The cost of an optimal $(T_v, \mathcal{D})$-coloring of T_v is denoted by* $\text{opt}(T_v, \mathcal{D})$.

If T_v has less than $|\mathcal{D}|$ vertices or $\mathcal{D} = \emptyset$, then no $(T_v, \mathcal{D})$-coloring exists, and we set $\text{opt}(T_v, \mathcal{D}) = \infty$. A $(T_v, \mathcal{D})$-coloring is a convex recoloring of T_v that uses exactly the colors from $\mathcal{D}$. Thus, the cost of an optimal convex recoloring of T can be calculated as $\min_{\mathcal{D} \subseteq \mathcal{C}} \text{opt}(T_r, \mathcal{D})$. To retrieve the recoloring that realizes this cost, we can use standard dynamic programming backtracing methods. It remains to describe how to fill in the dynamic programming table opt. For this, we need a second table opt_r.

Definition 2. *Let* $v \in V$, $\mathcal{D} \subseteq \mathcal{C}$ *and* $c \in \mathcal{C}$. *A recoloring* C' *is a* $(T_v, \mathcal{D}, c)$-coloring *if it is a* $(T_v, \mathcal{D})$*-coloring such that* $C'(v) = c$. *The cost of an optimal* $(T_v, \mathcal{D}, c)$*-coloring is denoted by* $\mathrm{opt}_r(T_v, \mathcal{D}, c)$.

We set $\mathrm{opt}_r(T_v, \mathcal{D}, c) = \infty$ if $c \notin \mathcal{D}$. It is easy to calculate opt from opt_r:

$$\mathrm{opt}(T_v, \mathcal{D}) = \min_{c \in \mathcal{D}} \mathrm{opt}_r(T_v, \mathcal{D}, c). \quad (1)$$

For a subtree T_v consisting of only the vertex v, we set $\mathrm{opt}_r(T_v, \{c\}, c) = w(v, c)$ and $\mathrm{opt}_r(T_v, \mathcal{D}, c) = \infty$ for $\mathcal{D} \neq \{c\}$. For an interior vertex v with children $w_1, \ldots, w_p$, we inductively assume that the values $\mathrm{opt}_r(T_{w_i}, \cdot, \cdot)$ for $1 \leq i \leq p$ are already calculated. We then iteratively calculate $\mathrm{opt}_r(T_{v,i+1}, \cdot, \cdot)$ for $i = 0, \ldots, p$, obtaining $\mathrm{opt}_r(T_v, \cdot, \cdot) = \mathrm{opt}_r(T_{v,p}, \cdot, \cdot)$. Thus, each iteration has to take into account the subtree $T_{w_{i+1}}$ in addition to the subtree $T_{v,i}$ considered in the previous iteration. In contrast, Moran and Snir [15] take into account all child subtrees at once. The childwise iterative approach to dynamic programming on trees has also been used e. g. to find minimum-weight subtrees of a tree [14, 4] or for unweighted CONVEX RECOLORING with a different parameter [5]. In our context, the technique allows to avoid the maximum vertex degree Δ in the base of the exponential part of the running time of Moran and Snir's algorithm.

For a simpler notation of the recurrence for $\mathrm{opt}_r(T_{v,i+1}, \cdot, \cdot)$, we define the function $\mathrm{opt}_c(T_{w_{i+1}}, \mathcal{D}, c)$ for the $(i+1)$th child w_{i+1} of v and $\mathcal{D} \subseteq \mathcal{C}, v \in \mathcal{C}$ as

$$\mathrm{opt}_c(T_{w_{i+1}}, \mathcal{D}, c) = \min\{\mathrm{opt}(T_{w_{i+1}}, \mathcal{D} \setminus \{c\}), \mathrm{opt}_r(T_{w_{i+1}}, \mathcal{D} \cup \{c\}, c)\}. \quad (2)$$

Thus, the value $\mathrm{opt}_c(T_{w_{i+1}}, \mathcal{D}, c)$ is the minimum cost of a convex recoloring C' of $T_{w_{i+1}}$ that uses every color in $\mathcal{D} \setminus \{c\}$, no color from $\mathcal{C} \setminus (\mathcal{D} \cup \{c\})$, and uses color c in $T_{w_{i+1}}$ only if $C'(w_{i+1}) = c$.

The following lemma describes the central recurrence for opt_r.

Lemma 1. *Let* v *be an interior vertex with children* $w_1, \ldots, w_p$. *For any color set* $\mathcal{D}$ *and any color* $c \in \mathcal{D}$ *it holds that*

$$\mathrm{opt}_r(T_{v,i+1}, \mathcal{D}, c) = \min_{\substack{\mathcal{D}_1 \cup \mathcal{D}_2 = \mathcal{D} \setminus \{c\} \\ \mathcal{D}_1 \cap \mathcal{D}_2 = \emptyset}} (\mathrm{opt}_r(T_{v,i}, \mathcal{D}_1 \cup \{c\}, c) + \mathrm{opt}_c(T_{w_{i+1}}, \mathcal{D}_2, c)). \quad (3)$$

Proof. "$\geq$": Let C' be an optimal $(T_{v,i+1}, \mathcal{D}, c)$-coloring. The weight of the recoloring C' is then $w(C') = \mathrm{opt}_r(T_{v,i+1}, \mathcal{D}, c)$. Let $\mathcal{D}'_1 = C'[T_{v,i}] \setminus \{c\}$ be the set of colors different from c that C' uses in the recoloring of $T_{v,i}$, and let $\mathcal{D}'_2 = C'[T_{w_{i+1}}] \setminus \{c\}$ be the set of colors different from c that C' uses in the recoloring of $T_{w_{i+1}}$. Given the fact that $C'(v) = c$ and the convexity of $(C', T_{v,i+1})$, it follows that $\mathcal{D}'_1 \cap \mathcal{D}'_2 = \emptyset$. By the definitions of the sets $\mathcal{D}'_1$, $\mathcal{D}'_2$, and $\mathcal{D}$, it holds that $\mathcal{D}'_1 \cup \mathcal{D}'_2 = \mathcal{D} \setminus \{c\}$.

For a subtree T' of a tree T and a coloring C of T, let $C|_{T'}$ be the restriction of C to the vertices of T'. Since $C'|_{T_{v,i}}$ is a $(T_{v,i}, \mathcal{D}'_1, c)$-coloring of $T_{v,i}$ and $C'|_{T_{w_{i+1}}}$ is a $(T_{w_{i+1}}, \mathcal{D}'_2)$- or $(T_{w_{i+1}}, \mathcal{D}'_2, c)$-coloring of $T_{w_{i+1}}$, it holds that $w(C'|_{T_{v,i}}) \geq \text{opt}_r(T_{v,i}, \mathcal{D}'_1 \cup \{c\}, c)$ and $w(C'|_{T_{w_{i+1}}}) \geq \text{opt}_c(T_{w_{i+1}}, \mathcal{D}'_2, c)$. Consequently, $w(C')$ is at least the right-hand side of (3).

"$\leq$": Consider $\mathcal{D}'_1$ and $\mathcal{D}'_2$ with $\mathcal{D}'_1 \cup \mathcal{D}'_2 = \mathcal{D} \setminus \{c\}$ and $\mathcal{D}'_1 \cap \mathcal{D}'_2 = \emptyset$ such that the sum $\text{opt}_r(T_{v,i}, \mathcal{D}'_1 \cup \{c\}, c) + \text{opt}_c(T_{w_{i+1}}, \mathcal{D}'_2, c)$ is minimized. Denote with $C'_{v,i}$ the recoloring of $T_{v,i}$ witnessing the cost $\text{opt}_r(T_{v,i}, \mathcal{D}'_1 \cup \{c\}, c)$ and with $C'_{w_{i+1}}$ the recoloring of $T_{w_{i+1}}$ witnessing the cost $\text{opt}_c(T_{w_{i+1}}, \mathcal{D}'_2, c)$. We can then combine $C'_{v,i}$ and $C'_{w_{i+1}}$ to obtain a coloring C' for $T_{v,i+1}$. The weight of C' equals the right-hand side of (3). By construction, C' is a convex recoloring of $T_{v,i+1}$ that uses exactly the colors in $\mathcal{D}$ and has $C'(v) = c$. Thus, $w(C')$ is at least $\text{opt}_r(T_{v,i+1}, \mathcal{D}, c)$. □

Theorem 1. *Non-uniformly weighted* CONVEX RECOLORING *can be solved in* $O(3^\kappa \cdot \kappa n)$ *time for a tree with* n *vertices and* κ *colors.*

Proof. We have shown how to solve non-uniformly weighted CONVEX RECOLORING by dynamic programming using the recurrences (1), (2), and (3). By visiting each vertex in postorder, it is possible to fill in opt, opt_c, and opt_r while only accessing already calculated entries. It remains to bound the running time. The bottleneck is clearly the calculation of (3). Since there are $O(n)$ edges in a tree, we have $O(n)$ values for the first component $T_{v,i+1}$. For fixed c and $T_{v,i+1}$, the computation of $\text{opt}_r(T_{v,i+1}, \mathcal{D}, c)$ effectively needs to examine all 3-ordered partitions of $\mathcal{C} \setminus \{c\}$ of the form $(\mathcal{C} \setminus \mathcal{D}, \mathcal{D}_1, \mathcal{D}_2)$; there are $3^{\kappa-1}$ such partitions. In total, we arrive at the claimed running time. □

2.2 Uniformly Weighted Convex Recoloring

In this section, we show that for the uniformly weighted case, the parameter κ (number of colors) can be replaced by β (number of bad colors) in the running time of Theorem 1. This is particularly attractive for scenarios where the input is already almost convex. Moran and Snir [15] have shown how to get an $O(\Delta^\beta \cdot \beta n)$ time algorithm from their $O(\Delta^\kappa \cdot \kappa n)$ time dynamic programming algorithm for the non-uniformly weighted case. We show that analogously, our $O(3^\kappa \cdot \kappa n)$ time algorithm (Theorem 1) can be improved to $O(3^\beta \cdot \beta n)$ time for the non-uniformly weighted case. The approach is similar to that of Moran and Snir, but we considerably simplify some concepts and proofs.

When recoloring, typically good colors are overwritten by bad colors, in order to connect different regions of a bad color. It is tempting to just restrict the search of alternative colors to bad colors, which would reduce the size of the dynamic programming tables defined in Sect. 2.1 and give the desired speedup of replacing κ by β in the base of the exponential factor. However, this is not correct: sometimes a bad color has to be overwritten with a good color in order to wipe out a region of this bad color. The central observation of Moran and Snir [15] is that when overwriting a color with a good color, we do not have to decide immediately which good color to use—the goal is after all only to get rid of the bad color of the vertex that is being recolored. We capture this in the notion of a *restricted* recoloring, which is a coloring $V \to \mathcal{C} \cup \{*\}$, where $* \notin \mathcal{C}$ serves to mark vertices that are *uncolored*.[2] It is easy to see that a standard recoloring is convex iff all vertices on a path between two vertices with the same color c also have color c. In analogy, we say that a restricted recoloring is convex iff all vertices on a path between two vertices with the same color $c \neq *$ have color c.

In the uniform cost model, we can assign a cost to a restricted recoloring by simply giving cost $w(v)$ to the recoloring of $v \in V$ with $*$ (this is not possible in the non-uniform model, where cost also depends on the actual color used in recoloring a vertex). The following lemma shows that to find an optimal convex recoloring, it suffices to look for optimal restricted recolorings.

Lemma 2. *In the uniformly weighted model, any convex restricted recoloring can be converted in linear time into a convex recoloring of the same weight and vice versa.*

Proof. Given a convex restricted recoloring, we can fill in the colors of the uncolored vertices by a depth-first search starting from some not uncolored vertex, where we recolor an uncolored vertex with the color of its predecessor in the search. This clearly produces a convex recoloring with the same weight.

The only way a convex recoloring $\hat{C}$ of a coloring C might not already be a restricted recoloring is that some vertex color was overwritten with a good color. We construct C' from $\hat{C}$ by recoloring these vertices by $*$ instead. Clearly, C' has the same weight as $\hat{C}$, and we claim that C' is also convex. For this, consider two vertices v_1, v_2 with $C'(v_1) = C'(v_2) = c \neq *$. By construction of C', then also $\hat{C}(v_1) = \hat{C}(v_2) = c$. Thus, every vertex on the path between v_1 and v_2 is colored c by $\hat{C}$. If c is a bad color, then also every vertex on the path between v_1 and v_2 is colored c by C', since only good colors are used differently between $\hat{C}$ and C'; if c is a good color, then $\hat{C}$ has left v_1 and v_2 unchanged from C, and because c is a good color, any vertex between v_1 and v_2 must also be colored c by C, and thus also by $\hat{C}$ and C'. In summary, every vertex on the path between v_1 and v_2 is colored c by C', and thus C' is convex. □

By Lemma 2, it suffices to find the weight of an optimal restricted recoloring to solve the uniformly weighted CONVEX RECOLORING problem. The dynamic programming from Sect. 2.1, based on the three tables opt, opt_r, and opt_c calculated by the recurrences (1), (2), and (3) in tree postorder remains almost unchanged. Therefore, we only point out the differences here.

Let $\mathcal{B}$ be the set of bad colors. The table opt now only covers restricted colorings, that is, $\text{opt}(T_v, \mathcal{D})$ is the weight of an optimal restricted coloring C' such that for each $c \in \mathcal{B}$ there is a vertex x in T_v with $C'(x) \neq C(x)$ and $C'(x) = c$ iff $c \in \mathcal{D}$. Table opt_r is adapted analogously, and we additionally allow $*$ as third argument. In the initialization, we also set $\text{opt}_r(T_v, \emptyset, *) = w(v)$

[2] Moran and Snir [15] use the more complicated notion of a *conservative recoloring*.

and $\text{opt}_r(T_v, \mathcal{D}, *) = \infty$ for all $\mathcal{D} \neq \emptyset$. For the case that $c = *$ in recurrence (3), we use

$$\text{opt}_r(T_{v,i+1}, \mathcal{D}, *) = \min_{\substack{\mathcal{D}_1 \cup \mathcal{D}_2 = \mathcal{D} \\ \mathcal{D}_1 \cap \mathcal{D}_2 = \emptyset}} (\text{opt}_r(T_{v,i}, \mathcal{D}_1, *) + \text{opt}(T_{w_{i+1}}, \mathcal{D}_2)). \quad (4)$$

We omit the proof, which is analogous to that in Sect. 2.1.

Theorem 2. *Uniformly weighted* Convex Recoloring *can be solved in* $O(3^\beta \cdot \beta n)$ *time for a tree with* n *vertices and* β *bad colors.*

3 Fast Subset Convolution

When we restrict the weights to be integers bounded by M, we can further improve the exponential part of the running time for Convex Recoloring by using *fast subset convolution.* This novel technique was developed by Björklund et al. [3], who used it to speed up several dynamic programming algorithms such as the classical Dreyfus–Wagner algorithm [10] for Steiner Tree in graphs. It was also used to improve the speed of subgraph homeomorphism algorithms [13].

Let f and g be functions defined on the power set of a finite set N with $|N| = p$, that is, $f, g : \mathcal{P}(N) \to I$. For any ring over I that defines addition and multiplication on elements of I, the *subset convolution* of f and g, denoted by $f * g$, is defined for each $S \subseteq N$ as

$$f * g : \mathcal{P}(N) \to I, \quad (f * g)(S) = \sum_{T \subseteq S} f(T) g(S \setminus T). \quad (5)$$

To calculate the subset convolution means to determine the value of $f * g$ for all 2^p possible inputs, assuming that f and g can be evaluated in constant time (typically by being stored in a table). A naive algorithm that calculates each value independently needs $O(\sum_{i=0}^{p} \binom{p}{i} 2^i) = O(3^p)$ ring operations. The following result shows a substantial improvement.

Theorem 3 (Björklund et al. [3]). *The subset convolution over an arbitrary ring can be computed with* $O(2^p \cdot p^2)$ *ring operations.*

Björklund et al. [3] showed how to apply Theorem 3 to also calculate the subset convolution for the integer min-sum semiring

$$f * g : \mathcal{P}(N) \to \mathbb{Z}, \quad (f * g)(S) = \min_{T \subseteq S} f(T) + g(S \setminus T) \quad (6)$$

by embedding it into the standard integer sum-product ring. Here, it is not appropriate to assume that addition and multiplication can be done in constant time, since the numbers involved can have up to n bits.[3] Björklund et al. [3] did not give a precise estimation, but it is not too hard to derive the following bound from their Theorem 3 [3].

Proposition 1. *The subset convolution over the integer min-sum ring with* $M := \max_{i \in (f(\mathcal{P}(N)) \cup g(\mathcal{P}(N)))} |i|$ *can be computed in* $O(2^p \cdot p^3 M \log^2(Mp))$ *time.*

[3] To avoid complicated terms, we assume a bound of $O(n \log^2 n)$ on the running time of integer multiplication of two n-bit numbers. Better bounds are known [12].

3.1 Convex Recoloring

We now use fast subset convolution over the integer min-sum to speed up the dynamic programming for Convex Recoloring. Recall that the bottleneck in deriving the running time of $O(3^\kappa \cdot \kappa n)$ (Theorem 1) comes from recurrence (3), which we recall here:

$$\mathrm{opt}_r(T_{v,i+1}, \mathcal{D}, c) = \min_{\substack{\mathcal{D}_1 \cup \mathcal{D}_2 = \mathcal{D} \setminus \{c\} \\ \mathcal{D}_1 \cap \mathcal{D}_2 = \emptyset}} (\mathrm{opt}_r(T_{v,i}, \mathcal{D}_1 \cup \{c\}, c) + \mathrm{opt}_c(T_{w_{i+1}}, \mathcal{D}_2, c)).$$

Consider fixed $T_{v,i+1}$ and c. Then (3) can be seen as a subset convolution over the integer min-sum semiring (like in (6)) by setting $f^c_{v,i}(\mathcal{D}) = \mathrm{opt}_r(T_{v,i}, \mathcal{D} \cup \{c\}, c)$ and $g^c_{w_i+1}(\mathcal{D}) = \mathrm{opt}_c(T_{w_{i+1}}, \mathcal{D}, c)$:

$$\mathrm{opt}_r(T_{v,i+1}, \mathcal{D}, c) = (f^c_{v,i} * g^c_{w_{i+1}})(\mathcal{D} \setminus \{c\}). \tag{7}$$

Theorem 4. *The non-uniformly weighted* Convex Recoloring *problem with integer weights bounded by M can be solved in $O(2^\kappa \cdot \kappa^4 n^2 M \log^2(nM))$ time for a tree with n vertices and κ colors.*

Proof. We solve non-uniformly weighted Convex Recoloring by dynamic programming using the recurrences (1), (2), and (3), where (3) is calculated by fast subset convolution as in (7). Using Proposition 1, for fixed $T_{v,i+1}$ and c, we can calculate (7) in $O(2^\kappa \cdot \kappa^3 n M \log^2(nM))$ time, because the values of $f^c_{v,i}$ and $g^c_{w_i+1}$ are bounded by nM, since they are weights of recolorings, and $\kappa \leq n$. The rest of the analysis is as in Theorem 1. □

In the same way as for Theorem 2, we obtain a running time of $O(2^\beta \cdot \beta^4 n^2 M \log^2(nM))$ for the uniformly weighted case.

Theorem 5. *The uniformly weighted* Convex Recoloring *problem with integer weights bounded by M can be solved in $O(2^\beta \cdot \beta^4 n^2 M \log^2(nM))$ time for a tree with n vertices and β bad colors.*

3.2 1-Connected Coloring Completion

Chor et al. [8] gave simple linear-time preprocessing rules that allow without loss of generality to assume $\kappa \leq k$, that is, there are at most as many colors as uncolored vertices. The data reduction also collapses each maximal connected monochromatic subgraph into a single vertex. Thus, the problem can be restated as finding a coloring of the set U of uncolored vertices such that each color induces a connected subgraph in U and each vertex in $V \setminus U$ with color c is adjacent to a vertex with color c in U.

Chor et al. [8] solved 1-Connected Coloring Completion by using a binary-valued dynamic programming table $T(\mathcal{C}', U')$ for $\mathcal{C}' \subseteq \mathcal{C}$ and $U' \subseteq U$ with the following semantics: $T(\mathcal{C}', U') = 1$ iff if it is possible to color U' with $\mathcal{C}'$ such that each color $c \in \mathcal{C}'$ induces a connected subgraph G_c in U that *dominates* the vertices colored c in $V \setminus U$, meaning that each such vertex is adjacent to at least one vertex in G_c. Thus, if $T(\mathcal{C}', U') = 0$, then it is not possible to solve the instance by assigning (exclusively) the colors from $\mathcal{C}'$ to the vertices in U'; but if $T(\mathcal{C}', U') = 1$, then solving the instance is still possible by finding a suitable allocation of the remaining colors $\mathcal{C} \setminus \mathcal{C}'$ to the remaining uncolored vertices $U \setminus U'$. Clearly, if $T(\mathcal{C}, U) = 1$, then the instance is solvable, and we can find the corresponding solution by backtracing.

Chor et al. [8] used the following recurrence to fill in T:

$$T(\mathcal{C}', U') = 1 \iff \exists c \in \mathcal{C}', U'' \subset U' : T(\mathcal{C}' \setminus \{c\}, U'') = 1 \tag{8}$$
$$\text{and } U' \setminus U'' \text{ induces a connected subgraph that dominates the vertices of color } c,$$

which can be simplified to

$$T(\mathcal{C}', U') = \bigvee_{U'' \subseteq U'} \left(T(\mathcal{C}' \setminus \{c\}, U'') \wedge T(\{c\}, U' \setminus U'') \right) \tag{9}$$

for some $c \in \mathcal{C}'$. To be able to calculate recurrence (9), we need all values of $T(\{c\}, U')$ for $c \in \mathcal{C}$ and $U' \subseteq U$. The calculation can clearly be done in $O(2^k \cdot kn)$ time, since there are $2^k \cdot k$ such entries, and each can be calculated in linear time. A straightforward calculation of (9) for an entry then takes $O(2^k)$ time, and there are 4^k table entries, thus giving a total running time of $O(8^k + 2^k \cdot kn)$.

To speed up the exponential part of the calculation of (9), we use fast subset convolution over the or-and semiring.

Proposition 2. *The subset convolution over the or-and semiring*

$$f \circ g : \mathcal{P}(N) \to \{0, 1\}, \quad (f \circ g)(S) = \bigvee_{T \subseteq S} f(T) \wedge g(S \setminus T) \tag{10}$$

with $|N| = p$ *can be calculated in* $O(2^p \cdot p^3 \log^2 p)$ *time.*

Proof. It holds that

$$(f \circ g)(S) = \bigvee_{T \subseteq S} f(T) \wedge g(S \setminus T) \tag{11}$$
$$= \begin{cases} 1 & \text{if } \max_{T \subseteq S}(f(T) + g(S \setminus T)) = 2 \\ 0 & \text{otherwise} \end{cases} \tag{12}$$
$$= \begin{cases} 1 & \text{if } (f * g)(S) = 2 \\ 0 & \text{otherwise,} \end{cases} \tag{13}$$

and the subset convolution "$*$" in the integer min-sum semiring can be calculated using Proposition 1 with $M = 1$. □

Next, we define

$$f_{\mathcal{C}'}(U') = T(\mathcal{C}' \setminus \{c\}, U') \tag{14}$$
$$g_{\mathcal{C}'}(U') = T(\{c\}, U'), \tag{15}$$

for some $c \in \mathcal{C}'$, which gives us

$$T(\mathcal{C}', U') = (f_{\mathcal{C}'} \circ g_{\mathcal{C}'})(U'). \tag{16}$$

Theorem 6. 1-Connected Coloring Completion *can be solved in* $O(4^k \cdot k^3 \log^2 k + 2^k \cdot kn)$ *time.*

Proof. By Proposition 2, for fixed $\mathcal{C}'$, we can calculate (16) in $O(2^k \cdot k^3 \log^2 k)$ time. There are 2^k subsets $\mathcal{C}' \subseteq \mathcal{C}$. Thus, together with the $O(2^k \cdot kn)$ time for the table initialization, we arrive at the claimed running time. □

4 Outlook

We improved known fixed-parameter tractability results based on dynamic programming for several NP-hard recoloring problems in trees and graphs. These problems are mainly motivated by applications in bioinformatics (particularly, phylogenetics). The running times now seeming practically feasible, so it would be desirable to experimentally test the algorithms on real-world data. In particular, it would be interesting to see how the improvements concerning the exponential factors that have been achieved due to fast subset convolution pay off in practice. Moreover, also the space consumption of our algorithms is exponential and so memory space could become the real bottleneck in applications—this invites further research on improvement strategies.

References

[1] Bachoore, E.H., Bodlaender, H.L.: Convex recoloring of leaf-colored trees. In: Proc. 3rd ACiD. Texts in Algorithmics, vol. 9, pp. 19–33. College Publications, London (2007)
[2] Bar-Yehuda, R., Feldman, I., Rawitz, D.: Improved approximation algorithm for convex recoloring of trees. Theory of Computing Systems, (to appear, 2007)
[3] Björklund, A., Husfeldt, T., Kaski, P., Koivisto, M.: Fourier meets Möbius: fast subset convolution. In: Proc. 39th STOC, pp. 67–74. ACM Press, New York (2007)
[4] Blum, C.: Revisiting dynamic programming for finding optimal subtrees in trees. European Journal of Operational Research 177(1), 102–115 (2007)
[5] Bodlaender, H.L., Weyer, M.: Convex and connected recolorings of trees and graphs (unpublished manuscript, 2005)
[6] Bodlaender, H.L., Fellows, M.R., Langston, M.A., Ragan, M.A., Rosamond, F.A., Weyer, M.: Kernelization for convex recoloring. In: Proc. 2nd ACiD. Texts in Algorithmics, vol. 7, pp. 23–35. College Publications, London (2006)
[7] Bodlaender, H.L., Fellows, M.R., Langston, M.A., Ragan, M.A., Rosamond, F.A., Weyer, M.: Quadratic kernelization for convex recoloring of trees. In: Lin, G. (ed.) COCOON. LNCS, vol. 4598, pp. 86–96. Springer, Heidelberg (2007)
[8] Chor, B., Fellows, M.R., Ragan, M.A., Razgon, I., Rosamond, F.A., Snir, S.: Connected coloring completion for general graphs: Algorithms and complexity. In: Lin, G. (ed.) COCOON. LNCS, vol. 4598, pp. 75–85. Springer, Heidelberg (2007)

[9] Downey, R.G., Fellows, M.R.: Parameterized Complexity. Springer, Heidelberg (1999)

[10] Dreyfus, S.E., Wagner, R.A.: The Steiner problem in graphs. Networks 1(3), 195–207 (1972)

[11] Flum, J., Grohe, M.: Parameterized Complexity Theory. Springer, Heidelberg (2006)

[12] Fürer, M.: Faster integer multiplication. In: Proc. 39th STOC, pp. 57–66. ACM Press, New York (2007)

[13] Lingas, A., Wahlen, M.: On exact complexity of subgraph homeomorphism. In: Cai, J.-Y., Cooper, S.B., Zhu, H. (eds.) TAMC 2007. LNCS, vol. 4484, pp. 256–261. Springer, Heidelberg (2007)

[14] Maffioli, F.: Finding a best subtree of a tree. Technical Report 91.041, Politecnico di Milano, Dipartimento di Elettronica, Italy (1991)

[15] Moran, S., Snir, S.: Convex recolorings of strings and trees: Definitions, hardness results and algorithms. In: Dehne, F., López-Ortiz, A., Sack, J.-R. (eds.) WADS 2005. LNCS, vol. 3608, pp. 218–232. Springer, Heidelberg (2005) (to appear in Journal of Computer and System Sciences)

[16] Moran, S., Snir, S.: Efficient approximation of convex recolorings. Journal of Computer and System Sciences 73(7), 1078–1089 (2007)

[17] Moran, S., Snir, S., Sung, W.-K.: Partial convex recolorings of trees and galled networks: Tight upper and lower bounds (February 2007) (manuscript)

[18] Niedermeier, R.: Invitation to Fixed-Parameter Algorithms. Oxford Lecture Series in Mathematics and Its Applications, vol. 31. Oxford University Press, Oxford (2006)

[19] Ponta, O.: The Fixed-Parameter Approach to the Convex Recoloring Problem. Diplomarbeit, Mathematisches Institut, Ruprecht-Karls-Universität. Springer, Heidelberg (2007)

[20] Razgon, I.: A $2^{O(k)}\text{poly}(n)$ algorithm for the parameterized convex recoloring problem. Information Processing Letters 104(2), 53–58 (2007)

A Linear-Time Algorithm for Finding All Door Locations That Make a Room Searchable
(Extended Abstract)

John Z. Zhang[1] and Tsunehiko Kameda[2]

[1] Department of Mathematics and Computer Science, University of Lethbridge
Lethbridge, AB, Canada T1K 3M4
zhang@cs.uleth.ca
[2] School of Computing Science, Simon Fraser University
Burnaby, BC, Canada V5A 1S6
tiko@cs.sfu.ca

Abstract. A *room* is a simple polygon with a prespecified point, called the *door*, on its boundary. Search may be conducted by two guards on the boundary who keep mutual visibility at all times, or by a single boundary searcher with a flashlight. Search starts at the door, and must detect any intruder that was in the room at the time the search started, preventing the intruder from escaping through the door. A room may or may not be searchable, depending on where the door is placed or no matter where the door is placed. We want to find all intervals on the boundary where the door can be placed for the resultant room to be searchable. It is known that this problem can be solved in $O(n \log n)$ time, if the given polygon has n sides. We improve this complexity to $O(n)$.

1 Introduction

Imagine *intruders* who move freely within a dark polygonal region. One or more *searchers*, who are equipped with flashlights, try to detect them. It is assumed that the intruders can move arbitrarily fast and try to escape detection. Suzuki and Yamashita [1] formulated *polygon search* by introducing the *k-searcher* whose vision is restricted to the beams from the k flashlights. The purpose of search is to detect the intruders by eventually illuminating all of them. A searcher who has 360° visibility and can illuminate an intruder in any direction is called an ∞-*searcher*.

In *street search* [2], two points on the polygon boundary, the *entrance* and the *exit*, are prespecified. Two guards start moving in the opposite directions from the entrance along the boundary, while maintaining mutual visibility. They may move backwards from time to time, as long as each of them does not move beyond the entrance and exit. The search completes when they meet at the exit. A street is said to be *walkable* if every intruder is detected by the time the search is completed. Heffernan [3] proposed a linear-time algorithm to check whether a street is walkable. Tseng *et al.* [4] considered the problem of finding all the

M. Agrawal et al. (Eds.): TAMC 2008, LNCS 4978, pp. 502–513, 2008.

pairs of entrance and exit to make the resultant street walkable. Bhattacharya *et al.* [5] came up with the optimal algorithm for the same problem. In [6], Crass *et al.* studied an ∞-searcher in an open-edge "corridor", which uses edges as the entrance and exit instead of vertices.

Another special case of polygon search is *room search*, which has been studied extensively [7,8,9,10,11]. A *room*, denoted by $P(d)$, is a simple polygon P with a designated point d on its boundary, called the *door*, which is like the entrance in street search. Unlike street search, however, no exit is prespecified. If two boundary guards are used, they move on the boundary as in street search and eventually meet somewhere on the boundary [9].

We may also use a *boundary 1-searcher* who can move only along the room boundary, starting at the door [7]. An intruder is detected if he is illuminated by the beam. We want to test if a given room $P(d)$ is searchable by two guards or a 1-searcher. We proposed a linear-time algorithm for checking whether a given room is searchable by two guards in [12]. In [13], we also presented an $O(n \log n)$ time algorithm for finding all the door locations for a given polygon to make the resultant room searchable by two guards or a 1-searcher, respectively, where n is the number of vertices of the polygon. In this paper, we show that this problem can be solved in $O(n)$ time. Other related work can be found in [8,10,11].

The paper is organized as follows. In Section 2, we introduce the notation used throughout the paper, and review some basic concepts related to room search. In particular, we review the *visibility diagram*. In Section 3, we discuss the relation between a room's searchability and its LR-visibility. In Section 4, we analyze and characterize searchable rooms using the visibility diagram. Section 5 presents our linear-time algorithm for identifying a set of possible positions for the door such that the resultant rooms are searchable by two guards. We then extend our discussions to the 1-searcher case. Finally in Section 6, we summarize our work. The full version of this paper, complete with all the proofs, is available as a technical report [14].

2 Preliminaries

2.1 Notation

A simple polygon P is defined by a clockwise sequence of distinct *vertices* numbered $0, 1, \cdots, n-1$, $(n \geq 3)$, and n *edges*, connecting adjacent vertices. The edge between vertices u and v is denoted by (u, v). The *boundary* of P, denoted by ∂P, consists of all its vertices and edges. The vertices immediately preceding and succeeding vertex v in the clockwise order are denoted by $Pred(v)$ and $Succ(v)$, respectively. For any two points $a, b \in \partial P$, the open and closed portions of ∂P clockwise from a to b are denoted by $\partial P(a, b)$ and $\partial P[a, b]$, respectively.

A vertex whose interior angle between its two incident edges in the polygon is more than 180° is called a *reflex vertex*. Consider reflex vertex r. Extend the edge $(Succ(r), r)$ toward the interior of P, and let $B(r) \in \partial P$ denote the *backward extension point*, where this extension leaves P for the first time. The polygonal area formed by $\partial P[r, B(r)]$ and the chord $\overline{rB(r)}$ is called the *clockwise component* associated with r, and is denoted by $C_{cw}(r)$. Similarly, the extension of ($Pred(r)$, r) determines *forward extension point*, $F(r)$, and the *counter-clockwise component*, $C_{ccw}(r)$, associated with r is bounded by $\partial P[F(r), r]$ and the chord $\overline{rF(r)}$. If $C_{cw}(r)$ (resp. $C_{ccw}(r)$) does not totally contain any other clockwise (resp., counter-clockwise) component, it is said to be a *non-redundant* component.

Since our result in this paper makes use of some previous results, which are based on the assumption that no three vertices of the polygon are collinear, and no three lines defined by edges intersect in a common point, we adopt the same assumption in our discussions.

2.2 Visibility Diagram and Skeleton V-Diagram

Two points u and v are said to be *mutually visible* if the line segment $\overline{uv}$ is completely contained inside P, where we consider that ∂P is inside P. Given a polygon P, we define a *configuration* to be an element in $\partial P \times \partial P$ [15]. We call $\langle p, q\rangle \in \partial P \times \partial P$ a *visible* (resp. *invisible*) configuration, if p and q are mutually visible (resp. invisible).

Let us pick an arbitrary point on the boundary ∂P as the origin, and measure all distances along ∂P clockwise from the origin. Let $|\partial P|$ denote the length of ∂P. For $x \in \mathbf{R}$,[1] x represents the point $p(x) \in \partial P$ which is at distance $x - k|\partial P|$ from the origin, where k is an integer such that $0 \leq x - k|\partial P| < |\partial P|$. We thus can consider any $x \in \mathbf{R}$ as representing point $p(x)$ on ∂P. Note that there are infinitely many real numbers $x \in \mathbf{R}$ that represent the same point $p(x)$ on ∂P. Let $x, y \in \mathbf{R}$. Without loss of generality, we maintain one side of the beam (or the line connecting two guards on the polygon boundary) clear of the intruders. As the search progresses, x and y, which are the points on ∂P where the cleared area ends, can be considered as real-valued functions $x, y : [0, 1] \rightarrow \mathbf{R}$.[2] A *schedule* is a pair $\sigma = (x(t), y(t))$. Since the length of the cleared part on the boundary is at most $|\partial P|$, we impose the constraint $x(t) - |\partial P| \leq y(t) \leq x(t)$ for any $t \in [0, 1]$.

The *visibility space*, denoted by $\mathcal{V}$, consists of the infinite area between and including the lines $y = x$ (the *start line* S) and $y = x - |\partial P|$ (the *goal line* G), as shown in Fig. 1 [16].[3] The *visibility diagram* (*V-diagram* for short) for a given polygon is drawn in the V-space by shading some areas in it gray as follows: point $(x, y) \in \mathcal{V}$ is gray if configuration $\langle x, y\rangle$ is invisible. For example, Fig. 2 (b) shows the V-diagram for the polygon in Fig. 2 (a). In the diagram, coordinate (x, y) on line S or G is labeled by vertex v if $x = \ell + k|\partial P|$ and $y = \ell + k'|\partial P|$ for some integers k and k', where ℓ is the distance of v from the origin on ∂P.

The straight-edge boundaries of the gray areas touching either the S line of G line constitute the *skeleton V-diagram* (*SV-diagram*, for short). See the highlighted line segments on the boundary of the gray areas in Fig. 2 (b). The maximal contiguous white area in an SV-diagram is called a *cell*. A cell that has either three or four sides is called a *simple cell*.

[1] $\mathbf{R}$ denotes the set of all real numbers.

[2] We shall see in Section 3 that the formulation here naturally maps to the room search by two guards.

[3] The two dashed lines originating from point (d, d) will be discussed in Section 4.1.

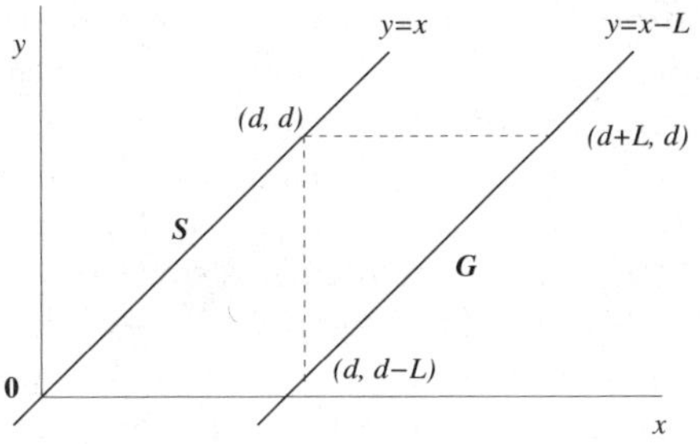

Fig. 1. Visibility space ($L = |\partial P|$)

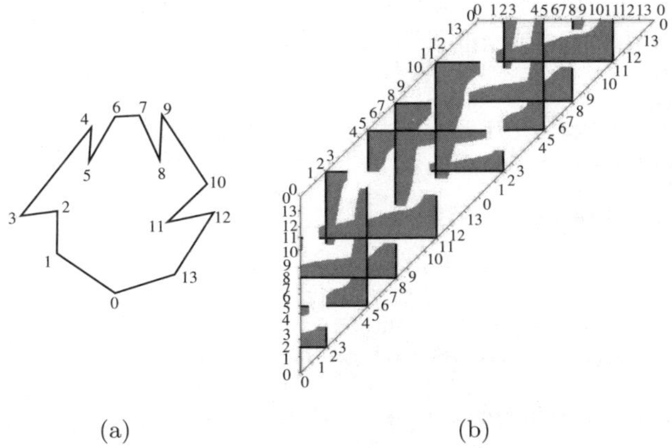

Fig. 2. (a) An example polygon; (b) Its V-diagram

3 Searchability and LR-Visibility

Given a polygon P, let $d \in \partial P$. Room $P(d)$ is *searchable* by two boundary guards if there exist two continuous functions $\ell : [0, 1] \to \partial P$ and $r : [0, 1] \to \partial P$ such that [9]

(a) $\ell(0) = r(0) = d$ and $\ell(1) = r(1) \in \partial P$;
(b) For any $t \in (0, 1)$, $\ell(t) \neq r(t)$ and $\ell(t)$ and $r(t)$ are mutually visible; and
(c) For any $t \in [0, 1]$, $d \in \partial P[r(t), \ell(t)]$ holds. □

Functions $\ell(t)$ and $r(t)$, both of which are continuous, represent the positions of the left guard and right guard[4] on ∂P at $t \in [0, 1]$, respectively. It is clear from the above definition that, if a room is searchable, for any $t \in [0, 1]$, the partition of P that lies on the door side of the line segment $\overline{\ell(t)r(t)}$ is always clear of intruders. If $P(d)$ is searchable, we call d a *safe* door.

[4] Viewed from door d.

A polygon P is *LR-visible* if there exists a pair of points u and v on ∂P such that $\partial P(u,v)$ and $\partial P(v,u)$ are mutually weakly visible, i.e., any point on $\partial P(u,v)$ is visible to at least one point on $\partial P(v,u)$ and vice versa [17]. If a polygon is LR-visible, there is an $O(n)$ time algorithm to determine all possible pairs of maximal boundary chains $\{A_i, B_i\}$, $i = 0, 1, \cdots, m$, such that for any $u \in A_i$ and any $v \in B_i$, P is LR-visible with respect to the pair $\{u, v\}$ [17].

It is shown in [1] that if a simple polygon P is 1-searchable, there exists a pair of vertices $\{u, v\}$ such that P is weakly visible from the shortest path connecting u and v, i.e., every point on the two sides of P partitioned by u and v can be seen at least by one point on the shortest path connecting u and v. In this case, it is clear that P is LR-visible with respect to u and v [18]. Similarly, there is a relation between the searchability of a room and its LR-visibility.

Lemma 1. [9,10] *Let v be a reflex vertex of a searchable room $P(d)$. If $d \notin C_{cw}(v)$ (resp. $d \notin C_{ccw}(v)$), the final meeting place of the two guards $p \in \partial P(v,d)$ (resp. $\partial P(d,v)$).* □

Lemma 2. *If a room $P(d)$ is searchable by two guards, P is LR-visible with respect to d and a point $p \in \partial P$.*

Proof. For the full proof, see [14]. □

Corollary 1. *If $d \in \partial P$ is a safe door, d is on some A_i or B_i.* □

4 Searchability Analysis of a Room

4.1 A Simple Characterization

We investigate room search by two guards using the V-diagram introduced in Section 2.2. Since no guard can move across the door at any time according to condition (c) of the definition of a searchable room given in Section 3, configuration $\langle l(t), r(t)\rangle$ is always within the V-space bounded by $x \geq d$ and $y \leq d$. See the triangular portion of the V-space bounded by the dashed lines in Fig. 1. The initial configuration for the two guards $\langle d, d\rangle$ is represented by the only point on S, called the *door point*, at the upper left corner of this triangular section. The portion of the SV-diagram for room $P(d)$ bounded by $x \geq d$ and $y \leq d$ is called the *visibility triangle* for $P(d)$ and denoted by $VT_P(d)$. Fig. 3 shows an example. By convention, we label point (x, x) on the diagonal by x.

By definition, in a searchable room $P(d)$, all the configurations in $\{\langle l(t), r(t)\rangle \mid t \in (0,1)\}$ correspond to white points in the V-diagram. Thus, the following proposition follows immediately [16].

Proposition 1. *A room is searchable by two boundary guards if and only if there exists a cell in $VT_P(d)$ that touches both the door point and the diagonal side.* □

A cell in $VT_P(d)$ that does *not* touch *both* the door point and the diagonal side is called a *blocking cell.* We call a blocking cell that has either three sides

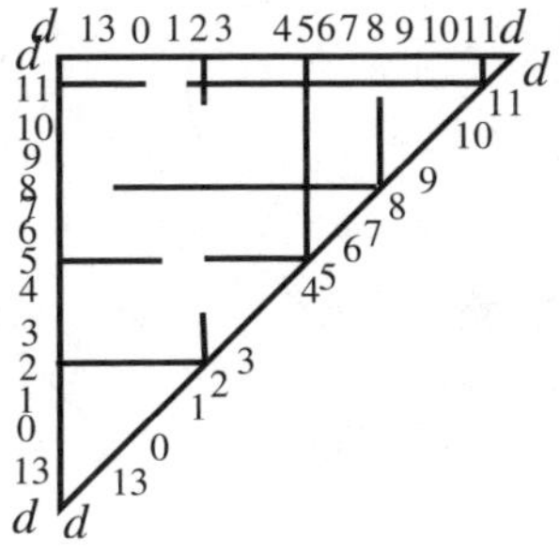

Fig. 3. The visibility triangle where vertex 12 is considered as the door for the polygon in Fig. 2

(a triangle) or four sides (a rectangle) a *simple blocking cell.*[5] Note that in Fig. 3, for example, the blocking cell to the right of the vertical line segment extending from vertex 5 on the diagonal G and bounded by the two horizontal line segments at vertices 8 and 11 does not touch G at 8, since the line segments have non-zero width.

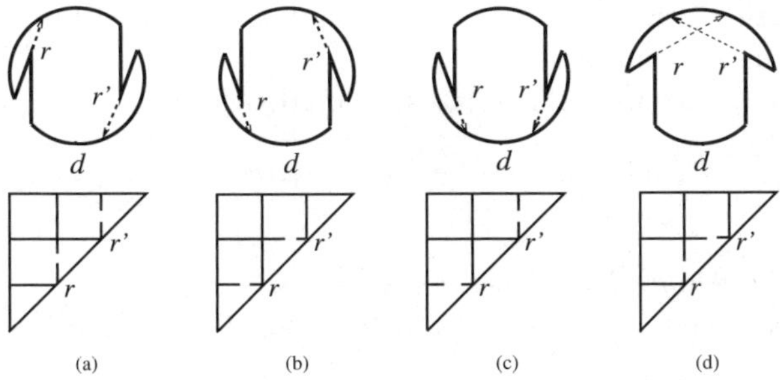

Fig. 4. Simple blocking cells that touch the door point

All the simple blocking cell patterns that touch the door point are shown in Fig. 4. In Fig. 4 (a) (resp. Fig. 4 (b)) there are two reflex vertices whose clockwise (resp. counter-clockwise) components are disjoint from each other. In Fig. 4 (c) there are two reflex vertices such that the clockwise component of one is disjoint from the counter-clockwise component of the other. The two arrows in Fig. 4 (d) need not intersect.

It follows immediately from Proposition 1 that if $P(d)$ contains any pattern in Fig. 4, $P(d)$ is not searchable by two boundary guards. However, Fig. 4 does not exhaust all the possible cases where $P(d)$ is not searchable. Let us call an interval on the diagonal that is touched by a simple blocking cell (a triangle) an

[5] There may be other "dangling" line segments inside a simple blocking cell.

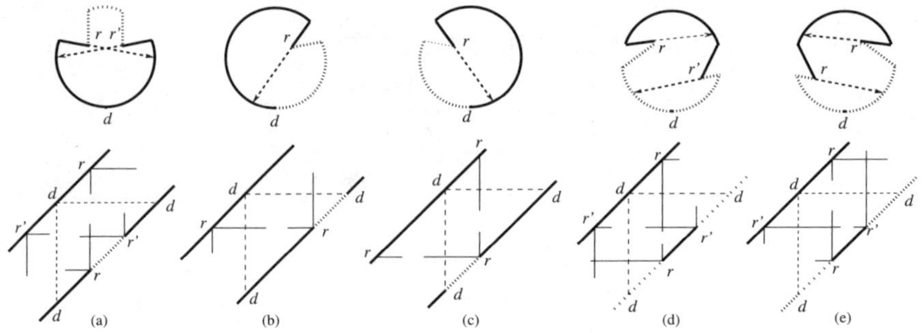

Fig. 5. Simple blocking cells that touch the diagonal

unreachable interval. An interval that is not unreachable is called *reachable*. Fig. 5 shows all the simple blocking cell patterns that cause unreachable intervals [13]. In the figure, the unreachable intervals are indicated by dashed line segments (in the bottom half of the figure), and the corresponding boundary portions are indicated by dotted curves (in the top half of the figure).

In Fig. 5 we show the visibility triangles embedded in V-diagrams. Note that the unreachable intervals of Fig. 5 (a) and (b) appear in (d) and (e), respectively, as well. In addition, Fig. 5 (a), (d), and (e) appear in Fig. 4 but the positions of d are different in the two figures.

The following theorem characterizes a safe door.

Theorem 1. [13] *Room $P(d)$ is searchable by two boundary guards if and only if*

(a) no simple blocking cell touches the door point, and
(b) there exists a point in a reachable interval on the diagonal of $VT_P(d)$. □

4.2 A Linear-Time Testing Algorithm

Suppose the given P is LR-visible with respect to some pair of vertices. Then all the pairs of maximal boundary chains $\{\{A_i, B_i\} \mid i = 1, 2, \cdots, m\}$, such that for any $s \in A_i$ and any $t \in B_i$, P is LR-visible with respect to the pair $\{s, t\}$, can be computed in $O(n)$ time [17]. In [12] we made use of the LR-visibility of a searchable room in our linear-time algorithm for checking if a given room is searchable by two guards. Below we present our algorithm as a procedure [13]. Its major goal is to detect the patterns in Fig. 5 (d) and (e) in linear time. We focus on the pattern in Fig. 5 (d), since the one in Fig. 5 (e) can be treated symmetrically. We will use this procedure in Section 5.

We first construct two lists in linear time: L_{cw} is a list of reflex vertices on $\partial P(d, p)$ that cause non-redundant clockwise components, ordered counter-clockwise from p to d, while R_{cw} is a list of all reflex vertices on $\partial P(p, d)$ ordered clockwise from p to d. We now modify R_{cw} to construct a new list $\overline{R}_{cw}$ by merging R_{cw} and $\{B(u) \mid u \in L_{cw}\}$ and appending d at the end. If L_{cw} could be augmented by $\{B(u) \mid u \in R_{cw}\}$, it would be easy to find the farthest r', but it would require $O(n \log n)$ time to do so [13], because R_{cw} may contain reflex vertices that cause redundant clockwise components. Our aim is to do away with the computation of $\{B(u) \mid u \in R_{cw}\}$, since we do not know how to do it in linear time. We use variables P_r and P_l, initialized to the first element of $\overline{R}_{cw}$ and to be null (Λ), respectively, in order to scan $\overline{R}_{cw}$ and L_{cw}, respectively. We maintain another variable, v, to remember the "current" reflex vertex in L_{cw}.

Procedure FARTHEST(d)

1. While $P_r = B(u)$ for some vertex u do $\{v = u$; advance $P_r\}$;
2. If $P_r = d$ no pattern in Fig. 5 (d) is present (stop);
3. If P_r and v are mutually visible, set $w = v$. Otherwise, go to step 6;
4. If $r = w$ and $r' = P_r$ form the pattern of Fig. 5 (d), P_r is the farthest vertex (stop);[6]
5. Advance P_r to the next element in $\overline{R}_{cw}$ and go to Step 1;
6. If $P_l = \Lambda$, set $P_l = v$;
7. While P_l and P_r are not mutually visible {advance P_l to the next element in L_{cw}};
8. Set $w = P_l$ and go to Step 4; □

5 A Linear-Time Algorithm for Finding all Safe Door Locations

We presented an $O(n \log n)$ solution to the problem of finding all safe doors in [13]. Here we present an optimal $O(n)$ algorithm in five steps, by reusing the same visibility information for all candidate door positions. If a polygon is not LR-visible, by Lemma 2 there can be no safe door. Therefore, we shall always assume that the given polygon is LR-visible.

5.1 Critical Points and Sections

We define a *section* to be the part of the polygon boundary between two adjacent critical points. Observe in Fig. 5 (d) that, as we slide an imaginary d clockwise from r', the interval $\partial P(r', r)$ remains unreachable as long as d is within $\partial P(r', r)$. In other words, $\partial P(r', r)$ remains unreachable for the range of d bounded by the two reflex vertices r' and r. This means that the safety of any door position on $\partial P(r', r)$ is not affected by the position of $B(r')$. Note that the clockwise component from r' could be redundant. A similar observation applies to r in Fig. 5 (e), due to its symmetry to Fig. 5 (d). Therefore, we do not need to consider the extension points from the reflex vertices whose components are redundant. Note that we do need to consider the reflex vertices themselves.

Proposition 2. *The points in a section are either all safe or all unsafe.* □

[6] To test this condition efficiently, determine if the extension of edge $(Succ(P_r), P_r)$ is to the left of line segment $\overline{P_r w}$ viewed from P_r and $P_r \neq B(w)$.

5.2 Data Structures

We maintain two lists, `Candidates` and `Sections`, and an array $A[\,]$. `Candidates` initially contains one representative door position for each section, and unsafe positions will be deleted from it as they become known. `Candidates` functions as an index for `Sections`. The entry of `Sections` corresponding to a candidate door d in `Candidates` contains at most two maximal unreachable intervals that are currently known for d.

The edge $(i, i+1)$ will be numbered i, where $i = 0, 1, \ldots, n-1$. $A[\,]$ is an array indexed by the *vertex* number. If edge i is in interval $\partial P(r, r')$ for any pair of reflex vertices r and r' in the pattern shown in Fig. 5 (a) components $A[i].cw$ and $A[i].ccw$ will store the clockwise and counter-clockwise end (reflex) vertices, respectively, of the maximal unreachable interval on which edge i lies. Otherwise, we set $A[i].cw = A[i].ccw = i$. We can construct $A[\,]$ in linear time.

5.3 Patterns in Fig. 4 and Fig. 5 (a)

By condition (a) of Theorem 1, point d in any of the figures in Fig. 4 is unsafe.

Step 1: Determine the critical points by computing the extension points of all the reflex vertices that cause non-redundant components. Construct `Candidates` by placing one representative door position for each section, excluding the sections that are unsafe, according to Fig. 4. □

Since we assume the given polygon is LR-visible, we can compute the A_is and B_is in $O(n)$ time. By deleting from `Candidates` all positions that do not belong to any A_i or B_i (see Corollary 1), we eliminate all candidates that are positioned as d in Figs. 4 (a), (b), or (c). As a result of Step 1, a safe d can only be in the clockwise component due to r or r'.

Step 2: Referring to the deadlock intervals in Fig. 5 (a), construct array $A[\,]$. □

5.4 Patterns in Fig. 5 (b)-(e)

Fig. 5 (b) and (c) are shown combined in Fig. 6 (a), and Fig. 5 (d) and (e) are shown combined in Fig. 6 (b).

We now construct the entry of `Sections` for each representative door d. It consists of at most two unreachable intervals on the diagonal of visibility triangle $VT_P(d)$. They are of the form $(r_H(d),\ d)$ or $(d,\ r_L(d)$, if any. In Fig. 6, $r_H(d)$ and $r_L(d)$ are the farthest reflex vertices from d that cause such patterns [13]. Thus the intervals $\partial P(d, r_L(d))$ and $\partial P(r_H(d), d)$ are unreachable. We update the unreachable intervals in the entry of `Sections` corresponding to d by Steps 3 and 4. Clearly, if the union of the unreachable intervals in the entry for position d and those in $A[\,]$ spans the entire boundary, d is unsafe.

Step 3: For each point d in `Candidates`, compute $r_L(d)$ and $r_H(d)$ of Fig. 6 (a), by determining the farthest (clockwise from d) non-redundant clockwise component and the farthest (counter-clockwise from d) non-redundant counter-clockwise component, respectively. □

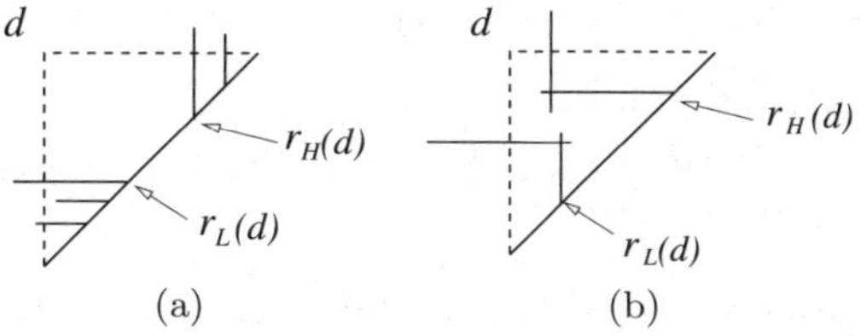

Fig. 6. (a) Unreachable interval of Type A; (b) Unreachable interval of Type B

We can partition `Candidates` into two groups: group G_1^d consists of those door positions belonging to the same clockwise components as d and group G_2^d consists of the rest. Note that all door positions in G_1^d (resp. G_2^d) share the same farthest r_H. This implies that we can invoke procedure FARTHEST(d) for one representative d from each group.

The pattern in Fig. 5 (e) can be treated similarly to determine $r_L(d)$ for each $d \in$ `Candidates`.

Step 4: Pick an element d from `Candidates` and determine the two groups, G_1^d and G_2^d, defined above. Update $r_L(d')$ and $r_H(d')$ for each d' in `Sections`. □

5.5 Final Step

All the information needed to determine the safety of the doors in `Candidates` is contained in $A[\,]$ and `Sections`. We now need to compute the union of the unreachable intervals on the diagonal, if they overlap. We then test if the resulting unreachable interval equals the entire boundary.

Step 5: For each entry (let it be the entry for representative door d) in `Sections` do the following:

1. If either $r_L(d) = d$ or $r_H(d) = d$, the diagonal is reachable and d is safe. Stop
2. Using $r_L(d)$ and $r_H(d)$ as indices into array $A[\ ]$, find $A[r_L(d)].cw$ and $A[r_H(d)].ccw$. If, starting clockwise from d, $A[r_H(d)].ccw$ is encountered before $A[r_L(d)].cw$, door d is unsafe and otherwise it is safe. □

Theorem 2. *Given polygon P with n vertices, we can identify in $O(n)$ time all the safe door locations when two boundary guards search.* □

5.6 Extension to 1-Searcher Case

A boundary 1-searcher always moves on the polygon boundary, starting from the door, aiming her flashlight at the other side of the door. Her vision is restricted to the beam (inside the polygon) from the flashlight, up to the point where it leaves the polygon for the first time. In the 1-searcher case, we need to consider the *beam head jump* [15,16]. The following results are immediate [14].

Theorem 3. *Room $P(d)$ is searchable by a boundary 1-searcher if and only if*

(a) no simple blocking cell touches the door point, and
(b) the unreachable intervals caused by the patterns in Figs. 5 (a)-(d) do not span the entire boundary ∂P. □

Theorem 4. *Given polygon P with n vertices, we can identify in $O(n)$ time all the safe door locations when a boundary 1-searcher searches.* □

6 Conclusion

The ultimate objective of our research on polygon search is to characterize the sets of polygons that can be searched by a given number of k-searchers or guards and generate search schedules in an efficient way. For some special cases where $k = 1$ or 2, or where there are other restrictions (such as streets or rooms), several interesting results are known, thanks to recent research effort by many researchers.

As for room search, it is now fairly well-understood, and an $O(n)$ time algorithm is known for testing if a given door on a given polygon is safe, i.e., it makes the room searchable by two guards. There is also an $O(n \log n)$ time algorithm to find all such safe door locations. We have shown in this paper that all the safe doors can be found in $O(n)$ time, in other words, we can do so in the most efficient way. In a sense this concludes research on the room search problem by two guards or a 1-searcher.

References

1. Suzuki, I., Yamashita, M.: Searching for a mobile intruder in a polygonal region. SIAM J. on Computing 21(5), 863–888 (1992)
2. Icking, C., Klein, R.: The two guards problem. Int'l J. of Computational Geometry and Applications 2(3), 257–285 (1992)
3. Heffernan, P.: An optimal algorithm for the two-guard problem. Int'l J. of Computational Geometry and Applications 6, 15–44 (1996)
4. Tseng, L.H., Heffernan, P.J., Lee, D.T.: Two-guard walkability of simple polygons. Int'l J. of Computational Geometry and Applications 8(1), 85–116 (1998)
5. Bhattacharya, B.K., Mukhopadhyay, A., Narasimhan, G.: Optimal algorithms for two-guard walkability of simple polygons. In: Proc. 7th Int'l Workshop on Algorithms and Data Structures, pp. 438–449 (2001)
6. Crass, D., Suzuki, I., Yamashita, M.: Searching for a mobile intruder in a corridor: the open edge variant of the polygon search problem. Int'l J. of Computational Geometry and Applications 5(4), 397–412 (1995)
7. Lee, J.H., Park, S.M., Chwa, K.Y.: Searching a polygonal room with one door by a 1-searcher. Int'l J. of Computational Geometry and Applications 10(2), 201–220 (2000)
8. Lee, J.H., Shin, S.Y., Chwa, K.Y.: Visibility-based pursuit-evasions in a polygonal room with a door. In: Proc. ACM Symp. on Computational Geometry, pp. 281–290 (1999)

9. Park, S.M., Lee, J.H., Chwa, K.Y.: Characterization of rooms searchable by two guards. In: Proc. Int'l Symp. on Algorithms and Computation, pp. 515–526 (2000)
10. Park, S.M., Lee, J.H., Chwa, K.Y.: Searching a room by two guards. Int'l J. of Computational Geometry and Applications 12(4), 339–352 (2002)
11. Tan, X.: Efficient algorithms for searching a polygonal room with a door. In: Japanese Conf. on Discrete and Computational Geometry, pp. 339–350 (2000)
12. Bhattacharya, B.K., Zhang, J.Z., Shi, Q.S., Kameda, T.: An optimal solution to room search problem. In: Proc. 18th Canadian Conf. on Computational Geometry, August 2006, pp. 55–58 (2006)
13. Zhang, J.Z., Kameda, T.: Where to build a door. In: Proc. IEEE/RSJ Int'l Conf. on Intelligent Robots and Systems, October 2006, pp. 4084–4090 (2006)
14. Zhang, J.Z., Kameda, T.: A linear-time algorithm for finding all door locations that make a room searchable. Technical Report CMPT TR 2007-27 (submitted to a journal), School of Computing Science, Simon Fraser University (November 2007), `http://www.cs.sfu.ca/research/publications/techreports/`
15. LaValle, S.M., Simov, B., Slutzki, G.: An algorithm for searching a polygonal region with a flashlight. Int'l J. of Computational Geometry and Applications 12(1-2), 87–113 (2002)
16. Kameda, T., Yamashita, M., Suzuki, I.: On-line polygon search by a seven-state boundary 1-searcher. IEEE Trans. on Robotics 22, 446–460 (2006)
17. Das, G., Heffernan, P.J., Narasimhan, G.: LR-visibility in polygons. Computational Geometry 7, 37–57 (1997)
18. Bhattacharya, B., Ghosh, S.K.: Characterizing LR-visibility polygons and related problems. Computational Geometry: Theory and Applications 18, 19–36 (2001)

Model Theoretic Complexity of Automatic Structures (Extended Abstract)

Bakhadyr Khoussainov[1] and Mia Minnes[2]

[1] Department of Computer Science
University of Auckland
Auckland, New Zealand
bmk@cs.auckland.ac.nz
[2] Mathematics Department
Cornell University
Ithaca, New York 14853
minnes@math.cornell.edu

Abstract. We study the complexity of automatic structures via well-established concepts from both logic and model theory, including ordinal heights (of well-founded relations), Scott ranks of structures, and Cantor-Bendixson ranks (of trees). We prove the following results: 1) The ordinal height of any automatic well-founded partial order is bounded by ω^ω; 2) The ordinal heights of automatic well-founded relations are unbounded below ω_1^{CK}; 3) For any infinite computable ordinal α, there is an automatic structure of Scott rank at least α. Moreover, there are automatic structures of Scott rank $\omega_1^{CK}, \omega_1^{CK} + 1$; 4) For any ordinal $\alpha < \omega_1^{CK}$, there is an automatic successor tree of Cantor-Bendixson rank α.

1 Introduction

In recent years, there has been increasing interest in the study of structures that can be presented by automata. The underlying idea is to apply techniques of automata theory to decision problems that arise in logic and applications such as databases and verification. A typical decision problem is the model checking problem: for a structure $\mathcal{A}$ (e.g. a graph), design an algorithm that, given a formula $\phi(\bar{x})$ in a formal system and a tuple $\bar{a}$ from the structure, decides if $\phi(\bar{a})$ is true in $\mathcal{A}$. In particular, when the formal system is the first order predicate logic or the monadic second order logic, we would like to know if the theory of the structure is decidable. Fundamental early results in this direction by Büchi ([4], [5]) and Rabin ([20]) proved the decidability of the monadic second order theories of the successor on the natural numbers and of the binary tree. There have been numerous applications and extensions of these results in logic, algebra, verification, model checking, and databases (see, for example, [9] [23] [24] and [25]). Moreover, automatic structures provide a theoretical framework for constraint databases over discrete domains such as strings and trees [1].

M. Agrawal et al. (Eds.): TAMC 2008, LNCS 4978, pp. 514–525, 2008.

A structure $\mathcal{A} = (A; R_0, \ldots, R_m)$ is **automatic** if the domain A and all the relations $R_0, \ldots, R_m$ of the structure are recognised by finite automata (precise definitions are in the next section). Independently, Hodgson [13] and later Khoussainov and Nerode [14] proved that for any given automatic structure there is an algorithm that solves the model checking problem for the first order logic. In particular, the first order theory of the structure is decidable. There is a body of work devoted to the study of resource-bounded complexity of the model checking problem for automatic structures. Most current results demonstrate that automatic structures are not complex in various concrete senses. However, in this paper we use well-established concepts from both logic and model theory to prove results in the opposite direction. We now briefly describe the measures of complexity we use (ordinal heights of well-founded relations, Scott ranks of structures, and Cantor-Bendixson ranks of trees) and connect them with the results of this paper.

A relation R is called **well-founded** if there is no infinite sequence $x_1, x_2, x_3, \ldots$ such that $(x_{i+1}, x_i) \in R$ for $i \in \omega$. In computer science, well-founded relations are of interest due to a natural connection between well-founded sets and terminating programs. We say that a program is **terminating** if every computation from an initial state is finite. This is equivalent to well-foundedness of the collection of states reachable from the initial state, under the reachability relation [3]. The **ordinal height** is a measure of the depth of well-founded relations. Since all automatic structures are computable, the obvious bound for ordinal heights of automatic well-founded relations is ω_1^{CK} (the first non-computable ordinal). Sections 3 and 4 study the sharpness of this bound. Theorem 1 characterizes automatic well-founded partial orders in terms of their ordinal heights, whereas Theorem 2 shows that ω_1^{CK} is the sharp bound in the general case.

Theorem 1. *For each ordinal α, α is the ordinal height of an automatic well-founded partial order if and only if $\alpha < \omega^\omega$.*

Theorem 2. *For each (computable) ordinal $\alpha < \omega_1^{CK}$, there is an automatic well-founded relation $\mathcal{A}$ whose ordinal height is greater than α.*

Section 5 is devoted to building automatic structures with high Scott ranks. The concept of Scott rank comes from a well-known theorem of Scott stating that for every countable structure $\mathcal{A}$ there exists a sentence ϕ in $L_{\omega_1,\omega}$-logic which characterizes $\mathcal{A}$ up to isomorphism [22]. The minimal quantifier rank of such a formula is called the Scott rank of $\mathcal{A}$. A known upper bound on the Scott rank of computable structures implies that the Scott rank of automatic structures is at most $\omega_1^{CK} + 1$. But, until now, all the known examples of automatic structures had low Scott ranks. Results in [19], [7], [17] suggest that the Scott ranks of automatic structures could be bounded by small ordinals. This intuition is falsified in Section 5 with the theorem:

Theorem 3. *For each infinite computable ordinal α there is an automatic structure of Scott rank at least α.*

In the last section, we investigate the Cantor-Bendixson ranks of automatic trees. A **partial order tree** is a partially ordered set $(T, \leq)$ such that there is a $\leq$-minimal element of T, and each subset $\{x \in T : x \leq y\}$ is finite and is linearly ordered under $\leq$. A **successor tree** is a pair (T, S) such that the reflexive and transitive closure $\leq_S$ of S produces a partial order tree $(T, \leq_S)$. The **derivative** of a tree $\mathcal{T}$ is obtained by removing all the nonbranching paths of the tree. One applies the derivative operation to $\mathcal{T}$ successively until a fixed point is reached. The minimal ordinal that is needed to reach the fixed point is called the **Cantor-Bendixson (CB) rank** of the tree. The CB rank plays an important role in logic, algebra, and topology. Informally, the CB rank tells us how far the structure is from algorithmically (or algebraically) simple structures. Again, the obvious bound on CB ranks of automatic successor trees is ω_1^{CK}. In [16], it is proved that the CB rank of any automatic partial order tree is finite and can be computed from the automaton for the $\leq$ relation on the tree. It has been an open question whether the CB ranks of automatic successor trees can be bounded by small ordinals. We answer this question in the following theorem.

Theorem 4. *For $\alpha < \omega_1^{CK}$ there is an automatic successor tree of CB rank α.*

The main tool we use to prove results about high ranks is the configuration spaces of Turing machines, considered as automatic graphs. It is important to note that graphs which arise as configuration spaces have very low model-theoretic complexity: their Scott ranks are at most 3, and if they are well-founded then their ordinal heights are at most ω (see Propositions 1 and 2). Hence, the configuration spaces serve merely as building blocks in the construction of automatic structures with high complexity, rather than contributing materially to the high complexity themselves.

2 Preliminaries

A (relational) **vocabulary** is a finite sequence $(P_1^{m_1}, \ldots, P_t^{m_t}, c_1, \ldots, c_s)$, where each $P_j^{m_j}$ is a predicate symbol of arity $m_j > 0$, and each c_k is a constant symbol. A **structure** with this vocabulary is a tuple $\mathcal{A} = (A; P_1^{\mathcal{A}}, \ldots, P_t^{\mathcal{A}}, c_1^{\mathcal{A}}, \ldots, c_s^{\mathcal{A}})$, where $P_j^{\mathcal{A}}$ and $c_k^{\mathcal{A}}$ are interpretations of the symbols of the vocabulary. When convenient, we may omit the superscripts $\mathcal{A}$. We only consider infinite structures.

A **finite automaton** $\mathcal{M}$ over an alphabet Σ is a tuple (S, ι, Δ, F), where S is a finite set of **states**, $\iota \in S$ is the **initial state**, $\Delta \subset S \times \Sigma \times S$ is the **transition table**, and $F \subset S$ is the set of **final states**. A **computation** of $\mathcal{A}$ on a word $\sigma_1\sigma_2\ldots\sigma_n$ ($\sigma_i \in \Sigma$) is a sequence of states, say $q_0, q_1, \ldots, q_n$, such that $q_0 = \iota$ and $(q_i, \sigma_{i+1}, q_{i+1}) \in \Delta$ for all $i \in \{0, \ldots, n-1\}$. If $q_n \in F$, then the computation is **successful** and we say that automaton $\mathcal{M}$ **accepts** the word $\sigma_1\sigma_2\ldots\sigma_n$. The **language** accepted by the automaton $\mathcal{M}$ is the set of all words accepted by $\mathcal{M}$. In general, $D \subset \Sigma^\star$ is **finite automaton recognisable**, or **regular**, if D is the language accepted by some finite automaton $\mathcal{M}$.

To define automaton recognisable relations, we use n-variable (or n-tape) automata. An n**–tape automaton** can be thought of as a one-way Turing machine with n input tapes [8]. Each tape is regarded as semi-infinite, having written on it a word over the alphabet Σ followed by an infinite succession of blanks (denoted by $\diamond$ symbols). The automaton starts in the initial state, reads simultaneously the first symbol of each tape, changes state, reads simultaneously the second symbol of each tape, changes state, etc., until it reads a blank on each tape. The automaton then stops and accepts the n–tuple of words if it is in a final state. The set of all n–tuples accepted by the automaton is the relation recognised by the automaton. For a formal definition see, for example, [14].

Definition 1. *A structure $\mathcal{A} = (A; R_0, R_1, \ldots, R_m)$ is* **automatic** *over Σ if its domain A and all relations R_0, R_1, ..., R_m are regular over Σ.*

The configuration graph of any Turing machine is an example of an automatic structure. The graph is defined by letting the configurations of the Turing machine be the vertices, and putting an edge from configuration c_1 to configuration c_2 if the machine can make an instantaneous move from c_1 to c_2. Many examples of automatic structures can be formed using the ω-fold disjoint union of a structure $\mathcal{A}$ (the disjoint union of ω many copies of $\mathcal{A}$).

Lemma 1. *[21] If $\mathcal{A}$ is automatic then its ω-fold disjoint union is isomorphic to an automatic structure.* □

The class of automatic structures is a proper subclass of the computable structures. In this paper, we will be coding computable structures into automatic ones. Good references for the theory of computable structures include [11], [15].

Definition 2. *A* **computable structure** *is a structure $\mathcal{A} = (A; R_1, \ldots, R_m)$ whose domain and relations are all computable.*

The domains of computable structures can always be identified with the set ω of natural numbers. Under this assumption, we introduce new constant symbols c_n for each $n \in \omega$ and interpret c_n as n. In this context, $\mathcal{A}$ is computable iff the **atomic diagram** of $\mathcal{A}$ (the set of Gödel numbers of all quantifier-free sentences in the extended vocabulary that are true in $\mathcal{A}$) is computable.

3 Ranks of Automatic Well-Founded Partial Orders

In this section we consider structures $\mathcal{A} = (A; R)$ with a single binary relation. An element x is said to be R**-minimal for a set** X if for each $y \in X$, $(y, x) \notin R$. The relation R is said to be **well-founded** if every non-empty subset of A has an R-minimal element. This is equivalent to saying that $(A; R)$ has no infinite chains $x_1, x_2, x_3, \ldots$ where $(x_{i+1}, x_i) \in R$ for all i.

A **ranking function** for $\mathcal{A}$ is an ordinal-valued function f such that $f(y) < f(x)$ whenever $(y, x) \in R$. For f a ranking function on $\mathcal{A}$, let $ord(f) = \sup\{f(x) : x \in A\}$. The structure $\mathcal{A}$ is well-founded if and only if $\mathcal{A}$ admits a ranking function. The **ordinal height** of $\mathcal{A}$, denoted $r(\mathcal{A})$, is the least ordinal α which is $ord(g)$ for some ranking function g on $\mathcal{A}$. For $B \subseteq A$, we write $r(B)$ for the ordinal height of the structure obtained by restricting R to B. Recall that if $\alpha < \omega_1^{CK}$ then α is a computable ordinal.

Lemma 2. *If $\alpha < \omega_1^{CK}$, there is a computable well-founded relation of ordinal height α.*

Lemma 2 amounts to taking a computable copy of any linear order of type α. The next lemma follows easily from well-foundedness of ordinals and of R.

Lemma 3. *For a structure $\mathcal{A} = (A; R)$ where R is well-founded, if $r(\mathcal{A}) = \alpha$ and $\beta < \alpha$ then there is an $x \in A$ such that $r_{\mathcal{A}}(x) = \beta$.* □

For the remainder of this section, we assume further that R is a partial order. For convenience, we write $\leq$ instead of R. Thus, we consider automatic well-founded partial orders $\mathcal{A} = (A, \leq)$. We will use the notion of **natural sum of ordinals**. The natural sum of ordinals α, β (denoted $\alpha +' \beta$) is defined recursively: $\alpha +' \beta$ is the least ordinal strictly greater than $\gamma +' \beta$ for all $\gamma < \alpha$ and strictly greater than $\alpha +' \gamma$ for all $\gamma < \beta$.

Lemma 4. *Let A_1 and A_2 be disjoint subsets of A such that $A = A_1 \cup A_2$. Consider the partially ordered sets $\mathcal{A}_1 = (A_1, \leq_1)$ and $\mathcal{A}_2 = (A_2, \leq_2)$ obtained by restricting $\leq$ to A_1 and A_2 respectively. Then, $r(\mathcal{A}) \leq \alpha_1 +' \alpha_2$, where $\alpha_i = r(\mathcal{A}_i)$.*

Proof. We will show that there is a ranking function on A whose range is contained in the ordinal $\alpha_1 +' \alpha_2$. For each $x \in A$ consider the partially ordered sets $\mathcal{A}_{1,x}$ and $\mathcal{A}_{2,x}$ obtained by restricting $\leq$ to $\{z \in A_1 \mid z < x\}$ and $\{z \in A_2 \mid z < x\}$, respectively. Define $f(x) = r(\mathcal{A}_{1,x}) +' r(\mathcal{A}_{2,x})$. It is not hard to see that f is the desired ranking function. □

Corollary 1. *If $r(\mathcal{A}) = \omega^n$ and $A = A_1 \cup A_2$, where $A_1 \cap A_2 = \emptyset$, then either $r(\mathcal{A}_1) = \omega^n$ or $r(\mathcal{A}_2) = \omega^n$.* □

Khoussainov and Nerode [14] show that, for each n, there is an automatic presentation of the ordinal ω^n. It is clear that such a presentation has ordinal height ω^n. The next theorem shows that ω^ω is the sharp bound on ranks of all automatic well-founded partial orders. Now that Corollary 1 has been established, the proof of Theorem 1 follows Delhommé [7] and Rubin [21].

Theorem 1. *For each ordinal α, α is the ordinal height of an automatic well-founded partial order if and only if $\alpha < \omega^\omega$.*

Proof. One direction of the proof is clear. For the other, assume for a contradiction that there is an automatic well-founded partial order $\mathcal{A} = (A, \leq)$ with $r(\mathcal{A}) = \alpha \geq \omega^\omega$. Let $(S_A, \iota_A, \Delta_A, F_A)$ and $(S_\leq, \iota_\leq, \Delta_\leq, F_\leq)$ be finite automata over Σ recognizing A and $\leq$ (respectively). By Lemma 3, for each $n > 0$ there is $u_n \in A$ such that $r_{\mathcal{A}}(u_n) = \omega^n$. For each $u \in A$ we define the set

$$u \downarrow = \{x \in A : x < u\}.$$

Note that if $r_{\mathcal{A}}(u)$ is a limit ordinal then $r_{\mathcal{A}}(u) = r(u \downarrow)$. We define a finite partition of $u \downarrow$ in order to apply Corollary 1. To do so, for $u, v \in \Sigma^\star$, define

$X_v^u = \{vw \in A : w \in \Sigma^\star \;\&\; vw < u\}$. Each set of the form $u\downarrow$ can then be partitioned based on the prefixes of words as follows:

$$u \downarrow = \{x \in A : |x| < |u| \;\&\; x < u\} \cup \bigcup_{v \in \Sigma^\star : |v| = |u|} X_v^u .$$

(All the unions above are finite and disjoint.) Hence, applying Corollary 1, for each u_n there exists a v_n such that $|u_n| = |v_n|$ and $r(X_{v_n}^{u_n}) = r(u_n \downarrow) = \omega^n$.

On the other hand, we use the automata to define the following equivalence relation on pairs of words of equal lengths:

$$(u,v) \sim (u',v') \iff \Delta_A(\iota_A, v) = \Delta_A(\iota_A, v') \;\&\; \Delta_{\leq}(\iota_{\leq}, \binom{v}{u}) = \Delta_{\leq}(\iota_{\leq}, \binom{v'}{u'})$$

There are at most $|S_A| \times |S_{\leq}|$ equivalence classes. Thus, the infinite sequence $(u_1, v_1), (u_2, v_2), \ldots$ contains m, n such that $m \neq n$ and $(u_m, v_m) \sim (u_n, v_n)$.

Lemma 5. *For any $u, v, u', v' \in \Sigma^\star$, if $(u,v) \sim (u',v')$ then $r(X_v^u) = r(X_{v'}^{u'})$.*

To prove the lemma, consider $g : X_v^u \to X_{v'}^{u'}$ defined as $g(vw) = v'w$. From the equivalence relation, we see that g is well-defined, bijective, and order preserving. Hence $X_v^u \cong X_{v'}^{u'}$ (as partial orders). Therefore, $r(X_v^u) = r(X_{v'}^{u'})$.

By Lemma 5, $\omega^m = r(X_{v_m}^{u_m}) = r(X_{v_n}^{u_n}) = \omega^n$, a contradiction with the assumption that $m \neq n$. Therefore, there is no automatic well-founded partial order of ordinal height greater than or equal to ω^ω. □

4 Ranks of Automatic Well-Founded Relations

4.1 Configuration Spaces of Turing Machines

In the following, we embed computable structures into automatic ones via configuration spaces of Turing machines. Let $\mathcal{M}$ be an n-tape deterministic Turing machine. The **configuration space** of $\mathcal{M}$, denoted by $Conf(\mathcal{M})$, is a directed graph whose nodes are configurations of $\mathcal{M}$. The nodes are n-tuples, each of whose coordinates represents the contents of a tape. Each tape is encoded as $(w\ q\ w')$, where $w, w' \in \Sigma^\star$ are the symbols on the tape before and after the location of the read/write head, and q is one of the states of $\mathcal{M}$. The edges of the graph are all the pairs of the form (c_1, c_2) such that there is an instruction of $\mathcal{M}$ that transforms c_1 to c_2. The configuration space is an automatic graph. The out-degree of every vertex in $Conf(\mathcal{M})$ is 1; the in-degree need not be 1.

Definition 3. *A deterministic Turing machine $\mathcal{M}$ is* **reversible** *if $Conf(\mathcal{M})$ consists only of finite chains and chains of type ω.*

Lemma 6. *[2] For any deterministic 1-tape Turing machine there is a reversible 3-tape Turing machine which accepts the same language.*

Proof. (Sketch) Given a deterministic Turing machine, define a 3-tape Turing machine with a modified set of instructions. The modified instructions have the property that neither the domains nor the ranges overlap. The first tape performs the computation exactly as the original machine would have done. As the new machine executes each instruction, it stores the index of the instruction on the second tape, forming a history. Once the machine enters a state which would have been halting for the original machine, the output of the computation is copied onto the third tape. Then, the machine runs the computation backwards and erases the history tape. The halting configuration contains the input on the first tape, blanks on the second tape, and the output on the third tape. □

We establish the following notation for a 3-tape reversible Turing machine $\mathcal{M}$ given by the construction in this lemma. A **valid initial configuration** of $\mathcal{M}$ is of the form $(\lambda\ \iota\ x, \lambda, \lambda)$, where x in the domain, λ is the empty string, and ι is the initial state of $\mathcal{M}$. From the proof above, observe that a **final (halting)** configuration is of the form $(x, \lambda, \lambda\ q_f\ y)$, with q_f a halting state of $\mathcal{M}$. Also, because of the reversibility assumption, all the chains in $Conf(\mathcal{M})$ are either finite or ω-chains (the order type of the natural numbers). In particular, this means that $Conf(\mathcal{M})$ is well-founded. We call an element of in-degree 0 a **base** (of a chain). The set of valid initial or final configurations is regular. We classify the components (chains) of $Conf(\mathcal{M})$ as follows:

- **Terminating computation chains:** finite chains whose base is a valid initial configuration; that is, one of the form $(\lambda\ \iota\ x, \lambda, \lambda)$, for $x \in \Sigma^\star$.
- **Non-terminating computation chains:** infinite chains whose base is a valid initial configuration.
- **Unproductive chains:** chains whose base is not a valid initial configuration.

Configuration spaces of reversible Turing machines are locally finite graphs (graphs of finite degree) and well-founded. Hence, the following proposition guarantees that their ordinal heights are small. The proof is left to the reader.

Proposition 1. *If $G = (A, E)$ is a locally finite graph then E is well-founded and the ordinal height of E is not above ω, or E has an infinite chain.* □

4.2 Automatic Well-Founded Relations of High Rank

Theorem 2. *For each computable ordinal $\alpha < \omega_1^{CK}$, there is an automatic well-founded relation $\mathcal{A}$ whose ordinal height is greater than α*

Proof. The proof of the theorem uses properties of Turing machines and their configuration spaces. We take a computable well-founded relation whose ordinal height is α, and "embed" it into an automatic well-founded relation with similar ordinal height.

By Lemma 2, let $\mathcal{C} = (C, L_\alpha)$ be a computable well-founded relation of ordinal height α. We assume without loss of generality that $C = \Sigma^\star$ for some finite alphabet Σ. Let $\mathcal{M}$ be the Turing machine computing the relation L_α. On each pair (x, y) from the domain, $\mathcal{M}$ halts and outputs "yes" or "no" . By Lemma 6, we can assume that $\mathcal{M}$ is reversible. Recall that $Conf(\mathcal{M}) = (D, E)$ is an automatic graph. We define the domain of our automatic structure to be $A = \Sigma^\star \cup D$. The binary relation of the automatic structure is:

$$R = E \cup \{(x, (\lambda\ \iota\ (x, y), \lambda, \lambda)) : x, y \in \Sigma^\star\} \cup \{(((x, y), \lambda, \lambda\ q_f \text{ "yes" }), y) : x, y \in \Sigma^\star\}.$$

Intuitively, the structure $(A; R)$ is a stretched out version of (C, L_α) with infinitely many finite pieces extending from elements of C, and with disjoint pieces which are either finite chains or chains of type ω. The structure $(A; R)$ is automatic because its domain is a regular set of words and the relation R is recognisable by a 2-tape automaton. We should verify, however, that R is well-founded. Let $Y \subset A$. If $Y \cap C \neq \emptyset$ then since (C, L_α) is well-founded, there is $x \in Y \cap C$ which is L_α-minimal. The only possible elements u in Y for which $(u, x) \in R$ are those which lie on computation chains connecting some $z \in C$ with x. Since each such computation chain is finite, there is an R-minimal u below x on each chain. Any such u is R-minimal for Y. On the other hand, if $Y \cap C = \emptyset$, then Y consists of disjoint finite chains and chains of type ω. Any such chain has a minimal element, and any of these elements are R-minimal for Y. Therefore, $(A; R)$ is an automatic well-founded structure.

We now consider the ordinal height of $(A; R)$. For each element $x \in C$, an easy induction on $r_C(x)$, shows that $r_C(x) \leq r_{\mathcal{A}}(x) \leq \omega + r_C(x)$. We denote by $\ell(a, b)$ the (finite) length of the computation chain of $\mathcal{M}$ with input (a, b). For any element $a_{x,y}$ in the computation chain which represents the computation of $\mathcal{M}$ determining whether $(x, y) \in R$, we have $r_{\mathcal{A}}(x) \leq r_{\mathcal{A}}(a_{x,y}) \leq r_{\mathcal{A}}(x) + \ell(x, y)$. For any element u in an unproductive chain of the configuration space, $0 \leq r_{\mathcal{A}}(u) < \omega$. Therefore, since $C \subset A$, $r(\mathcal{C}) \leq r(\mathcal{A}) \leq \omega + r(C)$. □

5 Automatic Structures and Scott Rank

The Scott rank of a structure is introduced in the proof of Scott's Isomorphism Theorem [22]. Here we follow the definition of Scott rank from [6].

Definition 4. *For structure $\mathcal{A}$ and tuples $\bar{a}, \bar{b} \in A^n$ (of equal length), define*

- $\bar{a} \equiv^0 \bar{b}$ *if $\bar{a}, \bar{b}$ satisfy the same quantifier-free formulas in the language of $\mathcal{A}$;*
- *For $\alpha > 0$, $\bar{a} \equiv^\alpha \bar{b}$ if for all $\beta < \alpha$, for each $\bar{c}$ (of arbitrary length) there is $\bar{d}$ such that $\bar{a}, \bar{c} \equiv^\beta \bar{b}, \bar{d}$; and for each $\bar{d}$ (of arbitrary length) there is $\bar{c}$ such that $\bar{a}, \bar{c} \equiv^\beta \bar{b}, \bar{d}$.*

Then, the **Scott rank** *of the tuple $\bar{a}$, denoted by $\mathcal{SR}(\bar{a})$, is the least β such that for all $\bar{b} \in A^n$, $\bar{a} \equiv^\beta \bar{b}$ implies that $(\mathcal{A}, \bar{a}) \cong (\mathcal{A}, \bar{b})$. Finally, the Scott rank of $\mathcal{A}$, denoted by $\mathcal{SR}(\mathcal{A})$, is the least α greater than the Scott ranks of all tuples of $\mathcal{A}$.*

Example 1. $\mathcal{SR}(\mathbb{Q}, \leq) = 1$, $\mathcal{SR}(\omega, \leq) = 2$, and $\mathcal{SR}(n \cdot \omega, \leq) = n + 1$.

Configuration spaces of reversible Turing machines are locally finite graphs. By the Proposition below, they all have low Scott Rank.

Proposition 2. *Let $G = (V, E)$ be a locally finite graph, then $SR(G) \leq 3$.*

Proof. The neighbourhood of diameter n of a subset U, denoted $B_n(U)$, is defined as follows: $B_0(U) = U$ and $B_n(U)$ is the set of $v \in V$ which can be reached from U by n or fewer edges. The proof of the proposition relies on two lemmas whose proofs are left to the reader.

Lemma 7. *Let $\bar{a}, \bar{b} \in V$ be such that $\bar{a} \equiv^2 \bar{b}$. Then for all n, there is a bijection of the n-neighbourhoods around $\bar{a}, \bar{b}$ which sends $\bar{a}$ to $\bar{b}$ and which respects E.*

Lemma 8. *Let $G = (V, E)$ be a graph. Suppose $\bar{a}, \bar{b} \in V$ are such that for all n, $(B_n(\bar{a}), E, \bar{a}) \cong (B_n(\bar{b}), E, \bar{b})$. Then there is an isomorphism between the component of G containing $\bar{a}$ and that containing $\bar{b}$ which sends $\bar{a}$ to $\bar{b}$.*

To prove the proposition, we note that for any $\bar{a}, \bar{b}$ in V such that $\bar{a} \equiv^2 \bar{b}$, Lemmas 7 and 8 yield an isomorphism from the component of $\bar{a}$ to the component of $\bar{b}$ that maps $\bar{a}$ to $\bar{b}$. Hence, if $\bar{a} \equiv^2 \bar{b}$, there is an automorphism of G that maps $\bar{a}$ to $\bar{b}$. Therefore, for each $\bar{a} \in V$, $SR(\bar{a}) \leq 2$, so $SR(G) \leq 3$. □

Let $\mathcal{C} = (C; R_1, \ldots, R_m)$ be a computable structure. We construct an automatic structure $\mathcal{A}$ whose Scott rank is (close to) the Scott rank of $\mathcal{C}$. Since the domain of $\mathcal{C}$ is computable, we assume that $C = \Sigma^\star$ for some finite Σ. The construction of $\mathcal{A}$ involves connecting the configuration spaces of Turing machines computing relations $R_1, \ldots, R_m$. Note that Proposition 2 suggests that the high Scott rank of the resulting automatic structure is the main part of the construction because it is not provided by the configuration spaces themselves. We detail the construction for R_i. Let $\mathcal{M}_i$ be a Turing machine for R_i. By a simple modification of the machine we assume that $\mathcal{M}_i$ halts if and only if its output is "yes" . By Lemma 6, we can also assume that $\mathcal{M}_i$ is reversible. We now modify the configuration space $Conf(\mathcal{M}_i)$ so as to respect the isomorphism type of $\mathcal{C}$. This will ensure that the construction (almost) preserves the Scott rank of $\mathcal{C}$. We use the terminology from Subsection 4.1.

Smoothing out unproductive parts. The length and number of unproductive chains is determined by the machine $\mathcal{M}_i$ and hence may differ even for Turing machines computing the same set. In this stage, we standardize the format of this unproductive part of the configuration space. We add ω-many chains of length n (for each n) and ω-many copies of ω. This ensures that the (smoothed) unproductive section of the configuration space of any Turing machine will be isomorphic and preserves automaticity.

Smoothing out lengths of computation chains. We turn our attention to the chains which have valid initial configurations at their base. The length of each finite chain denotes the length of computation required to return a "yes" answer. We will smooth out these chains by adding "fans" to each base. For this, we connect to each base of a computation chain a structure which consists of ω many chains of each finite length. To do so we follow Rubin [21]: consider the structure whose domain is $0^\star 01^\star$ and whose relation is given by xEy if and only if $|x| = |y|$ and y is the least lexicographic successor of x. This structure has a finite chain of every finite length. As in Lemma 1, we take the ω-fold disjoint union of the structure and identify the bases of all the finite chains. We get a "fan" with infinitely many chains of each finite size whose base can be identified with a valid initial computation state. Also, the fan has an infinite component if and only if R_i does not hold of the input tuple corresponding to the base. The result is an automatic graph, $Smooth(R_i) = (D_i, E_i)$, which extends $Conf(\mathcal{M}_i)$.

Connecting domain symbols to the computations of the relation. We apply the construction above to each R_i in the signature of $\mathcal{C}$. Taking the union of the resulting automatic graphs and adding vertices for the domain, we have the structure $(\Sigma^\star \cup \cup_i D_i, E_1, \ldots, E_n)$ (where we assume that the D_i are disjoint). Assume that each $\mathcal{M}_i$ has a different initial state, and denote it by ι_i. We add n predicates F_i to the signature of the automatic structure connecting the elements of the domain of $\mathcal{C}$ with the computations of the relations R_i:

$$F_i = \{(x_0, \ldots, x_{m_i-1}, (\lambda\ \iota_i\ (x_0, \ldots, x_{m_i-1}), \lambda, \lambda)) \mid x_0, \ldots, x_{m_i-1} \in \Sigma^\star\}.$$

Note that for $\bar{x} \in \Sigma^\star$, $R_i(\bar{x})$ if and only if $F_i(\bar{x}, (\lambda\ \iota_i\ \bar{x}, \lambda, \lambda))$ holds and all E_i chains emanating from $(\lambda\ \iota_i\ \bar{x}, \lambda, \lambda)$ are finite. We have built the automatic structure

$$\mathcal{A} = (\Sigma^\star \cup \cup_i D_i, E_1, \ldots, E_n, F_1, \ldots, F_n).$$

Two technical lemmas are used to show that the Scott rank of $\mathcal{A}$ is close to α:

Lemma 9. *For $\bar{x}, \bar{y}$ in the domain of $\mathcal{C}$ and for ordinal α, if $\bar{x} \equiv^{\alpha}_{\mathcal{C}} \bar{y}$ then $\bar{x} \equiv^{\alpha}_{\mathcal{A}} \bar{y}$.*

Lemma 10. *If $\bar{x} \in \Sigma^\star \cup \cup_i D_i$, there is $\bar{y} \in \Sigma^\star$ with $\mathcal{SR}_{\mathcal{A}}(\bar{x}\bar{x}'\bar{u}) \leq 2 + \mathcal{SR}_{\mathcal{C}}(\bar{y})$.*

Putting these together, we conclude that $\mathcal{SR}(\mathcal{C}) \leq \mathcal{SR}(\mathcal{A}) \leq 2 + \mathcal{SR}(\mathcal{C})$. Applying the above construction to the computable structures of Scott rank ω_1^{CK} and $\omega_1^{CK} + 1$ built by Harrison [12] and Knight and Millar [18], we get automatic structures of Scott rank $\omega_1^{CK}, \omega_1^{CK} + 1$. We also apply the construction to [10], where it is proved that there are computable structures with Scott ranks above each computable ordinal. In this case, we get the following theorem.

Theorem 3. *For each infinite computable ordinal α, there is an automatic structure of Scott rank at least α.*

6 Cantor-Bendixson Rank of Automatic Successor Trees

In this section we show that there are automatic successor trees of high Cantor-Bendixson (CB) rank. Recall the definitions of partial order trees and successor trees from Section 1. Note that if $(T, \leq)$ is an automatic partial order tree then the successor tree (T, S), where the relation S is defined by $S(x,y) \iff (x < y)\ \&\ \neg\exists z(x < z < y)$, is automatic.

Definition 5. *The* **derivative** *of a tree T, $d(T)$, is the subtree of T whose domain is $\{x \in T :\ x$ lies on at least two infinite paths in $T\}$. By induction, $d^0(T) = T$, $d^{\alpha+1}(T) = d(d^{\alpha}(T))$, and for γ a limit ordinal, $d^{\gamma}(T) = \cap_{\beta<\gamma} d^{\beta}(T)$. The* **CB rank** *of the tree, $CB(T)$, is the least α such that $d^{\alpha}(T) = d^{\alpha+1}(T)$.*

The CB ranks of automatic partial order trees are finite [16]. This is not true of automatic successor trees. The main theorem of this section provides a general technique for building trees of given CB ranks. It uses the fact that for each $\alpha < \omega_1^{CK}$ there is a computable successor tree of CB rank α. This fact can be proven by recursively coding up computable trees of increasing CB rank.

Theorem 4. *For $\alpha < \omega_1^{CK}$ there is an automatic successor tree of CB rank α.*

Proof. Suppose we are given $\alpha < \omega_1^{CK}$. Take a computable tree R_α of CB rank α. We use the same construction as in the case of well-founded relations (see the proof of Theorem 2). The result is a stretched out version of the tree R_α, where between each two elements of the original tree we have a coding of their computation. In addition, extending from each $x \in \Sigma^\star$ we have infinitely many finite computation chains. Those chains which correspond to output "no" are not connected to any other part of the automatic structure. Finally, there is a disjoint part of the structure consisting of chains whose bases are not valid initial configurations. By the reversibility assumption, each unproductive component of the configuration space is isomorphic either to a finite chain or to an ω-chain. Moreover, the set of invalid initial configurations which are the base of such an unproductive chain is regular. We connect all such bases of unproductive chains to the root and get an automatic successor tree, T_α.

We now consider the CB rank of T_α. Note that the first derivative removes all the subtrees whose roots are at distance 1 from the root and are invalid initial computations. This occurs because each of the invalid computation chains has no branching and is not connected to any other element of the tree. Next, if we consider the subtree of T_α rooted at some $x \in \Sigma^\star$, we see that all the paths which correspond to computations whose output is "no" vanish after the first derivative. Moreover, $x \in d(T_\alpha)$ if and only if $x \in d(R_\alpha)$ because the construction did not add any new infinite paths. Therefore, after one derivative, the structure is exactly a stretched out version of $d(R_\alpha)$. Likewise, for all $\beta < \alpha$, $d^\beta(T_\alpha)$ is a stretched out version of $d^\beta(R_\alpha)$. Hence, $CB(T_\alpha) = CB(R_\alpha) = \alpha$. □

Acknowledgement

We thank Moshe Vardi who posed the question about ranks of automatic well-founded relations. We also thank Anil Nerode and Frank Stephan with whom we discussed Scott and Cantor-Bendixson ranks of automatic structures.

References

1. Benedikt, M., Libkin, L.: Tree extension algebras: Logics, automata, and query languages. In: Proceedings of LICS 2002, pp. 203–212 (2002)
2. Bennett, C.H.: Logical Reversibility of Computation. IBM Journal of Research and Development, 525–532 (1973)
3. Blaas, A., Gurevich, Y.: Program Termination and Well Partial Orderings. ACM Transactions on Computational Logic, 1–25 (2006)

4. Büchi, J.R.: Weak second-order arithmetic and finite automata. Zeitschrift Math. Logik und Grundlagen det Mathematik, 66–92 (1960)
5. Büchi, J.R.: On a decision method in restricted second-order arithmetic. In: Nagel, E., Suppes, P., Tarski, A. (eds.) Proc. International Congress on Logic, Methodology and Philosophy of Science, 1960, pp. 1–12. Stanford University Press (1962)
6. Calvert, W., Goncharov, S.S., Knight, J.F.: Computable structures of Scott rank ω_1^{CK} in familiar classes. In: Advances in Logic (Proceedings of the North Texas Logic Conference). Contemporary Mathematics, vol. 425, pp. 49–66. American Mathematical Society (2007)
7. Delhommé, C.: Automaticité des ordinaux et des graphes homogènes. C.R. Acadèmie des sciences Paris, Ser. I 339, 5–10 (2004)
8. Eilenberg, S.: Automata, Languages, and Machines (Vol. A). Academic Press, New York (1974)
9. Epstein, D.B.A., et al.: Word Processing in Groups, A.K. Peters Ltd. (Natick, Massachusetts) (1992)
10. Goncharov, S.S., Knight, J.F.: Computable structure and non-structure theorems. Algebra and Logic 41, 351–373 (2002)
11. Harizanov, V.S.: Pure Computable Model Theory. In: Yu. Ershov, S., Goncharov, A., Nerode, J. (eds.) Handbook of Recursive Mathematics, pp. 3–114. North-Holland, Amsterdam (1998)
12. Harrison, J.: Recursive Pseudo Well-Orderings. Transactions of the American Mathematical Society 131(2), 526–543 (1968)
13. Hodgson, B.R.: On Direct Products of Automaton Decidable Theories. Theoretical Computer Science 19, 331–335 (1982)
14. Khoussainov, B., Nerode, A.: Automatic presentations of structures. In: Leivant, D. (ed.) LCC 1994. LNCS, vol. 960, pp. 367–392. Springer, Heidelberg (1995)
15. Khoussainov, B., Shore, R.A.: Effective Model Theory: The Number of Models and Their Complexity. In: Cooper, S.B., Truss, J.K. (eds.) Models and Computability, Invited Papers from LC 1997. LMSLNS, vol. 259, pp. 193–240. Cambridge University Press, Cambridge, England (1999)
16. Khoussainov, B., Rubin, S., Stephan, F.: On automatic partial orders. In: Proceedings of 18th LICS, pp. 168–177 (2003)
17. Khoussainov, B., Rubin, S., Stephan, F.: Automatic linear orders and trees. ACM Transactions on Computational Logic 6(4), 675–700 (2005)
18. Knight, J.F., Millar, J.: Computable Structures of Rank ω_1^{CK}. Journal of Mathematical Logic; posted on arXiv 25 (August 2005) (submitted)
19. Lohrey, M.: Automatic structures of bounded degree. In: Y. Vardi, M., Voronkov, A. (eds.) LPAR 2003. LNCS (LNAI), vol. 2850, pp. 344–358. Springer, Heidelberg (2003)
20. Rabin, M.O.: Decidability of Second-Order Theories and Automata on Infinite Trees. Transactions of the American Mathematical Society 141, 1–35 (1969)
21. Rubin, S.: Automatic Structures, PhD Thesis, University of Auckland (2004)
22. Scott, D.: Logic with Denumerably Long Formulas and Finite Strings of Quantifiers. In: Addison, J., Henkin, L., Tarski, A. (eds.) The Theory of Models, pp. 329–341. North-Holland, Amsterdam (1965)
23. Vardi, M.Y., Wolper, P.: Automata-Theoretic Techniques for Modal Logics of Programs. In: Proceedings of 16th STOC, pp. 446–456 (1984)
24. Vardi, M.Y.: An Automata-Theoretic Approach to Linear Temporal Logic. In: Moller, F., Birtwistle, G. (eds.) Logics for Concurrency. LNCS, vol. 1043, pp. 238–266. Springer, Heidelberg (1996)
25. Vardi, M.Y.: Model Checking for Database Theoreticians. In: Proceedings of ICDL, vol. 5 (2005)

A Separation between Divergence and Holevo Information for Ensembles

Rahul Jain[1,*], Ashwin Nayak[2,**], and Yi Su[3,***]

[1] University of Waterloo
rjain@cs.uwaterloo.ca
[2] University of Waterloo & Perimeter Institute
anayak@math.uwaterloo.ca
[3] University of Waterloo
y6su@student.math.uwaterloo.ca

Abstract. The notion of *divergence information* of an ensemble of probability distributions was introduced by Jain, Radhakrishnan, and Sen [1,2] in the context of the "substate theorem". Since then, divergence has been recognized as a more natural measure of information in several situations in quantum and classical communication.

We construct ensembles of probability distributions for which divergence information may be significantly smaller than the more standard Holevo information. As a result, we establish that lower bounds previously shown for Holevo information are weaker than similar ones shown for divergence information.

1 Introduction

In this article, we study the relationship between two different measures of information contained in an ensemble of probability distributions. The first measure, *Holevo information*, is a standard notion from information theory, and is equivalent to the notion of *mutual information* between two random variables. Consider jointly distributed random variables XY, with X taking values in a sample space $\mathcal{X}$. Consider the ensemble of distributions $\mathcal{E} = \{(\lambda_i, Y_i) \; : \; i \in \mathcal{X}\}$,

* School of Computer Science, and Institute for Quantum Computing, University of Waterloo, 200 University Ave. W., Waterloo, ON N2L 3G1, Canada. Research supported in part by ARO/NSA USA.

** Department of Combinatorics and Optimization, and Institute for Quantum Computing, University of Waterloo, 200 University Ave. W., Waterloo, ON N2L 3G1, Canada. Research supported in part by NSERC Canada, CIFAR, MITACS, QuantumWorks, and an ERA from the Province of Ontario. A.N. is also Associate Member, Perimeter Institute for Theoretical Physics, Waterloo, Canada. Research at Perimeter Institute for Theoretical Physics is supported in part by the Government of Canada through NSERC and by the Province of Ontario through MRI.

*** Department of Pure Mathematics, University of Waterloo, 200 University Ave. W., Waterloo, ON N2L 3G1, Canada. Research supported in part by an NSERC Canada Undergraduate Research Award.

M. Agrawal et al. (Eds.): TAMC 2008, LNCS 4978, pp. 526–541, 2008.

where $\lambda_i = \Pr(X = i)$, and $Y_i = Y|(X = i)$, obtained by conditioning on values assumed by X. The Holevo information of the ensemble is given by $\chi(\mathcal{E}) = \mathrm{I}(X : Y) = \mathbb{E}_{i \sim X} \mathrm{S}(Y_i \| Y)$, where $\mathrm{S}(\cdot \| \cdot)$ measures the relative entropy of a random variable (equivalently, distribution) with respect to another. This notion may be extended to ensembles of quantum states (see, e.g., the text [3]), and the term 'Holevo information' is derived from the literature in quantum information theory.

The second measure, *divergence information*, was introduced by Jain, Radhakrishnan, and Sen [1,2]. It arises in the study of relative entropy, and its connection with a "substate property". The *observational divergence* (or simply *divergence*) of two classical distributions P, Q on the same finite sample space is $\max_E P(E) \log_2(P(E)/Q(E))$, where E ranges over all events. We may view this as a (scaled) measure of the factor by which P may exceed Q for an event of interest. The notion of *divergence information* is derived from this as $\mathrm{D}(\mathcal{E}) = \mathbb{E}_{i \sim X} \mathrm{D}(Y_i \| Y)$, in analogy with Holevo information. A quantum generalisation of this measure may also be defined [2].

Relative entropy and Holevo (or mutual) information have been studied extensively in communication theory and beyond (see, e.g, [4]) as they arise in a variety of applications. Since the discovery of the substate theorem [1], divergence is being recognized as a more natural measure of information in a growing number of applications [2, Section 1]. The applications include privacy trade-offs in communicatioin protocols for computing relations [5] and *bit-string commitment* [6], and the communication complexity of *remote state preparation* [7]. In particular, divergence captures, up to a constant factor, the substate property for probability distributions. It thus becomes relevant in every application where the substate theorem is used.

We construct ensembles of probability distributions (equivalently, jointly distributed random variables) for which the Holevo and divergence information are quantitatively different.

Theorem 1. *For every positive integer N, and real number $k \geq 1$ such that $N > 2^{36k^2}$, there is an ensemble $\mathcal{E}$ of distributions over a sample space of size N such that $\mathrm{D}(\mathcal{E}) = k$ and $\chi(\mathcal{E}) = \Theta(k \log \log N)$.*

A more precise statement of this theorem (Theorem 4) and related results may be found in Section 3.

The ensembles we construct satisfy the property that the ensemble average (i.e., the distribution of the random variable Y in the description above) is uniform. We show that the above separation is essentially the best possible whenever the ensemble average is uniform (Theorem 6). The result also applies to ensembles of quantum states, where the ensemble average is the completely mixed state (Theorem 7). We leave open the possibility of larger separations for classical or quantum ensembles with non-uniform averages.

The difference between the two measures demonstrated by Theorem 1 shows that in certain applications, divergence is quantitatively a more relevant measure of information. In Appendix A, we describe two applications where functionally similar lower bounds have been established in terms of both measures. This article shows that the lower bounds in terms of divergence information are, in fact, stronger.

In prior work on the subject, Jain *et al.* [2, Appendix A] compare relative entropy and divergence for classical as well as quantum states. For pairs of distributions P, Q over a sample space of size N, they show that $\mathrm{D}(P\|Q) \leq \mathrm{S}(P\|Q)+1$, and $\mathrm{S}(P\|Q) \leq \mathrm{D}(P\|Q)\cdot(N-1)$. This extends to the corresponding measures of information in an ensemble: $\mathrm{D}(\mathcal{E}) \leq \chi(\mathcal{E})+1$ and $\chi(\mathcal{E}) \leq \mathrm{D}(\mathcal{E})\cdot(N-1)$. They show qualitatively similar relations for ensembles of quantum states. In addition, they construct a pair of distributions P, Q such that $\mathrm{S}(P\|Q) = \Omega(\mathrm{D}(P\|Q)\cdot N)$. However, they do not translate their construction to a similar separation for *ensembles* of probability distributions. Our work fills this gap for ensembles (of classical or quantum states) with a uniform average.

2 Preliminaries

Here, we summarise our notation and the information-theoretic concepts we encounter in this work. We refer the reader to the text by Cover and Thomas [4] for a deeper treatment of (classical) information theory. While the bulk of this article pertains to classical information theory, as mentioned in Section 1, it is motivated by studies in (and has implications for) quantum information. We refer the reader to the text [3] for an introduction to quantum information.

For a positive integer N, let $[N]$ represent the set $\{1,\ldots,N\}$. We view probability distributions over $[N]$ as vectors in $\mathbb{R}^N$. The probability assigned by distribution P to a sample point $i \in [N]$ is denoted by p_i (i.e., with the same letter in small case). We denote by $P^{\downarrow}$ the distribution obtained from P by composing it with a permutation π on $[N]$ so that $p_i^{\downarrow} = p_{\pi(i)}$ and $p_1^{\downarrow} \geq p_2^{\downarrow} \geq \cdots \geq p_N^{\downarrow}$. For an event $E \subseteq [N]$, let $P(E) = \sum_{i\in E} p_i$ denote the probability of that event. We denote the uniform distribution over $[N]$ by U_N. The expected value of a function $f : [N] \to \mathbb{R}$ with respect to the distribution P over $[N]$ is abbreviated as $\mathbb{E}_P f$.

We appeal to the *majorisation* relation for some of our arguments. The relation tells us which of two given distributions is "more random".

Definition 1 (Majorisation). *Let P, Q be distributions over $[N]$. We say that* P majorises Q, *denoted as $P \succeq Q$, if*

$$\sum_{j=1}^{i} p_j^{\downarrow} \geq \sum_{j=1}^{i} q_j^{\downarrow},$$

for all $i \in [N]$.

The following is straightforward.

Fact 2. *Any probability distribution P on $[N]$ majorises U_N, the uniform distribution over $[N]$.*

Throughout this article, we use 'log' to denote the logarithm with base 2, and 'ln' to denote the logarithm with base e.

Definition 2 (Entropy, relative entropy). *Let P, Q be probability distributions on $[N]$. The* entropy *of P is defined as* $\mathrm{H}(P) \stackrel{\text{def}}{=} -\sum_{i=1}^{N} p_i \log p_i$. *The* relative entropy *between P, Q, denoted $\mathrm{S}(P\|Q)$, is defined as*

$$\mathrm{S}(P\|Q) \quad \stackrel{\text{def}}{=} \quad \sum_{i=1}^{N} p_i \log \frac{p_i}{q_i} \ .$$

Note that the relative entropy with respect to the uniform distribution is connected to entropy as $\mathrm{S}(P\|\mathrm{U}_N) = \log N - \mathrm{H}(P)$.

We can formalise the connection between majorisation and randomness through the following fact.

Fact 3. *If P, Q are distributions over $[N]$ such that P majorises Q, i.e. $P \succeq Q$, then $\mathrm{H}(P) \leq \mathrm{H}(Q)$.*

The notion of *observational divergence* was defined by Jain, Radhakrishnan, and Sen [1] in the context of the "substate theorem".

Definition 3 (Observational divergence). *Let P, Q be probability distributions on $[N]$. Then the* observational divergence *between them, denoted $\mathrm{D}(P\|Q)$, is defined as*

$$\mathrm{D}(P\|Q) \quad \stackrel{\text{def}}{=} \quad \max_{f:[N]\to[0,1]} \ (\mathbb{E}_P f) \ \log \frac{\mathbb{E}_P f}{\mathbb{E}_Q f} \ .$$

Note that we allow the quantity to take the value $+\infty$. Throughout the paper we refer to 'observational divergence' as simply 'divergence'.

Divergence $\mathrm{D}(P\|Q)$ is always non-negative, and it is finite precisely when the support of P is contained in the support of Q [1]. Due to convexity, the divergence between two distributions is attained by the characteristic function of an event.

Lemma 1. $\mathrm{D}(P\|Q) \quad = \quad \max_{E\subseteq[N]} \ P(E) \log \frac{P(E)}{Q(E)}$.

Proof. Let $\mathcal{F}$ denote the (convex) set of functions from $[N]$ to $[0,1]$. The extreme points of $\mathcal{F}$ are precisely the characteristic functions of events in $[N]$. For an extreme point, say the characteristic function f_E of the event $E \subseteq [N]$, we have $\mathbb{E}_P f_E = P(E)$.

If the divergence is $+\infty$, then there is an event for which the right hand side also takes the value $+\infty$. So assume that the divergence is finite. In this case, the right hand side also is finite, as the support of P is contained in the support of Q. By restricting $f : [N] \to [0,1]$ to characteristic functions of events, we see that $\mathrm{D}(P\|Q)$ is at least the expression on the right hand side above.

For the inequality in the other direction, we note that the function

$$g(x) \quad = \quad (ax+b) \log \left(\frac{ax+b}{cx+d} \right)$$

defined on $[0,1]$ is convex in x, for any $a, b, c, d \in \mathbb{R}$ such that $ax + b \geq 0$ and $cx + d > 0$ when $x \in [0,1]$. Therefore, the function $g(x)$ attains its maximum at either $x = 0$ or at $x = 1$.

The convexity of $g(x)$ implies that for any $\alpha \in [0,1]$, and functions $f, f' \in \mathcal{F}$, we have

$$\begin{aligned}
&(\mathbb{E}_P(\alpha f + (1-\alpha) f'))\ \log \frac{\mathbb{E}_P(\alpha f + (1-\alpha) f')}{\mathbb{E}_Q(\alpha f + (1-\alpha) f')} \\
&= (\alpha(\mathbb{E}_P f - \mathbb{E}_P f') + \mathbb{E}_P f')\ \log \frac{\alpha(\mathbb{E}_P f - \mathbb{E}_P f') + \mathbb{E}_P f'}{\alpha(\mathbb{E}_Q f - \mathbb{E}_Q f') + \mathbb{E}_Q f'} \\
&\leq \max\left\{ (\mathbb{E}_P f) \log \frac{\mathbb{E}_P f}{\mathbb{E}_Q f}, \quad (\mathbb{E}_P f') \log \frac{\mathbb{E}_P f'}{\mathbb{E}_Q f'} \right\}.
\end{aligned}$$

Thus, the divergence is attained at an extreme point of $\mathcal{F}$. This proves the claim.

Henceforth, we only use the equivalent definition of divergence given by this lemma.

The divergence of any distribution with respect to the uniform distribution is bounded.

Lemma 2. *For any probability distribution P on $[N]$, we have $0 \leq \mathrm{D}(P \| \mathrm{U}_N) \leq \log N$.*

Proof. Consider the event E which achieves the divergence between P and U_N. W.l.o.g., the event E is non-empty. Therefore $P(E) \geq \mathrm{U}_N(E) \geq 1/N$, and

$$0 \quad \leq \quad \mathrm{D}(P \| U_N) \quad \leq \quad P(E) \log P(E) N \quad \leq \quad \log N.$$

We observe that we need only maximise over N events to calculate divergence with respect to the uniform distribution.

Lemma 3. *For any probability distribution P on $[N]$ such that $P^{\downarrow} = P$, i.e., $p_1 \geq p_2 \geq \cdots \geq p_N$, we have*

$$\mathrm{D}(P \| \mathrm{U}_N) \quad = \quad \max_{i \in [N]} P([i]) \log \frac{N \cdot P([i])}{i} .$$

Proof. By definition of observational divergence, the RHS above is bounded by $\mathrm{D}(P \| \mathrm{U}_N)$. For the inequality in the other direction, we note that the probability $P(E)$ of any event E with size $n_E = |E|$ is bounded by $P([n_E])$, the probability of the first n_E elements in $[N]$. We thus have

$$\begin{aligned}
\mathrm{D}(P \| Q) &= \max_{E \subseteq [N]} P(E) \log \frac{N \cdot P(E)}{n_E} \\
&\leq \max_{E \subseteq [N]} P(E) \log \frac{N \cdot P([n_E])}{n_E} \\
&\leq \max_{E \subseteq [N]} P([n_E]) \log \frac{N \cdot P([n_E])}{n_E},
\end{aligned}$$

since P majorises U_N (Fact 2) and $P([n_E]) \geq \frac{n_E}{N}$. This is equivalent to the RHS in the statement of the lemma.

Definition 4 (Ensemble). *An* ensemble *is a sequence of pairs* $\{(\lambda_j, Q_j): j \in [M]\}$, *for some integer* M, *where* $\Lambda = (\lambda_j) \in \mathbb{R}^M$ *is a probability distribution on* $[M]$ *and* Q_j *are probability distributions over the same sample space.*

Definition 5 (Holevo information). *The* Holevo information *of an ensemble* $\mathcal{E} = \{(\lambda_j, Q_j) : j \in [M]\}$, *denoted as* $\chi(\mathcal{E})$, *is defined as*

$$\chi(\mathcal{E}) \quad \stackrel{\text{def}}{=} \quad \sum_{j=1}^{M} \lambda_j \, \mathrm{S}(Q_j \| Q),$$

where $Q = \sum_{j=1}^{M} \lambda_j Q_j$ *is the* ensemble average.

Definition 6 (Divergence information). *The* divergence information *of an ensemble* $\mathcal{E} = \{(\lambda_j, Q_j) : j \in [M]\}$, *denoted as* $\mathrm{D}(\mathcal{E})$ *is defined as*

$$\mathrm{D}(\mathcal{E}) \quad \stackrel{\text{def}}{=} \quad \sum_{j=1}^{M} \lambda_j \, \mathrm{D}(Q_j \| Q),$$

where $Q = \sum_{j=1}^{M} \lambda_j Q_j$ *is the* ensemble average.

3 Divergence Versus Relative Entropy

In this section, we describe the construction of an ensemble for which there is a large separation between divergence and Holevo information. The ensemble has the property that the ensemble average is uniform. As a by-product of our construction, we also obtain a bound on the maximum possible separation for ensembles with a uniform average.

We begin with the construction of the ensemble. Let $f_{\mathrm{L}}(k, N) = k(\ln \log(kN) - \ln(6k) + 1) - \log(1 + k \ln 2) - 1 - \frac{1}{\ln 2}$ on point in the positive orthant in $\mathbb{R}^2$ with $Nk > 1$.

Theorem 4. *For every integer* $N > 1$, *and every positive real number* $\frac{16}{N} \leq k < \log N$, *there is an ensemble* $\mathcal{E} = \left\{(\frac{1}{N}, Q_i) : i \in [N]\right\}$ *with* $\frac{1}{N} \sum_i Q_i = \mathrm{U}_N$, *the uniform distribution over* $[N]$, *with* $\mathrm{D}(\mathcal{E}) \leq k$, *and*

$$\chi(\mathcal{E}) \quad \geq \quad f_{\mathrm{L}}(k, N).$$

To construct the ensemble described in the theorem above, we first construct a probability distribution P on $[N]$ with observational divergence $\mathrm{D}(P \| \mathrm{U}_N) \leq k$ such that its relative entropy $\mathrm{S}(P \| \mathrm{U}_N)$ is large as compared with k. Let $f_{\mathrm{U}} = k(\ln \log(Nk) - \ln k + 1)$ be defined on points in the positive orthant of $\mathbb{R}^2$ with $kN > 1$.

Theorem 5. *For every integer* $N > 1$, *and every positive real number* $\frac{16}{N} \leq k < \log N$, *there is a probability distribution* P *with* $\mathrm{D}(P \| \mathrm{U}_N) = k$, *and*

$$f_{\mathrm{L}}(k,N) \quad \leq \quad \mathrm{S}(P\|\mathrm{U}_N) \quad \leq \quad f_{\mathrm{U}}(k,N).$$

The construction of the ensemble is now immediate.

Proof of Theorem 4: Let $Q_j = P \circ \pi_j$, where π_j is the cyclic permutation of $[N]$ by $j-1$ places. We endow the set of the N cyclic permutations $\{Q_j : j \in [N]\}$ of P with the uniform distribution. By construction, the ensemble average is U_N. Since both observational divergence and relative entropy with respect to the uniform distribution are invariant under permutations of the sample space, $\mathrm{D}(\mathcal{E}) = \mathrm{D}(P\|\mathrm{U}_N) \leq k$, and $\chi(\mathcal{E}) = \mathrm{S}(P\|\mathrm{U}_N) \geq f_{\mathrm{L}}(k,N)$. □

We turn to the construction of the distribution P. Our construction is such that $P^{\downarrow} = P$, i.e., $p_1 \geq p_2 \geq \cdots \geq p_N$. Lemma 3 tells us that we need only ensure that

$$P([i]) \log \frac{N \cdot P([i])}{i} \leq k, \quad \forall\, i \in [N], \tag{1}$$

to ensure $\mathrm{D}(P\|Q) \leq k$. Since $\mathrm{S}(P\|\mathrm{U}_N) = \log N - \mathrm{H}(P)$, we wish to minimise the entropy of P subject to the constraints in Eq. (1). This is equivalent to successively maximising $p_1, p_2, \ldots$, and motivates the following definitions.

Define the function $g(y,x) = y \log(Ny/x) - k$ on the positive orthant of $\mathbb{R}^2$. Consider the function $h : \mathbb{R}^+ \to \mathbb{R}^+$ implicitly defined by the equation $g(h(x), x) = 0$.

Lemma 4. *The function $h : \mathbb{R}^+ \to \mathbb{R}^+$ is well-defined, strictly increasing, and concave.*

Proof. Fix an $x \in \mathbb{R}^+$, and consider the function $g_x(y) = g(y,x)$. This function is continuous on $\mathbb{R}^+$, tends to $-k < 0$ as $y \to 0^+$, and tends to ∞ as $y \to \infty$. By Intermediate Value Theorem, for some $y > 0$, we have $g_x(y) = 0$. Moreover, $g_x(y) < -k$ for $0 < y \leq x/N$, and is strictly increasing for $y > x/N\mathrm{e}$ (its derivative is $g'_x(y) = \log \frac{\mathrm{e}Ny}{x}$). Therefore there is a unique y such that $g_x(y) = 0$ and $h(x)$ is well-defined.

The function h satisfies the equation $h \log \frac{Nh}{x} = k$, and therefore the identity

$$x \quad = \quad Nh \exp\left(-\tfrac{k \ln 2}{h}\right).$$

Differentiating with respect to h, we see that

$$\frac{dx}{dh} = N\left(1 + \frac{k \ln 2}{h}\right) \exp\left(-\tfrac{k \ln 2}{h}\right), \text{ and}$$

$$\frac{d^2x}{dh^2} = \frac{N(k \ln 2)^2}{h^3} \exp\left(-\tfrac{k \ln 2}{h}\right).$$

So $\frac{dh}{dx} > 0$ for all $x > 0$, and h is a strictly increasing function. Note also that $\frac{d^2x}{dh^2} > 0$ for all $h > 0$, so x is a convex function of h. Since h is an increasing function, convexity of $x(h)$ implies concavity of $h(x)$.

Let $v_0 = 0$. For $i \in [N]$, let $v_i = h(i)$, i.e., $v_i \log \frac{Nv_i}{i} = k$. Let $s_i \stackrel{\text{def}}{=} \min\{1, v_i\}$, for $i \in [N]$. Let $p_1 = s_1$, and $p_i = s_i - s_{i-1}$ for all $2 \le i \le N$. Lemma 4 guarantees that these numbers are well-defined. We claim that

Lemma 5. *The vector $P = (p_i) \in \mathbb{R}^N$ defined above is a probability distribution, and $P^{\downarrow} = P$, i.e., $p_1 \ge p_2 \ge \cdots \ge p_N$.*

Proof. By definition, we have $v_i > 0$ for all $i \in [N]$. Therefore $s_1 = \min\{1, v_1\} > 0$. Since $h(x)$ is an increasing function in x, the sequence (v_i) is also increasing, so (s_i) is non-decreasing. Therefore $p_i = s_i - s_{i-1} \ge 0$ for $i > 1$.

Now $v_N \log v_N = k > 0$. Since $x \log x \le 0$ for $x \in (0, 1]$, we have $v_N > 1$. So $s_N = \min\{1, v_N\} = 1$. Therefore $\sum_{i=1}^{N} p_i = s_N = 1$. So P is a probability distribution on $[N]$.

Note that $(v_2/2)\log(Nv_2/2) = k/2 < k$, so $v_1 > v_2/2$. So $s_1 \ge s_2/2$, i.e., $p_1 \ge p_2$. For $i \ge 2$, we have $p_i - p_{i+1} = (s_i - s_{i-1}) - (s_{i+1} - s_i) = 2s_i - s_{i-1} - s_{i+1}$. Since $h(x)$ is concave, so is the function $\min\{1, h(x)\}$. Therefore, $s_i \ge (s_{i-1} + s_{i+1})/2$, and the sequence (p_i) is non-decreasing.

The vector $S = (s_i) \in \mathbb{R}^N$ thus represents the (cumulative) distribution function corresponding to P.

Proof of Theorem 5: We claim that the probability distribution P constructed above satisfies the properties stated in the theorem.

Since $P^{\downarrow} = P$, by Lemma 3, we need only verify that $s_i \log(Ns_i/i) \le k$ for $i \in [N]$. If $s_i = v_i$, then the condition is satisfied with equality. (Note that since $k < \log N$, we have $s_1 = v_1 < 1$.) Else, $s_i = 1 < v_i$, so $s_i \log(Ns_i/i) < v_i \log(Nv_i/i) = k$.

We now bound the relative entropy $\mathrm{S}(P\|\mathrm{U}_N)$ from below. Let n be the smallest positive integer such that $v_{n-1} \le 1$ and $v_n > 1$. Note that $n > 1$. We also have $n \le N$, since $v_N > 1$ (as $v_N \log v_N = k > 0$). Therefore, we have $s_i = v_i$ (equivalently, $Ns_i = i2^{k/s_i}$) for $i \in [n-1]$, and $s_n = 1 < v_n$. Thus, for $1 < i < n$,

$$
\begin{aligned}
Np_i &= i2^{\frac{k}{s_i}} - (i-1)2^{\frac{k}{s_{i-1}}} \\
&= 2^{\frac{k}{s_i}} + (i-1)(2^{\frac{k}{s_i}} - 2^{\frac{k}{s_{i-1}}}) \\
&= 2^{\frac{k}{s_i}} + (i-1)2^{\frac{k}{s_{i-1}}}(2^{\frac{k}{s_i} - \frac{k}{s_{i-1}}} - 1) \\
&= 2^{\frac{k}{s_i}} + Ns_{i-1}(2^{\frac{k}{s_i} - \frac{k}{s_{i-1}}} - 1) \\
&\ge 2^{\frac{k}{s_i}} + Ns_{i-1}\left(\frac{k}{s_i} - \frac{k}{s_{i-1}}\right)\ln 2 \\
&= 2^{\frac{k}{s_i}} - \frac{Np_i k}{s_i}\ln 2.
\end{aligned}
$$

The penultimate line follows from the inequality $2^x \ge 1 + x\ln 2$ for all $x \in \mathbb{R}$. Thus we have

$$
Np_i \quad \ge \quad \frac{2^{\frac{k}{s_i}}}{1 + \frac{k}{s_i}\ln 2}. \tag{2}
$$

Since $Np_1 = Ns_1 = 2^{\frac{k}{s_1}}$, this also holds for $i = 1$.

We bound the relative entropy using Eq. (2).

$$\begin{aligned} \mathrm{S}(P\|\mathrm{U}_N) = \sum_{i=1}^{N} p_i \log Np_i &= \sum_{i=1}^{n} p_i \log Np_i \\ &\geq \sum_{i=1}^{n-1} p_i \log \frac{2^{\frac{k}{s_i}}}{1 + \frac{k}{s_i}\ln 2} + p_n \log Np_n \\ &\geq \sum_{i=1}^{n-1} \frac{p_i k}{s_i} - \sum_{i=1}^{n-1} p_i \log\left(1 + \frac{k \ln 2}{s_i}\right) + p_n \log Np_n. \end{aligned} \quad (3)$$

We bound each of the three terms in the RHS of Eq. (3) separately.

We start with $\sum_{i=1}^{n-1} \frac{p_i k}{s_i}$. Let $p = p_1$, and let $m = \left\lfloor \frac{1}{p} \right\rfloor$. For every $j \in [m]$, there is an $i \in [n]$, say $i = i_j$, such that $jp \leq s_{i_j} \leq (j+1)p$. (Otherwise, for some $i > 1$, the probability $p_i = s_i - s_{i-1}$ is strictly larger than p, an impossibility.)

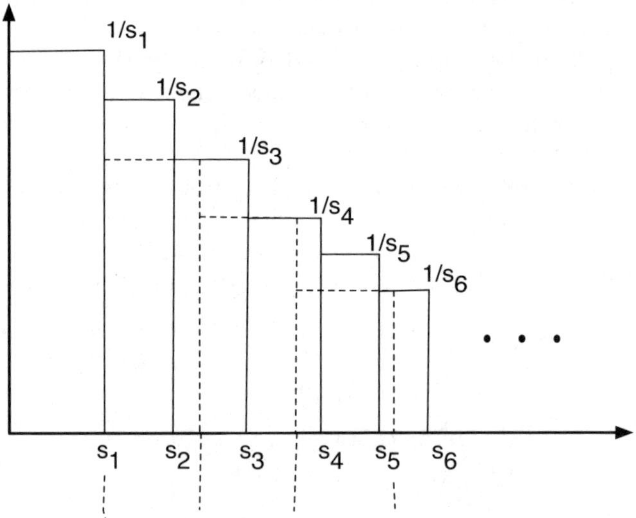

We interpret the sum $\sum_{i=2}^{n-1} \frac{p_i}{s_i} = \sum_{i=2}^{n-1} \frac{s_i - s_{i-1}}{s_i}$ as a Riemann sum approximating the area under the curve $1/x$ between s_1 and s_{n-1} with the area under the solid lines in Figure 3. This area is bounded from below by the area under the dashed lines, which corresponds to the area of rectangles of uniform width p and height $1/s_{j+1}$ for the jth interval. Thus,

$$\begin{aligned}
\sum_{i=1}^{n-1} \frac{p_i k}{s_i} &\geq k + k \sum_{j=1}^{m} p \cdot \frac{1}{s_{i_{j+1}}} \\
&\geq k + k \sum_{j=1}^{m} p \cdot \frac{1}{(j+2)p} \\
&= k + k \sum_{j=1}^{m} \frac{1}{j+2} \\
&\geq k + k \int_{3}^{m+3} \frac{1}{x} dx \\
&= k + k \ln \frac{m+3}{3}. \qquad (4)
\end{aligned}$$

We lower bound $m = \left\lfloor \frac{1}{p} \right\rfloor$ next. Recall that $g_1(y) = y \log(Ny)$ is an increasing function for $y > \frac{1}{eN}$, and $p = p_1 \geq 1/N$. Consider the value of $g_1(y)$ at the point $q = \frac{2k}{\log kN}$:

$$g_1(q) \quad = \quad \frac{2k}{\log kN} \log \frac{2Nk}{\log kN} \quad > \quad 2k\left(1 - \frac{\log \log kN}{\log kN}\right) \quad \geq \quad k,$$

since $kN \geq 16$. As $g_1(q) > g_1(p) > 0$, we have $q > p$. Therefore, $m \geq \frac{1}{p} - 1 \geq \frac{\log kN}{2k} - 1$. Together with Eq. (4), we get

$$\sum_{i=1}^{n-1} \frac{p_i k}{s_i} \geq k(\ln \log kN - \ln 6k + 1). \qquad (5)$$

Next, we derive a lower bound for the second term in Eq. (3).

$$\begin{aligned}
-\sum_{i=1}^{n-1} p_i \log\left(1 + \frac{k \ln 2}{s_i}\right) &= -\sum_{i=1}^{n-1} p_i \log(s_i + k \ln 2) + \sum_{i=1}^{n-1} p_i \log s_i \\
&\geq -\log(1 + k \ln 2) + \sum_{i=1}^{n-1} p_i \log s_i. \qquad (6)
\end{aligned}$$

Viewing the second term above as a Riemann sum, we get

$$\begin{aligned}
\sum_{i=1}^{n-1} p_i \log s_i &\geq \int_{0}^{s_{n-1}} \log x \, dx \\
&\geq \int_{0}^{1} \log x \, dx \\
&= -\frac{1}{\ln 2}. \qquad (7)
\end{aligned}$$

Combining Eq. (6) and (7), we get

$$-\sum_{i=1}^{n-1} p_i \log\left(1+\frac{k\ln 2}{s_i}\right) \geq -\log(1+k\ln 2)-\frac{1}{\ln 2}. \tag{8}$$

We bound the third term in Eq. (3) crudely as $p_n \log N p_n \geq -1$. Along with the bounds for the previous two terms, Eq. (5), (8), this shows that

$$\mathrm{S}(P\|\mathrm{U}_N) \quad \geq \quad f_{\mathrm{L}}(k,N) \quad \stackrel{\text{def}}{=} \quad k(\ln\log kN - \ln 6k + 1) - \log(1+k\ln 2) - 1 - \frac{1}{\ln 2}. \tag{9}$$

This proves the lower bound on the relative entropy.

Moving to an upper bound, we have for $i \geq 2$,

$$\begin{aligned} Np_i &= i2^{\frac{k}{s_i}} - (i-1)2^{\frac{k}{s_{i-1}}} \\ &= 2^{\frac{k}{s_i}} + (i-1)(2^{\frac{k}{s_i}} - 2^{\frac{k}{s_{i-1}}}) \\ &\leq 2^{\frac{k}{s_i}}, \end{aligned}$$

since the second term is negative. This also holds for $i = 1$, since $p_1 = s_1$ and $s_1 \log N s_1 = k$. Therefore,

$$\begin{aligned} \mathrm{S}(P\|\mathrm{U}_N) &= \sum_{i=1}^{n} p_i \log N p_i \\ &\leq \sum_{i=1}^{n} \frac{kp_i}{s_i} \\ &\leq k + k\int_{s_1}^{1} \frac{1}{s} ds \\ &= k - k\ln s_1 \\ &\leq k + k\ln\left(\frac{\log Nk}{k}\right) \\ &= k(1 - \ln k + \ln(\log Nk)). \end{aligned}$$

In the last inequality, we used the lower bound $s_1 \geq k/\log Nk$. □

The upper and lower bounds on the relative entropy of P with respect to the uniform distribution both behave as $k \log\log Nk$ up to constant factors.

Proof of Theorem 1: The dominating term in both of lower bound and upper bound on the relative entropy $\mathrm{S}(P\|\mathrm{U}_N)$, with P as in Theorem 5, is $k\ln\log Nk$ when N is large as compared with k. Specifically, when $N > 2^{36k^2}$, we have

$$\frac{1}{2}k\log\log Nk \quad \leq \quad \mathrm{S}(P\|\mathrm{U}_N) \quad \leq \quad 2k\log\log Nk.$$

By hypothesis, $1 \leq k$ and by Lemma 2, we have $k \leq \log N$. Thus,

$$\mathrm{S}(P\|\mathrm{U}_N) \quad = \quad \Theta(\mathrm{D}(P\|\mathrm{U}_N)\log\log N).$$

The same holds for the ensembles constructed in Theorem 4. □

The separation we demonstrated above is the best possible for ensembles of distributions that have a uniform average distribution.

Theorem 6. *For any positive integer* N*, and any ensemble* $\mathcal{E}=\{(\lambda_j, Q_j) : j \in [M]\}$ *of distributions over* $[N]$ *such that* $\sum_{j=1}^{M} \lambda_j Q_j = \mathrm{U}_N$*, we have*

$$\chi(\mathcal{E}) \leq K(2 \ln \log N - \ln K + 1) + 16,$$

where $K = \mathrm{D}(\mathcal{E})$*.*

Proof. Let $\mathrm{D}(Q_j \| \mathrm{U}_N) = k_j$. We show that $\mathrm{S}(Q_j \| \mathrm{U}_N) \leq k_j(2 \ln \log N - \ln k_j + 1)$ when $k_j \geq \frac{16}{N}$. When $k_j < \frac{16}{N}$, we have $\mathrm{S}(Q_j \| \mathrm{U}_N) < 16$. Since $k(2 \ln \log N - \ln k + 1)$ is a concave function in k, averaging over j with respect to the distribution $\Lambda = (\lambda_j)$ gives the claimed bound.

Fix an j such that $k_j > \frac{16}{N}$. Let $R = Q_j^{\downarrow}$. Note that $\mathrm{D}(R \| \mathrm{U}_N) = k_j$ and $\mathrm{S}(R \| \mathrm{U}_N) = \mathrm{S}(Q_j \| \mathrm{U}_N)$. Consider the distribution P constructed as in Section 3 with $k = k_j$. Using the notation of that section, we have $s_i \log(N s_i / i) = k_j$ for all $i < n$, and $s_n = 1$. Let $t_i = \sum_{l=1}^{i} r_l$, where $r_l \stackrel{\text{def}}{=} \Pr(R = l)$. By definition, we have $t_i \log(N t_i / i) \leq k_j = s_i \log(N s_i / i)$. Since the function $g_i(y) = y \log(N y / i)$ is strictly increasing for $y \geq i / N\mathrm{e}$, and $t_i \geq i/N$ (Fact 2), we have $t_i \leq s_i$ for $i < n$. Since $s_i = 1$ for $i \geq n$, we have $t_i \leq s_i$ for these i as well. In other words, $P \succeq R$. By Fact 3, we have $\mathrm{H}(P) \leq \mathrm{H}(R)$. This is equivalent to $\mathrm{S}(R \| \mathrm{U}_N) \leq \mathrm{S}(P \| \mathrm{U}_N)$. By Theorem 5, $\mathrm{S}(P \| \mathrm{U}_N) \leq k_j(\ln \log(N k_j) - \ln k_j + 1)$. Since $k_j \leq \log N$, this is at most $k_j(2 \ln \log N - \ln k_j + 1)$.

Finally, we observe that this is also the best separation possible for an ensemble of quantum states with a completely mixed ensemble average.

Theorem 7. *For any positive integer* N*, and any ensemble* $\mathcal{E} = \{(\lambda_j, \rho_j) : j \in [M]\}$ *of quantum states* ρ_j *over a Hilbert space of dimension* N *such that* $\sum_{j=1}^{M} \lambda_j \rho_j = \frac{\mathrm{I}}{N}$*, the completely mixed state of dimension* N*, we have*

$$\chi(\mathcal{E}) \leq K(2 \ln \log N - \ln K + 1) + 16,$$

where $K = \mathrm{D}(\mathcal{E})$*.*

Proof. Let Q_j be the probability distribution on $[N]$ corresponding to the eigenvalues of ρ_j. By definition of observational divergence for quantum states, $\mathrm{D}(Q_j \| \mathrm{U}_N) \leq \mathrm{D}(\rho_j \| \frac{\mathrm{I}}{N})$. Further, we have $\mathrm{S}(\rho_j \| \frac{\mathrm{I}}{N}) = \mathrm{S}(Q_j \| \mathrm{U}_N)$. We now apply the same reasoning as in the proof of Theorem 6, note that the divergence of the ensemble $\{(\lambda_j, Q_j) : j \in [M]\}$ is bounded by $\mathrm{D}(\mathcal{E})$, and that the RHS in the statement is a non-decreasing function of K. This gives us the stated bound. (Note that we do not need $\sum_{j=1}^{M} \lambda_j Q_j = \mathrm{U}_N$ to use the reasoning in Theorem 6.)

References

1. Jain, R., Radhakrishnan, J., Sen, P.: Privacy and interaction in quantum communication complexity and a theorem about the relative entropy of quantum states. In: Proceedings of the 43rd Annual IEEE Symposium on Foundations of Computer Science, 2002, pp. 429–438. IEEE Computer Society Press, Los Alamitos (2002) (A more complete version appears as In: [2])
2. Jain, R., Radhakrishnan, J., Sen, P.: A theorem about relative entropy of quantum states with an application to privacy in quantum communication. Technical Report arXiv:0705.2437v1, ArXiv.org Preprint Archive (2007), `http://www.arxiv.org/`
3. Nielsen, M.A., Chuang, I.L.: Quantum Computation and Quantum Information. Cambridge University Press, Cambridge (2000)
4. Cover, T.M., Thomas, J.A.: Elements of Information Theory. Wiley Series in Telecommunications. John Wiley & Sons, New York, NY, USA (1991)
5. Jain, R., Radhakrishnan, J., Sen, P.: Prior entanglement, message compression and privacy in quantum communication. In: Proceedings of the 20th Annual IEEE Conference on Computational Complexity, 2005, pp. 285–296. IEEE Computer Society Press, Los Alamitos (2005)
6. Jain, R.: Stronger impossibility results for quantum string commitment. Technical Report arXiv:quant-ph/0506001v4, ArXiv.org Preprint Archive (2005), `http://www.arxiv.org/`
7. Jain, R.: Communication complexity of remote state preparation with entanglement. Quantum Information and Computation 6, 461–464 (2006)
8. Buhrman, H., Christandl, M., Hayden, P., Lo, H.-K., Wehner, S.: Security of quantum bit string commitment depends on the information measure. Physical Review Letters 97 (2006) (Article no. 250501)
9. Mayers, D.: Unconditionally secure quantum bit commitment is impossible. Physical Review Letters 78(17), 3414–3417 (1997)
10. Lo, H.-K., Chau, H.F.: Is quantum bit commitment really possible? Physical Review Letters 78, 3410–3413 (1997)
11. Kent, A.: Quantum bit string commitment (Article no. 237901). Physical Review Letters 90 (2003) (Article no. 237901)

A Implications for Quantum Protocols

A.1 Quantum String Commitment

A *string commitment* scheme is an extension of the well-studied and powerful cryptographic primitive of *bit commitment*. In such schemes, one party, Alice, wishes to commit an entire string $x \in \{0,1\}^n$ to another party, Bob. The protocol is required to be such that Bob not be able to identify the string until it is revealed by Alice. In turn, Alice should not be able to renege on her commitment at the time of revelation. Formally, quantum string commitment protocols are defined as follows [8,6].

Definition 7 (Quantum string commitment (QSC)). *Let $P = \{p_x : x \in \{0,1\}^n\}$ be a probability distribution and let B be a measure of information contained in an ensemble of quantum states. A (n,a,b)-B-***QSC** *protocol for P is a quantum communication protocol between two parties, Alice and Bob.*

Alice gets an input $x \in \{0,1\}^n$ chosen according to the distribution P. The starting joint state of the qubits of Alice and Bob is some pure state independent of x. The protocol runs in two phases: the commit phase, followed by the reveal phase. There are no intermediate measurements during the protocol. At the end of the reveal phase, Bob measures his qubits according to a POVM $\{M_y : y \in \{0,1\}^n\} \cup \{I - \sum_y M_y\}$ to determine the value of the committed string by Alice or to detect cheating. The protocol satisfies the following properties.

1. **(Correctness)** *Suppose Alice and Bob act honestly. Let ρ_x be the state of Bob's qubits at the end of the reveal phase of the protocol, when Alice gets input x. Then $(\forall x, y)$ $\mathrm{Tr}\, M_y \rho_x = 1$ iff $x = y$, and 0 otherwise.*
2. **(Concealing property)** *Suppose Alice acts honestly, and Bob possibly cheats, i.e., deviates from the protocol in his local operations. Let σ_x be the state of Bob's qubits after the commit phase when Alice gets input x. Then the B information $B(\mathcal{E})$ of the ensemble $\mathcal{E} = \{p_x, \sigma_x\}$ is at most b. In particular, this also holds when both Alice and Bob follow the protocol honestly.*
3. **(Binding property)** *Suppose Bob acts honestly , and Alice possibly cheats. Let $c \in \{0,1\}^n$ be a string in a special cheating register C with Alice that she keeps independent of the rest of the registers till the end of the commit phase. Let τ_c be the state of Bob's qubits at the end of the reveal phase when Alice has c in the cheating register. Let $q_c \stackrel{\mathrm{def}}{=} \mathrm{Tr}\, M_c \tau_c$. Then*

$$\sum_{c \in \{0,1\}^n} p_c q_c \leq 2^{a-n}$$

The idea behind the above definition is as follows. At the end of the reveal phase of an honest run of the protocol Bob identifies x from ρ_x by performing the POVM measurement $\{M_y\}_y \cup \{I - \sum_y M_y\}$. He accepts the committed string to be x iff the observed outcome $y = x$; this happens with probability $\mathrm{Tr}\, M_x \rho_x$. He declares that Alice is cheating if outcome $I - \sum_x M_x$ is observed. Thus, at the end of an honest run of the protocol, with probability 1, Bob accepts the committed string as being exactly Alice's input string. The concealing property ensures that the amount of B information about x that a possibly cheating Bob gets is bounded by b. In *bit*-commitment protocols, the concealing property is quantified in terms of the probability with which Bob can guess Alice's bit. Here we instead use different notions of information contained in the corresponding ensemble. The binding property ensures that when a cheating Alice wishes to postpone committing to a string string until after the commit phase, then she succeeds in forcing an honest Bob to accept her choice with bounded probability (in expectation).

Strong string commitment, in which both parameters a, b above are required to be 0, is impossible for the same reason that of *strong* bit-commitment protocols are impossible [9,10]. Weaker versions are nonetheless possible, and exhibit a trade-off between the concealing and binding properties. The trade-off between the parameters a and b has been studied by several researchers [11,8,6]. Buhrman, Christandl, Hayden, Lo, and Wehner [8] study this trade-off both in the scenario of a single execution of the protocol and also in the asymptotic regime, with an unbounded number of parallel executions of the protocol. In the asymptotic scenario, they show the following result in terms of Holevo information (which is denoted by χ).

Theorem 8 ([8]). *Let Π be an (n, a_1, b)-χ-**QSC** scheme. Let Π_m represent m independent, parallel executions of Π (so $\Pi_1 = \Pi$). Let a_m represent the binding parameter of Π_m and let $a \stackrel{\text{def}}{=} \lim_{m\to\infty} a_m/m$. Then, $a + b \geq n$.*

Jain [6] shows a similar trade-off result regarding **QSC**s, in terms of the divergence information of an ensemble (denoted by D).

Theorem 9 ([6]). *For single execution of the protocol of an (n, a, b)-D-**QSC** scheme,*

$$a + b + 8\sqrt{b+1} + 16 \geq n.$$

As mentioned before, for any ensemble $\mathcal{E}$, divergence information is bounded by the Holevo χ-information $\mathrm{D}(\mathcal{E}) \leq \chi(\mathcal{E}) + 1$. This immediately implies:

Theorem 10 ([6]). *For single execution of the protocol of a (n, a, b)-χ-**QSC** scheme*

$$a + b + 8\sqrt{b+2} + 17 \geq n.$$

As Jain shows, this implies the asymptotic result due to Buhrman *et al.* (Theorem 8).

The separation that we demonstrate between divergence and Holevo information (Theorem 1) shows that for some ensembles over n qubits, $\mathrm{D}(\mathcal{E})$ may be a $\log n$ larger than $\chi(\mathcal{E})$. For such ensembles the binding-concealing trade-off of Theorem 9 is stronger than that of Theorem 8.

A.2 Privacy Trade-Off for Two-Party Protocols for Relations

Let us consider two-party protocols between Alice and Bob for computing a relation $f \subseteq \mathcal{X} \times \mathcal{Y} \times \mathcal{Z}$. The goal here is to find a $z \in \mathcal{Z}$ such that $(x, y, z) \in f$, when Alice and Bob are given $x \in \mathcal{X}$ and $y \in \mathcal{Y}$, respectively. Jain, Radhakrishnan, and Sen [1] studied to what extent the two parties may solve f while keeping their respective inputs hidden from the other party. They showed the following:

Result 11 ([5], informal statement). *Let μ be a product distribution on $\mathcal{X} \times \mathcal{Y}$. Let
$Q^{\mu, A\to B}_{1/3}(f)$ represent the one-way distributional complexity of f with a single communication from Alice to Bob and distributional error under μ at most $1/3$. Let X and Y represent the random variables corresponding to Alice and Bob's inputs respectively. If there is a quantum communication protocol for f where Bob* leaks *divergence information at most b about his input Y, then Alice leaks divergence information at least $\Omega(Q^{\mu, A\to B}_{1/3}(f)/2^{O(b)})$ about her input X. A similar statement also holds with the roles of Alice and Bob interchanged.*

From the upper bound on the divergence information in terms of Holevo information this immediately implies the following.

Result 12 ([5], informal statement). *Let μ be a product distribution on $\mathcal{X} \times \mathcal{Y}$. Let $Q_{1/3}^{\mu, A \to B}(f)$ represent the one-way distributional complexity of f with a single communication from Alice to Bob and distributional error under μ at most $1/3$. Let X and Y represent the random variables corresponding to Alice and Bob's inputs respectively. If there is a quantum communication protocol for f where Bob* leaks *Holevo information at most b about his input Y, then Alice leaks Holevo information at least $\Omega(Q_{1/3}^{\mu, A \to B}(f)/2^{O(b)})$ about her input X. A similar statement also holds with the roles of Alice and Bob interchanged.*

It follows from Theorem 1 that Result 11 is much stronger than the second, Result 12 in case the ensembles arising in the protocol between Alice and Bob has divergence information much smaller than its Holevo information.

Unary Automatic Graphs: An Algorithmic Perspective

Bakhadyr Khoussainov[1], Jiamou Liu[1], and Mia Minnes[2]

[1] Department of Computer Science
University of Auckland, New Zealand
[2] Department of Mathematics
Cornell University, USA

Abstract. This paper studies infinite graphs produced from a natural unfolding operation applied to finite graphs. Graphs produced via such operations are of finite degree and can be described by finite automata over the unary alphabet. We investigate algorithmic properties of such unfolded graphs given their finite presentations. In particular, we ask whether a given node belongs to an infinite component, whether two given nodes in the graph are reachable from one another, and whether the graph is connected. We give polynomial time algorithms for each of these questions. Hence, we improve on previous work, in which non-elementary or non-uniform algorithms were found.

1 Introduction

The underlying idea of automatic structures consists of using automata to represent structures and then to study the logical and algorithmic consequences of such presentations. For example, there are descriptions of automatic linear orders and trees in model theoretic terms such as Cantor-Bendixson ranks [13], [10]. Thomas and Oliver gave a full description of finitely generated automatic groups [12]. Khoussainov, Nies, Rubin and Stephan have characterized the isomorphism types of automatic Boolean algebras [8]. These results give the decidability of the isomorphism problems for automatic ordinals and Boolean algebras [13].

The complexity of the first-order theories of automatic structures has also been studied. Grädel and Blumensath constructed examples of automatic structures whose first-order theories are non-elementary [2]. Lohrey, on the other hand, proved that the first-order theory of any automatic graph of bounded degree is elementary [11]. This paper continues this line of research and investigates computational properties of unary automatic graphs of finite degree. We use a fundamental algorithmic property of automatic structures proved by Khoussainov and Nerode: the first-order theory of any automatic graph is decidable [7]. In particular, for a fixed first-order formula $\phi(\bar{x})$ and an automatic graph $\mathcal{G}$, determining if a tuple $\bar{a}$ from $\mathcal{G}$ satisfies $\phi(\bar{x})$ can be done in linear time. Refining this, we find polynomial time algorithms for natural graph theoretic questions in the class of unary automatic graphs of finite degrees. Since all such graphs can be obtained by an unfolding operation applied to finite graphs (see Theorem 2),

M. Agrawal et al. (Eds.): TAMC 2008, LNCS 4978, pp. 542–553, 2008.

we measure complexity based on the input size of the finite graphs. Specifically, we are interested in the following decision problems for the graph $\mathcal{G}$ determined by the pair of finite graphs $(\mathcal{D}, \mathcal{F})$:

- **Connectivity Problem.** Is the graph $\mathcal{G}$ connected?
- **Reachability Problem.** Given vertices x, y, is there a path from x to y?
- **Infinite Component Problem.** Does $\mathcal{G}$ have an infinite component?
- **Infinity Testing Problem.** Given a vertex x, is it in an infinite component?

For finite graphs, the first two problems can be solved in linear time and the last two have obvious answers. However, for infinite graphs, much more work is needed to investigate these problems. In the class of all automatic graphs, all of these problems are undecidable (see [13]). Since all unary automatic graphs are first-order definable in $S1S$ (the monadic second-order logic of the successor function), it is not hard to prove that all the problems above are decidable ([1], [13]). However, the constructions which appeal to $S1S$ yield algorithms with non-elementary time complexity, since one needs to transform $S1S$ formulas into automata ([4]). The reachability problem has been studied in [3], [5], and [14] via pushdown graphs. A pushdown graph is the configuration space of a pushdown automaton. Unary automatic graphs are examples of pushdown graphs [14]. In [3], [5], [14] it is proved that for a given node v in a pushdown graph, there is an automaton that recognizes all nodes reachable from v. The size of this automaton depends on the input node v. Moreover, the automata constructed by this algorithm are not uniform (different automata are built for different vertices v). It is therefore interesting to see for which classes of graphs the reachability problem has a uniform solution (an automaton that tells whether any two nodes belong to the same component). The practical advantage of a uniform solution is that, once the automaton that recognizes reachability relation is built, deciding whether node v is reachable from u by a path takes only linear time In this paper, we show that for unary automatic graphs of finite degree, all the problems above can be solved in polynomial time. Moreover, the reachability problem has a uniform solution.

We now outline the rest of the paper. Section 2 introduces the main definitions needed and recalls a characterization theorem (Theorem 1) for unary automatic graphs. Section 3 introduces unary automatic graphs of finite degree; Theorem 2 explicitly provides a method for building these graphs and is used throughout the paper. Section 4 and Section 5 solve the infinite component problem and infinity testing problem, respectively. For easy reference, we list the main results below. $\mathcal{G}$ is a given unary automatic graph of finite degree, $\mathcal{A}$ is the unary automaton recognizing $\mathcal{G}$, and n is the number of states of $\mathcal{A}$.

Theorem 3. *The infinite component problem for $\mathcal{G}$ is solved in $O(n^{\frac{3}{2}})$.*

Theorem 4. *The infinity testing problem for $\mathcal{G}$ is solved in $O(n^{\frac{5}{2}})$. When $\mathcal{A}$ is fixed, a constant time algorithm decides the infinity testing problem on $\mathcal{G}$.*

Section 6 gives a polynomial time algorithm constructing uniform automata that solve the reachability problem. This algorithm also yields a solution to the connectivity problem for unary automatic graphs of finite degree.

Theorem 5. *A polynomial time algorithm solves the reachability problem on* $\mathcal{G}$. *For inputs* u, v, *the running time of the algorithm is* $O(|u| + |v| + n^{\frac{5}{2}})$.

Theorem 6. *The connectivity problem for* $\mathcal{G}$ *is solved in* $O(n^3)$.

2 Preliminaries

A **finite automaton** $\mathcal{A}$ over Σ is a tuple (Q, ι, Δ, F), where Q is a finite set of **states**, $\iota \in Q$ is the **initial state**, $\Delta \subset Q \times \Sigma \times Q$ is the **transition table**, and $F \subset Q$ is the set of **final states**. A **run** of $\mathcal{A}$ on a word $\sigma_1 \ldots \sigma_n \in \Sigma^\star$ is a sequence $q_0, \ldots, q_n$ such that $q_0 = \iota$ and $(q_i, \sigma_{i+1}, q_{i+1}) \in \Delta$ for all $i \leq n-1$. If $q_n \in F$ then the run is **successful** and we say that the automaton $\mathcal{A}$ **accepts** the word. The **language** accepted by the automaton $\mathcal{A}$ is the set of all words accepted by $\mathcal{A}$. A set $D \subset \Sigma^\star$ is **FA recognizable** if D is the language accepted by some finite automaton. For two states q_0, q_1, the **distance** from q_0 to q_1 is the minimum number of transitions required for $\mathcal{A}$ to go from q_0 to q_1. If $|\Sigma| = 1$, we call $\mathcal{A}$ a **unary automaton**. A **2-tape automaton** is a one-way Turing machine with two semi-infinite input tapes. Each tape has written on it a word from $\Sigma^\star$ followed by a succession of $\diamond$ symbols. The automaton starts in the initial state, reads simultaneously the first symbol of each tape, changes state, reads simultaneously the second symbol of each tape, changes state, etc., until it reads $\diamond$ on each tape. The automaton then stops and accepts the 2-tuple of words on its input tapes if it is in a final state. Formally, set $\Sigma_\diamond = \Sigma \cup \{\diamond\}$ where $\diamond \notin \Sigma$. The **convolution** of a tuple $(w_1, w_2) \in \Sigma^{\star 2}$ is the string $w_1 \otimes w_2$ of length $\max_i |w_i|$ over the alphabet $(\Sigma_\diamond)^2$ which is defined as follows: the kth symbol is (σ_1, σ_2) where σ_i is the kth symbol of w_i if $k \leq |w_i|$, and is $\diamond$ otherwise. The **convolution** of a relation $E \subset \Sigma^{\star 2}$ is the language $\otimes E = \{w_1 \otimes w_2 \mid (w_1, w_2) \in E\}$. The relation $E \subset \Sigma^{\star 2}$ is **FA recognizable** if $\otimes E$ is recognizable by a 2-tape automaton.

A graph $\mathcal{G} = (V, E)$ is **automatic** over Σ if its vertex set $V \subset \Sigma^\star$ and the edge relation E are FA recognizable. The binary tree $(\{0,1\}^\star, E)$, where $E = \{(x, y) \mid y = x0 \text{ or } y = x1\}$, is an automatic graph. We are interested in the following class of automatic graphs:

Definition 1. *A* **unary automatic graph** *is a graph* (V, E) *whose domain is a regular subset of* $\{1\}^\star$ *and whose edge relation* E *is regular.*

Convention. To eliminate bulky exposition, we fix the following assumptions: 1) By "automatic graph", we always mean "unary automatic graph". 2) All graphs are infinite unless explicitly specified otherwise. 3) The domains of automatic graphs coincide with the set $1^\star$ of all unary strings $\{\lambda, 1, 11, 111, \ldots\}$. Hence, the automaton recognizing the edge relation is sufficient for describing the graph. 4) The graphs are undirected. All the notions and results below can be adapted to the case when the domains are regular subsets of $1^\star$ and when the graphs are directed without materially changing the complexity of the algorithms.

Let $\mathcal{G} = (V, E)$ be an automatic graph. Let $\mathcal{A}$ be a unary automaton recognizing E with n states. The general shape of $\mathcal{A}$ is given in Figure 1. All the states reachable from the initial state by reading input $(1, 1)$ are called **$(1, 1)$-states**. A **tail** in $\mathcal{A}$ is a sequence of states linked by transitions without repetition. A **loop** is a sequence of states linked by transitions such that the last state coincides with the first one, and with no repetition in the middle. The set of $(1, 1)$-states is a disjoint union of a tail and a loop, called the **$(1, 1)$-tail** and the **$(1, 1)$-loop**. Let q be a $(1, 1)$-state. All the states reachable from q by reading inputs $(1, \diamond)$ are called **$(1, \diamond)$-states**. This collection of $(1, \diamond)$-states is also a disjoint union of a tail and a loop (see the figure), called the **$(1, \diamond)$-tail** and the **$(1, \diamond)$-loop**. The **$(\diamond, 1)$-tails** and **$(\diamond, 1)$-loops** are defined in a similar way. Since we consider undirected graphs, we simplify the general shape of the automaton by only considering edges labelled by $(\diamond, 1)$ and $(1, 1)$. An automaton is **standard** if the lengths of all its loops and tails equal some number p, called the **loop constant**.

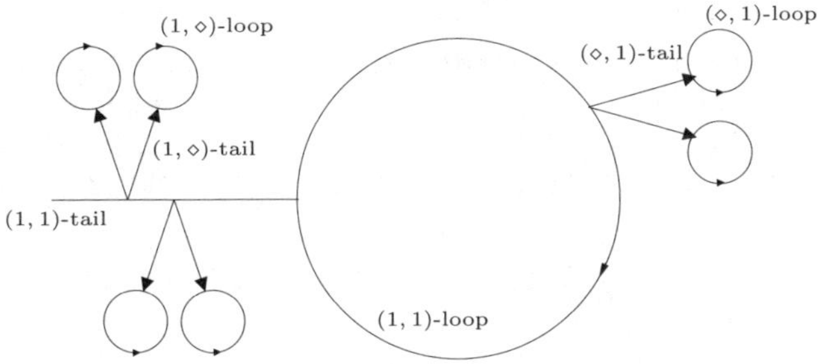

Fig. 1. A Typical Unary Graph Automaton

We recall a characterization theorem of unary automatic graphs from [13]. Let $\mathcal{D} = (D, E_D)$ and $\mathcal{F} = (F, E_F)$ be finite graphs. Let R_1, R_2 be subsets of $D \times F$, and R_3, R_4 be subsets of $F \times F$. Consider the graph $\mathcal{D}$ followed by ω many copies of $\mathcal{F}$, ordered as $\mathcal{F}^0, \mathcal{F}^1, \mathcal{F}^2, \ldots$. Formally, the vertex set of $\mathcal{F}^i$ is $F \times \{i\}$ and we write $f^i = (f, i)$ for $f \in F$ and $i \in \omega$. The edge set E^i of $\mathcal{F}^i$ consists of all pairs (a^i, b^i) such that $(a, b) \in E_F$. We define the infinite graph, *unwind*$(\mathcal{D}, \mathcal{F}, \bar{R})$, as follows: the vertex set is $D \cup F^0 \cup F^1 \cup F^2 \cup \ldots$; the edge set contains $E_D \cup E^0 \cup E^1 \cup \ldots$ as well as the following edges, for all $a, b \in F$, $d \in D$, and $i, j \in \omega$:

- (d, b^0) when $(d, b) \in R_1$, and (d, b^{i+1}) when $(d, b) \in R_2$,
- (a^i, b^{i+1}) when $(a, b) \in R_3$, and (a^i, b^{i+2+j}) when $(a, b) \in R_4$.

Theorem 1. *[9] A graph $\mathcal{G}$ has a unary automaton presentation if and only if it is isomorphic to unwind$(\mathcal{D}, \mathcal{F}, \bar{R})$ for some parameters $\mathcal{D}$, $\mathcal{F}$, and $\bar{R}$. Moreover, if $\mathcal{A}$ is a standard automaton representing $\mathcal{G}$ then the parameters $\mathcal{D}, \mathcal{F}, \bar{R}$ can be extracted in $O(n^2)$; otherwise, the parameters can be extracted in $O(n^{2n})$, where n is the number of states in $\mathcal{A}$.*

3 Unary Automatic Graphs of Finite Degree

A graph is of **finite degree** if there are finitely many edges connected to each vertex v. A unary automaton $\mathcal{A}$ recognizing a binary relation is a **one-loop automaton** if its transition diagram contains exactly one loop, the $(1,1)$-loop. The following is an easy proposition:

Proposition 1. *Let $\mathcal{G} = (V, E)$ be a unary automatic graph, then $\mathcal{G}$ is of finite degree if and only if there is a one-loop unary automaton $\mathcal{A}$ recognizing E.* □

Each unary automaton has an equivalent standard unary automaton. In general, the standard automaton may have exponentially more states. However, if $\mathcal{A}$ is a one-loop automaton with n states, the $(1,1)$-loop of the equivalent standard one-loop automaton has at most n states, so the automaton itself has at most $4n^2$ states. Below, we assume the input automaton $\mathcal{A}$ is standard. Let p be the loop constant of $\mathcal{A}$, then $\mathcal{A}$ has exactly $4p^2$ states. In the following, we state all results in terms of p rather than n, the number of states of the input automaton.

Definition 2 (Unfolding Operation). *Let $\mathcal{D} = (V_\mathcal{D}, E_\mathcal{D})$ and $\mathcal{F} = (V_\mathcal{F}, E_\mathcal{F})$ be finite graphs. The finite sets $\Sigma_{\mathcal{D},\mathcal{F}}, \Sigma_\mathcal{F}$ contain all mappings $\eta : V_\mathcal{D} \to P(V_\mathcal{F})$ and $\sigma : V_\mathcal{F} \to P(V_\mathcal{F})$ (respectively). The sequence $\alpha = \eta\sigma_0\sigma_1\dots$ where $\eta \in \Sigma_{\mathcal{D},\mathcal{F}}$ and $\sigma_i \in \Sigma_\mathcal{F}$ for each i yields the infinite graph $\mathcal{G}_\alpha = (V_\alpha, E_\alpha)$ as follows:*

- $V_\alpha = V_\mathcal{D} \cup \{(v, i) \mid v \in V_\mathcal{F}, i \in \omega\}$.
- $E_\alpha = E_\mathcal{D} \cup \{(d, (v, 0)) \mid v \in \eta(d)\} \cup \{((v, i), (v', i)) \mid (v, v') \in E_\mathcal{F}, i \in \omega\} \cup \{((v, i), (v', i + 1)) \mid v' \in \sigma_i(v), i \in \omega\}$.

Figure 2 illustrates the general shape of a unary automatic graph of finite degree built from $\mathcal{D}$, $\mathcal{F}$, η, and σ^ω (σ^ω is the infinite word $\sigma\sigma\sigma\cdots$). We use Definition 2 to recast Theorem 1 for graphs of finite degree. The proof is omitted.

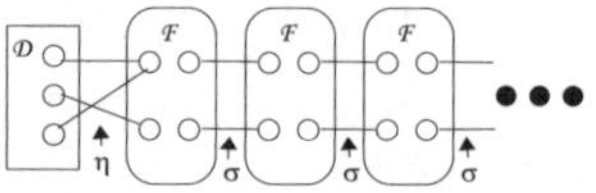

Fig. 2. Unary automatic graph of finite degree $\mathcal{G}_{\eta\sigma^\omega}$

Theorem 2. *A graph of finite degree $\mathcal{G} = (V, E)$ possesses a unary automatic presentation if and only if there exist finite graphs $\mathcal{D}, \mathcal{F}$ and mappings $\eta : V_\mathcal{D} \to P(V_\mathcal{F})$ and $\sigma : V_\mathcal{F} \to P(V_\mathcal{F})$ such that $\mathcal{G}$ is isomorphic to $\mathcal{G}_{\eta\sigma^\omega}$.* □

If $\mathcal{G}$ is a unary automatic graph of finite degree, the parameters $\mathcal{D}$, $\mathcal{F}$, σ and η can be extracted in $O(p^2)$ time, where p is the loop constant of the one-loop automaton representing the graph. Furthermore, $|V_\mathcal{F}| = |V_\mathcal{D}| = p$.

4 Deciding the Infinite Component Problem

A **component** of a graph is the transitive closure of a vertex under the edge relation. The **infinite component problem** asks whether a graph $\mathcal{G}$ has an infinite component.

Theorem 3. *The infinite component problem for unary automatic graphs of finite degree $\mathcal{G}$ is solved in $O(p^3)$, where p is the loop constant of the unary automaton recognizing $\mathcal{G}$.*

By Theorem 2, it suffices to consider the case when $\mathcal{G} = \mathcal{G}_{\sigma^\omega}$ since $\mathcal{G}_{\eta\sigma^\omega}$ has an infinite component if and only if $\mathcal{G}_{\sigma^\omega}$ has one. Let $\mathcal{F}^i$ be the ith copy of $\mathcal{F}$ in $\mathcal{G}$ and x^i be the copy of vertex x in $\mathcal{F}^i$. The finite directed graph $\mathcal{F}^\sigma = (V^\sigma, E^\sigma)$ is defined as follows. Nodes in V^σ are the distinct connected components of $\mathcal{F}$. For simplicity, we assume that $|V^\sigma| = |V_\mathcal{F}|$ and use x to denote its own component in $\mathcal{F}$. The case in which $|V^\sigma| < |V_\mathcal{F}|$ is similar. For $x, y \in V_\mathcal{F}$, put $(x, y) \in E^\sigma$ if and only if $y' \in \sigma(x')$ for some x' and y' that are in the same component as x and y, respectively. Constructing $\mathcal{F}^\sigma$ requires finding connected components of $\mathcal{F}$ and hence takes time $O(p^2)$. To prove Theorem 3, we make essential use of the following definition which is taken from [6].

Definition 3. *An **oriented walk** in a directed graph G is a subgraph $\mathcal{P}$ of G that consists of a sequence of nodes $v_0, \ldots, v_k$ such that for $1 \le i \le k$, either (v_{i-1}, v_i) or (v_i, v_{i-1}) is an arc in G, and for each $1 \le i \le k$, exactly one of (v_{i-1}, v_i) and (v_i, v_{i-1}) belongs to $\mathcal{P}$. An oriented walk is an **oriented cycle** if $v_0 = v_k$ and there are no repeated nodes in $v_1, \ldots, v_k$.*

In an oriented walk $\mathcal{P}$, an arc (v_i, v_{i+1}) is called a **forward arc** and (v_{i+1}, v_i) is called a **backward arc**. The **net length** of $\mathcal{P}$, denoted $disp(\mathcal{P})$, is the difference between the number of forward arcs and backward arcs. Note that the net length can be negative. Given an oriented walk $\mathcal{P} = v_0, \ldots, v_m$, we define the **low point** of $\mathcal{P}$ as $\min\{disp(v_0 \ldots v_\ell) \mid 0 \le \ell \le m\}$. The low point of the oriented walk $\mathcal{P}$ is at most $\min\{0, disp(\mathcal{P})\}$, and hence is not positive. The next lemma establishes a connection between oriented walks in $\mathcal{F}^\sigma$ and paths in G.

Lemma 1. *Let $\mathcal{P}$ be an oriented walk from x to y whose net length is d and low point is $-\ell$. For every $i \ge \ell$, the oriented walk $\mathcal{P}$ defines a path P^i in $\mathcal{G}$ from x^i to y^{i+d}. Moreover, the smallest j such that $P^i \cap \mathcal{F}^j \ne \emptyset$ is equal to $i - \ell$.* □

Lemma 2. *There is an infinite component in $\mathcal{G}$ if and only if there is an oriented cycle in $\mathcal{F}^\sigma$ with positive net length.*

Proof. We prove one direction; the other is left to the reader. Suppose there is an infinite component D in $\mathcal{G}$. Since $\mathcal{F}$ is finite, there must be some x in $V_\mathcal{F}$ such that there are infinitely many copies of x in D. Let x^i and x^j be two copies of x in D with $i < j$. Consider a path between x^i and x^j. We can assume that on this path there is at most one copy of any vertex $y \in V_\mathcal{F}$ apart from x (otherwise, there is another vertex in $V_\mathcal{F}$ having an infinite number of copies in the infinite component with these properties). By definition of $\mathcal{G}_{\sigma^\omega}$ and $\mathcal{F}^\sigma$, the node x must be on an oriented cycle of $\mathcal{F}^\sigma$ with net length $j - i$. □

Proof (Theorem 3). By Lemma 2, it suffices to decide if $\mathcal{F}^\sigma$ contains an oriented cycle with positive net length. Such an oriented cycle exists if and only if there is an oriented cycle with negative net length. Therefore, the following algorithm searches for oriented cycles with non-zero net length.

`ALG:Oriented-Cycle`

1. Pick the first node $x \in \mathcal{F}^\sigma$ for which a queue has not been built. Initialize the queue Q_x to be empty. Let $d(x) = 0$, and put x into Q_x marked as *unprocessed*. If there is no such $x \in \mathcal{F}^\sigma$, stop the process and return *NO*.
2. Define y to be the first *unprocessed* node in the queue Q_x. If there are no *unprocessed* nodes in Q_x, return to (1).
3. For each node z in the set $\{z \mid (y,z) \in E^\sigma \text{ or } (z,y) \in E^\sigma\}$, do the following:
   (a) If $(y,z) \in E^\sigma$, set $d'(z) = d(y) + 1$; if $(z,y) \in E^\sigma$, set $d'(z) = d(y) - 1$. (If both hold, do steps (a), (b), (c) first for (z,y) and then for (y,z).)
   (b) If $z \notin Q_x$, set $d(z) = d'(z)$, put z into Q_x, and mark z as *unprocessed.*
   (c) If $z \in Q_x$ then if $d(z) = d'(z)$, move to next z; if $d(z) \neq d'(z)$, stop the process and return *YES*.
4. Mark y as *processed* and go back to (2).

We claim that the algorithm returns *YES* if and only if there is an oriented cycle in $\mathcal{F}^\sigma$ with non-zero net length. Suppose the algorithm returns *YES*. Then, there is a base node x and a node z such that $d(z) \neq d'(z)$. Thus, there is an oriented walk $\mathcal{P}$ from x to z with net length $d(z)$ and there is an oriented walk $\mathcal{P}'$ from x to z with net length $d'(z)$. Let $(\mathcal{P}')^-$ be the oriented walk $\mathcal{P}'$ in reverse direction. Consider the oriented walk $\mathcal{P}(\mathcal{P}')^-$: it is an oriented walk from x to x with net length $d(z) - d'(z) \neq 0$. If there are no repeated nodes in $\mathcal{P}(\mathcal{P}')^-$, it is the required oriented cycle. Otherwise, let y be a repeated node in $\mathcal{P}(\mathcal{P}')^-$ such that no nodes between the two occurrences of y are repeated. Consider the oriented walk between these two occurrences of y; if it has a non-zero net length it is our required oriented cycle and otherwise we can make the oriented walk $\mathcal{P}(\mathcal{P}')^-$ shorter without altering its net length.

Conversely, suppose there is an oriented cycle $\mathcal{P} = x_0, \ldots, x_m$ of non-zero net length where $x_0 = x_m$. We assume for a contradiction that the algorithm returns *NO*. Consider how the algorithm acts when we pick x_0 at step (1). For each $0 \leq i \leq m$, the following statements hold (by induction on i).

($\star$) x_i gets a label $d(x_i)$
($\star\star$) $d(x_i)$ equals the net length of the oriented walk from x_0 to x_i in $\mathcal{P}$.

These statements suffice to yield a contradiction, and hence prove the correctness of `Oriented-Cycle`.

Putting these pieces together, the following algorithm solves the infinite component problem. Suppose we are given a unary automaton (with loop constant p) which recognizes the unary automatic graph of finite degree $\mathcal{G}$. Recall that $p = |V_\mathcal{F}|$. We compute $\mathcal{F}^\sigma$ in time $O(p^2)$. Then we run `Oriented-Cycle` to decide if $\mathcal{F}^\sigma$ contains an oriented cycle with positive net length. For each node x in $\mathcal{F}^\sigma$, the run time is $O(p^2)$. Since $\mathcal{F}^\sigma$ contains p nodes, this takes time $O(p^3)$. □

5 Deciding the Infinity Testing Problem

The **infinity testing problem** asks for an algorithm that, given a vertex v and graph $\mathcal{G}$, decides if the vertex belongs to an infinite component of the graph $\mathcal{G}$.

Theorem 4. *The infinity testing problem for $\mathcal{G}$, a unary automatic graph of finite degree with loop constant p, is solved in $O(p^5)$. When $\mathcal{A}$ is fixed, there is a constant time algorithm that decides the infinity testing problem on $\mathcal{G}$.*

To prove Theorem 4 we outline several lemmas, the more difficult of which we prove. The set $\mathcal{C}$ is defined as all nodes x in $\mathcal{F}^\sigma$ for which there exists an oriented cycle from x with positive net length and low point 0. We call an oriented walk **simple** if it contains no repeated nodes. For any $k \geq 0$, let $\mathcal{C}[k]$ be the set of all nodes $x \notin \mathcal{C}[0] \cup \ldots \cup \mathcal{C}[k-1]$ that can reach $\mathcal{C}$ via a simple oriented walk with low point $-k$. Note that $\mathcal{C} \subseteq \mathcal{C}[0]$. Moreover, since $|\mathcal{F}^\sigma| \leq p$, a simple oriented walk may have at most p steps and hence $\mathcal{C}[k] = \emptyset$ for $k > p-1$.

Lemma 3. *Let $x \in V_\mathcal{F}$. If x^i belongs to an infinite component of $\mathcal{G}$ then for all $j > 0$, x^{i+j} also belongs to an infinite component of $\mathcal{G}$.* □

Lemma 4. *If $x \in \mathcal{C}$, then x^i is in an infinite component for all $i \in \omega$.* □

Lemma 5. *For each vertex x^i, x^i belongs to an infinite component in $\mathcal{G}$ if and only if node $x \in \mathcal{C}[k]$ for some $0 \leq k \leq \min\{i, p-1\}$.*

Proof. We prove the harder direction: if x^i is in an infinite component, there is an oriented walk from x to $\mathcal{C}$ with low point $-k$, where $0 \leq k \leq \min\{i, p-1\}$. Let D be the infinite component of x^i. Since $\mathcal{F}$ is finite, there must be y in $V_\mathcal{F}$ such that D contains infinitely many copies of y. Let y^s and y^t be two copies of y in D with $s < t$. Take a path P in $\mathcal{G}$ between y^s and y^t such that P contains no more than one copy of each vertex in $V_\mathcal{F}$ apart from y. (If there is no such path P, choose another vertex y in $V_\mathcal{F}$ with these properties). Let ℓ be the least number such that $P \cap \mathcal{F}^\ell \neq \emptyset$. Let z^ℓ be a vertex in P. Then P is divided into two paths P_1 and P_2, where P_1 goes from y^s to z^ℓ and P_2 goes from z^ℓ to y^t. Hence there is a path P_3 from y^t to $z^{\ell+t-s}$. By joining P_2 and P_3 together we obtain a path between z^ℓ and $z^{\ell+t-s}$. We have defined an oriented cycle in $\mathcal{F}^\sigma$ with positive net length and low point 0. Hence, $z \in \mathcal{C}$. Take a path in $\mathcal{G}$ between x^i and a copy of z in D containing no more than one copy of each vertex in $\mathcal{F}$. This is an oriented walk in $\mathcal{F}^\sigma$ from x to z with low point not more than $\min\{i, p-1\}$. □

Lemma 6. *If $\mathcal{G}$ is a unary automatic graph of finite degree presented by $\mathcal{A}$ with loop constant p, the set $\mathcal{C}$ for $\mathcal{G}$ can be computed in time $O(p^4)$.*

Proof. For each $x \in \mathcal{F}^\sigma$, do a breadth-first search through $\mathcal{F}^\sigma$ for oriented walks starting at x. To compute the path $\mathcal{P}$, put (y, d) in a queue, where y is the incremental destination of $\mathcal{P}$ and d is its net length. We keep track of the following properties of the pair (y, d):

1. *level*(y, d) is the length of the oriented walk $\mathcal{P}$ from x to y; and

2. $path(y,d)$ is a tuple of pairs $(x_0, d_0) \ldots (x_{level(y,d)}, d_{level(y,d)})$ coding the initial segment of $\mathcal{P}$. Note: $(x_0, d_0) = (x, 0)$ and $(x_{level(y,d)}, d_{level(y,d)}) = (y, d)$.

Given input $x \in \mathcal{F}^\sigma$, the following algorithm checks membership in $\mathcal{C}$.
`ALG:C-Membership`

1. Put $(x, 0)$ into the (initially empty) queue Q. Mark $(x, 0)$ as *unprocessed* and set $level(x, 0) = 0$, $path(x, 0) = (x, 0)$.
2. If no *unprocessed* pair is left in the queue, stop and output *NO*. Otherwise, take the first *unprocessed* (y, d) in Q.
3. If $level(y, d) \geq p$, stop and output *NO*.
4. For arcs e of the form (y, z) or (z, y) in E^σ do the following:
   (a) If $e = (y, z)$, set $j = d + 1$; if $e = (z, y)$, set $j = d - 1$.
   (b) If $z = x$ and $j > 0$, stop the process and return *YES*.
   (c) If (z, d') is not in $path(y, d)$ for any d', and if $j \geq 0$ and $(z, j) \notin Q$, then put (z, j) into Q, mark (z, j) as *unprocessed*, set $level(z, j) = level(y, d) + 1$, set $path(z, j) = path(y, d) \cdot (z, j)$.
5. Mark (y, d) as *processed* and go back to (2).

We claim that `C-Membership` on input x returns *YES* if and only if $x \in \mathcal{C}$. Suppose the algorithm returns *YES*. Then there is a simple oriented walk $\mathcal{P}$ from x to x with positive net length. Let $\mathcal{P}$ be $x_0, \ldots, x_m$ such that $x_0 = x_m = x$. The algorithm ensures that the net length of the (sub)oriented walk in $\mathcal{P}$ from x_0 to each x_i is non-negative. Thus, the low point of $\mathcal{P}$ is no less than 0 and $x \in \mathcal{C}$. For the other direction, suppose $\mathcal{P} = x_0, \ldots, x_m$ is an oriented cycle of positive net length and zero low point. Assume the algorithm does not return *YES*. Run the algorithm from x_0. For all x_i, the following statements hold by induction.

($\star$) There exists $d_i \geq 0$ such that $(x_i, d_i) \in Q$.
($\star\star$) d_i equals the net length of the oriented walk from x_0 to x_i in $\mathcal{P}$.

Note that $level(y, d) \leq p$ for all $(y, d) \in Q$. Moreover, every time the level is incremented by 1 the net length either goes up or down by 1, hence d must also be no more than p. Thus, the cardinality of Q is bounded above by p^2 and so for each $x \in \mathcal{F}^\sigma$, the algorithm takes time $O(p^3)$. To compute $\mathcal{C}$, we need to run `C-Membership` on every x in $\mathcal{F}^\sigma$, taking time $O(p^4)$. □

Using Lemma 6, we iteratively compute $\mathcal{C}[k]$ for any $0 \leq k \leq p - 1$ as follows. First, compute the set $\mathcal{C}$ in time $O(p^4)$. For each $x \notin \mathcal{C}[0] \cup \cdots \cup \mathcal{C}[k-1]$ we run operations similar to the ones described above, except that at step (4)(b), the process stops and returns *YES* whenever $z \in \mathcal{C}$ and $j \geq -k$, and at step (4)(c), the process puts a pair (z, j) into the queue Q if $j \geq -k$ and $(z, j) \notin Q$. The proof of correctness is like that of `C-Membership`. This algorithm runs in $O(p^4)$.

Proof (Theorem 4). We assume the input vertex x^i is given by tuple (x, i). By Lemma 5, to check if x^i is in an infinite component, the algorithm needs to compute $\mathcal{C}[0], \ldots, \mathcal{C}[\min\{i, p-1\}]$. As a consequence of Lemma 6, this takes time $O(p^5)$. The algorithm then checks whether $x \in \mathcal{C}[k]$ for some $0 \leq k \leq \min\{i, p-1\}$. Once the sets $\mathcal{C}[0], \ldots, \mathcal{C}[p-1]$ are found, checking whether x^i belongs to an infinite component takes constant time. □

6 Deciding the Reachability and Connectivity Problems

The **reachability problem** asks whether two given vertices u and v in a unary automatic graph of finite degree belong to the same component.

Theorem 5. *Suppose $\mathcal{G}$ is a unary automatic graph of finite degree represented by unary automaton $\mathcal{A}$ of loop constant p. A polynomial time algorithm solves the reachability problem on $\mathcal{G}$. For inputs x^i, y^j, the algorithm runs in $O(i+j+p^5)$.*

We restrict to the case $\mathcal{G} = \mathcal{G}_{\sigma^\omega}$ (the general case requires few changes). The infinity testing algorithm checks if x^i is in a finite component in $O(p^5)$ time, and leads to two possible cases. First, suppose that x^i is in a finite component.

Lemma 7. *If x^i is in a finite component, then x^i and y^j are in the same component only if $i - p < j < i + p$.* □

To check if x^i and y^j are in the same component, we run a breadth first search in $\mathcal{G}$ starting from x^i visiting all vertices in $\mathcal{F}^{i-i'}, \ldots, \mathcal{F}^{i+p}$ $(i' = \min\{p, i\})$.
`ALG: FiniteReach`

1. Put $(x, 0)$ into the (initially empty) queue Q, marked as *unprocessed.*
2. If there are no *unprocessed* pairs in Q, stop the process. Otherwise, let (y, d) be the first *unprocessed* pair. For arcs e of the form (y, z) or (z, y) in E^σ:
   (a) If $e = (y, z)$, let $d' = d + 1$; if $e = (z, y)$, let $d' = d - 1$.
   (b) If $-i' \le d' \le p$ and $(z, d') \notin Q$, put (z, d') into Q marked as *unprocessed.*
3. Mark (y, d) as *processed*, and go to (2).

Then, x^i and y^j are in the same (finite) component if and only if after running `FiniteReach` on the input x^i, the pair $(y, j - i)$ is in Q. The running time is bounded by the number of edges in $\mathcal{G}$ restricted to $\mathcal{F}^0, \ldots, \mathcal{F}^{2p}$, hence is $O(p^3)$.

Corollary 1. *If all components of $\mathcal{G}$ are finite and if we represent (x^i, y^j) by $(x^i, y^j, j - i)$, an $O(p^3)$-algorithm checks reachability for x^i and y^j.* □

On the other hand, suppose that x^i is in an infinite component. We begin with an algorithm that computes all vertices $y \in V_\mathcal{F}$ whose i^{th} copy lies in the same component as x^i. The algorithm is identical to `FiniteReach`, except that Line (2b) in `FiniteReach` is changed to the following: (**2b'**) If $|d'| \le p$ and $(z, d') \notin Q$, then put (z, d') into Q and mark (z, d') as *unprocessed.* We use this modified algorithm to define the set $Reach(x) = \{y \mid (y, 0) \in Q\}$. Intuitively, we can think of the algorithm as a breadth first search through $\mathcal{F}^0 \cup \cdots \cup \mathcal{F}^{2p}$ which originates at x^p. Therefore, $y \in Reach(x)$ if and only if there exists a path from x^p to y^p in $\mathcal{G}$, restricted to $\mathcal{F}^0 \cup \cdots \cup \mathcal{F}^{2p}$.

Lemma 8. *If x^i, y^i are both in infinite components, they are in the same component iff $y \in Reach(x)$.*

Proof. Assume x^i, y^i are in infinite components. Suppose $y \in Reach(x)$. There is a path P in $\mathcal{G}$ from x^p to y^p. Let ℓ be least such that $\mathcal{F}^\ell \cap P \neq \emptyset$. If $i \ge p - \ell$, then x^i and y^i are in the same component. Thus, suppose $i < p - \ell$. Let z be such that $z^\ell \in P$. Then P is P_1P_2 where P_1 is a path from x^p to z^ℓ and P_2 is a path from z^ℓ to y^p. By Lemma 3, since x^i is in an infinite component, so is x^p. There is $r > 0$ such that the set $\{x^{p+rm} \mid m \in \omega\}$ is contained in a single component. Likewise, there is an $r' > 0$ such that $\{y^{p+r'm} \mid m \in \omega\}$ is in one component. Consider $x^{p+rr'}$ and $y^{p+rr'}$. There is a path $P_1'P_2'$ from $x^{p+rr'}$ to $y^{p+rr'}$. A second path P' from x^p to y^p goes from x^p to $x^{p+rr'}$, then along $P_1'P_2'$ from $x^{p+rr'}$ to $y^{p+rr'}$, and finally to y^p. The least ℓ' such that $\mathcal{F}^{\ell'} \cap P' \neq \emptyset$ is larger than ℓ. Iteratively lengthening the path between x^p and y^p until $i < p - \ell'$ brings us to the previous case.

To prove the implication in the other direction, we assume that x^i and y^i are in the same infinite component. We want to prove that $y \in Reach(x)$. Let $i' = \min\{p, i\}$. Let P be a path in G from x^i to y^i. We use P to construct a path which stays in $\mathcal{F}^{i-i'} \cup \cdots \cup \mathcal{F}^{i+p}$. Let $\ell(P)$ be largest such that $P \cap \mathcal{F}^{\ell(P)} \neq \emptyset$; let $\ell'(P)$ be least such that $P \cap \mathcal{F}^{\ell'(P)} \neq \emptyset$. If $i - i' \leq \ell'(P)$ and $\ell(P) \leq i + p$, we are done. Otherwise, let $P_1, \ldots, P_k$ be a sequence of subpaths of P, each beginning and ending in $\mathcal{F}^i$, such that $P = P_1 \cdots P_k$ and for each $1 \leq j \leq k$, $\ell(P_j) = i$ or $\ell'(P_j) = i$. It is not hard to see that each P_j can be replaced by a path P_j' with the same start and end points and which satisfies $i - i' \leq \ell'(P_j') \leq \ell(P_j') \leq i + p$. This new path witnesses that $y \in Reach(x)$. □

We inductively define a sequence $Cl_0(x), Cl_1(x), \ldots$ such that each $Cl_k(x)$ is a subset of $V_{\mathcal{F}}$. Set $Cl_0(x) = Reach(x)$. For $k > 0$, set $Cl_k(x) = Reach(\sigma(Cl_{k-1}(x)))$.

Lemma 9. *Suppose $j \geq i$ and x^i, y^j are both in infinite components. x^i and y^j are in the same component if and only if $y \in Cl_{j-i}(x)$.* □

The following algorithm uses the lemma to solve the reachability problem.
`ALG: NaïveReach`

1. Check if each of x^i, y^j are in an infinite component of $\mathcal{G}$ (see Theorem 4).
2. If exactly one of x^i and y^j is in a finite component, then return *NO*.
3. If both x^i, y^j are in finite components, run `FiniteReach` on x^i and check if $(y, j - i) \in Q$.
4. If both x^i and y^j are in infinite components, check if $y \in Cl_{j-i}(x)$.

`Naïve Reach` computes $Cl_0(x)$ in time $O(p^3)$. Given $Cl_{k-1}(x)$, we can compute $Cl_k(x)$ in time $O(p^4)$. Hence, on input x^i, y^j, `NaïveReach` takes time $O((j - i) \cdot p^4)$. We will now improve this bound. From Lemma 5, x^i is in an infinite component in $\mathcal{G}$ if and only if there is an oriented cycle $\mathcal{C}$ with positive net length, zero low point, and reachable from x by a simple oriented walk with low point $\geq -i$. Assume x^i is in an infinite component. The algorithm for the infinity testing problem finds such an oriented cycle $\mathcal{C}$. And, it can compute the net length r of $\mathcal{C}$. All vertices in $\{x^{i+mr} \mid m \in \omega\}$ belong to the same component.

Lemma 10. $Cl_0(x) = Cl_r(x)$. □

We give a new algorithm, `Reach`, by replacing line (4) in `NaïveReach` with: (4') If x^i and y^j belong to infinite components, compute $Cl_0(x)$, ..., $Cl_{r-1}(x)$. If $y \in Cl_k(x)$ for $k < r$ with $j - i = k \mod r$, return *YES*; otherwise, return *NO*.

Proof (Theorem 5). By Lemma 9 and Lemma 10, `Reach` returns *YES* iff x^i and y^j are in the same component. Calculating $Cl_0(x), \ldots, Cl_{r-1}(x)$ requires time $O(p^5)$. Therefore the running `Reach` on x^i, y^j takes $O(i + j + p^5)$. □

In fact, the algorithm produces $k < p$ such that to check if x^i, y^j $(j > i)$ are in the same component, we need to test if $j - i < p$ and if $j - i = k \mod p$. If $\mathcal{G}$ is fixed, we may pre-compute $Cl_0(x), \ldots, Cl_{r_x-1}(x)$ for all x, so deciding if two vertices u, v belong to the same component takes linear time. The above proof can also be used to build a unary automaton that decides reachability uniformly.

Corollary 2. *With $\mathcal{G}$ as above, there is a deterministic automaton with at most $2p^4 + p^3$ states that solves the reachability problem on $\mathcal{G}$. The time required to construct this automaton is $O(p^6)$.* □

This corollary can be applied to solve the **connectivity problem**.

Theorem 6. *The connectivity problem for unary automatic graphs of finite degree is solved in time $O(p^6)$, where p is the loop constant of the unary automaton.* □

References

1. Blumensath, A.: Automatic Structures. Diploma Thesis, RWTH Aachen (1999)
2. Blumensath, A., Grädel, E.: Finite presentations of infinite structures: Automata and interpretations. Theory of Computing Systems 37, 642–674 (2004)
3. Bouajjani, A., Esparza, J., Maler, O.: Reachability analysis of pushdown automata: Application to model-checking. In: Mazurkiewicz, A., Winkowski, J. (eds.) CONCUR 1997. LNCS, vol. 1243, pp. 135–150. Springer, Heidelberg (1997)
4. Büchi, J.R.: On a decision method in restricted second-order arithmetic. In: Proc. CLMPS, pp. 1–11. Stanford University Press (1960)
5. Esparza, J., Hansel, D., Rossmanith, P., Schwoon, S.: Efficient algorithms for model checking pushdown systems. In: Emerson, E.A., Sistla, A.P. (eds.) CAV 2000. LNCS, vol. 1855, pp. 232–247. Springer, Heidelberg (2000)
6. Hell, P., Nešetřil, J.: Graphs and Homomorphisms. Oxford University Press, Oxford (2004)
7. Khoussainov, B., Nerode, A.: Automatic presentation of structures. In: Leivant, D. (ed.) LCC 1994. LNCS, vol. 960, pp. 367–392. Springer, Heidelberg (1995)
8. Khoussainov, B., Nies, A., Rubin, S., Stephan, F.: Automatic structures: richness and limitations. In: Proc. LICS, pp. 44–53 (2004)
9. Khoussainov, B., Rubin, S.: Graphs with automatic presentations over a unary alphabet. J. of Automata, Languages and Combinatorics 6(4), 467–480 (2001)
10. Khoussainov, B., Rubin, S., Stephan, F.: Automatic linear orders and trees. ACM Trans. Comput. Log. 6(4), 675–700 (2005)
11. Lohrey, M.: Automatic structures of bounded degree. In: Vardi, M.Y., Voronkov, A. (eds.) LPAR 2003. LNCS (LNAI), vol. 2850, pp. 344–358. Springer, Heidelberg (2003)
12. Oliver, G.P., Thomas, R.M.: Automatic presentations for finitely generated groups. In: Diekert, V., Durand, B. (eds.) STACS 2005. LNCS, vol. 3404, pp. 693–704. Springer, Heidelberg (2005)
13. Rubin, S.: Automatic Structures, PhD Thesis, University of Auckland (2004)
14. Thomas, W.: A short introduction to infinite automata. In: Kuich, W., Rozenberg, G., Salomaa, A. (eds.) DLT 2001. LNCS, vol. 2295, pp. 130–144. Springer, Heidelberg (2002)

Search Space Reductions for Nearest-Neighbor Queries

Micah Adler[1] and Brent Heeringa[2]

[1] Department of Computer Science, University of Massachusetts, Amherst
140 Governors Drive Amherst, MA 01003
[2] Department of Computer Science, Williams College, Williamstown, MA, 01267
micah@cs.umass.edu, heeringa@cs.williams.edu

Abstract. The vast number of applications featuring multimedia and geometric data has made the R-tree a ubiquitous data structure in databases. A popular and fundamental operation on R-trees is nearest neighbor search. While nearest neighbor on R-trees has received considerable experimental attention, it has received somewhat less theoretical consideration. We study pruning heuristics for nearest neighbor queries on R-trees. Our primary result is the construction of non-trivial families of R-trees where k-nearest neighbor queries based on pessimistic (i.e. min-max) distance estimates provide exponential speedup over queries based solely on optimistic (i.e. min) distance estimates. The exponential speedup holds even when $k = 1$. This result provides strong theoretical evidence that min-max distance heuristics are an essential component to depth-first nearest-neighbor queries. In light of this, we also consider the time-space tradeoffs of depth-first versus best-first nearest neighbor queries and construct a family of R-trees where best-first search performs exponentially better than depth-first search even when depth-first employs min-max distance heuristics.

1 Introduction

Nearest neighbor queries on the R-tree play an integral role in many modern database applications. This is due in large part to the prevalence and popularity of multimedia data indexed geometrically by a vector of features. It is also because nearest neighbor search is a common primitive operation in more complex queries [1].

Although the performance of nearest neighbor search on R-trees has received some theoretical consideration (e.g., [2,3]), its increasing prominence in todays computing world warrants even further investigation. The authors of [1] note that three issues affect the performance of nearest neighbors on R-trees:

- the order in which children are visited,
- the traversal type, and
- the pruning heuristics.

We show that at least two of these — traversal type and pruning heuristics — have a quantitatively profound impact on efficiency. In particular we prove the following:

M. Agrawal et al. (Eds.): TAMC 2008, LNCS 4978, pp. 554–567, 2008.

1. There exists a family of R-trees where depth-first k-nearest neighbor search with pessimistic (i.e. min-max) distance pruning performs exponentially better than optimistic (i.e. min) distance pruning alone. This result holds even when $k = 1$.
2. There exists a family of R-trees where best-first k-nearest neighbor queries perform exponentially better than depth-first nearest neighbor queries even when the depth-first search uses both optimistic and pessimistic pruning heuristics. This result also holds when $k = 1$.

Our first result provides strong theoretical evidence that pruning strategies based on pessimistic distance estimates are valuable in depth-first nearest neighbor queries. These results rely on subtle changes to existing algorithms. In fact, without these nuanced changes, the exponential speedup may completely disappear.

Our second result deals with the known time efficiency benefits of best-first nearest neighbor algorithms over depth-first nearest neighbor algorithms. Given our first result, it is natural to ask whether pessimistic distance pruning closes part of the time efficiency gap. We answer this question in the negative through several general constructions. Still, the benefit of pessimistic pruning strategies should not be overlooked. They provably enhance the time-efficiency of the already space-efficient depth-first nearest neighbor queries. Such algorithms still play an increasingly prominent role in computing given the frequency and demand for operations on massive data sets.

The outline of this paper is as follows: In sections 2 and 3 we briefly review definitions for R-trees, MinDist, MinMaxDist, and the three common pruning heuristics employed in depth-first nearest neighbor queries. Sections 4 and 5 describe our R-tree constructions and prove the power of pessimistic pruning. Section 6 discusses the time-space tradeoffs of best-first versus depth-first nearest neighbor queries. We conclude in Section 7.

2 Background

R-trees [4] and their variants (e.g. [5,6]. See [1] for a list of others) are data structures for organizing spatial objects in Euclidean space. They support dynamic insertion and deletion operations. Internal and leaf nodes contain records. A record r belonging to an internal node is a tuple $\langle M, \mu\rangle$ where μ is a pointer to the child node of r and M is an n-dimensional minimum bounding rectangle (MBR). M tightly bounds the spatial objects located in the subtree of r. For example, given the points $(1, 2)$, $(4, 5)$, and $(3, 7)$ in 2-space, the MBR would be $\langle(1, 4), (2, 7)\rangle$. The records of leaf nodes are also tuples but have the form $\langle M, o\rangle$ where o is either the actual spatial object or a reference to it.

The number of records in a node is its branching factor. Every node of an R-tree contains between b and B records where both b and B are positive integers and $b \leq \lfloor \frac{B}{2} \rfloor$. One exception is the root node which must have at least two records. R-trees are completely balanced—all leaf nodes have the same depth. Figure 1 depicts an example collection of spatial objects, their MBRs, and their R-tree. More details on R-trees are available in [1].

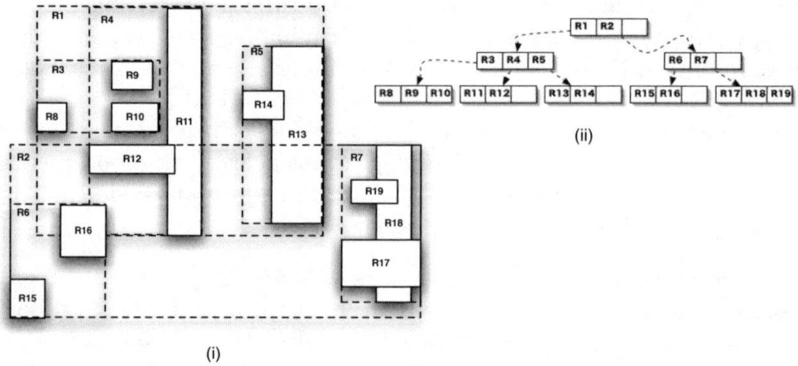

Fig. 1. (i) A collection of spatial objects (solid lines) and their hierarchy of minimum bounding rectangles (dashed lines). (ii) An R-tree for the objects in (i).

We consider nearest neighbor searches on R-trees. In all cases these searches have the form: *Given a query point q and an R-tree T with spatial objects of matching dimension to q, find the k-nearest objects to q in T.* Nearest here and throughout the rest of the paper is defined by *Euclidean distance.*

3 Nearest Neighbors

There are two dominant nearest neighbor algorithms for the R-tree. The first is a best-first search algorithm (denoted HS) due to Hjatlson and Samet [7]. HS is optimal in the sense that it only searches nodes with bounding boxes intersecting the k-nearest neighbor hypersphere [7,8]. However, it has worst case space complexity that is linear in the total number of tree nodes. With large data sets this cost may become prohibitive [3].

The second algorithm due to Roussopoulos et al. [9] (denoted here by RKV) is a branch and bound depth-first search. RKV employs several heuristics to prune away branches of the tree. We define and discuss the subtlety of these heuristics below. While RKV may search more nodes than HS, it has worst-case space complexity that is only logarithmic in the number of tree nodes. In addition, the authors of [10] note that statically constructed indices map all pages on a branch to contiguous regions on disk, so a depth-first search may "yield fewer disk head movements than the distance-driven search of the HS algorithm." In these cases RKV may be preferable to HS for performance reasons beyond space complexity.

3.1 Distances

RKV uses three different strategies (here called H1, H2, and H3 respectively and defined formally below) to prune branches. H3 is based on a measure called MinDist which gives the actual distance between a node and a query point. In

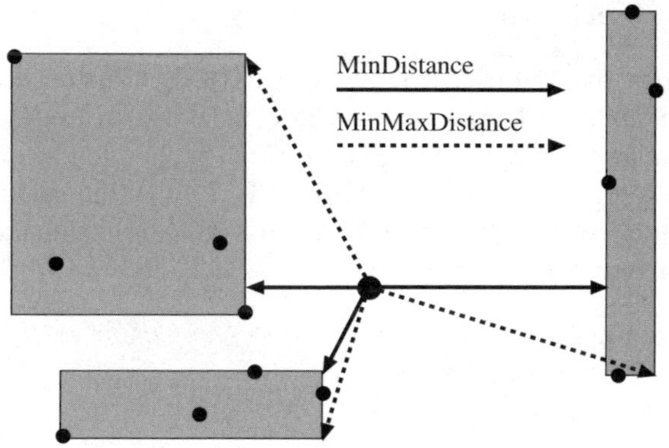

Fig. 2. A visual explanation of MinDist and MinMaxDist in two dimensions

other words, MinDist(q, M) is the length of the shortest line between the query point q and the nearest face of the MBR M. When the query point lies within the MBR, MinDist is 0. Figure 2 shows the MinDist values for a query point and three minimum bounding rectangles. Because an MBR tightly encapsulates the spatial objects within it, each face of the MBR must touch at least one of the objects it encloses [9]. This is called the MBR face property.

Figure 2 shows that MinDist is a lower bound or optimistic estimate of the distance between the query point and some spatial object inside its MBR. However, the actual distance between a query point and the closest object may be much larger.

H1 and H2 use a second measure called MinMaxDist which provides an upper bound on the distance between an actual object in a node and a query point. In other words, MinMaxDist provides a pessimistic estimate of the distance between the query point and some spatial object within its MBR. Figure 2 depicts these distances for a query point and three MBRs. From its name one can see MinMaxDist is calculated by finding the minimal distance from a set of maximal distances. This set of maximal distances is formed as follows: Suppose we have an n-dimensional minimum bounding rectangle. If we fix one of the dimensions, we are left with two $n-1$ dimensional hyperplanes; one representing the MBR's lower bounds, the other representing its upper bounds. We know from the MBR face property that at least one spatial object touches each of these hyperplanes. However, given only the MBR, we cannot identify this location. But, given a query point, we can say that an object is at least as close as the distance from that point to the farthest point on the closest hyperplane. This distance is an upper bound on the distance between the query point and a spatial object located within the MBR. By iteratively fixing each dimension of an MBR and finding the upper bound, we can form the set of maximal distances. Since each maximal distance is an upper bound, it follows that the minimum of these is also an upper bound. This minimum distance is what we call the MinMaxDist(q, M) of a query point q and an MBR M.

3.2   Pruning Heuristics

Pruning strategies based on MINDIST and MINMAXDIST potentially remove large portions of the search space. The following three strategies were originally defined in [9] for use in RKV. All assume a query point q and a list of MBRs $\mathcal{M}$ (to potentially prune) sorted by MINDIST. The latter assumption is based on empirical results from both [9] and [7]. In addition, two strategies, H2 and H3, assume a current nearest object o.

Definition 1 (H1). *Discard any MBR $M_i \in \mathcal{M}$ if there exists $M_j \in \mathcal{M}$ with* MINDIST$(q, M_i) >$ MINMAXDIST(q, M_j)

Definition 2 (H2). *Discard o if there exists $M_i \in \mathcal{M}$ such that* MINMAXDIST $(q, M_i) <$ DIST(q, o).

Definition 3 (H3). *Discard any minimum bounding rectangle $M \in \mathcal{M}$ if* MINDIST$(q, M) >$ DIST(q, o)

Both Cheung et al. [11] and Hjaltason et al. [7] show that any node pruned by H1 is also pruned by H3. Furthermore, they note that H2 serves little purpose since it does not perform any pruning. This has led to the development of simpler but behaviorally-identical versions of RKV that rely exclusively on H3 for pruning. As a result, *we take* RKV *to mean the original* RKV *without H1 and H2.*

Algorithm 1. 1NN(q, n, e)

Require: A query point q, a node n and a nearest neighbor estimate e. e may be a distance estimate object or a (pointer to a) spatial object.

```
 1: if LEAFNODE(n) then
 2:    for ⟨M, o⟩ in records[n] do {M is a MBR, o is a (pointer to a) spatial object}
 3:      if DIST(q, M) ≤ DIST(q, e) then
 4:         e ← o
 5:      end if
 6:    end for
 7: else
 8:    ABL ← SORT(records[n]) {Sort records by MINDIST}
 9:    for ⟨M, μ⟩ in ABL do {M is an MBR, μ points to a child node}
10:      if MINMAXDIST(q, M) ≤ e then {H2* Pruning}
11:         e ← MINMAXDIST(q, M)))
12:      end if
13:    end for
14:    for ⟨M, μ⟩ in ABL do {M is an MBR, μ points to a child node}
15:      if MINDIST(q, M) < DIST(q, e) then {H3 Pruning}
16:         1NN(q, μ, e)
17:      end if
18:    end for
19: end if
```

The benefits of MinMaxDist, however, should not be overlooked — it can provide very useful information about unexplored areas of the tree. The key is to replace an actual object o with a distance estimate e (some call this the closest point candidate) and then adjust H2 so that we *replace* the estimate with the MinMaxDist instead of *discarding* the object. This gives us a new definition of H2 which we call H2*.

Definition 4 (H2*). *Replace e with* MinMaxDist(q, M_i) *if there exist $M_i \in \mathcal{M}$ such that*
MinMaxDist$(q, M_i) < e$.

This definition is not new. In fact, the authors of [2] use *replace* instead of *discard* in their description of H2. However, updating the definition of H2 to H2* in RKV does not yield the full pruning power of MinMaxDist. We need to apply H2* early in the search process. This variation on RKV yields Algorithm 1 which we refer to it as 1NN. Note that the RKV traditionally applies H2 after line 1. This diminishes the power of pessimistic pruning. In fact, our exponential speedup results in Sections 4 and 5 hold even when H2* replaces H2 in the original algorithm.

3.3 k-Nearest Neighbors

Correctly generalizing H2* to k-nearest neighbor queries is essential in light of the potential power of pessimistic pruning. However, as Böhm et. al [10] point out, such an extension takes some care.

We begin by replacing e with a priority queue L of k-closest neighbors estimates. Note that H2* doesn't perform direct pruning, but instead, updates the neighbor estimate when distance guarantees can be made. If the MinMaxDist of a node is less than the current distance estimate, then we can update the estimate because the future descent into that node is guaranteed to contain an object with actual distance at most MinMaxDist. We call estimates in updates of this form *promises* because they are not actual distances, but are upper bounds on distances. Moreover each estimate is a promise of, or place holder for, a spatial object that is as least as good the promise's prediction. A natural but incorrect generalization of H2 places a promise in the priority queue whenever the maximum-distance element in L is farther away than the MinMaxDist. This leads to two problems. First, multiple promises may end up referring to the same spatial object. Second, a promise may persist past its time and eventually refer to a spatial object already in the queue. These problems are depicted visually in Figure 3. The key to avoiding both problems is to always remove a promise from the queue before searching the node which generated it; it will always be replaced by an equal or better estimate or by an actual object. This leads us to the following generalization of H2 which we call Promise-Pruning:

Definition 5 (Promise-Pruning). *If there exists $M_i \in \mathcal{M}$ such that $\delta(q, M_i) =$* MinMaxDist$(q, M_i) < Max(L)$*, then add a promise with distance $\delta(q, M_i)$ to L. Additionally, replace any promise with distance $\delta(q, M_i)$ from L with ∞ before searching M_i.*

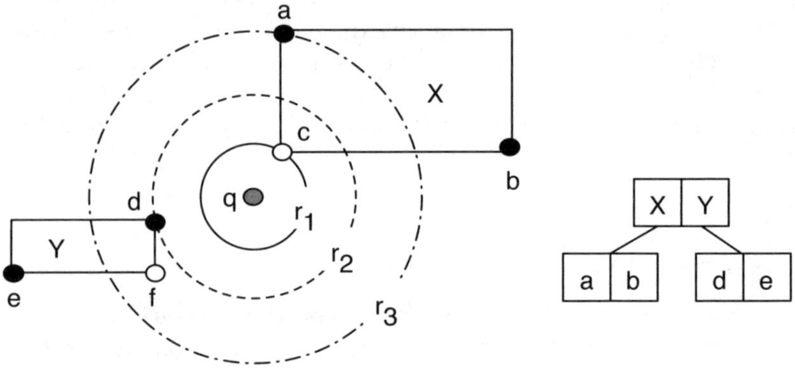

Fig. 3. Blindly inserting promises into the queue, without removing them correctly, wreaks havoc on the results. For example, when performing a 2-nearest neighbor search on the tree above, a promise with distance f is placed in the queue at the root. After investigating X, the queue retains f and a. However, if f is not removed before descending into Y, the final distances in the queue are e and f — an incorrect result.

This generalization is tantamount to an extension suggested by Böhm et al. in [10]. Our primary contribution is to show that this extension, when performed at the right point, may provide an exponential performance speedup on depth-first nearest neighbor queries.

For completeness, we also provide generalizations of H1 and H3 which we call K1 and K3 respectively. We also prove that K3 dominates K1, just as it did in the 1-nearest neighbor case. This proof is a simple extension of those appearing in [11,7].

Definition 6 (K1). *Discard any minimum bounding rectangle $M_i \in \mathcal{M}$ if there exists $\mathcal{M}' \subseteq \mathcal{M}$ such that $|\mathcal{M}'| \geq k$ and for every $M_j \in \mathcal{M}'$ it is the case that* $\text{MinDist}(q, M_i) > \text{MinMaxDist}(q, M_j)$

Definition 7 (K3). *Discard any minimum bounding rectangle $M_i \in \mathcal{M}$ if* $\text{MinDist}(q, M_i) > Max(L)$ *where Max returns the largest estimate in the priority queue.*

Theorem 1. *Given a query point q, a list of MBRs $\mathcal{M}$, and a priority queue of k closest neighbor estimates L, any MBR pruned by K1 in the depth-first RKV algorithm is also pruned by K3.*

Proof. Suppose we are performing a k nearest neighbor search with query point q and K1 prunes MBR M from $\mathcal{M}$. From Definition 6 there exists $\mathcal{M}' \subset \mathcal{M}$ such that $|\mathcal{M}'| \geq k$ and every M' in $\mathcal{M}'$ has $\text{MinMaxDist}(q, M') < \text{MinDist}(q, M)$. Since for any MBR N, $\text{MinDist}(q, N) \leq \text{MinMaxDist}(q, N)$, each M' in $\mathcal{M}'$ will be searched before M because $\mathcal{M}$ is sorted by MinDist. Because each M' in $\mathcal{M}'$ is guaranteed to contain a spatial object with actual distance at most that of than an object found in M and since we have $|\mathcal{M}'| \geq k$ we know $\text{Max}(L) < \delta(q, M)$. Therefore, from Definition 7, M would also be pruned using K3.

Algorithm 2. KNN(k, q, n, L)

Require: An integer k, a query point q, a node n and a priority queue L of fixed size k. L initially contains k neighbor estimates with distance ∞.

```
if LeafNode(n) then
  for ⟨M, o⟩ in records[n] do {M is a MBR, o is a (pointer to a) spatial object}
    if Dist(q, M) < max(L) then {Here max returns the object or estimate of
    greatest distance}
      insert(L, o) {Inserting o into L replaces some other estimate or object.}
    end if
  end for
else
  ABL ← Sort(records[n]) {Sort records by MinDist }
  for ⟨M, μ⟩ in ABL do {M is an MBR, μ points to a child node}
    if MinMaxDist(q, M) < max(L) then {Promise-Pruning}
      insert(L, Promise(MinMaxDist(q, M)))
    end if
  end for
  for ⟨M, μ⟩ in ABL do {M is an MBR, μ points to a child node}
    if MinDist(q, M) < max(L) then {K3 Pruning}
      if L contains a promise generated from M then
        remove(L, Promise(MinMaxDist(q, M)))
      end if
      KNN(k, q, μ, L)
    end if
  end for
end if
```

Theorem 1 means K1 is redundant with respect to K3, so we do not use it in our depth-first k-nearest neighbor procedure outlined in Algorithm 2. We call this procedure KNN.

4 The Power of Pessimism

In this section we show that 1NN may perform exponentially faster than RKV despite the fact that it differs only slightly from the original definition. As we noted earlier, the original RKV is equivalent to RKV with only H3 pruning. Thus, RKV is Algorithm 1 without lines 9-13.

Theorem 2. *There exists a family of R-tree and query point pairs* $\mathcal{T} = \{(T_1, q_n), \ldots, (T_m, q_m)\}$ *such that for any* (T_i, q_i)*, RKV examines* $O(n)$ *nodes and* 1NN *examines* $O(\log n)$ *nodes on a 1-nearest neighbor query.*

Proof. For simplicity, we restrict our attention to R-trees composed of points in $\mathbb{R}^2$ so that all the MBRs are rectangles. Also, we only construct complete binary trees where each node has two records. The construction follows the illustration in Figure 4. Let $\delta(i, j)$ be the Euclidean distance from point i to point j and let (i, j) be their rectangle. Let q be the query point. Choose three points a,

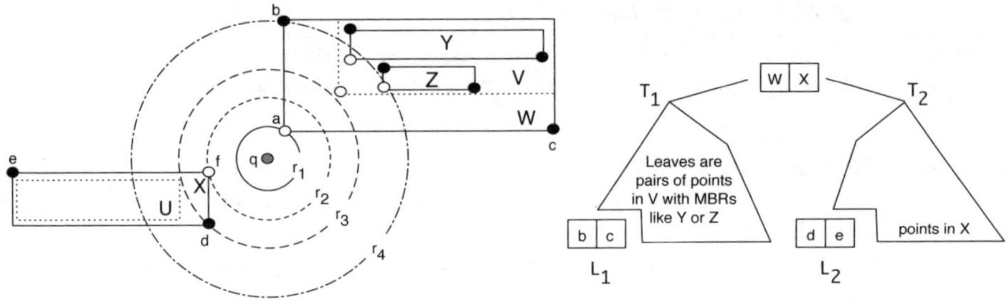

Fig. 4. A visual explanation of the tree construction for Theorem 2

b, and c such that $\delta(q,a) = r_1 < \delta(q,b) = r_4 < \delta(q,c)$ and (b,c) forms a rectangle W with a corner a. Similarly, choose three points d, e, and f such $r_1 < \delta(q,f) = r_2 < \delta(q,d) = r_3 < r_4 < \delta(q,e)$ and (d,e) forms a rectangle X with corner f. Let T be a complete binary tree over n leaves where each node has two records. Let T_1 be the left child of T and let T_2 be the right child of T. Let L_1 be the far left leaf of T_1 and let L_2 be the far left leaf of T_2. Place b and c in L_1, and d and e in L_2. In the remaining leaves of T_1, place pairs of points (p_i, p_j) such that p_i and p_j are interior to V, $\delta(q,p_i) > r_4$ and $\delta(q,p_j) > r_4$ but (p_i,p_j) form a rectangle with corner p' such that $r_3 < \delta(q,p') < r_4$. Rectangles Y and Z in Figure 4 are examples of this family of point pairs. In the remaining leaves of T_2 place pairs of points (p_k,p_l) such that p_k and p_l are interior to U (*i.e.*, so that $\text{MINDIST}(q,(p_k,p_l)) > r_3$). The construction yields a valid R-tree because we place pairs of points at the leaves and build up the MBRs of the internal nodes accordingly.

Claim. Given a tree T and query point q as constructed above, both RKV and 1NN prune away all of T_2 save the left branch down to L_2 on a 1-nearest neighbor query.

Proof. Note that d is the nearest neighbor to q in T so both algorithms will search T_2. L_2, by construction, has the least MINDIST of any subset of points in T_2, so both algorithms, when initially searching T_2, will descend to it first. Since $\delta(q,d)$ is the realization of this MINDIST and no other pair of points in X has MINDIST $<$ $\delta(q,d)$, both algorithms will prune away the remaining nodes using H3.

Now we'll show that RKV must examine all the nodes in T_1 while 1NN can use information from X to prune away all of T_1 save the left branch down to L_1.

Lemma 1. *Given an R-tree T and query point q as constructed above, RKV examines every node in T_1 on a 1-nearest neighbor query.*

Proof. Since $\text{MINDIST}(q,W) = \delta(q,a)$, RKV descends to L_1 first and claims b as its nearest neighbor. However, RKV is unable to prune away any of the remaining leaves of T_1. To see this, let $L_i = (p_{i1}, p_{i2})$ and $L_j = (p_{j1}, p_{j2})$ be distinct leaves of

T_1 (but not L_1). Note that $\text{MinDist}(q,(p_{i1},p_{i2})) < r_4 < \min(\delta(q,p_{j1}),\delta(q,p_{j2}))$ and $\text{MinDist}(q,(p_{j1},p_{j2})) < r_4 < \min(\delta(q,p_{i1}),\delta(q,p_{i2}))$. This means that the MinDist of any leaf node is at most r_4 but every point is at least r_4 so RKV must probe every leaf. As a result, it cannot prune away any branches.

Lemma 2. *Given a tree T and query point q as constructed above,* 1NN *prunes away all nodes in T_1 except those on the branch leading to L_1 in a 1-nearest neighbor query.*

Proof. 1NN uses the MinMaxDist information from X as an indirect means of pruning. Before descending into T_1, the algorithm updates its neighbor estimate with $\delta(q,d)$. Like RKV, 1NN descends into T_1 directly down to L_1 since $\delta(q,W) < \delta(q,d)$ and W has the smallest MinDist of all the nodes. Unlike RKV, it reaches and ignores b because $\delta(q,d) < \delta(q,b)$. In fact, the promise granted by X allows us to prune away all other branches of the tree since the remaining nodes are all interior to V and $\delta(q,d) < \text{MinDist}(q,V)$.

The theorem follows from Lemma 1 and Lemma 2. RKV searches all of T_1 ($O(n)$ nodes) while 1NN searches only the paths leading to L_1 and L_2 ($O(\log n)$ nodes). As a consequence, 1NN can reduce the search space exponentially over RKV. □

In the original RKV, H2 pruning appears on line 1. Our results hold even if we replace H2 with H2*. This is because all the pruning in T_1 relies on the MinMaxDist found at the root node. Hence the promotion of pessimistic pruning in Algorithm 1 plays a crucial role in the performance of depth-first nearest neighbor queries.

5 Search Space Reductions with K-Nearest Neighbors

Here we show that the benefits 1NN reaps from MinMaxDist extend to KNN when H2* is properly generalized to Promise-Pruning. In particular, we construct a class of R-trees where KNN reduces the number of nodes visited exponentially when compared with RKV. As in Section 4, we take RKV to mean Algorithm 2 without lines 9-13 (and additionally lines 16-18).

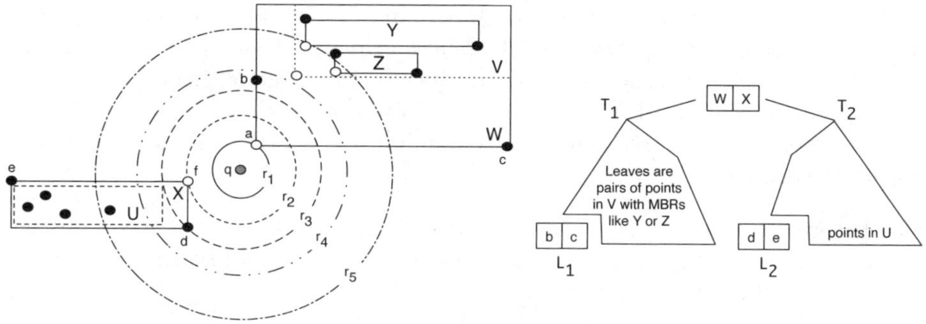

Fig. 5. A visual explanation of the tree construction used in Theorem 3

Theorem 3. *There exists a family of R-trees and query point pairs* $\mathcal{T} = \{(T_1, q_n), \ldots, (T_m, q_m)\}$ *such that for any* (T_i, q_i), *RKV examines* $O(n)$ *nodes and KNN examines* $O(\log n)$ *nodes on a 2-nearest neighbor query.*

Proof. The proof follows the outline of Theorem 2. The construction is similar to Figure 4 except that b shifts slightly down and V shifts slightly up so that $\delta(q, b) < \textsc{MinDist}(q, V)$. We illustrate this in Figure 5 and give details where the two proofs differ.

Claim. Given a tree T and query point q as constructed in Figure 5, both KNN and RKV prune away all of T_2 save the left branch down to L_2 on a 2-nearest neighbor query.

Proof. Note that b and d are the two nearest-neighbors to q in T. Both algorithms will search T_1 first because $\textsc{MinDist}(q, W) < \textsc{MinDist}(q, X)$. Since both algorithms are depth-first searches, b is in tow by the time T_2 is searched. Because d has yet to be realized, both algorithms will search T_2 and prune away the remaining nodes just as in 4.

Just as before, we'll show that RKV must examine all the nodes in T_1 while KNN can use information from X to prune away all of T_1 save the left branch down to L_1.

Lemma 3. *Given an R-tree* T *and query point* q *as constructed above, RKV examines every node in* T_1 *in a 2-nearest neighbor search.*

Proof. Since $\textsc{MinDist}(q, W) = \delta(q, a)$, RKV descends to L_1 first and inserts b (and c) into its 2-best priority queue. However, RKV is unable to prune away any of the remaining leaves of T_1 because every pair of leaf points have $\textsc{MinDist}$ at most r_5 but all points in T_1 (besides b) lie outside r_5 . As a result RKV must probe every leaf.

Lemma 4. *Given a tree* T *and query point* q *as constructed above, KNN prunes away all nodes in* T_1 *except those on the branch leading to* L_1 *in a 2-nearest neighbor search.*

Proof. Before descending into T_1, KNN inserts a promise with distance $\textsc{MinMaxDist}(q, X) = \delta(q, d)$ into its 2-best priority queue. The algorithm descends into T_1 directly down to L_1, finding b and inserting it into its 2-best priority queue. Unlike RKV, the promise granted by X allows us to prune away all other nodes of the tree since the remaining nodes are all interior to V and $\delta(q, d) < \textsc{MinDist}(q, V)$.

The theorem follows from Lemma 3 and Lemma 4. Note that if KNN did not remove the promise granted by d at X the final result would be the point d and its promise – an error. □

We can generalize the construction given in Theorem 3 so that the exponential search space reduction holds for any k nearest neighbor query. In particular,

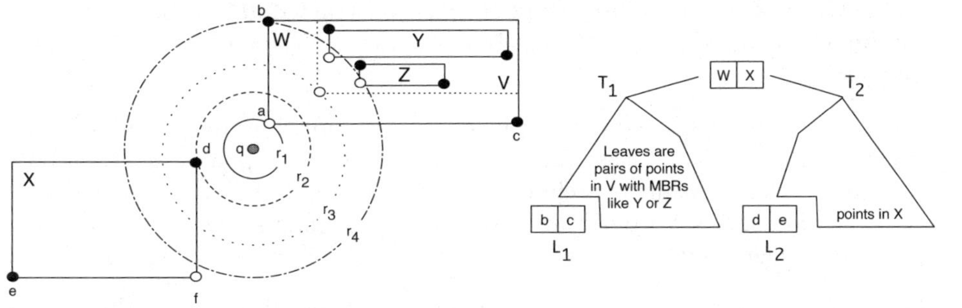

Fig. 6. A visual explanation of the tree construction used in Theorem 4

for any constant $k > 0$ there exists a class of R-tree and query point pairs for which KNN reduces the search space exponentially over RKV. A simple way of accomplishing this is to place $k-1$ points at b and insert them into the far left leaves of T_1.

6 Time / Space Tradeoffs

Given the search space reductions offered by 1NN and KNN, it is natural to ask if the space-efficient depth-first algorithms can approach the time efficiency of the best-first algorithms. In other words, how does KNN stack up against HS? We answer this question here in the negative by constructing a family of R-trees where HS performs exponentially better than KNN. The HS algorithm uses a priority queue to order the nodes by MinDist. It then performs a best-first search by MinDist while pruning away nodes using K3. We direct the reader to [7] for more details.

Theorem 4. *There exists a family of R-tree and query point pairs $\mathcal{T} = \{(T_1, q_n), \ldots, (T_m, q_m)\}$ such that for any (T_i, q_i), HS examines $O(\log n)$ nodes and* 1NN *examines $O(n)$ nodes on a 1-nearest neighbor query.*

Proof. This construction resembles the construction in Theorem 2 and is depicted visually in Figure 6. We organize T_1 in exactly the same way as Theorem 2: we choose three points a, b, and c such that $\delta(q, a) = r_1 < \delta(q, b) = r_4 < \delta(q, c)$ and (b, c) forms a rectangle W with a corner a. Next we choose three points d, e, and f such $r_1 < \delta(q, d) = r_2 < r_4 < \delta(q, f) < \delta(q, e)$ and (d, e) forms a rectangle X with corner f. Let T be a complete binary tree over n leaves where each node has two records. Let T_1, T_2, L_1, and L_2 be as in Theorem 2. Place b and c in L_1, and d and e in L_2. In the remaining leaves of T_1, place pairs of points (p_i, p_j) such that p_i and p_j are interior to V, $\delta(q, p_i) > r_4$ and $\delta(q, p_j) > r_4$ but (p_i, p_j) form a rectangle with corner p' such that $r_3 < \delta(q, p') < r_4$. Rectangles Y and Z in Figure 6 are examples of this family of point pairs. In the remaining leaves of T_2 place pairs of points (p_k, p_l) such that p_k and p_l are interior to X.

Claim. Given a tree T and query point q as constructed above, both 1NN and HS prune away all of T_2 save the left branch down to L_2 on a 1-nearest neighbor query.

Proof. d is the nearest neighbor to q in T so both algorithms must search T_2. L_2, by construction, has the least MINDIST of any subset of points in T_2, so both algorithms, when initially searching T_2, will descend to it first. Since $\delta(q, d)$ is the realization of this MINDIST and no other pair of points in X has MINDIST $<$ $\delta(q, d)$, both algorithms will prune away the remaining nodes using H3.

Now we'll show that 1NN must examine all the nodes in T_1 while HS can use the MINDIST of X to bypass searching all of T_1 save the path down to L_1.

Lemma 5. *Given an R-tree T and query point q as constructed above,* 1NN *examines every node in T_1 on a 1-nearest neighbor query.*

Proof. 1NN descends into T_1 before T_2 since $\text{MINDIST}(q, W) < \text{MINDIST}(q, X)$. Note that the distance estimate delivered by f is useless given that every pair of leaf points in T_1 has MINDIST less than $\delta(q, f)$. 1NN will descend to L_1 and find b but it cannot rule out the rest of T_1 because every pair of leaf points forms a rectangle with MINDIST smaller than r_4. To see this, let $L_i = (p_{i1}, p_{i2})$ and $L_j = (p_{j1}, p_{j2})$ be distinct leaves of T_1 (but not L_1). Note that $\text{MINDIST}(q, (p_{i1}, p_{i2})) < r_4 < \min(\delta(q, p_{j1}), \delta(q, p_{j2}))$ and $\text{MINDIST}(q, (p_{j1}, p_{j2})) < r_4 < \min(\delta(q, p_{i1}), \delta(q, p_{i2}))$. This means that the MINDIST of any leaf node is at most r_4 but every point is beyond r_4 so 1NN cannot use H3 pruning. Furthermore, since MINMAXDIST is always an upper bound on the points, it can never use H2* pruning. Thus, 1NN searches all of T_1.

Lemma 6. *Given a tree T and query point q as constructed above, HS prunes away all nodes in T_1 except those on the branch leading to L_1 in a 1-nearest neighbor query.*

Proof. Like 1NN, HS descends directly to L_1, however, once b is in tow, it immediately jumps back to X since $\text{MINDIST}(q, X) < \text{MINDIST}(q, V)$. Since d is the 1-nearest neighbor and since $\text{MINDIST}(q, X) = \delta(q, d)$, it immediately descends to L_2 to find d. Since $\delta(q, d) < \text{MINDIST}(q, V)$ it can use H3 to prune away the rest of T_1.

The theorem follows from Lemma 5 and Lemma 6. 1NN searches all of T_1 ($O(n)$ nodes) while HS searches only the paths leading to L_1 and L_2 ($O(\log n)$ nodes). As a consequence, HS can prune the search space exponentially over 1NN even when 1NN has the advantages of H2*. □

Extending Theorem 4 to k-nearest neighbors is fairly straight-forward. Add $k-1$ points to X between r_2 and r_3 and place these points in leaves as adjacent to L_2 as possible. Since this set of points forms a rectangle with MINDIST smaller than r_3 and since this rectangle is encountered en route to L_2, HS will find the points immediately and then use H3 to prune away the rest of T_1 and T_2. This gives us the following theorem:

Theorem 5. *There exists a family of R-tree and query point pairs* $\mathcal{T} = \{(T_1, q_n), \ldots, (T_m, q_m)\}$ *such that for any* (T_i, q_i)*, HS examines* $O(\log n)$ *nodes and* KNN *examines* $O(n)$ *nodes on a* k*-nearest neighbor query.*

7 Open Problems and Future Work

The most natural open problem is quantifying the time/space trade-off of depth-first versus best-first nearest-neighbor algorithms on the R-tree. One line of future work might explore hybrid algorithms that combine the space-efficiency of depth-first search along with the time-efficiency of best-first search.

References

1. Manolopoulos, Y., Nanopoulos, A., Papadopoulos, A.N., Theodoridis, Y.: R-Trees: Theory and Applications. In: Advanced Information and Knowledge Processing, 1st edn., Springer, Heidelberg (2006)
2. Papadopoulos, A., Manolopoulos, Y.: Performance of nearest neighbor queries in r-trees. In: Proceedings of the 6th International Conference on Database Theory, pp. 394–408 (1997)
3. Berchtold, S., Böhm, C., Keim, D.A., Krebs, F., Kriegel, H.P.: On optimizing nearest neighbor queries in high-dimensional data spaces. In: Van den Bussche, J., Vianu, V. (eds.) ICDT 2001. LNCS, vol. 1973, pp. 435–449. Springer, Heidelberg (2000)
4. Guttman, A.: R-trees: A dynamic index structure for spatial searching. In: Proceedings of the ACM SIGMOD International Conference on Management of Data, pp. 47–57 (1984)
5. Sellis, T., Roussopoulos, N., Faloutsos, C.: R+-tree: A dynamic index for multidimensional objects. In: Proceedings of the 13th International Conference on Very Large Databases, pp. 507–518 (1988)
6. Beckmann, N., Kriegel, H., Schneider, R., Seeger, B.: R*-tree: An efficient and robust access method for points and rectangles. In: Proceedings of the ACM SIGMOD International Conference on Management of Data, pp. 322–331 (1990)
7. Hjaltason, G.R., Samet, H.: Distance browsing in spatial databases. ACM Transactions on Database Systems 24, 265–318 (1999)
8. Berchtold, S., Böhm, C., Keim, D.A., Kriegel, H.P.: A cost model for nearest neighbor search in high-dimensional data space. In: Proceedings of the Sixteenth ACM Symposium on Principles of Database Systems, pp. 78–86. ACM Press, New York (1997)
9. Roussopoulos, N., Kelley, S., Vincent, F.: Nearest neighbor queries. In: Proceedings ACM SIGMOD Internaiontal Conference on the Management of Data, pp. 71–79 (1995)
10. Böhm, C., Berchtold, S., Keim, D.A.: Searching in high-dimensional spaces: Index structures for improving the performance of multimedia databases. ACM Computing Surveys (CSUR) 33, 322–373 (2001)
11. Cheung, K.L., Fu, A.W.C.: Enhanced nearest neighbour search on the r-tree. SIGMOD Record 27, 16–21 (1998)

Total Degrees and Nonsplitting Properties of Σ_2^0 Enumeration Degrees

M.M. Arslanov[1,⋆], S.B. Cooper[2], I.Sh. Kalimullin[1], and M.I. Soskova[2,⋆⋆]

[1] Department of Mathematics, Kazan State University, Kazan 420008, Russia
[2] School of Mathematics, University of Leeds, Leeds LS2 9JT, U.K.

This paper continues the project, initiated in [ACK], of describing general conditions under which relative splittings are derivable in the local structure of the enumeration degrees.

The main results below include a proof that any high total e-degree below $\mathbf{0}_e'$ is splittable over any low e-degree below it, and a construction of a Π_1^0 e-degree unsplittable over a Δ_2 e-degree below it.

In [ACK] it was shown that using semirecursive sets one can construct minimal pairs of e-degrees by both effective and uniform ways, following which new results concerning the local distribution of total e-degrees and of the degrees of semirecursive sets enabled one to proceed, via the natural embedding of the Turing degrees in the enumeration degrees, to results concerning embeddings of the diamond lattice in the e-degrees. A particularly striking application of these techniques was a relatively simple derivation of a strong generalisation of the Ahmad Diamond Theorem.

This paper extends the known constraints on further progress in this direction, such as the result of Ahmad and Lachlan [AL98] showing the existence of a nonsplitting Δ_2^0 e-degree $> \mathbf{0}_e$, and the recent result of Soskova [Sos07] showing that $\mathbf{0}_e'$ is unsplittable in the Σ_2^0 e-degrees above some Σ_2^0 e-degree $< \mathbf{0}_e'$. This work also relates to results (e.g. Cooper and Copestake [CC88]) limiting the local distribution of total e-degrees.

For further background concerning enumeration reducibility and its degree structure, the reader is referred to Cooper [Co90], Sorbi [Sor97] or Cooper [Co04], chapter 11.

Theorem 1. *If* $\mathbf{a} < \mathbf{h} \leq \mathbf{0}'$, $\mathbf{a}$ *is low and* $\mathbf{h}$ *is total and high then there is a low total e-degree* $\mathbf{b}$ *such that* $\mathbf{a} \leq \mathbf{b} < \mathbf{h}$.

Corollary 2. *Let* $\mathbf{a} < \mathbf{h} \leq \mathbf{0}'$, $\mathbf{h}$ *be a high total e-degree,* $\mathbf{a}$ *be a low e-degree. Then there are* Δ_2^0 *e-degrees* $\mathbf{b}_0 < \mathbf{h}$ *and* $\mathbf{b}_1 < \mathbf{h}$ *such that* $\mathbf{a} = \mathbf{b}_0 \cap \mathbf{b}_1$ *and* $\mathbf{h} = \mathbf{b}_0 \cup \mathbf{b}_1$.

Proof. Immediately follows from Theorem 1, and Theorem 6 of [ACK]. □

Proof of Theorem 1. Assume A has low e-degree, $H \oplus \overline{H}$ has high e-degree (i.e., H has high Turing degree) and $A \leq_e H \oplus \overline{H}$.

⋆ The first and the third authors are partially supported by RFBR grant 05-01-00605.
⋆⋆ The fourth author is partially supported by Marie Curie Early Training Site MATHLOGAPS(MEST-CT-2004-504029).

M. Agrawal et al. (Eds.): TAMC 2008, LNCS 4978, pp. 568–578, 2008.

We want to construct an H-computable increasing sequence of initial segments $\{\sigma_s\}_{s\in\omega}$ such that the set $B = \cup_s \sigma_s$ satisfies the requirements

$$P_n : n \in A \iff (\exists y)[\langle n, y\rangle \in B]$$

and

$$R_n : (\exists \sigma \subset B)[n \in W_n^\sigma \vee (\forall \tau \supset \sigma)[\tau \in S^A \implies n \notin W_n^\tau]]$$

for each $n \in \omega$, where

$$S^A = \{\tau : (\forall x)(\forall y)[\tau(\langle x, y\rangle) \downarrow= 1 \implies x \in A]\}.$$

Note that P_n-requirements guarantee that $A \leq_e B$, and hence $A \leq_e B \oplus \overline{B}$. To prove that the R_n-requirements provide $B' \equiv_T \emptyset'$, first note that $S^A \equiv_e A$, which has low e-degree, and

$$X = \{\langle \sigma, n\rangle : (\exists \tau \supset \sigma)[\tau \in S^A \ \&\ n \in W_n^\tau]\} \leq_e S^A.$$

Then $X \in \Delta_2^0$ and

$$n \notin B' \iff (\exists \sigma \subset B)[\langle \sigma, n\rangle \notin X],$$

so that B' is co-c.e. in $B \oplus \emptyset' \equiv_T \emptyset'$. Thus $B' \leq_T \emptyset'$ by Post's Theorem.

Since the set B will be computable in H, the set

$$Q = \{n : (\forall \sigma \subset B)(\exists \tau \supset \sigma)[\tau \in S^A \ \&\ n \in W_n^\tau]\}$$

will be computable in $(H \oplus \emptyset')' \equiv_T H'$ – indeed, we have $n \in Q \iff (\forall \sigma \subset B)[\langle \sigma, n\rangle \in X]$, so that Q is co-c.e. in $H \oplus \emptyset'$. Now to construct the desired set B we can apply the Recursion Theorem and fix an H-computable function g such that $Q(x) = \lim_s g(x, s)$.

Let $\{A_s\}_{s\in\omega}$ and $\{S_s^A\}_{s\in\omega}$ be respective H-computable enumerations of A and S^A.

Construction

Stage $s = 0$. $\sigma_0 = \lambda$.

Stage $s + 1 = 2\langle n, z\rangle$ (to satisfy P_n). Given σ_s define $l = |\sigma_s|$.

If $n \notin A_s$, then let $\sigma_{s+1} = \sigma_s{}^\frown 0$.

If $n \in A_s$, then choose the least $k \geq l$ such that $k = \langle n, y\rangle$ for some $y \in \omega$ and define $\sigma_{s+1} = \sigma_s{}^\frown 0^{k-l}{}^\frown 1$ (so that $\sigma_{s+1}(k) = 1$).

Stage $s + 1 = 2\langle n, z\rangle + 1$ (to satisfy R_n). H-computably find the least stage $t \geq s$ such that either $g(n, t) = 0$, or $n \in W_{n,t}^\tau$ for some τ satisfying $\tau \in S_t^A$ and $\tau \supset \sigma_s$. (Such stage t exists since if $\lim_s g(n, s) = 1$ then $n \in Q$, and hence there exists some $\tau \supset \sigma_s$ such that $n \in W_n^\tau$ and $\tau \in S^A$.)

If $g(n, t) = 0$ then define $\sigma_{s+1} = \sigma_s{}^\frown 0$.

Otherwise, choose the first $\tau \supset \sigma_s$ such that $\tau \in S_t^A$ and $n \in W_{n,t}^\tau$. Define $\sigma_{s+1} = \tau$.

This completes the description of the construction.

Let $B = \cup_s \sigma_s$. Clearly $B \leq_T H$ since each σ_s is obtained effectively in H. Each P_n-requirement is satisfied by the even stages of the construction since $\sigma_s \in S^A$ for any $s \in \omega$.

To prove that each R_n-requirement is met suppose that

$$(\forall \sigma \subset B)(\exists \tau \supseteq \sigma)[\tau \in S^A \ \& \ n \in W_n^{\tau}]$$

for some n. This means that $n \in Q$. Choose any odd stage $s = 2\langle n, z\rangle + 1$ such that $g(n,t) = 1$ for all $t \geq s$. Then by the construction $n \in W_n^{\sigma_s}$.

Hence $A \leq_e B \oplus \overline{B} \leq_e H \oplus \overline{H}$, and $\deg_e(B \oplus \overline{B})$ is low. □

Theorem 3. *There is a* Π_1^0 *e-degree* **a** *and a* 3*-c.e. e-degree* **b** < **a** *such that* **a** *is not splittable over* **b**.

Proof. We construct a Π_1^0 set A and 3-c.e. set B satisfying both the global requirement:

$G : B = \Omega(A)$,

and the requirements

$$R_{\Xi,\Psi,\Theta} : A = \Xi(\Psi(A) \oplus \Theta(A)) \Longrightarrow (\exists \text{ e-operator } \Gamma) A = \Gamma(\Psi(A) \oplus B) \vee$$
$$(\exists \text{ e-operator } \Lambda) A = \Lambda(\Theta(A) \oplus B)$$

for each triple of e-operators Ξ, Ψ, Θ, and

$$N_\Phi : A \neq \Phi(B)$$

for each e-operator Φ.

In fact A will be constructed as a 2-c.e. set. Note that the e-degrees of Π_1 sets coincide with the e-degrees of 2-c.e. sets. Hence this will still produce the desired enumeration degrees.

Basic Strategies

Suppose we have an effective listing of all requirements $R_1, R_2, \ldots$ and $N_1, N_2, \ldots$ The requirements will then be arranged by priority in the following way: $G < R_1 < N_1 < R_2 < N_2 < \ldots$

To satisfy the requirement G we will make sure that every time we enumerate an element into the set B, we enumerate a corresponding axiom into the set Ω; and every time we extract an element from B, we make the corresponding axiom invalid by extracting elements from A. More precisely every element y that enters B will have a corresponding marker m in A and an axiom $\langle y, \{m\}\rangle$ in Ω. If y is extracted from B then we extract m from A. If y is later re-enumerated into B – this can happen since B is 3-c.e. – then we will just enumerate the axiom $\langle y, \emptyset\rangle$ into Ω.

To satisfy the requirements R_i we will initially try to construct an operator Γ using information from both of sets B and $\Psi(A)$. Again, enumeration of elements into A is always accompanied by enumeration of axioms into Γ, and extraction of elements from A can be rectified via B-extractions.

The N-strategies follow a variant of the Friedberg -Muchnik strategy while at the same time respecting the Γ-rectification, so we will call them (N_Φ, Γ)-strategies. They choose a follower x, enumerate it in A, then wait until $x \in \Phi(B)$. If this happens - they extract the element x from A while restraining $B \upharpoonright \varphi(x)$ in B. The need to rectify Γ after the extraction of the follower x from A can be in conflict with the restraint on B. To resolve this conflict we try to obtain a change in the set $\Psi(A)$ which would enable us to rectify Γ without any extraction from the set B. To do this we monitor the length of agreement

$$l_{\Xi,\Psi,\Theta}(s) = \max\{y : (\forall y < x)[y \in A[s] \iff y \in \Xi(\Psi(A) \oplus \Theta(A))[s]]\}.$$

We only proceed with actions directed at a particular follower once it is below the length of agreement. This ensures that the extraction of x from A will have one of the following consequences

1. The length of agreement will never return so long as at least one of the axioms that ensure $x \in \Xi(\Psi(A) \oplus \Theta(A))$ remains valid.
2. There is a useful change in the set $\Psi(A)$.
3. There is a useful change in the set $\Theta(A)$.

We will initially assume that it is the case that the third consequence is true and commence a backup strategy (N_Φ, Λ) which is devoted to building an enumeration operator Λ with information from A and $\Theta(A)$. This is a new copy of the N-strategy working with the same follower. It will try to make use of this change in $\Theta(A)$ to satisfy the requirement. Only when we are provided with evidence that our assumption is wrong will we return to the initial strategy (N_Φ, Γ).

Basic module for an N_Φ-strategy below one $R_{\Xi,\Psi,\Theta}$-strategy

We will first consider the simple case involving just two requirements. Assume we have N_Φ, which we refer to as the N-requirement, below $R_{\Xi,\Psi,\Theta}$, which we refer to as the R-requirement.

At the root we have the R-strategy denoted by (R, Γ). It will have two outcomes $e <_L gw$. The R-strategy will monitor all elements $x \notin A$. In the case in which there is an element $x \notin A$ such that $x \in \Gamma(\Psi(A) \oplus B)$ the operator Γ cannot be rectified. The (R, Γ)-strategy will then have outcome gw, and we will be able to argue that $x \in \Xi(\Psi(A) \oplus \Theta(A))$, which indicates a global win for the R-requirement. Strategies working below this outcome will follow a simple Friedberg-Muchnik strategy and preserve the difference at x by using followers of big enough value. In case there is no such x the operator, Γ can be rectified and the (R, Γ)-strategy will have outcome e.

Below e we will try to meet N satisfying $A = \Gamma(\Psi(A) \oplus B)$. The (N, Γ)-strategy will have four outcomes: three finitary outcomes, f, w and l, and one infinitary outcome λ. The outcomes are arranged in the following way: $\lambda <_L f <_L w <_L l$. Outcome l indicates that at that node the R-requirement is globally satisfied since the follower x enumerated in A is not in $\Xi(\Psi(A) \oplus \Theta(A))$.

Outcome w indicates that Γ is correct on x and the N-requirement is satisfied as $x \in A - \Phi(B)$. Outcome f is only accessible once a follower x has been returned. It will indicate that Γ is again correct on x and the N-requirement is satisfied via $x \in \Phi(B) - A$.

Below outcome λ strategies will be devoted to constructing an operator Λ with $A = \Lambda(\Theta(A) \oplus B)$ where they will receive their followers from (N, Γ). Again we have a controlling strategy (R, Λ) with only one outcome e which makes sure that the operator Λ can be rectified at all times. In case it sees an element $x \notin A$ for which the axiom enumerated in Λ is valid, it will send x back to (N, Γ). We will be able to argue that x has provided evidence of a useful change in $\Psi(A)$.

Below (R, Λ)'s only outcome e we try to meet N by (N, Λ) with $A = \Lambda_\Phi(\Theta(A) \oplus B)$. The strategy below the outcome λ acts only when the (N, Γ)-strategy *sends* its follower x. It performs similar actions with regard to (N, Γ) and has two outcome $f <_L w$ both indicating that the N-requirement is satisfied and the operator Λ remains intact.

The R strategy

1. Scan all followers $x \notin A$ defined up to the current stage.
2. If $x \in \Gamma(\Psi(A) \oplus B)$, then let the outcome be $o = gw$.
3. If all followers are scanned and none has produced an outcome $o = gw$, then let the outcome be $o = e$.

The (N, Γ) strategy

At stage s the strategy will start its work at the step of the module indicated at the previous stage.

Setup 1) Choose a new follower x as a fresh number (bigger than any previously set up restraint). Enumerate it into A_s.
2) If there are finite sets $G(x), H(x), L(x)$ with $x \in \Xi(G(x) \oplus H(x))$, $G(x) \subset \Psi(L(x))$, $H(x) \subset \Theta(L(x))$ and $L(x) \subset A$ then restrain A on $\max(L(x))$ and go to *Setup 3*. Otherwise let the outcome be $o = l$ and return to *Setup*2) at the next stage.
3) Define $x's$ B-marker $y(x)$, along with its corresponding A-marker $m(x)$, as fresh numbers bigger than any previously set restraint on A or B. Enumerate $y(x)$ in B_s and $m(x)$ in A_s. Define a new axiom $\langle y(x), \{m(x)\}\rangle$ for Ω_s.
Enumerate each $\langle z, G_x \oplus B \restriction y(x)\rangle$ into Γ where z is either x, or $m(x)$, or a follower $z \in A$ from a previous cycle of the strategy. Note that we enumerate axioms for previous followers as well. So at this point the operator Γ is rectified. Let the outcome be $o = w$. Go to *Wait* at the next stage.

Wait If $x \in \Phi(B_s)$ then go to *Attack*. Otherwise let the outcome be $o = w$ and return to *Wait* at the next stage.

Attack 1) Check if any previously sent follower has been returned. If so go to *Result*. Otherwise go to *Attack*2.

2)Let $v(x) = \max(\varphi(x), y(x))$ and restrain B on $v(x)$. Extract $y(x)$ from B_s and $m(x)$ from A_s, noting that x is still in $\Xi(\Psi(A) \oplus \Theta(A))$ as the marker $m(x)$ is chosen as a fresh number after $G(x)$ and $H(x)$ are already defined.

Send x. Let the outcome be $o = \lambda$. At the next stage start from *Setup*1, choosing a new current follower. The strategy working below outcome λ will believe B only below a right boundary $R_s = y(x)$. Note that the next follower will choose its B-marker of greater value. So if the outcome λ is visited infinitely often then the right boundary R will grow unboundedly.

Result Let the returned follower be x. Put $y(x)$ into B_s and $\langle y(x), \emptyset \rangle$ into Ω_s. For each follower z of this strategy such that $z \in A$ put the axiom $\langle z, \emptyset \rangle$ into Γ_s.

1) For the returned follower we know that $x \notin A_s$ and $H(x) \subset \Theta(A_s)$. The outcome λ will not be accessible anymore so we can preserve $H(x) \subseteq \Theta(A_t)$ at further stages t. Also if $G(x) \subseteq \Psi(A_s)$ then the (R, Γ)-strategy would have outcome gw preserving the difference and satisfying R globally. The (N, Γ)-strategy would not be accessible any longer. Otherwise $G(x) \nsubseteq A$ and the outcome is $o = f$. Return at *Result*1 at the next stage.

The (R, Λ)-strategy below outcome λ

1. Scan all followers $x \notin A$.
2. If $x \in \Lambda(\Theta(A) \oplus B)$ then return x. End this stage.
3. If all followers are scanned and none have been returned then let the outcome be e.

The (N, Λ)-strategy below outcome λ

Setup 1) Let $x \in A$ be a new integer which was sent by the (N, Γ)-strategy. Now x becomes the *follower* of the (N, Λ)-strategy. Go to *Setup*2.
2) Put $\langle x, H_x \oplus B \restriction v(x) \rangle$ into Λ. Go to *Wait*.

Wait If $x \in \Phi(B)$ with use $\varphi(x) < R_s$ then go to *Attack*. Otherwise the outcome is $o = w$, return to *Wait* at the next stage.

Attack Extract x from A. Go to result.

Result Let the outcome be $o = f$. Return to *Result* at the next stage.

The (N, FM)-strategy below outcome l or gw

Setup Choose a new follower x bigger than any previously set restraint on A and enumerate it into A. Go to *Wait*.

Wait If $x \in \Phi(B)$ go to *Attack*. Otherwise the outcome is $o = w$, return to *Wait* at the next stage.

Attack Extract x from A and go to *Result*.

Result Let the outcome be $o = f$. Return to *Result* at the next stage.

Now the (N, FM) strategy below outcome l will also be changing A. To keep Γ and Λ rectified, every time we initialise the (N, FM)-strategy and cancel its follower x, if $x \in A$ we will add the axiom $\langle x, \emptyset\rangle$ in Γ and Λ.

If the (R, Γ)-strategy has outcome gw on stage s for the first time, then the (N, FM)-strategy working below will will be initialised on the previous stage and will choose its follower x anew, respecting the restraint on A that (N, Γ) has set up. So (R, Γ) will have outcome gw on all further stages and B will not be modified any longer. The (N, FM)-strategy will be able to satisfy its requirement.

Suppose that (R, Γ)-strategy never has outcome gw. We will analyse all possible outcomes of the N-strategies and see that in each case the requirements are satisfied.

Consider first the possible outcomes of the strategy (N, Γ). If one of the cycles stops at $Setup2$, i.e. on all stages $t > s$ the strategy has outcome l, then the true outcome will be $(o = l)$. The length of agreement $l_{\Xi,\Psi,\Theta}(s) = \max\{y : (\forall y < x)[y \in A[s] \iff y \in \Xi(\Psi(A) \oplus \Theta(A))[s]]\}$ is bounded and hence the requirement R is trivially satisfied.

The set B is not modified after stage s and the simple strategy (N, FM), active on all stages $t \geq s$ succeeds to satisfy the requirement N.

Suppose now that no cycle of the (N, Γ)-strategy stops at $Setup2$. In this case the (N, FM)-strategy may be activated infinitely many times and will be initialised every time the $(N\Gamma)$-strategy moves on to $Wait$. The current follower x of the (N, FM)-strategy will be cancelled and if it is not yet extracted from A the corresponding axiom $\langle x, \emptyset\rangle$ will be enumerated in Γ and Λ. This ensures that both operators will be correct at x for all cancelled followers x of the strategy (N, FM).

We first consider the case when the (N, Γ)-strategy during its work sends only finitely many integers. Then some cycle with a follower x stops either at $Wait$ or reaches $Result$. If the cycle stops at $Wait$ then the outcome is $o = w$ and $x \in A - \Phi(B)$, hence the N-requirement is satisfied. On the other hand for all followers z we have $z \in A \iff z \in \Gamma(\Psi(A) \oplus B)$ and $m(z) \in A \iff m(z) \in \Gamma(\Psi(A) \oplus B)$ since $y(z) \in B \iff z = x$. Hence Γ is correct at all followers z.

If the cycle reaches $Result$ then we have $y(x) \in B$ and hence $x \in \Phi(B) - A$, so N is satisfied. Also $H_x \subseteq \Theta(A)$ via some finite set $P_x \subset A$. If $G_x \subseteq \Psi(A)$ then this will be apparent at some finite stage s, i.e. on stage s we will see a finite set $Q_x \subset A$ such that $G_x \subseteq \Psi(Q_x)$. Then from stage s on the (R, Γ)-strategy will have outcome $o = gw$, contradicting our assumption. So $G_x \nsubseteq \Psi(A)$ giving $x \notin \Gamma(\Psi(A) \oplus B)$. Since again $y(z) \in B \iff z = x$ we have $z \in A \iff z \in \Gamma(\Psi(A) \oplus B)$ and $m(z) \in A \iff m(z) \in \Gamma(\Psi(A) \oplus B)$ for any follower z. Hence the operator Γ remains correct at all further stages.

Suppose now that the (N, Γ)-strategy during its work sends infinitely many integers. In particular, no x is returned to (N, Γ). Then the true outcome is $o = \lambda$ and we will see that the (N, Λ)-strategy is successful.

If the (N, Λ)- strategy stops at $Wait$ then $x \in A - \Phi(B)$. Indeed if we assume that $x \in \Phi(B)$ then there is some finite $M_x \subset B$ such that $x \in \Phi(M_x)$. The right boundary R grows unboundedly, so eventually there will be a stage s with $R_s > \max(M_x)$ and the strategy will move on to *Attack*.

The second case is if the strategy reaches *Result*. Then $x \in \Phi(B) - A$ because at some stage s we found a set $M_x \subset B_s$ with $\max M_x < R$ such that $x \in \Phi(M_x)$. The strategy (N, Γ) will not extract any more markers from B after stage s that are below the right boundary R_s, hence $x \in \Phi(B)$.

At this stage of the construction we can only prove that Λ will be correct at the follower x and all cancelled followers of the strategy (N_Φ, FM). To prove that the operator is correct at the rest of the followers enumerated in A by the (N, Γ)-strategy we will need to consider how all N-strategies will work together.

Basic module for many N_Φ-strategies under one $R_{\Xi,\Psi,\Theta}$-strategy

We will try to meet all requirements $N_{\Phi_1}, N_{\Phi_2}, \ldots$. Each requirement N_{Φ_j} will be denoted by N_j and met by one of the following strategies:

1. (N_j, Γ) with outcomes λ, f, w and l;
2. (N_j, FM) with outcomes f and w and situated in the subtree of the strategy (N_i, Γ) with outcome l, where $i \leq j$.
3. (N_j, Λ) with outcomes f and w and situated in the subtree of the strategy (N_i, Γ) with outcome λ where where $i \leq j$.

We now need to be more careful as more strategies will enumerate and extract markers from A and B. We will have to ensure that the operator constructed on the true path is correct and manages to satisfy the R-requirement.

The first rule that we will implement in order to achieve this follows the idea of cancelling followers of the (N, FM)-strategy from the previous section. Namely, whenever we initialise a strategy (N_j, S) on an node α in the tree of strategies whose follower x is in A we will enumerate an axiom $\langle x, \emptyset \rangle$ into all operators Γ and Λ that are constructed on nodes $\beta < \alpha$. If $m(x)$ is in A we will also enumerate an axiom $\langle m(x), \emptyset \rangle$ into these operators.

Secondly we will be more careful when enumerating axioms in the corresponding operators. Instead of just using the sets $G(x)$ and $H(x)$, we will use the information from all axioms defined up until now. More precisely we will modify the modules of the strategies from the previous section in the following way:

The (N_j, Γ)**-strategy** is the same as the as the (N_Φ, Γ)-strategy with the exception of step *Setup*3, which is now as follows:

*Setup*3) Enumerate all $\langle z, G_x \oplus B \restriction y_x \cup U \rangle$ into Γ where z is either x, or m_x, or a follower $z \in A$ from a previous cycle of the strategy and U is the union of all sets D such that $\langle v, D \rangle$ is a valid axiom in Γ, where $v \in A$ is a follower of the strategy (N_i, Γ) with $i < j$.

The (N_{Φ_j}, Λ)**-strategy** is the same as the (N_Φ, Λ)-strategy with the exception of *Setup*2), which is now as follows:

*Setup*2) Enumerate $\langle x, (H_x \oplus B \restriction v(x)) \cup U \rangle$ into Λ where U is the union of all finite sets D such that $\langle v, D \rangle \in \Lambda$ for some follower $v \in A$ of an (N_k, Λ)-strategy with $k < j$.

The main idea behind the added sets U in the axioms is that a strategy α working below another strategy β where α and β construct the same operator O believes that $\beta's$ work is final and the axioms enumerated in O by β will remain true. In the case that β changes its mind and invalidates one of these axioms α will be initialised as β will have an outcome to the left of α. If $\alpha's$ followers are still in A then an axioms for them will be enumerated in the operator as stated in above. But if $\alpha's$ follower is not in A, then we need to ensure that there isn't a valid axiom in O for it. α will not be able to monitor this follower any longer, so the job is going to be transferred to β automatically via the set U which includes an axiom for $\beta's$ follower, which β observes and makes sure is invalid.

Two *R*-requirements

Now we need to consider the case when there are two R-requirements. Corresponding to them there are nodes on the tree: an (R_1, Γ_1)-strategy and an (R_2, Γ_2)-strategy along each path, scanning for an appropriate global win for the R-requirements. Below outcome gw for an R_i-strategy the N-requirements simply ignore the requirement R_i and act as in the previous section.

There now more possibilities for an N-strategy working below outcomes e of both (R_i, Γ_i)-strategies depending on how it believes the R_i-requirements will be satisfied.

The main strategy will be again the one that deals with operators Γ_1 and Γ_2. It will try to obtain the necessary changes in the sets $\Psi_1(A)$ and $\Psi_2(A)$ using backup strategies that try to satisfy the R requirements in a different manner. The requirement R_1 is of higher priority. The method for satisfying the lower priority requirement R_2 will be decided after we have established the method for satisfying R_1 unless we have already evidence that the R_2-requirement is trivially satisfied. The N-strategy starts off assuming that the requirements will be satisfied via operators Γ_1 and Γ_2. It will be denoted by (N, Γ_1, Γ_2). Its outcomes are $\lambda_2 <_L f <_L w <_L l_2 <_L l_1$. Outcomes w and f will represent the fact that the strategy has succeeded in satisfying its requirement while keeping both operators rectified.

Outcome l_1 will represent a global win for R_1. The price we pay for it is that the operator Γ_2 will not be rectified. Below this outcome there will be a backup (N, FM_1, Γ_2')-strategy. It will construct a new operator Γ_2' and meet the requirement N. Its outcomes are $\lambda_2 <_L f <_L w <_L l_2$ and it acts just as the (N, Γ)-strategy from the previous section.

Outcome l_2 will represent a global win for R_2. Below it we have a strategy (N, Γ_1, FM_2) which continues to construct the same operator Γ_1 as the (N, Γ_1, Γ_2)-strategy. Strategies below will simply treat R_2 as satisfied - that is, this requirement will be invisible to them.

Below outcome λ_2 is the (R_2, Λ_2)-strategy followed by a backup strategy (N, Γ_1, Λ_2). It continues to construct the same operator for the first strategy Γ_1 but switches the method for the second strategy to Λ_2. Its outcomes are $\lambda_1 <_L f <_L w$.

Below outcome λ_1 is the (R_1, Λ_1)-strategy a backup strategy that changes the method for satisfying the requirement R_1. As a consequence the method for

R_2 must be decided again. The strategy is $(N, \Lambda_2, \Gamma_2'')$ with outcomes $\lambda_2 <_L f <_L <_w <_L l_2$. The method for satisfying R_1 cannot be switched anymore. The method for R_2 can be further switched via (N, Λ_1, FM_2) below l_2 and to $(N, \Lambda_1, \Lambda_2'')$ below outcome λ_2.

In this way all possible combinations of methods for satisfying the two R-requirements are distributed through the tree.

The modules for each of the described strategies above follow the basic steps as outlined in the previous section. The (N, Γ_1, Γ_2) strategy chooses a follower x. It tries to define the parameters for R_1 - $H_1(x)$, $G_1(x)$, $y_1(x)$ and $m_1(x)$ and rectifies Γ_1. Then it focuses on the second requirement R_2. Once R_2's parameters are defined a new element $m_2(x)$ will be enumerated in A. The new point is that this new change in A must be reflected in the definition of Γ_1. So an axiom $\langle m_2(x), G_1(x) \oplus \{y_2(x)\}\rangle$ is enumerated in Γ_1. If $m_2(x)$ is extracted from A then we will extract $y_2(x)$ from B and this axiom will not be valid. We will enumerate $y_2(x)$ back in B only if x has been returned in which case $G_1(x) \not\subseteq \Psi_1(A)$.

The axioms enumerated in Γ_2 will have to include additionally $m_1(x)$ and all $m_1(z)$ for previously defined followers of this strategy from previous cycles, that are still in A.

Once we have established that $x \in \Phi(B)$, we start the attack by sending the follower x with defined $v(x) = \max(\varphi(x), y_1(x), y_2(x))$ to (N, Γ_1, Λ_2). This strategy will need to get further permission from Γ_1. An axiom $\langle z, H_2(x) \oplus B \upharpoonright v(x)\rangle$ will be enumerated for each z which is a follower from a previous cycle, x or $m_1(x)$. This strategy also starts an attack by sending x to $(N, \Lambda_1, \Gamma_2'')$ and extracting $y_1(x)$ and $m_1(x)$ from A once it has observed that $x \in \Phi(B)$. Note that this will make the axiom for x in Λ_2 invalid.

The $(N, \Lambda_1, \Gamma_2'')$-strategy now must define parameters $G_2''(x)$ and $H_2''(x)$, markers $y_2''(x)$ and $m_2''(x)$. And then it will initiate the last attack sending x to $(N, \Lambda_1, \Lambda_2'')$.

Once the follower is extracted from A it can climb back up these strategies. (R_2, Λ_2'') will send it back to $(N, \Lambda_1, \Gamma_2'')$ in case $H_2''(x) \subset \Theta_2(A)$.

(R_1, Λ_1) will send the follower x back to (N, Γ_1, Λ_2) in case $H_1(x) \subset \Theta_1(A)$.

Then (R, Λ_2) will send it back to $(N, \Gamma_1\Gamma_2)$ in case $H_2(x) \subset \Theta_2(A)$.

When the $(N, \Gamma_1\Gamma_2)$-strategy re-receives x it will have proof that $H_1(x) \subseteq \Theta_1(A)$, so that $G_1(x) \not\subseteq \Psi_1(A)$ and Γ_1 is rectified and $H_2(x) \subset \Theta_2(A)$, so $G_2(x) \not\subseteq \Psi_2(A)$ and Γ_2 is rectified.

Considering two requirements we can justify the need for the (R_i, Λ_i)-strategies. Suppose $\alpha\hat{\ }l_2 \subset \beta$ and β is sharing the same method Λ_1 as α. If a follower x of β is extracted from A we must ensure that the axioms for x defined in the operator Λ_1 are invalid. It could be the case that α moves on to outcome w and initialises β. The follower x will not be observed any longer. But as $\Theta_1(A)$ is not in our control it is possible that $H_1(x) \subset \Theta_1(A)$ and this is revealed at a later stage after x has been cancelled. If x is not sent back, then Λ_1 will not be correct. This is why we need the (R_1, Λ_1) strategy which observes all followers. It will return x even after x is cancelled.

The (R, Γ_1) strategy plays a similar role. Suppose that $\alpha\hat{}l_2 \subset \beta$. Now β is sharing the same method Γ_1 as α. If a follower x of β is extracted from A we must ensure that the axioms for x defined in the operator Γ_1 are invalid. If α moves on to outcome w thereby initialising β we lose control on x and it could happen that $G_1(x) \subset \Psi_1(A)$ at a later stage. We will be able to argue that if the axiom for x in Γ_1 is valid, then $H_1(x) \subset \Theta_1(A)$ and (R, Γ_1) will have outcome gw at all further stages.

In [ACKS] we combine the ideas from the above description to obtain the construction that meets all requirements. □

References

[Ah91] Ahmad, S.: Embedding the diamond in the Σ_2 enumeration degrees. J. Symbolic Logic 50, 195–212 (1991)
[AL98] Ahmad, S., Lachlan, A.H.: Some special pairs of Σ^0_2 e-degrees properties of Σ^0_2- enumeration degrees. Math. Logic Quarterly 44, 431–449 (1998)
[ACK] Arslanov, M.M., Cooper, S.B., Sh. Kalimullin, I.: Splitting properties of total enumeration degrees. Algebra and Logic 42(1) (2003)
[ACKS] Arslanov, M.M., Cooper, S.B., Sh. Kalimullin, I., Soskova, M.: Splitting and nonsplitting in the Σ^0_2 enumeration degrees (to appear)
[AS99] Arslanov, M.M., Sorbi, A.: Relative splittings of $\mathbf{0}'_e$ in the Δ^0_2 enumeration degrees. In: Buss, S., Pudlak, P. (eds.) Logic Colloquium 1998, Springer, Heidelberg (1999)
[Co90] Cooper, S.B.: Enumeration reducibility, nondeterministic computations and relative computability of partial functions. In: Ambos-Spies, K., Müller, G., Sacks, G.E. (eds.) Recursion Theory Week, Oberwolfach 1989. Lecture Notes in Mathematics, vol. 1432, pp. 57–110. Springer, Heidelberg (1990)
[Co04] Cooper, S.B.: Computability Theory. Chapman & Hall/CRC, Boca Raton, London, New York, Washington, D.C (2004)
[CC88] Cooper, S.B., Copestake, C.S.: Properly Σ_2 enumeration degrees. Z. Math. Logik Grundlag. Math. 34, 491–522 (1988)
[MC85] McEvoy, K., Cooper, S.B.: On minimal pairs of enumeration degrees. J. Symbolic Logic 50, 983–1001 (1985)
[La75] Lachlan, A.H.: A recursively enumerable degree which will not split over all lesser ones. Ann. Math. Logic 9, 307–365 (1975)
[NS00] Nies, A., Sorbi, A.: Branching in the enumeration degrees of the Σ^0_2 sets. J. Symb. Log. 65(1), 285–292 (2000)
[Sor97] Sorbi, A.: The enumeration degrees of the Σ^0_2 sets. In: Sorbi, A. (ed.) Complexity, Logic and Recursion Theory, pp. 303–330. Marcel Dekker, New York (1997)
[Sos07] Soskova, M.I.: A non-splitting theorem in the enumeration degrees. Annals of Pure and Applied Logic (to appear)

s-Degrees within e-Degrees

Thomas F. Kent*

Dipartimento di Scienze Matematiche ed Informatiche "Roberto Magari"
University of Siena, Italy
kent@unisi.it

Abstract. For any enumeration degree $\mathbf{a}$ let $D^s_{\mathbf{a}}$ be the set of s-degrees contained in $\mathbf{a}$. We answer an open question of Watson by showing that if $\mathbf{a}$ is a nontrivial Σ^0_2-enumeration degree, then $D^s_{\mathbf{a}}$ has no least element. We also show that every countable partial order embeds into $D^s_{\mathbf{a}}$.

1 Introduction

Positive reducibilities formalize models of relative computability which use only "positive " oracle information. The most comprehensive positive reducibility is *enumeration reducibility*, denoted by $\leq_e$. Intuitively a set A is enumeration reducible to a set B if there is some effective procedure for enumerating A given any enumeration of B. This is made mathematically precise by defining $A \leq_e B$ if there exists a c.e. set Φ such that

$$A = \{x : (\exists \text{ finite } D)[\langle x, D\rangle \in \Phi \;\&\; D \subseteq B]\}$$

(often denoted by $A = \Phi^B$) where finite sets are identified with their canonical indices. In this context a c.e. set Φ is also called an *enumeration operator.*

It is clear that given a set B, an enumeration operator Φ, and a given x, there is no bound to the number n of oracle questions which are needed to enumerate x in Φ^B, i.e. to the cardinality of a finite set D for which we need $D \subseteq B$, in order to have $x \in \Phi^B$. One can therefore introduce restricted, or strong, versions of enumeration reducibility by requesting instead that there be such a bound, such as is done in [1].

Although extreme, the case $n = 1$, in which for any given x we need at most *one* oracle question, is particularly interesting, and occurs often in practical applications of enumeration reducibility. This suggests the following definition:

Definition 1.1. An enumeration operator Φ is called an *s-operator* if for every $\langle x, D\rangle \in \Phi$, we have that D has at most one element.

It is straightforward to see that the s-operators (s stands for singleton) can be effectively listed, and give rise to a reducibility (called *s-reducibility*), denoted by $\leq_s$. The corresponding degree structure, denoted by $\mathcal{D}_s$, consists of the

* The author has been supported by a Marie Curie Incoming International Fellowship of the European Community FP6 Program under contract number MIFI-CT-2006-021702.

M. Agrawal et al. (Eds.): TAMC 2008, LNCS 4978, pp. 579–587, 2008.

equivalence classes, called *s-degrees*, of the subsets of ω under the equivalence relation $\equiv_s$ generated by $\leq_s$. The s-degree of a set A will be denoted by $\deg_s(A)$. The structure $\mathcal{D}_s$ is an upper semilattice with least element $\mathbf{0}_s = \deg_s(\emptyset)$ consisting of the c.e. sets, and the operation of least upper bound is given by $\deg_s(A) \cup \deg_s(B) = \deg_s(A \oplus B)$, where $\oplus$ denotes the usual disjoint union of sets.

It is clear that if $A \leq_s B$ then $A \leq_e B$ and so it is natural to ask questions about the structure of the s-degrees contained within a single enumeration degree. Given a set A, define $A^* = \{n : D_n \in A\}$ where n is the canonical index of the finite set D_n. It follows that $A^* \equiv_e A$ and if $B \leq_e A$ then $B \leq_s A^*$, and so $\deg_s(A^*)$ is the greatest s-degree in $\deg_e(A)$. Zacharov [10] showed that $\leq_s$ is properly contained in $\leq_e$ by showing that every nonzero enumeration degree contains at least two s-degrees. Zacharov's proof is a unique gem of its kind and will be sketched in Section 3. Watson [9] showed that no nonzero Δ^0_2- or Σ^0_2-high enumeration degree contains a minimal s-degree, thus showing that every such enumeration degree contains infinitely many s-degrees. Based also on Copestake's result in [3] that every 1-generic enumeration degree contains infinitely many s-degrees (in fact an ω-chain of s-degrees), Watson then raised the question ([9, p. 90]) as whether every nontrivial Σ^0_2-enumeration degree contains infinitely many s-degrees, conjecturing that this is so. In this paper we give a positive answer to Watson's question, by giving in Theorem 4.1 a uniform priority-free proof of the fact that no nonzero Σ^0_2-enumeration degree contains a minimal s-degree. That every nontrivial Σ^0_2-enumeration degree contains infinitely many s-degrees is also implied by Theorem 5.1, in which we show that one can embed any countable partial order in any nontrivial Σ^0_2 enumeration degree. This theorem generalizes also Copestake's result on ω-chains within any 1-generic enumeration degree. For the reader who is interested in restricted versions of enumeration reducibility, we finally observe that Theorem 4.1 still holds if we replace enumeration reducibility with computably bounded enumeration reducibility, where we say that a set A is *computably bounded enumeration reducible to* a set B, if $A \leq_e B$ via an enumeration operator Φ such that there exists a computable function f satisfying

$$(\forall x, D)[\langle x, D\rangle \in \Phi \Rightarrow |D| \leq f(x)],$$

where $|D|$ denotes the cardinality of the finite set D.

The reader is referred to the papers [1], [2], and [8] for a survey of results on s-reducibility.

2 Conventions

We give some conventions and notation that we will use throughout the article.

Definition 2.1. Given a Σ^0_2-approximation $\langle A_s\rangle_{s\in\omega}$ to a set A, we say that the stage s is *true* if $A_s \subseteq A$, and we say that the approximation is *good* if it contains infinitely many true stages.

Even though we are using a good Σ^0_2-approximation to A, it is possible that a strategy on the true path of the construction will be active only during finitely many true stages. To circumvent this problem, each strategy in the tree of strategies will use its own relativized version of the approximation to A defined by $\langle A^\alpha_s \rangle_{s\in\omega} = \langle \bigcap_{s_0<t\leq s} A_t \rangle_{s\in\omega}$, where $s_0 < s$ is the last stage at which α was active prior to stage s, or -1 if no such stage exists.

Clearly, if α is active infinitely often, then $\langle A^\alpha_s \rangle$ is a good Σ^0_2-approximation to A. During the construction, instead of referring to A^α_s and $B^\alpha_s = \Theta_s^{A^\alpha_s}$ (for some given enumeration operator Θ approximated by $\langle \Theta_s \rangle_{s\in\omega}$), we will just refer to these sets as A and B, with the understanding that we are actually using the relativized approximations.

3 Sketch of Zacharov's Proof

Let A be a non-c.e. set. We outline the main steps of Zacharov's proof to show that there exists a set B such that $A \equiv_e B$, but $A \not\leq_s B$. One of the key points is the following observation: If a set X is not c.e. and for every c.e. subset $W \subseteq X$ there exists a computable set V such that $W \subseteq V \subseteq X$ (examples of such sets X include the immune sets and the non-c.e. semirecursive sets), then for every set Y containing a simple set, we have that $X \not\leq_s Y$.

1. Define $J(A) = \{x : x \in \Phi^A_x\}$ where $\langle \Phi_e \rangle_{e\in\omega}$ denotes here an effective listing of all enumeration operators. It is easy to see that $A \equiv_e J(A)$, and for every set B, if $B \leq_e J(A)$ then $B \leq_1 J(A)$, where $\leq_1$ denotes $1-1$-reducibility.
2. Assume now that S is Post's simple set, and let $\{F_n\}_{n\in\omega}$ be a strong array of finite sets (i.e. $F_n = D_{f(n)}$ for some computable function f), which partition ω and such that $F_n \cap \overline{S} \neq \emptyset$, for every n. Define
$$B = S \cup \bigcup_{n\in J(A)} F_n,$$
so that B contains a simple set. It is not difficult to see that $J(A) \equiv_e B$.
3. We now claim that $J(A) \not\leq_s B$, thus showing that the enumeration degree of A contains two distinct s-degrees. Suppose that $J(A) \leq_s B$. By [4, Theorem 3.6] let R be a semirecursive set such that $R \leq_e J(A)$ and $J(A) \leq_T R$. Hence $R \leq_1 J(A)$, which implies that $R \leq_s B$. But, as observed at the beginning of this section, since R is semirecursive this is possible only if R is c.e.. It follows that $J(A) \in \Delta^0_2$. It is now easy to see that there exists an immune set C such that $C \leq_e J(A)$: In fact any non-c.e. Σ^0_2 set is enumeration equivalent to a hyperimmune set (see [6]). Therefore $C \leq_1 J(A)$, which implies that $C \leq_s B$, a contradiction since B contains a simple set, and C is immune.

4 There Is No Minimal Element

Watson [9] proves that if A is Δ^0_2 and non-c.e., or A is Σ^0_2-high, then there exists a set B such that $B \equiv_e A$ and $B <_s A$. He actually gives two distinct proofs depending on whether A lies in Δ_2^0 or A is Σ_2^0-high. The following theorem gives a uniform and priority-free proof of Watson's result that works for every non-c.e. Σ_2^0 set.

Theorem 4.1. *Let A be a non-c.e. Σ_2^0-set. There exists a set $B \equiv_e A$ such that $B <_s A$.*

Corollary 4.1. *If* **a** *is a nontrivial Σ_2^0-enumeration degree, then there is no minimal s-degree within* **a**.

Proof (of Corollary 4.1). Immediate.

We now give the construction, and a sketch of the verification, for Theorem 4.1.

4.1 The Requirements

Let A be a non-c.e. Σ_2^0-set. We build an enumeration operator Γ and an s-operator Λ such that for each s-operator Φ we meet the following requirements:

$$\begin{aligned} &\mathcal{R}\colon\ B = \Lambda^A \\ &\mathcal{Q}\colon\ A = \Gamma^B \\ &\mathcal{P}_\Phi\colon\ A = \Phi^B \Rightarrow A \text{ is c.e.} \end{aligned}$$

Order the $\mathcal{P}_\Phi$-requirements as $\langle P_i \rangle_{i\in\omega}$.

4.2 The Strategies

Partition ω as $\{F_n : n \in \omega\}$ where $|F_n| = (n+2)^3$ and $\max(F_n) < \min(F_{n+1})$. For each x, define $\sigma(x)$ to be such that $x \in F_{\sigma(x)}$. Define

$$\Gamma = \{\langle n, F_n\rangle : n \in \omega\}$$

The construction of Λ is by stages. At stage s define a finite approximation Λ_s to Λ, and let of course $B_s = \Lambda_s^{A_s}$. Initially define

$$\Lambda_0 = \{\langle y, \{n\}\rangle : n \in \omega \text{ and } y \in F_n\}\,.$$

Notice that if we defined $B = \Lambda_0^A$, we would immediately have

$$n \in A \Leftrightarrow F_n \subseteq B \Leftrightarrow n \in \Gamma^B,$$

i.e. $A = \Gamma^B$. Unfortunately, strategies attempting to meet $\mathcal{P}_\Phi$-requirements may enumerate additional axioms into Λ as needed during the construction, while Γ is never changed. These additional axioms will have the form $\langle z, \emptyset\rangle \in \Lambda$, in which case we say that we *dump* z into B. However we will guarantee that $A = \Gamma^B$ for the eventual set $B = \Lambda^A$, by ensuring that for every n, not all the elements of F_n are dumped into B. It follows from the definition of the operator Γ that if n belongs to A then F_n is a subset of B. The fact that not all the elements of F_n are dumped into B guarantees that if n does not belong to A then F_n will not be a subset of B and hence n will not belong to Γ^B.

The strategy for $\mathcal{P}_\Phi$. The basic strategy to meet requirement $\mathcal{P}_\Phi = P_t$ tries to permanently restrain in Φ^B the elements of $A \cap \Phi^B$: If at stage s we see a number $y \in A \cap \Phi^B$ which has not been so far restrained, then we do so, by either automatic restraint provided by an axiom $\langle y, \emptyset \rangle \in \Phi$, or otherwise by taking an axiom $\langle y, \{z\} \rangle \in \Phi$, with currently $z \in B$, and dumping z into B. If $A = \Phi^B$ then we can show that A is c.e. by arguing that A coincides, modulo a c.e. set, with the elements restrained in Φ^B. Extra care will be taken in order to achieve that the strategy only dumps elements of sets F_n with $t \leq n$, and for each such n, $\mathcal{P}_\Phi$ dumps at most $n+1$ elements of F_n. This, together with the fact that $|F_n| = (n+2)^3$, will guarantee that not all the elements of F_n are dumped into B: In fact, for each n, $\left|F_n \cap \Lambda^\emptyset\right| \leq (n+1)^3$.

4.3 The Construction

Each stage s of the construction consists of substages $t < s$, where strategy P_t is eligible to act. Let strategy $P_t = P_\Phi$ be eligible to act at substage t of stage s. Define $\widetilde{A}_s = \bigcup_{i \in \omega} A_s^i$ where $A_s^0 = A_s \restriction t$, $B_s^{i+1} = \Lambda_s^{A_s^i}$ and $A_s^{i+1} = \Phi_s^{B_s^{i+1}}$. Let $y \in A_s \cap \Phi_s^{B_s}$ be least such that $y \notin \widetilde{A}_s$. If y exists, let $z \in B_s$ be least such that $\langle y, \{z\} \rangle \in \Phi_s$ and $\sigma(z) \in \Phi_s^{B_s}$. If z exists, dump z into B by enumerating the axiom $\langle z, \emptyset \rangle$ into Λ. (Intuitively, $\widetilde{A}_s$ is the smallest c.e. set such that, for $\widetilde{B}_s = \Lambda_s^{\widetilde{A}_s}$, we have $\widetilde{A}_s = \Phi_s^{\widetilde{B}_s}$, and $A_s \restriction t \subseteq \widetilde{A}_s$.)

The reader is referred to [5] for a thorough verification that the construction works.

5 Embedding Countable Partial Orders

A further easy consequence of Theorem 4.1 is the following corollary. Let $\mathbb{Z}^-$ denote the partial order of the nonpositive integers:

Corollary 5.1. *If* $\mathbf{a}$ *is any nontrivial* Σ_2^0 *enumeration degree, then* $\mathbb{Z}^-$ *embeds into the s-degrees within* $\mathbf{a}$.

Proof. If A is a non-c.e. Σ_2^0 set, define $B_0 = A$. Having defined B_i then apply Theorem 4.1 to B_i to get $B_{i+1} \equiv_e B_i$ (hence $B_{i+1} \equiv_e A$), and $B_{i+1} <_s B_i$.

We improve on the previous embedding result by showing that in fact every countable partial order embeds in the s-degrees within any nontrivial Σ_2^0 enumeration degree. Theorem 5.1 generalizes also Copestake's result in [3] stating that every 1-generic enumeration degree contains a chain of s-degrees order-isomorphic to ω.

We say that a family $\{\mathbf{b}_i\}_{i \in \omega}$ of s-degrees is *independent* if for every $i \in \omega$, and any computable set J with $i \notin J$, one has

$$\mathbf{b}_i \not\leq \bigoplus_{j \in J} \mathbf{b}_j$$

where the join is defined by taking suitable representatives $B_j \in \mathbf{b}_j$ and letting

$$\bigoplus_{j \in J} \mathbf{b}_j = \deg_s(\bigcup_{j \in J}(\{j\} \times B_j))$$

Theorem 5.1. *Let* $\mathbf{a}$ *be a non-trivial* Σ_2^0*-enumeration degree. Then there exists an independent family of s-degrees* $\{\mathbf{b}_i\}_{i \in \omega}$ *such that for each* i, $\mathbf{b}_i \subseteq \mathbf{a}$.

Corollary 5.2. *For every non-trivial* Σ_2^0*-enumeration degree* $\mathbf{a}$ *and countable partial order* $\mathcal{P} = \langle P, \leq \rangle$, *we have that* $\mathcal{P}$ *embeds into the s-degrees within* $\mathbf{a}$, *i.e. there exists a mapping* F, *associating with any* $p \in \mathcal{P}$ *an s degree* $F(p) \subseteq \mathbf{a}$, *and satisfying, for all* $p, q \in P$,

$$p \leq q \Leftrightarrow F(p) \leq_s F(q).$$

Proof (of Corollary 5.2). It is a standard argument (see for instance [7, Theorem V.2.9]) to show how to embed any countable partial order in a class of degrees which contains a computably independent collection of degrees, and is closed under computable joins.

We now give the construction, and a sketch of the verification, for Theorem 5.1.

5.1 The Requirements

Fix a non-c.e. Σ_2^0-set A. We build enumeration operators Γ_i and Λ_i, such that for all s-operators Φ and $i \in \omega$, we meet the following requirements:

$$\begin{aligned}
&\mathcal{R}_i\colon\ B_i = \Lambda_i^A \\
&\mathcal{Q}_i\colon\ A = \Gamma_i^{B_i} \\
&\mathcal{P}_{i,\Phi}\colon\ B_i = \Phi^{\bigoplus_{j \neq i} B_j} \Rightarrow \exists \Delta\ (A = \Delta)
\end{aligned}$$

where Δ is a c.e. set constructed by us.

5.2 The Tree of Strategies

Let our set of outcomes be $0 < 1 < \cdots < \infty$ and define $\mathbf{T} = (\omega \cup \{\infty\})^{<\omega}$. Order the $\mathcal{P}$-requirements as $\langle P_i \rangle_{i \in \omega}$ and assign to each $\alpha \in \mathbf{T}$ the requirement $P_{|\alpha|}$. In the rest of the proof, notations and terminology about trees are standard.

5.3 The Construction

Let $\langle A_s \rangle_{s \in \omega}$ be a good approximation to A. We define $B_i \subseteq \omega^{[i]}$, and so we can take $\bigoplus_{j \neq i} B_j = \bigcup_{j \neq i} B_j$. Using the convention that $\max(\emptyset) = \min(\emptyset) = -1$, partition ω as $\{F_n : n \in \omega\}$ where $|F_n^{[i]}| = (n+2)^2$ if $i \leq n$ and $F_n^{[i]} = \emptyset$ otherwise, and such that $\max(F_n^{[i]}) \leq \min(F_{n+1}^{[i]})$. For each x define $\sigma(x)$ to be such that $x \in F_{\sigma(x)}$. For each $i \in \omega$, define

$$\Lambda_i = \left\{ \langle x, A_n \rangle : n \geq i \text{ and } x \in F_n^{[i]} \right\}, \text{and}$$

$$\Gamma_i = \left\{ \langle x, F_n^{[i]} \rangle : n \geq i \text{ and } x \in A_n \right\}.$$

We will enumerate additional axioms into Λ_i as needed during the construction, and so at each stage of the construction, for $i, x \in \omega$ and $\alpha \in \mathbf{T}$, define

$$G(i,x) = \bigcap \{F : \langle x, F\rangle \in \Lambda_i\}, \text{ and}$$

$$G^\alpha = \bigcup \{G(i,x) : x \in B_i \text{ and } x = b^\alpha_m \text{ for some } m < n^\alpha\},$$

where n^α and the b^α_m are parameters defined by α during the construction.

5.4 The Strategies

As in the proof of the previous theorem, it is immediate to see that if we let $B_i = \Lambda_i^A$, with Λ_j as is initially defined, then it would follow that $A = \Gamma_i^{B_i}$. In defining the additional axioms for Λ_i (as needed by the $\mathcal{P}_{i,\Phi}$-strategies) we must therefore ensure that for every k, there is at least one element $y \in F_k^{[i]}$ for which we do not define additional axioms.

The strategy α for $\mathcal{P}_{i,\Phi}$ can be visualized as working on cycles, one cycle for each natural number n:

Cycle n:

1. If $b_m \in \Phi^{\bigoplus_{j\neq i} B_j} - B_i$, for some $m < n$, then stop all cycles $k > m$. If later $b_m \in \Phi^{\bigoplus_{j\neq i} B_j} \cap B_i$ again then resume all existing cycles;
2. otherwise, wait for a number $b \in \Phi^{\bigoplus_{j\neq i} B_j} \cap B_i$, with $b \notin \{b_0, \ldots, b_{n-1}\}$, such that there is either an axiom $\langle b, \emptyset\rangle \in \Phi$, or an axiom $\langle b, \{y\}\rangle \in \Phi$ where $y \in B_j$, some $j \neq i$, and y can be *conditionally* dumped into B_j, by adding an axiom $\langle y, G\rangle \in \Lambda_j$, where G is such that if $b_0, \ldots, b_{n-1} \in B_i$ then $G \subseteq \oplus_{k\neq i} B_k$, forcing $b \in \Phi^{\bigoplus_{j\neq i} B_j}$. In doing so we must not violate the constraints imposed by higher priority strategies, and care should be taken to guarantee that for every z the strategy conditionally dumps at most one element of $F^{[j]}_{\sigma(z)}$.
3. Once such a number b appears, appoint $b_n = b$, conditionally dump the corresponding y into B_j (if $\langle y, \emptyset\rangle \notin \Phi$), impose a restraint on b_n so that lower priority strategies are not allowed to add additional Λ-axioms for b_n, enumerate the elements of the finite set G, as defined in Step 2, into an auxiliary c.e. set Δ, and go to Cycle $n+1$.

We briefly describe the outcomes of this strategy, and how they are recorded on the true path if α is on the true path. If there is a least m such that outcome (1) holds infinitely often, than the requirement is satisfied since in this case $b_m \in \Phi^{\bigoplus_{j\neq i} B_j} - B_i$ and so $\alpha^\frown\langle m\rangle$ is on the true path. Outcome (2) would imply that $\Phi^{\bigoplus_{j\neq i} B_j} \cap B_i$ is c.e.; therefore if $\Phi^{\bigoplus_{j\neq i} B_j} = B_i$ then B_i is c.e., and so $A = \Gamma_i^{B_i}$ is c.e., a contradiction. On the other hand, each time we pass through (2) we let $\alpha^\frown\langle\infty\rangle$ be eligible to act next. Finally if all cycles are defined and there is no $b_m \in \Phi^{\bigoplus_{j\neq i} B_j} - B_i$, then we can argue that $A = \Delta$, i.e. A is c.e., again a contradiction. In this last case, the construction ends the current stage. However, since A is not c.e., we can conclude that there is indeed an outcome o

such that $\alpha^\frown\langle o\rangle$ is on the true path, and this outcome is of the form $o = m$, for some $b_m \in \Phi^{\bigoplus_{j\neq i} B_j} - B_i$.

Before proceeding with the details of the construction let us spend a few more words on how different strategies interact with each other. Let α_1 have higher priority than α_2. By initialization, we may assume that $\alpha_1 \subset \alpha_2$, and by the above informal remarks we consider only the case of some m such that $\alpha_1^\frown\langle m\rangle \subseteq \alpha_2$. The restraints imposed by α_1 on α_2 are that α_2 is not allowed to add additional Λ-axioms for numbers b_n that α_1 conditionally dumps in (3) of the above module. We can argue that these restraints guarantee that for each i and $n \geq i$, there is a $y \in F_n^{[i]}$ such that the only axiom of the form $\langle y, F\rangle \in \Lambda_i$ is $\langle y, A_n\rangle$, thus letting the $\mathcal{Q}_i$- and $\mathcal{R}_i$-requirements be satisfied. On the other hand these restraints, being finitary by the above discussion, do not prevent α_2 from being satisfied.

Each stage s of the construction consists of substages $t < s$, where a strategy $\alpha \in \mathbf{T}$, with $|\alpha| = t$ is eligible to act. Let α be a $\mathcal{P}_{i,\Phi}$-strategy eligible to act at substage t of stage s and let s_0 be the first stage at which α was eligible to act after its last initialization. If this is the first time that α has been eligible to act, set $n = 0$. Choose the first case which applies.

Case 1. There is an $m < n$ such that $b_m \in \Phi^{\bigoplus_{j\neq i} B_j} - B_i$: Let m_0 be the least such m and end the current substage by letting $\alpha^\frown\langle m_0\rangle$ be eligible to act next.

Case 2. For all $m < n$, $b_m \in \Phi^{\bigoplus_{j\neq i} B_j} \cap B_i$:

Case 2.1. There is a $b \in \Phi^{\bigoplus_{j\neq i} B_j} \cap B_i - \{b_m : m < n\}$ with $\sigma(b) > s_0$ and $G^\alpha \subseteq G(i, b)$ such that either $\langle b, \emptyset\rangle \in \Phi$ or $\langle b, \{y\}\rangle \in \Phi$ with $y \in \bigoplus_{j\neq i} B_j$, $\sigma(y) > s_0$, $y \neq b_m^\beta$ for all $\beta \subseteq \alpha$ and $m < n^\beta$, and $y \notin F_{\sigma(y_m)}$ for all $m < n$: Let b_n to be the least such b. If $\langle b_n, \emptyset\rangle \notin \Phi$, let y_n be the least such y for b_n, and enumerate $\langle y_n, \bigcup_{\beta\subseteq\alpha} G^\beta\rangle$ into Λ_j for the j such that $y_n \in B_j$. Enumerate $G(i, b_n)$ into Δ. Set $n = n + 1$, and end the current stage.

Case 2.2. Else: Let $\alpha^\frown\langle\infty\rangle$ be eligible to act next.

Ending the stage s: For any $\alpha > f_s$ cancel all local parameters and set $\Delta = \emptyset$.

Again, the reader is referred to [5] for more details on the construction and its verification.

References

1. Cooper, S.B.: Enumeration reducibility using bounded information: counting minimal covers. Z. Math. Logik Grundlag. Math. 33(6), 537–560 (1987)
2. Cooper, S.B.: Enumeration reducibility, nondeterministic computations and relative computability of partial functions. In: Ambos-Spies, K., Müller, G., Sacks, G.E. (eds.) Recursion theory week (Oberwolfach, 1989). Lecture Notes in Math, vol. 1432, pp. 57–110. Springer, Heidelberg (1990)
3. Copestake, C.S.: Enumeration Degrees of Σ_2^0-sets. PhD thesis, Department of Pure Mathematics, University of Leeds (1987)

4. Jockusch Jr., C.G.: Semirecursive sets and positive reducibility. 131, 420–436 (1968)
5. Kent, T.F.: On a conjecture of Watson (to appear, 2008)
6. McEvoy, K.: The Structure of the Enumeration Degrees. PhD thesis, School of Mathematics, University of Leeds (1984)
7. Odifreddi, P.: Classical Recursion Theory (vol. I), Amsterdam (1989)
8. Sh. Omanadze, R., Sorbi, A.: Strong enumeration reducibilities. Arch. Math. Logic 45(7), 869–912 (2006)
9. Watson, P.: On restricted forms of enumeration reducibility. Ann. Pure Appl. Logic 49(1), 75–96 (1990)
10. Zakharov, S.D.: e- and s-degrees. Algebra i Logika 23(4), 395–406, 478 (1984)

The Non-isolating Degrees Are Upwards Dense in the Computably Enumerable Degrees

S. Barry Cooper[1], Matthew C. Salts, and Guohua Wu[2,*]

[1] School of Mathematics, University of Leeds, Leeds LS2 9JT, U.K.
pmt6sbc@leeds.ac.uk
http://www.amsta.leeds.ac.uk/ pmt6sbc/

[2] School of Physical and Mathematical Sciences, Nanyang Technological University, Singapore 639798, Republic of Singapore
guohua@ntu.edu.sg
http://www.ntu.edu.sg/home/guohua/

Abstract. The existence of isolated degrees was proved by Cooper and Yi in 1995 in [7], where a d.c.e. degree $\mathbf{d}$ is isolated by a c.e. degree $\mathbf{a}$ if $\mathbf{a} < \mathbf{d}$ is the greatest c.e. degree below $\mathbf{d}$. A computably enumerable degree $\mathbf{c}$ is non-isolating if no d.c.e. degree above $\mathbf{c}$ is isolated by $\mathbf{c}$. Obviously, $\mathbf{0}$ is a non-isolating degree. Cooper and Yi asked in [7] whether there is a nonzero non-isolating degree. Arslanov et al. showed in [3] that nonzero non-isolating degrees exist and that these degrees are downwards dense in the c.e. degrees and can also occur in every jump class. In [11], Salts proved that there is an interval of computably enumerable degrees, each of which isolates a d.c.e. degree. Recently, Cenzer et al. [4] proved that such intervals are dense in the computably enumerable degrees, and hence the non-isolating degrees are nowhere dense in the computably enumerable degrees. In this paper, using a different type of construction to that of [3], we prove that the non-isolating degrees are upwards dense in the computably enumerable degrees. In the context of [4], this is the best possible such result.

1 Introduction

The existence of isolated degrees was proved by Cooper and Yi in 1995 in [7], where a d.c.e. degree $\mathbf{d}$ is isolated by a c.e. degree $\mathbf{a}$ if $\mathbf{a} < \mathbf{d}$ is the greatest c.e. degree below $\mathbf{d}$. Ding and Qian [8], LaForte [10] independently, proved that the isolated degrees, and hence the isolating degrees, are dense in the computably enumerable degrees. In [3], Arslanov, Lempp and Shore proved that the non-isolated degrees are also dense in the computably enumerable degrees. In [13], Wu use the isolation phenomenon an alternative proof of Downey's diamond embedding theorem. Ishmukhametov and Wu

* This research was partially supported by a London Mathematical Society collaborative small grant. The first author was supported by EPSRC grant No. GR /S28730/01, and by the NSFC Grand International Joint Project, No. 60310213, *New Directions in the Theory and Applications of Models of Computation,* the second author by an EPSRC Research Studentship and the EU Human Capital and Mobility Network *Complexity, Logic and Recursion Theory* (COLORET), and the third author is partially supported by a research grant No. RG58/06 from Nanyang Technological University.

M. Agrawal et al. (Eds.): TAMC 2008, LNCS 4978, pp. 588–596, 2008.

[12] proved that sometimes, the isolated degrees can be far from the corresponding isolating degrees. That is, there is a high d.c.e. degree isolated by a low c.e. degree. This too has had an interesting application (see [1]), related to Post's problem for the d.c.e. degrees.

In this paper, we are mainly concerned with the non-isolating degrees, where a computably enumerable degree $\mathbf{c}$ is non-isolating if no d.c.e. degree above $\mathbf{c}$ is isolated by $\mathbf{c}$. Obviously, $\mathbf{0}$ is a non-isolating degree. Cooper and Yi asked in [7] whether there is a nonzero non-isolating degree. Arslanov et al. showed in [3] that nonzero non-isolating degrees exist and that these degrees are downwards dense in the c.e. degrees and can also occur in every jump class. Arslanov et al. actually proved a stronger result. They first pointed out that for any c.e. degree $\mathbf{c}$ and d.c.e. degree $\mathbf{d} > \mathbf{c}$, there is a degree $\mathbf{a}$ c.e. in $\mathbf{c}$ such that $\mathbf{c} < \mathbf{a} < \mathbf{d}$, and then proved that there is a c.e. degree failing to isolate any 2-CEA degree in it. In [11], Salts proved that there is an interval of computably enumerable degrees, each of which isolates a d.c.e. degree. Recently, Cenzer et al. [4] proved that such intervals are dense in the computably enumerable degrees, and hence the non-isolating degrees are nowhere dense in the computably enumerable degrees. In this paper, we prove that the non-isolating degrees are upwards dense in the computably enumerable degrees, so completing a near comprehensive characterisation of the situation.

Theorem 1. *For any incomplete c.e. degree* $\mathbf{a} < \mathbf{0}'$, *there is an incomplete non-isolating degree* $\mathbf{c}$ *above* $\mathbf{a}$.

Our construction of the non-isolating degrees is direct, and is different from the one given by Arslanov et al. in [3]. In section 2, we show how to construct a non-isolating degree. In section 3, we describe how to combine our construction with the upwards density to prove Theorem 1.

2 Constructing a Non-isolating Degree

In this section, we present a new construction of non-isolating degrees. We will construct c.e. sets A, C satisfying the following requirements:

$$\mathcal{P}_e: \quad A \neq \Phi_e;$$
$$\mathcal{Q}_e: \quad C \neq \Phi_e^A;$$
$$\mathcal{R}_e: \quad \widetilde{D}_e = \Phi_e^A \Rightarrow \exists B_e \leq_T D_e \oplus A(B_e \not\leq_T A) \vee D_e \leq_T A;$$

where $\{(D_e, \Phi_e) : d \in \omega\}$ is an effective list of pairs (D, Φ), where D is a d.c.e. set and Φ is a partial computable functional. Here, $\widetilde{D}$ is the Lachlan set of D, with respect to an effective (d.c.e.) approximation $\{D_s : s \in \omega\}$. That is:

$$\widetilde{D} = \{\langle x, s\rangle : x \in D_s \;\&\; \exists t > s(x \notin D_t)\}.$$

Obviously, $\widetilde{D} \leq_T D$ and it is a c.e. set. From the approximation $\{D_s : s \in \omega\}$, we have the following effective enumeration of $\widetilde{D}$: $\langle x, s\rangle$ is enumerated into $\widetilde{D}$ at stage t if t is the least stage such $x \notin D_t$.

The strategy for satisfying the $\mathcal{P}$ and $\mathcal{Q}$ requirements is the standard Friedberg-Muchnik one, and we assume the readers are familiar with it. The $\mathcal{R}$-requirements are non-isolating requirements. That is, for a d.c.e. set D, if D is not reducible to A, then we want to find a c.e. set reducible to $A \oplus D$, but not reducible to A, so that A does not isolate $A \oplus D$. This c.e. set may *either* be the natural candidate $\widetilde{D}$, *or* some other, B, which we will need to construct. On the other hand, if D itself is reducible to A (where, of course, $\widetilde{D}$ is reducible to A), then we will need to show this fact.

To satisfy $\mathcal{R}_e$, we need to construct a c.e. set B_e, and a p.c. functional Γ_e for which $B_e = \Gamma_e^{A \oplus D}$. At the same time, we also want to ensure that $B_e \not\leq_T A$, provided that D is not reducible to A. That is, the following requirements should be also satisfied:

$$\mathcal{S}_{e,i}: \quad B_e \neq \Phi_i^A \text{ or there is a p.c. functional } \Delta_{e,i} \text{ such that } D = \Delta_{e,i}^A.$$

An $\mathcal{R}_e$ strategy defines Γ_e at sufficiently large, that is "big", expansionary stages. Here we say that an expansionary stage is *big* if the length of agreement between $\widetilde{D}$ and Φ_e^A is bigger than any number specified by a substrategy $\mathcal{S}$. Obviously, if there are infinitely many expansionary stages, there are also infinitely many big expansionary stages. An $\mathcal{R}_e$ strategy has two outcomes: f for finitely many expansionary stages, and ∞ for infinitely many expansionary stages, with $\infty <_L f$. Below outcome ∞, we will list substrategies $\mathcal{S}_{e,i}$, $i \in \omega$, which will work together to construct B_e, and to satisfy requirement $\mathcal{R}_e$.

An $\mathcal{S}_{e,i}$-module consists of (infinitely many) steps, where each step n tries to find a number x_n such that either $\Phi_i^A(x_n) \neq B_e(n)$ or $\Delta_{e,i}^A(n)$ is defined. Step n works as follows:

1. Choose x_n as a big number.
2. Wait for $\Phi_i^A(x_n) \downarrow= 0$.
3. For $\Phi_i^A(x_n) \downarrow= 0$ at stage s –

   We define $\Delta_{e,i}^A(n) = D(n)$ with use $\delta_{e,i}(n) = \varphi_i(x_n)$. *Here, again, when we see $\Phi_i^A(x_n)[s] \downarrow= 0$, we do not put restraint on A to preserve this computation.* Whenever A changes below $\delta_{e,i}(n)$, go back to (2), in which case, $\Delta_{e,i}^A(n)$ is undefined by this A-change.

   [Here, n can be in D_s or not in in D_s.
   If n is currently not in D_s, then n can enter D later, and leave at a further stage. When n enters D, at stage s', we will use the assumption that $\widetilde{D}$ is equal to Φ_e^A, as at stage s', $\langle n, s\rangle$ is not in $\widetilde{D}$, and we want to restrain A from changing to preserve this computation. If later n leaves D, then $\langle n, s\rangle$ will enter $\widetilde{D}$, making $\widetilde{D}$ and Φ_e^A disagree at $\langle n, s\rangle$.
   If n is in D_s, we do nothing here, since if n leaves D later, this change will enable us to undefine $\Gamma^{A \oplus D}(x_n)$, and allow us to put $x_{j,n}$ into B_e. Note that axioms enumerated into Γ_e before n enters D are all invalidated by the A-changes. Otherwise we will be in the situation described in the last paragraph, where n leaving D causes a disagreement between $\widetilde{D}$ and Φ_e^A.]

   Wait for $D(n)$ to change, and simultaneously start the step $n+1$.

4. Say $D(n)$ changes at stage $t > s$. There are two possibilities.

   (a) n enters D at stage t. In this case, the $D(n)$ change undefines $\Gamma_e^{A\oplus D}(x_n)$, and instead of putting x_n into B_e immediately at this stage, we wait for $\widetilde{D}$ and Φ_e^A to agree on (all numbers $\leq$) $\langle n, t\rangle$.

   [*We delay the enumeration of x_n into B_e mainly because n may leave D later, making $\Gamma_e^{A\oplus D}$ correct, forcing the enumeration of a number into A to change $\Gamma_e^{A\oplus D}(x_n)$, resulting in no real progress.*]

   (b) n leaves D at stage t. Then $\langle n, s\rangle$ enters $\widetilde{D}$ at this stage. Since the computation $\Phi_e^A(\langle n, s\rangle) = 0$ is preserved at stage s' by restraining A, and we get a global win via

$$\Phi_e^A(\langle n, s\rangle) = 0 \neq 1 = \widetilde{D}(\langle n, s\rangle).$$

5. At stage $t' > t$, $\widetilde{D}$ and Φ_e^A agree on (all numbers $\leq$) $\langle n, t\rangle$. We enumerate x_n into B_e, and put restraint on A to preserve both computations $\Phi_i^A(x_n) = 0$, and $\Phi_e^A(\langle n, t\rangle) = 0$.

   [*Here, at stage t, n enters D, which undefines $\Gamma_e^{A\oplus D}(x_n)$. From now on, we wait for the agreement between $\widetilde{D}$ and Φ_e^A to exceed $\langle n, t\rangle$, and during this period, we do not define $\Gamma_e^{A\oplus D}(x_n)$. Thus, at stage t', we can enumerate x_n into B_e directly, as $\Gamma_e^{A\oplus D}(x_n)$ is undefined at this stage. If there is no such a stage t', then $\mathcal{R}_e$ will have outcome f, and below this outcome, no substrategies $\mathcal{S}_{e,i}$ are listed. We consider the case when a strategy $\mathcal{X}$ is between $\mathcal{R}_e$ and $\mathcal{S}_{e,i}$. W.o.l.g., suppose that $\mathcal{X}$ is an $\mathcal{S}_{e',j}$-strategy with $e' < e$. Then under outcome f, another strategy is arranged to satisfy the $\mathcal{S}_{e',j}$-requirement. A nontrivial case is that after $\mathcal{X}$ puts a number into $B_{e'}$, $D_{e'}$ changes now and the permission from $D_{e'}$ used for the enumeration of the number into $B_{e'}$ goes away — (6) is reached. In this case, $\mathcal{R}_{e'}$ is satisfied permanently because a disagreement between $\widetilde{D}_{e'}$ and $\Phi_{e'}^A$ is found. We will also see that in the whole construction, only $\mathcal{P}$ strategies put numbers into A. This explains why* **0** *is non-isolating and why the nonzero non-isolating degrees are downwards dense in the c.e. degrees. We will see in the next section that it will not be like this when we prove the upwards density, where a change of K is required.*]

6. At stage $t'' > t'$, n leaves D. Then as in 4(b), $\langle n, t\rangle$ enters $\widetilde{D}$ at this stage. As the computation $\Phi_e^A(\langle n, t\rangle) = 0$ is preserved at stage t', by keeping this restraint on A, we get a global win via

$$\Phi_e^A(\langle n, t\rangle) = 0 \neq 1 = \widetilde{D}(\langle n, t\rangle).$$

We now consider the outcomes of $\mathcal{S}_{e,i}$, which can run finitely many or infinitely many steps. Notice that if step n_2 is in progress, and now we have an A-change so that we come back to 2 of step $n_1 < n_2$, then $\Delta_{e,i}^A(n_2)$ is undefined automatically by this A-change. After this, we need to start step n_2 from 1. That is, we choose a new x_{n_2}.

If some step n passes 3, then we win either by 4(b) or 6, which is a global win via $\Phi_e^A \neq \widetilde{D}$, where this disagreement is preserved forever. This means that there will be no more $\mathcal{R}_e$-expansionary stages, or we win by 5, where we get a D-change at n, where this n remains in D, ensuring that the previous axioms enumerating x_n into $\Gamma^{A,D}$ are invalid forever.

If no step passes 3, then notice that we put no restraint on A. It can happen that there is a least n such that $\Phi_e^A(x_n)$ does not converge to 0, or for each n, $\Phi_e^A(x_n)$ converges to 0. $\mathcal{S}_{e,i}$ is satisfied in both cases: In the former case, $B_e(x_n) = 0 \neq \Phi_e^A(x_n)$; and in the latter case, each step remains at 3, and hence $\Delta_{e,i}^A(n)$ is defined and equal to $D(n)$. So $\Delta_{e,i}^A = D$.

So this $\mathcal{S}_{e,i}$ module has two outcomes: w and s with $w <_L s$. Here w denotes the case in which no step passes 3, and s the case in which some step passes 3. In the latter case, we put restraint on A to preserve computations.

3 Upwards Density

We are now ready to present the strategy for constructing an incomplete non-isolating degree above any incomplete c.e. degree.

Fix U as an incomplete c.e. set. We will construct c.e. sets A and C satisfying the following requirements:

$$\mathcal{P}_e: \quad C \neq \Phi_e^{A,U};$$

$$\mathcal{R}_e: \quad \widetilde{D}_e = \Phi_e^{A,U} \Rightarrow (\exists B_e \leq_T A \oplus D_e \oplus U)(B_e \not\leq_T A \oplus U) \vee D_e \leq_T A \oplus U;$$

where $\{(D_e, \Phi_e) : d \in \omega\}$ is a standard list of pairs (D, Φ), where D is a d.c.e. set and Φ is a partial computable functional, and $\widetilde{D}$ is the Lachlan set of D. Here we did not require that A is not reducible to U, as we can obtain this by applying Sacks' density theorem first.

We apply the Sacks preservation strategy to satisfy the $\mathcal{P}$ requirements by running (infinitely many) cycles to threaten the assumption that U is incomplete via a p.c. functional Θ. Cycle n behaves as follows:

1. Choose a big number x_n.
2. Wait for $\Phi_e^{A,U}(x_n) \downarrow= 0$.
3. Say $\Phi_e^{A,U}(x_n)$ converges to 0 at stage s. Define $\Theta^U(n) = K_s(n)$ with use $\theta(n) = \varphi(x_n)$. Restrain A from changing below $\varphi(x_n)$.

   Wait for a change of $K(n)$ or a change of U below $\varphi(n)$, and simultaneously start the next cycle.
4. Say U changes below $\varphi(x_n)$ first. Then go back to 2. Note that $\Theta^U(n')$, each $n' \geq n$, is undefined by this U change.
5. Say $K(n)$ changes first. Then we put x_n into C, and wait for a change of U below $\varphi(x_n)$.

   [*If there is no such a U-change, then $\Theta^U(n)$ differs from $K(n)$. But as U below $\varphi(x_n)$ is fixed, we satisfy $\mathcal{P}$ because $\Phi_e^{A,U}(x_n) \downarrow= 0$ is preserved forever, and $C(x_n) = 1$. Otherwise, as above, a U-change undefines $\Theta^U(n')$, each $n' \geq n$, allowing us to redefine $\Theta^U(n) = K(n) = 1$.*]

6. Undefine $\Theta^U(n')$ for $n' \geq n$, and redefine $\Theta^U(n) = K(n) = 1$ with use 0. Start the next cycle.

The $\mathcal{P}$ module has the following outcomes:

$\langle n, w\rangle$: cycle n stops at 2. $\mathcal{P}$ is satisfied since $\Phi_e^{A,U}(x_n)$ does not converge to 0 and $C(x_n) = 0$.
$\langle n, s\rangle$: cycle n stops at 2. $\mathcal{P}$ is satisfied since $\Phi_e^{A,U}(x_n)$ converges to 0 and $C(x_n) = 1$.
$\langle n, u\rangle$: cycle n runs through the loop 2-3-4-2 infinitely often. $\mathcal{P}$ is satisfied since $\Phi_e^{A,U}(x_n)$ does not converge at all, and $C(x_n) = 0$.

These outcomes are ordered:

$$\langle 0, u\rangle <_L \langle 0, w\rangle <_L \langle 0, s\rangle <_L \langle 1, u\rangle <_L \langle 1, w\rangle <_L \langle 1, s\rangle <_L \langle 2, u\rangle$$

$$<_L \langle 2, w\rangle <_L \langle 2, s\rangle <_L \cdots <_L \langle n, u\rangle <_L \langle n, w\rangle <_L \langle n, s\rangle <_L \cdots .$$

As U is incomplete, there is a least n such that one of $\langle n, u\rangle, \langle n, w\rangle, \langle n, s\rangle$ is the true outcome for $\mathcal{P}$, since otherwise, Θ^A will be totally defined, and will compute K correctly, which is impossible.

We now consider how to satisfy the $\mathcal{R}$-requirements, with U included. Again, for any d.c.e. set D, if D is not reducible to $A \oplus U$, we need to find a c.e. set reducible to $A \oplus D \oplus U$, but not reducible to $A \oplus U$, so that $A \oplus U$ does not isolate $A \oplus D \oplus U$. Again, this c.e. set can be $\widetilde{D}$, or another set B which we will construct. On the other hand, if D itself is reducible to $A \oplus U$, then we need to construct a p.c. functional Δ to reduce D to $A \oplus U$. The following are the details.

We will construct a c.e. set B_e and a p.c. functional Γ_e such that $B_e = \Gamma_e^{A \oplus D \oplus U}$. At the same time, if D is not reducible to $A \oplus U$, we need to ensure that $B_e \not\leq_T A \oplus U$. That is, the following requirements should also be satisfied:

$$\mathcal{S}_{e,i}: \quad B_e \neq \Phi_i^{A \oplus U} \text{ or there is a p.c. functional } \Delta_{e,i} \text{ such that } D = \Delta_{e,i}^{A \oplus U}.$$

As in section 2, a $\mathcal{R}_e$ strategy defines Γ_e at big expansionary stages, and has two outcomes: f for finitely many expansionary stages, and ∞ for infinitely many expansionary stages, with $\infty <_L f$. Below outcome ∞, we will list substrategies $\mathcal{S}_{e,i}$, $i \in \omega$.

An $\mathcal{S}_{e,i}$-module consists of infinitely many cycles and each cycle consists of infinitely many steps. All the cycles are devoted to defining a p.c. functional $\Theta_{e,i}$, and each cycle j tries to find some some x such that $B_e(x) \neq \Phi_i^{A \oplus U}(x)$, or to define a p.c. functional $\Delta_{e,i,j}$ such that $D = \Delta_{e,i,j}^{A \oplus U}$, or to define $\Theta_{e,i}^{A \oplus U}(j) = K(j)$. This task will be realized by cycle j's steps, where as in section 2, each step n of cycle j, denoted by $\langle j, n\rangle$, tries to find a number $x_{j,n}$ such that either $\Phi_i^A(x_{j,n}) \neq B_e(x_{j,n})$ or $\Delta_{e,i,j}^A(n)$ is defined. Step $\langle j, n\rangle$ proceeds as follows:

1. Choose $x_{j,n}$ as a big number.
2. Wait for $\Phi_i^{A,U}(x_{j,n}) \downarrow = 0$.
3. If $\Phi_i^{A,U}(x_{j,n}) \downarrow = 0$ at stage s —

   We define $\Delta_{e,i}^{A,U}(n) = D(n)$ with use $\delta_{e,i}(n) = \varphi_i(x_n)$. *Here, again, when we see $\Phi_i^{A,U}(x_n)[s] \downarrow = 0$, we do not put restraint on A to preserve this computation.*

Whenever A or U changes below $\delta_{e,i}(n)$, we go back to 2, in which case, $\Delta_{e,i}^{A,U}(n)$ is undefined.

[*Note that here, n can be in D_s or not in D_s.*
If n is currently not in D_s, then n can enter D later, and leave at a further stage. When n enters D, at stage s', we will use the assumption that $\widetilde{D}$ is equal to $\Phi_e^{A\oplus U}$, as then $\langle n, s\rangle$ is not in $\widetilde{D}$, and we want to restrain A from changing to retain this computation. If n leaves D later ($\Gamma^{A\oplus D\oplus U}(x_{j,n})$ reverts to a previous value, which is equal to 0), then $\langle n, s\rangle$ will enter $\widetilde{D}$, making $\widetilde{D}$ and $\Phi_e^{A\oplus U}$ disagree at $\langle n, s\rangle$. This disagreement can now fail only when U changes, which will undefine $\Gamma^{A\oplus D\oplus U}(x_{j,n})$.

If n is in D_s, then we do nothing here, as if n leaves D later, this change will allow us to undefine $\Gamma^{A\oplus D\oplus U}(x_n)$, and we can put $x_{j,n}$ into B_e. Notice that axioms enumerated into Γ_e before n enters D are all invalid due to the A or U-changes. Since otherwise we will be in the situation described in the last paragraph, where n leaving D will cause a disagreement between $\widetilde{D}$ and $\Phi_e^{A\oplus U}$.]

Wait for $D(n)$ to change, and simultaneously commence the step $\langle j, n+1\rangle$.

4. Say $D(n)$ changes at stage $t > s$. There are two possibilities:
   (a) n enters D at stage t. In this case, the $D(n)$ change makes $\Gamma_e^{A\oplus D\oplus U}(x_{j,n})$ undefined, and instead of putting $x_{j,n}$ into B_e immediately at this stage, we wait for $\widetilde{D}$ and Φ_e^A to agree on (all numbers $\leq$) $\langle n, t\rangle$.
   (b) n leaves D at stage t. Then this $D(n)$ change necessarily results in $\Gamma_e^{A\oplus D\oplus U}$ $(x_{j,n})$ becoming undefined, and we define $\Theta^U(j) = K(j)$ with use $\theta(j) = \varphi_i(x_{j,n})$, and wait for U to change below $\theta(j)$ or $K(n)$ to change. Of course, a restraint is put on A to preserve the computation $\Phi_e^{A\oplus U}(x_{j,n})$.
   If U changes first, go back to (2), and the restraint is canceled. If $K(n)$ changes first, go to (6).

5. At stage $t' > t$, $\widetilde{D}$ and Φ_e^A agree on (all numbers $\leq$) $\langle n, t\rangle$. We put restraint on A to preserve both computations $\Phi_i^{A,U}(x_{j,n}) = 0$, and $\Phi_e^{A,U}(\langle n, t\rangle) = 0$. Create a link between $\mathcal{S}_{e,i}$ and $\mathcal{R}_e$, the mother node. *This link can be traveled and canceled when we find at $\mathcal{R}_e$ that U has changes below $\varphi_e(\langle n, t\rangle)$. If there is no such U-change, then this link can survive for ever.*
   We define $\Theta^U(j) = K(j)$ with use $\theta(j) = \max\{\varphi_i(x_{j,n}), \phi_e(\langle x, s\rangle)\}$, and wait for U to change below $\theta(j)$, or $K(n)$ to change, or n to leave D. Simultaneously, start cycle $j+1$.
   If $K(j)$ changes first, then go to (6). If n leaves D first, go to (7).
   If U changes first, see whether U has a change below $\varphi_i(x_{j,n})$. If yes, then go back to (2). Of course, the restraint on A established at stage t' will be canceled. Otherwise, go back to (4) and wait for $\widetilde{D}$ and $\Phi_e^{A\oplus D}$ to agree on (all numbers $\leq$) $\langle n, t\rangle$.

6. $K(j)$ changes.

We enumerate $x_{j,n}$ into B_e, and the current $\gamma_e(x_{j,n})$ into A. *Note that this new $\gamma_e(x_{j,n})$ is bigger than $\theta(j)$, since it is defined after stage t'.*

Wait for U to change below $\theta(j)$, and also wait for n to leave D. If n leaves D first, then go to (7). If U changes first, go to (8).

7. At stage $t'' > t'$, n leaves D. Then $\langle n, t\rangle$ enters $\widetilde{D}$ at this stage. As the computation $\Phi_e^A(\langle n, t\rangle) = 0$ is preserved at stage t', we get a global win via
$$\Phi_e^{A,U}(\langle n, t\rangle) = 0 \neq 1 = \widetilde{D}(\langle n, t\rangle),$$
provided that U does not change change below $\varphi_e(\langle n, t\rangle)$.

   Wait for U to change below $\theta(j)$.
8. Put $\gamma_e(x_{j,n})$ into A and redefine $\Theta^U(j) = K(j)$, with use 0. Start cycle $j + 1$.

The outcomes for $\mathcal{S}_{e,i}$ are rather more complicated. As in the $\mathcal{P}$-module, we will have outcomes for each cycle, and each cycle has outcomes different from those given in section 2. Below, we only describe the outcome for a particular cycle j.

$\langle j, w\rangle$: Cycle j runs many steps, and each one stays at (2) or (3), or returns to (2) infinitely often from (3) or (4b). In this case, no restraint is put on A, and $\mathcal{S}_{e,i}$ is satisfied by either $\Phi_e^{A,U}(x_{j,n})$ not converging for some $x_{j,n}$ (*some step stops at (2) or loops between (2) and (3), or (4b)*), or by $\Delta_{e,i}^{A,U}$ becoming totally defined and computing D correctly. [*Note that if this outcome is true, a step can return to (2) from other point beyond (5) at most once, via a U-change.*]

No restraint is put on A in this outcome.

$\langle j, n, u\rangle$: Step n of cycle j reaches (5) and later returns to (4) to wait for $\widetilde{D}$ and $\Pi_e^{A\oplus U}$ to agree on number $\leq \langle n, t\rangle$, infinitely often by U-changes, where t is the stage at which n enters D. That is, infinitely many links to the mother node $\mathcal{R}_e$ are created and canceled in the construction. If this outcome is true, then $\Phi_e^{A\oplus U}(\langle n, t\rangle)$ diverges, leading to a global win for $\mathcal{R}_e$.

No restraint is put on A in this outcome.

$\langle j, n, d\rangle$: Step n of cycle j reaches (7) because after (5), n leaves D, no matter whether $K(j)$ has changed or not. As we put a restraint on A at (5), if U does not change below $\varphi_e(\langle n, t\rangle)$, then we will have $\Phi_e^{A\oplus D}(\langle n, t\rangle) = 0$ and $\langle n, t\rangle \in \widetilde{D}$ — again, a global win for $\mathcal{R}_e$. If this outcome applies, then there are only finitely many expansionary stages. That is, $\mathcal{R}_e$ will have f as its outcome, and we do not put this outcome below $\mathcal{S}_{e,i}$.

[*Notice that if this outcome applies, $\Gamma_e^{A\oplus D\oplus U}(x_{j,n})$ can be wrong in the case that step n reaches (6), and $x_{j,n}$ is put into B_e when n is in D. If later n leaves D, then $\Gamma_e^{A\oplus D\oplus U}(x_{j,n})$ can be reinstated to the one before n enters D, which is currently defined as 0. If so, then either cycle j reaches (7) and stays at (7), or it eventually reaches (8). If the former case, as indicated above, we have a global win for $\mathcal{R}_e$. If cycle j eventually reaches (8), then cycle j also wins as $\Theta^U(j)$ is defined, and is equal to $K(j)$. To keep $\Gamma_e^{A\oplus D\oplus U}(x_{j,n})$ correct, at (8), we also put $\gamma_e(x_{j,n})$ (the old one) into A. This enumeration does not injure cycle j.*]

$\langle j, s\rangle$: Some step n of cycle j reaches and stays at (6). Then $\mathcal{S}_{e,i}$ is satisfied, via $B_e(x_{j,n}) = 1 \neq 0 = \Phi_i^{A\oplus U}(x_{j,n})$. Notice that since n remains in D (otherwise step n will go to (7), which cannot be true by our assumption), the enumeration of $\gamma_e(x_{j,n})$ (the new one) into A will not change the computation $\Phi_i^{A\oplus U}(x_{j,n})$. The enumeration of this $\gamma_e(x_{j,n})$ makes $\Gamma_e^{A\oplus D\oplus U}$ well-defined and correct at $x_{j,n}$. Also note that U will not change so as to injure this computation, since otherwise step n will go to (8), which again cannot be true.

We arrange the outcomes of cycle j as:

$$\langle j, w\rangle <_L \langle j, 0, u\rangle <_L \langle j, 1, u\rangle <_L \cdots <_L \langle j, n, u\rangle <_L \cdots <_L \langle j, s\rangle,$$

where the outcomes for j_1 are always to the left of those for j_2, whenever $j_1 < j_2$. Again, since U is incomplete, there is a least j such that one of the outcomes for cycle j is the true outcome relative to $\mathcal{S}_{e,i}$.

The whole construction turns out to be a $0'''$ argument, in a quite standard way. The details will appear in [6].

References

1. Afshari, B., Barmpalias, G., Cooper, S.B., Stephan, F.: Post's Programme for the Ershov Hierarchy. J. of Logic and Computation (to appear)
2. Arslanov, M.M., Lempp, S., Shore, R.A.: Interpolating d-r.e. and REA degrees between r.e. degrees. Ann. Pure. Appl. Logic 78, 29–56 (1994)
3. Arslanov, M.M., Lempp, S., Shore, R.A.: On isolating r.e. and isolated d-r.e. degrees. In: Computability, enumerability, unsolvability. London Math. Soc. Lecture Note Ser, vol. 224, pp. 61–80. Cambridge Univ. Press, Cambridge (1996)
4. Cenzer, D., LaForte, G., Wu, G.: The nonisolated degrees are nowhere dense (in preparation)
5. Cooper, S.B.: Computability Theory. Chapman & Hall/CRC, Boca Raton, London, New York, Washington, D.C (2004)
6. Cooper, S.B., Salts, M., Wu, G.: The non-isolating degrees are upwards dense in the computably enumerable degrees (to appear)
7. Cooper, S.B., Yi, X.: Isolated d.r.e. degrees. University of Leeds, Dept. of Pure Math. 17 (1995) (preprint series)
8. Ding, D., Qian, L.: Isolated d.r.e. degrees are dense in r.e. degree structure. Arch. Math. Logic 36, 1–10 (1996)
9. Downey, R.G.: D.r.e. degrees and the nondiamond theorem. Bull. London Math. Soc. 21, 43–50 (1989)
10. LaForte, G.: The isolated d.r.e. degrees are dense in the r.e. degrees. Math. Logic Quart. 42, 83–103 (1996)
11. Salts, M.: An interval of computably enumerable isolating degrees. Math. Logic Quart. 45, 59–72 (1999)
12. Ishmukhametov, S., Wu, G.: Isolation and the high/low hierarchy. Arch. Math. Logic 41, 259–266 (2002)
13. Wu, G.: Isolation and lattice embeddings. Jour. Symb. Logic 67, 1055–1064 (2002)

Author Index

Printing: Mercedes-Druck, Berlin
Binding: Stein+Lehmann, Berlin

Lecture Notes in Computer Science

Sublibrary 1: Theoretical Computer Science and General Issues

For information about Vols. 1–4664
please contact your bookseller or Springer

Vol. 4988: R. Berghammer, B. Möller, G. Struth (Eds.), Relations and Kleene Algebra in Computer Science. X, 397 pages. 2008.

Vol. 4981: M. Egerstedt, B. Mishra (Eds.), Hybrid Systems: Computation and Control. XV, 680 pages. 2008.

Vol. 4978: M. Agrawal, D. Du, Z. Duan, A. Li (Eds.), Theory and Applications of Models of Computation. XII, 598 pages. 2008.

Vol. 4974: M. Giacobini, A. Brabazon, S. Cagnoni, G.A. Di Caro, R. Drechsler, A. Ekárt, A.I. Esparcia-Alcázar, M. Farooq, A. Fink, J. McCormack, M. O'Neill, J. Romero, F. Rothlauf, G. Squillero, A.Ş. Uyar, S. Yang (Eds.), Applications of Evolutionary Computing. XXV, 701 pages. 2008.

Vol. 4973: E. Marchiori, J.H. Moore (Eds.), Evolutionary Computation, Machine Learning and Data Mining in Bioinformatics. X, 213 pages. 2008.

Vol. 4972: J. van Hemert, C. Cotta (Eds.), Evolutionary Computation in Combinatorial Optimization. XII, 289 pages. 2008.

Vol. 4971: M. O'Neill, L. Vanneschi, S. Gustafson, A.I. Esparcia Alcázar, I. De Falco, A. Della Cioppa, E. Tarantino (Eds.), Genetic Programming. XI, 375 pages. 2008.

Vol. 4963: C.R. Ramakrishnan, J. Rehof (Eds.), Tools and Algorithms for the Construction and Analysis of Systems. XVI, 518 pages. 2008.

Vol. 4962: R. Amadio (Ed.), Foundations of Software Science and Computational Structures. XV, 505 pages. 2008.

Vol. 4961: J.L. Fiadeiro, P. Inverardi (Eds.), Fundamental Approaches to Software Engineering. XIII, 430 pages. 2008.

Vol. 4960: S. Drossopoulou (Ed.), Programming Languages and Systems. XIII, 399 pages. 2008.

Vol. 4959: L. Hendren (Ed.), Compiler Construction. XII, 307 pages. 2008.

Vol. 4957: E.S. Laber, C. Bornstein, L.T. Nogueira, L. Faria (Eds.), LATIN 2008: Theoretical Informatics. XVII, 794 pages. 2008.

Vol. 4943: R. Woods, K. Compton, C. Bouganis, P.C. Diniz (Eds.), Reconfigurable Computing: Architectures, Tools and Applications. XIV, 344 pages. 2008.

Vol. 4934: U. Brinkschulte, T. Ungerer, C. Hochberger, R.G. Spallek (Eds.), Architecture of Computing Systems - ARCS 2008. XI, 287 pages. 2008.

Vol. 4927: C. Kaklamanis, M. Skutella (Eds.), Approximation and Online Algorithms. X, 289 pages. 2008.

Vol. 4921: S.-i. Nakano, M.. S. Rahman (Eds.), WALCOM: Algorithms and Computation. XII, 241 pages. 2008.

Vol. 4919: A. Gelbukh (Ed.), Computational Linguistics and Intelligent Text Processing. XVIII, 666 pages. 2008.

Vol. 4917: P. Stenström, M. Dubois, M. Katevenis, R. Gupta, T. Ungerer (Eds.), High Performance Embedded Architectures and Compilers. XIII, 400 pages. 2008.

Vol. 4915: A. King (Ed.), Logic-Based Program Synthesis and Transformation. X, 219 pages. 2008.

Vol. 4912: G. Barthe, C. Fournet (Eds.), Trustworthy Global Computing. XI, 401 pages. 2008.

Vol. 4910: V. Geffert, J. Karhumäki, A. Bertoni, B. Preneel, P. Návrat, M. Bieliková (Eds.), SOFSEM 2008: Theory and Practice of Computer Science. XV, 792 pages. 2008.

Vol. 4905: F. Logozzo, D.A. Peled, L.D. Zuck (Eds.), Verification, Model Checking, and Abstract Interpretation. X, 325 pages. 2008.

Vol. 4904: S. Rao, M. Chatterjee, P. Jayanti, C.S.R. Murthy, S.K. Saha (Eds.), Distributed Computing and Networking. XVIII, 588 pages. 2008.

Vol. 4878: E. Tovar, P. Tsigas, H. Fouchal (Eds.), Principles of Distributed Systems. XIII, 457 pages. 2007.

Vol. 4875: S.-H. Hong, T. Nishizeki, W. Quan (Eds.), Graph Drawing. XIII, 402 pages. 2008.

Vol. 4873: S. Aluru, M. Parashar, R. Badrinath, V.K. Prasanna (Eds.), High Performance Computing – HiPC 2007. XXIV, 663 pages. 2007.

Vol. 4863: A. Bonato, F.R.K. Chung (Eds.), Algorithms and Models for the Web-Graph. X, 217 pages. 2007.

Vol. 4860: G. Eleftherakis, P. Kefalas, G. Păun, G. Rozenberg, A. Salomaa (Eds.), Membrane Computing. IX, 453 pages. 2007.

Vol. 4855: V. Arvind, S. Prasad (Eds.), FSTTCS 2007: Foundations of Software Technology and Theoretical Computer Science. XIV, 558 pages. 2007.

Vol. 4854: L. Bougé, M. Forsell, J.L. Träff, A. Streit, W. Ziegler, M. Alexander, S. Childs (Eds.), Euro-Par 2007 Workshops: Parallel Processing. XVII, 236 pages. 2008.

Vol. 4851: S. Boztaş, H.-F.(F.) Lu (Eds.), Applied Algebra, Algebraic Algorithms and Error-Correcting Codes. XII, 368 pages. 2007.

Vol. 4848: M.H. Garzon, H. Yan (Eds.), DNA Computing. XI, 292 pages. 2008.

Vol. 4847: M. Xu, Y. Zhan, J. Cao, Y. Liu (Eds.), Advanced Parallel Processing Technologies. XIX, 767 pages. 2007.

Vol. 4846: I. Cervesato (Ed.), Advances in Computer Science – ASIAN 2007. XI, 313 pages. 2007.

Vol. 4838: T. Masuzawa, S. Tixeuil (Eds.), Stabilization, Safety, and Security of Distributed Systems. XIII, 409 pages. 2007.

Vol. 4835: T. Tokuyama (Ed.), Algorithms and Computation. XVII, 929 pages. 2007.

Vol. 4818: I. Lirkov, S. Margenov, J. Waśniewski (Eds.), Large-Scale Scientific Computing. XIV, 755 pages. 2008.

Vol. 4800: A. Avron, N. Dershowitz, A. Rabinovich (Eds.), Pillars of Computer Science. XXI, 683 pages. 2008.

Vol. 4783: J. Holub, J. Žďárek (Eds.), Implementation and Application of Automata. XIII, 324 pages. 2007.

Vol. 4782: R. Perrott, B.M. Chapman, J. Subhlok, R.F. de Mello, L.T. Yang (Eds.), High Performance Computing and Communications. XIX, 823 pages. 2007.

Vol. 4771: T. Bartz-Beielstein, M.J. Blesa Aguilera, C. Blum, B. Naujoks, A. Roli, G. Rudolph, M. Sampels (Eds.), Hybrid Metaheuristics. X, 202 pages. 2007.

Vol. 4770: V.G. Ganzha, E.W. Mayr, E.V. Vorozhtsov (Eds.), Computer Algebra in Scientific Computing. XIII, 460 pages. 2007.

Vol. 4769: A. Brandstädt, D. Kratsch, H. Müller (Eds.), Graph-Theoretic Concepts in Computer Science. XIII, 341 pages. 2007.

Vol. 4763: J.-F. Raskin, P.S. Thiagarajan (Eds.), Formal Modeling and Analysis of Timed Systems. X, 369 pages. 2007.

Vol. 4759: J. Labarta, K. Joe, T. Sato (Eds.), High-Performance Computing. XV, 524 pages. 2008.

Vol. 4746: A. Bondavalli, F. Brasileiro, S. Rajsbaum (Eds.), Dependable Computing. XV, 239 pages. 2007.

Vol. 4743: P. Thulasiraman, X. He, T.L. Xu, M.K. Denko, R.K. Thulasiram, L.T. Yang (Eds.), Frontiers of High Performance Computing and Networking ISPA 2007 Workshops. XXIX, 536 pages. 2007.

Vol. 4742: I. Stojmenovic, R.K. Thulasiram, L.T. Yang, W. Jia, M. Guo, R.F. de Mello (Eds.), Parallel and Distributed Processing and Applications. XX, 995 pages. 2007.

Vol. 4739: R. Moreno Díaz, F. Pichler, A. Quesada Arencibia (Eds.), Computer Aided Systems Theory – EUROCAST 2007. XIX, 1233 pages. 2007.

Vol. 4736: S. Winter, M. Duckham, L. Kulik, B. Kuipers (Eds.), Spatial Information Theory. XV, 455 pages. 2007.

Vol. 4732: K. Schneider, J. Brandt (Eds.), Theorem Proving in Higher Order Logics. IX, 401 pages. 2007.

Vol. 4731: A. Pelc (Ed.), Distributed Computing. XVI, 510 pages. 2007.

Vol. 4728: S. Bozapalidis, G. Rahonis (Eds.), Algebraic Informatics. VIII, 291 pages. 2007.

Vol. 4726: N. Ziviani, R. Baeza-Yates (Eds.), String Processing and Information Retrieval. XII, 311 pages. 2007.

Vol. 4719: R. Backhouse, J. Gibbons, R. Hinze, J. Jeuring (Eds.), Datatype-Generic Programming. XI, 369 pages. 2007.

Vol. 4711: C.B. Jones, Z. Liu, J. Woodcock (Eds.), Theoretical Aspects of Computing – ICTAC 2007. XI, 483 pages. 2007.

Vol. 4710: C.W. George, Z. Liu, J. Woodcock (Eds.), Domain Modeling and the Duration Calculus. XI, 237 pages. 2007.

Vol. 4708: L. Kučera, A. Kučera (Eds.), Mathematical Foundations of Computer Science 2007. XVIII, 764 pages. 2007.

Vol. 4707: O. Gervasi, M.L. Gavrilova (Eds.), Computational Science and Its Applications – ICCSA 2007, Part III. XXIV, 1205 pages. 2007.

Vol. 4706: O. Gervasi, M.L. Gavrilova (Eds.), Computational Science and Its Applications – ICCSA 2007, Part II. XXIII, 1129 pages. 2007.

Vol. 4705: O. Gervasi, M.L. Gavrilova (Eds.), Computational Science and Its Applications – ICCSA 2007, Part I. XLIV, 1169 pages. 2007.

Vol. 4703: L. Caires, V.T. Vasconcelos (Eds.), CONCUR 2007 – Concurrency Theory. XIII, 507 pages. 2007.

Vol. 4700: C.B. Jones, Z. Liu, J. Woodcock (Eds.), Formal Methods and Hybrid Real-Time Systems. XVI, 539 pages. 2007.

Vol. 4699: B. Kågström, E. Elmroth, J. Dongarra, J. Waśniewski (Eds.), Applied Parallel Computing. XXIX, 1192 pages. 2007.

Vol. 4698: L. Arge, M. Hoffmann, E. Welzl (Eds.), Algorithms – ESA 2007. XV, 769 pages. 2007.

Vol. 4697: L. Choi, Y. Paek, S. Cho (Eds.), Advances in Computer Systems Architecture. XIII, 400 pages. 2007.

Vol. 4688: K. Li, M. Fei, G.W. Irwin, S. Ma (Eds.), Bio-Inspired Computational Intelligence and Applications. XIX, 805 pages. 2007.

Vol. 4684: L. Kang, Y. Liu, S. Zeng (Eds.), Evolvable Systems: From Biology to Hardware. XIV, 446 pages. 2007.

Vol. 4683: L. Kang, Y. Liu, S. Zeng (Eds.), Advances in Computation and Intelligence. XVII, 663 pages. 2007.

Vol. 4681: D.-S. Huang, L. Heutte, M. Loog (Eds.), Advanced Intelligent Computing Theories and Applications. XXVI, 1379 pages. 2007.

Vol. 4672: K. Li, C. Jesshope, H. Jin, J.-L. Gaudiot (Eds.), Network and Parallel Computing. XVIII, 558 pages. 2007.

Vol. 4671: V.E. Malyshkin (Ed.), Parallel Computing Technologies. XIV, 635 pages. 2007.

Vol. 4669: J.M. de Sá, L.A. Alexandre, W. Duch, D.P. Mandic (Eds.), Artificial Neural Networks – ICANN 2007, Part II. XXXI, 990 pages. 2007.

Vol. 4668: J.M. de Sá, L.A. Alexandre, W. Duch, D.P. Mandic (Eds.), Artificial Neural Networks – ICANN 2007, Part I. XXXI, 978 pages. 2007.

Vol. 4666: M.E. Davies, C.J. James, S.A. Abdallah, M.D. Plumbley (Eds.), Independent Component Analysis and Signal Separation. XIX, 847 pages. 2007.

Vol. 4665: J. Hromkovič, R. Královič, M. Nunkesser, P. Widmayer (Eds.), Stochastic Algorithms: Foundations and Applications. X, 167 pages. 2007.